EXPLORING EARTH

EXPLORING EARTH

AN INTRODUCTION TO PHYSICAL GEOLOGY

JON P. DAVIDSON
UNIVERSITY OF CALIFORNIA, LOS ANGELES

WALTER E. REED
UNIVERSITY OF CALIFORNIA, LOS ANGELES

PAUL M. DAVIS
UNIVERSITY OF CALIFORNIA, LOS ANGELES

Prentice Hall
Upper Saddle River, NJ 07458

Library of Congress Cataloging-in-Publication Data

Davidson, Jon P.
Exploring earth: an introduction to physical geology / Jon P. Davidson, Walter E. Reed, Paul M. Davis.
p. cm.
Includes index.
ISBN 0-13-463936-7
1. Physical geology. I. Reed, Walter E. II. Davis, Paul M. III. Title.
QE28.2.D375 1997
551—dc20 96-32914
CIP

Acquisitions Editor: *Daniel Kaveney*
Editor in Chief: *Paul F. Corey*
Editorial Director: *Tim Bozik*
Assistant Vice President, Production and Manufacturing: *David W. Riccardi*
Development Editor: *Carol Trueheart*
Editor in Chief of Development: *Ray Mullaney*
Production Editors: *Edward Thomas, Debra Wechsler, Fay Ahuja*
Executive Managing Editor, Production: *Kathleen Schiaparelli*
Assistant Managing Editor, Production: *Shari Toron*
Marketing Manager: *Leslie Cavaliere*
Creative Director: *Paula Maylahn*
Art Director: *Joseph Sengotta*
Art Manager: *Gus Vibal*
Cover Designer: *Joseph Sengotta*
Interior Designer: *Kevin Kall*
Assistant Editor: *Wendy Rivers*
Editorial Assistant: *Betsy Williams*
Summer Intern: *Adam Velthaus*
Manufacturing Manager: *Trudy Pisciotti*
Photo Editor: *Lorinda Morris-Nantz*
Photo Researcher: *Yvonne Gerin*
Page Layout: *David Tay*
Illustrations: *Biographics/Visible Productions, Maryland Cartographics, Inc.*
Artists: *Charles Pelletreau, Patrice Van Acker*
Cover Photo: *Volcan Villarrica, Active Volcano, Lake District Southern Chile—
Galen Rowell/Mountain Light Photography, Inc.*

10 9 8 7 6 5 4 3 2 1

ISBN 0-13-463936-7

Prentice-Hall International (UK) Limited, *London*
Prentice-Hall of Australia Pty. Limited, *Sydney*
Prentice-Hall Canada Inc. *Toronto*
Prentice-Hall Hispanoamericana, S.A., *Mexico*
Prentice-Hall of India Private Limited, *New Delhi*
Prentice-Hall of Japan, Inc., *Toyko*
Simon and Schuster Asia Pte. Ltd., *Singapore*
Editora Prentice-Hall do Brasil, Ltda., *Rio de Janeiro*

Dedication

To Donna, Lauri, and Cecily

Contents

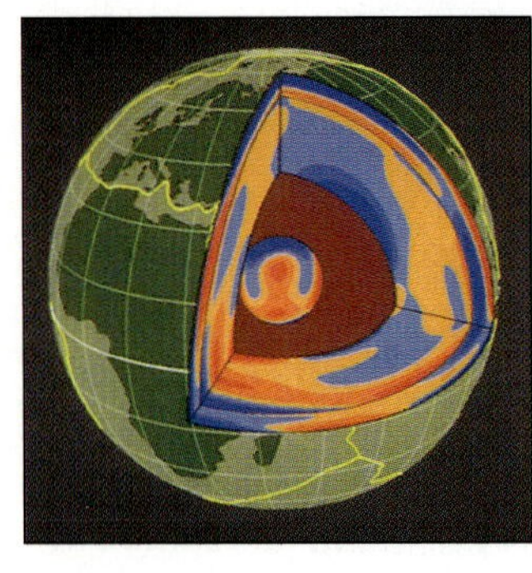

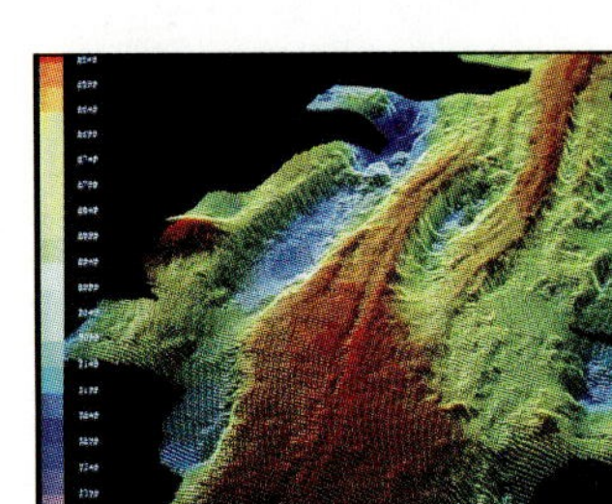

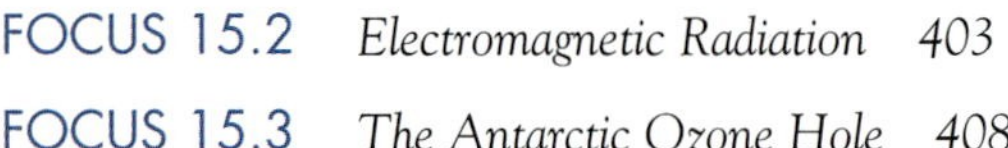

Preface

Preface

With this book, we hope to introduce readers to the world of physical geology and to share with them the excitement of exploring Earth and the processes that formed it. We hope that readers of this book will gain a better understanding of Earth and an increased awareness of our planet. When traveling by airplane, a reader will appreciate how mountains, rivers, and deserts were formed, and what governs their locations, shapes, and textures. A hiker will understand how multicolored pebbles in a stream formed and where they came from. Earthquakes, volcanic eruptions, tidal waves, and other geologic hazards will be viewed with a greater understanding of the processes that give rise to such events.

We have taught the introductory geology and geophysics courses at the University of California at Los Angeles for a cumulative total of 40 years. Students have traveled with us on field trips to every continent. Because our special areas of study include not only geology, but also geophysics and geochemistry, we seized the opportunity to present an interdisciplinary approach to physical geology. We felt the need for a book that presents geology in the framework of plate tectonics with a strong emphasis on geologic processes. We feel strongly that this approach presents the field of geology in its full richness while making the subject both more interesting and easier to learn.

We wrote this book for students with little or no scientific background. Toward this end, we have avoided unnecessary use of jargon, introducing geologic terms in context and only as needed. We describe geologic time using millions and billions of years rather than using the traditional time scale, which we have found can cause unnecessary difficulty for students taking their first geology course. We have limited the use of mathematical expressions, but those that we have included serve to introduce students to a few of the powerful and concise tools geologists use to quantify the processes and characteristics of Earth.

We think this book will give pleasure to students by increasing their awareness and understanding of their surroundings. Indeed, we hope that some will continue on to become geologists and share in the excitement that we have experienced working in this field. In any case, it has been our intention here to make the subject easy to learn and one of never-ending fascination.

Organization of the Book

After a brief introduction to plate tectonics and geologic time in Chapter 1, the order of material presented in the book follows the evolution of Earth from its formation to the development of plate tectonics and present-day geology. Chapter 2 presents the formation of Earth and the elements, and the partition of Earth into crust, mantle, and core. Throughout, we emphasize how scientists arrived at our present understanding of Earth's internal composition and processes. Students can then appreciate the origin of Earth's huge internal energy source, and understand how that energy drives plate tectonics. In Chapters 3 and 4, we turn to the material of Earth itself, and explore rocks and the minerals from which rocks are formed.

Chapter 5 introduces the approaches that gave rise to our present understanding of Earth's interior and internal processes. These include measurements of seismic waves, gravity and magnetism of Earth. It also provides the background for the development of the plate-tectonic theory, which is described in Chapter 6.

The subsequent chapters concentrate on processes that occur at plate margins, starting with earthquakes and deformation in Chapter 7. Chapters 8 through 10 explore the lithospheric plates, their relative movements, and how interaction of these plates gives rise to earthquakes, volcanoes, and mountain ranges. Many of Earth's features are most conveniently and logically discussed in the context of the three types of plate margins—divergent margins, convergent margins and transform margins. We point out the context in which igneous, metamorphic, and sedimentary rocks form and how they differ from margin to margin.

Chapter 11 presents plate interiors, discussing the plate-tectonic role in continental accretion. The following chapters deal with surficial processes, examining the processes of weathering and erosion that occur at Earth's surface, and the transport and deposition of the sedimentary material formed in this way (Chapters 12 and 13). A natural consequence of these processes is the modification of landscapes, and we next explore the characteristics of different landforms in the geomorphology chapter (Chapter 14). Students will discover that different combinations of climate, geology, and environment will produce distinct landform characteristics. Finally, we address some of the more practical aspects of the earth sciences. From an understanding of how Earth works, we take a look at the environment from a geologic perspective in Chapter 15 and at how natural resources are formed in Chapter 16.

Pedagogical Features

Table of Contents This text integrates topics in context rather than isolating a topic in a given chapter; therefore some topics are found in more than one location. For example, volcanoes are introduced in Chapter 4 during our discussion of rocks. Chapter 8 discusses the vast regions of volcanic activity along the mid-ocean ridges, which, until 30 years ago, were largely unknown and unexplored. Chapter 9 presents volcanoes in the context of convergent margins, while Chapter 11 explores intraplate magmatism and hot spots. Finally, Chapter 15 examines volcanic hazards and the climatic effects of volcanoes.

Speedbumps Brief interim summaries are incorporated throughout each chapter, especially after discussion of an important concept, allowing students to pause, review, remember, and then continue on to the next concept.

Cross references Cross references are included to help students connect concepts. They are indicated by the icon ∞, and indicate where the referenced material was introduced or previously mentioned.

Key Terms and Glossary Important vocabulary terms are boldfaced in the text and defined in the glossary at the back of the text.

Focus Boxes Each chapter contains several boxed topics; this boxed text either applies a concept that is being discussed in a given chapter or takes a closer, more in-depth look at a topic being presented.

Living With Geology Several chapters include a final section called "Living With Geology." These sections focus on helping students understand their surrounding environment and how it can affect our lives. Some of the topics include "Can Earthquakes Be Controlled?" "Acid Rain," and "Damming a Wild River."

Art Program The art program was carefully designed both to facilitate student understanding of difficult concepts, and to present geologic processes and place these processes in context. Believing strongly in the adage that a picture is worth a thousand words, we have included a rich variety of photographs and illustrations. Examination of these illustrations will help students realize the naturalist, or observational nature, of geology as a science.

Appendices Appendices include a conversion table for metric units, a geologic time scale, a periodic table of the elements, a table of common minerals and their properties, and a list of Earth statistics.

Supplements

The supplement package includes the traditional supplements that students and professors have come to expect from authors and publishers, as well as some new kinds of supplements that involve electronic media.

Study Guide: Written by William McLoda at Mountain View College, the study guide includes a chapter overview, chapter objectives, key terms, and an extensive selection of study questions for each chapter.

Instructor's Resource Manual: Also written by William McLoda, the instructor's manual contains a variety of lecture outlines, teaching tips, and advice on how to integrate visual supplements. The instructor's manual is meant to be both a guide for less experienced instructors and a resource and source of new ideas for experienced instructors.

Test Item File: Written by Shannon O'Dunn, this file includes over 1000 questions keyed directly to the text. ISBN: 0-13-464082-9.

Prentice Hall Custom Test: Based on the powerful testing technology developed by Engineering Software Associates, Inc., this supplement allows instructors to tailor exams to their own needs. With the On-line Testing option, exams can be administered on-line so data can be automatically transferred for evaluation. A comprehensive desk reference guide is included, along with on-line assistance. The Custom Test is available for both MAC and Windows platforms.

Transparencies: Over 100 full-color acetates of illustrations from the text.

The New York Times **Themes of the Times—The Changing Earth:** This unique newspaper-format supplement features recent articles on dynamic geologic applications from the pages of **The New York Times.** This free supplement is available in quantity, and encourages students to make connections between the classroom and the world around them.

Life on the Internet: Geosciences: Written by Andrew Stull of California State University, Fullerton and Duane Griffin of the University of Wisconsin, Madison, this guide helps earth science students use the internet to enrich their studies. It includes general information on the internet and WWW basics, as well as specific material about how to use the internet for the geosciences. *Life on the Internet* is free to qualified adopters of *Exploring Earth*.

World Wide Web Resources: Designed to give the students an interactive way to enhance their studies, the Exploring Earth WWW site reaches out into cyberspace and brings the myriad of geologic resources available there closer to you and your students. Check out our interactive study resources, interesting links, and items for further thought at *http://www.prenhall.com/davidson*.

Prentice Hall Geodisc: This laser disk allows instructors to bring a wealth of images in to the classroom. It contains over 50 minutes of motion video, and over 900 still images and animations. The disk is accompanied by a bar code manual to access images easily. ISBN: 0-13-304168-3.

Acknowledgments

We would like to thank the reviewers of this text, who provided extremely valuable input and many helpful suggestions:

E. Calvin Alexander, Jr. *(University of Minnesota)*, Don L. Anderson *(California Institute of Technology)*, Duwayne M. Anderson *(Texas A&M University)*, Thomas W. Broadhead *(University of Tennessee)*, George H. Davis *(University of Arizona)*, Wakefield Dort *(University of Kansas)*, Karen Goodman *(Broome Community College)*, Barbara E. Grandstaff *(New Jersey State Museum)*, Bryce M. Hand *(Syracuse University)*, Vicki Hansen *(Southern Methodist University)*, W. Burleigh Harris *(University of North Carolina—Wilmington)*, Richard B. Hathaway *(State University of New York—College at Geneseo)*, John R. Huntsman *(University of North Carolina at Wilmington)*, Karl E. Karlstrom *(University of New Mexico)*, David King, Jr. *(Auburn University)*, Ronald Krauth *(Middlesex County College)*, Lawrence Krissek *(The Ohio State University)*, Albert M. Kudo *(University of New Mexico)*, Ralph L. Langenheim, Jr. *(University of Illinois at Urbana-Champaign)*, Douglas Levin *(Bryant College)*, Lawrence Lundgren *(University of Rochester)*, Greg Mack *(New Mexico State University)*, John A. Madsen *(University of Delaware)*, Bart S. Martin *(Ohio Wesleyan)*, Richard L. Mauger, William S. McLoda *(Mountain View College)*, James M. McWhorter *(Miami-Dade Community College)*, Donald S. Miller *(Rensselaer Polytechnic Institute)*, Kula C. Misra *(University of Tennessee)*, Paul Morgan *(Northern Arizona University)*, David B. Nash *(University of Cincinnati)*, David A. Nellis *(University of Massachusetts at Boston)*, Hallan C. Noltimier *(Ohio State University)*, Nilgun Okay *(Hunter College of the City University of New York)*, Bruce C. Panuska *(Mississippi State University)*, Richard Pardi *(William Paterson College)*, Robert W. Pinker *(Johnson County Community College)*, Nicholas Rast *(University of Kentucky)*, Gregory J. Retallack *(University of Oregon)*, Justin Revenaugh *(University of California - Santa Cruz)*, Richard Robinson *(Santa Monica College)*, Barbara L. Ruff *(University of Georgia)*, Douglas L. Smith *(University of Florida)*, William A. Smith *(Western Michigan University)*, Bryan Tapp *(University of Tulsa)*, Paul Tayler *(Utah Valley State College)*, Jack W. Travis *(University of Wisconsin—Whitewater)*, Kenneth L. White *(Texas A&M University)*, Monte D. Wilson *(Boise State University)*, William H. Wright *(Sonomo State University)*, Michael Wysession *(Washington University)*.

We also extend a special thanks to the author of the Instructor's Resource Manual, William S. McLoda.

We thank many folks at Prentice Hall for their efforts in bringing this book into existence. Lee Englander roped us into writing it in the first place, and badgered us relentlessly. We thank her for overestimating our abilities and never giving up on us. We also thank Holly Hodder, David Brake, Karen Karlin, Ray Henderson, Deirdre Cavanaugh, and Dan Kaveney, all of whom deserve credit for their patience and enthusiasm for our project. Paul Corey has been truly motivating in guiding our book home. Ed Thomas and Debra Wechsler were instrumental in assembling the product at short notice, while Yvonne Gerin worked incredibly hard to provide the rich and diverse collection of photographs. Finally, Carol Trueheart, our development editor, was the real heroine in producing this book. Her tireless efforts, her endurance of the frustrations we presented on a daily basis, and her cheerful disposition are appreciated enormously. She deserves six month's vacation in the Caribbean.

We wish to thank our colleagues at UCLA for supporting the project and providing keen insights and advice. A particular mention goes to Art Montana, who was chair of the Department when the project was initiated and gave it his strongest blessing and support. We thank Frank and Joy Stacey for editing chapters dealing with geophysical topics, Bob Tilling and Wendell Duffield of the United States Geological Survey for freely providing information and illustrations for plate tectonics and volcanism, and Paul Davis thanks the Guggenheim Foundation for partial support.

Finally, we would like to thank our wives for putting up with us throughout this project and our students from whom we are still constantly learning about this planet.

Jon Davidson
Walter E. Reed
Paul M. Davis

About the Authors

Jon Davidson

Jon Davidson received his undergraduate degree in Geology from the University of Durham and a Ph.D. in Geology from the University of Leeds. He has held a Visiting Assistant Professor position at both the Southern Methodist University in Texas and at the University of Michigan. He was appointed Assistant Professor of Geology and Geochemistry at the University of California, Los Angeles in 1988, where he is currently an Associate Professor. He has taught courses in Earth Science, Historical Geology, Igneous Petrology, Isotope Geochemistry, Volcanology, and the Regional Geology of Britain. Professor Davidson has led field trips to the Cascades, the Mojave Desert, Eastern California, Hawaii, and Britain. He received the UCLA Harriet and Charles Luckman Outstanding Teaching award in 1994, and is currently an associate editor of the *Journal of Petrology*.

Davidson is an igneous petrologist and geochemist, with a keen interest in volcanology. His work focuses primarily on volcanoes in the Caribbean, the Andes, Ascension Island in the south Atlantic, and Kamchatka in Russia. In his spare time, he enjoys travel, photography, cricket, football (both types), and music.

Walter E. Reed

Walter Reed received his Ph.D. in 1972 from the University of California, Berkeley and joined the faculty of the Department of Earth and Space Sciences at the University of California, Los Angeles, in 1973. Prior to his arrival at UCLA, he worked in an oil company research laboratory for four years and worked for two years for the Department of Defense on the Nuclear Test Site and at the National Reactor Test Range. He has won two "best paper" awards, one in organic geochemistry and one (with his graduate student) in structural geology. Professor Reed has taught continuously since arriving at UCLA, and his courses include Introductory Geology, Sedimentology and Sedimentary Petrology, and Field Geology, spending six to eight weeks each summer with students in the Sierra Nevadas with the latter course.

Professor Reed is a field geologist with experience throughout the western United States, the Aleutian Islands, Spitsbergen, Norway, and Israel. His recent work focuses on California's western Transverse Ranges, and on a tectonically emplaced metamorphic-plutonic complex in the Sierra Nevada Mountains. Professor Reed's hobbies include trout fishing, skiing, ice climbing, and building and riding Harley Davidson motorcycles.

Paul M. Davis

Paul Davis is a Professor of Geophysics at the University of California, Los Angeles. He received his Ph.D. in Physics at the University of Queensland, followed by postdoctoral studies at both the Institute of Geophysics and Planetary Physics, University of Alberta and the Department of Geodesy and Geophysics at the University of Cambridge. He joined the faculty at UCLA in 1980 and has recently served as the Vice-Chair and Chair of the Department of Earth and Space Sciences. He teaches Seismology and Applied Geophysics.

Professor Davis received a Guggenheim Fellowship in 1995 to conduct research in the Department of Earth Sciences at the University of Oxford. Upon his return to UCLA, he assumed the position of senior editor of the American Geophysical Union *Journal of Geophysical Research (Solid Earth)*. His research uses geophysical experiments to study lithospheric dynamics. He has installed magnetometer arrays on volcanoes on Kilauea, Hawaii and Washington's Mount St. Helens and has carried out seismic array studies of the Mount Etna volcano and the Rio Grande, East African, and Baikal rifts. Professor Davis' interests include racquetball, sailing, hiking, and backpacking with his family.

1 The Changing Planet

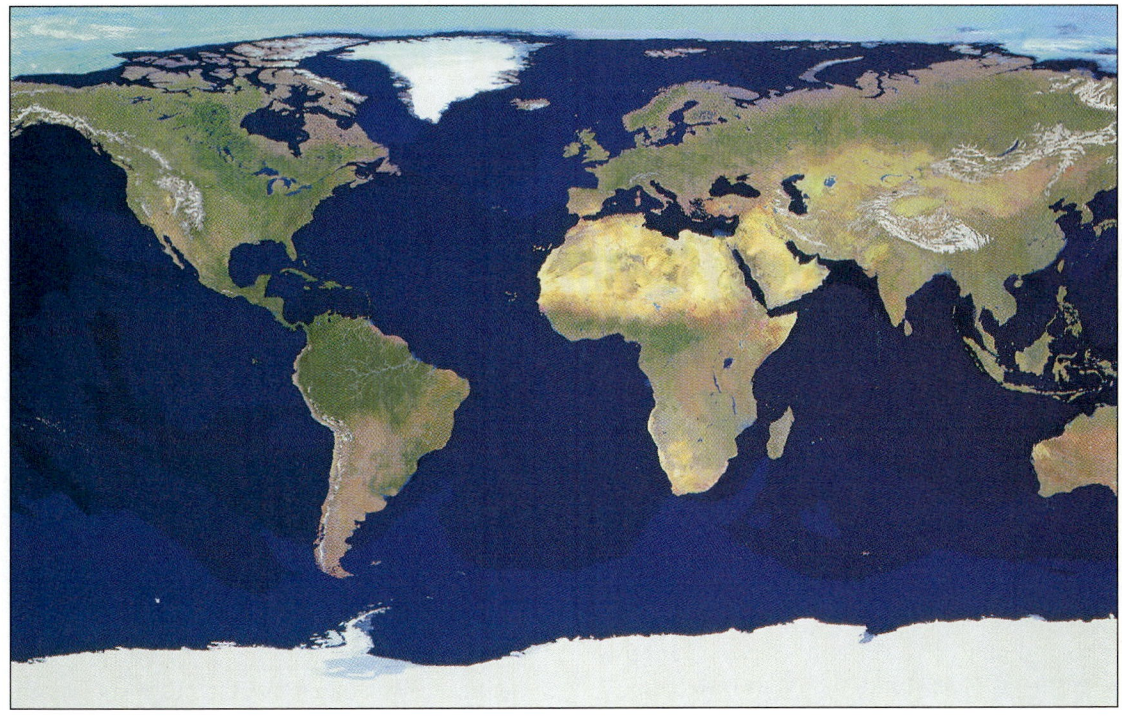

A composite cloud-free image of Earth created from thousands of individual satellite images.

Overview

- Earth is a dynamic planet, with diverse climates and landscapes.
- The planet is powered by two energy sources: solar energy and internal heat.
- Plate tectonics is the system of interaction between fragments (plates) of Earth's outer shell, which is driven by internal heat.
- Plate tectonics gives rise to volcanoes, earthquakes, and deformation of Earth's crust.
- Solar energy drives the planet's climate system, which transfers water between the ocean and land in what is known as the hydrologic cycle.
- The hydrologic cycle modifies Earth's surface, wearing down high land and depositing the residue of this activity on lower land or in oceans.

Introduction

Earth's surface is a very diverse place. There are many different landscapes—from islands to continents, from flat plains to high, jagged mountains (Fig. 1.1). Earthquakes and volcanic eruptions are common in specific regions of the planet, whereas other areas are geologically tranquil and apparently inactive. Varied climates give rise to deserts, arctic tundras, and tropical rain forests. Two thirds of the globe is covered by oceans that conceal mountain ranges, plains, and volcanoes.

The natural world around us can be explained in terms of the interactions between two main driving forces that shape our planet: the Sun and the internal energy of the planet itself. Solar heat is the major external source of energy for Earth and the driving force behind Earth's climate. Solar energy is responsible for clouds, rain, wind, rivers, and even glaciers. These agents help shape Earth's surface into the landscapes that we see today.

The second source of Earth's energy is the planet's internal energy. The hot interior of the planet is the result of heat left over from the huge amount of energy released during the formation of Earth over 4 billion years ago, coupled with the added energy of natural radioactivity. The existence of such heat is evident when molten lava pours out of a volcano (Fig. 1.2); in fact, this internal heat has been escaping from the planet ever since Earth's formation. But the internal heat does more than simply drive volcanic eruptions. It causes ocean basins to form and mountains to rise. It unleashes enormous amounts of energy in the form of earthquakes and volcanoes. This internal energy drives the system of **plate tectonics**.

FIGURE 1.1 A view of North America from space, illustrating a diversity of landscapes and climate.

FIGURE 1.2 Lava erupting from a volcano in Hawaii. The temperature of the lava (about 1100°C) is just one line of evidence that Earth's interior is hot.

Plate tectonics has become a unifying theory in geology. It explains spectacular natural phenomena, and it also allows us to explain the origins of oceans and continents, inland seas, and mountain belts. Plate tectonic processes control both the distribution of landmasses relative to oceans, and the locations of landmasses relative to climate belts. These processes also exert a significant influence on Earth's climate. Similar processes are thought to have acted on Earth for at least 2.5 billion years (more than half the life of the planet); thus, we have a comprehensive theoretical framework in which to interpret the record of evolving continents and oceans preserved in the rocks of Earth's crust. Evidence shows that entire ocean basins have opened and closed, mountain ranges have been uplifted and worn down to flat plains, and the remains of ancient jungles are now preserved beneath the Antarctic ice sheets. The effects of past glaciations are now found in the deserts of Africa. All these observations can be explained in the context of plate tectonics. Therefore, we must first gain a basic understanding of this important fundamental theory.

FIGURE 1.3 A view of Earth from space. Earth's lithosphere forms the solid surface of the planet, which is visible here in the form of the continental landmasses. The hydrosphere is represented by the oceans, which cover two-thirds of the planet's surface. The clouds are part of the planet's outer envelope of gas—the atmosphere.

FIGURE 1.4 The Moon, Earth's satellite. The Moon does not have a hydrosphere or atmosphere; it has an ancient rocky surface bearing the scars of meteorite bombardments, most of which occurred more than 3 billion years ago.

An Introduction to Plate Tectonics

Earth is the only planet in the Solar System that is known to possess active plate tectonics. This, combined with the presence of abundant liquid water, makes Earth a dynamic planet (Fig. 1.3) and, as far as we know, the only planet in our Solar System that supports life. Earth's uniqueness can be appreciated by a simple comparison with our nearest planetary body: the Moon (Fig. 1.4).

The Moon has a lifeless surface, pockmarked with craters most of which are the result of meteorite impacts over 3 billion years ago. Earth's surface, although probably affected by similar meteorite bombardment, is constantly changing—with the result that few ancient impact features remain. Plate tectonics constantly reshapes the continents and oceans of Earth, while the land surfaces are constantly subjected to the action of wind, ice, and running water. In this section, we will take a brief look at the basic principles of the theory of plate tectonics, which form an organizing framework for the chapters that follow.

Earth is divided into compositional layers called the **crust**, **mantle**, and **core** (Fig. 1.5). The crust and mantle are made of rocky material, and the core is largely composed of iron. The Earth is also divided into layers based on strength; these layers do not correspond to the compositional layers but rather to how they behave. Earth's outer shell is a strong layer, about 100 to 200 km thick, called the **lithosphere**. The lithosphere overlies a weaker layer—the **asthenosphere** (Fig. 1.5). The lithospheric shell is not a continuous layer; rather, it is formed from a number of fragments, or **plates**. There are about eight large plates and many smaller ones distributed over Earth's surface (Fig. 1.6). These plates have changed their positions, shapes, and sizes through geologic time. Rates of plate movement are as much as 10 cm per year, almost imperceptible on the scale of a human lifetime. However, the geologic record underscores the significance of such plate movements. The geology of Africa and South America, for instance, suggests that they were joined together 150 million years ago and that the Atlantic Ocean has opened between them since that time. Recent, very precise satellite measurements of plate motions have confirmed that the plates are indeed moving relative to one other.

In Figure 1.6, you will notice that most of the volcanic activity, earthquakes, and mountain belts on Earth's surface are located at the boundaries of the different plates. These observations suggest that the processes taking place at these plate margins—where lithospheric plates are pulling apart, pushing together, or sliding past each other—may be responsible for volcanoes, earthquakes, and deformation of the crust. There are three different types of plate margins: **divergent margins**, **convergent margins**, and **transform margins** (Fig. 1.7).

Divergent Plate Margins As the name implies, lithospheric plates move away from each other at a divergent plate margin (Fig. 1.7a). Hot rock, which is more buoy-

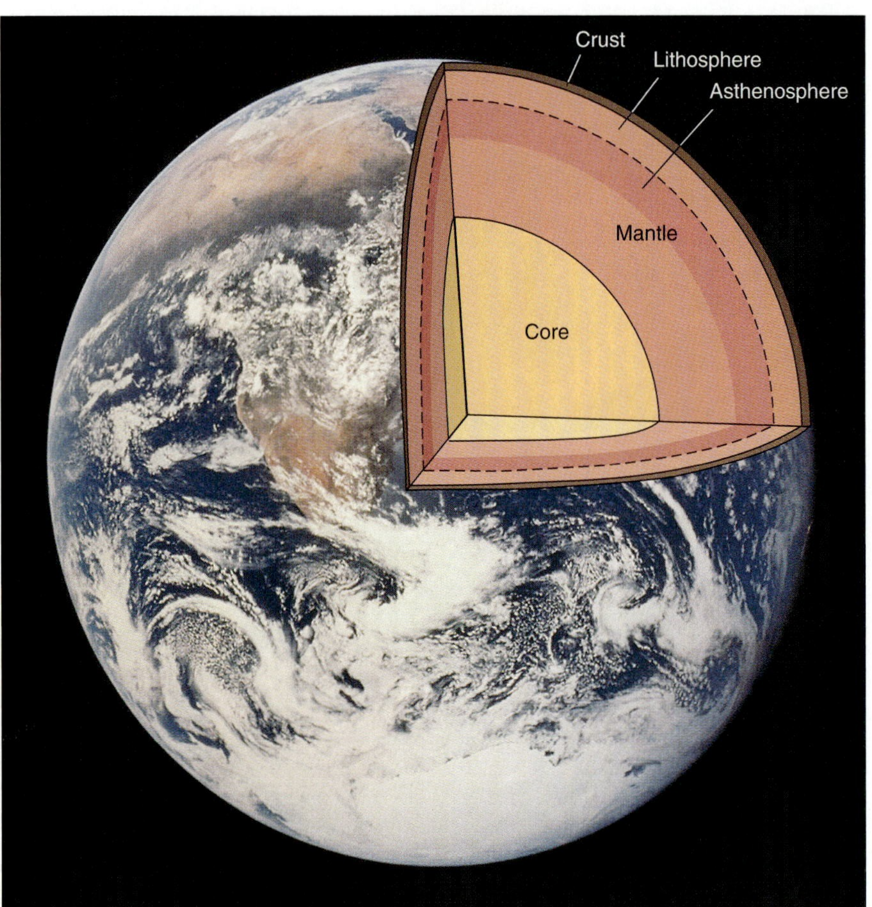

FIGURE 1.5 A cutaway view of Earth showing the planet's internal layering (not to scale).

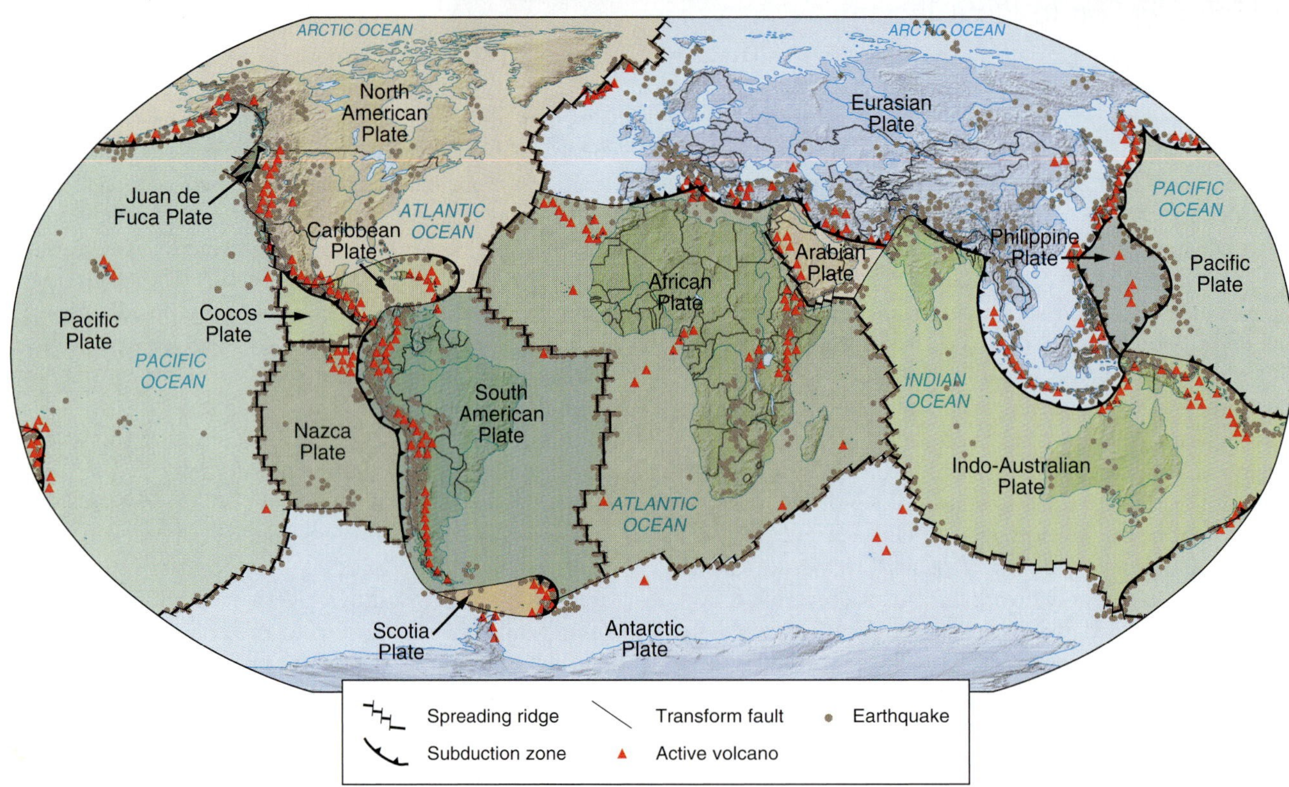

FIGURE 1.6 Map of the world indicating the distribution of plate boundaries. Note the correspondence of the plate margins to the locations of volcanoes and earthquakes.

ant than the surrounding material, wells up from the mantle below and melts as it moves to the surface, filling the gap created as the plates move apart. New lithosphere is generated in this way, and at Earth's surface, this process is observed as volcanic activity along a plate boundary, or **spreading ridge** (Fig. 1.8). The gradual separation of the plates and consequent volcanic activity causes many small earthquakes (see Fig. 1.6). A spreading ridge runs down the middle of the Atlantic Ocean between the North American and the Eurasian plates. Satellite measurements indicate that the cities of New York and London—located on the North American and Eurasian plates, respectively—are moving away from each other by about 2 cm annually.

Convergent Plate Margins If new lithosphere is created at spreading ridges as plates pull apart from each other, then the total area of lithosphere on Earth's surface should be increasing. This, in turn, would suggest that the planet is expanding. Because there is no evidence that the planet is growing (satellite measurements prove that Earth is not expanding today, and the cooling of Earth over the past 4.55 billion years is not consistent with expansion during the past), there must be lithosphere destroyed to balance the new lithosphere that is generated. This destruction of lithosphere occurs at convergent plate margins, or **subduction zones**. Two plates move toward one another at a convergent margin. One plate sinks beneath the other as the two converge, subducting into the mantle (Fig. 1.7b). This process causes earthquakes and gives rise to volcanoes on the overriding plate (Fig. 1.9). Subduction zones surround most of the Pacific Ocean. Convergence at these plate boundaries means that the Pacific Plate (Fig. 1.6) is being consumed. The island of Hawaii, located on the Pacific Plate, is moving toward Japan at a rate of about 7 cm per year because there is a subduction zone between them.

Transform Plate Margins At the third type of plate boundary—a transform boundary—lithosphere is neither created nor destroyed. The two plates simply slide laterally past each other (Fig. 1.7c). Perhaps the best-known example of a transform plate boundary is California's San Andreas fault, which lies between the North American and Pacific plates (Fig. 1.10). Along this boundary, the city of Los Angeles on the Pacific Plate and the city of Oakland on the North American Plate are moving toward each other at about 5 cm annually. As the plates slide past one another, their margins rub together and trigger many of the earthquakes that are all too familiar to Californians (see Fig. 1.6). There is, however, very little volcanic activity associated with transform boundaries.

The plate boundaries shown in Figure 1.6 are not fixed, but have varied through time. It is changes in the geometry and distribution of lithospheric plates that have given rise to the rich variety of landscapes and rock types that now characterize the surface of our planet.

Climate and the Hydrologic Cycle

Conditions on Earth are unique in the Solar System in allowing liquid water to exist on Earth's surface. Earth possesses a **hydrosphere**, which is an envelope of water

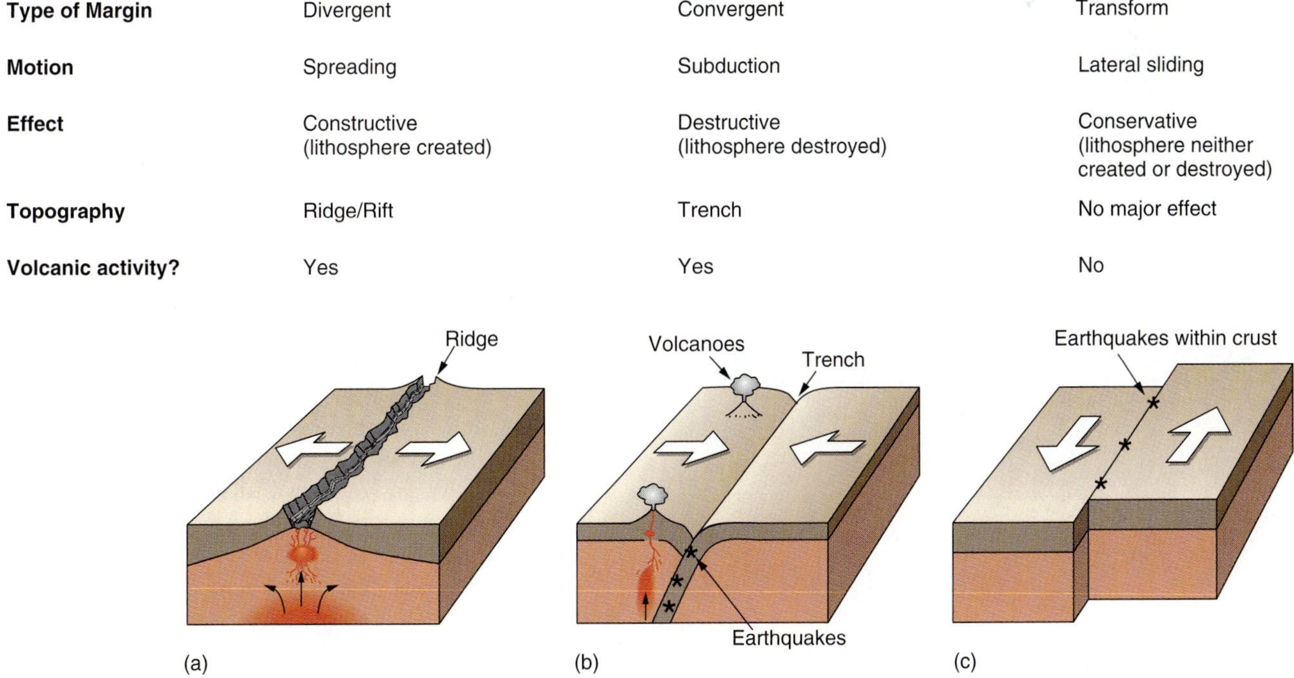

Type of Margin	Divergent	Convergent	Transform
Motion	Spreading	Subduction	Lateral sliding
Effect	Constructive (lithosphere created)	Destructive (lithosphere destroyed)	Conservative (lithosphere neither created or destroyed)
Topography	Ridge/Rift	Trench	No major effect
Volcanic activity?	Yes	Yes	No

FIGURE 1.7 The three different types of plate margins: (a) divergent, (b) convergent, and (c) transform.

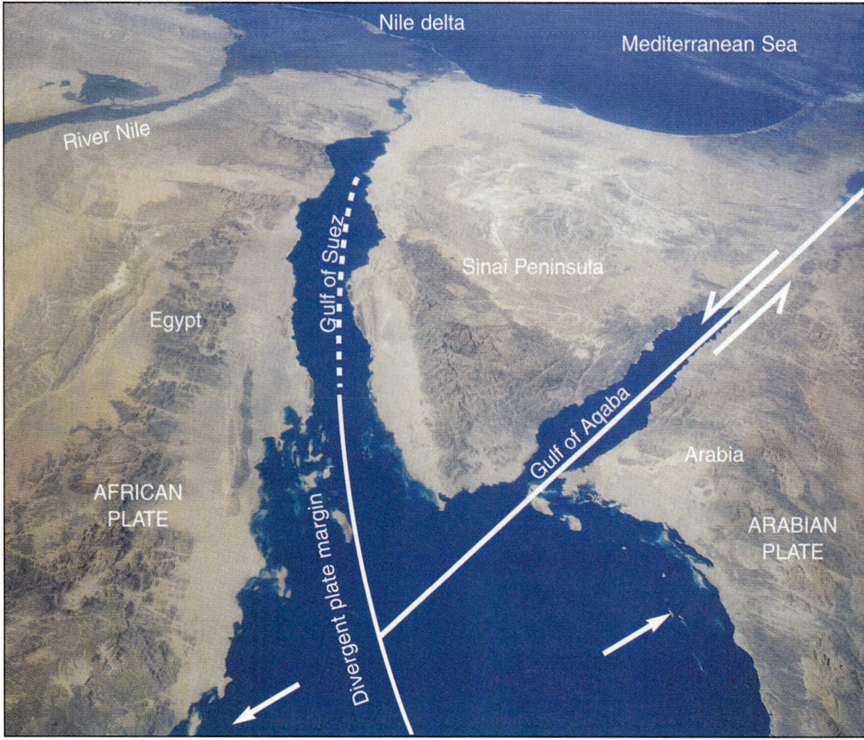

FIGURE 1.8 A satellite view of an oceanic spreading center: the East Africa/Arabian peninsula region. Here, the Arabian Plate is moving away from the African Plate, and new lithosphere is being formed at a spreading center beneath the Red Sea. The spreading center has moved up the Gulf of Suez and the Gulf of Aqaba, rather like a zipper that acts to separate Arabia from Africa.

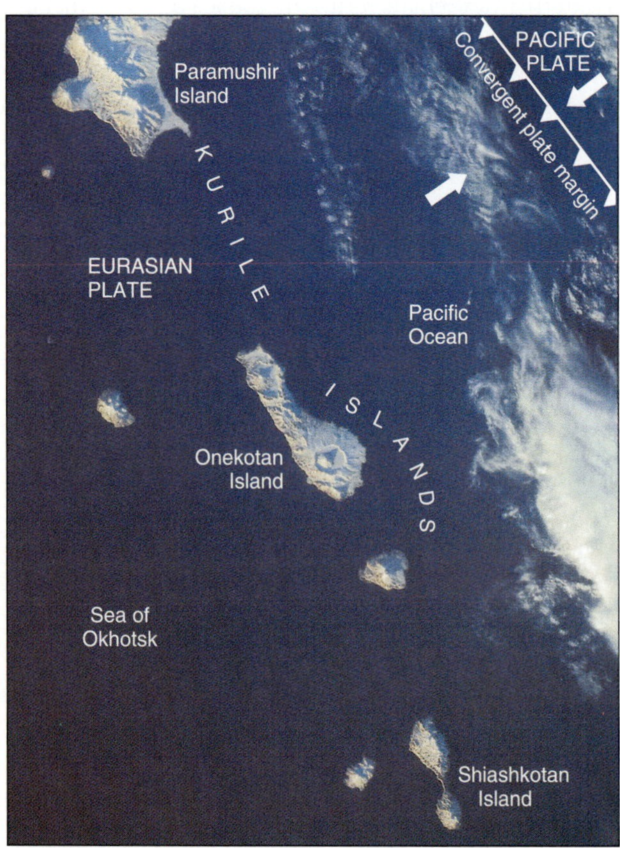

FIGURE 1.9 Satellite view of a convergent margin: Kurile Islands north of Japan. The most obvious expression of the subduction zone is the chain of volcanoes.

contained in the oceans as well as rivers, lakes, groundwater, and glaciers (Fig. 1.11). It is surrounded by an atmosphere of gas that includes water vapor. Condensed water vapor in the atmosphere forms clouds. Solar energy heats the atmosphere and the hydrosphere, driving what is known as the **hydrologic cycle** (Fig. 1.12). This cycle refers to the flux of water at Earth's surface; water is continually transferred among the oceans, clouds, rain, groundwater, lakes, rivers, and glaciers.

Briefly, the Sun heats the surface of the oceans and causes water to evaporate. Water vapor moves into the atmosphere and is transported by winds. Atmospheric circulation—wind—is itself driven by solar energy. Variations in temperature due to solar heating cause gas in the atmosphere to circulate—hot air moves up and cooler air sinks. Water vapor may eventually precipitate as rain or snow, depending on the temperatures. If the precipitation falls on land, it feeds rivers or glaciers and eventually makes its way back to the oceans, flowing downhill under the influence of gravity. A fraction of this water may soak into the ground, adding to the underground reservoirs of water called **groundwater**. Groundwater also moves downhill through pore spaces in the soil and rock, albeit slowly.

The continual cycling of water from the oceans to the continents and back again modifies Earth's surface (Fig. 1.13). Wind, rain, rivers, and other agents of weather act on the surface to wear it down. Two processes occur: **weathering** (the disintegration and decomposition of solid rock into small fragments) and **erosion** (the

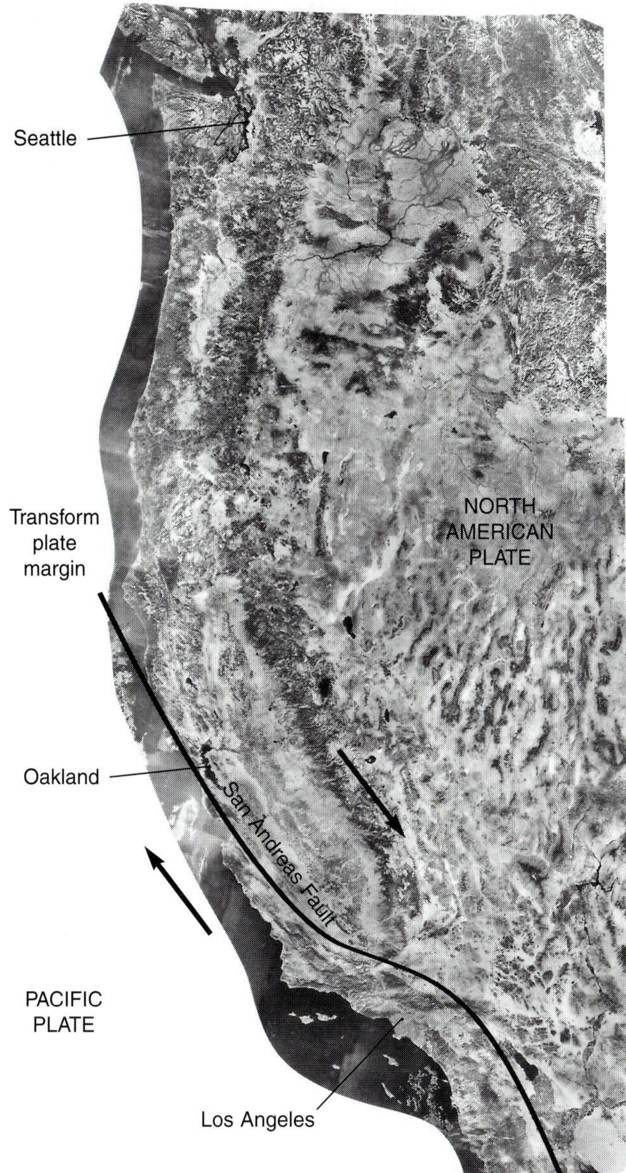

FIGURE 1.10 Satellite view of a transform margin: the San Andreas fault, California. The Pacific Plate (left) is moving north relative to the North American Plate. The plate boundary is clearly visible here, although many boundaries (such as divergent and convergent boundaries) are actually located beneath the oceans.

removal of weathered material). If there were no plate tectonics, the surface of the planet would be uniformly flat, with the mountains worn down nearly to sea level. The interaction between externally driven climate and internally driven plate tectonics is therefore a dynamic battle—a continuous cycle of uplift, upheaval, and deformation tempered by weathering and erosion.

The map of the globe is not fixed in time. Continents may shift their positions, break apart, or join together in huge collisions. The fundamental climatic zonation of the planet, however, is relatively fixed through time. This is controlled largely by the orientation of Earth relative to the Sun, resulting in cold climates at the poles and hot climates near the equator.

We can therefore appreciate the interplay between climate and plate tectonics. As continents move across the surface of the globe, they may pass through different climate zones, which will leave a distinct signature in the rock record. For instance, the now-arid regions of South Africa and southern Australia were located close to the South Pole some 400 million years ago. The rocks of that age in South Africa and southern Australia bear the scars of ancient glaciations, which tell us that those regions have moved from a polar location to the more tropical locations that they now occupy.

Furthermore, the topography of the continents—which itself is a function of plate tectonics—will influence the climate, as demonstrated today in Central Asia. Mongolia and Tibet are dry and desertlike regions of the Asian interior. To the south, the Indian subcontinent is visited annually by monsoon rains, but this moisture never reaches Tibet because the path of the monsoon weather system is obstructed by the Himalayan Mountain range lying between India and Central Asia (Fig. 1.14). The Himalayas, in turn, were formed by the collision between India and Asia and did not even exist 50 million years ago.

Throughout this book, we will emphasize the dynamic roles of plate tectonics and weathering in shaping our planet (Figs. 1.14, 1.15). You will understand what gives rise to the rich tapestry of landscapes that make up the surface of our planet. We will explore the causes of earthquakes and volcanoes and the reasons why these phenomena only occur in certain places. We will discover where our natural resources (such as coal, oil, and metals) come from, and how they got there. We will try to understand the environment in which we live, enabling us to evaluate how best to interact with it—what natural aspects should be protected and how. Consider the questions that are posed in Figure 1.15 (page 10).

Before analyzing these questions in detail, we must first discuss the perspective from which we will examine Earth—the sizes and distances that we encounter in the natural world, and the enormity of time over which the planet has been changing.

Understanding Earth: The Question of Scale

With our everyday experiences, we are familiar with certain scales of measurement and we can relate easily to them. If you want to drive a few kilometers across town to visit a friend, you can estimate how long the trip might take and you would have a good understanding of the distance involved. If you want to drive from Los Angeles to New York City, you might be less confident

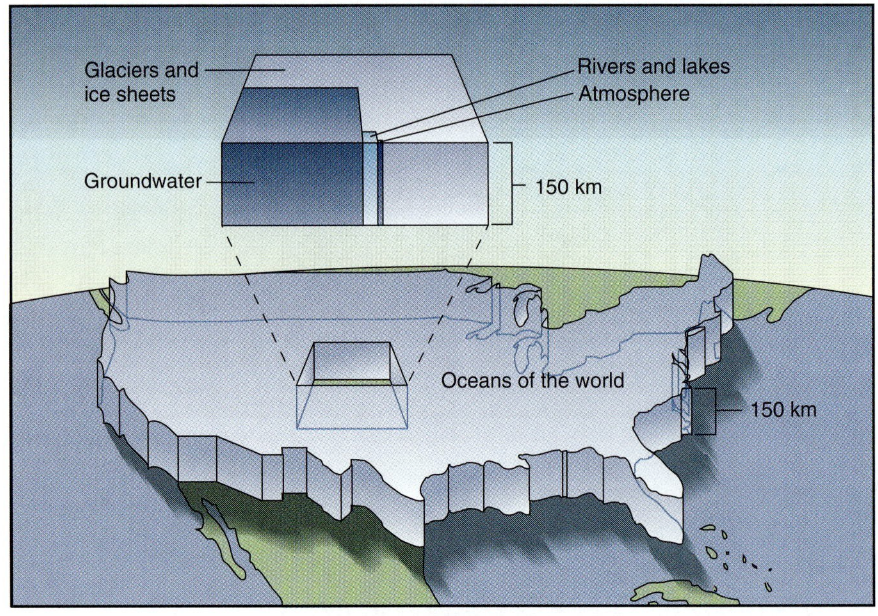

FIGURE 1.11 Relative volumes of the different reservoirs of water in the hydrosphere. The entire volume of the hydrosphere is represented by a volume with the area of the United States, to 150 km depth. The greatest volume is contained in the oceans, while freshwater reservoirs are represented by slices with relative volumes as indicated.

in your estimate of travel time. Although it is outside the realm of "everyday" experience for most of us, we have some idea of the meaning of a few thousand kilometers in distance. If you want to travel to the Moon, however, it is nearly 400,000 km away, and a trip to the nearest star outside of our Solar System would involve a distance of some 40 trillion (40×10^{12}) km (Focus 1.1). These distances are difficult to appreciate, and analogies are frequently used to translate imponderable numbers into everyday terms.

The vital statistics of the Solar System may seem difficult to comprehend: The Sun has a diameter of

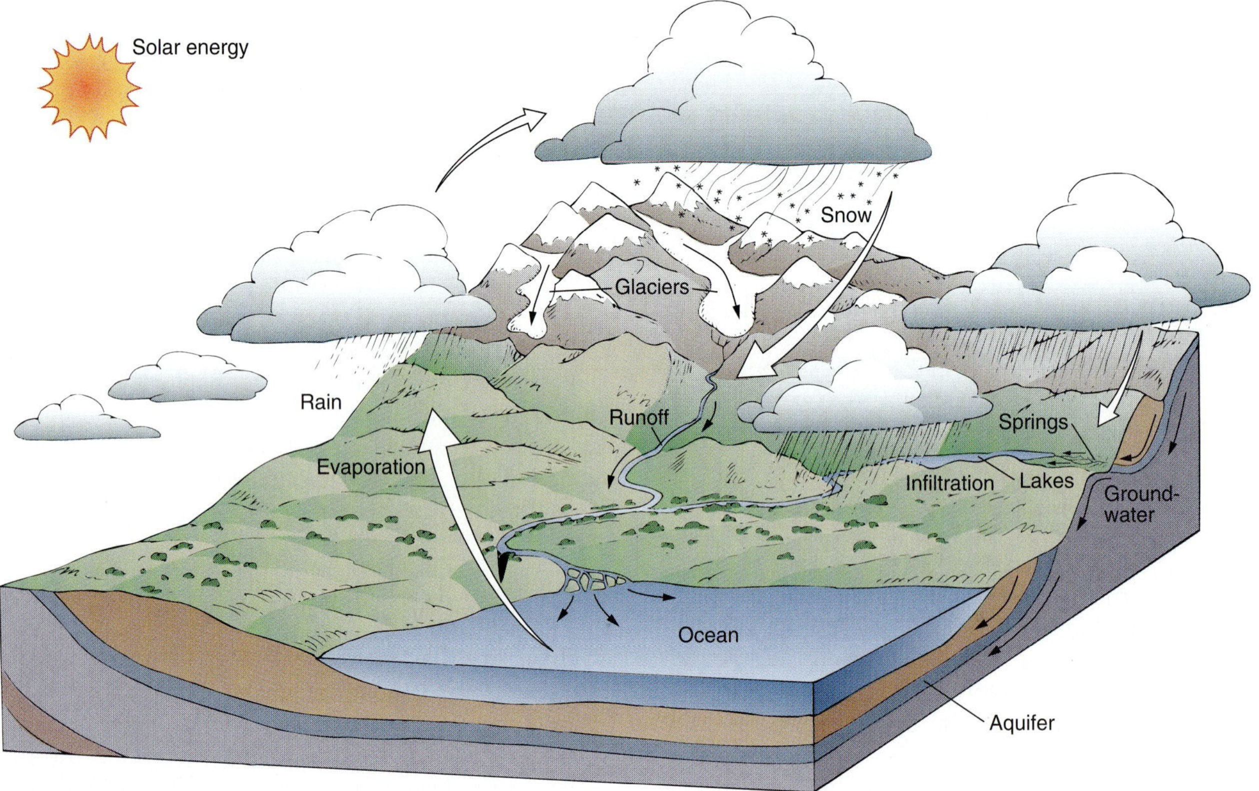

FIGURE 1.12 The hydrologic cycle. This cycle represents the constant flux of water between the main reservoirs at Earth's surface (oceans, lakes, rivers, glaciers, and groundwater).

FIGURE 1.13 A satellite view of the Grand Canyon in Arizona. This image emphasizes the effect that the hydrologic cycle has on modifying Earth's surface. The Colorado River has cut a deep gorge (about 2000 m) into the Colorado Plateau. Sediment produced by cutting down into the plateau is carried by the river out to sea in the Gulf of California.

1,392,000 km, and it is 5,908,000,000 km from the center of the Sun to the outermost planet, Pluto. Translated into the more familiar dimensions of the football field (Fig. 1.16), we can appreciate such enormous distances: Let the 100-yard length of the football field represent the distance from the middle of the Sun to Pluto. If you were to stand on the goal line representing the center of the Sun, Earth would be 2.5 yards downfield. Jupiter, the largest planet in our Solar System, would be 13 yards from the goal line. The nearest star, Proxima Centauri, would still be 385 miles away!

The smallest scale of the natural world—that of atoms and subatomic particles—is equally imponderable. The diameter of an atom is on the order of 0.0000000001 m (10^{-10} m). Most of the atom consists of empty space; nearly all of the atom's mass is concentrated in the central nucleus, which is only 10^{-14} m in diam-

FIGURE 1.14 An example of how plate tectonics affects climate. This is a Space Shuttle photograph of the Himalayan Mountain range, looking west. The high mountaintops are snow-capped. Note the strong contrast between the arid dry highlands of the Tibetan Plateau to the north (right) and the vegetated lowlands of northern India to the south (left). The Himalayas were gradually uplifted in response to the enormous tectonic forces that resulted from the Indian continent colliding with the Asian continent. The uplift of this huge mountain range has changed the entire weather system of the planet; at over 4 km in height, the mountains act as a barrier to winds circulating in the atmosphere, preventing rain-carrying clouds from moving over the Tibetan Plateau.

QUESTION:

Why are volcanoes located in the northwestern United States and none are found on the eastern coast or in the midwest?

(a) Mount St. Helens volcano, erupting in 1980.

EXPLANATION:

Washington and Oregon are located at a convergent plate margin (Fig. 1.6). It is the action of a plate moving eastward *beneath* Washington and Oregon that gives rise to these volcanoes (see Chapter 9).

QUESTION:

How is oil formed and why is it only found in certain places?

(b) Oil field in North Dakota.

EXPLANATION:

Oil is formed from the natural breakdown of organic material as it is buried. The material must be buried and subjected to certain pressures and temperatures. Oil is found in geologic features called "traps" (see Chapter 16).

QUESTION:

What caused this earthquake, and why does California have so many earthquakes?

(c) In 1994, the Northridge, California earthquake caused extensive damage.

EXPLANATION:

Figure 1-6 shows that a transform plate boundary runs along the western edge of California. The two plates slide horizontally past each other, rubbing together. The movement is jerky rather than smooth. Each jerk—as the plates move—causes an earthquake (see Chapters 7, 10 and 15).

QUESTION:

Why do mountain belts form?

(d) The Alps, southern Europe.

EXPLANATION:

Most mountain belts are formed as a result of plate convergence, which may result in the collision of two landmasses. In this case, Africa collided with Europe, thrusting up the Alps. Such plate-tectonic procresses will clearly affect the local topography but, as we saw in Figure 1.14, they can also affect climate (see Chapters 6, 9, 11).

FIGURE 1.15 Some examples of the interplay between plate tectonics and processes at Earth's surface.

eter. To help make sense of these small distances, we can imagine an atom's nucleus to be represented by a baseball suspended in the middle of a baseball stadium. In this analogy, the atom's electrons are mainly located out in the grandstands.

Why must we consider such large or small dimensions when studying geology? Shouldn't we just concern ourselves with materials that are as small as sand grains or as large as mountains? Actually, a true appreciation of the natural processes of our planet requires a basic knowledge of many branches of science—from astrophysics through biology, oceanography, physics, and chemistry. The Earth sciences integrate all science disciplines at all scales of magnitude. To understand Earth's composition, we must understand the processes that form the elements in stars, and the subsequent formation of planets from these materials. To understand the behavior of minerals—the fundamental building blocks of rocks—we must examine the atomic structure of a mineral because it is this arrangement of atoms in a mineral that determines its properties.

In the next few chapters, we will consider processes operating at scales ranging from the scale of the Universe to the scale of atoms. We will also come to appreciate that many geologic processes occur very slowly, over vast periods of time.

Geologic Time: An Important Perspective

To understand Earth and its processes, we must be able to appreciate the important additional dimension of time. Many geologic processes occur through immense stretches of time. As humans, our experience of time is very limited—seconds, minutes, days, and years. The human lifetime is a matter of several decades, and human history is recorded in terms of centuries and millennia. Ancient wonders, such as the 4000-year-old Egyptian pyramids, are commonly described as "timeless." Although it is difficult for us to appreciate the age of such artifacts, these time scales are merely a blink of an eye in geologic time.

Earth is 4.55 billion (4.55×10^9) years old. The record of organic life goes back more than 3 billion years. The rocks in the walls of the Grand Canyon record more than 1.5 billion years of geologic history, and the dinosaurs lived 100 million years ago. An analogy cast in terms of our everyday experience can serve to put these incalcu-

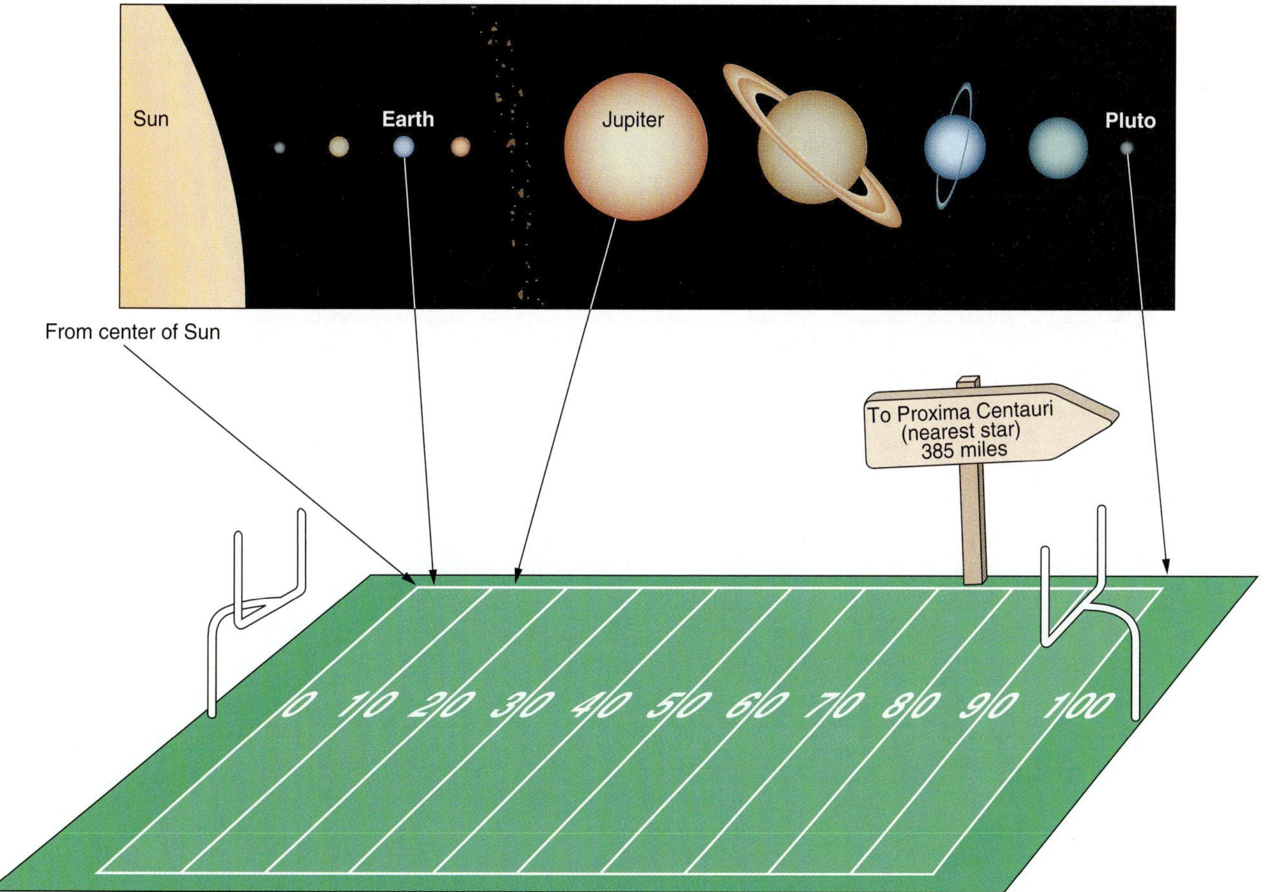

FIGURE 1.16 A matter-of-distance scale: Our Solar System is represented by a football field 100 yards long. On this scale, Earth is 2.5 yards from the center of the Sun.

lable numbers into perspective. Consider the entire history of the planet Earth condensed into a period of just one year—beginning on January 1 with the formation of Earth (Fig. 1.17). The oldest rocks that we have dated were formed in early February (4 billion years ago). The first signs of life—simple one-celled organisms—appeared in late March (3.5 billion years ago). Complex organisms with shells or skeletons first appeared in mid-November (600 million years ago), and the first land animals—early amphibians—crawled onto the shore in late November. December was a relatively busy month: The dinosaurs became extinct late on Christmas day, and humans appeared in the afternoon of December 31. The last Ice Age began later that evening. The pyramids of Egypt were built 30 seconds before midnight, the Declaration of Independence was signed at about 2 seconds before midnight.

Measuring Geologic Time

How do we know the magnitude of these very old ages? Various ways exist for establishing the relative ages of rocks. Three centuries ago, early geologists realized that layers of rock represent progressive deposition through time in oceans, lakes, rivers, or deserts. Thus, when we see a thick succession of relatively undeformed rock layers, such as those exposed in the walls of the Grand Canyon (Fig. 1.18), we surmise that the lowest layer was deposited first and the top layer was deposited last. If you were to walk down into the Grand Canyon, you would be taking a trip back through geologic time as you passed progressively older rock layers. But how can we tell whether a particular layer of rock in the Grand Canyon is older or younger than a rock elsewhere on the continent, and how can we tell the relative ages of rock layers that are strongly deformed rather than stacked in simple horizontal layers?

We need a way to correlate rocks of similar ages. Perhaps the most widely used method is the use of fossils. A **fossil** is preserved evidence of a preexisting organism; literally, something "dug up." Organic evolution gives us a natural scale to distinguish the relative ages of rocks by the types of fossils contained within them. Evolution is a slow process, and organisms that exhibit a very primitive level of evolution are probably very old. We can tell, for instance, that rocks containing fossils of a primitive type of armor-plated fish are older than those containing more evolved (and perhaps more familiar looking) fish fossils. These methods, however, provide relative ages; they are not absolute or quantitative.

Moreover, correlation of rocks using fossils is only useful in rock layers where complex life forms existed, and this represents only the last 600 million years—less than 15 percent—of Earth history (Fig. 1.17). To determine the actual ages of different specimens, we use the natural process of radioactive decay to date the rocks in which they are found.

Focus On 1.1 A MATTER OF PERSPECTIVE: BIG AND SMALL NUMBERS

Whether we discuss distance, time, or any physical quantity, we must be able to relate these quantities to each other. Normally, we use quantities that are relatively easy to understand. Meters (abbreviated m) are a good unit of measurement for length; most people are 1.5 to 2 m in height. Scientists use the metric system of units, and thus use units of meters rather than feet, and kilograms or grams rather than pounds or ounces. See Appendix I for a conversion table.

In the earth sciences, we deal with phenomena ranging from the very small to the very large. The distance between two atoms in a crystal is approximately 0.000000001 m, whereas Earth's circumference is some 40 million m. To make these numbers easier to communicate, we use scientific notation, or "powers of ten." In this notation, 0.000000001 m equals 10^{-9} m ("ten to the minus nine meters"), and 1,000,000 m is 10^6 m ("ten to the six meters"). Just as 2^2 means 2×2, 10^6 represents $10 \times 10 \times 10 \times 10 \times 10 \times 10$, and 10^{-9} means $1/(10 \times 10 \times 10 \times 10 \times 10 \times 10 \times 10 \times 10 \times 10)$. This notation is illustrated in Figure 1, where we start with an immense distance—the distance across the galaxy—and then focus on increasingly smaller dimensions, ending with atomic diameters.

We also use large numbers in the earth sciences in the context of time. Earth is 4,550,000,000, or 4.55×10^9, years old. Because using millions of years is a more helpful measure in this context, geologists use the unit Ma to represent "million years old." Ma stands for million years, from the Greek prefix "mega" meaning million and the Latin "annus" meaning year. Thus, Earth is 4550 Ma old. Similarly, geologists use the prefix *Giga* (Ga), where 1 Ga equals 10^9 or 1 billion years. We can therefore also say that Earth is 4.55 Ga old.

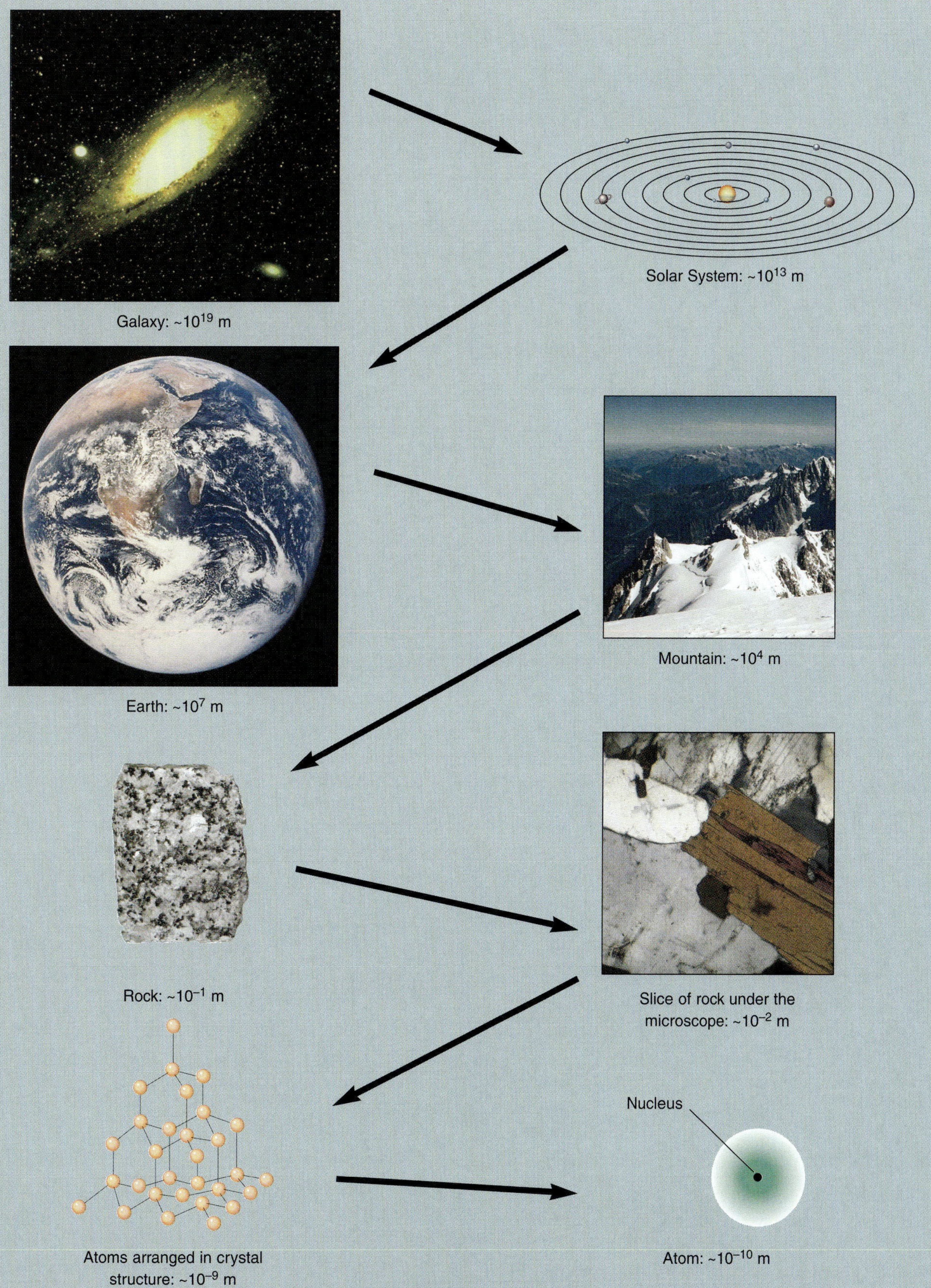

FIGURE 1 Exploring distances: from large to small, to illustrate the use of scientific notation. All distances give approximate width of the object and are in units of meters.

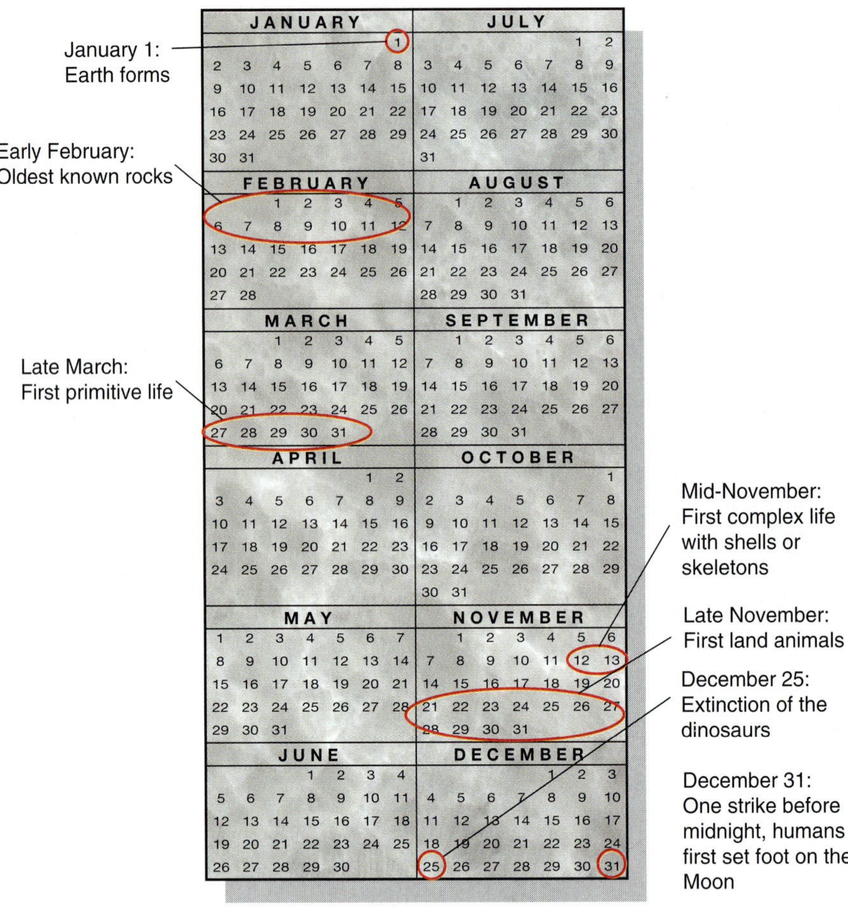

FIGURE 1.17 A matter-of-time scale: The calendar represents the age of Earth—4.55 billion years (4.55×10^9) of geologic time condensed into one year. On this scale, humans evolved on New Year's Eve.

Radioactive *isotopes* (see Chapter 2) are atoms that spontaneously break down, by radioactive decay, into atoms of another element. The products of this disintegration are referred to as *daughter isotopes*, whereas the original isotope is called the *parent isotope*. The disintegration, or *decay*, rate of radioactive parent isotopes into daughter isotopes is a constant for a given isotope. The time required for half of the parent isotopes to decay is defined as the **half-life**. When molten rock cools, the isotopes may be trapped in crystallizing minerals so that the parent and daughter atoms are retained and their abundances can be measured. By knowing the decay rate of a given isotope and by measuring the amount of daughter and parent isotopes in a rock, scientists can estimate the time of formation of that rock.

For example, suppose that isotope X decays to isotope Y with a half-life of 1 million years (Fig. 1.19); it takes 1 million years for 100 percent X to decay to 50 percent X and 50 percent Y. If a rock was formed with 100 atoms of parent isotope X (and no Y), and we now measure 50 atoms of X and 50 atoms of the daughter isotope Y, then half of the parent atoms have decayed, which means that the rock has experienced one half-life and therefore must be 1 million years old. After 2 million years (one more half-life), the remaining 50 atoms of parent isotope X will have decayed to 25 atoms of daughter Y, giving a total of 75 atoms of Y, with 25 atoms of X remaining.

The Present Is the Key to the Past

Before this type of **isotope dating** was developed, geologists had inferred that Earth was many millions of years old. It was recognized that sediments—rock debris created by the weathering of Earth's surface—settled in layers on Earth's surface until the pressure of the successive layers became so high that sediments were compressed together to form **sedimentary rock**. Using the rates at which this process occurred—which can be readily measured—geologists concluded that the rocks of the Grand Canyon must have taken millions of years to form (see Fig. 1.18). Radioactive dating has provided spectacular corroborative evidence for this reasoning (proving, for instance, that the oldest rocks at the bottom of the Grand Canyon are over 1.5 billion years old) and has strengthened what has become known as the "uniformitarian" approach to geology. This approach was originally proposed in the eighteenth century by James Hutton, known as the father of modern geology. The theory of uniformitarianism suggests that *the past history of Earth can be explained by what can be seen to be happening now*. In other words, "the present is the key to the past."

FIGURE 1.18 The Grand Canyon, Arizona. The Colorado river has cut a deep valley through layers of rock (compare with the satellite image of Fig. 1.13). These layers become increasingly older, and the rocks at the bottom of the valley are about 1.5 billion (1.5×10^9) years old.

In our time scale, Earth seems a very stable place. Of course, we are familiar with transient natural events like volcanic eruptions, landslides, floods, and earthquakes, which can change the local landscape to some extent. This is evident in the addition of new land onto the island of Hawaii as a result of lava pouring into the sea from Kilauea volcano. Nevertheless, it is difficult to imagine the area where the Himalayan Mountains now stand being covered by an ocean, or to imagine a time when the Atlantic Ocean did not exist. These changes are not merely figments of imagination; if our interpretations are correct, they are part of geologic history and can only be appreciated in the context of the great stretch of geologic time. The Himalayas are still rising at a rate of 5 to 10 mm annually. At this rate, Mt. Everest could have risen from sea level to its present lofty elevation of nearly 9000 m in less than 2 million years—a perfectly reasonable feat given the magnitude of the geologic time scale. The Atlantic Ocean, currently widening at the rate of 2 cm per year, has been expanding for the last 150 million years (since mid-December in our calendar analogy of Fig. 1.17).

Earth's history is recorded in the rocks that we study. However, much of the time over which Earth has existed is actually represented by *gaps* in the record; that is, the *absence* of rocks. In a series of layered rocks, each layer is separated from the next by a bedding plane; these

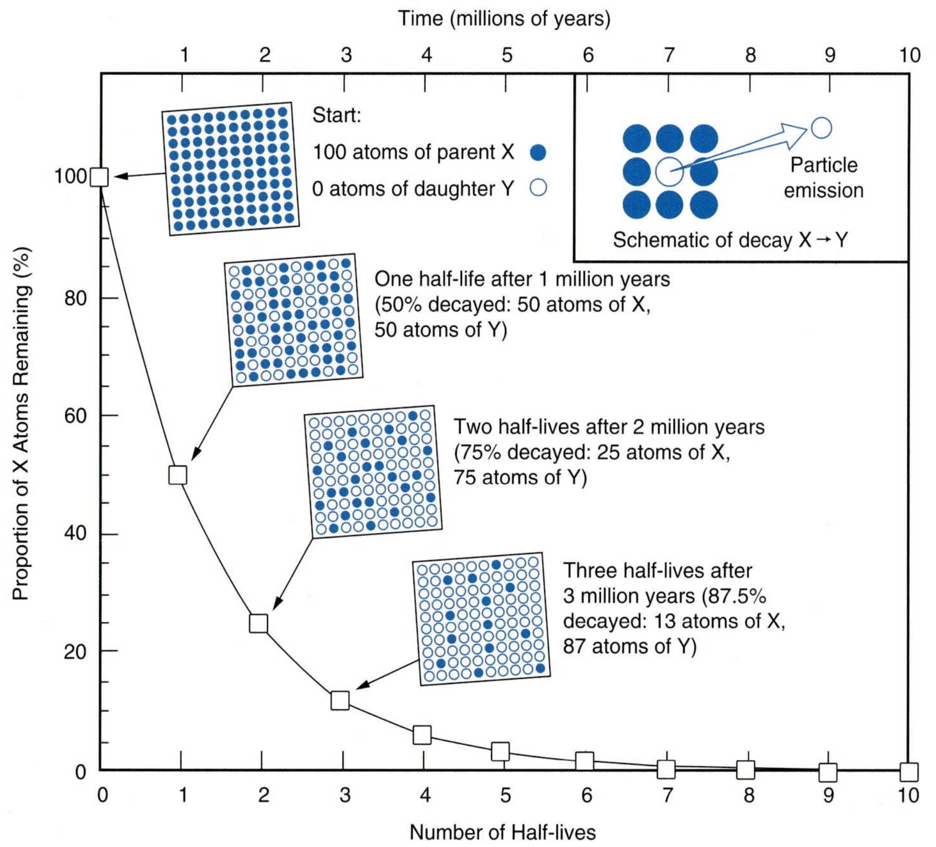

FIGURE 1.19 The process of radioactive decay. If isotope X is present in a rock or a mineral sample, we can determine the age of that sample in the following way: If the atoms of an element X undergo radioactive decay to a daughter isotope Y, and we know that the half-life of X is 1 million years, then we can measure the relative amounts of X atoms and Y atoms. If there is 50 percent X and 50 percent Y, the sample is 1 million years old.

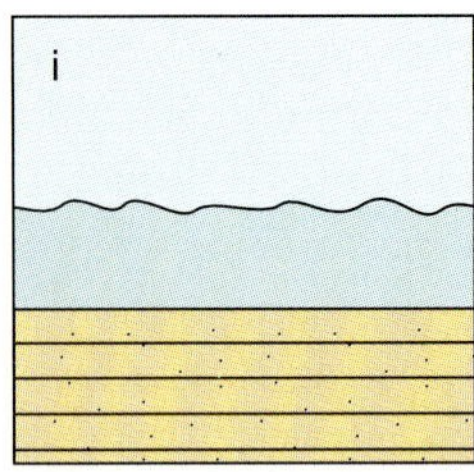

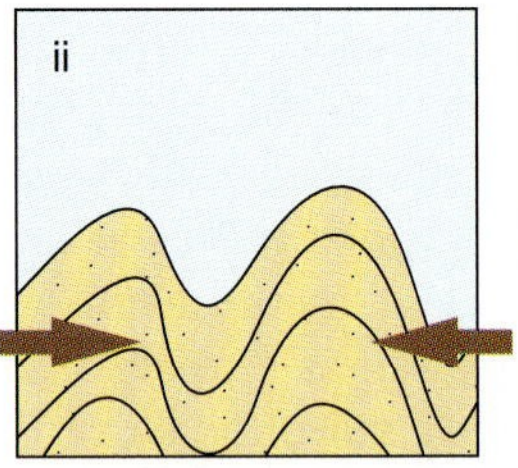

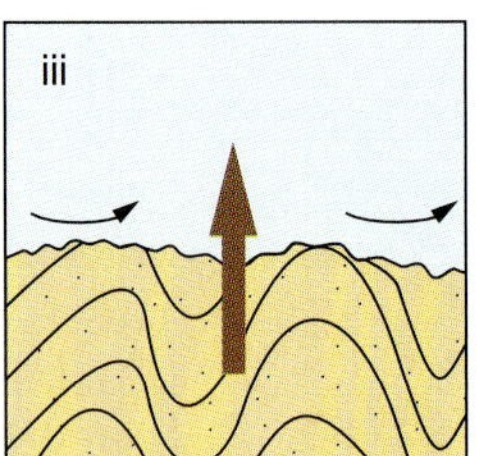

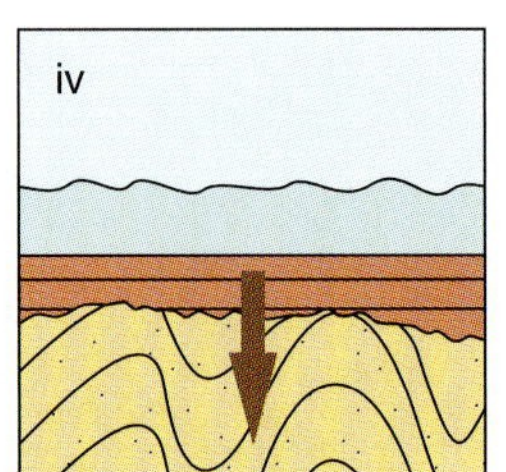

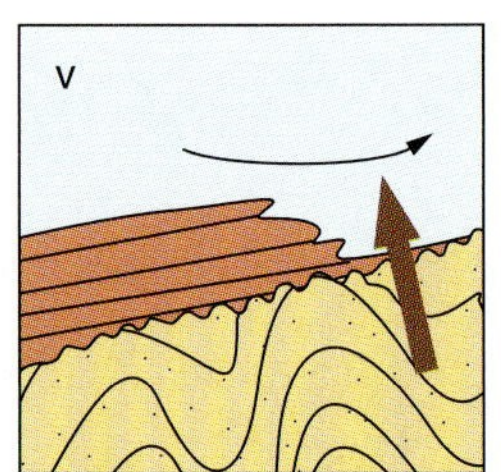

FIGURE 1.20 The unconformity—known as "Hutton's Unconformity"—at Siccar Point, east of Edinburgh, Scotland. The geologic history represented by this outcrop is shown schematically. (i) Deposition of older rock layers; (ii) deformation of older rock layers; (iii) uplift and erosion of older rock layers, forming an ancient land surface; (iv) subsidence of old land surface and deposition of younger rock layers over it; and (v) uplift, tilting, and erosion. Hutton recognized that this outcrop required an enormous stretch of time to undergo all of these processes if the processes occurred at rates comparable with those observed today.

represent small gaps in time. Larger breaks in time are reflected by **unconformities**; they are interruptions in what is normally a continuous process.

James Hutton observed one particular unconformity at Siccar Point, just east of Edinburgh, Scotland (Fig. 1.20), which played an important role in the development of geology. This unconformity exhibits a high angularity between the underlying and overlying layers. Hutton had not seen it before, but he immediately recognized the significance of the outcrop. He surmised that the older beds below the unconformity must have been deposited, consolidated, deformed, uplifted, and eroded before the younger, upper succession of rocks were deposited on the eroded surface (Fig. 1.20). The unconformity must therefore represent a significant length of time—actually about 50 million years. The record between the two series of rocks is incomplete even though geologic processes were still ongoing.

Clearly, many geologic processes can only be understood in the context of long periods of time. In fact, a *geologic time scale* has been established, comprising eons and periods. These are divided mainly on the basis of significant changes in fossil life-forms and on the occurrence of widespread unconformities, which tell us of important changes in sea level. This time scale is given for reference in Table 1.1 and Appendix II. Note that for simplicity, we will use time (in millions of years) rather than these period and eon names in the text when discussing events in the geologic past. Among geologists though, these names are as familiar as the names of the elements are to a chemist. It is implicit that the matter of scale—including time—will be important in our understanding of Earth.

A Process Approach to Physical Geology

It often helps to examine a map before setting out on a journey; thus, let's take a moment to explain the organizing framework of this book and the philosophy behind it.

We begin in Chapter 2 from the logical standpoint of the formation of Earth. This is perhaps the least accessible and most poorly known aspect of the earth sciences, but it is critical to view Earth in the context of the cosmos rather than as an isolated entity. From an understanding of Earth formation, we can draw some conclusions about the composition of Earth; for example, why Earth contains a core of iron, and why it has such an abundance of water at its surface—in contrast to the other oceanless planets. In thinking about our planet's formation, we get our first inkling of the origin of Earth's huge internal energy source and how this energy source is the driving mechanism for plate tectonics.

Next, we turn to the material of Earth itself: the rocks and the minerals from which rocks are formed (∞ see Chapters 3 and 4). We will learn that rocks and minerals are in a continual state of flux, influenced by both plate tectonics and weathering at Earth's surface. Molten rock from deep within Earth is buoyant and moves toward the surface where it may solidify at depth or erupt at volcanoes. As the molten rock cools, it crystallizes into minerals. At Earth's surface, rocks are broken up into miner-

TABLE 1.1 The Geologic Time Scale

EON	ERA	PERIOD	EPOCH	MILLION YEARS	APPEARANCE OF...
Phanerozoic	Cenozoic	Quaternary	Holocene	0.01	Humans
			Pleistocene	1.64	
		Tertiary	Pliocene	5.2	
			Miocene	23.3	
			Oligocene	35.4	Mammals
			Eocene	56.5	
			Paleocene	65	Extinction of dinosaurs
	Mesozoic	Cretaceous		145.6	
		Jurassic		208	
		Triassic		245	
	Paleozoic	Permian		290	Reptiles
		Carboniferous		362.5	
		Devonian		417	Amphibians
		Silurian		443	Land plants, fishes
		Ordovician		495	
		Cambrian		545	Shelled organisms
Proterozoic				2500	
Archean				4000	First primitive life
				4550	Age of Earth

al grains, and minerals change composition through chemical reactions. The mineral grains (sediment) may be washed into the sea and deposited. If these sediments are buried to a sufficient depth, the heat and pressure of burial will turn the sediment back into rock. The rock—and the minerals that form it—may change composition and texture as the pressure and temperature change; in extreme cases, the rock may even melt again. At any time, the forces of plate tectonics can lift buried rocks to the surface where they are once again weathered and washed away. Thus, the cycle continues.

After examining the rocks and minerals that make up Earth, we take a broad look at the planet. The physical phenomena that help us understand the way Earth works—gravitational attraction, magnetism, seismic wave behavior and so on—are investigated in Chapter 5. This will help us to determine Earth's internal structure and explain the first order differences between continents and oceanic crust (not least of which is why the latter is submerged beneath the oceans). It also helps us understand the principles of plate tectonics and appreciate how the theory was developed.

We then return to a more detailed examination of plate tectonics—processes driven by the internal energy of Earth. Here, we have just introduced the concept that the outer layer of Earth is divided into plates that move relative to one another. In Chapters 6 through 11, we will revisit this concept in greater detail because it is the relative movement and interaction of these plates that give rise to earthquakes, volcanoes, and mountain ranges. Many of the features of Earth are most conveniently discussed (∞ see Chapters 8, 9, 10) in terms of the three types of plate margins that we introduced in this chapter.

Following the discussion of plate tectonics, we turn to the features and processes that result from Earth's external source of energy. In this context, the hydrologic cycle is of primary importance. We examine the processes of weathering and erosion that occur at Earth's surface (∞ see Chapter 12), and the transport and deposition of the sedimentary material formed in this way (∞ see Chapter 13). A natural consequence of these processes is the modification of landscapes, and we survey the characteristics of different landforms in a chapter (∞ see Chapter 14) on geomorphology (the study of landforms). We discover that different combinations of climate, geology, and environment—for instance, deserts, shorelines, river valleys—produce distinct landform characteristics.

Finally, we address some practical aspects of the earth sciences (Chapters 15 and 16). From an understanding of how Earth works, we can determine how natural resources are formed. Not least of these in the present world of ever-

THE PHILOSOPHY OF SCIENCE

What is a scientist? Scientists are people who approach a given problem in a specific way—a critical, systematic approach often referred to as the "scientific method."

The basis of science is observation, which can take the form of a simple observation of natural phenomena, such as the temperature of a lava as it erupts from a volcano or the thickness of a rock layer in a roadcut, or it can take the form of data collected from experiments, such as measuring the force it takes to break a slab of rock in the laboratory. Although geologists can perform many useful experiments, we rely more heavily on observations of the natural world. By now, you may well appreciate that the time scales and magnitudes of geologic processes are usually far too great to simulate in the laboratory.

Careful observation allows us to characterize the phenomena we are studying. A well-defined behavior that appears to be unchanging can be expressed as a scientific law. One thermodynamic law states that heat always flows from hot to cold. You can appreciate this infallible law every time you pour a cup of coffee and leave it on the table; it will become progressively cooler—it will never spontaneously heat up! Laws are so well defined that they can often be expressed in the form of mathematical equations. Although a law describes a phenomenon precisely, it does not indicate how the phenomenon works, or why. When an explanation is proposed on the basis of observation, this is known as a *hypothesis*, or model. On occasion, more than one explanation or hypothesis is proposed, and we talk about "multiple working hypotheses." Experiments or observations can be performed to test the model or to choose between multiple working hypotheses. If the results do not fit the model, the model is modified or perhaps abandoned in favor of an alternative one. If a model "stands the test of time" and is able to explain new sets of data or observations consistently, then it is elevated to the status of a theory. Plate tectonics is an excellent example of a theory. As we will see in the following chapters, plate-tectonics theory accounts for and interrelates many observations of the natural world, observations previously enigmatic or inadequately explained.

To illustrate the scientific philosophy, consider the hypothesis proposed in the Middle Ages that Earth is flat. At that time, no specific observations suggested otherwise. It looked flat, and travelers, despite having to climb and descend hills, considered, overall, that they were traversing a flat surface. This, however, is a testable hypothesis. If Earth is flat, it should have edges (unless it extends to infinity); if it is spherical, then it will not have edges but be a continuous, curved surface. A simple experiment, albeit demanding considerable trepidation, would be to travel in a given direction until you either reached the edge of Earth, or returned back to where you started. This was what Portuguese explorer Ferdinand Magellan effectively achieved in 1522, when his expedition completed the first circumnavigation of Earth, proving it to be a globe. Today, with the development of space travel, we can directly observe the spherical shape of Earth, as is shown spectacularly in Figure 1.3.

increasing population is our resource of fresh water—most of which is obtained from beneath the ground (groundwater). We have realized that we must manage the way in which we interact with our planet, and everyone must protect the environment. Earth scientists are well qualified to understand the problems of toxic-waste disposal, water pollution, and excessive exploitation of natural resources (Fig. 1.21).

After each chapter, we will pose some questions to help you think about the material. In geology—as with other sciences—there is a necessary amount of terminology to learn. We have tried to keep this to a minimum—only introducing a term in context when it is necessary to understand a particular concept. We believe it is more exciting to learn about Earth processes than to become mired in the details of naming and classifying rocks, minerals, and landforms. Nevertheless, some knowledge of important terms is needed, and these terms are boldfaced and defined both in the text and in the glossary at the end of the book. We have already introduced some terms in this first chapter because they are essential as we progress through the book.

Throughout the text, you will encounter boxed material. These Focus boxes either discuss a topic in greater detail or present specific geographic or historical examples of phenomena examined in the text. You will run

FIGURE 1.21 The oil fires of Kuwait following the 1991 Gulf War. This satellite image shows black smoke streaming from the burning oil wells. This image underscores the problems encountered in recovery and responsible management of our natural resources (in this case, oil) and the potential environmental hazards that we must address.

across short, interim summaries of the most important material covered in the preceding section. These are also used as chapter summaries.

We end this chapter on a philosophical note: Earth science is fundamentally a naturalist science. Most of what we know about Earth is the result of observation. We can perform experiments, of course, but the time scales and length scales of many Earth processes simply cannot be reproduced in the laboratory. In a sense, then, Earth is a gigantic natural laboratory. In this context, we mention the concept of a *model* throughout the text. A model is a description or an explanation based on theory, experimentation, and observation. Models are proposed—like a hypothesis—and subsequently tested by further observation, by new experiments, or by theoretical calculations based on measured data (Focus 1.2). Earth scientists have established a number of models constituting our present way of thinking about Earth. A good model will stand the test of time, but 30 years from now, many of our current models may be greatly modified or even abandoned as we obtain more data. It is worth mentioning that the plate-tectonics model of how Earth works did not even exist 50 years ago.

KEY TERMS

plate tectonics, 2
crust, 3
mantle, 3
core, 3
lithosphere, 3
asthenosphere, 3
plate, 3
divergent margin, 3
convergent margin, 3
transform margin, 3
spreading ridge, 5
subduction zone, 5
hydrosphere, 5
hydrologic cycle, 6
groundwater, 6
weathering, 6
erosion, 6
fossil, 12
half-life, 14
isotope dating, 14
sedimentary rock, 14
unconformity, 16

2 Earth: Origin and Composition

A column of interstellar hydrogen gas and dust in the Eagle Nebula, recorded by the Hubble Space Telescope.

Overview

- The universe is thought to have formed of material from the "big bang."
- Hydrogen and helium are the principal components of stars. Fusion processes within stars produce elements ranging in mass from helium to iron. Supernova explosions of stars form the remaining elements.
- Meteorites, moon rocks, and the properties of comets and stars tell us about the composition of the material that formed Earth, which is thought to be about 30 percent iron and 70 percent silicate rocky material.
- Gravitational energy released at Earth's formation was sufficient to have melted it. The planet then separated into the core and mantle.
- Earth's rocky material contains radioactive elements that generate heat. The planet would be entirely molten if this heat did not escape.
- As heat is generated within Earth, hot material rises to the surface and cools, allowing the heat to escape. The cool material then sinks. This process, called convection, drives plate tectonics.

Introduction

Why is Earth the only planet in the Solar System fit for life as we know it? Why do we have an oxygen-rich atmosphere when our sister planet Venus has a carbon dioxide–rich atmosphere, Mars has hardly any, and Mercury has none at all? Why is Earth the only terrestrial planet with oceans and a temperate climate? Are there planets with life in other parts of the universe? People have long sought answers to these questions. In this chapter, we investigate the origins of Earth's uniqueness, its elemental composition, its internal structure, and the factors that make Earth a dynamic planet.

Our view of Earth today can be thought of as a snapshot of the planet in the span of geologic time. If you pick up a handful of sand on the beach, you are looking at the end products of a process that began about 15 billion years ago with the **big bang**—a gigantic explosion thought to be the origin of our present universe (Fig. 2.1). The big bang was followed by star formation, supernova explosions, planetary formation, and subsequent volcanic eruptions, earthquakes, hurricanes, and floods. The fundamental particles that make up those sand grains have been in existence through all of these cataclysmic processes and now simply lie on the beach, washed by the ocean waves. To gain a better understanding of Earth and its history, we will take a closer look at what these fundamental particles are, how they were formed, and what geologic processes have affected them.

FIGURE 2.1 Most scientists believe that the universe began in a fiery explosion some 15 billion years ago called the "big bang." Out of this explosion emerged the energy and fundamental atomic particles that would later form the galaxies stars, and planets. Recently acquired photographs of distant galaxies show light emitted from what is the most distant galaxy known—possibly as far away as 10 million light-years.

Piecing together the geologic history of these sand grains is a giant puzzle for which we have many clues. Fifteen billion years of history must be traced through scraps of evidence from Earth's surface, from meteorites, from samples recovered from the Moon, and from our understanding of the basic laws of physics and chemistry. We will construct our best estimate of the events that may have taken place throughout Earth's history—its formation, its separation into crust, mantle and core, and the onset of plate tectonics. As we project back into the dim recesses of time, our understanding of geologic events necessarily becomes more uncertain in comparison with our understanding of the geologic events of the present, which we can frequently observe going on around us.

The Composition of the Universe

Matter consists of fundamental components called **atoms**. The three basic constituents of an atom are the electron, the proton, and the neutron. An **electron** is a negatively charged particle that orbits the center, or nucleus, of the atom. A **proton** is a positively charged particle that resides in the atomic nucleus and has a substantially larger mass than does an electron. The **neutron**, also found in the nucleus, has approximately the same mass as a proton, but is electrically neutral.

Different atoms have different numbers of protons and electrons. An **element** is composed of just one type of atom; the atoms of an element all have the same number of protons and electrons (Fig. 2.2). The number of protons in an atom is known as its **atomic number**. Each element is defined by a unique atomic number.

However, a given element may have different numbers of neutrons, which give rise to different **isotopes.** **Atomic mass** is equal to the total number of protons and neutrons in the nucleus, so the isotopes of a given element have the same atomic number but different atomic mass. The simplest atom is the hydrogen atom (atomic number 1), which has a single proton and a single electron. The helium atom (atomic number 2) contains two protons and two neutrons in its nucleus with two orbiting electrons. **Molecules** are combinations of atoms bonded together—either atoms of the same type or atoms of different elements (∞ see Chapter 3). All stars emit light that is consistent with their composition,

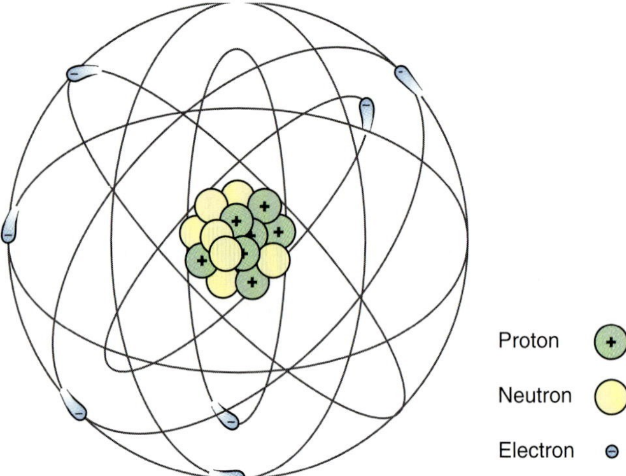

FIGURE 2.2 A simple model of an atom. Protons and neutrons lie at its center, or nucleus, and electrons orbit this nucleus in zones called *shells*, which are shown schematically.

which is mainly hydrogen and helium. There is a remarkable uniformity to the universe in that it appears to be made up of very common elements that are found in abundance on Earth rather than some exotic set of elements completely unfamiliar to us. These elements do not appear to vary from star to star; we do not get a nitrogen star here, a neon star there.

How can we tell that the light emitted from stars is consistent with this composition? We have never traveled to a star, and we are not likely to do so in the foreseeable future. The nearest star is Proxima Centauri, which is 40 trillion kilometers away, or 4.2 light-years (see Fig. 1.16). The spacecraft Voyager travels at 1/10,000 the speed of light (or 67,000 mph). If such a state-of-the-art spacecraft were to travel to Proxima Centauri, it would take 42,000 years. Despite our inability to travel to the stars, we still know a good deal about them from the light they emit.

Light travels in the form of waves that resemble ripples moving across a pond. The number of wave crests, or peaks, in a light wave passing by a given point per unit time is called the wave's **frequency**. We know that light is composed of electromagnetic waves of different frequencies, with each frequency corresponding to a different color. Given the appropriate weather conditions, we can see the Sun's white light separated into its different-colored components in the form of a rainbow. The same effect can be achieved by passing white light through a prism (Fig. 2.3). The prism causes the different frequencies of light to bend at different angles: The blue light is bent more than the yellow light which, in turn, is bent more than the red light, resulting in separation into the familiar colors of the rainbow. A representation of light separated into its constituent colors, as shown in Figure 2.3, is called a **spectrum**.

An element heated to a high temperature will emit light. We are all familiar with the example of a tungsten filament of a lightbulb. When the emitted light from a

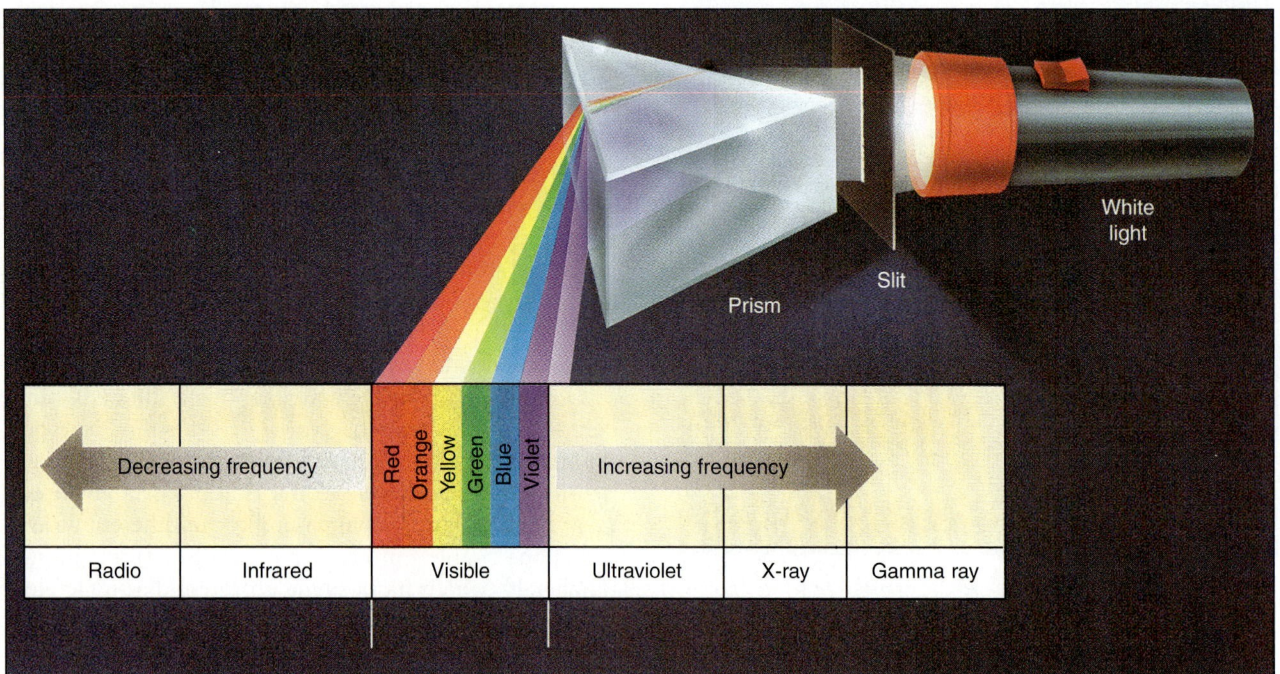

FIGURE 2.3 White light passing though a prism separates into different colors like the pattern of a rainbow. The colors from red to blue merge into each other with no intervening gaps, which is why this type of spectrum is termed a *continuous spectrum*. Light is the visible part of the electromagnetic spectrum, which also includes radio, infrared, ultraviolet, X, and gamma rays.

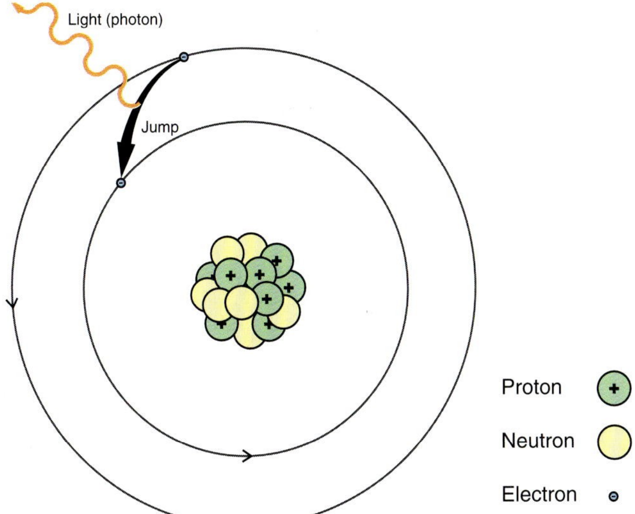

FIGURE 2.4 Electrons that jump from one orbital shell to a lower-energy shell emit light photons at a frequency dictated by the size of the jump. The spectra have energy only at frequencies corresponding to the possible orbital jumps; this gives rise to an emission-line spectrum.

given element is closely examined, it is found to have a unique spectrum that can be used as a fingerprint to identify the element. The spectrum is unique because it depends on the distribution of negatively charged electrons orbiting the nuclei of the particular type of atom making up that element. Electrons orbit the nucleus of an atom in orbital shells, and can move from shell to shell. Electrons with the lowest energies reside in the innermost shell, closest to the nucleus. Electrons with increasingly higher energies are found in shells that lie at increasing distances from the nucleus.

In a hot environment, such as a star, electrons can gain or lose energy and thus be knocked from one shell to another. When an electron jumps from an outer shell to an inner shell of an atom—from higher energy to lower energy—it releases a burst of light that has a frequency (color) proportional to the energy jump between the shells (Fig. 2.4). This burst of light is referred to as a **photon**. For a particular atom, some shell jumps might occur much more readily than others, and more light is emitted at those frequencies. The corresponding spectral colors will be the brightest in the spectrum and will show up as distinct lines.

Suppose that light is being emitted from the electrons of the atoms of hydrogen gas. We can pass the emitted light through a prism and look at the brightness of the colors. We find a series of colored lines, which correspond to the electron jumps specific to hydrogen. These electron jumps emit light at only those frequencies. Gaps occur between the frequencies because no corresponding jumps exist (Fig. 2.5).

When light is emitted from hot hydrogen atoms, the resulting spectrum is called an *emission* line spectrum. Conversely, when white light passes through hydrogen gas, we obtain an *absorption* line spectrum for hydrogen (Fig. 2.5). The hydrogen atoms absorb the light energy at frequencies corresponding to the line spectrum of hydro-

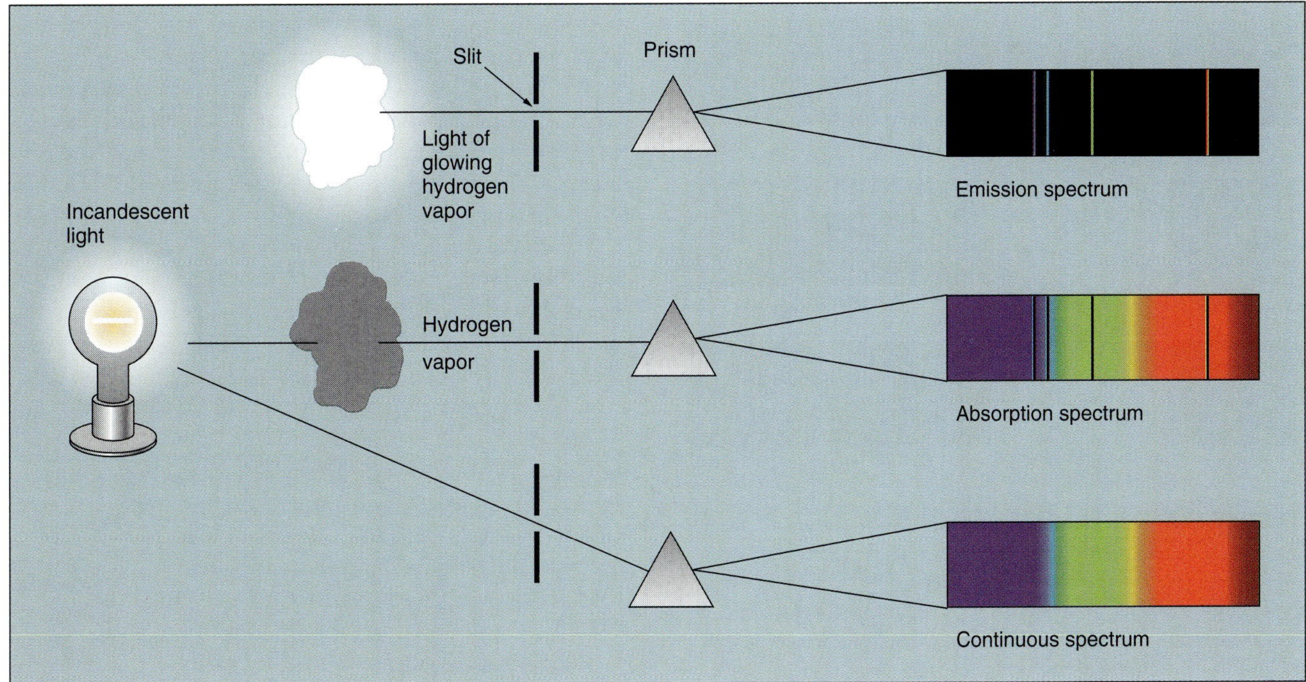

FIGURE 2.5 Three types of spectra for hydrogen: An emission spectrum for hydrogen, an absorption spectrum for hydrogen, and a continuous spectrum. The first two spectra can be used to identify the compositions of stars.

gen. In practice, the spectra we observe from stars are absorption spectra because white light from the interior of a star must pass through gases of the star's exterior, where light energy is absorbed.

Atoms have unique spectra that appear as colored lines of different brightness in a spectrum. The spectra of stars reveal their compositions.

Even though astronomers noticed that the absorption spectra from stars of distant galaxies nearly correspond to the spectra expected for the elements hydrogen and helium, the lines obtained from the starlight were not at quite the same frequencies as those measured for hydrogen and helium in the laboratory. The general features of the spectrum were the same, but the lines were shifted to lower frequencies. The red lines of the hydrogen spectrum from the stars appeared darker red; the yellow lines were shifted in the direction of the red lines. The observed shift in frequency increased with a star's increased distance from Earth. This phenomenon is called the **red shift**; it tells us that stars of distant galaxies are moving away from Earth and results from the Doppler effect (∞ see Focus 2.1). The red shift is the primary evidence that tells us that the universe is expanding (Fig. 2.6).

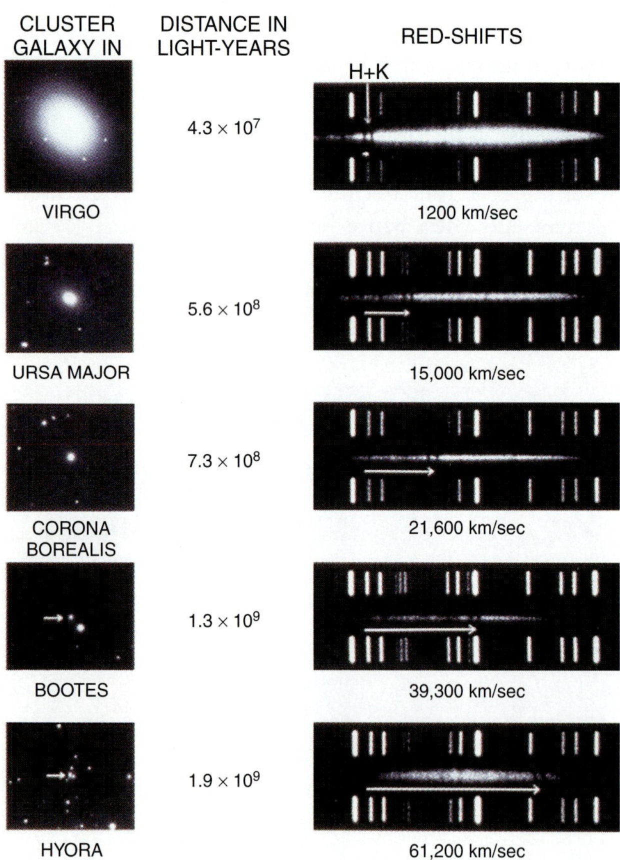

FIGURE 2.6 Red-shifted light from stars in distant receding galaxies. The absorption spectrum is shown in the middle of the right-hand side panels relative to reference-line spectrum above and below. Reductions in intensity correspond to two absorption lines marked H and K. The arrow shows how much they are red-shifted. The number below each spectrum gives the velocity in km/sec causing the red shift. The data indicate that distant galaxies travel faster, which is expected if the universe is expanding.

The Building Blocks of the Universe

Formation of the Elements

The spectra of light from stars indicate that stars are composed almost entirely of hydrogen and helium. Hydrogen and helium are the first two elements listed in the periodic table of the elements (∞ see Focus 3.1 and Appendix III). The periodic table lists the elements in order of the increasing number of protons in atoms, and it groups elements together that share similar properties and behave similarly in chemical reactions (∞ see Chapter 3). If the stars are predominantly filled with hydrogen and helium, how were the remaining elements formed and why are they so prevalent on Earth?

Nuclear Fusion and Fission Understanding these stellar processes requires a basic knowledge of the nuclear energy driving the universe. The fundamental matter spewed into space by the big bang is presumed to have condensed into clumps of greater concentration, driven by the forces of gravitational attraction. As this matter became more and more concentrated, it generated heat by a nuclear process called **nuclear fusion.**

We know that two positive electric charges (or two negative ones) repel each other, whereas charges of opposite sign attract each other; the closer the two charges, the greater the force of repulsion or attraction. Nuclear fusion is a process that occurs when two positively charged nuclei are brought so close together that the short-range nuclear forces of attraction are able to overcome the electrical forces of repulsion. The two nuclei fuse together to form a single, larger nucleus, releasing a tremendous amount of energy in the process. The fused nucleus has a slightly lower mass than the sum of the two separated nuclei. This change in mass is reflected in the release of energy and is represented by Albert Einstein's famous equation $E = mc^2$, where E equals the energy, c is the velocity of light (3×10^8 m/sec in a vacuum), and m represents the change in mass. The velocity of light is so great that a very small change in mass generates an enormous amount of energy.

Because of the formidable difficulty in overcoming the repulsion between nuclei, such fusions are rare occurrences on Earth. For example, the hydrogen bomb—which brings hydrogen nuclei into the required proximi-

Focus On 2.1 THE DOPPLER SHIFT: EVIDENCE FOR THE EXPANDING UNIVERSE

Astronomers proposed the big bang theory for the origin of the universe because they realized that most stars and galaxies are moving away from each other. Thus the universe is expanding. Astronomer Edwin Powell Hubble estimated the rate of this expansion using the frequency shift in the spectra observed from stars.

Because the universe expands, the frequencies emitted by starlight appear to be reduced. The observed reduction in the frequency of waves emitted by an object moving away from an observer is known as the *Doppler effect.* You probably have experienced this effect: An ambulance siren has a higher pitch as it approaches you than it does as it passes you and moves away. As the ambulance approaches, the wave peaks are bunched together to produce more waves passing by you per second (higher frequency, or pitch). When the ambulance speeds away from you, the peaks are spread out, which means that fewer sound waves pass by you per second and you hear a lower pitch.

In an analogous manner, Hubble found that the light frequencies of stars in distant galaxies were shifted to values that are lower than you would expect if the galaxies were stationary with respect to Earth, which suggests that they are all moving away from Earth (Fig. 2.6). Given the present dimensions of the universe and the rate of expansion that Hubble calculated, it can be shown that all matter in the universe was concentrated at a single point at the time of the big bang, 15 billion years ago.

ty for fusion—uses the force of an atomic bomb to overcome the electrostatic repulsion forces. In research apparatus such as a particle accelerator, the appropriate nuclear proximity is achieved by accelerating nuclei toward each other at enormous speeds. The quest for fusion energy on Earth is pursued using giant pulsed lasers, or other means, to generate the necessary force. Gravity provides the force in stars. Within a star, the pressure is so great and the temperature is so high that nuclei fuse to produce fusion reactions equivalent to the detonation of trillions of hydrogen bombs every second.

The fundamental nuclear reaction that occurs in stars involves collisions between hydrogen nuclei, forming helium as a product of reaction (Fig. 2.7). About 75 percent of the matter in most stars consists of hydrogen nuclei; most of the remaining 25 percent is made up of helium nuclei. Other elements comprise just a few percent and depend on nuclear reactions that are less favored, such as the fusing of two helium nuclei to form the element beryllium or a hydrogen and a helium fusing to form lithium. The process of fusion becomes much less efficient as the atomic number increases.

Experiments in particle accelerators and spectroscopic observations of stars have indicated that stellar fusion creates elements with atomic numbers up to and including that of iron, which has atomic number 26. For elements with atomic numbers greater than 26, the mass of the fused nucleus is greater than the sum of the individual nuclear masses. It thus requires an *input* of energy to fuse them. Breaking apart the nuclei of elements with

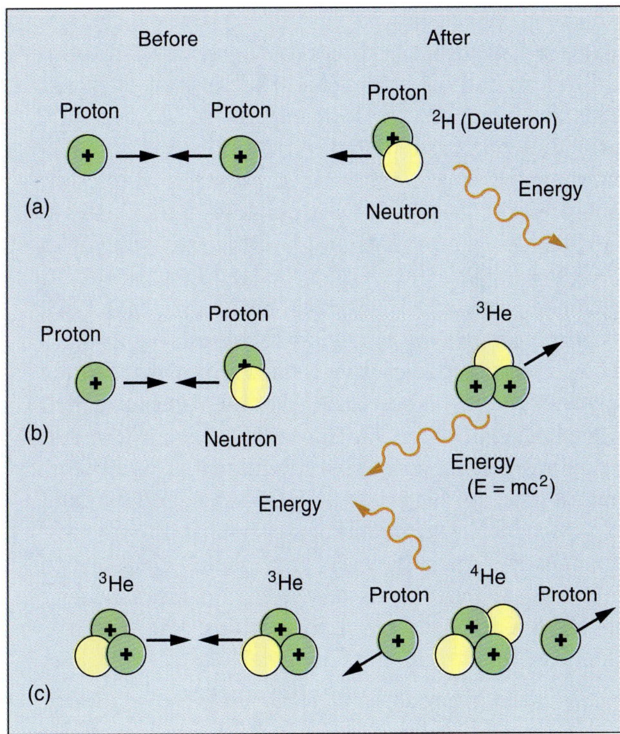

FIGURE 2.7 Nuclear fusion: The proton-proton cycle of a nuclear-fusion reaction, in which the fusion of atomic nuclei releases energy. (a) Protons are fused together to form an isotope of hydrogen ^{2}H called a deuteron (the superscript denotes the mass of the isotope). (b) A proton and a deuteron combine to form an isotope of helium ^{3}He. (c) Two ^{3}He combine to form ^{4}He, releasing two protons for further reactions.

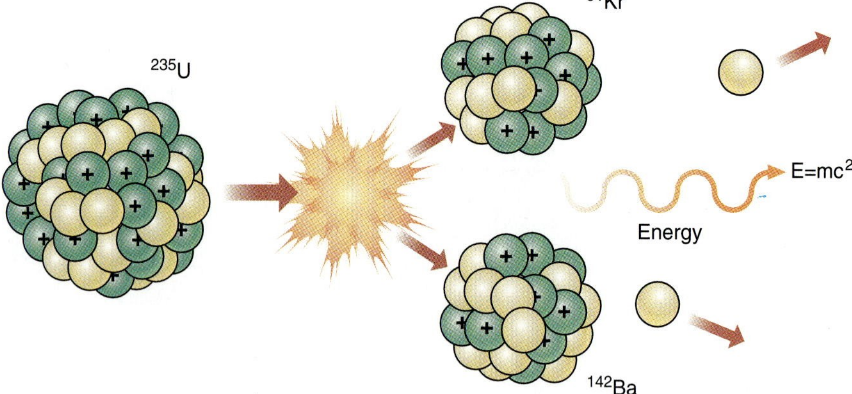

FIGURE 2.8 Nuclear fission: The nucleus of an atom of uranium breaks into two nuclei, thereby releasing energy.

atomic numbers greater than 26 releases energy in a process called **nuclear fission** (Fig. 2.8). In contrast to fusion, fission is the basic principle utilized in the atomic bomb. The decrease in nuclear mass due to radioactive fission of elements such as uranium is converted to radiation energy, once again given by $E=mc^2$.

Nuclear energy drives the universe. Nuclear fusion occurs when nuclei bind together to form new atoms, releasing energy in the process. Nuclear fission occurs when nuclei split apart to release energy.

Supernova Explosions Forces even greater than those found in a star are needed to fuse the heaviest elements together to complete the formation of the known elements. The appropriate conditions are found in a **supernova**, the explosion of a massive star. Such an explosion causes the core of the star to collapse and releases a tremendous amount of energy. This process is thought to provide the full array of elements as we know them.

Elements with low atomic numbers that are found in a star can be thought of as nuclear fuel. Fusion consumes this fuel, creating less reactive elements of higher atomic number. As the fuel is consumed, a point arrives at which the gravitational force from the outer layers of the star can no longer be supported by the thermal pressure of its fusing interior. For large stars, a cataclysmic gravitational collapse occurs. At the center of the star, the intense pressure causes the space between protons and electrons to decrease to such an extent that the short-range nuclear forces fuse them together to form neutrons. The center, or core, thus becomes an impenetrable neutron star. The exterior matter collapsing onto this impenetrable center ricochets off the neutron star and explodes outward, sending a cascade of elements out into space—including heavy elements of large atomic number (Fig. 2.9). Elements of high atomic number such as silver, gold, lead, platinum, mercury, and uranium are formed from the extremely high temperatures and pressures at the center of the star. All this is thought to take place within a matter of seconds. The elements with atomic numbers up to that of iron took billions of years to make; in a few seconds, the rest of the elements that give our planet its varied composition are assembled. Thus, it is not surprising that these heavier elements are found only in traces on Earth. In fact, 99.9 percent of Earth is made up of elements that are no heavier than iron.

Elements up to atomic number 26 (iron) in the periodic table are made by nuclear fusion reactions in stars. The remaining elements are formed from supernovae—giant star explosions that occur when the nuclear fuel at the star's center is consumed and gravitational collapse occurs, followed by outward explosion.

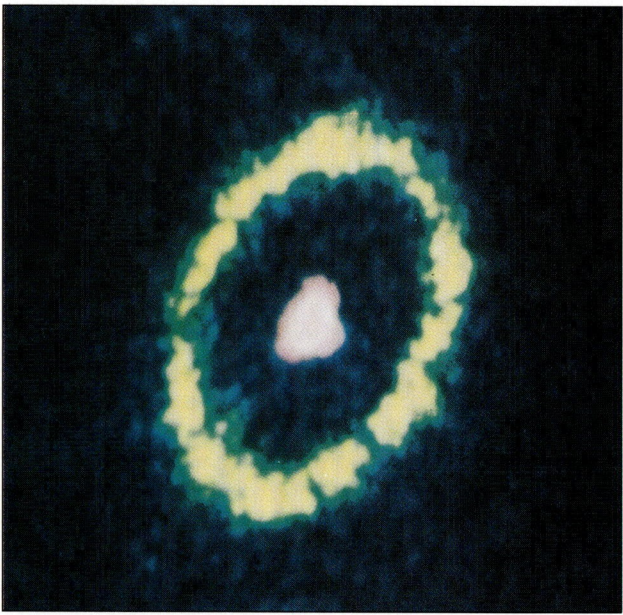

FIGURE 2.9 A Hubble telescope image of Supernova 1987A taken in 1990. The remnant of the star is seen at the center. Interaction of radiation from the explosion with surrounding gas gave rise to the glowing shell shown in the figure. This well-documented explosion provided unprecedented data on stellar evolution and the formation of the elements of the universe.

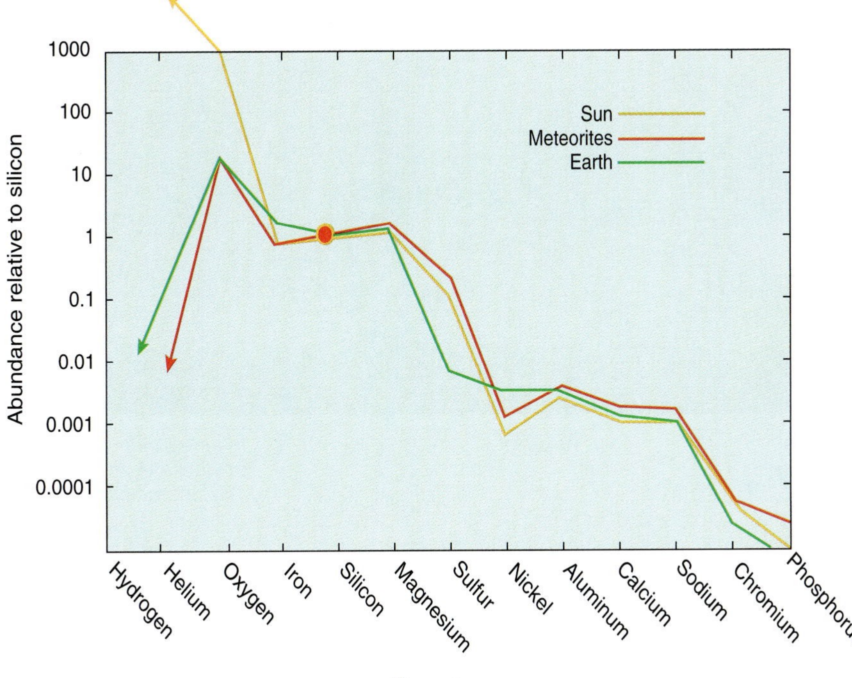

FIGURE 2.10 Relative abundance of the number of atoms of common elements of the Sun, meteorites, and Earth with respect to the number of silicon atoms in each (large dot). Except for hydrogen and helium, the similarities suggest that all condensed from the solar nebula, which is the primeval material of the Solar System.

Our Solar System

Composition of the Solar System

Throughout the billions of years since the big bang, many cycles of star formation and supernovae have occurred. Our Solar System is in such a cycle now. In each cycle, exploded matter from previous supernovae, along with primordial (originating from the big bang) hydrogen and helium, is concentrated by gravity to form new stars. The result is the present distribution of elements in the universe.

Figure 2.10 shows the relative abundances of the common elements in stars (such as the Sun), meteorites, and Earth. Data are taken from spectroscopic observations of starlight (including sunlight), analyses of meteorites, and analyses of rocks from the Moon and Earth. Note that the average compositions of Earth and meteorites are very similar. The main difference between the solar abundance of elements and the elemental composition of planetary material (meteorites or Earth) is the paucity of helium and hydrogen in the planetary material. At the time of condensation, the Sun was very active and a solar wind—a constant stream of particles emitted from the Sun—of high-speed particles may have scoured gaseous helium and hydrogen from the inner Solar System. In addition, the heat from the Sun may have been so intense upon the closest inner planets that it heated the molecules of the primordial atmospheres, giving them such high velocities that they escaped the gravitational attraction of the growing planets (Fig. 2.11).

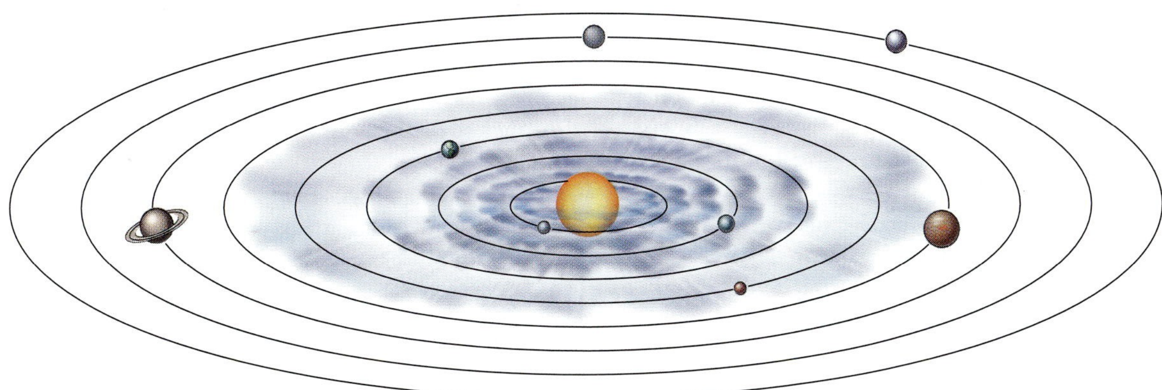

FIGURE 2.11 Exposure of the inner planets to a violent solar wind swept these planets free of volatiles, leaving only the outer giant planets with their original gaseous components of hydrogen and helium.

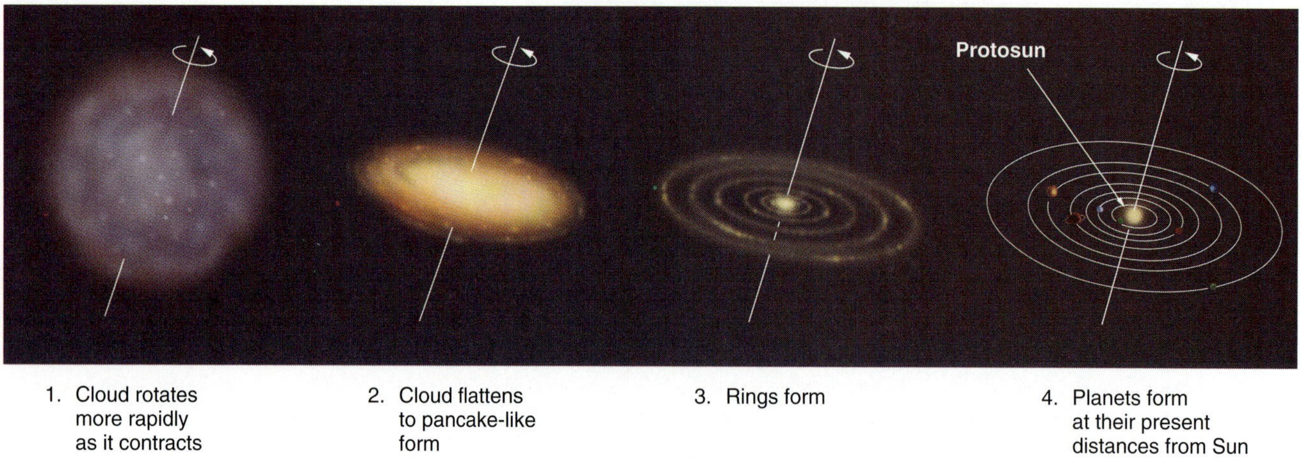

1. Cloud rotates more rapidly as it contracts
2. Cloud flattens to pancake-like form
3. Rings form
4. Planets form at their present distances from Sun

Protosun

FIGURE 2.12 Steps in the formation of the Solar System: (1) Solar nebular cloud forms from previous supernovae; (2) as the spinning cloud contracts due to gravity, it flattens to a disk shape; (3) rings such as those seen about Saturn form, and matter collects at the center to form the Sun; (4) the Sun forms and rings condense to produce planets.

The Age and Origin of the Solar System

The time scale over which Earth and other planets in our Solar System have evolved is incredibly large compared with our everyday experience (see Fig. 1.17). Nevertheless, we possess a precise measure of the time involved owing to the natural "clocks" provided by radioactive isotopes (∞ see Chapter 1). It is estimated that our Solar System is about 4.55 billion years old; that is, 4.55 billion years ago, a vast volume of hydrogen and helium remaining from the big bang and remnants of supernovae (interplanetary dust) had collapsed together because of their combined gravity. This system of gases, called the *solar nebula*, was probably spinning slightly, which might be expected given its explosive history (Fig. 2.12).

As the gaseous nebula contracted, the Sun formed at its center, but not all particles collapsed into the Sun. Some particles may have had greater orbital velocities and thus took longer to fall into the center; these particles may have become trapped in stable orbits around the Sun, forming a set of rings—rather like those we currently see around the planet Saturn. Eventually, the high gravitational attraction associated with denser parts of these rings started to attract other matter in their vicinity as they orbited. First, it is thought that small clumps gathered, called *planetesimals*. Then planetesimals coalesced and captured additional matter to form the planets.

The compositions of Earth and of meteorites are very similar. The Sun has similar proportions of heavy elements but contains much more hydrogen and helium. Earth's original hydrogen and helium were lost when Earth formed.

The Planets

We can divide the planets in our Solar System into two groups: the inner terrestrial planets—Mercury, Venus, Earth, and Mars, and the giant planets—Jupiter, Saturn, Uranus, and Neptune (Fig. 2.13). Pluto does not fit easily into either category. The giant planets are thought to be composed of the original constituents of the Solar System: an abundance of hydrogen and helium, and smaller amounts of methane, ammonia, and water. The terrestrial (Earth-like) planets are rocky and have retained very little of the primeval atmosphere. It has been estimated that if we were to take a giant planet such as Saturn—which is 95 times the mass of Earth—and blow on it as with a dandelion puff, blowing away all hydrogen and helium, we would end up with an Earth-like planet.

Escape Velocity and Planetary Degassing The mass and temperature of a planet are key factors in determining whether or not that planet will retain an atmosphere. The stream of particles due to the solar wind may also affect the potential for an atmosphere to exist, particularly on the inner planets. Molecules in a gas are in continual motion—flying through space and bouncing off each other when they collide; the hotter the gas, the higher the velocity of the molecules. The **escape velocity** of a planet is defined as the minimum velocity needed to escape the gravitational pull of that planet. If the planet is hot, gas molecules will have high velocities and will be able to escape the planet's atmosphere. If the mass of the planet is small, the escape velocity will be low and, once again, the gas molecules will be able to escape into outer space. The escape velocity for Earth is 11.6 km/sec.

In addition to scouring of gases by the early solar wind, the radiated heat of the early Sun warmed the inner planets, causing them to start to lose their atmospheres as the gas molecules gained sufficient velocity to escape their gravitational pull. The process gradually accelerated because the masses of the planets decreased as they lost their atmospheres. This, in turn, allowed the remaining atmosphere to escape more easily until no original atmosphere was left. As

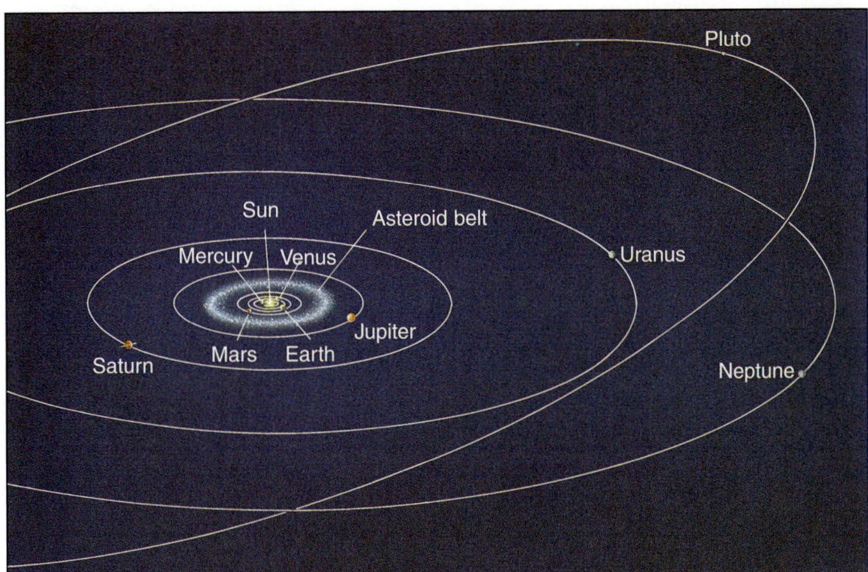

FIGURE 2.13 The Solar System comprises the terrestrial planets—Mercury, Venus, Earth, Mars, and the giant planets—Jupiter, Saturn, Uranus, and Neptune. Pluto does not fit into either category. The asteroid belt is either the remains of a planet that has broken up or the components of a planet that never completely assembled.

a result, the present atmospheres of the terrestrial planets are the products of ongoing internal degassing: the steady escape of gas molecules from the interior of a planet where the gas was trapped when the planet formed.

Earth's internal degassing generates our nitrogen- and oxygen-rich atmosphere faster than the gaseous constituents can be lost into space. At a given temperature, molecules of the atmosphere have a wide range of velocities with an average that depends on their mass; light molecules travel fastest. If significant numbers have velocities greater than the escape velocity, the population gradually diminishes as they leak off into space. In Earth's atmosphere, nitrogen and oxygen have velocities of about 0.5 km/sec, much slower than the escape velocity of 11.6 km/sec. Negligible numbers of molecules have velocities greater than the escape velocity. The time needed to severely deplete the atmosphere of these constituents is greater than the age of Earth. In contrast, the lighter molecules of hydrogen have an average velocity of about 2 km/sec and many hydrogen molecules travel faster than the escape velocity. Thus, most hydrogen is lost from the atmosphere by the escape process. The same is true for helium. Smaller bodies with low escape velocities, such as Mercury, the Moon, and most of the moons of other planets, have completely lost their atmospheres. The giant planets are different from the terrestrial ones in that they retained the original gases of the Solar System because they were farther away from the Sun's winds and heat at the time of formation and have higher escape velocities.

Ever since our planet's formation, its materials have undergone severe heating and pressure changes. This means that Earth's present material is chemically different from its original material. To model the chemical composition of Earth and how it may have changed with time, we must estimate the original composition of the solar nebula and the planetesimals that condensed out of it. Comets and meteorites provide samples of primitive material that may have remained unchanged since this condensation.

Comets

Not all of the original nebula condensed to form the planets and the Solar System. Some of the material in the farthest reaches of the Solar System appears to have condensed independently to form ice and rock accretions called **comets**. Comets, which have been described as mountains of dirty snow, come in a range of sizes and have elliptical orbits that may extend out from the Sun more than a thousand times the radius of the Solar System (Fig. 2.14). The "snow" is composed of ordinary frozen water plus carbon dioxide and other frozen gases; the dirt consists of grains of rocky material.

As the comet approaches the Sun, gas and dust are swept away from the comet by the solar wind (Fig. 2.15), leaving a spectacular tail that stretches millions of kilometers. When comets disintegrate, they leave trails of debris that continue their solar orbit along the path of the original comet. At times in its orbit, Earth intersects these trails. The cometary debris falls into Earth's atmosphere, burning up and producing the bright streaks of light in the night sky referred to as shooting stars or **meteors**. Analysis

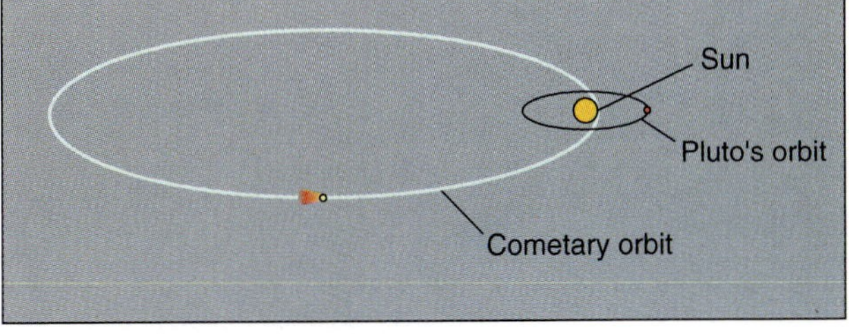

FIGURE 2.14 The orbit of a comet relative to the Solar System.

FIGURE 2.15 Halley's comet approaches Earth every 76 years during its orbit around the Sun.

of this light using spectrometers enables us to determine the composition of the comet, and by inference the composition of primitive Solar System material.

The approach of Halley's comet in 1986—an event that happens only once every 76 years—provided a rare opportunity for scientists to study the source material of the Solar System (Fig. 2.16). The spacecraft Giotto navigated to within 600 km of Halley's nucleus, a hazardous accomplishment since the comet and its surrounding dust cloud were traveling at 70 km/sec relative to the spacecraft. Dust particles at this speed impacting the spacecraft damaged the camera, but not before many images were relayed back to Earth. The nucleus of the comet was found to be a potato-shaped object 10 to 15 km in diameter. It is thought to consist of dust particles trapped in a cold mixture of methane, ammonia, and water ice, blanketed with an opaque layer of dust. The comet's composition was determined by measuring the spectra of light emitted from it. These data provided new information on the composition of the gases present in the comet and, thus, in the primordial solar nebula at the time of formation of the Solar System.

Asteroids and Meteorites

The distance from the Sun to each planet in the Solar System is found to increase in a uniform pattern, roughly doubling from planet to planet. This observation is known as Bode's law, after astronomer Johann Bode, and has never been satisfactorily explained. Between Mars and Jupiter, however, this law breaks down; the planet that is predicted by Bode's law to exist between the terrestrial planets and the giant planets appears to have either broken up or never to have formed. What remains is the asteroid belt, which contains rocky objects ranging in size from millimeters to several hundreds of kilometers across. Bode's law also breaks down badly for Pluto, which may be a large chunk of debris left over from the formation of the Solar System.

Collisions in the asteroid belt can cause fragments of asteroids to be ejected into orbits that follow a collision course with Earth. On entering Earth's atmosphere, most burn up, forming meteors or shooting stars; at this point, they are indistinguishable from incoming cometary debris (Fig. 2.17). Those fragments that do reach Earth's surface are called **meteorites** (Focus 2.2).

FIGURE 2.16 This tapestry (the Bayeux Tapestry) records Halley's comet passing Earth in the year 1066. Examination of Halley's comet by spacecraft in 1986 provided scientists with a rare opportunity to analyze material of the early Solar System.

FIGURE 2.17 A near miss. This meteor was so large it created a fireball over the Grand Teton mountains, Wyoming, that was visible in daylight. Referred to as a daylight fireball, had it struck Earth, the explosion might have been enormous.

If geologists want to know what is inside a rock, they might crack it open and examine its contents. On a planetary scale, asteroids might be the remains of a planet (or planets) similar to Earth that was cracked apart by bombardment by other planet-sized bodies in the early stages of the Solar System. Alternatively, they might just be the components of a terrestrial planet that was unable to completely assemble. In either case, meteorites provide us with samples of the asteroids that allow us to discover what may be inside a planet like Earth.

The Planet Earth

Determining Earth's Composition

We can directly measure the composition of crustal rocks as well as the composition of Earth's hydrosphere (the oceans) and atmosphere. We have even obtained samples of upper-mantle rocks brought to the surface by geologic processes. Some rare types of volcanoes (Fig. 2.18) may yield samples from depths of up to 200 km via kimberlite pipes—giving us an idea of Earth's composition at those depths. But these data sample only the outer 10 percent of the planet's volume. Scientists must use their ingenuity, clues from the cosmos, laboratory observations, and the laws of chemistry and physics to work out the composition of the remaining 90 percent of the planet, which includes most of the mantle and core.

Similarities Between Meteorites and Earth Many meteorites contain small, rounded inclusions with elements in different abundances from the surrounding material. The inclusions are called **chondrules** (Fig. 2.19, page 34) and meteorites containing them are called **chondrites**. Chondrules are thought to be droplets that condensed from the original solar nebula from which the Solar System derives. They have been preserved in their original form and composition ever since, protected by the surrounding shell of the meteorite that formed around them. Of the chondrites, the *carbonaceous chondrites* are of particular interest because up to 5 percent of their mass consists of various carbon compounds. This suggests that the building blocks for life exist both in outer space and on Earth.

Given that chondritic meteorites formed directly from the solar nebula, we expect that the relative abundances of elements within the bulk of Earth are similar to those of chondrites. The elements in chondrules are uniformly distributed throughout their volume in contrast to the distri-

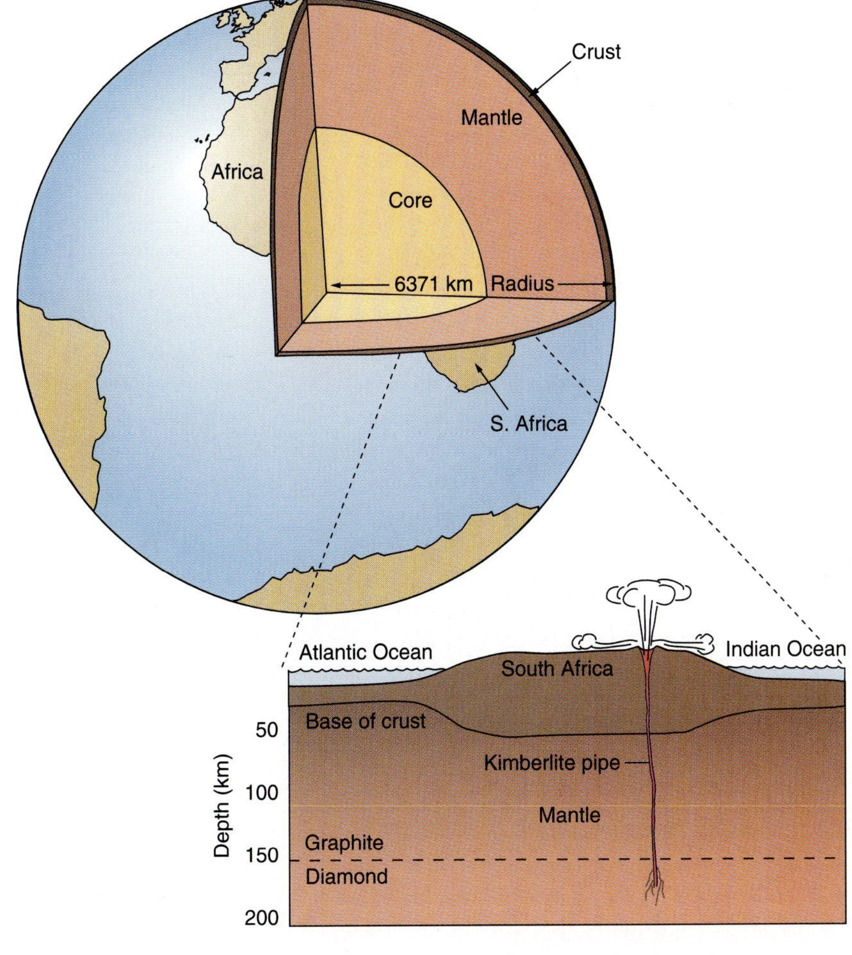

FIGURE 2.18 Volcanoes with deep roots bring samples of material to the surface from depths of up to 200 km. This is a small fraction of the radius of Earth, however, which is 6371 km. Geophysical methods, such as seismology, must be used to interpret Earth's composition at greater depths. At depths greater than 150 km, the stable form of carbon transforms from graphite to diamond. Kimberlite pipes are the deep conduits that transport rocks containing diamonds to the surface.

Focus On 2.2 IMPACTS AND THEIR EFFECTS ON EARTH

Impacts from asteroids and comets that have collided with Earth have left their record both on the surface in the form of craters and in the fossil record because of their effect on life. Giant impacts are currently favored as having caused a number of phenomena in the Solar System. The Moon is thought to have originated from the impact of a Mars-sized body with Earth when the Earth was forming. This would explain why the Moon appears to be formed from differentiated Earth material. If the impact occurred after the ion-rich core had settled to the center of Earth, the ejected material from the impact would come primarily from Earth's mantle, and would be relatively deficient in iron. It is thought that the Moon formed by re-accretion from this ejected material. Impacts can explain why the spin axis of Uranus is on its side, why Venus rotates backward, and why Neptune's spin is also highly tilted. These observations point to some large impacts that, depending on size and impact angle, changed the spin properties of these planets from the average.

Examination of the terrestrial planets and the Moon reveals a range of crater sizes generated by impacts (e.g., ∞ Fig. 1.4). On Earth, lava flows, erosion, and plate tectonics have just about eradicated the evidence of past impacts. There are, however, spectacular exceptions such as Meteor Crater in Arizona (Fig. 1). More than 30 tons of iron fragments from a meteorite have been found, mostly around the crater and rarely inside it. This was taken as evidence that the falling meteorite was made of iron. It was presumed that the bulk of this meteorite had become buried beneath the crater floor on impact. Early in the twentieth century, the crater floor was drilled to locate the parent body and to mine it for iron. No iron was discovered. Subsequently, gravity and magnetic surveys in the crater have found no evidence of a buried meteorite.

The lack of iron in the crater is now understood. The meteorite would have entered Earth's atmosphere at supersonic speed (14 km/sec or 31,500 mph). The tremendous stresses from passage through the atmosphere caused it to break up, and smaller fragments would have been slowed down significantly by atmospheric drag. The main body, however, probably hit Earth at enormous initial speed. The smaller fragments that scattered around the crater for distances up to 25 km are still being collected today. The large body would have speared into Earth and compressed elastically until its motion was arrested, perhaps several kilometers down. Once the motion stopped, the meteorite would have expanded in a giant explosion. It is estimated that there was enough energy in this process to vaporize the main body and excavate Meteor Crater, peeling back the geologic layering like a flower. As evidence for this, we can see a reversal of the sequence of geologic layering in the crater walls along the rim.

Other examples of craters are found worldwide. Evidence for very old craters, though, has been largely erased by erosion. Recently, magnetic anomalies have been used to uncover evidence for a truly giant crater in Africa. For many years it was recognized that one of the most significant magnetic anomalies on the continents is the Bangui anomaly in central Africa (Fig. 2). Evidence that it was caused by a giant meteorite impact came from topographic maps derived from space altimeters on satellites. The entire region was found to be surrounded by a perfectly circular crater rim with a diameter of 810 km. Such symmetry is most consistent with an enormous impact thousands of times larger than the one in Arizona. It has been estimated that the impact occurred more than a billion years ago.

In the geologic record, there are many examples of species that have become extinct. The dinosaurs, which became extinct about 65 million years ago, are a well-known example. In many cases, extinctions of many species are found to occur simultaneously. For example, at the time of extinction of the dinosaurs many other species also became extinct, ranging from the size of plankton to that of the largest dinosaur. Such extinction episodes are called "mass extinctions," and geologists seek to explain what major change in the environment might have been responsible. Understanding the origins of mass extinctions is of great import for the human species. Some possible causes include major changes in climate or intense volcanism (∞ see Chapter 15). Recently, extraterrestrial phenomena, such as the effects of giant impacts, have also been considered possible causes of extinctions. In the rock layer deposited at the time of the dinosaur extinction 65 million years ago, traces of a metal are found that is most likely to have come from a meteorite. Scientists suggest that a 10-km-diameter asteroid impacted Earth 65 million years ago, causing enormous widespread firestorms and a giant dust cloud. Debris in the atmosphere over a significant period of time would block out sunlight

FIGURE 1 Meteor crater, near Winslow, Arizona has a diameter of 1.2 km and is 0.2 km deep. A large iron meteorite weighing 300,000 tons is thought to have impacted the ground here 25,000 years ago at a speed of 10 km/sec. Although fragments are found outside the crater, none of the original body is thought to have remained inside, having vaporized at the time of impact.

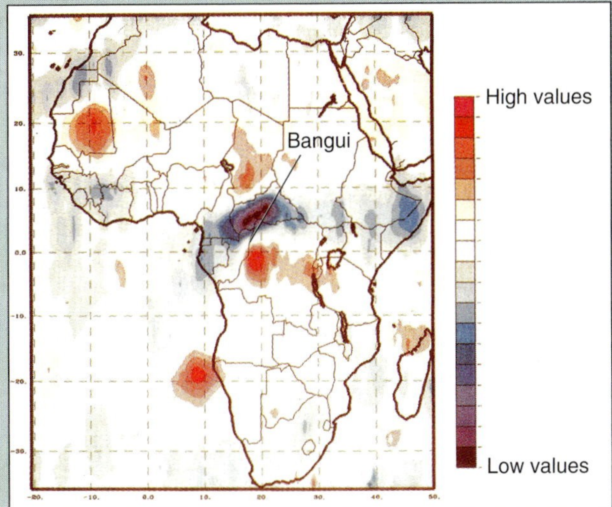

FIGURE 2 The magnetic field in Africa shows that one of the largest magnetic anomalies on the continent is associated with the Bangui impact site.

and inhibit photosynthesis, causing mass extinctions in the biosphere. A crater in the Yucatan Peninsula, Mexico, is currently thought to be the most likely site for this impact.

Some researchers have suggested that such impact-generated extinctions have occurred many times in the past, with a periodicity of about 26 million years. They hypothesize that Earth is bombarded periodically in response to the gravitational effects of a body (sometimes fancifully called "The Death Star") that has an orbit which brings it into proximity with Earth with this periodicity. Such a body has never been seen astronomically. Also, there is considerable debate as to the reliability of dating the extinctions and whether, as in the case of the dinosaur episode, they are even associated with extraterrestrial material. Nonetheless, NASA conducts a comet-watch program, and deflecting a 10-km comet from an Earth collision course may not be out of the question. Is the human species at risk or even under threat of extinction from an impact? A recent event reveals that the planets are still vulnerable. In July 1994, chunks of comet (named Shoemaker-Levy after its discoverers) collided with the planet Jupiter at a speed of 60 km/sec. The largest fragment generated an explosion greater than that of a billion nuclear bombs, making it the largest explosion on a planet in the Solar System since telescopic observations began. It has been estimated that it was larger than the explosion thought to have caused the extinction of the dinosaurs. Spectral analysis of the light from the fireball confirmed its cometary composition—that of a dust-impregnated snowball. The comet disintegrated in Jupiter's upper atmosphere, before reaching Jupiter's thick cloud banks.

On a day-to-day basis, however, the chance of being hit by a meteorite is negligible—although some near misses have occurred. On October 9, 1992, a fireball brighter than the full Moon appeared over West Virginia. It occurred at a time when many Friday night football games were being played and was recorded by a number of television camera operators. At the end of its 700-km trajectory, a 12-kg fragment passed right through a parked car at Peekskill, New York. Scientists retrieved many of the television records and pieced together its history, finding it to be consistent with an asteroid fragment having an Earth-intersecting orbit. The car was destroyed, but its owner bought a new one with the proceeds from selling the meteorite to a museum.

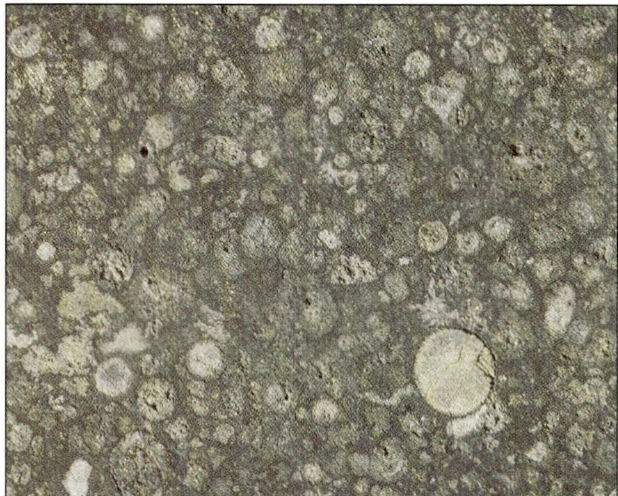

FIGURE 2.19 The early material of the Solar System condensed in droplets called *chondrules*. Chondrules are preserved in meteorites called *chondrites*, which have not undergone subsequent heating.

bution of elements found within Earth. If we propose that Earth formed as one giant chondrite, or agglomeration of chondrites, we must determine how the separation of elements into the iron core, the silicate mantle, and crust occurred; that is, how the planet became differentiated.

Evidence from Meteorites The chondrite meteorites are about 35 percent iron in composition. Rocks on Earth's surface contain an average of just 6 percent iron. This is a puzzle because the meteorites are thought to represent material typical of the terrestrial planets, so we would expect to find similar amounts of iron in the two rock types. Did Earth once contain more iron, or are terrestrial planets so different that we cannot use meteorites to tell us anything about Earth?

FIGURE 2.20 A core–mantle boundary as seen preserved in a stony-iron meteorite, which indicates that core differentiation occurred in asteroidal planetesimals.

Once again, we turn to the meteorites for clues as to what might have happened. Many meteorites other than chondrites show evidence of melting. Most are composed of nearly pure iron; the remainder are mainly composed of rocky material similar to Earth's mantle. A few are rich in both iron and rocky material (Fig. 2.20). The existence of these different types of meteorites can be explained if they are the result of melting of chondritic material. It is inferred that the meteorites originated from a parent body of chondritic composition, which was sufficiently large that the heat of formation caused them to melt.

As planetesimals coalesced to form Earth, the bombarding impacts generated frictional heat. We can demonstrate this effect by pounding on a piece of metal with a hammer and observing that the metal becomes warm. This generation of heat would have led to two significant effects: First, this heat would have boiled off gases; second, the heat would have melted the planet. At this stage, the planet would have consisted of a mixture of molten iron and rock. A volume of molten iron weighs more than the same volume of molten rock. The property that describes this difference is called **density**, which is defined as the amount of mass in a given volume (∞ see Focus 2.3). Materials with high densities will sink in materials of lower density; conversely, low-density materials will float on those of higher density. When Earth melted, denser materials such as iron (that were not chemically bound to stony materials) sank in toward Earth's center. Gravitational settling of the iron core is thought to have occurred rapidly, at least in the first 2 percent of Earth's history (i.e., the first 100 million years). Separation of the core from the rocky part of the planet explains the low concentrations of iron found in Earth's rocks. This was the first stage in the partitioning of Earth into the distinct compositional zones that we see today. This partitioning, known as **differentiation** is largely the result of differences in density. Evidence of this process is seen in the stony-iron meteorites (Fig. 2.20).

Heating of primitive chondritic material that formed Earth caused separation between iron and rock, resulting in an iron core and a rocky mantle. The low concentration of iron near Earth's surface can be explained if the iron has separated from the silicate mantle and settled into the core.

Differentiation of Earth

Over the 4.55 billion years since its formation, the material forming Earth has partitioned into five major zones: the core, mantle, crust, hydrosphere, and atmosphere (Fig. 2.21). The compositions of these divisions are quite different. Separation into core and mantle is believed to have occurred during and shortly after the time of Earth's formation. In contrast, the crust, hydrosphere, and atmosphere have evolved over a much longer time.

Focus On 2.3 CALCULATION OF DENSITY

Throughout this text, we use the Standard International (SI) system of units whenever we substitute into a formula. This system uses the units of kilograms, meters, and seconds. The density of water is 1000 kg/m^3 in this system. However, the units of grams/cm^3 (g/cm^3) are still used in many earth science reports because densities in this system are quite simple numbers. For example, the density of water is 1 g/cm^3, which is equivalent to 1000 kg/m^3. The density of crustal rock is, on average, 2.67 g/cm^3. The density of mantle rock is about 3.3 g/cm^3. To convert to SI units, multiply g/cm^3 by 1000 to obtain kg/m^3.

GUIDE TO SUBSTITUTION IN FORMULAS

1. Make sure all values are in SI units.
2. For ease in evaluation, express each value as a power of ten; that is, as a number between 1 and 10 multiplied by 10 raised to an integer power. For example, 5976×10^{21} kg becomes 5.976×10^{24} kg.
3. Collect powers of 10.
4. Make a mental estimate of the result by rounding off values to the nearest integer and calculating a rough answer in your head.
5. Use a calculator to obtain the final result, and make sure it agrees reasonably with your estimate.
6. Include the units in your answer.

AN EXAMPLE

Calculation of the density of Earth:

$$\begin{aligned} \text{mass} &= 5.976 \times 10^{24}\ \text{kg} \\ \text{volume} &= 1.083 \times 10^{21}\ \text{m}^3 \\ \text{density} &= \text{mass/volume} \\ &= \frac{5.976 \times 10^{24}}{1.083 \times 10^{21}} \\ &= 5.52 \times 10^3\ \text{kg/m}^3 \\ &= 5.52\ \text{g/cm}^3 \end{aligned}$$

Check: 5.97/1.08 is approximately 6/1 = 6. Note that this is an *average* density. As we will see in Chapter 5, density actually varies enormously.

It has been estimated that the amount of energy released from the impacting planetesimals that made up Earth, together with the gravitational energy released when iron sank into the center to form the core, was more than enough to bring the entire planet to a molten state. We might imagine a planet that would have appeared to space travelers as Io, a moon of Jupiter, appears to us now (Fig. 2.22). A crust might have formed at Earth's surface, which was cooled as heat radiated into space but was continuously broken by turbulent motion in the red-hot molten rock below it—much as we see today in a lake of molten lava (Fig. 2.23). Incandescent eruptions of red and white hot magmas spewing out into space released gases into the early atmosphere. The thin crust would have been formed and devoured by the boiling cauldron and re-formed again. As far as we know, none of these early rocks are preserved today (see Fig. 1.17). At some stage in this early chaotic part of Earth's history, the planet is thought to have been subjected to an enormous impact, which ejected a fraction of the mantle material. This material settled in orbit around Earth and consolidated as our Moon (Focus 2.4).

The oldest rock yet discovered on Earth was found in northwestern Canada and is nearly 4 billion years old.

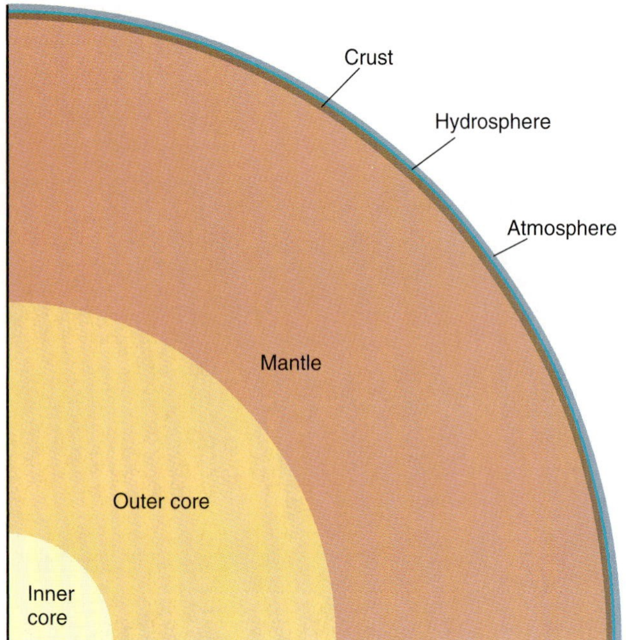

FIGURE 2.21 The five zones of the differentiated planet Earth: (1) inner and outer core, (2) mantle, (3) crust, (4) hydrosphere, and (5) atmosphere. (Compare with Figure 1.5)

This rare find suggests that, during the first 500 million years of Earth history, the surface material was recycled back into the planet's interior. Eventually, rocks formed that were so buoyant—had such low densities—they resisted being recycled.

Earth formed 4.55 billion years ago. It took 500 million years for Earth to cool sufficiently to allow formation and preservation of rocks.

FIGURE 2.23 Mauna Ulu lava lake in Hawaii may resemble the surface of primordial Earth.

Core, Mantle, Crust, Hydrosphere, and Atmosphere

Our planet's *hydrosphere*—the portion of Earth that is water—and the gaseous *atmosphere* have been released from the rocky interior and have accumulated at the planet's surface. Scientists estimate that the present-day oceans and atmosphere largely formed in the first 500 million years of Earth's history. Nevertheless, material is still being lost as light elements escape Earth's gravitational field; at the same time, the oceans and atmosphere are being replenished as the planet continues to degas—mainly as a result of volcanic eruptions.

Some of the constituents of the oceans and atmosphere are recycled into the mantle by plate tectonics. Water reacts with ocean floor basalts to form minerals that are carried into Earth at subduction zones. The water is then released back into the atmosphere/hydrosphere

FIGURE 2.22 A Voyager picture of a volcanic eruption of Jupiter's moon, Io; this is how early Earth might have once looked.

system at arc volcanoes or, if completely entrained in the convection system, can be recycled at hot spots or mid-ocean ridges. Carbon dioxide and oxygen are absorbed from the atmosphere during weathering and recycled back into Earth as sediments are subducted.

Earth's early atmosphere was made up of gases vented at volcanoes. Volcanic gases are rich in water vapor, methane, carbon dioxide, sulfur dioxide, and gases containing nitrogen. Carbon dioxide and sulfur dioxide reacted with rocks. Radiation from the Sun decomposed water and the nitrogen gases into hydrogen, oxygen, and nitrogen. The lighter hydrogen escaped Earth leaving free nitrogen and oxygen. Because oxygen reacts readily with rocks to form oxides, any free oxygen was removed from the atmosphere as soon as it formed. Once life appeared on Earth more than 3 billion years ago, photosynthesis converted carbon dioxide to oxygen. The production of oxygen by life organisms became faster than loss from the atmosphere. Thus the free oxygen in Earth's atmosphere, so important for the sustenance of life, relies on life itself.

Earth's *crust* is the outer rocky layer extending to a depth of about 30 km beneath the continents and to an average depth of 7 km beneath the ocean floor. The crust is largely made up of minerals containing the elements calcium, aluminum, magnesium, iron, silicon, sodium, potassium, and oxygen. Between the base of the crust and the core lies the *mantle*, which is formed from minerals containing mostly magnesium, iron, silicon, and oxygen. The *core* occupies the inner 3480 km of the planet and is divided into a solid *inner core*, which extends out to a radius of about 1220 km, and a liquid *outer core* that is approximately 2260 km in thickness. Based on analogy with meteorites discussed earlier, the core is believed to be composed mainly of iron.

Earth is thought to have formed from material similar to that of chondritic meteorites. The planet then became differentiated into core, mantle, crust, hydrosphere, and atmosphere.

Earth's Core: Is It Iron?

We will spend much of the rest of the text discussing geologic processes affecting the mantle and the crust, but we pay far less attention to the core. We have never sampled the core. Yet the core has great importance: It generates Earth's magnetic field, which has been used for centuries to navigate by compass. The core contains most of Earth's iron, and is important in understanding how Earth's geochemical reservoirs are distributed.

Let us take this opportunity to review some of the most important indirect observations bearing on the nature of Earth's core.

Evidence from Density We can investigate the composition of a planet by calculating its density. The planet's density can then be compared with the densities of known rocks, and we can draw some tentative conclusions about the planet's composition.

How do we accomplish this? If we know both the mass of a planet and its volume, we can calculate the planet's density by simply dividing the mass by the volume (see Focus 2.3). The **mass** of a planet is a measure of the amount of matter it contains, and the mass can be estimated from the orbital periods of its moons or satellites by applying Newton's laws of motion. An orbital period is the time it takes a moon or satellite to complete one circuit of an orbit. The volume of a planet is measured by surveying techniques—either on Earth or from space.

We know that typical rocks found on Earth's surface have densities of about 2.5–3.0 g/cm^3, yet Earth has an average density of 5.5 g/cm^3. How can we reconcile these values? With increasing depth, the pressure within a planet becomes immense. The weight of the overlying rock causes the volume of the interior to be compressed, which increases the density. We can measure this effect in the laboratory; however, if we compress surface rocks to pressures equivalent to Earth's interior, we do not obtain an overall average density as high as 5.5 g/cm^3. We conclude, therefore, that Earth is not made entirely of rock, and that the material deep within Earth must have a much higher density than simply that of compressed rock. If we perform our calculations assuming that Earth's core is composed of iron (plus a small amount of nickel), the overall density then nearly matches the observed value.

Evidence from Seismic Waves We cannot drill to Earth's core; the deepest hole is currently being bored in the Kola Peninsula in Russia—at 15 km, it would still have to penetrate 2900 km before reaching the core! We therefore rely on indirect evidence such as analysis of seismic waves to ascertain the composition of the core. An earthquake generates seismic waves that travel through the interior of the planet. By analyzing these waves as they travel along different paths through Earth, seismologists have constructed a model of Earth's interior (∞ see Chapter 5). A sudden increase in Earth's density has been found at a radial distance of 3480 km from the planet's center, which is the core–mantle boundary. The density of the core calculated from these seismic data fits with what is expected for the density of iron at those enormous pressures. (In fact, the density is a bit less than expected, pointing to a presence of a second, lighter element. There is much debate as to the identity of this lighter element—sulfur or oxygen being currently favored.)

Evidence from Planetary Precession Another line of evidence that the interior of Earth is much denser than its exterior comes from observations of Earth's motion with respect to the Sun, the Moon, and the stars. Imagine Earth as a spinning top with the spin axis tilted (23.5°) from the axis of its orbit around the Sun (Fig.

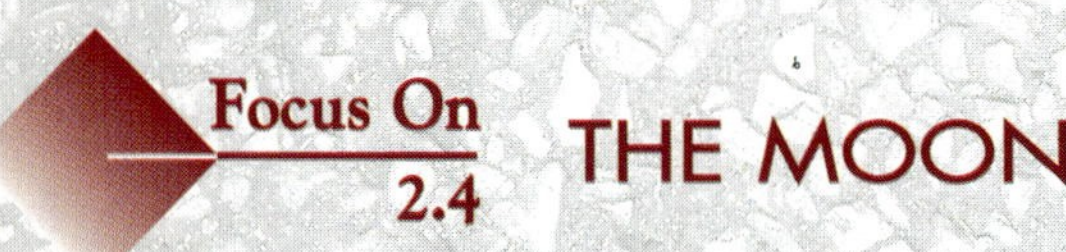

Focus On 2.4 THE MOON

Rocks returned to Earth from the Apollo space program showed that the Moon's composition is similar to that of Earth's mantle; in particular, they are both deficient in iron relative to meteorites. The low density of the Moon compared with the densities of the terrestrial planets is also consistent with a lack of iron. The Moon is also deficient in compounds such as water, indicating that it probably formed hot, but cooled rapidly. Because the Moon is a cool and water-deficient body, its geologic history is completely different from that of Earth.

The favored explanation for the Moon's formation is that it is the result of a Mars-sized body impacting Earth after Earth had settled out its core. This violent origin would have projected mantle material into space, which agglomerated to form the Moon. Then the iron deficit can be explained because much of the ejecta would have arisen from Earth's mantle. Moon rocks are nearly as old as the age of Earth; thus, its formation occurred at a time when large-scale impacts were common in the early Solar System. Other suggested explanations for lunar formation such as capture or simultaneous accretion with Earth must answer why the Moon is so different in iron content from meteorites, Earth, Mercury, Venus, and Mars.

The first 1.5 billion years of the Moon's history were violent. The crust differentiated from the mantle. Massive bombardment of the surface by meteorites left giant craters that persist today. Volcanism was widespread. It has even been suggested that, upon condensation, the Moon was covered with a magma ocean. The first billion years included volcanism that poured basalts onto the crust, filling giant impact craters to form the lunar maria, which stretch over hundreds of kilometers. These dark basalt flows can be seen by the naked eye as the darker regions on the surface of a full Moon (see Fig. 1.4). Maria (singular *mare*) means "seas," named by Galileo, who, using his low-resolution telescope, thought they might be lunar seas. The youngest volcanic rocks are about 3 billion years old. Since this time, the Moon has apparently been tectonically inactive. Its small size allows the heat of radioactivity to escape by conduction, without the violent effects of convection on a larger planet such as Earth.

Examination of the surface rocks and seismological investigations indicate that the Moon is a highly differentiated body with a crust, mantle, and a small core. The radius of the Moon is 1738 km. The lunar crust is varied in thickness and, on average, is thought to be about 100 km thick. The core is thought to begin at a depth of 1000 km. The Moon's surface is covered by the lunar *regolith*: a 1-km layer of pulverized rock created by bombardment of meteorites and other cosmic particles. The upper several meters of the regolith are a fine powder in which the Apollo astronauts left their footprints.

Sedimentary processes on the Moon differ markedly from those on Earth, owing to the lack of water. Without an atmosphere or a hydrosphere to weather rocks and deposit sediments, solar energy—which plays such a major role in shaping Earth's surface—is ineffective on the Moon. Another consequence of the Moon's being a cool body is the absence of plate tectonics to cause uplift. Weathering and erosion occur by meteorite impacts, fragmenting surface rocks and distributing them across the surface to form the regolith. Some impacts are thought to have been large enough to eject lunar fragments from the Moon to be captured by Earth. Meteorites found on Earth with lunar geochemical signatures are thought to have originated in this way.

The Sun and the Moon cause tides on Earth. The energy dissipated in the tides is taken from the Moon's orbit. Early in its history, the Moon's spin was decreased by tidal forces to the point that it now presents a constant face to Earth. Thus, we never see the far side of the Moon from Earth. Tidal forces also have caused the lunar orbital distance to increase with time by about 4 cm/century. The Moon was significantly closer to the Earth in the past and would have generated much larger tides. The increase in the orbital distance has been measured directly using lasers to track distance as a function of time. In addition, tidal forces from the Moon cause Earth's spin to slow down. Using growth rings in fossils, it has been estimated that, 650 million years ago, the length of a day was just 22 hours.

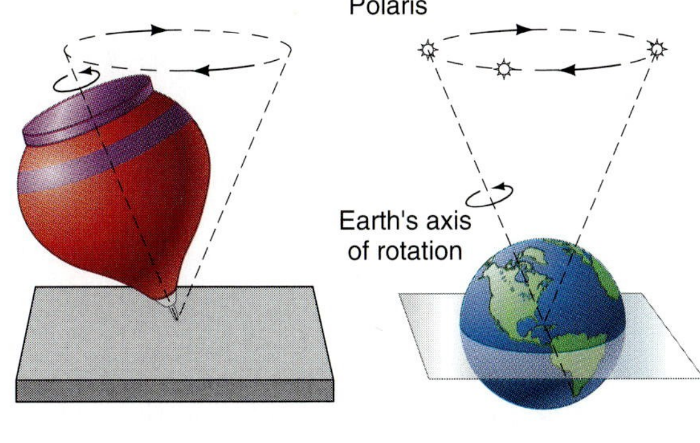

(a)

(b)

FIGURE 2.24 (a) A top precesses because gravity tries to change the direction of its spin. In an analogous manner, the Sun (and Moon) pull on Earth and try to change its direction of spin. (b) Earth's axis currently points toward the North Star (Polaris). About 12,000 years from now it will point toward a star named Vega, and in 26,000 years it will complete the loop to point again at the North Star. Earth's precession takes 26,000 years to complete one loop. This photo depicts a map that traces the path of the axis on a star chart.

2.24a). Presently, Earth's spin axis points at the North Star (Polaris) but it varies a little from year to year. If this variation were tracked over 26,000 years, the direction in which the axis points would leave the North Star, make a loop up to 47° away, then return to the North Star (Fig. 2.24b). This motion is called the **precession** of Earth, which refers to the slow movement of Earth's rotation axis (Fig. 2.24a). Scientists can now make very precise astronomical measurements of this process.

Why does Earth precess? Earth's spin causes it to bulge slightly (21 km) at the equator and to contract a little at the poles. As a result, Earth's shape, although approximately spherical, is more strictly described as an "oblate spheroid." Because Earth's spin axis is tilted, the gravity fields of both the Sun and the Moon pull on the tilted bulge of Earth and try to line up the axis. Earth resists this gravitational pull. The overall effect is that the spin axis precesses about the orbit axis, making a complete loop once in 26,000 years. This is similar to the precession of a top when it is spun on the floor (Fig. 2.24a). If the top is perfectly vertical, it spins smoothly and its axis stays in one place. If it is set spinning at an angle to the vertical, rather than just falling over, it precesses. Its spin axis rotates at an angle around the vertical at a much slower rate than the rate at which the top is spinning. Precession is commonly observed in spinning systems when an external force attempts to change the direction of spin. Simply put, if you try to tilt a top sideways, it will start going around in circles.

By measuring Earth's precession, Earth's bulge, and estimating the forces of gravity of the Sun and the Moon, it has been shown that Earth's interior must contain material of a much higher density than that which exists on its surface. For example, if Earth had the same density throughout, its precession period would be 31,500 years rather than the 26,000 years observed. Iron in Earth's core explains the precessional observations.

In summary, three reasons exist for proposing the existence of iron at Earth's core: (1) the density of the planet, (2) the passage of seismic waves through the planet, and (3) the precession of the planet. Given evidence from meteorites indicating that terrestrial planets should have much more iron than is found in Earth's surface rocks, the argument for iron in the core is a persuasive one.

A dense iron core explains the overall density of Earth, the passage of seismic waves through the planet, and Earth's precession.

How Earth Works

Primordial Heat and Radioactive Heat Earth is a giant heat engine. The heat energy within it causes upwelling motions in the mantle that drive plate tectonics—creating earthquakes, volcanoes, mountains, and rifts. Earth's internal energy comes from two main sources. The first is primordial heat accumulated during the process of Earth's formation: Heat from impacting planetesimals, and the release of gravitational energy as molten iron metal sank inward to form the core. If primordial heat were the only source of energy, Earth should have cooled much more rapidly over the 4.55 billion years since its formation; in fact, it should have cooled to the point where it would be tectonically "dead," with an unchanging surface like that of the Moon. Earth, however, is clearly a dynamic planet with a surface that is in a continual state of change as a result of plate tectonics. Another source of heat must be present; this second source of heat is the heat of **radioactivity**.

During radioactive decay, the nucleus of a radioactive isotope spontaneously disintegrates into another element (∞ see Chapter 1) while emitting radiation. When this occurs, the recoiling particles of the disintegration are brought to rest by the surrounding rock, thus generating heat. The main heat-producing elements within Earth are uranium, thorium, and potassium. Each disintegration produces a very small amount of heat. When summed over the huge bulk of Earth, however, the heat produced is significant.

Heat Transfer One of the keys to understanding how Earth works is a basic understanding of **convection**, a process by which heat is transferred within Earth. Convection is based on the same principle of flotation. Imagine a blob of material deep within Earth that becomes heated. The material expands, which reduces its density, and it becomes lighter than its surroundings. It floats toward the surface, where it may cool to such a degree that it now becomes denser and heavier; it then sinks back into the interior. This cycle of floating and sinking is called convection (Fig. 2.25a). It occurs in both Earth's mantle and core, and it transports heat efficiently from the planet's interior to the surface.

A much slower process for transporting heat is called **conduction** (Fig. 2.25b). In this case, the material is strong enough to resist rising to the surface; instead, the heat is carried internally from atom to atom. The heat moves from regions of high temperature to regions of lower temperature. The amount of heat flow depends on the temperature difference divided by the distance the heat must travel. Conduction is the most common method of heat transport in the lithosphere.

Heating of a material that can flow may cause convection where hot expanded material rises and cooler material sinks. Heat can be transported within a strong material by conduction.

The Effects of Convection upon Earth The broad picture of convective motions within Earth consists of lithospheric plates moving fairly rapidly across a slowly convecting mantle. Beneath the mantle, complicated and vigorous motions are taking place in the core. The core–mantle boundary is a zone of decoupling between these two regimes, where heat from the fast-moving core feeds into the slow-moving mantle. Because the bulk of the mantle probably moves too slowly to rid the core of heat efficiently, it has been proposed that **plumes** of rising hot material are generated at the core–mantle interface, heated at their base by uprising convection in the core. This heated mantle material rises in columns with a broad return flow of material to replace the lost mass (Fig. 2.26).

The total heat energy that flows out of Earth today is estimated to be 42×10^{12} watts. For comparison, the total use of electrical energy in the world is 10^{12} watts. A model of Earth's heat budget based on estimates of the distribution of radioactive elements throughout Earth finds that 20 percent of this heat comes from crustal radioactivity, 50 percent from mantle radioactivity, and 25 percent from the cooling of Earth—releasing the residual heat from accretion (Fig. 2.27a). The rest of the heat results from other internal processes beyond the scope of this text.

Because the number of radioactive atoms within Earth decreases with time, heat generated by radioactivity must have been much greater in the past. We know the decay rates and the present concentration of radioactive ele-

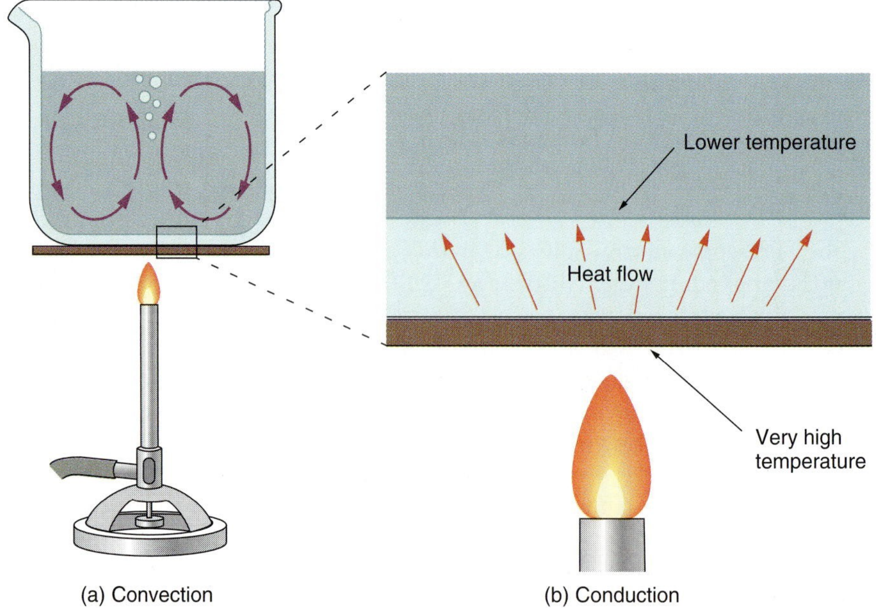

FIGURE 2.25 (a) A familiar example of convection, in which soup is heated in a pot. Hot soup at the bottom of the pot above the flame expands and rises because of its buoyancy. It cools upon reaching the surface and sinks, completing the cycle. Within Earth, the heat is mainly supplied by radioactivity. (b) When the material through which heat flows is strong, like the base of a pot, the heat is transported by conduction. Conduction is the mechanism by which heat is transported across the lithosphere—the strong outer shell of Earth.

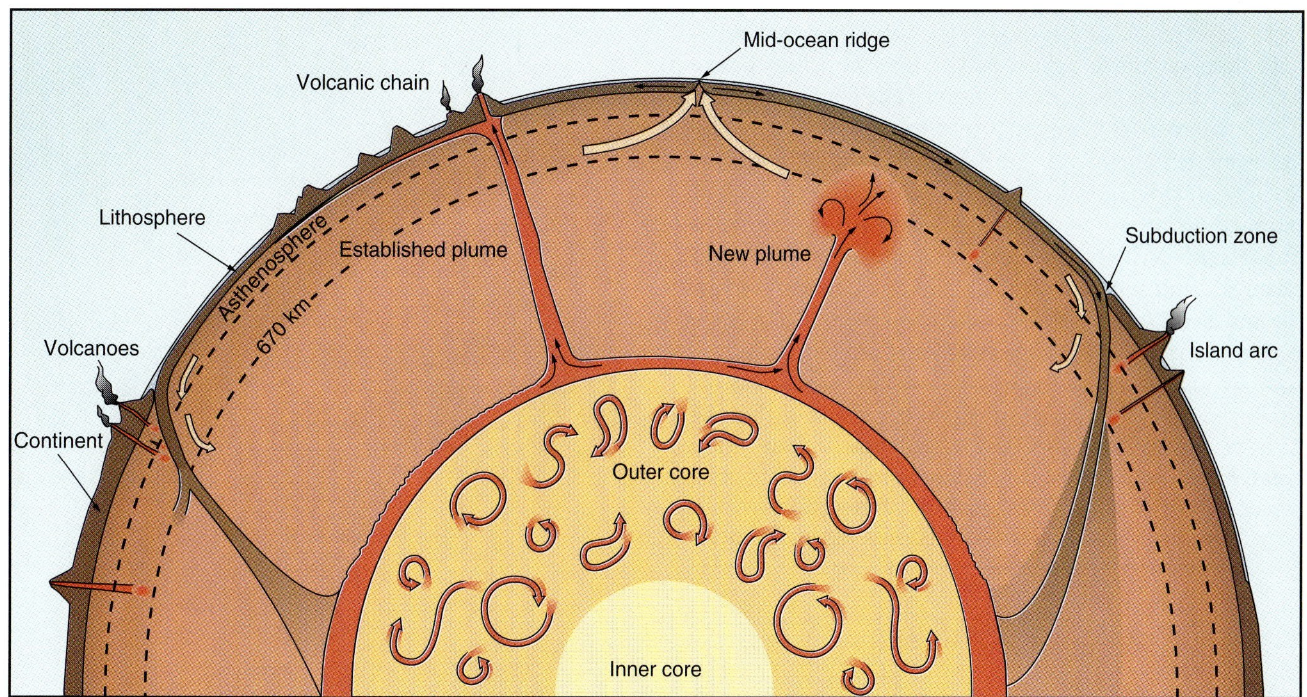

FIGURE 2.26 General circulation within Earth's interior. Rapid convection in the outer core drives Earth's geomagnetic field. Plates slide above the asthenosphere, and the mantle flows slowly into gaps located at mid-ocean ridges. Plumes feed hot spots at a rate that is faster than the overall mantle motion.

ments. Using these, we can estimate that, at the time of Earth's formation, the heat generated by radioactivity was nearly four times greater than it is today. Earth was much hotter and was cooling more rapidly.

Figure 2.27b illustrates one model for the thermal history of the planet, which estimates that Earth was heated 4.55 billion years ago to its average mantle melting temperature of about 3000°C. Because of vigorous convection early in formation, this model proposes that the average mantle temperature has dropped to its present value of about 2200°C. Today, there is too much internal heat generated by radioactivity for it to be conducted out passively, as is the case with both Mercury and the Moon. Because heat is conducted so slowly, the radioactivity will cause the temperature to rise, causing the rock to flow, and the heat will be carried away by convection. This has had a stabilizing effect on temperatures within Earth, which have dropped by only 26 percent since its formation

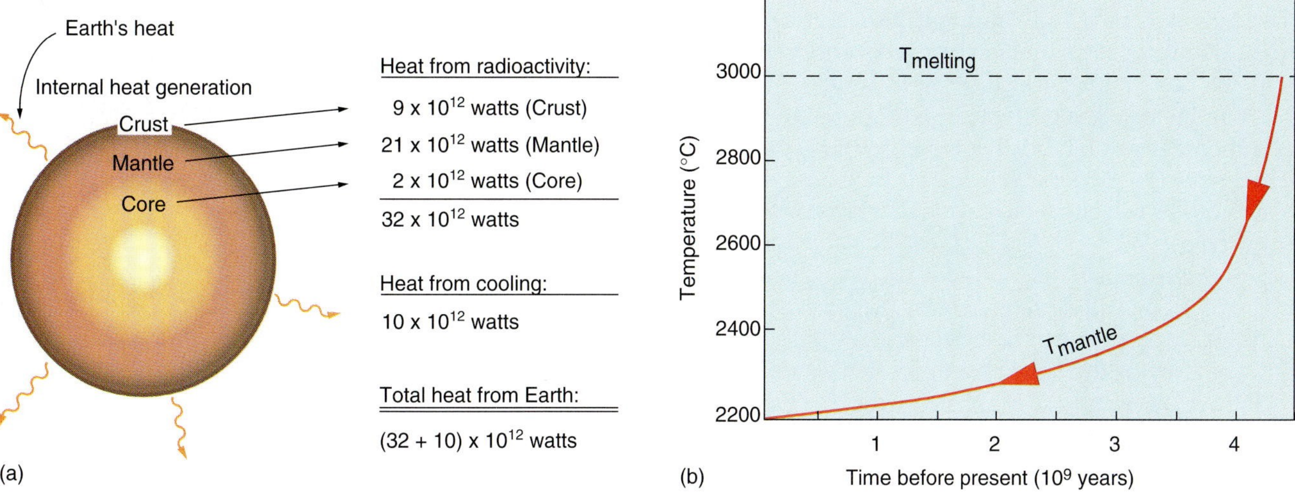

FIGURE 2.27 (a) Heat budget for Earth. (b) The decrease in the mantle's internal temperature since formation of Earth. Based on a model by F.D. Stacey (1992).

(3000° to 2200°C) even though the internal heat sources have dropped by a factor of 4. The most significant change has been an alteration in the vigor of the convection.

We will see in Chapter 6 that the movement of the lithospheric plates across Earth's surface, their creation at mid-ocean ridges, and their sinking into the interior at subduction zones all provide evidence that Earth is indeed convecting. Plates move at a range of speeds up to about 10 cm each year. As they sink back into the mantle and the material spreads out, however, the average flow in the mantle is much less than 1 cm per year.

At the time of Earth's formation, it is thought that the heat generated by radioactivity was nearly four times greater than it is today. Earth has been steadily cooling since its formation. Consequently, convection within Earth's interior has become less vigorous through time.

The Future of Earth and Our Solar System

The Big Crunch

We can only guess as to what might have happened before the big bang. Perhaps a previous universe—created by its own big bang—had gravitationally collapsed, resulting in the explosion that began our universe. A major question in predicting the future of the universe is whether there is enough matter in it that its gravitational attraction will eventually overcome the expansion caused by the big bang. The expanding movement would be replaced with a contracting motion, ending in what has sometimes been called the "big crunch." Perhaps this would then set up conditions for another cycle as nuclear forces, resisting the gravitational implosion, build up and explode, generating a new big bang and, in turn, a new universe.

The End of the Sun and Earth

The Sun is thought to be nearly halfway through its life span. During the next 1 billion years, it has been estimated that the Sun's output will increase by about 10 percent. Such an increase has been predicted to trigger a runaway greenhouse effect on Earth that would boil away the oceans and destroy life. About 8 billion years from now, the Sun's hydrogen fuel will burn out, and its outer layers will expand to encompass the inner planets of the Solar System. It will have become a giant star called a *red giant*. In its dying phases, the Sun will explode and may extend even as far as Earth's orbit. It then will become a *white dwarf* (a burned-out star), leaving the Solar System a frozen wasteland.

If Earth survives the dying phases of the Sun's activity—which may be quite violent—it will necessarily change quite dramatically. Internal convection would continue, but at a slower pace as the amount of radioactivity diminishes. Further degassing of the interior will continue.

(a)

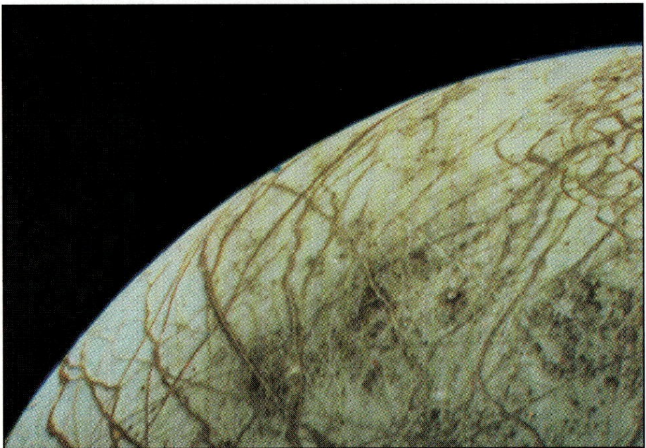

(b)

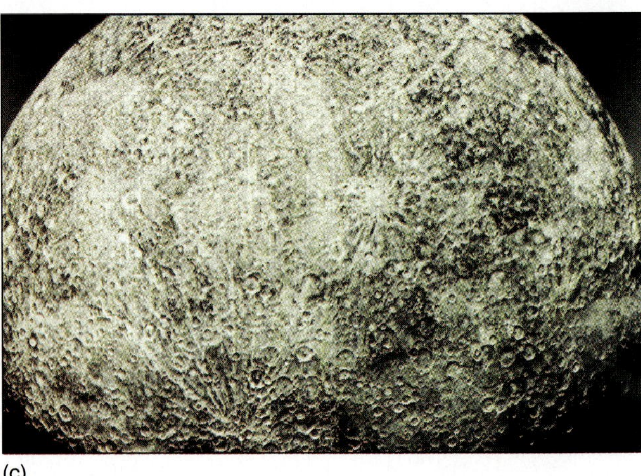

(c)

FIGURE 2.28 The present and future of Earth? (a) The present; (b) possibly frozen in the future and ice-covered, similar to the moon Europa; and (c) in the distant future, cratered, similar to Mercury.

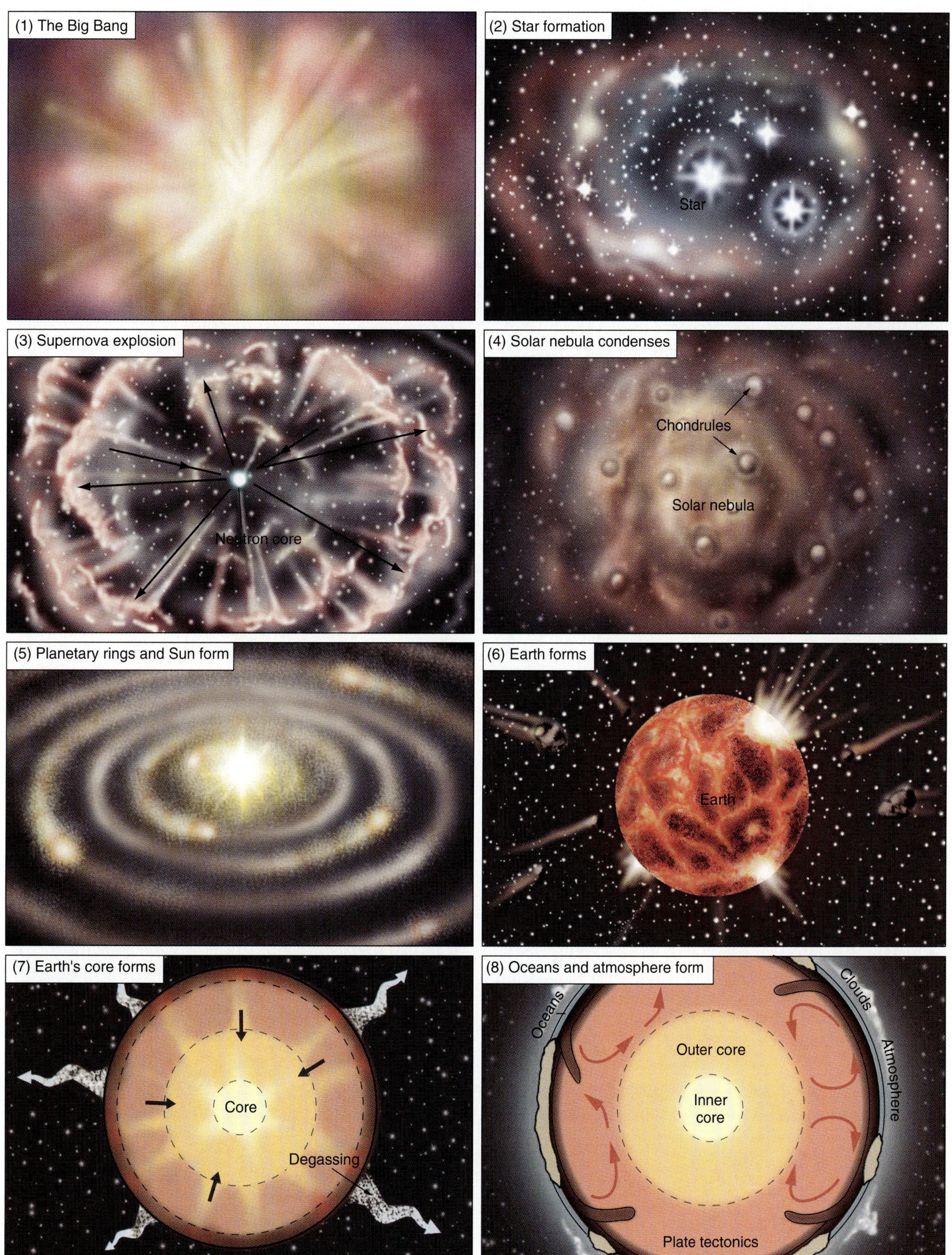

FIGURE 2.29 Eight steps in the history of Earth.

Deprived of the Sun's heat, any surface water will freeze. Earth might appear like Europa, one of the moons of Jupiter that is covered in ice (Fig. 2.28b). Eventually, a time will come when radioactive heat sources have diminished to such an extent that the heat can only be passively carried out of the interior by conduction. Then plate tectonics will grind to a halt. Volcanism and earthquakes will cease. Vigorous degassing and replenishment of the atmosphere will stop, and the atmosphere and frozen oceans will escape into outer space. Life will long since have ceased to exist. Without plate tectonics to generate new surface, Earth will start changing from a planet dominated by erosion features to a surface pock-marked by meteor craters, such as we see today on the Moon or Mercury (Fig. 2.28c). Eventually, tidal effects from the Sun will cause Earth to spiral inward toward the Sun and collide with it.

For the human species to survive, it will require the technology to venture to another solar system. A habitable planet in that system must possess a number of unique features. It must be at just the right distance from its own sun to produce a temperate climate (0° to 100°C so that water can exist in its liquid form). It should have generated plate tectonics so that it forms and constantly replenishes an atmosphere required to sustain life and to protect the planet from meteorite impacts. A magnetic field is needed for additional protection from tissue-damaging cosmic rays that constantly rain in from space. The plate tectonics, however, should not be too vigorous, otherwise humans must constantly deal with violent earthquakes and volcanic eruptions. This means the planet, like our present Earth, should be in its middle age—not too young and boisterous nor too old and "out of gas."

We have seen how the elements of the grains of sand mentioned in the introduction were formed somewhere out in space in stars and supernova explosions. We have followed their separation from the iron in Earth by differentiation and their possible transport to the surface by mantle convection. This long and checkered history is illustrated in Figure 2.29. These elements still have a long way to go from this point, however, and many geologic processes to experience before they eventually form into the sand grains and reach their resting place on the beach.

SUMMARY

- Atoms have unique spectra that appear as colored lines of different brightness in a spectrum. The spectra of stars reveal their compositions.
- Nuclear energy drives the universe. Nuclear fusion occurs when nuclei bind together to form new atoms, releasing energy in the process. Nuclear fission occurs when nuclei split apart to release energy.
- Elements up to atomic number 26 (iron) in the periodic table are made by nuclear fusion reactions in stars. The remaining elements are formed from supernovae—giant star explosions that occur when the nuclear fuel at the star's center is consumed and gravitational collapse occurs, followed by outward explosion.
- The average compositions of Earth and of most meteorites are very similar. The Sun has similar proportions of heavy elements but contains much more hydrogen and helium. Earth's original hydrogen and helium were lost when Earth formed.
- Heating of primitive chondritic material that formed Earth caused separation between iron and rock, resulting in an iron core and a rocky mantle. The low concentration of iron near Earth's surface can be explained if the iron has separated from the silicate mantle and settled into the core.
- Earth formed 4.55 billion years ago. It took 500 million years for Earth to cool sufficiently to allow formation and preservation of rocks.
- Earth is thought to have formed from material similar to that of chondritic meteorites. The planet then became differentiated into core, mantle, crust, hydrosphere, and atmosphere.
- A dense iron core explains the average density of Earth, the passage of seismic waves through the planet, and Earth's precession.
- Heating of a material that can flow may cause convection where hot expanded material rises and cooler material sinks. Heat can be transported within a strong material by conduction.
- At the time of Earth's formation, it is thought that the heat generated by radioactivity was nearly four times greater than it is today. Earth has been steadily cooling since its formation. Consequently, convection within Earth's interior has become less vigorous through time.

KEY TERMS

big bang, 21
atom, 21
electron, 21
proton, 21
neutron, 21
element, 21
atomic number, 21
isotope, 21
atomic mass, 21
molecule, 21
frequency, 22
spectrum, 22
photon, 23
red shift, 24
nuclear fusion, 24
nuclear fission, 26
supernova, 26
escape velocity, 28
comet, 29
meteor, 29

meteorite, 30
chondrule, 31
chondrite, 31
density, 34
differentiation, 34
mass, 37
precession, 39
radioactivity, 40
convection, 40
conduction, 40
plume, 40

QUESTIONS FOR REVIEW AND FURTHER THOUGHT

1. What is the evidence that the universe started with a big bang?
2. Describe a supernova explosion of a star.
3. Why are supernovae important processes in the formation of the elements?
4. Why does Earth have an atmosphere, whereas the Moon and Mercury do not?
5. What do comets tell us about the Solar System?
6. Some meteorites are made of iron, some are made of stone, and some are a combination of the two. Explain why this may have happened and why this is relevant to the formation of Earth.
7. The average iron content of crustal rocks is 6 percent, whereas the average iron content of chondrite meteorites is 35 percent. If the Earth, on average, has a chondrite composition, how can we reconcile these differences?
8. Give three reasons why we think that Earth's core is made of iron.
9. Describe the convection in Earth's core and in the mantle.

3 Minerals: The Building Blocks of Rocks

Crystals of the mineral fluorophyllite.

Overview

- Minerals are naturally occurring inorganic compounds with a definite composition and a definite crystal structure.
- Minerals can be distinguished from each other by properties such as crystal shape, color, and hardness.
- By far the most abundant minerals (the "rock-forming" minerals) are silicate minerals.
- Although they are less abundant than the silicate minerals, carbonate, oxide, halide, sulfide, and sulfate minerals are important sources of many raw materials.
- Earth is divided into three layers of very different composition: the crust (oceanic and continental), the mantle, and the core.
- The different chemical compositions of these layers result in the formation of different minerals, and therefore, different rocks.

Introduction

To understand the behavior of Earth, we must know something about the material from which it is made, namely **rocks**. Most rocks are complex mixtures of **minerals**, which, in turn, are made of atoms arranged in a regular structure. In this chapter, we will briefly review atoms and their properties and behavior: for example, how they combine with each other. We will see how atoms combine in regular geometrical arrangements to form **crystals**, and we will discover how the crystal shape and composition are used to define a specific mineral. Finally, we will return to the structure and composition of Earth and learn that minerals are not randomly distributed. Their occurrence is controlled by the chemical composition of the local environment and by conditions that exist there, such as pressure and temperature.

FIGURE 3.1 (a) The stable form of silica (SiO_2) at Earth's surface is the mineral quartz, shown here as well-formed crystals. (b) Snowflake obsidian. The white "snowflakes" are clusters of microscopic crystals that have formed in a natural volcanic glass.

What Is a Mineral?

Nearly all rocks—both at the surface and deep within Earth—are made up of one or more minerals. There are a few exceptions, such as rocks made of natural glass, which include the volcanic rock *obsidian* (glass is not a true mineral because it does not have a regular crystal structure). Some minerals are very common: Quartz is the dominant component of most beach sand. Other minerals are rare and valuable, such as diamond and sapphire. Still others are important to society and are used in the manufacture of materials from fertilizers to automobiles.

The broad definition of a mineral is *a naturally occurring inorganic solid with a definite (fixed) composition and an orderly arrangement of fundamental components (or atoms), called a crystal structure*. It is this orderly arrangement that gives a mineral its specific recognizable form. Note that, although minerals are defined as inorganic solids, they may *originate* by either inorganic or organic processes. Many minerals crystallize from chemical solutions or melts. But certain organisms can also generate minerals, such as those that form shells and corals. By the definition above, not all solids are minerals. In glass, which is mainly composed of an oxide of silicon (SiO_2, or silica), silicon and oxygen atoms are arranged rather randomly, much as they are in the liquid state. The liquid state is energetically unstable compared with that of quartz, which is the mineral form of silica (Fig. 3.1a). Over long periods of time, glass may become reordered into a microcrystalline solid formed from tiny quartz crystals, giving old glass a "frosted" appearance. This process of devitrification can be seen in natural glasses such as obsidian (Fig. 3.1b). In contrast to the beautiful quartz crystals of Figure 3.1, natural glass and material such as agate (Fig. 3.2) can be described as **amorphous,** meaning "without form."

The study of minerals (mineralogy) is a subdiscipline of the earth sciences overlapping into the realms of chemistry, physics, and material sciences. Mineralogists are particularly interested in the structure of minerals and their properties. They examine the effects of pressure, temperature, and composition on minerals. Amateur mineralogists take pleasure in collecting the best possible specimens they can find—either the largest or the finest crystal shapes. You need only look over the mineral collection of a natural history museum to appreciate the diversity of minerals, their different colors, forms, and modes of occurrence. These beautiful specimens are rare; special conditions are required to grow perfectly shaped minerals. A good specimen of any mineral can be very valuable. Although the grains of quartz you pick up on the beach are nearly worthless, the same mineral occurring as perfectly formed crystals a meter long may be worth well over $1000!

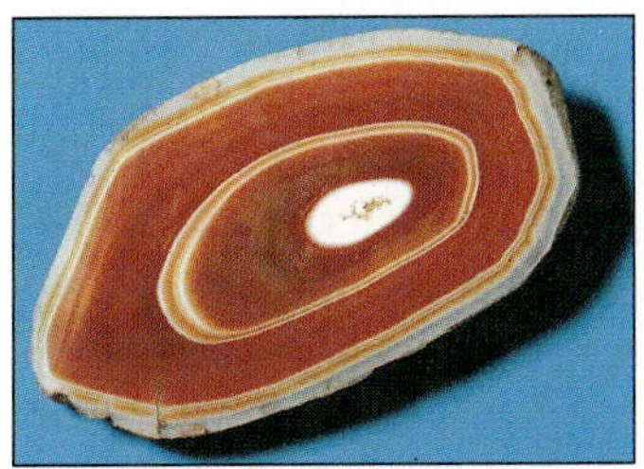

FIGURE 3.2 Agate, a naturally occurring form of silica. It has the same chemical composition as the quartz crystals in Figure 3.1a.

(a)

(b)

(c)

FIGURE 3.3 (a) A geologist samples rocks by breaking a sample off an outcrop using a geologic hammer to expose crystals unaffected by weathering. (b) Visual examination of crystals. (c) A hand lens reveals crystals not clearly visible to the naked eye.

After visiting a museum, you may start looking for beautiful mineral specimens similar to the ones you saw there—and you may become quickly discouraged. The rocks that are found at roadcuts or sea cliffs do indeed contain minerals, but they are typically very small grains and you may need a hand lens to see them (Fig. 3.3). Furthermore, they tend to have poorly defined crystal shapes. It is difficult to make the connection between minerals and rocks if we think in terms of museum-quality specimens, but if we look closely we can still identify the minerals present in any given rock. This is an important objective because the minerals in a rock reflect its chemical composition and tell us something about how the rock was formed.

A mineral is a naturally occurring inorganic solid with a definite composition and an orderly arrangement of atoms that give rise to a recognizable crystal form. The minerals in a rock reflect the rock's chemical composition and its conditions of formation.

Atoms: The Building Blocks of Minerals

Minerals are composed of atoms. We saw that atoms, in turn, are composed of even smaller particles: protons, neutrons, and electrons (∞ see Chapter 2). The nucleus, or central core, of an atom is made of protons, which are particles that have one unit of positive electric charge, and neutrons, which are particles with no charge. We also saw how protons and neutrons are successively combined to form the elements. A given element is composed only of atoms that have the same number of protons and electrons. Each element is thus distinguished by the number of protons in its nucleus—its atomic number.

The nucleus, with a net positive charge, is surrounded by one or more electrons. Each electron has one unit of negative electric charge and is therefore attracted to the positively charged nucleus. For an atom to be electrically neutral, then, it must have an equal number of protons and electrons. The number of protons in an atom is important because it dictates the number of electrons required to make a neutral atom of a particular element. An electrically charged atom—one that has either more protons than electrons, or vice versa—is called an **ion**. More specifically, an ion with a net negative charge (more electrons than protons) is called an **anion**, and an ion with a positive charge (more protons than electrons) is called a **cation**.

For simplicity, electrons can be thought of as particles orbiting the nucleus in hypothetical shells. Each orbital shell can accommodate a limited number of electrons. If an atom interacts with neighboring atoms, the electrons residing in the atom's outermost electron shells are the ones involved in that reaction. Atoms with a full, or complete, outer electron shell are stable and do not tend to react with other atoms. Atoms with an incomplete outer shell tend to be more reactive and will give up those electrons or obtain extra ones to fill the shell; accordingly, these atoms then acquire a charge imbalance. When dominated by the positive charge of the protons, they become cations; when dominated by the negative charge, they become anions.

Atoms with different atomic numbers possess different numbers of electrons, and therefore each type of atom has a unique arrangement of electrons orbiting its nucleus. As a result, atoms of different elements behave differently and that is how we are able to distinguish them as distinct elements. The different elements have been systematized in the **periodic table** (Appendix III), which is arranged according to the electron distribution of atoms.

The periodic table was originally devised on the basis of chemical behavior of the elements; thus, elements that share similar properties and behave similarly in chemical reactions are grouped together (Focus 3.1).

Atoms are made of a nucleus of protons and neutrons and surrounding shells of electrons. The atomic number—the number of protons in a nucleus—equals the number of electrons in the shells of the atom and serves to define each chemical element.

Building Minerals from Atoms

Most minerals are **compounds**, which are combinations of two or more elements. Plagioclase feldspar, for instance, is a compound of calcium, sodium, aluminum, silicon, and oxygen. Some minerals are composed of just one element: examples include diamond (carbon), gold, and sulfur.

Minerals are formed from atoms or ions stacked in regular geometrical arrangements called *lattices* (Focus 3.2; Fig. 3.4a and b) that give rise to the characteristic shapes observed in museum-quality crystals. For a given mineral, the geometrical arrangement of atoms imparts a certain symmetry to the crystals and is a defining property of that mineral.

Mineral Formation: Chemical Bonding

For atoms or ions to be held together in a regular arrangement, there must be some means of "gluing" together adjacent atoms or ions. This is accomplished by **chemical bonds**, which are attractive forces between atoms. These bonds are of two principal varieties: ionic bonds and covalent bonds.

Ionic Bonds The *ionic bond* is an electrical force of attraction between positively and negatively charged ions. A good example of a mineral in which ions are held together by ionic bonds is halite—better known as common table salt. Halite is the mineral name for the compound sodium chloride (NaCl) (Fig. 3.4). Sodium has 1 electron in its outermost shell and tends to give up that electron in a chemical reaction, thus becoming a positive ion (Na^+) (Fig. 3.4c). The outermost shell of chlorine has 7 electrons, one electron short of this shell's full complement of 8 electrons. A chlorine atom will readily gain an electron to complete this shell and will thereby become a negative ion (Cl^-). The ions of sodium and chlorine are bonded by the electrical attraction between them (Fig. 3.4c). Halite is electrically neutral, so it must have an equal number of Na^+ and Cl^- ions. Ionically bonded compounds share several characteristics and properties. For instance, they are soluble in water and form solutions that tend to be good electrical conductors.

Covalent Bonds *Covalent bonds* are typical of compounds containing atoms that have partially filled shells; however, it takes too much energy for these atoms to give

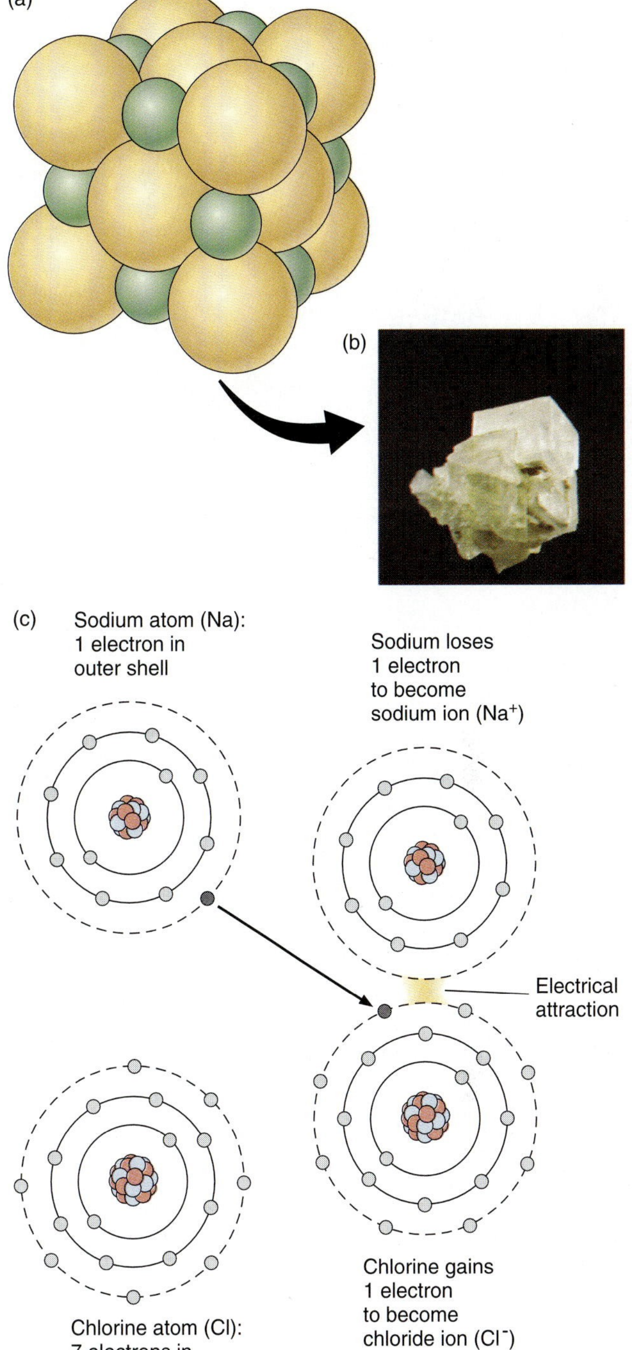

FIGURE 3.4 (a) The arrangement of sodium and chlorine ions in halite (table salt). The charge on the ion is represented by the superscript + (positive) or – (negative). If the charge is more than one electron unit, then the number is also given in the superscript—so that a Mg ion that has given up two electrons from its outer shell will be designated Mg^{2+}. (b) This arrangement is reflected in the cubic symmetry of the mineral. (c) As indicated by the arrow, the outer shell of sodium gives up one electron and the outer shell of chlorine gains one (although not necessarily the same one). The result is a positively charged sodium ion and a negatively charged chlorine ion, which are attracted to each other via an ionic bond.

Focus On 3.1 THE PERIODIC TABLE

The properties of different elements have been systematized in the *periodic table* (Fig. 1). This chart received its name because the electrons that make up the shells of atoms behave in a predictable or "periodic" pattern as their numbers increase and as successive shells are filled with electrons. The first shell of an atom—the shell closest to the nucleus—can accommodate a maximum of 2 electrons and the next shell accommodates 8 electrons. Each row, or *period*, reading from left to right across the periodic table, represents the successive filling of electron shells. Each column, or *group*, comprises elements that have the same number of electrons in their outermost shell, and they will therefore behave in a similar chemical fashion.

The behavior of atoms depends on the status of their outermost electron shells. The most stable atoms—and therefore the least chemically reactive—have complete outermost electron shells. The *inert gases* (helium, neon, argon, krypton, xenon, and radon) all have full outermost electron shells. They do not share electrons with other atoms; hence their designation as "inert." As a result, they behave similarly and are classified in the same group (VIII) of the periodic table.

Atoms with a nearly full outermost shell tend to gain an electron in order to complete that shell, and such atoms attain a more stable electron arrangement. Once an atom has gained an extra electron, it possesses a net negative electrical charge. The *halogens* (fluorine, chlorine, bromine, iodine, and astatine) fall into this category, and again, they are grouped together (Group VII) in the periodic table.

Atoms with only one electron in their outer shell tend to give up those electrons; accordingly, these atoms have a charge imbalance dominated by the positive charge of the protons, so they form cations. The *alkali metals* (lithium, sodium, potassium, rubidium, cesium, and francium) all share this property and thus occupy the same group (I) in the periodic table.

In fact, as shown in the periodic table, we can make a general subdivision of all the elements into those that easily give up electrons and tend to be on the left-hand side of the periodic table (metals), and those on the right-hand side of the periodic table that readily accept electrons (nonmetals). Thus, the periodic table is arranged according to the electron structure of atoms, and elements in the same group have similar atomic properties and behave similarly in chemical reactions.

up or accept the additional electrons needed to achieve a complete shell structure. Instead, these atoms *share* their electrons. The electrons orbit the nuclei of both atoms, in effect tying them together (Fig. 3.5). Diamond is a good example of a covalently bonded mineral (Fig. 3.5a). There is only one type of atom in diamond (carbon), and thus there is no tendency to form ions—the way there is with halite, where two different elements are combined. Covalently bonded compounds tend to be poor electrical conductors and are much less soluble in water than are compounds with ionic bonds.

Many carbon compounds are covalently bonded. Most common rock-forming minerals have covalent bonds because such minerals are dominated by the elements silicon and oxygen, which tend to share their electrons rather than to exchange them and form ions. We should point out, however, that many bonds are actually of an intermediate type—they are not completely ionic or completely covalent; further, more than one type of bond may be present in a given mineral.

Van der Waals and Metallic Bonds *Van der Waals bonds* are weak forces of electrostatic attraction. They are particularly important in minerals where the main structure takes the form of continuous chains or sheets. Within these chains or sheets, atoms are connected by ionic or covalent bonds, but the chains or sheets may, in turn, be held together by Van der Waals bonds.

In metals, the individual atoms are bonded by *metallic bonds*. This involves sharing electrons between all the atoms in the solid (rather than electrons being shared by specific atoms, as is the case with covalent bonds). The shared electrons can flow between the atoms, which makes metals good electrical conductors.

Atoms in solids are linked together by bonds. Ionic bonds form from the electrical attraction between oppositely charged ions. Covalent bonds form when two or more atoms share their electrons.

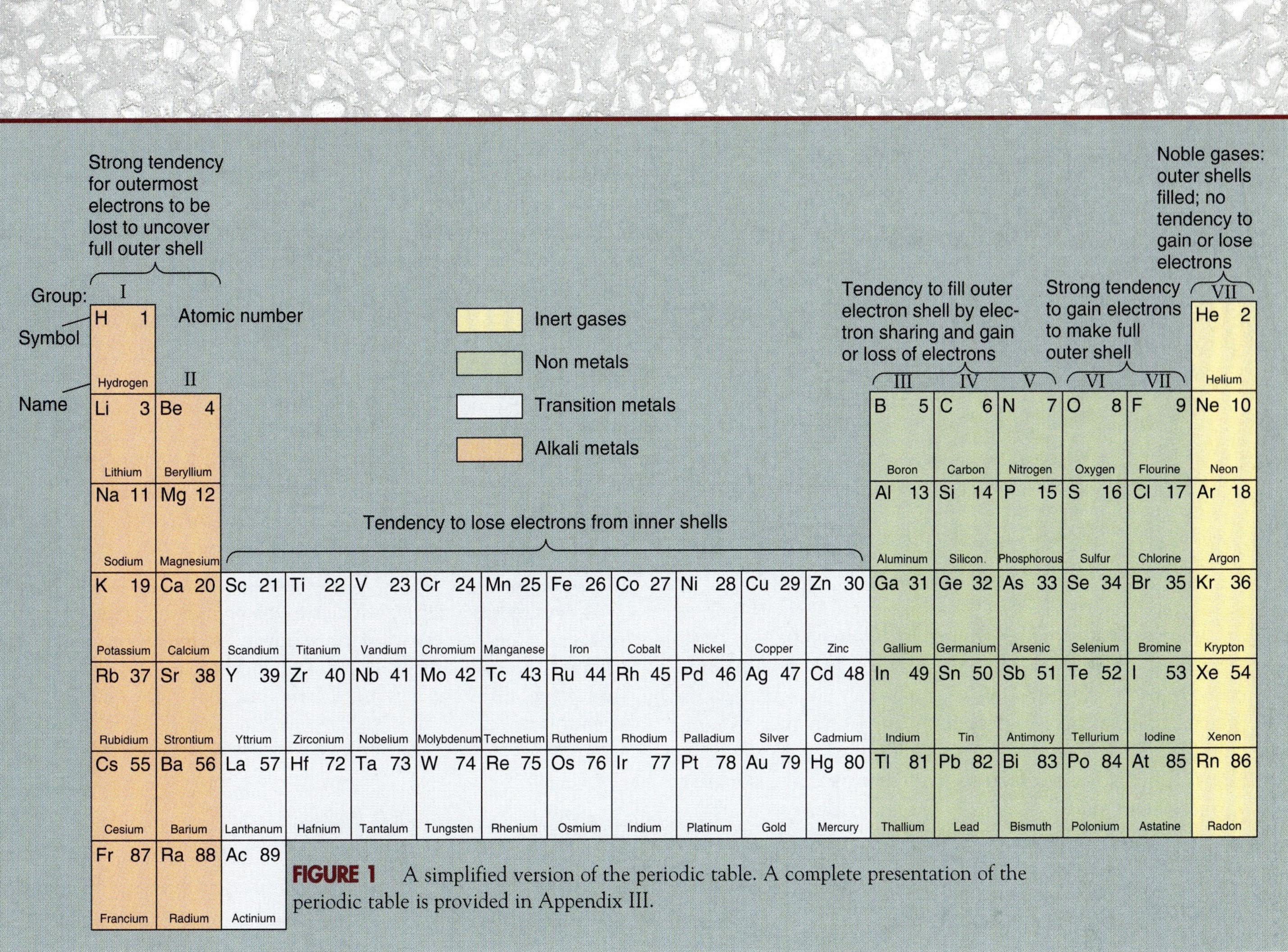

FIGURE 1 A simplified version of the periodic table. A complete presentation of the periodic table is provided in Appendix III.

CRYSTALS

Crystal Structures

One of the most striking features of a collection of beautiful minerals is their **crystal form**, which is the outward appearance of a mineral based on its internal arrangement of constituent atoms. The formal definition of a crystal is *a body bounded by surfaces, usually planar, arranged in a definite plan that is an expression of the internal arrangements of the atoms*. This means that crystals may differ widely in shape, but crystals of any given mineral always have the same crystal form.

Let us reconsider the combination of sodium ions and chlorine ions in halite (Fig. 3.4a). They stack in a regular geometrical arrangement that is determined by the energy required to hold the ions at certain distances from each other, which, in turn, depends on the sizes and charges of the constituent ions. You can visualize the effect of size on packing by trying to stack a few layers of tennis balls on top of each other in a box (Fig. 3.6a). You can arrange the bottom layer so that all the balls are touching. The next layer can be placed directly on top of the underlying layer (arrangement I), or it could be packed even more efficiently by placing each ball of the top layer into the gap between the underlying balls (arrangement II). Notice how arrangement II takes up less space; the box does not need to be as tall as it does in arrangement I.

In addition, the forms of arrangements I and II are different. In two dimensions, they form a rectangle and a parallelogram, respectively. If these tennis balls were atoms of crystals, arrangements I and II would represent two different crystal forms.

Now if there were a much smaller sphere present in addition to the tennis balls—we can use a golf ball or a marble to represent a smaller atom or ion—this "atom" could be placed into the gaps between the tennis balls. This organization will only work with arrangement I, though, where the gap between the tennis balls is large enough to fit a marble (Fig. 3.6b).

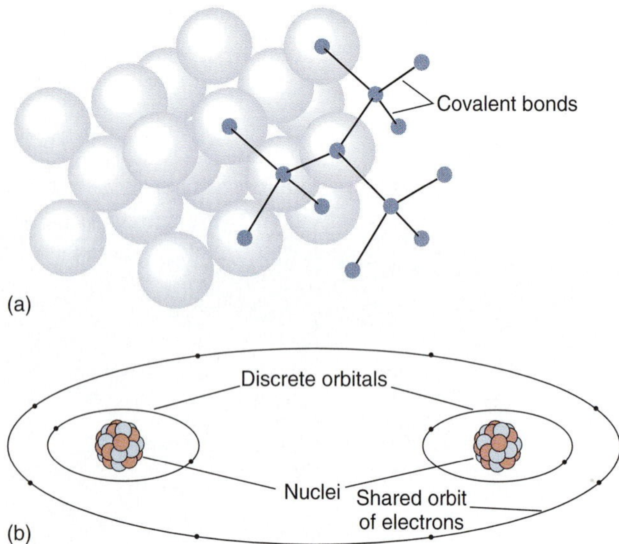

FIGURE 3.5 (a) The arrangement of carbon atoms in diamond. (b) The outer shells of the atoms share electrons, forming covalent bonds. Compare this arrangement with Figure 3.4c.

In the arrangements shown in Figure 3.6, the balls remain in their stack owing to gravity; in a real crystal, the atoms or ions are bonded together by electrical forces that attract and repel, depending on which type of ions are placed together and on the amount of charge on the ions. In halite, these factors give rise to a cubic stacking arrangement (Fig. 3.4a) rather like that shown in arrangement I; thus, the crystals have cubic shapes (Fig. 3.4b, Fig. 3.6b). The way atoms combine in a given mineral therefore gives rise to a consistent crystal form or *symmetry*, which serves as a means of identifying that mineral (Focus 3.3).

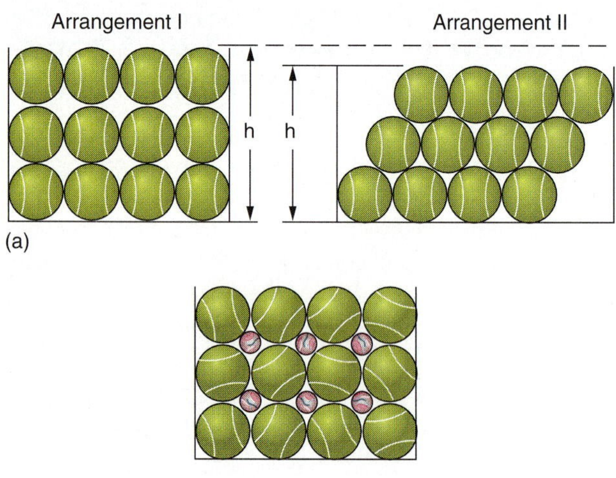

FIGURE 3.6 An analogy to show how the stacking of atoms controls crystal structure. Successive layers of tennis balls can be placed directly on top of each other (arrangement I) or in staggered layers (arrangement II). Arrangement II is more closely packed and takes up less space. The different arrangements result in different shapes for the stacked balls. (b) In the more loosely packed arrangement (I), smaller balls can fit into the spaces between the large balls.

Some compounds exist in more than one crystal form. Because each mineral has a fixed crystal form, these different forms correspond to different minerals that have the same chemical composition. Such minerals are called **polymorphs**. A change from one polymorph to another is known as a *phase change*. Diamond and graphite have the same chemical composition—but different crystal symmetries (see Focus 3.3) and different properties. Diamond has a cubic crystal form, whereas graphite is hexagonal (Fig. 3.7). Diamond is the hardest known mineral, a precious gem. The carbon atoms in diamond are linked by strong covalent bonds (Fig. 3.5; Fig. 3.7a). Graphite is black, relatively soft, and cheap—the material from which pencil "lead" is made. In graphite, carbon atoms are bonded together to form sheets; the sheets, in turn, are held together by weak Van der Waals bonds (Fig. 3.7b).

A crystal has a definite structure, is bounded by planar surfaces, and its outward appearance is based on its internal arrangement of atoms. Polymorphs are minerals that have different crystal forms but have the same chemical composition.

Controls on Crystal Structure

Why does carbon occur in two polymorphs as graphite and diamond? The answer lies in the influence that external factors, such as pressure and temperature, have on the internal arrangement of atoms in a mineral. At low pressures, material of a given composition will crystallize as a mineral that is a stable structure under low-pressure conditions. The term *stable* means that the mineral is at equilibrium with its environment; it will not tend to change its composition or its crystal structure.

At higher pressures, different minerals may form that are stable at these pressures. Typically, crystal structures that are more closely packed (denser) are more stable at high pressures.

Let us consider what we mean by mineral *stability*. You can think of it in terms of an energy balance that is analogous to the situation sketched in Figure 3.8. A ball sitting at the top of a hill is unstable. Gravity will cause it to roll down the hill into the valley. In contrast, the ball sitting in the valley is energetically stable. It will not move unless it is pushed uphill—unless energy is added to it. Figure 3.8 also illustrates an intermediate condition, represented by the ball sitting on the ledge halfway up the hill. This ball is unstable compared with the ball in the valley, but it will not spontaneously roll downhill. It requires a little push.

Focus On 3.2 DETERMINING CRYSTAL STRUCTURES BY X-RAY DIFFRACTION

How do we know the arrangement of atoms or ions in a crystal structure? The most powerful and common technique used to address this question is X-ray diffraction (Fig. 1). In this approach, X-rays (very short-wavelength, high-energy radiation) are passed through a solid crystal. As they pass through the crystal, the rays are *diffracted*, or scattered. For a given wavelength of X-ray, the amount and direction of diffraction is dependent both on the orientation of a particular layer of atoms in the crystal structure and on the spacing between atoms forming successive layers; thus different wavelengths will be diffracted differently. These diffracted X-rays are detected either on photographic film (Fig. 1) or by an electronic detector. The pattern made on photographic film is like a fingerprint; the distribution of the dots (where X-rays have hit the film) can be analyzed and measured; and this distribution will correspond to a specific crystal structure. This method is very precise and can be used to distinguish small variations in mineral composition such as the substitution of one type of ion for another, which might slightly change the spacing between layers of ions.

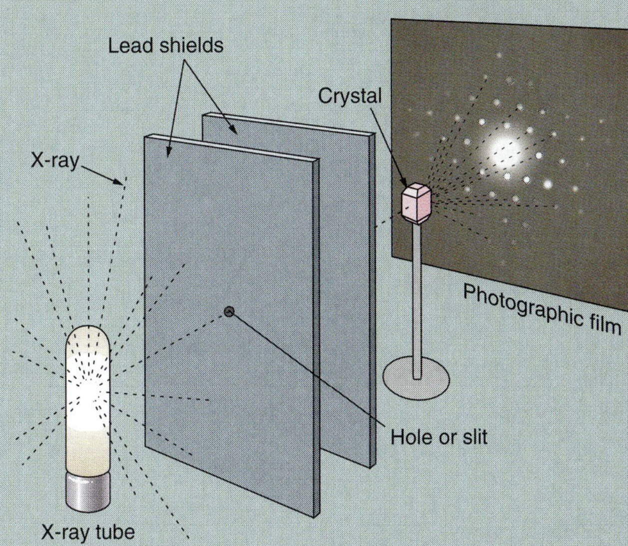

FIGURE 1 The technique of X-ray diffraction. X-rays are passed through a crystal and diffracted to form a pattern on photographic film. The pattern on the film is specific to the type of mineral that forms the crystal.

Mineral Stability: Thermodynamics A study of the relationship between heat and work and the balance, or equilibrium, between the two is called *thermodynamics*. It governs the effects of temperature and pressure on materials and can tell us what mineral should be stable under any conditions and the direction in which any chemical reaction should go. For geologic materials, heat and pressure, rather than the gravitational energy of Figure 3.8, dictate which materials are stable under given conditions.

Again, consider the example of diamond and graphite. Diamond represents a closely packed arrangement of carbon atoms that are stable at pressures equivalent to depths within Earth of more than 150 km. This closely packed crystal structure, which gives the mineral a higher density, is a typical response to increasing pressure. Graphite has a much looser crystal structure and is stable at low pressures (Fig. 3.9, page 56). In terms of our hill and valley example, diamond occupies the energy valley at high pressures, whereas graphite is in the valley at low pressures.

Geologists experiment to determine the behavior of minerals under different pressure and temperature conditions. Results they obtain can be summarized in a chart called a **phase diagram**. Focus 3.4 gives a more detailed explanation of the principles of phase diagrams.

In the earth sciences, we use phase diagrams to describe the stability of rocks and minerals. Figure 3.9 is a phase diagram of the polymorphs of carbon. In this figure, the two phases represented are both solids; they can be distinguished from each other because they have distinct crystal structures. We use such a diagram to tell which polymorph is more stable under a given set of pres-

Focus On 3.3 CRYSTAL SYMMETRY

The positions of crystal faces and edges and the angles between them exhibit a striking regularity in form. This regularity is referred to as *symmetry*. The symmetry properties of minerals are used to distinguish and identify them. We can describe symmetry in terms of *planes of symmetry*, which are like mirrors, and *axes of symmetry*.

Consider the square shown in Figure 1a. Four separate planes of symmetry can be drawn on it: two from corner to corner and two from the middles of opposite sides. If you picture a line running from corner to corner or from the middle of one side of the square to the middle of the opposite side, and stand a mirror up on this line, then the reflection of one side will not change the original shape—the reflection shows a square. This holds true for all four planes of symmetry. We can say then that the square has four planes of symmetry.

Now take a look at the rectangle sketched in Figure 1b. Only two planes of symmetry can be drawn on this shape: between the middles of the opposite sides. If a line is drawn from corner to corner, the reflection will alter the original shape from a rectangle to a "kite" shape.

We can now consider *axes* of symmetry. Imagine cutting out a new square that fits over the square shown in Figure 1a. Now put a pin through the center of this new square and rotate this top square (Fig. 1c). After you have rotated it by 90°, it will fit exactly over the original square again (Fig. 1c). It will also fit over the square after rotating it either 180°, 270°, or 360°. We say that the square has a *fourfold axis of rotation*.

Try the same exercise with the rectangle. You must rotate it a full 180° before it fits exactly over the original rectangle again (Fig. 1d). Because it fits perfectly over the underlying rectangle only twice during a full rotation, we say that it possesses a *twofold axis of rotation*. The square is "more symmetrical" than the rectangle and is said to have a higher *degree of symmetry*. A hexagon would have an even higher degree of symmetry—a sixfold axis of rotation. Mineralogists use such properties of symmetry to distinguish minerals from each other.

(a) Reflection in diagonals results in a square like the original shape

(b) Reflection in halfway plane gives the same rectangle, but reflection in diagonal does not

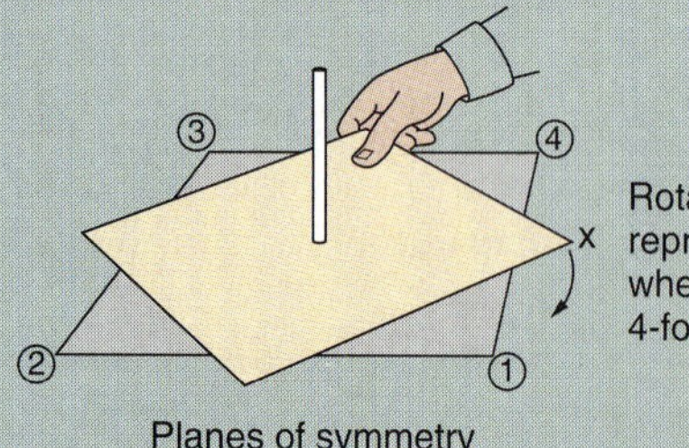

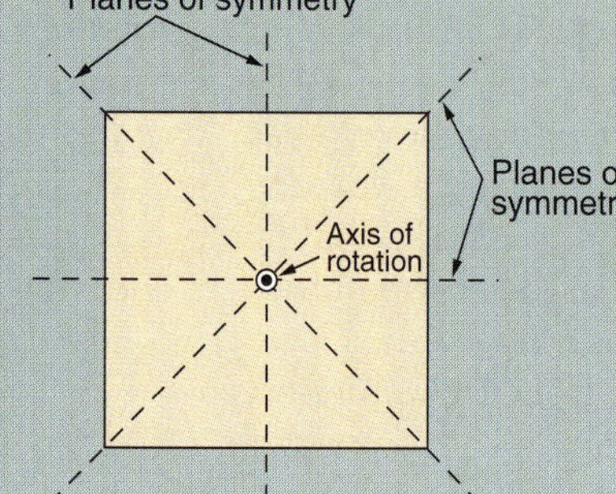

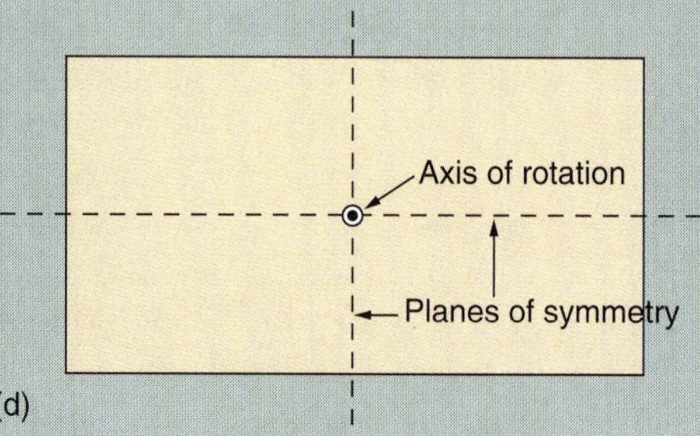

FIGURE 1 Principles of symmetry. The square has four planes of symmetry (a) compared with the rectangle (b), which has two. The square has a fourfold axis of rotation (c) compared with the rectangle (d), which has a twofold axis of rotation. The square has a higher *degree* of symmetry than does the rectangles.

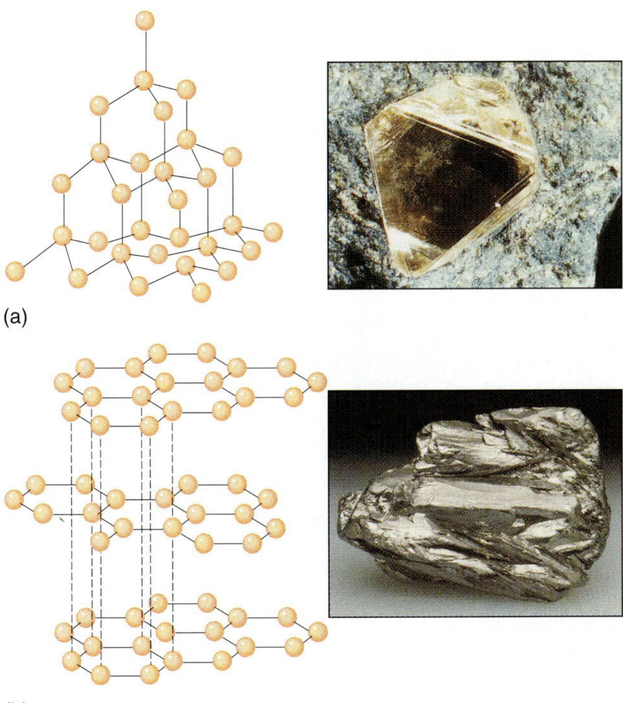

FIGURE 3.7 Two different minerals with the same chemical composition. Diamond (a) and graphite (b) are both formed from pure carbon. The arrangement of atoms in the crystal structure is indicated.

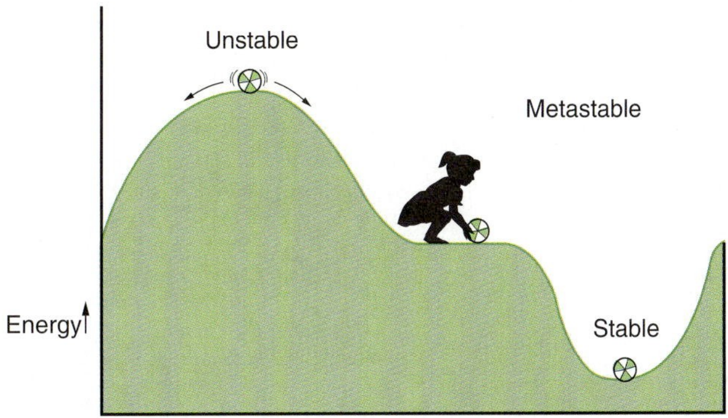

FIGURE 3.8 Principles of equilibrium and stability, as illustrated by a ball on a hill. At the top of the hill, the ball is unstable, or out of equilibrium. At the bottom (in the valley), it is stable, or at equilibrium. The ball is in equilibrium on a ledge halfway up the hill, provided it is not pushed.

sure and temperature conditions, and therefore we can determine the conditions under which a given rock formed. If we discover diamond in a rock, not only are we extremely lucky, but we are also aware that the rock must have originated from depths of more than 150 km, where the pressure is very high.

Rates of Reaction: Kinetics Most of the rocks that we collect are on Earth's surface. How, then, can we retrieve minerals such as diamond, which are stable only at significant depths within the planet? When we talk about mineral stability, we are referring to equilibrium conditions. To respond to a change in these conditions, one polymorph may change into another or minerals may react together to form a different set of minerals. If a mineral is truly at equilibrium, there will be no tendency for it to react with, or transform into, other minerals. The rate at which any reaction takes place, however, is not usually instantaneous (although nearly instantaneous chemical reactions do occur, such as explosions).

Why doesn't the diamond in a piece of jewelry transform into graphite—the polymorph of carbon that is stable at Earth's surface? Consider the example of the hill and valley of Figure 3.8 one more time. At low pressures, we might expect diamond to be at the top of the energy hill, or unstable. In reality, though, it sits on the ledge; it is not really stable, but it will sit there until it is given a push. That "push" could be heat. If diamond is heated, it will transform into graphite—or roll down the hill in our analogy—because heating adds energy, which speeds up the reaction that causes diamond to become graphite at low pressure.

The study of reaction rates is called *kinetics*. For most geologic materials, the rates at which reactions occur are extremely slow, such as the transformation of diamond to graphite at Earth's surface. Relative to human lifetimes, diamond is stable at Earth's surface even though it is not in equilibrium. We mentioned earlier that glass may become reordered into a microcrystalline solid formed from tiny quartz crystals over a time scale of a few hundred years. In fact, the kinetics of a reaction such as a mineral transformation depend on a number of factors, including mineral composition, temperature, and pressure. Appreciation of these principles is important if we are to gain information from minerals about the rocks in which these minerals occur.

In geology, thermodynamics is used to study the stability of different minerals under different conditions. The rates at which mineral reactions may occur under different conditions are governed by kinetics.

A Survey of Mineral Properties

A mineral collector will use a variety of properties to distinguish different minerals. The properties by which you can identify minerals include crystal shape, specific gravity, color, streak, cleavage, fracture, luster, opacity, and hardness (Table 3.1, page 58). One word of caution: Some of the mineral properties described are easily observable only on rocks with large crystals. Rocks are typically composed of minerals the size of sand grains, which may be difficult to see without a hand lens or microscope.

A catalog of common minerals and their properties is provided in Appendix IV.

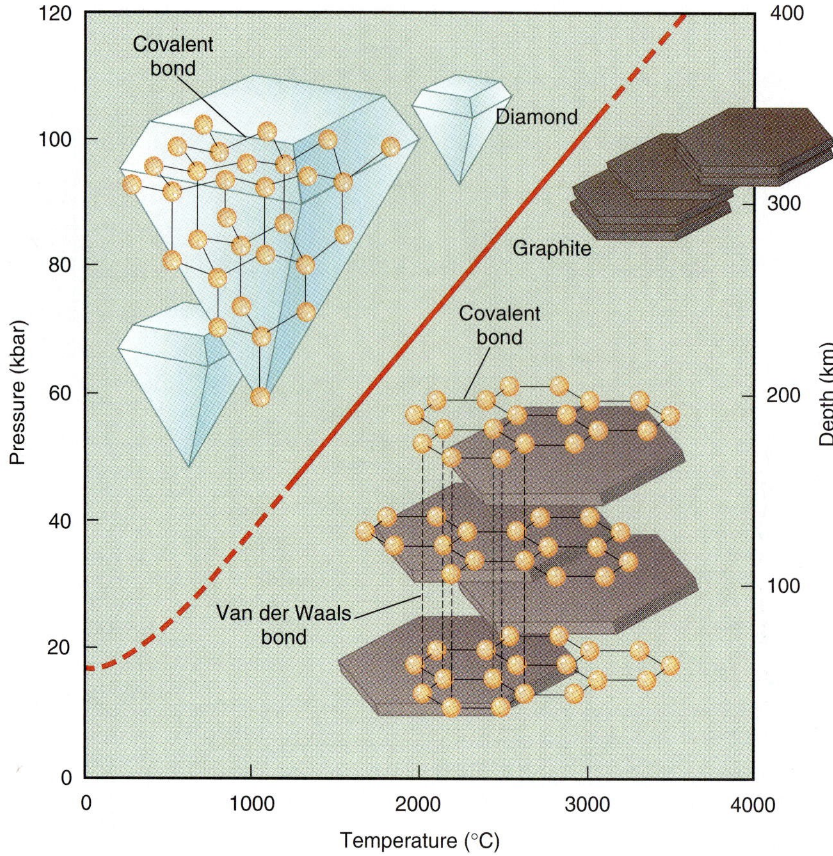

FIGURE 3.9 A pressure versus temperature diagram (phase diagram) showing the stability ranges of diamond and graphite, which are polymorphs of carbon. Diamond is stable at high pressures; graphite is stable at low pressures.

Mineral Groups

Minerals can be divided into a number of groups, based on the anions involved in the compounds. For example, we can distinguish carbonates (one or more metal cations bonded to a carbonate (CO_3^{2-}) ion and sulfides (one or more metals bonded to sulfur ions). The most important group is the silicates, in which metals are bonded to silicon-oxygen anion groups.

Silicate Minerals

Although thousands of minerals occur naturally within Earth, only a handful make up the vast majority of rocks exposed at Earth's surface. Because silicon and oxygen are the two most abundant elements, most rocks are made of *silicate minerals*: compounds of silicon and oxygen. These compounds generally contain one or more metals, such as calcium, magnesium, aluminum, or iron. Of particular importance are the silicate minerals olivine, pyroxene, amphibole, mica, feldspar, and quartz. These common minerals are referred to as *rock-forming minerals*. We will see in Chapter 4 that the minerals found in different rocks are a reflection of the bulk composition of each rock and are an important consideration in naming and classifying rocks.

The fundamental building block of any silicate mineral is the *silica tetrahedron*, which is a silicon atom bonded to four oxygen atoms. This compound is represented by the formula SiO_4 and the atoms are arranged in the shape of a four-cornered pyramid (Fig. 3.10). You can build a simple model of a silica tetrahedron by placing three tennis balls together on a flat surface to represent oxygen atoms. Then place a small marble in the center

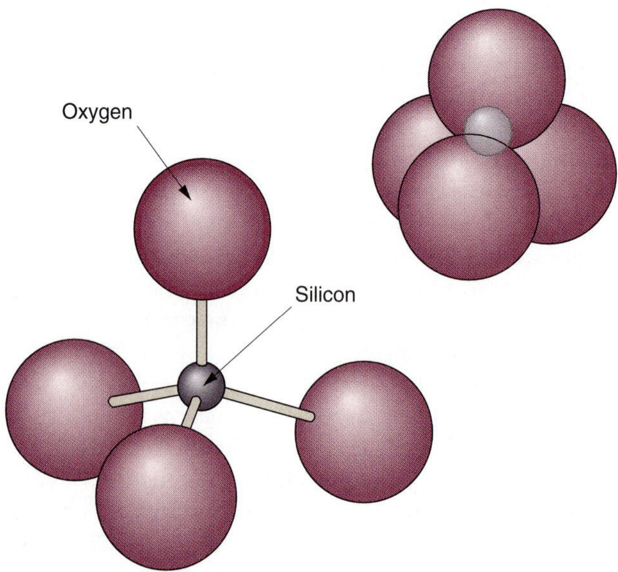

FIGURE 3.10 Atomic structure of the silica tetrahedron. This representation is expanded to show the small silicon atom in the center of the much larger oxygen atoms.

PHASE DIAGRAMS

By performing experiments (see Focus 3.1), geologists can determine how minerals behave, and the results are then summarized in a phase diagram. The term "phase" refers to a homogeneous, physically distinct portion of a heterogeneous system. For example, ice is a solid phase of the system ice, water, and steam; diamond and graphite are solid phases of carbon. A phase can be a liquid, a solid, or a gas. A phase diagram is essentially a chart; for any set of pressures, temperatures, and compositions, it gives the stable phase of a given element or compound.

For a simple element, or for a compound such as water, we can construct a phase diagram using pressure and temperature as the axes. Figure 1 shows a phase diagram for water (H_2O). At a pressure of 1 atmosphere or 0.1 megapascals—which is equivalent to conditions at Earth's surface—water boils at 100°C and freezes at 0°C. You are familiar with these temperatures and you may also realize that the boiling point of water decreases as pressure decreases—as it would on a mountaintop. If we were to measure how the freezing point and boiling point of water change with changing pressure, we could join these points in a line on a phase diagram; such a line is called a *phase boundary*. You can see the phase boundaries in Figure 1 that divide the diagram into three fields: solid (ice), liquid (water), and gas (steam). Each field indicates the ranges of pressures and temperatures for which a particular phase is stable.

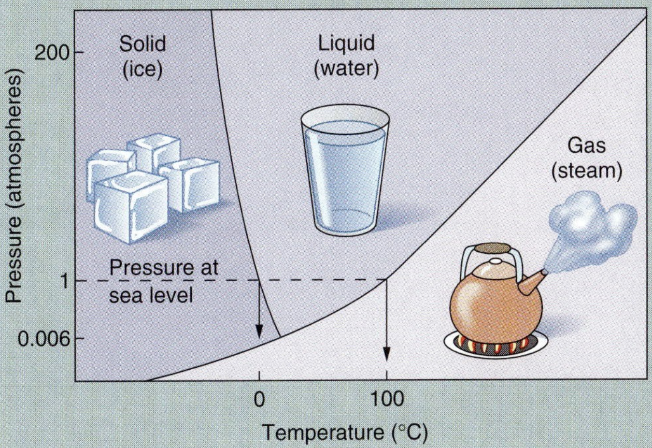

FIGURE 1 A phase diagram for water (H_2O). From this diagram, we can determine the phase that will be stable at a given temperature and pressure. (Note that the scales are not linear.)

of the tennis balls to represent a silicon atom. Finally, put a fourth tennis ball on top to complete the pyramid. This particular structure is possible only because the silicon atom (the marble) is small enough to fit into the hole between the four oxygen atoms (tennis balls) that form the corners of the tetrahedron. If you had a relatively larger ion—a golf ball in our model rather than a marble—then the stacking would not work.

This example underscores the importance of the relative *sizes* of atoms and ions in defining mineral compositions. Within a given mineral structure, there are spaces of limited dimensions that will only accommodate atoms or ions of a given size (see Fig. 3.6). An ion that is the correct size to fit into a given mineral structure is said to be compatible with that mineral. If an ion is too large (or too small in the sense that it does not comfortably fill a particular site in a crystal) to fit into the mineral structure, it is said to be incompatible. In the case of minerals crystallizing from molten rock, for instance, the compatible elements will be preferentially included in the minerals, while the incompatible elements will concentrate into the remaining melt.

Various metal cations may be bonded into a network of silica tetrahedra to form other silicate minerals (Table 3.3). The bonding between silicon atoms and oxygen atoms is primarily covalent, and the silica tetrahedra tend to act as coherent units. The tetrahedra may be linked together in one direction to form a long chain, linked in two dimensions to form planar structures or sheets, or linked in three directions to form a three-dimensional framework. The silica tetrahedra form anionic (negative-

CRYSTAL SHAPE

A crystal's shape reflects the way its atoms are packed together. This, in turn, is controlled by the mineral's composition. Halite generally occurs as cubic crystals, whereas quartz forms elongate crystals with hexagonal cross section and pointed (pyramidal) ends (compare Figs. 3.4 and 3.1). Under ideal circumstances, crystals grow from a liquid by the addition of successive layers of atoms onto previously formed crystal faces. The size of a crystal and the perfection of its shape are both related to how quickly it grows and how much space is available to it. The largest single crystals may even grow to be the size of a house! Certain mineral types also exhibit a property known as *twinning,* in which the crystal form is repeated one or more times at different orientations, to give a complex shape to the specimen.

We cannot rely on crystal shape for mineral identification because the individual crystals are small and poorly formed in most rocks. Geologists often use hand lenses or microscopes to identify minerals (see Focus 4.1). Minerals form perfect crystals only when they grow in a relatively large volume of liquid. This situation occurs in a *geode,* which is a cavity in a rock that is lined with well-formed crystals projecting toward the center of the cavity (Fig. 1). A geode forms when a liquid containing dissolved elements or compounds penetrates into the cavity and minerals slowly crystallize from this liquid.

The first minerals to crystallize in a cooling lava tend to have well-defined crystal shapes because they grow unrestricted. Figure 2a shows feldspar in a lava, where the crystals have an obvious, characteristic shape because they crystallized before other minerals. If feldspar was one of the last crystallizing minerals in a rock, however, there would not be much space left at that point. In that case, the feldspar would occupy whatever space remained available and would probably crystallize as an irregular shape (Fig. 2b). We can use crystal shapes to tell us which minerals crystallized first (these will have regular crystal shapes) and which crystallized later (and have irregular shapes).

Crystalline rocks may disintegrate at Earth's surface into individual mineral grains. If you look through a microscope or hand lens at the tiny sand grains that make up ordinary beach sand, you will probably see rounded grains instead of well-shaped crystals. These crystals have been worn down during transport by rivers, wind, and waves on their way to the beach (Fig. 2c).

FIGURE 1 A geode. The well-formed crystals in the middle grew inward from the walls of a cavity in a rock.

(a)

(b)

(c)

FIGURE 2 (a) Feldspar crystals in lava. (b) Poorly formed (pink and white) feldspar crystals in granite. (c) Sand grains (feldspar and quartz) from a beach, rounded and smoothed as they were carried by wind, rivers, and waves.

CRYSTAL HABIT

In addition to variations in how *well* a crystal is formed, we can also distinguish a different characteristic *appearance,* or *crystal habit* (Fig. 3). Minerals grow in a variety of different ways: Many crystals grow as simple prisms, which is a habit referred to as columnar or prismatic (Fig. 3a). Individual crystals may also be long and needle-like (acicular; Fig. 3b), long and flat like a knife blade (bladed), like flat sheets (platy or foliated; Fig. 3c), or like fine threads (fibrous; Fig. 3d). Crystals often grow as groups or aggregates. In such cases, aggregates may be described as dendritic (diverging network of crystals like the branches of a tree; Fig. 3e), radiating (where elongate crystals grow out in different directions from a common point, Fig. 3f), or botryoidal (groups of fine radiating clumps that form spherical shapes resembling a bunch of grapes).

(a) (b) (c) (d) (e) (f)

FIGURE 3 Examples of different crystal habits, with minerals that show these characteristics. (a) Prismatic (kyanite); (b) acicular (actinolite); (c) platy (mica); (d) fibrous (asbestos); (e) dendritic (pyrolusite); and (f) radiating (pyrophyllite).

SPECIFIC GRAVITY

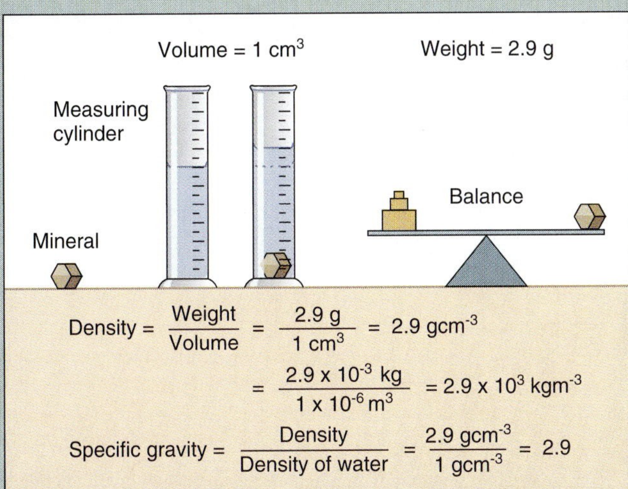

FIGURE 4 Specific gravity can be determined using a measuring cylinder and a balance. A mineral specimen is placed in the cylinder, displacing a volume of water that is equal to the volume of the mineral. The mineral is then weighed; its weight divided by the volume gives the mineral's density. This, divided by the density of water (1 g/cm^3), yields the specific gravity.

The *specific gravity* of a mineral is the ratio of the mineral's density to the density of water. As the density of water is 1 g/cm^3, the specific gravity of a mineral is numerically equal to its density. Specific gravity, however, is a dimensionless quantity; unlike density, it has no units. You can measure these quantities with simple instruments such as a balance and a measuring cylinder (Fig. 4).

You can, however, get a general sense of a mineral's specific gravity simply by holding mineral specimens in your hand and comparing their weights. This test, sometimes loosely referred to as "heft," can distinguish minerals with the same or similar volume that have significantly different specific gravities. Many iron- or lead-bearing minerals have high specific gravities (the specific gravity of magnetite [Fe_3O_4] is 5.18 and galena [PbS] is 7.5) compared with common minerals such as feldspar (2.6–2.7) and quartz (2.65).

COLOR

FIGURE 5 Varieties of the mineral quartz are distinguished by color: "Normal" transparent, or colorless, quartz; rose quartz; smoky quartz; and purple amethyst.

One of the great attractions of minerals is the wide variety of colors in which they occur. A mineral's color depends on the way in which light waves interact with the electrons of the elements that make up the mineral; simply put, color is a function of composition. Some minerals have a single, distinctive color. Sulfur, for instance, is characterized by a yellow color, whereas malachite (copper carbonate) is always green. Among the common rock-forming minerals, those containing iron and magnesium (olivine, pyroxene, amphibole, and biotite) are dark in color (typically black, brown, or green) and are referred to as *mafic* minerals. The light-colored rock-forming minerals (feldspar and quartz), which are typically white, pink or colorless, are referred to as *felsic*.

Other minerals can occur in a range of different colors, generally reflecting different "impurities" in the mineral. These impurities are small quantities of an element not normally included in the pure form of the mineral. The gemstones sapphire and ruby are blue and red, respectively, yet both are varieties of the mineral corundum (aluminum oxide, which normally occurs with a dull gray color). The blue and red colors result from the presence of trace quantities of titanium and chromium, respectively. Quartz, which typically occurs as a white or colorless mineral, may occur as pink rose quartz (owing to the presence of titanium and iron), gray smoky quartz (due to aluminum), or purple amethyst (from manganese impurities) (Fig. 5). This example illustrates that caution should be exercised when color alone is used to identify minerals.

STREAK

FIGURE 6 Metallic hematite and the blood-red streak produced when a specimen is scraped on a streak plate.

For some minerals, a powder made from a sample of the mineral has a different color from the mineral itself. The color of the powder is known as the mineral's *streak*. The mineral hematite might be silvery and metallic in a hand sample but can be recognized by its characteristic red streak. The powdered form is usually made by scraping the mineral on a piece of unglazed porcelain or frosted glass, called a *streak plate* (Fig. 6).

CLEAVAGE

(a)

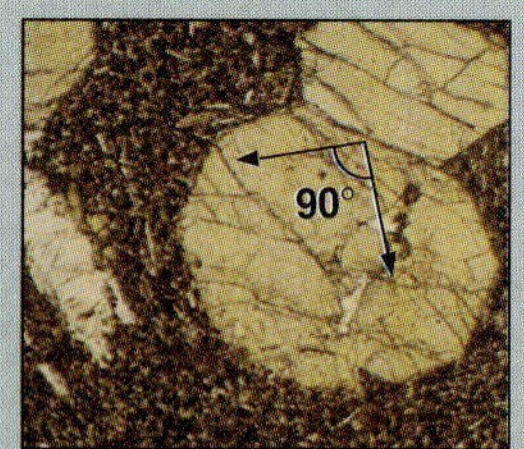

(b)

FIGURE 7 Examples of cleavage. (a) Mica displays perfect cleavage in one direction. (b) Pyroxene (upper photo) (with two sets of cleavage planes that intersect at 90°) and amphibole (lower photo) (with two sets of cleavage planes that intersect at 120°).

The *cleavage* of a mineral is its tendency to split apart along regular, well-defined planes where bonding is weakest. Cleavages are controlled by crystal structure. One or more sets may be present, always oriented at a particular angle with respect to the crystal structure for a given mineral. To appreciate cleavage, we can take a look at the mineral mica (Fig. 7a). The crystal structure of mica consists of sheets of covalently bonded silicon and oxygen clusters stacked on top of each other like the leaves of a book. The sheets are, in turn, held together by weaker Van der Waals bonds. The sheets can therefore be easily broken apart, or cleaved. Mica has one set of perfect cleavages.

In contrast, the minerals *amphibole* and *pyroxene* both have two sets of cleavages. In pyroxene, the two sets of cleavage planes intersect at nearly right angles (90°) (Fig. 7b); in amphibole, they intersect at an angle of about 120°. The properties of amphibole and pyroxene are very similar, but the difference in cleavage orientation is one of the main ways by which we can tell them apart. Both minerals are photographed in thin sections—ultra-thin slices of rock.

FRACTURE

(a)

(b)

FIGURE 8 (a) Conchoidal fracture of quartz. (b) Hackly fracture of chalcopyrite.

Fracture describes the way in which a mineral breaks when the break is not directed along a cleavage. Like cleavage, fracture reflects the arrangement of atoms in a crystal and the way in which these atoms are bonded together. Fracture surfaces can be conchoidal (curved or "shell-shaped," like broken glass) or hackly (rough, jagged). Quartz has a distinctive conchoidal fracture (Fig. 8a). Native copper and many other metallic minerals have a hackly fracture (Fig. 8b).

LUSTER

(a)

(b)

(c)

FIGURE 9 Examples of luster, transparency, translucency, and opacity. (a) Diamond has an adamantine luster and is transparent. (b) Galena has a metallic luster and is opaque. (c) Garnet has a vitreous luster and is translucent.

The *luster* of a mineral is the way in which the mineral reflects light. The capacity to reflect light depends on how smooth a crystal surface is at the atomic scale. Diamonds, for instance, have a brilliant sparkle and reflect most of the incident light. We refer to this as adamantine luster (Fig. 9a). Some minerals have metallic luster (Fig. 9b), especially sulfides of either iron (pyrite), copper and iron (chalcopyrite), or lead (galena). Minerals such as quartz and garnet (Fig. 9c) have vitreous (glassy) luster, and minerals such as sphalerite (zinc sulfide) have resinous luster.

TRANSPARENCY, TRANSLUCENCY, AND OPACITY

In contrast to luster, which describes the ability to *reflect* light, transparency, translucency, and opacity describe the capacity of a mineral to *transmit* light. If light can travel through a mineral—as it does through window glass—then the mineral is transparent (Fig. 9a). If light cannot pass through a mineral, then that mineral is opaque (Fig. 9b). Between these extremes lie translucent minerals; light passes through such minerals, but you cannot see clearly through them (Fig. 9c). The minerals diamond, galena, and garnet are transparent, opaque, and translucent, respectively.

TABLE 3.1 Mineral Properties (continued)

HARDNESS

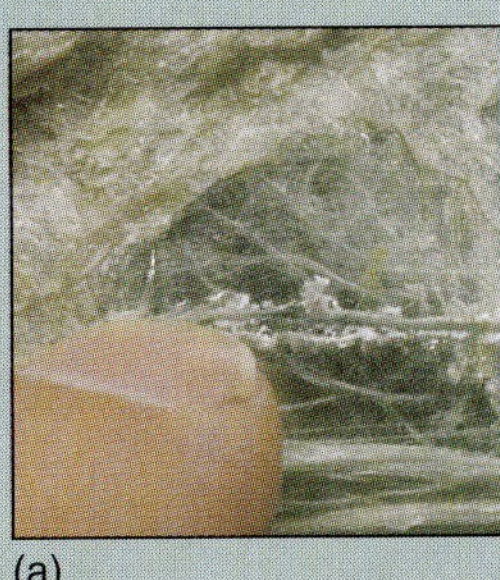
(a)

(b)

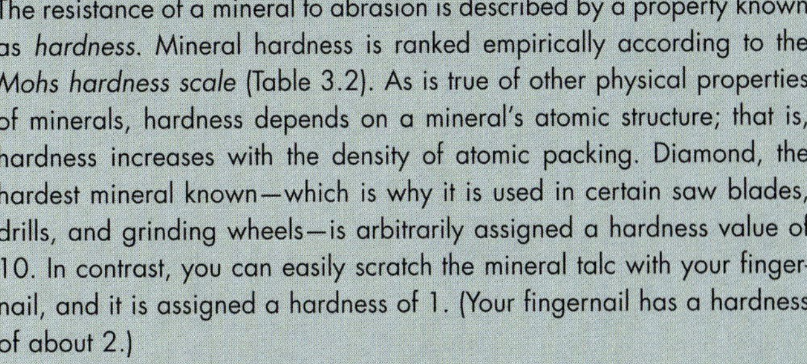
The resistance of a mineral to abrasion is described by a property known as *hardness*. Mineral hardness is ranked empirically according to the *Mohs hardness scale* (Table 3.2). As is true of other physical properties of minerals, hardness depends on a mineral's atomic structure; that is, hardness increases with the density of atomic packing. Diamond, the hardest mineral known—which is why it is used in certain saw blades, drills, and grinding wheels—is arbitrarily assigned a hardness value of 10. In contrast, you can easily scratch the mineral talc with your fingernail, and it is assigned a hardness of 1. (Your fingernail has a hardness of about 2.)

Hardness is a useful way to identify some minerals (Fig. 10). For example, quartz and calcite both occur in forms that are colorless and transparent. It can be difficult to distinguish between them if they do not exhibit good crystal forms. But you can try to scratch a mineral sample with a steel pocketknife, which has a hardness of 5.5 to 6. Steel will easily scratch calcite (hardness 3) but will have no effect on quartz (hardness 7) (Fig. 10). The hardest minerals are generally held together by strong covalent bonds. Ionically bonded solids, such as halite, are softer.

(c)

FIGURE 10 Testing the property of hardness (resistance to abrasion). (a) Talc, a very soft mineral (see Table 3.2), can be scratched by a fingernail. (b) Calcite can be scratched by a pocket-knife. (c) Quartz, which looks similar to calcite in many respects, cannot be scratched by the same pocketknife.

TABLE 3.2 The Mohs Hardness Scale

HARDNESS	MINERAL	TEST
1	Talc	Fingernail
2	Gypsum	Fingernail
3	Calcite	Copper coin
4	Fluorite	Glass plate
5	Apatite	
6	Orthoclase	Knife blade
7	Quartz	
8	Topaz	Steel file
9	Corundum	
10	Diamond	

ly charged) groups, which bond with metal cations primarily through ionic bonding. The types of cations (positively charged ions) occurring in a given silicate mineral depend on both the size of the ion (it must fit into the space available in the silica tetrahedra network; Fig. 3.11) and the charge of the ion (the mineral must have a net charge of zero, so the positive charges of the cations must balance the negative charge of the anionic network).

The silicate minerals represent structures in which progressively more silica tetrahedra are linked in one-, two-, or three-dimensional networks. Table 3.3 shows the structures of the silicate rock-forming minerals schematically. The silica tetrahedra are represented by discrete triangles connected to each other via their corners. In reality, the structures are three-dimensional, and the cations that are bonded between the silica tetrahedra are not shown.

Silicon and oxygen are the two most abundant elements within Earth. The basic building block of silicate minerals is the silica tetrahedron. Metal cations are combined with linked silica tetrahedra in the rock-forming minerals.

Olivines Olivines are silicate minerals formed from discrete, individual tetrahedra held together by ionic bonds with iron or magnesium cations. Olivine is a simple mineral; it varies in composition only by the relative proportions of iron or magnesium occupying the available cation sites. Olivine can contain either all iron or all magnesium cations or any intermediate ratio of these two cations. Ions of these two elements are similar in charge and size, so they substitute easily for each other in the crystal structure (Fig. 3.11).

As a result, olivine is characterized by **solid solution**, which is a variation in composition that controls many of the mineral properties and is the result of one solid (iron-bearing olivine or *fayalite*) dissolved in another (magnesium-bearing olivine or *forsterite*) (Focus 3.5). Olivine is the most abundant mineral in Earth's mantle; it is present as high-pressure polymorphs in the deep

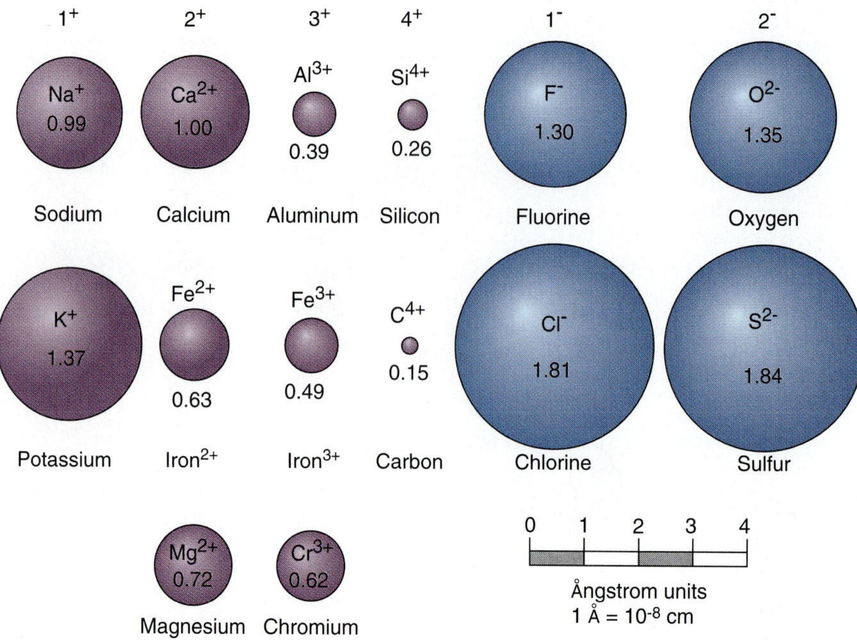

FIGURE 3.11 Relative sizes and charges of metal ions (cations) and nonmetal ions (anions) that are common in silicate minerals. The number associated with each ion is its radius in Angstrom units (1 Angstrom = 1 hundred millionth of a centimeter—or 10^{-10}m). The size of the ion determines whether it will fit into the spaces that exist between silica tetrahedra. The pyroxene structure has small spaces and will incorporate only small ions, such as iron (Fe^{+2}) and magnesium (Mg^{+2}). Relatively larger ions, such as potassium (K^+), cannot fit; thus, no pyroxenes contain potassium.

mantle. Olivine is also a common mineral in basalt; it is recognized by its light green color and the absence of cleavage (Fig. 3.12).

Pyroxenes Pyroxenes are silicate minerals formed from one-dimensional chains of silica tetrahedra. In this case, the tetrahedra along the chains are linked together by ionic bonds with cations of iron, magnesium, and calcium. Pyroxenes form solid-solution series among magnesium pyroxenes, iron pyroxenes, and calcium pyroxenes. Calcium is a larger cation than iron and magnesium (see Fig. 3.11) and thus does not readily fit into the spaces between the chains of silica tetrahedra. It distorts the regular crystal structure, so calcium-rich pyroxenes have a different symmetry from the magnesium-iron varieties (see Focus 3.3 for a detailed discussion of crystal symmetry). Like olivine, pyroxene is abundant in the mantle and in basaltic lavas (Fig. 3.12). Pyroxenes are dark green or brown and characteristically have two cleavages at roughly 90° to each other.

FIGURE 3.12 Silicate minerals identified in a rock. Olivine, pyroxene, and plagioclase feldspar are common in basalt.

Amphiboles Amphiboles form when two chains of silica tetrahedra are linked side-by-side. This structure is more complex than that of olivines or pyroxenes and has a wider range of holes, or sites, into which cations may fit. In addition to magnesium, iron, and calcium, the elements aluminum, sodium, potassium, and the hydroxyl ion (OH^-) commonly occur in amphiboles, giving rise to a wide range of compositions and solid-solution series (see Focus 3.5 for a closer look at solid solutions). Larger ions tend to fit into the big "holes" or rings of silica tetrahedra that result when two chains are linked (see Table 3.3).

Amphiboles are found in many crustal rocks. Common amphiboles, such as hornblende, resemble pyroxene in their mineral properties, although amphiboles can be distinguished because their cleavage planes intersect at roughly 120° rather than 90°. Like pyroxene, amphiboles are typically dark green or brown to black, although rarer varieties may be light green, blue, or even colorless. Amphibole typically forms elongate crystals and is common in rocks such as dacite and andesite (Fig. 3.13).

Micas Micas are silicate minerals in which the tetrahedra are linked with cations to form a two-dimensional planar sheet (Table 3.3). Larger ions, such as the hydroxyl ion and potassium, can fit into the spaces between the sheets. The sheet structure of micas is reflected in the perfect cleavage characteristic of these minerals (see Fig. 7a of Table 3.1). The two most common micas are *biotite*, a dark, iron-bearing mica, and *muscovite*, which is colorless and contains

TABLE 3.3 Structures of Silicate Minerals

	Silica tetrahedra (central Si^{4+} not shown)	Composition of a single unit	Mineral example
Isolated groups	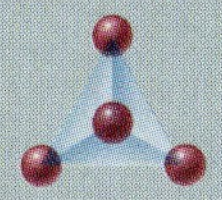	$(SiO_4)^{4-}$	Olivine, $(Mg, Fe)_2SiO_4$
One-dimensional chains — Single	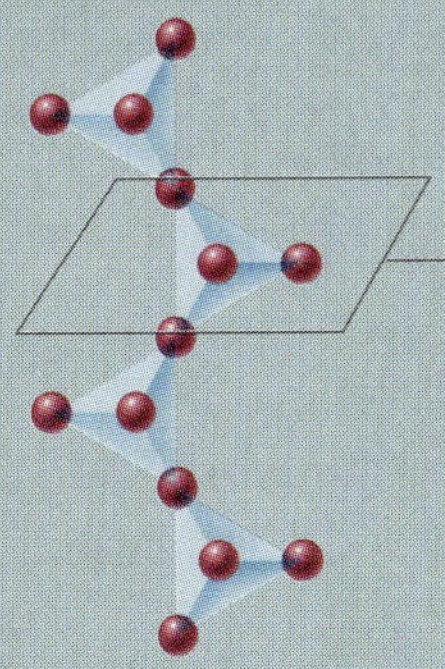	$(SiO_3)^{2-}$	Pyroxene e.g., Enstatite, $MgSiO_3$
One-dimensional chains — Double	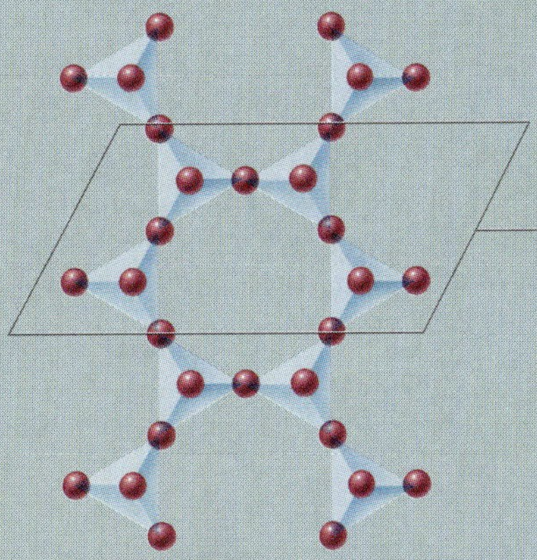	$(Si_4O_{11})^{6-}$	Amphibole e.g., Anthophyllite, $Mg_7Si_8O_{22}(OH)_2$
Two-dimensional sheets		$(Si_2O_5)^{2-}$	Mica e.g., Phlogopite, $KMg_3(AlSi_3O_{10})(OH)_2$
Three-dimensional framework	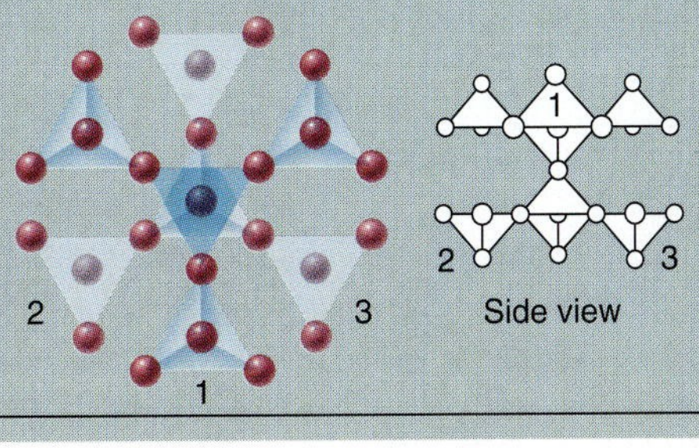 Side view	$(Si_3O_8)^{4-}$ and $(SiO_2)^0$	Feldspar e.g., Albite, $NaAlSi_3O_8$ Quartz, SiO_2

Focus On 3.5 SOLID SOLUTIONS

Many of the important silicate minerals are characterized by *solid solution*. The *composition* of such minerals can vary, although their fundamental properties (including crystal structure) are constant. The variation in composition generally takes the form of substitution between ions of similar size and charge. For instance, iron and magnesium are similar in charge and size and thus may substitute for each other in cation sites. Aluminum ions are sufficiently small that they can substitute for silicon in tetrahedral structures (see Fig. 3.11).

Consider the solid solution series of olivine, which we can represent in a phase diagram (Fig. 1). In this diagram, the composition or variations in the relative amounts of magnesium and iron are plotted along the horizontal axis, and temperature is plotted on the vertical axis. On the left and right ends of the graph are the *end members* of a solid solution series; for olivines, the end members are represented by *forsterite* (with 100 percent magnesium in the cation sites) and *fayalite* (with 100 percent iron in these sites). The melting point of forsterite is higher than that of fayalite. Above the liquidus—the temperature above which a substance is liquid—olivine is completely melted. Below the solidus—the temperature at which a substance begins to melt—it is completely crystallized. For solid solutions, or compositions other than those of the pure end members, there is a field between the solidus and liquidus in which crystals and melt exist together; they both are stable.

By reading the compositions off a horizontal line at a given temperature that intersects the solidus and liquidus, we can determine the crystals and liquid that are in *equilibrium* with each other at that temperature. At 1400°C, for example, crystals of 55 percent fayalite are in equilibrium with a liquid that has a composition of 83 percent fayalite. A liquid of this composition, when cooled to 1400°C, will crystallize olivine with a composition of 55 percent fayalite (45 percent forsterite).

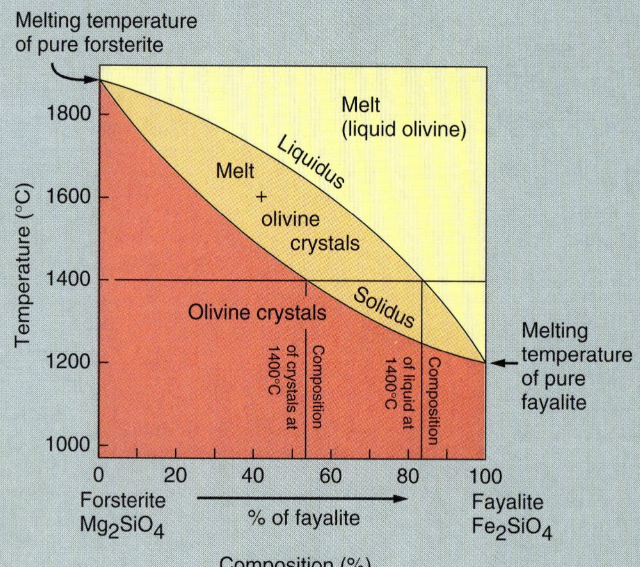

FIGURE 1 Phase diagram for the olivine solid-solution series. The composition of olivine varies between the end members forsterite (Mg_2SiO_4) and fayalite (Fe_2SiO_4).

potassium and aluminum. The micas are common in rocks of the continental crust, especially granite (Fig. 3.14).

Feldspars Feldspars are even more complex than micas; the silica tetrahedra of feldspars form a three-dimensional framework. Cation sites in the feldspar structure accommodate calcium, sodium, and potassium. The proportions of these elements allow us to distinguish among different varieties of feldspar. The common varieties are the *plagioclase feldspars*, which contain sodium and calcium, and the *alkali feldspars*, which contain sodium and potassium. Plagioclase feldspar can be distinguished from

FIGURE 3.13 Silicate minerals identified in a rock. Amphibole and plagioclase feldspar are common in andesite.

FIGURE 3.14 Silicate minerals identified in a rock. Biotite, alkali feldspar, and quartz are common in granite.

alkali feldspar by the presence of fine *striations* on the surfaces of the crystals. These fine parallel lines mark the contacts between twinned crystals of plagioclase feldspar.

Plagioclase feldspars offer another example of solid solution, analogous to the olivines described in Focus 3.5. In this case, solid solution exists between the high-temperature calcium plagioclase (anorthite) and low-temperature sodium plagioclase (albite).

Feldspars are characteristically white (although they may alter to a slightly pinkish tone), and have two cleavage directions. The feldspars are one of the most common minerals in Earth's crust; high-temperature, calcium-bearing plagioclase is found in basalts (see Fig. 3.12), and the feldspars become progressively more sodium-rich as rocks become richer in silica. Thus, granites are characterized by sodium-rich plagioclase feldspar and potassium-rich alkali feldspar (orthoclase) (see Fig. 3.14).

Quartz Quartz is a continuous framework of connected, covalently bonded silica tetrahedra without additional cations. Quartz is found in rocks such as granites (see Fig. 3.14), and is a common constituent of sandstones. It is typically colorless except for impurities, such as manganese, which gives rise to the purple color of the variety amethyst (see Fig. 5 of Table 3.1). It lacks cleavage and it has a conchoidal fracture. Furthermore, it is one of the hardest of the silicate minerals; most silicate minerals are about 5 to 6.5 on the Mohs hardness scale (Table 3.2), whereas quartz is rated 7.

Other Silicate Minerals Less abundant, but still important silicate minerals in some rocks include garnet, epidote, chlorite, and the clays. Garnet is commonly found in rocks formed at high pressures and temperatures (Fig. 3.15). It forms a range of compositions through solid solution, but the most common varieties are purple to red and generally occur as well-formed crystals. Garnet can be distinguished by its relatively high specific gravity. Epidote is broadly similar in composition to plagioclase feldspar, and commonly occurs as a result of feldspar alteration at high pressure and temperature. Chlorite is a soft green mineral that occurs largely as a result of low-temperature alteration of mafic minerals like olivine and pyroxene. The clay minerals also typically form as a result of alteration, and they resemble the micas in many respects. As you might expect from the minerals that form clay, however, the mineral grains are very small in size and usually cannot be viewed without the aid of a microscope.

You can see there is a gradual increase in the degree of linkage between the silica tetrahedra shown in Table 3.3 from olivine through pyroxenes, amphiboles, micas, and feldspar. The linking of individual tetrahedra is sometimes referred to as *polymerization*. Such polymerization corresponds to an increasing dominance of covalent bonds, which form the silica tetrahedra, over ionic bonds, which bond the metal cations into the structure. Recall that compounds formed from ionic bonds tend to be more soluble in water than are covalently bonded solids. As a consequence, minerals like quartz and

FIGURE 3.15 Silicate minerals identified in a rock. Garnet is the well-formed red mineral seen in this gneiss.

feldspar, which are largely held together by covalent bonds, tend to be more resistant to chemical weathering (reacting with water) at Earth's surface. In contrast, olivine, which contains more ionic bonds, is more soluble in water and thus more easily weathered.

Silicate minerals are by far the most important minerals that make up Earth's crust and mantle. The silicates are subdivided according to the way the silica tetrahedra are arranged and the types of cations that combine with them.

Nonsilicate Minerals

Most nonsilicate minerals are not as abundant as silicates but may be very important as ores: valuable sources for many of our raw materials. For instance, although most of the iron at Earth's surface is contained in silicate minerals, it is only economical to recover the element from nonsilicate minerals such as hematite and magnetite (iron oxides). This is largely because individual elements are difficult to separate from silicate minerals. We will deal with the nature and origin of economically important minerals in more detail (∞ see Chapter 16). Some nonsilicate minerals are of sufficient local abundance that they can be classified as rock-forming minerals, along with the silicate minerals previously listed. We can classify nonsilicate minerals according to their chemical composition.

Oxides Oxides are formed when metal ions bond directly with negative oxygen ions rather than with a silicon-oxygen complex. Some oxides, such as magnetite (Fe_3O_4), are common in ordinary igneous rocks. Others, like the oxide of aluminum (Al_2O_3), may occur as gemstones (ruby and sapphire). Oxides of iron such as magnetite and hematite (Fe_2O_3) are important economic sources of this metal.

Carbonates Carbonates contain the negative carbonate ion (CO_3^{2-}). The most important carbonate mineral is calcite ($CaCO_3$), which is the main component in limestone. Calcium carbonate is extracted from seawater by marine organisms, and secreted as small crystals of calcite, or of aragonite—another polymorph of $CaCO_3$—to form a protective exoskeleton. Most shells and corals, for instance, are composed of carbonates (Fig. 3.16). Limestones are used to make cement and are also used as one of the chemical agents in the process of iron ore smelting.

Sulfides and Sulfates Sulfides form when metal ions bond with sulfur ions, usually in oxygen-poor environments. Metals such as copper, zinc, and lead have a strong affinity for sulfur. Sulfides such as chalcopyrite ($CuFeS_2$), galena (PbS), and sphalerite (ZnS) are important sources of these metals, which are used in many everyday applications. For instance, copper is a good electrical conductor and is therefore used in

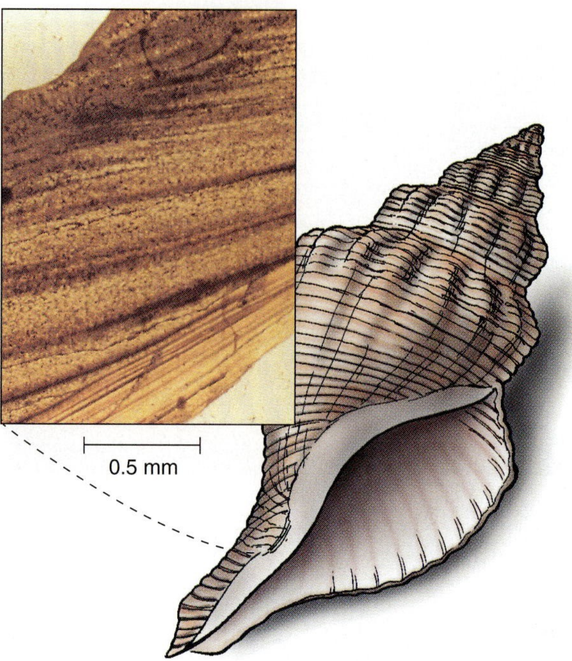

FIGURE 3.16 Layers of microcrystalline calcium carbonate form this shell.

wiring, and zinc is an important element in many metal alloys. Lead is a malleable, dense, and low-melting point metal that has been used to make water pipes and as an additive in gasoline. The realization that lead is toxic has discontinued its use in these applications. It is, however, still used extensively in lead-acid batteries for automobiles, where the lead is sealed within the casing.

If excess oxygen is present with sulfur, then sulfates, in which the negative ion is SO_4^{2-}, may form. Probably the most important sulfates are gypsum ($CaSO_4 \cdot 2H_2O$) and anhydrite ($CaSO_4$), which are widely used in the chemical industry and as components of fertilizers. Gypsum is also used to make plaster and Sheetrock® for the construction industry. Both gypsum and anhydrite may crystallize when seawater evaporates.

Halides Elements such as sodium and potassium combine readily with the halogen elements such as chlorine (Cl), and form soluble salts called halides. The most well-known halide is sodium chloride, or halite (see Fig. 3.4), better known as common table salt (NaCl). This mineral, dissolved in seawater, gives the sea its salty taste. When saltwater lakes or seas evaporate, halide minerals such as halite and sylvite (potassium chloride, KCl) may precipitate from solution.

Native Elements Although most metals are sufficiently reactive to form compounds such as silicates, carbonates, and oxides, some elements occur naturally in an uncombined or "native" state. Copper and gold are two such native elements; they may be found as naturally occurring pure metals, which form a characteristic crystal structure, so they can be considered as minerals by our

FIGURE 3.17 Sulfur (yellow) forming at the vent of a volcano.

definition. Under certain conditions, sulfur and carbon may also occur as native elements rather than combined in sulfides, sulfates, or carbonates. Sulfur is common around volcanic vents or hot springs (Fig. 3.17); carbon can occur as diamond in rocks that come from deep in Earth (see Fig. 3.7).

Distribution of Elements and Minerals in Earth

By now you have a feeling for the diverse range of minerals that form in and on planet Earth. You should also have some understanding of how minerals are identified; that is, which properties distinguish one mineral from another. For our purposes, though, knowing something about minerals is only a first step in understanding the formation of rocks in different parts of our Earth.

We must also look at the distribution of the elements throughout the planet. The elements that are locally available determine the chemical composition of (and thereby control the type of) minerals that can form at a given place. The local pressure and temperature also influence the type of mineral that might form. Minerals that crystallize deep within Earth differ from those that form at the surface.

The characteristics of the rocks at Earth's surface and their mineral constituents result from the way in which the planet has differentiated (∞ see Chapter 2). Recall that Earth is composed of distinct layers of different compositions. The origin of this layering (Fig. 3.18) is related to the way the planet formed and separated into compositionally distinct horizons; that is, how it differentiated.

Compositional Layering Within Earth

The Core Earth's core is thought to have formed within 100 million years of the formation of Earth itself and is made largely of iron and nickel (∞ see Chapter 2). The calculated density of the core (10 to 14 g/cm^3) does not agree well with the known density of iron and nickel under pressures appropriate to the depth of the core. Therefore, we interpret this to mean that the core must also contain one or more light elements such as sulfur or oxygen in order for its density to match the calculated value (see Figure 3.18). We know that the outer core is liquid and the inner core is solid because certain types of earthquake waves will not pass through the outer core. We do not know, however, what mineral structure the solid iron core possesses. We cannot sample the core, and it is very difficult to simulate the immense pressures and temperatures present in the core in the laboratory.

Earth's core is inferred to be mainly liquid and solid iron and nickel, with small amounts of one or more lighter elements such as sulfur or oxygen.

The Mantle Earth's mantle is made almost entirely of silicate minerals. The rock that forms most of the upper mantle is referred to as peridotite. A peridotite is a rock that contains abundant olivine. The term "peridotite" is derived from *peridot*, which is gem-quality olivine. The other minerals present in the mantle are pyroxene and a mineral that contains aluminum, such as garnet or spinel. These minerals are characteristically rich in magnesium and, to a lesser extent, iron, so the mantle is therefore magnesium-rich (see Fig. 3.18). Deeper in the mantle, olivine and pyroxene are no longer stable and undergo a phase change to high-pressure polymorphs (∞ see Chapter 5). The transformation to minerals that are stable deep in the mantle is analogous to the transformation of graphite into diamond at high pressures (see Fig. 3.9). Note that the chemical composition is probably similar throughout the mantle even though the mineralogy varies with depth.

Earth's mantle is rich in magnesium and is made of olivine and pyroxene (or high-pressure equivalents) together with an aluminum-bearing mineral.

We know a good deal more about the composition of Earth's uppermost mantle than we do about the core. We use three main lines of evidence to determine the composition of the mantle: slivers of mantle trapped in plate collisions, fragments of mantle (**xenoliths**) carried to the surface in lavas, and lavas derived from the mantle. In other words, we can actually obtain samples of the mantle, which tell us about its composition. Bits of mantle are pushed up to the surface by the immense forces involved in a collision between two continents. In this process,

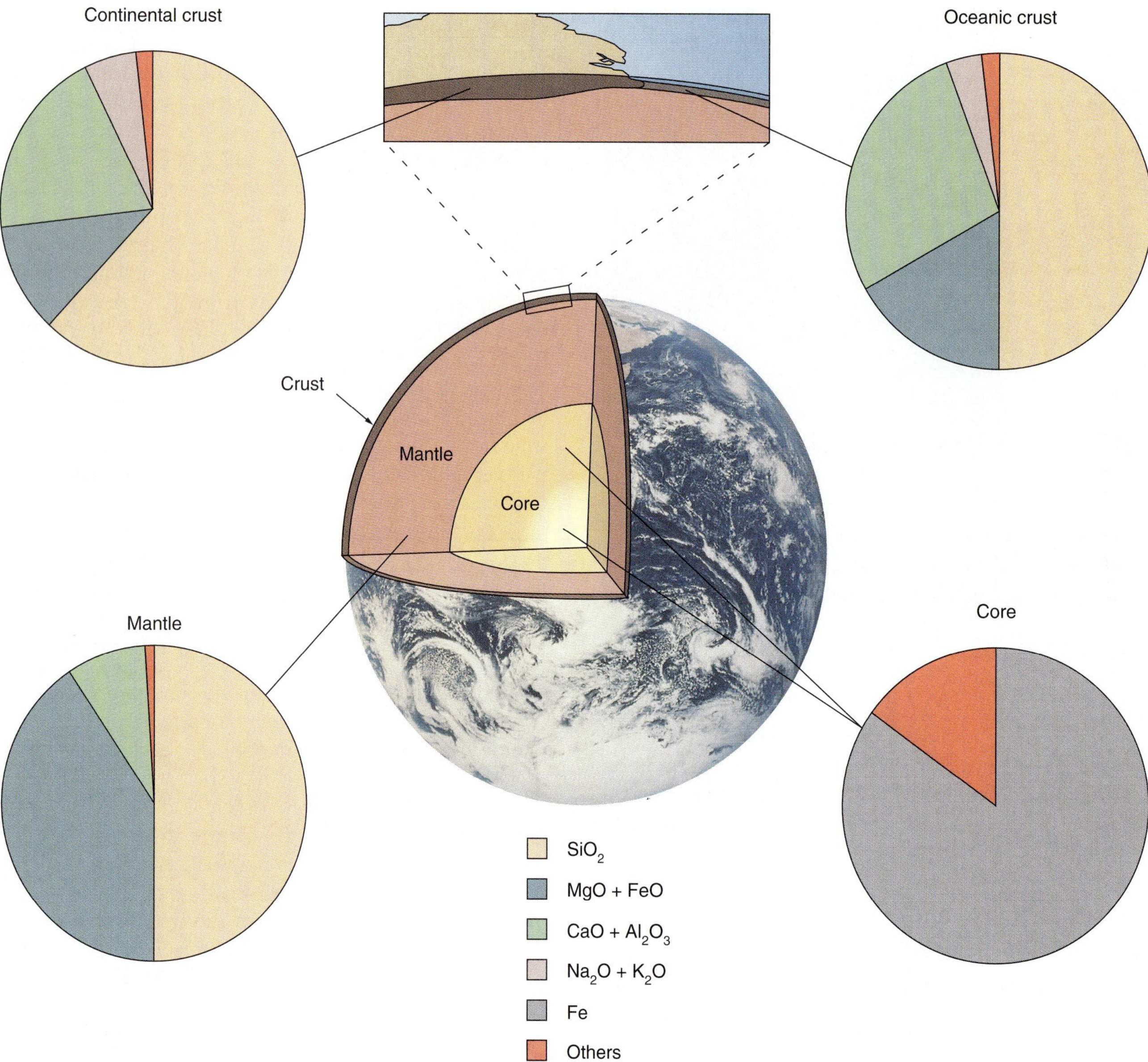

FIGURE 3.18 Schematic cutaway of Earth, showing the distribution of elements in the different compositional layers.

remnants of the oceanic lithosphere, which include both the basaltic oceanic crust and the underlying peridotitic mantle, may be trapped between colliding continents during the closing of an ocean basin and thrust up onto the land surface. These uplifted remnants of the ocean floor, called **ophiolites**, are extremely important rocks for geologists to study. After all, the only other access we have to the rocks of the ocean floor is through deep-sea drilling or observations made in deep-sea submersibles.

Another way to obtain samples of the mantle is from volcanic eruptions, when chunks of solid mantle are ripped up and carried to the surface as xenoliths in lavas (Fig. 3.19). Xenoliths are also common in rocks called *kimberlites*, which are brought to the planet's surface from great depths (>200 km) in a very short time. Kimberlites are well known as the source of diamonds (∞ see Fig. 2.18).

We gain information about the composition of Earth's mantle in a third way by studying melts that come from the mantle. The composition of a melt from the mantle is not the same as the composition of the mantle from which the melt was produced, because the mantle does not melt completely when it heats up; it undergoes partial melting. The mantle is actually a mixture of minerals, and the presence of certain minerals will lower melting temperatures so that they melt sooner than other minerals.

We have seen that many minerals are themselves a complex mixture of different elements and show a range of compositions. As these minerals start to melt, the composition of the first melts formed is different from that of

FIGURE 3.19 Mantle xenoliths containing mainly olivine brought to the surface in basaltic lava.

the original, solid minerals. This process is broadly similar to distillation, in which a distillate of a particular composition can be produced and separated from a mixture of several compounds. Distillation is how we separate gasoline from thick, sticky oil. The overall result of partial melting of the mantle is the production of a melt called **basalt**, which is richer in silica, calcium, and aluminum and poorer in magnesium than is the mantle. We will learn more about melting of the mantle in Chapter 4.

The Crust Like the mantle, Earth's crust is made predominantly from silicate minerals. The crust comprises two distinct subdivisions characterized by their chemical composition: oceanic crust and continental crust. The oceanic crust is basaltic in composition. The basalts of the oceanic crust are produced by melting of the mantle beneath mid-ocean ridges. Basalt typically contains the minerals olivine, pyroxene, and plagioclase feldspar (see Fig. 3.12). These minerals, and therefore the basaltic crust formed from them, are rich in the elements iron, magnesium, calcium, and aluminum (see Fig. 3.18).

Oceanic crust is made of basalt, which is rich in magnesium, iron, calcium, and aluminum and therefore contains the minerals olivine, pyroxene, and plagioclase feldspar.

The continental crust is rich in so-called incompatible elements such as potassium and sodium. These elements do not easily fit into the crystal structures of common silicate minerals of the mantle such as olivine and pyroxene. The composition of the continental crust is very diverse—just look at the huge variety of rocks that outcrop at Earth's surface. On average, the continental crust is richer in silicon than is the basalt of oceanic crust (see Fig. 3.18) and has a mineralogy characterized by plagioclase feldspar, potassium feldspar, and quartz.

Continental crust is rich in elements such as potassium, sodium, aluminum, and silicon. These elements are concentrated in low-density minerals such as quartz and feldspar.

SUMMARY

- A mineral is a naturally occurring inorganic solid with a definite composition and an orderly arrangement of atoms that give rise to a recognizable crystal form. The minerals in a rock reflect the rock's chemical composition and its conditions of formation.
- Atoms are made of a nucleus of protons and neutrons and surrounding shells of electrons. The atomic number—the number of protons in a nucleus—equals the number of electrons in the shells of the atom and serves to define each chemical element.
- Atoms in solids are linked together by bonds. Ionic bonds form from the electrical attraction between oppositely charged ions. Covalent bonds form when two or more atoms share their electrons.
- A crystal has a definite structure, is bounded by planar surfaces, and its outward appearance is based on its internal arrangement of atoms. Polymorphs are minerals that have different crystal forms but have the same chemical composition.
- In geology, thermodynamics is used to study the stability of different minerals under different conditions. The rates at which mineral reactions may occur under different conditions are governed by kinetics.
- Silicon and oxygen are the two most abundant elements within Earth. The basic building block of silicate minerals is the silica tetrahedron. Metal cations are combined with linked silica tetrahedra in the rock-forming minerals.
- Silicate minerals are by far the most important minerals that make up Earth's crust and mantle. The silicates are subdivided according to the way the silica tetrahedra are arranged and the types of cations that combine with them.
- Earth's core is inferred to be mainly liquid and solid iron and nickel, with small amounts of one or more lighter elements such as sulfur or oxygen.
- Earth's mantle is rich in magnesium and is made of olivine and pyroxene (or high-pressure equivalents) together with an aluminum-bearing mineral.
- Oceanic crust is made of basalt, which is rich in magnesium, iron, calcium, and aluminum and therefore contains the minerals olivine, pyroxene, and plagioclase feldspar.
- Continental crust is rich in elements such as potassium, sodium, aluminum, and silicon. These elements are concentrated in low-density minerals such as quartz and feldspar.

KEY TERMS

rock, 47
mineral, 47
crystal, 47
amorphous, 47
ion, 48
anion, 48
cation, 48
periodic table, 48
compound, 49
chemical bond, 49
crystal form, 51
polymorph, 52
phase diagram, 53
solid solution, 61
xenolith, 67
ophiolite, 68
basalt, 69

QUESTIONS FOR REVIEW AND FURTHER THOUGHT

1. Based on the definition given for a mineral, which of the following materials is truly a mineral and why (or why not)?
   a. Wood
   b. Table salt
   c. Sugar
   d. Coal
   e. The mercury in a thermometer
   f. The calcite that makes up some sea shells
2. What specific *property* makes the following minerals useful for the purposes listed here?
   (a) The use of diamonds to improve the cutting ability of industrial saws and drills; (b) the use of mica to make small windows on analytical instruments such as X-ray tubes.
3. The mineral rutile (TiO_2) is hard, insoluble in water, and does not conduct electricity. Would you expect the Ti–O bonds in this mineral to be ionic or covalent? Why?
4. Given that diamond and graphite are high- and low-pressure polymorphs of the element carbon, respectively, which would you expect to have a higher specific gravity? Why?
5. The mineral orthopyroxene has the chemical formula $(MgFe)_2Si_2O_6$. From Figure 3.18, which of the main subdivisions of Earth (crust, mantle, or core) would you expect to contain the most orthopyroxene, and why?
6. Most mineral grains that are present in rocks do not demonstrate a characteristic crystal form, even though they have an orderly internal arrangement of atoms. Why is this?
7. In some minerals, aluminum ions (Al^{3+}) may take the place of silicon ions (Si^{4+}) in the silica tetrahedra. From Figure 3.11, can you suggest a reason why this might be? Why might ions such as potassium (K^+) and calcium (Ca^{2+}) never substitute for silicon?
8. If you heat table salt (NaCl or sodium chloride), it will eventually melt. If we were to compare liquid (molten) salt and solid salt, which one of the following statements do you think would be true?
   (a) The liquid and solid have different compositions (different amounts of Na relative to Cl).
   (b) The liquid and solid have the same composition, but different arrangements of the atoms; the atoms are more ordered in the solid.
   (c) The Na^+ and Cl^- ions are bonded to each other in the solid, but not in the liquid.
   (d) The liquid and solid are both polymorphs of sodium chloride.

4 The Cornerstones of Geology: Rocks!

Spectacular outcrops of granite rocks in Yosemite National Park, California.

Overview

- Three main types of rock can be defined: igneous, sedimentary, and metamorphic.
- Igneous rocks form when molten rock cools and solidifies.
- Sedimentary rocks form from mineral and rock fragments and material of organic origin that is eroded or dissolved and then deposited at Earth's surface.
- Metamorphic rocks form when the composition, mineralogy, or texture of existing rocks changes in response to high temperatures and/or pressures or by reacting with fluids in the crust.
- Rocks within Earth are constantly changing in composition and location through processes such as melting and metamorphism.
- Near Earth's surface, tectonic forces cause deformation, while weathering and erosion recycle rocks at the surface.
- The continual cycling of rocks in the interior of Earth and at Earth's surface is known as the rock cycle.

INTRODUCTION

Chapter 3 introduced the basic building blocks of Earth: minerals. Just as there is a whole field defined as the study of minerals (mineralogy), there is a related field dealing with the study of rocks: petrology. Most rocks are aggregates of one or more minerals, but there are exceptions, such as volcanic glass and coal. Rocks can form by the crystallization of minerals from molten or dissolved material, by the accumulation and consolidation of loose mineral grains, or by the alteration of existing rocks in response to high temperature or pressure.

Geologists recognize three main types of rocks: **Igneous rocks** (from the Latin *ignis*, for "fire"), formed from the direct cooling and solidification of molten material; **sedimentary rocks** (from the Latin *sedimentum*, for "settling"), formed near Earth's surface by the hardening of sediments; and **metamorphic rocks** (from the Greek *meta*, for "change,"and *morphe*, for "form"), formed when a rock of any origin (sedimentary, igneous, or another metamorphic rock) is subjected to high pressures and temperatures or reacts with fluid in Earth's crust.

Igneous and metamorphic rocks make up 95 percent of Earth's crust by volume, whereas sedimentary rocks account for only 5 percent. Sedimentary rocks, however, cover much of Earth's surface and are the source of many important natural resources, including oil and coal.

This chapter serves as an introduction to the main rock groups. More detailed discussions of rocks are left until later in the text, when the processes leading to their formation can be explained in context.

IGNEOUS ROCKS

When we see molten rock erupting from a volcano (Fig. 4.1), we are looking at the end product of processes that began at great depths within the planet beneath that volcano. The volcano erupts melted silicate material that is generally derived from the mantle, although the crust may melt at times. This molten silicate material, together with gases dissolved in it (volatiles), is called **magma**. When magma reaches Earth's surface, much of the dissolved gas is released. Magma reaching the surface in liquid form is called **lava**.

The liquid form of a given material is usually less dense than its solid form. Molten magma is less dense than the surrounding material from which it has melted and so it rises. It may erupt at the surface from a volcano as lava and cool to form an **extrusive** igneous rock. Alternatively, it may crystallize deep in the crust below the surface as an intrusion formed from **intrusive** igneous rock.

Igneous rocks may form either by eruption of lava at the surface to form extrusive rocks or by solidification of magma below ground to form intrusive rocks.

FIGURE 4.1 Lava erupting from a volcano at Kilauea, Hawaii.

Intrusive and Extrusive Rocks

There are many different types of igneous rocks (Fig. 4.2). Names are based on texture and composition. Both properties reflect the mode of origin of the rock. *Texture* refers to both the size and the size distribution of the mineral grains (crystals) in the rock (Fig. 4.3). A rock in which all the crystals are easily seen with the naked eye is termed **phaneritic** (from the Greek word *phanes*, for "appearing"). This texture forms when the rock cools slowly, as it would if it were intruded into the crust. Alternatively, a magma that erupts at the surface may cool rapidly and form a fine-grained igneous rock where no distinct crystals are visible to the naked eye. This type of igneous rock is termed **aphanitic** (Fig. 4.3). If cooling is very rapid (chill or quench), a glassy texture may result. Rocks such as obsidian are truly glass, which means that they are noncrystalline. When broken, fragments of glassy rocks show a characteristic conchoidal fracture, which results in sharp cutting edges—ideal material from which to make arrows, spears, and knives.

Many extrusive igneous rocks have what is called a **porphyritic** texture. The bulk of the rock (the groundmass) is fine-grained, but it has a few large crystals, called **phenocrysts**, that are easily visible by eye. Such a texture usually results from a two-stage cooling process: First, magma intrudes into the crust, cools slowly, and begins to crystallize large, well-formed crystals of one or more minerals; second, the magma erupts at the surface, carrying the already formed phenocrysts, and the remaining liquid magma cools quickly into a fine-grained texture (see Fig. 4.3). In a sense, a porphyritic texture is a mixture of phaneritic and aphanitic textures that would be expected from the magma being subjected to both slow and rapid cooling. The identity of phenocryst minerals is commonly used to qualify the names of igneous rocks.

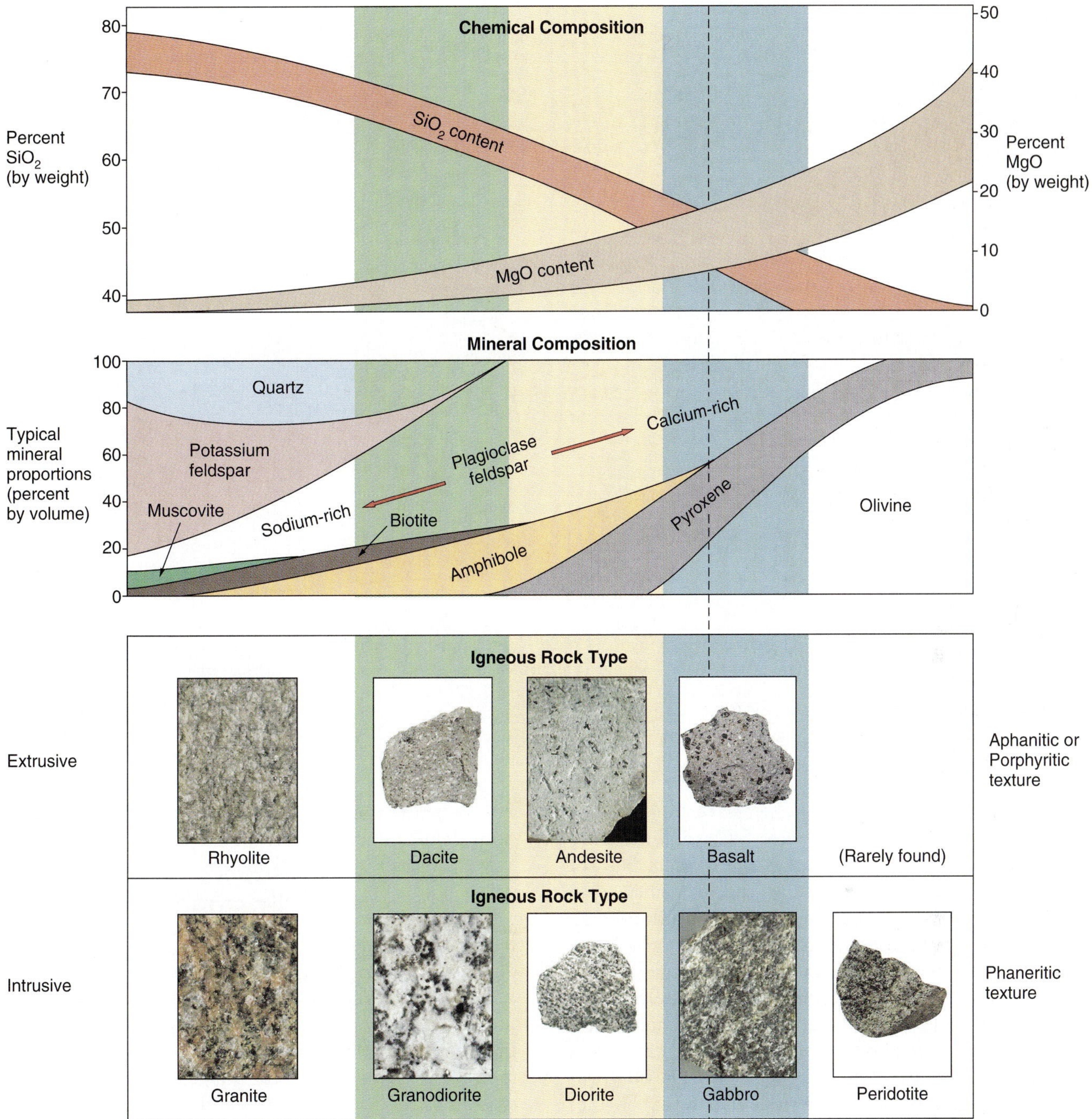

FIGURE 4.2 A simple classification of common igneous rocks; the boundaries between rock types are gradational. Two characteristics are used for classification: *composition*, which defines the mineralogy, and *texture*, which is a function of the cooling rate. These photographs are *representative*, but not all samples with the same name look alike. The dashed line drawn through basalt and gabbro is an example of the classification discussed in the text.

Thus, a rock with intermediate SiO_2 content and conspicuous hornblende phenocrysts could be referred to as a "hornblende andesite" (see Fig. 4.2).

The rock texture and types of minerals are sometimes difficult to determine simply by looking at a sample of rock—even using a hand lens. Geologists can get much more information by looking at a thin section, an ultra-thin slice of the rock, under a microscope. Entire undergraduate courses are designed to teach students the use of a microscope in geology and the identification of different mineral types. In Focus 4.1, we offer a brief introduction to the subject.

Figure 4.2 shows a simple classification scheme for igneous rocks. We can classify igneous rocks as intrusive or extrusive types by their texture. In reality, there is a gradation between the two groups. Similarly, a continuum in

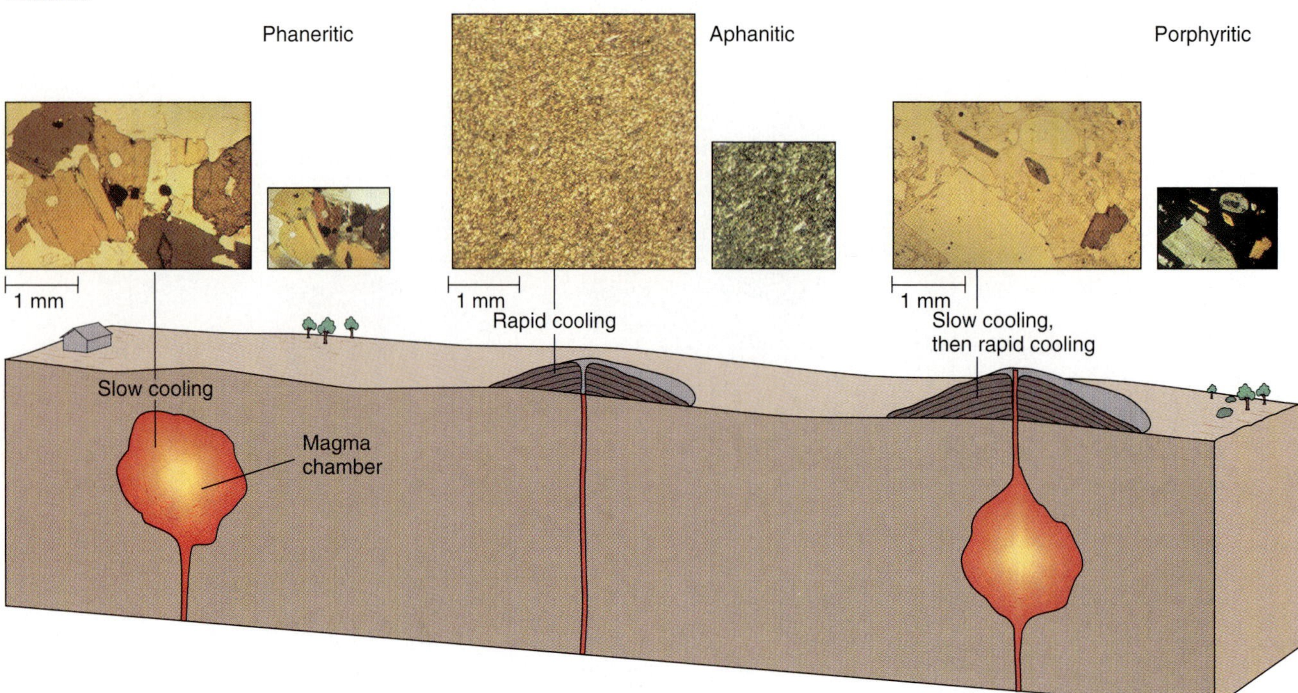

FIGURE 4.3 The formation of different textures in igneous rocks is determined primarily by cooling rate. If cooling is slow, deep in the crust, then a *phaneritic* texture is produced. Rapid cooling, as occurs when a magma erupts directly at the surface, results in *aphanitic* or glassy texture. Porphyritic textures are produced when a magma that has cooled slowly and partly crystallized then erupts at the surface. The photographs show thin sections of rocks with different textures as viewed through a microscope (Focus 4.1). The smaller photograph in each example shows the same view, but using crossed polarized light.

compositions exists as well. The chemical composition, as shown by the abundances of silica (SiO_2) and magnesia (MgO), varies continuously across the diagram but, for convenience, we classify rocks using specific names.

It is important to understand that these categories are arbitrary and that, in nature, there are gradations across all categories. For example, as indicated in Figure 4.2, a basalt or gabbro typically contains approximately 30 percent olivine, 30 percent pyroxene, and 40 percent calcium-rich plagioclase. If these minerals grew slowly to form a coarse-grained texture, then the rock would be an intrusive gabbro. If they grew quickly, the rock would be fine-grained and therefore would be called a basalt.

The composition of an igneous rock can be described in either mineralogical or chemical terms. The mineralogical composition depends on the types of minerals that have crystallized. Minerals that crystallize in a gabbro (plagioclase feldspar, olivine, and pyroxene) are different from those that crystallize in a granite (potassium feldspar, quartz, plagioclase feldspar, and biotite). If we look at the fine-grained extrusive equivalents of gabbro and granite (basalt and rhyolite, respectively), we might not see any crystals—the rock may simply be too fine-grained. The chemical composition, or bulk composition, can be determined by crushing the rocks into powders and then analyzing the powders. We would then find that the bulk composition of rhyolite is the same as that of granite and the bulk composition of basalt is equivalent to that of gabbro (see Fig. 4.2).

The type of mineral that crystallizes from a magma depends on the chemical composition of the magma. Whether minerals crystallize at all and the crystal size both depend on the cooling rate.

Differentiation of Igneous Rocks Looking at Figure 4.2, it becomes evident that the range of compositions of igneous rocks is vast. Consequently, Earth must be able to generate magmas of different chemical compositions. The processes that do this are described under the general heading of magmatic **differentiation.** Indeed, the separation of the crust and mantle from the chondritic material that formed the early Earth (∞ see Chapter 2) is largely the result of magmatic differentiation. Some of the specific processes are worth discussing here briefly.

Partial melting To understand melting or crystallization of a silicate melt, we must look carefully at the concept of melting and revisit the terms "solidus" and "liquidus." (You are encouraged to refer to Focus 3.5 for an in-depth example of the use of these terms.) The **solidus** is the temperature below which a material is completely solid. The **liquidus** is the temperature above which a material is completely liquid.

Focus On 4.1 THIN SECTIONS

To look more closely at rocks and the minerals they contain, geologists examine *thin sections* under the microscope. A thin section is a slice of rock mounted on a glass base or *slide*. It is made by cutting a slab of the rock with a diamond saw. One side is polished and glued onto the glass slide; the other side is then ground down until the slice is only 0.03 mm thick. The thinness of the slice enables us to see through it. When a thin section is placed beneath the microscope, light shines through the thin section, and the textures of the minerals can be seen clearly. In addition to the properties that can be seen in a hand specimen as described in the text, other properties can be viewed with the aid of the microscope.

In Figure 1a, some minerals stand out on the thin section more than others do; they have high *relief*. One particularly helpful tool used with petrographic microscopes is a *polarizer*. Light from objects such as the Sun or a lightbulb is made up of electromagnetic waves oscillating in all directions at right angles to the direction of travel. Any given direction of oscillation is called the polarization. A polarizer is a filter that only passes light waves with one direction of polarization, which is determined by the crystal structure of the polarizer. Polaroid sunglasses filter sunlight using this principle. Light viewed through two polarizers oriented at right angles (crossed polarizers) will be extinguished completely. If a mineral placed between two polarizers rotates the polarization direction of the incident light, by an effect called birefringence, some of the rotated component can then pass through the second polarizer. These minerals will stand out in contrast to those not possessing this property. The artificial colors of minerals under crossed polarizers (Fig. 1b) can be used for mineral identification. Compare Figures 1a and 1b. Under crossed polarizers, minerals such as olivine show up as brightly colored. Feldspars, however, are dull gray, but they often have a striped appearance due to twinning.

The rock shown in this thin section (Fig. 1) is a basalt. It contains large crystals (phenocrysts) of olivine, pyroxene, and smaller crystals of feldspar set in a fine-grained groundmass. This is a good example of porphyritic texture. The individual minerals can be identified easily by a trained geologist based on characteristics such as relief, color, cleavage, and birefringence.

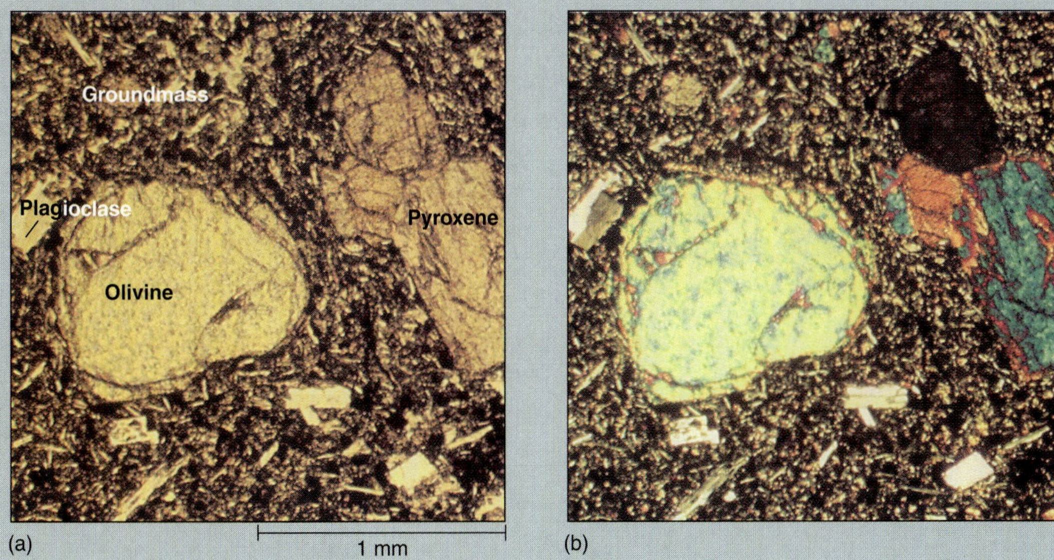

FIGURE 1 (a) A thin slice of rock, or *thin section*, seen with a petrographic microscope. The different types of minerals are labeled. (b) The same thin section viewed through crossed polarizers.

When the mantle or crust melts, it does not do so completely. If the mantle were a simple element or compound, then its solidus and liquidus would be the same. Water, for instance, freezes to a solid or melts to a liquid at one temperature (0°C at sea level). Rocks are complex mixtures of many minerals and therefore do not melt or crystallize at a single temperature but rather over a range of temperatures. For any given silicate rock, there is a temperature *interval* over which it is partly molten, with solid (crystals) and liquid (magma) existing together. Some minerals melt at lower temperatures than do others. When the mantle melts, the composition of the melt will therefore not be the same as that of the solid that is melting, because some minerals melt before others.

The melt (magma) is less dense than the solid mantle or crust. The relative buoyancy of the liquid produced allows it to rise away from the region of melting toward the surface. In effect, the solid is only *partially* melted and we say that the solid material has undergone **partial melting**. Here then, in the very process of generating magma, is a way to generate a range in compositions. Experiments have shown that partial melting of the peridotite mantle produces melts of basaltic composition, and that partial melting of basalts yields more silica-rich and magnesium-poor melts.

Fractional crystallization Very few partial melts erupt at volcanoes or intrude Earth's crust without undergoing some modification. Magma rising through the crust will cool and begin to crystallize. In a general sense, the process of crystallization is the reverse of melting. If an isolated volume of magma crystallizes completely, its bulk composition will not change. Within Earth's crust, however, a given volume of magma is seldom isolated.

Magma may collect in a subterranean reservoir or **magma chamber** and begin to crystallize. Again, because the magma is a complex solution of many elements, the individual minerals that crystallize are different in composition from the bulk composition of the liquid magma. The minerals that crystallize at a given time depend largely on the composition of the magma and on the temperature and pressure in the magma chamber (see Focus 4.2). Most minerals that crystallize are denser than the magma

(a)

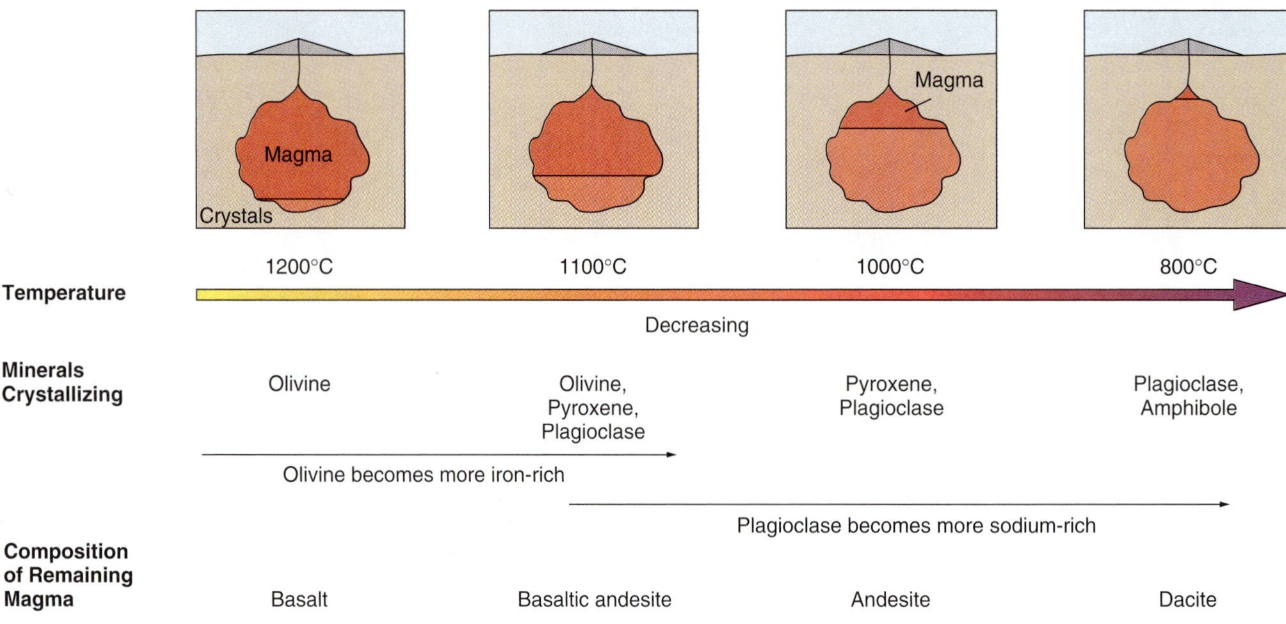

(b)

FIGURE 4.4 (a) Layers of crystals that accumulated at the bottom of a magma chamber, Skaergaard Intrusion, Greenland. The denser dark-colored minerals (olivine and pyroxene) settled out first, followed by less dense lighter-colored feldspars. (b) Typical sequence of progressive crystallization within a magma chamber. As crystals form, they sink to the bottom of the chamber and progressively change the composition of the remaining magma from basalt to dacite.

from which they form and may sink to the bottom of the chamber (Fig. 4.4a). Thus, there is a *fractionation*, or physical separation, of the liquid and crystals. The liquid becomes progressively less dense as denser crystals settle out; in fact, the liquid may become buoyant enough to rise from the magma chamber and perhaps erupt at the surface at a volcano. We will see later that gas pressure may also assist the process of eruption.

Crystallization and removal of minerals such as olivine and pyroxene from a basaltic magma, perhaps by gravitational sinking, give rise to andesitic liquids. We can see how this works by examining Figure 4.4b. We saw in Table 3.3 (∞ see Chapter 3) that olivine and pyroxene are rich in the elements magnesium (Mg), calcium (Ca), and iron (Fe). Removal of olivine and pyroxene from the basalt results in a liquid that contains correspondingly less magnesium, calcium, and iron. If a basaltic magma is allowed to crystallize for a while, various minerals will crystallize one after the other as the temperature drops (Focus 4.2). If these minerals are progressively removed from the remaining liquid, then this process is known as **fractional crystallization**.

The exact sequence of mineral crystallization varies according to magma composition, pressure, and temperature. For a basalt at low pressures, such as those in a magma chamber beneath a volcano, a typical crystallization series with decreasing temperature (increasing crystallization) would be olivine, pyroxene, plagioclase feldspar, and amphibole. This sequence would give rise to a corresponding change in the composition of the remaining liquid magma from basalt to andesite to dacite (Fig. 4.4).

More than one mineral may crystallize at any time. During the crystallization interval of a given mineral type, its composition can change as crystallization progresses. Plagioclase feldspar, for instance, has a range of compositions (called **solid solution;** see Focus 3.5). Thus, at higher temperatures, the plagioclase that crystallizes is calcium-rich but it is sodium-rich at lower temperatures. The plagioclase feldspar changes its composition gradually as temperature decreases because it is continuously reacting with the remaining magma. Discontinuous reaction can also occur—the replacement of olivine by pyroxene in a crystallizing sequence can be due to olivine actually reacting with the magma to form a new mineral, pyroxene. The sequence of minerals crystallizing from a magma as temperature decreases is commonly referred to as *Bowen's reaction series*, named after the famous American petrologist N. L. Bowen.

Contamination Magmas can change composition by melting the surrounding preexisting rock and subsequently becoming a mixture of different melt compositions. This process is commonly called *contamination* because of the incorporation of surrounding rock into the magma, and it is particularly important when basalt intrudes continental crust because the temperature of the basalt exceeds the melting temperature of the crust. As we previously learned (∞ see Chapter 3), the continental crust is andesitic to granitic in composition, which is very different from a basalt. Mixing melted continental crust into basaltic melts from the mantle may have a marked effect on the composition of the magma. Sometimes we see physical evidence of the process of contamination when xenoliths (fragments of wall rock) are trapped and may be partly melted in the magma (Fig. 4.5).

FIGURE 4.5 Contamination (melting of crustal material into a magma) may change the composition of a magma. Here a xenolith of granite has been melted by basalt lava.

A wide range of igneous rock compositions is generated by various processes of differentiation such as partial melting, fractional crystallization, and contamination.

Types of Intrusions

We know a good deal about extrusive igneous rocks—how they are erupted, how they cool, and so on—simply because we can observe them directly at volcanoes. In contrast, the formation of intrusive rocks cannot be observed directly. As is true of metamorphic rocks and many types of geologic structures, we must wait millions of years until the overlying crustal rock is eroded away to expose intrusions at the surface. We then examine the relationships between the intrusive rock and the wall rock to determine how the magma was intruded.

Enormous volumes of magma—hundreds to thousands of cubic kilometers—have intruded the crust at certain times and places. The largest of all igneous intrusions are called **batholiths** (from the Greek word *bathos*, meaning deep). These intrusive bodies tend to be elongate and are perhaps hundreds of kilometers long. Close examination shows that each batholith is actually a combination of many smaller intrusions, or **plutons** (from *Plutos*, the Greek god of the underworld), each of which may be only a few kilometers across and may vary slightly in composition. Most batholiths are formed of plutons

Focus On 4.2 CRYSTALLIZATION OF LAVA: THE HAWAIIAN LAVA LAKE

We can perform crystallization and melting experiments in the laboratory (see Focus 4.3). Under exceptional circumstances, we can also observe crystallization of lavas in nature. Figures 1 through 4 show lava collected from the lava lake of Kilauea volcano in Hawaii during the spring of 1965. The samples were collected from surface flows, from cores drilled into the solidified crust of the lava lake, and from lavas obtained at progressively deeper (hotter) levels. Temperatures were measured accurately by using a thermocouple—a device that distinguishes temperatures by electric current. If the lava is cooled rapidly (quenched), the minerals that had crystallized at the measured temperature are preserved and the still-molten material solidifies to a glass. The sequence of photographs shows progressive crystallization with decreasing temperature.

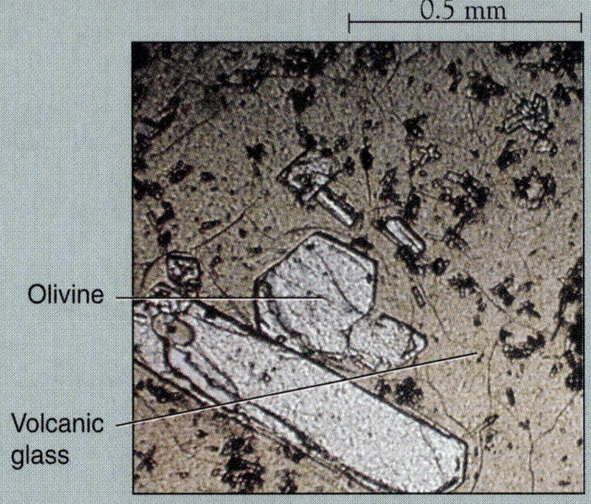

FIGURE 1 At 1170°C, large crystals of olivine form, enclosed in brown volcanic glass. The large size and the well-formed crystal faces indicate that the olivine had been growing slowly in the lava.

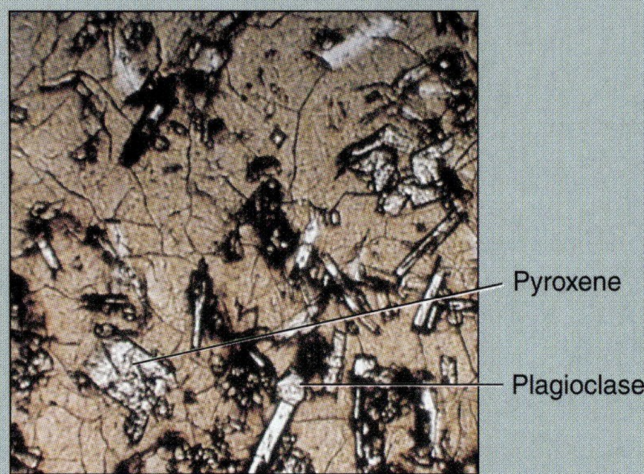

FIGURE 2 At 1130°C, the lava has crystallized to a greater extent. Plagioclase and pyroxene crystals are much larger now.

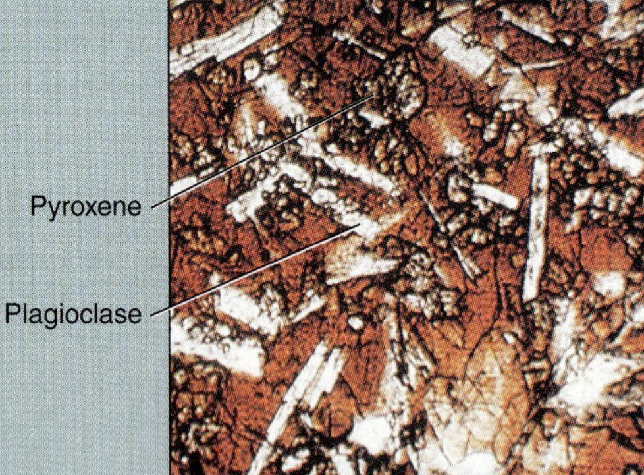

FIGURE 3 At 1075°C, the lava is about 70 percent crystalline. The brown glass is the solidified liquid remaining at this temperature. The deep brown color compared with Figures 1 and 2 is due to an increase in the amount of iron and titanium in the glass.

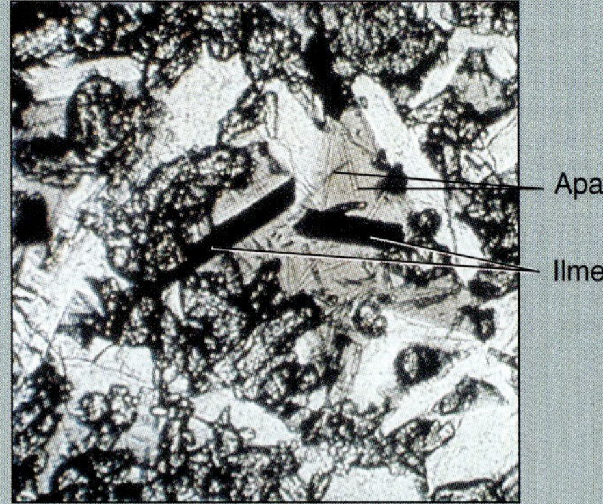

FIGURE 4 At 1020°C, the lava is almost totally crystalline. Two new minerals are present: ilmenite (black crystals) and apatite (small, needlelike crystals). These minerals must have crystallized at temperatures below 1075°C because they are not present in the sample collected at that temperature (Figure 3).

that are granitic to dioritic in composition. Figure 4.6a is a map of the batholith distribution in western North America. The Sierra Nevada range in California exposes a batholith composed of a huge mass of granitic plutons intruded some 100 million to 200 million years ago at depths of 10 to 25 km in the crust. Subsequent uplift and erosion of the overlying crust have exposed this enormous intrusion as the spectacular mountain range we see today (Fig. 4.6). We do not know the details of how such large masses of magma are intruded into the crust, but we can get an idea of what processes may be operating by looking at the relationships between the igneous rocks and the crustal rocks they intrude.

The main driving force behind the movement of magma is buoyancy. When a portion of the crust or mantle melts, the liquid that forms is usually less dense (lighter per unit volume) than the surrounding solid. As a result, magma tends to rise. Rocks in the upper part of the crust are brittle and may contain cracks that allow magma from below to rise to the surface, where it may eventually erupt at a volcano. Some of the magma may solidify in these cracks as shallow igneous intrusions (Fig. 4.7). Sheetlike intrusions that crosscut preexisting rocks are called **dikes**. Dikes are commonly vertical or steeply inclined. Intrusions that follow near-horizontal cracks parallel to the layers of near-surface rocks, rather than cutting across them, are called **sills**. At times, magma rises beneath a volcano along a simple cylindrical channel and solidifies to form a volcanic neck (Fig. 4.7).

The volumes of most intrusions that solidify at moderate depths in the crust are generally small, so they cool rapidly. The outer margins of these bodies in contact with the relatively cold, surrounding wall rock actually "chill" to a fine-grained or glassy texture. The shapes of dikes and sills are the result of the brittle behavior of the crust through which the magma ascends. The crust fractures, allowing the magma to fill the cracks (Fig. 4.8a). At greater depths, the crust is not so brittle and will not crack.

Deeper in the crust, the rise of buoyant magma is resisted by the overlying crust, which acts like a cap. There are no huge holes in the crust for magma to fill. At

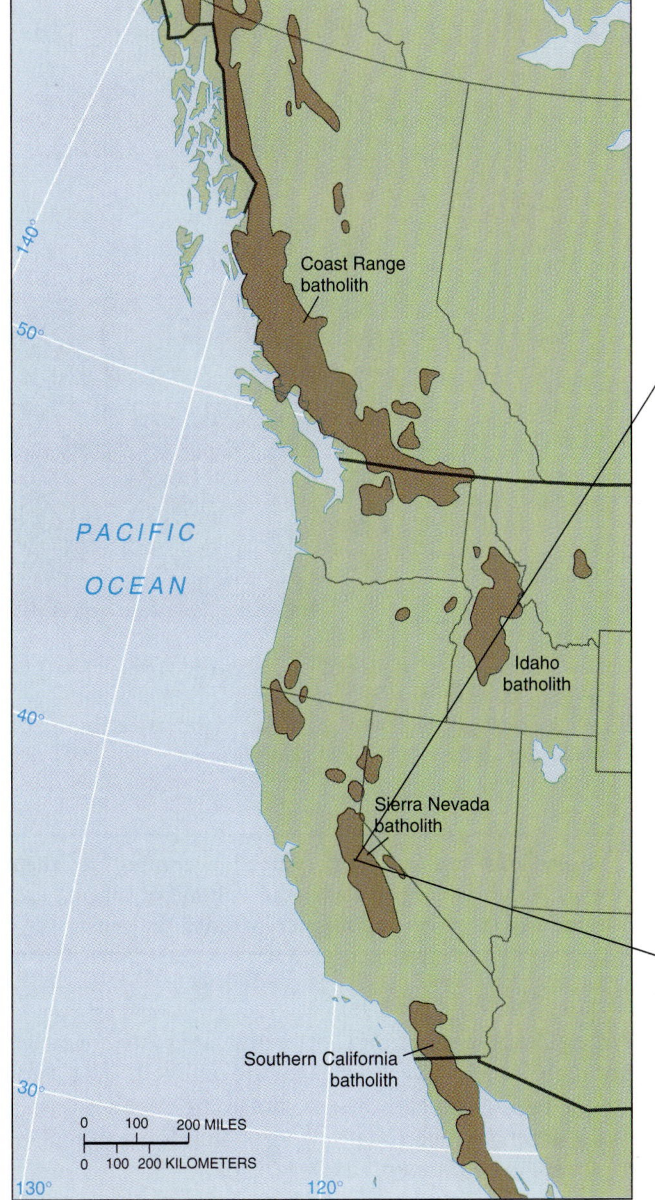

(b)

FIGURE 4.6 (a) Distribution of batholiths in western North America. Each batholith comprises many individual plutons, intruded over several millions to tens of millions of years. (b) Granitic outcrops in the Sierra Nevada at Yosemite National Park, California, are shown. The smoothness and massiveness of the outcrops are typical of granites but are accentuated in Yosemite by the effects of glacial erosion.

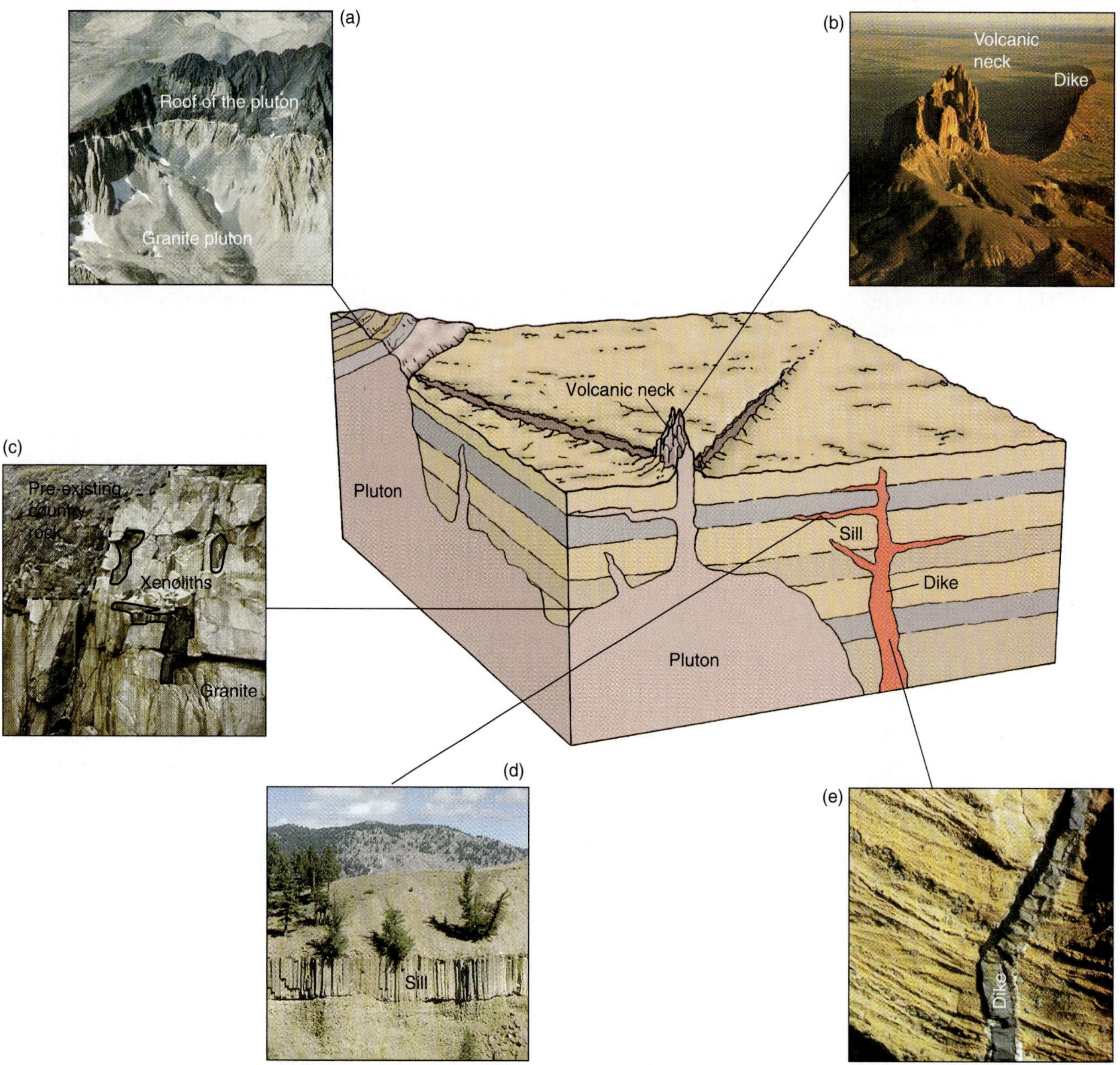

FIGURE 4.7 Igneous structures formed when magma solidifies. (a) Granite pluton exposed at the surface and eroded. Yosemite Valley, California. (b) Volcanic neck and dike, exposed by erosion. Shiprock, New Mexico. (c) Assimilation of preexisting (older) rock at the margin of a pluton. The chunks of older rock are known as xenoliths. Field of view 5 m. (d) Sill intruding parallel to rock layers. In this case it has moved from one layer to another. Sill is about 30 m thick. (e) Dike cutting through layers of cinders (Azores).

depth in the crust, the upward movement of magma takes place by *diapiric rise*. The magma may rise as a buoyant mass, or **diapir**, inflating the surrounding crust like a balloon and physically pushing it aside (Fig. 4.8b). Alternatively, the magma may "eat" its way up, melting and incorporating the overlying crust in its path, a process called **assimilation** (Fig. 4.8c). The magma loses heat to the wall rock, which both raises the temperature of this surrounding rock and melts it, contaminating the magma. A large amount of heat is required to convert solid wall rock at its melting temperature to a liquid at that same temperature. This heat is supplied by the intruding magma, which consequently loses heat and solidifies.

Intrusions are generally named according to their shape, which is controlled by the volume of magma and by the way in which the magma intrudes the crust.

Types of Volcanoes

Volcanoes are openings or vents at Earth's surface through which lava and gases are erupted. The vent might be an elongate fissure, but most commonly is a hole, around which the products of eruption accumulate to form a hill or mountain.

If you have ever watched a film of a volcanic eruption, you have probably seen red-hot lava rivers pouring down the slopes of a volcano such as Kilauea in Hawaii (Fig. 4.9a). These eruptions, although spectacular, are not considered dangerous. Geologists equipped with heat shields can walk right up to the molten lava and take samples. The volcanoes of Hawaii tend to have gentle slopes and are commonly termed **shield volcanoes** because they resemble a shield laid on the ground (Fig. 4.9b).

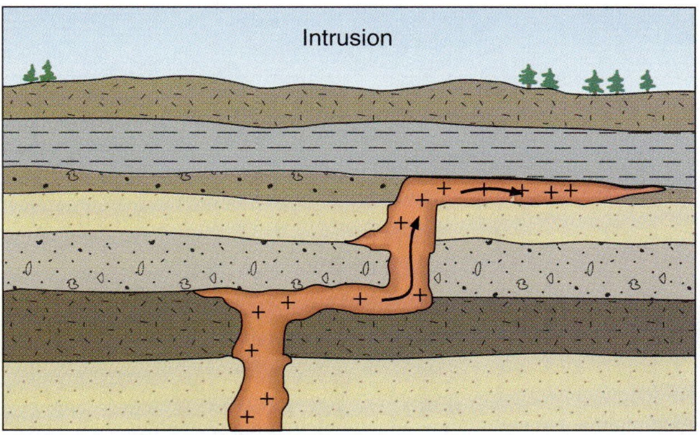

(a)

Diapiric Rise

(b)

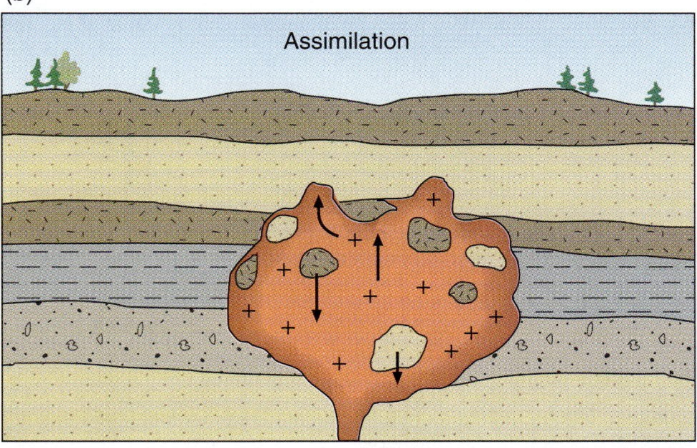

(c)

FIGURE 4.8 Different ways in which a magma can ascend through the crust, solidifying later as an intrusive rock. (a) *Intrusion into brittle fractures*: Magma fills cracks and forces its way up, forming dikes and sills when it solidifies. (b) *Diapiric rise*: A hot blob of magma melts its way up through the crust like a hot knife cutting through butter. (c) *Assimilation*: The surrounding rock is digested, allowing the magma to ascend by literally eating its way up through the crust.

You may have also seen a film of the 1980 Mount St. Helens eruption in the state of Washington (see Fig. 1.15a). There were no rivers of molten lava in this eruption. Instead, a very dangerous and explosive eruption threw volcanic ash high into the atmosphere and triggered flows of hot, ashy material and volcanic mud down former river valleys. Mount St. Helens is a very steep-sided volcano. A similar type of volcano, Mount Pinatubo in the Philippines, erupted in 1991 and poured more than 20 million tons of volcanic ash into the upper atmosphere—a volume that may rival that of the infamous Krakatoa eruption of 1883. Such large, steep-sided, explosive volcanoes are known as **composite volcanoes** (Fig. 4.9c,d). In contrast to the shield volcanoes that are formed from flow upon flow of basalt lava, composite volcanoes are composed of a range of different lava types together with *pyroclastic* deposits (ash, pumice, and lava fragments).

Why are these volcanoes so different in shape and in the type of eruption? The answer lies in the *composition* of the magma being fed to the volcano, which, in turn, controls the magma's *viscosity*. The more silica in the magma, the more it tends to form long, chainlike molecules (polymerization), and the more viscous the magma becomes. **Viscosity** is the measure of a fluid's resistance to flow. In Hawaii, the lavas are generally basalts. They are hot, low in silica (SiO_2) content (see Fig. 4.2), and tend to have a low viscosity. As a result, they flow readily and spread out from the volcano, building a gently sloping shield volcano. If the basalt contains gas bubbles, it may be thrown up in the air as frothy droplets of lava (Fig. 4.9a) that cool as they fall back around the vent to form cinders. Such a process forms small volcanoes known as **cinder cones** (Fig. 4.10a). Cinder cones may simply be constructions of loose cinders or may have one or two lava flows that issue from the same vent by breaking through the weak cone walls.

The magma at Mount St. Helens, which is dacitic in composition, erupts at a lower temperature, contains more silica (see Fig. 4.2), and is much more viscous than Hawaiian lava. The viscous magma squeezes out like thick tar, forming steep-sided domes such as the one now forming on the crater floor of Mount St. Helens. The magma that feeds Mount St. Helens contains large amounts of dissolved gas that is mainly water vapor. The solubility of this gas in the magma decreases as the pressure on the magma decreases—that is, as the magma moves up to the surface. For fluid lavas such as the basalts of Hawaii, the pressure of gases may shoot the lava into the air, forming spectacular fire fountains over 600 m high. In magma that is more viscous, however, such as andesites and dacites, the bubbles that form in the rising magma cannot escape or grow easily. The pressure that builds in the gases therefore is much greater. The magma explodes violently upon eruption, breaking up into frothy blocks (**pumice**) or fine-grained particles of ash. (Fig. 4.9c). This type of eruption produces pyroclastic deposits rather than lava flows.

Among the most viscous types of magma are rhyolites, which contain both the highest silica content of all common volcanic rocks (see Fig. 4.2) and the most dissolved water vapor. If erupted as a lava, rhyolites are so viscous that they flow at rates of only a few centimeters per day and will not flow far from the vent, forming steep-sided lava domes (Fig. 4.10b). More commonly, though, the gases disolved in rhyolites, dacites, and andesites escape violently on eruption, and the volcanoes vigorously explode a gas-laden froth that roars straight up into the air. Much of the solid debris falls straight back down immediately outside the tower of expanding gas and ash, cascading down the slope as **pyroclastic flows**.

A simple experiment demonstrates the influence of dissolved gases upon the production of an explosive eruption. Consider an unopened bottle of soda. The soda

(a)

(b)

(c)

(d)

FIGURE 4.9 Different types of eruptions produce different volcano forms. (a) Eruption of fluid basaltic lava builds up shield volcanoes, such as Kilauea, Hawaii. (b) Shield volcano, formed by layer upon layer of fluid basaltic lava. (Fernandina, Galapagos Islands) (c) More violent eruptions, such as Kluychevskoy Volcano, Kamchatka, Russia, are characteristic of magmas that are richer in silica and water. (d) The resulting volcanoes are typically steep-sided composite volcanoes. (Osorno, Chile)

(a)

(b)

FIGURE 4.10 Minor volcanic features: (a) A cinder cone; a small volcano formed by a pyroclastic eruption north of Flagstaff, Arizona. (b) A steep-sided dacite/rhyolite dome. (Wilson Butte, California)

is under pressure; the gases (carbonation) are dissolved in it so that the soda does not appear to have many bubbles. Now if the bottle were shaken then opened quickly, there would be a sudden eruption of froth. These bubbles are derived from gas that can no longer remain dissolved in the soda because of the decrease in pressure caused by opening the bottle cap. This is analogous to a pyroclastic eruption; gases escape explosively when the magma rises rapidly from depth and is consequently subjected to lower pressures. At Mount St. Helens, increasing pressure in the magma chamber caused the side of the volcano to break away in a giant landslide, allowing pyroclastic material to erupt violently (see Focus 9.3).

Pyroclastic eruptions are typically very violent and dangerous and may produce large volumes of ash and other volcanic fragments. Large eruptions, such as the 1991 eruption of Mt. Pinatubo in the Philippines, can actually affect Earth's climate by hurling ash and gases high into the upper atmosphere where they may remain for several years, blocking out a fraction of the incoming energy from the Sun. The significance of this will be discussed in Chapter 15.

The shapes of volcanoes are determined in part by the viscosity of the lava they produce. The viscosity, in turn, is controlled by chemical composition, temperature, and water content.

A strong connection exists between volcanic type and plate-tectonic setting (Fig. 4.11). We have just shown that the type of volcano is determined by the composition of magma produced. In turn, the type of magma is controlled by the plate-tectonic setting. Mount St. Helens is typical of volcanoes located at zones of plate convergence (subduction zones). The Hawaiian volcanoes are typical of intraplate volcanism, which is usually remote from plate boundaries. Much of the magma that comes to Earth's surface does so at fissures that mark divergent plate boundaries and erupts underwater at ocean ridges. When lava erupts underwater, it characteristically congeals in a bulbous pocket resembling a pillow. Magma progressively breaks through the crusts of these structures to form more pillows, and the result is referred to as *pillow lava*.

Melting and Magmatism

We have seen that the portion of the mantle beneath the lithosphere is convecting, or circulating, like soup heating on a stove (∞ see Chapter 2). In reality, even though the mantle is convecting, it is still solid. In terms of our everyday experience, it may seem like a paradox that a solid can "flow," but given the long time scales involved (millions of years) and the rather slow rates of mantle convection, it is possible. The effect is similar to ice in a glacier, which may flow slowly down a mountain valley over a period of many weeks, but will break or shatter if struck sharply with a hammer.

Over long periods of time, a solid such as glass can actually deform and flow. The rock that constitutes the mantle flows slowly, deforming plastically, but it is not molten (∞ see Chapter 5 for a discussion of plastic deformation). If we could bring a piece of the mantle to the surface, maintaining its properties as it exists at depth, we would find it hot to be sure, but it is also solid and brittle on our human time scale. It only behaves plastically over a very long period of time.

We know that the mantle does melt; however, it does not melt randomly across Earth's surface but only at very special sites. The volcanoes of the world are not randomly distributed; they are arranged systematically, and

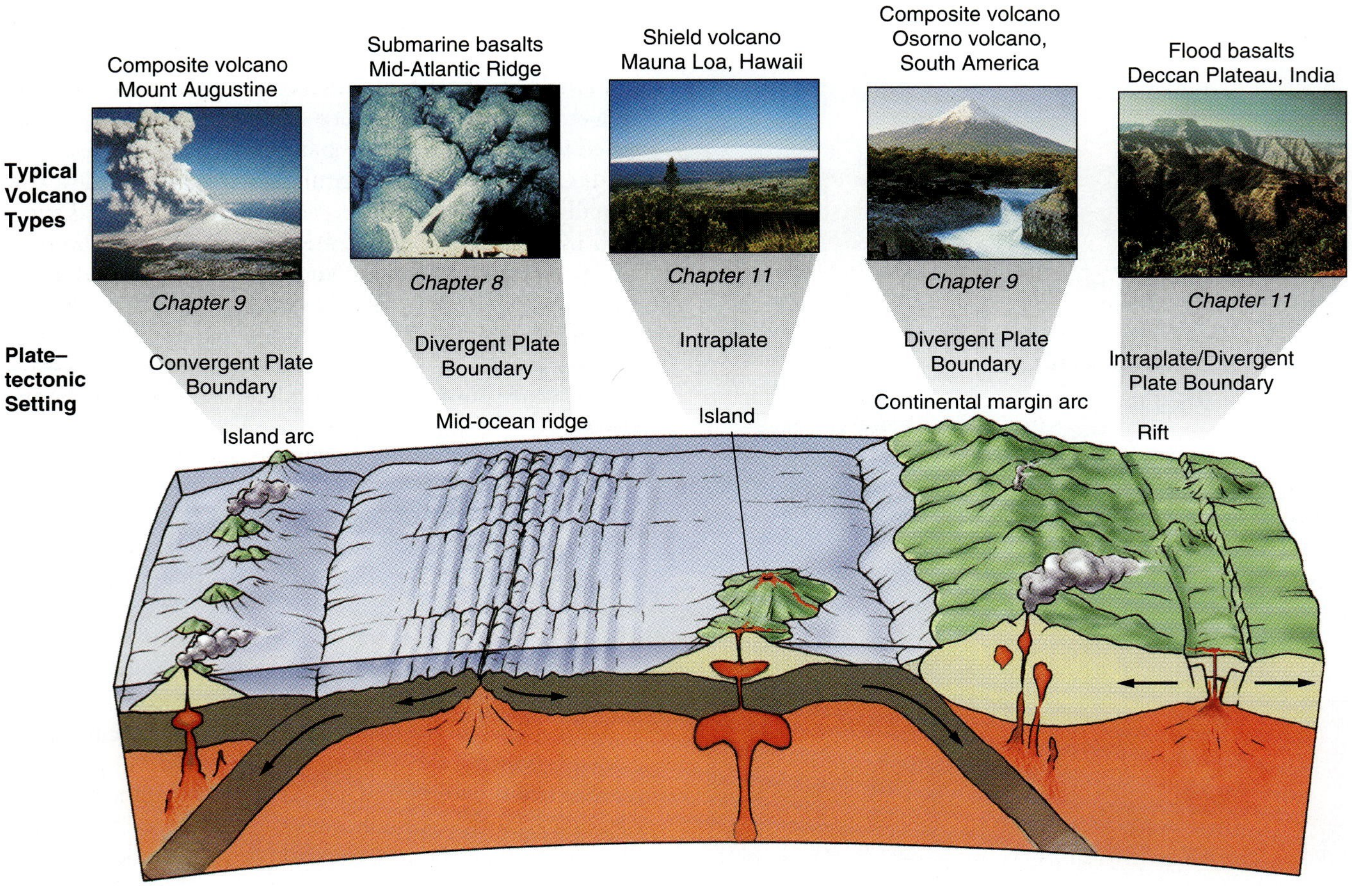

Typical Volcano Types
Composite volcano
Mount Augustine
Submarine basalts
Mid-Atlantic Ridge
Shield volcano
Mauna Loa, Hawaii
Composite volcano
Osorno volcano,
South America
Flood basalts
Deccan Plateau, India
Chapter 9
Chapter 8
Chapter 11
Chapter 9
Chapter 11
Plate–tectonic Setting
Convergent Plate Boundary
Divergent Plate Boundary
Intraplate
Divergent Plate Boundary
Intraplate/Divergent Plate Boundary
Island arc
Mid-ocean ridge
Island
Continental margin arc
Rift

FIGURE 4.11 Different types of volcanoes and how they correspond to different plate-tectonic settings.

Continent
y
Crust
Ocean
x
Lithosphere
Lithosphere
Mantle
Asthenosphere
Temperature (°C)
500
1000
1500
0
50
100
150
200
Depth
(km)
Partially molten
Liquid
Oceanic geotherm (x)
Continental geotherm (y)
Solid mantle
Solidus
Liquidus

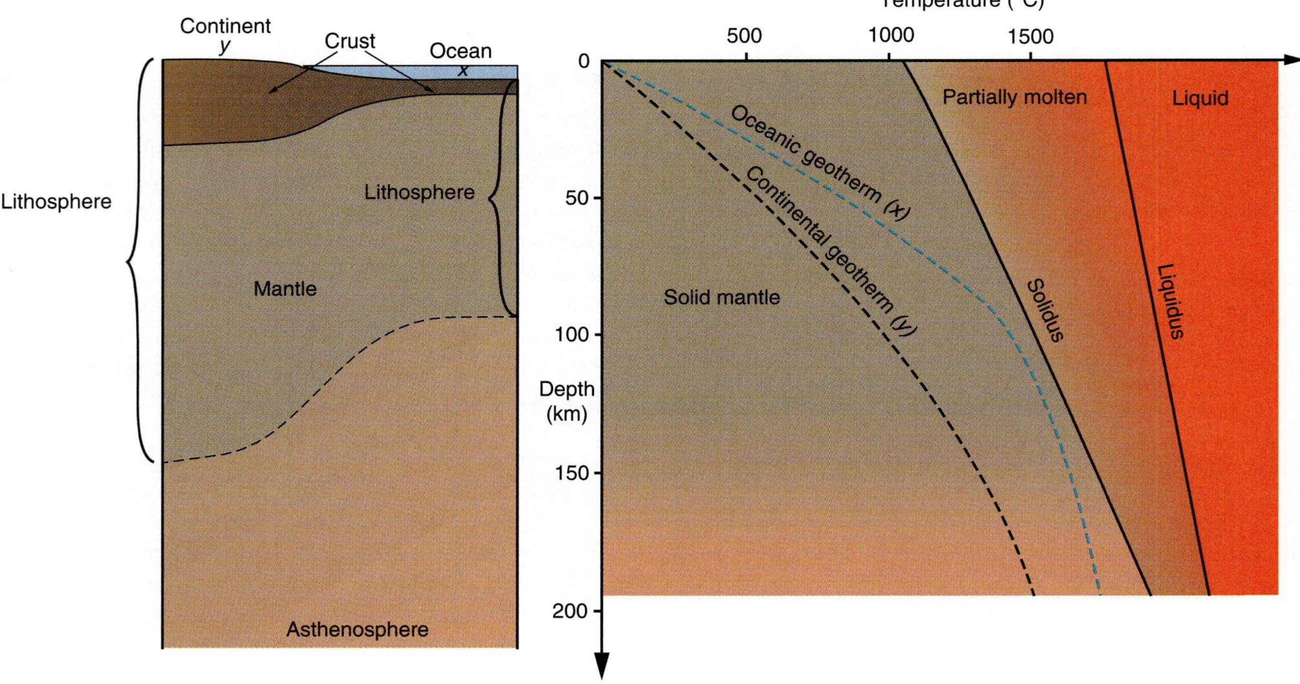

FIGURE 4.12 Pressure-temperature diagram for mantle material (peridotite) showing variation in melting temperature with depth (pressure), which is represented by the solidus. The dashed lines are typical variations of temperature with depth within Earth. On the left, a cross section of the upper part of Earth to a depth of 200 km. The oceanic and continental geotherms would be the temperatures measured at each depth below points *x* and *y*, respectively.

they show a strong correspondence with the locations of plate boundaries (see Fig. 1.6).

Melting and the Geothermal Gradient Both the solidus and liquidus temperatures of any rock differ with pressure, or depth, within Earth (Fig 4.12). For the material that constitutes the mantle (peridotite), the solidus temperature increases with increasing depth. In other words, deeper portions of the solid mantle will start to melt at higher temperatures than will portions of the mantle that are closer to the surface. Further, the solidus temperature of the mantle decreases as the percentage of water dissolved in the mantle increases (Fig. 4.13). The solid mantle, therefore, will start to melt at lower temperatures as the amount of water dissolved in it increases.

We know that Earth's temperature increases with increasing depth. One has only to go down into a deep mine or take temperature readings in a deep drill hole to confirm this. A plot of temperature vs. depth is called a *geotherm* (see Fig. 4.12). The rate of temperature change with depth is called the **geothermal gradient**. The geothermal gradient varies from location to location. If you were to drill a hole in oceanic crust, you would find that the temperature in that hole increases more for each kilometer drilled than if you drilled a hole in most parts of the continental crust (Fig. 4.12). In areas where Earth's crust is ancient (old and cold), the geothermal gradient is small, and the temperature increases slowly with depth. In areas of young crust, such as the ocean basins, the gradient is high.

Notice from Figure 4.12 that the geotherm does not actually cross the solidus; for any given depth, the geotherm is at a lower temperature than the solidus, and the mantle is therefore wholly solid. The closest approach between the geotherm and the solidus occurs at a depth of approximately 100 to 200 km and corresponds to the base of the lithosphere. It is at this depth that the mantle is closest to melting, and, therefore, this depth marks a major change in strength between the easily deformed, ductile material of the asthenosphere and the rigid lithospheric plates that move above it.

The geothermal gradient is the rate of temperature change with depth in Earth.

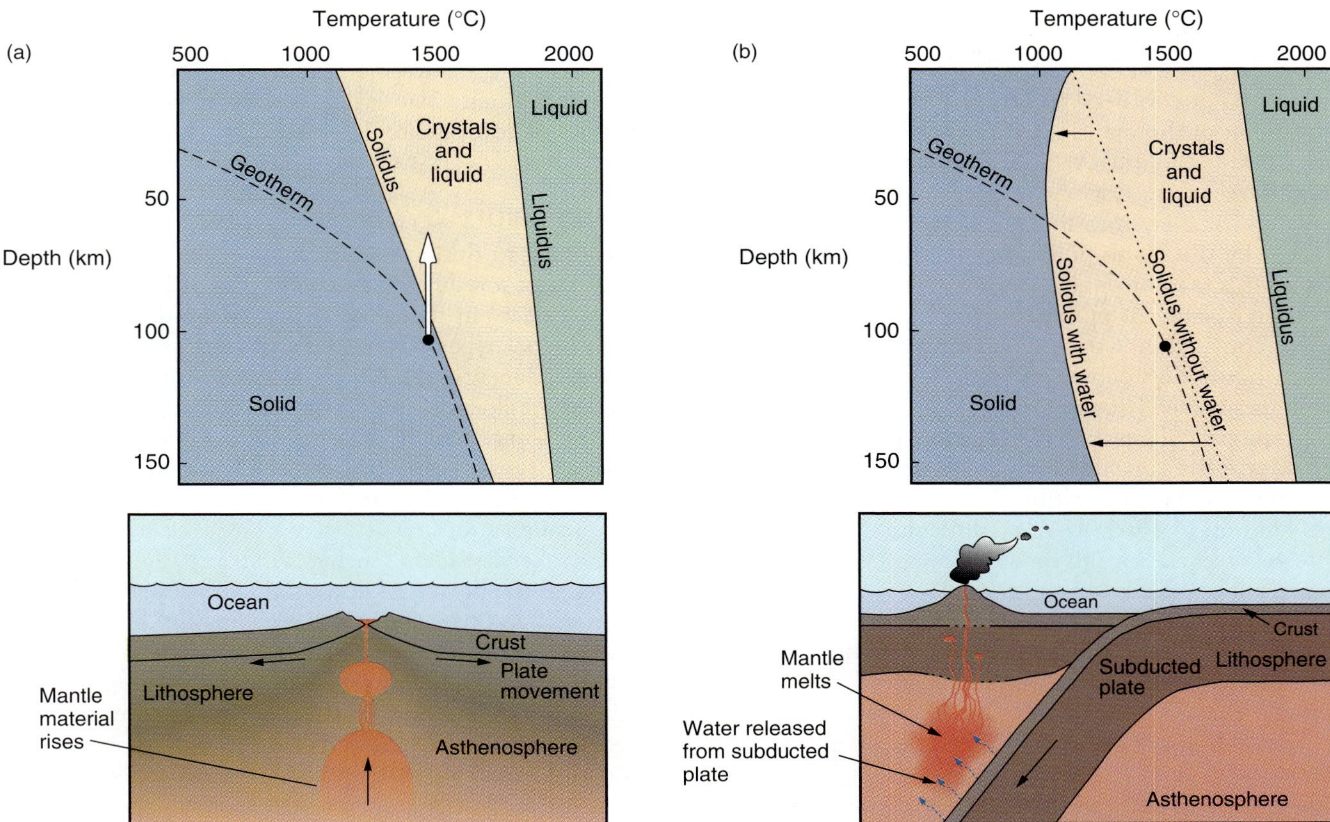

FIGURE 4.13 The two main ways in which melting occurs in the mantle and examples of such processes. (a) *Decompression*: Mantle material (represented by the black dot), originally on the geotherm, moves up toward the surface as indicated by the arrow. As this material rises, its path crosses the solidus so the material will start to melt. (b) *Hydration* (addition of water): Mantle material remains on the geotherm, but addition of water to the mantle lowers the solidus as shown by the two arrows. The material will be at a higher temperature than the new hydrous solidus and will start to melt.

Melting of the Mantle If the mantle is not molten under "normal" conditions, then it follows that some disturbance of these conditions is required to cause the mantle to melt, generating the magmas that later solidify into igneous rocks. The two principal ways in which melting can occur in the mantle are decompression and hydration (Fig. 4.13).

We will consider decompression first. As warm convecting mantle rises, it is under progressively less and less pressure; in other words, it decompresses. At a particular depth, its temperature-pressure path crosses the solidus for mantle material and this material will then start to melt.

The black dot in Figure 4.13a represents a piece of the mantle at a temperature that lies on the geotherm at that depth. You can see that the temperature of our sample is below its melting point. If this piece of the mantle rises, it will follow a path of decreasing pressure; it will, however, undergo little cooling if its ascent is sufficiently rapid. Such a path is represented by the upward-pointing arrow. At a depth of about 90 km, this arrow crosses the solidus into the field labeled "crystals and liquid." This means that, at 90 km, melting will begin and a small amount of liquid magma will be produced, although most of the mantle at that point will remain solid and crystalline. As the material continues to ascend, the amount of melting will increase, and the amount of liquid relative to crystals will become greater.

Melting due to decompression as a result of upward movement of the mantle occurs at divergent plate margins (mid-ocean ridges) and at ocean islands such as Hawaii. At ocean ridges, the mantle beneath the ridges wells up passively into the gap that is constantly being created by the plates pulling apart (Fig. 4.13). We will discuss this in more detail (∞ see Chapter 8). In contrast, at ocean islands, the upward moving mantle is part of the active convection system circulating in the mantle, in which hotter, less dense solid mantle is moving upward in plumes (∞ see Chapter 11).

In the case of melting by hydration, the mantle material stays at the same depth and temperature but the position of its solidus (its melting temperature) is changed by the addition of water. This is a difficult concept to visualize; it is like adding salt to ice to lower its melting point and turn the ice into water, as we do on icy roads in the winter. Similarly, antifreeze is added to the water in the radiator of a car to prevent it from freezing. Like salt, antifreeze prevents water from freezing until a lower temperature is reached.

The addition of water to the mantle moves the position of the solidus to lower temperatures—to the left on Figure 4.13b. The black dot that represents the mantle composition of our sample now lies at a temperature above the new solidus. The material is therefore above the melting point of the mantle. Without moving at all, it will begin to melt. Such a process may occur at convergent margins (subduction zones). There, water contained within the subducted lithosphere, which heats up as it descends, is released and added to the overlying mantle, lowering its melting point.

Geologists can simulate the pressures and temperatures that occur within Earth and investigate the processes of melting and crystallization, and also the mineral phases that are stable. Focus 4.3 offers a brief discussion.

Melting can be achieved by decompression or by lowering the melting point of the mantle by adding water. These conditions are encountered at divergent and convergent plate margins, respectively.

Sedimentary Rocks

Although sedimentary rocks form only 5 percent of Earth's crust by volume, they are the most common rock type encountered at Earth's surface, covering some two-thirds of the continents and probably most of the ocean floor. This is perhaps not surprising, as sediments form only at and near the surface. Most sedimentary rocks are formed from fragmented rock and mineral material such as gravel, sand, and mud. Some sedimentary rocks also form by the accumulation of fragments of organically generated material (such as shells), and by precipitation directly from solutions.

Sedimentary rocks typically form horizontal, flat layers until subjected to deformation. The folding and faulting of originally flat layers is easy to appreciate and enables us to evaluate deformation processes that have affected Earth's crust (∞ see Chapter 7).

Sedimentary Systems

It is useful to think of sedimentary rocks within a larger context, or a sedimentary system (Fig. 4.14). Sediments are secondary in the sense that they are derived from a range of rock types exposed at Earth's surface. The heritage of the sediment—the *source rock types* from which it is derived—influences the mineralogy of the final product.

Weathering, which will be discussed thoroughly in Chapter 12, is the process by which rocks are disaggregated and the constituent fragments or minerals undergo chemical reactions.

Transport mechanisms such as wind, ice, water, or gravity, as well as the process of deposition, influence the sorting, grain size, and type of layering (bedding) of the sediment. The *site of deposition*—whether river, lake, ocean, or desert—gives rise to differences in color, bedding, mineralogy, grain size, and grain-size distribution. Finally, the *plate-tectonic setting* of sediment deposition plays a significant role in the final product (∞ see Chapter 13). Sedimentary rocks formed at convergent margins are different in many ways from sedimentary rocks formed at plate interiors.

Types of Sediments

The term *sediment* encompasses *detrital* or *clastic sediments*, which consist of loose grains produced by weathering at Earth's surface; *biochemical sediments*, composed

Focus On 4.3 SIMULATING EARTH'S INTERIOR

We know from measuring the temperatures of lava flows that magma erupts at Earth's surface at temperatures commonly in excess of 1000°C. But as depth within Earth increases, the pressure increases (due to the weight of rock above). This increase in pressure will affect the melting temperature of rock, but how? We know also that a number of different minerals commonly crystallize from magmas. We described these minerals earlier (see Chapter 3); most are silicate minerals such as olivine, feldspar, pyroxene, amphibole, mica, and quartz. But what minerals crystallize from a given magma, and how are the types of minerals that crystallize from a given magma affected by the temperature and pressure at which crystallization occurs?

To answer these questions geologists subject rocks to high pressures and temperatures such as those that exist within Earth. Such experiments, however, are difficult and require very specialized equipment with precise temperature and pressure controls. The first geologist to attempt such experiments was Sir James Hall, who, in the early nineteenth century, sealed rocks in a cannon barrel and placed the end of the barrel in a furnace. Although he showed that rocks do indeed change in composition or even melt at these extreme conditions, Hall was unable to calibrate his experiments well; he could not easily measure the temperatures and pressures at which the experiment was occurring.

More modern equipment allows us to simulate conditions deep within Earth. Figure 1 shows a press that can reproduce the temperatures and pressures that exist to a depth of 200 km. These conditions correspond to the upper mantle, where most magmas are originally generated. Even more recently, presses have been built with "diamond anvils" that can exert pressures corresponding to thousands of kilometers within Earth (recall that diamond is the hardest mineral known). Thus we can even duplicate the extreme pressures that exist in Earth's core.

FIGURE 1 A piston cylinder apparatus. The sample is placed under a piston and heated. The gauges indicate temperature and pressure.

of material of organic origin, such as shell and coral fragments; and *chemical sediments*, made up of inorganic material such as salt precipitated from solution.

Weathering includes both the mechanical breakdown of minerals and rocks at Earth's surface and the chemical reaction of minerals, formerly at equilibrium within the crust, to conditions at Earth's surface. Surface conditions differ markedly from those at depth. Minerals are exposed to an atmosphere rich in water and oxygen, and plants and animals can accelerate the mechanical and chemical breakdown of rocks and minerals. The sediments are transported by water, wind, or ice and deposited elsewhere. Sediments eventually settle from the transporting medium owing to the force of gravity. As the grains are deposited, layering in the sediments naturally occurs. This layering is caused by variations in grain size, mineralogy, or grain orientation. When originally deposited, these layers, known as **strata** or *beds*, tend to be horizontal.

FIGURE 4.14 A sedimentary system. Satellite photograph of the coast of northern Chile, South America. To the northeast (top right) are the Andean mountains, which are weathered and eroded to provide sediment. Sediment is transported downhill toward the ocean, carried mainly by rivers, and is ultimately deposited there. Note that sediment may also be deposited on its way to the ocean in lakes, deserts, and riverbeds, and along the coast.

A familiar example is the huge "layer cake" effect of sedimentary strata exposed in the Grand Canyon or elsewhere in the southwestern U.S. (Fig. 4.15a). Sedimentary rocks can be faulted and folded subsequent to deposition, and their original layering serves as an important reference to the nature of the deformation (Fig. 4.15b).

(a)

(b)

FIGURE 4.15 (a) A thick succession of horizontal sedimentary strata exposed in Canyonlands National Park, Utah. (b) Strata (presumed to have been originally horizontal) that have been deformed and folded (Lulworth Cove, England).

Clastic Sediments The term "clastic" means "made of **clasts**" (particles or grains) and is derived from the Greek word *klastos*, for "broken."

The simplest classification of clastic sedimentary rocks is based on the sizes of the grains they contain (Fig. 4.16). Sedimentary rocks containing extremely large grains are called *conglomerates* if the clasts are rounded, and *breccias* if the clasts are angular. The large grains may be *pebbles*, *cobbles*, or *boulders*. Although specific definitions exist for each of these size terms, you may find the following explanation of sediment sizes easier to appreciate: If you can throw it easily, it is a pebble; if it is too big to throw far, but you can pick it up and carry it, it's a cobble; and if it's too big to pick up, it is a boulder.

Similarly, there is a size classification for smaller grains. *Sand* is a size term for grains that range from those that can be picked up easily between your index finger and thumb down to those grains for which forceps are required but can still be easily seen without the aid of a magnifying lens. *Silt* is the next smaller size term, ranging from the finest sand down to something that looks like a fine powder, but when rubbed between your fingers or

Rock type	Grain size	Grain type
Conglomerate	gravel, pebbles, cobbles (> 2mm)	Rock fragments  Conglomerate (rounded)  / Breccia (angular)
Sandstone	Sand (1/16mm to 2mm)	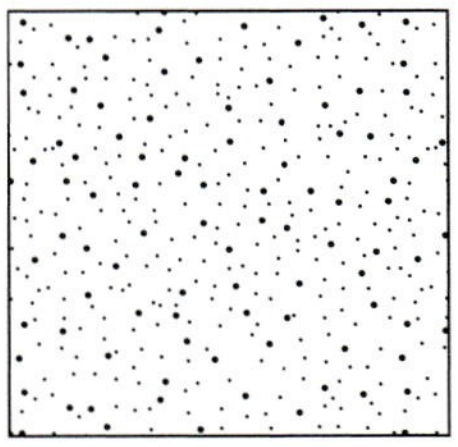 Grains are usually single mineral fragments. *Quartz* is generally most abundant. A sandstone made mostly from quartz is a *quartzite.*
Mudstone	Mud or silt (< 1/16 mm)	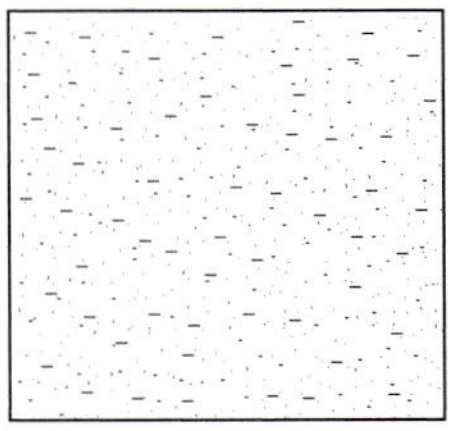 Very fine mineral grains; mostly *clay.* If clay minerals are aligned so rock splits into sheets, it is known as *shale.*

FIGURE 4.16 A simple scheme for naming clastic sedimentary rocks according to the types of clasts from which they are made.

chewed between your teeth (don't make a habit of this!), feels gritty. If you chew a *mud* or *clay* or rub wet mud between your fingers, it feels smooth rather than gritty.

The designation "-stone" (as in sandstone or mudstone) indicates that the sediments have been consolidated into rocks. The size grade of the sediment—sand, silt, mud, or clay—becomes the modifier, indicating the nature of the sedimentary rock. In clastic rocks, each grain is a single fragment derived from the disintegration of preexisting rocks. The fragment may be a piece of the rock or it may be a single mineral grain.

Conglomerates are sedimentary rocks made up of pebbles, cobbles, or boulders embedded in a matrix of mud, sand, or carbonate; the clasts are from preexisting rocks—igneous, metamorphic, or other sedimentary rocks. If the fragments are angular rather than rounded, the rock is more correctly referred to as a *breccia*.

Sandstones are sedimentary rocks made up of sand-sized grains, many of which are individual minerals; however, some grains may be sand-sized rock fragments.

Siltstones are fine-grained equivalents to sandstones, made of silt. *Mudstones* may contain tiny grains of clay minerals or micas or even calcium carbonate mud. During compaction, the clays and micas in mudstones might develop an orientation that is perpendicular to the compaction force (gravity) and more or less parallel to the bedding. This preferred orientation can cause the rock to split apart along closely spaced planar surfaces. In this case, we refer to the rock as *shale*.

The characteristics of a clastic sediment or sedimentary rock are a reflection of what has happened to the sediment from the time the detritus was first produced by weathering, to the time that it is finally deposited. Sediment processing, as this is called, will affect the size of the clasts, the shape of the clasts (rounding), the distribution of grain sizes (sorting), and even the compositions of the minerals found in the sediment.

Grain Size The grain size of a clastic sediment largely reflects the amount of processing to which that sediment has been subjected. Consider a mountain range being weathered and eroded by water, wind, and ice (Fig. 4.17). The riverbeds in the mountain canyons are littered with many large boulders, but there is little sand, silt, or mud deposited by the vigorously flowing streams. The finer material is carried out of the canyons and transported into the neighboring lowlands, where it is deposited farther along the stream course.

As the eroded rock material is carried farther and farther from its source (the mountain range), these fragments bang together repeatedly and are ground down into smaller and smaller sizes. All sharp edges and points on the boulders or cobbles are quickly removed by bumping into one another during transport, and the rocks become more and more rounded. A long way from the mountains, the sediment grains in the riverbed are mainly sand-sized or smaller and rounded rather than angular. The rock fragments that were eroded in the mountains mostly have been broken down into individual mineral grains.

Sorting Other changes occur in the sediment during its transit downstream; the grain size tends to become finer and the *range* in grain sizes decreases. Rather than a great array in the sizes of sediment particles deposited along the stream course, sediment grain sizes become more uniform. Pebbles may remain upstream because the stream did not have sufficient capacity to move them farther downstream, while the mud is carried downstream in the water. Thus, sediment deposited along the river becomes more uniform in size; that is, better **sorted**, farther and farther downstream.

Mineral Composition One final change that accompanies increased processing of sediment is an alteration in the types of minerals comprising the sediment. Of course, the rocks or minerals present in a sediment deposit must ultimately reflect the rocks and minerals being weathered and eroded in the source area. For instance, it is not possible to produce a quartz-rich sand by eroding limestone, which is made of pure calcium carbonate. Because of chemical weathering and some sorting, the sediment will not have exactly the same mineral composition as the source rock.

Certain minerals are more resistant to weathering and abrasion during transport and tend to survive in a sedimentary system longer. Let us compare three minerals: quartz, plagioclase feldspar, and olivine. We can use the mineral properties that we learned about (∞ see Chapter 3) to understand how these minerals behave in a sedimentary system. Quartz is one of the hardest of common rock-forming minerals (H = 7 on the Mohs scale; see Table 3.2) and it has no cleavage. It is composed of a three-dimensional network of silica tetrahedra, which are entirely covalently bonded; thus, quartz is nearly insoluble in water. Plagioclase feldspar is slightly softer (H = 6) and has two nearly perfect cleavages along which the mineral grain will preferentially break. Plagioclase is also made of a framework of silicate tetrahedra with calcium and sodium cations and is nearly insoluble in water. Olivine has an intermediate hardness (H = 6.5) and has no cleavage. The chemical bonding in the olivine crystal is ionic between the silica tetrahedra and magnesium (Mg); thus olivine is more soluble in water.

During weathering, olivine dissolves almost completely, whereas quartz and, to a lesser extent, plagioclase remain largely unaffected. During erosion and transport, plagioclase breaks into progressively smaller grains because of its tendency to break along the two cleavages. Quartz survives nearly unaltered because of its hardness and chemical stability.

Other minerals may exhibit behavior that falls between the extremes used in this illustration. Pyroxene and amphibole rapidly lose iron and magnesium by dissolution during

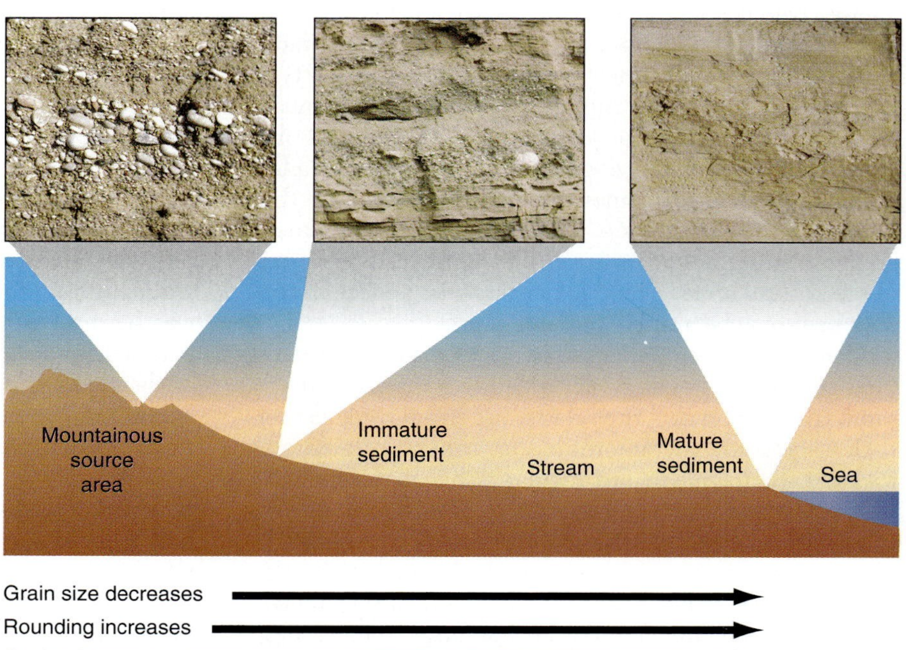

FIGURE 4.17 Schematic profile of a river, showing changes in sediments with progressive erosion and transportation. Note the decrease in overall grain size and the increase in rounding of individual grains.

weathering. Both also have two cleavages that allow them to break easily into smaller particles, enhancing chemical reactivity. Elements such as iron and magnesium are carried away in solution, and the remaining components of the original mineral may transform into more stable alteration products, such as clay minerals (∞ see Chapter 12). Thus, these minerals have a relatively low survival rate in sedimentary systems. With increased processing (increased distance from the source), the sediment will therefore become progressively depleted in minerals with a low survivability, such as olivine, pyroxene, and feldspar. It will become dominated by resistant minerals like quartz.

Maturity Taking all of these characteristics into account, we can describe a sediment (or a sedimentary rock) in terms of its **maturity**. A mature sediment is well sorted, relatively fine-grained (sandstone rather than conglomerate), and composed of rounded grains, most of which are quartz. Maturity signifies that a sediment has been well processed (Fig. 4.17). Unlike an immature sediment, which has not undergone much processing, a mature sediment might have been carried for long distances in a river or might have been recycled—that is, processed more than once, as would occur when sedimentary rocks are eroded a second time.

As the maturity of a sediment increases, average grain size decreases, the grains become more rounded and better sorted, and only the most resistant mineral types, such as quartz, survive.

Chemical and Biochemical Sediments Although most sedimentary rocks are formed from fragments of other rocks, some are formed by direct precipitation of minerals from seawater *(chemical sediments)* or by the actions of organisms *(biochemical sediments)*.

Chemical Sediments Water, particularly seawater, contains dissolved materials, called salts. Note that the term "salt" in this context refers to soluble ionic compounds in general, rather than just sodium chloride. Minerals crystallize directly from water when the concentration of dissolved ions exceeds their **solubility**; that is, when the ions exceed the maximum concentration that can remain dissolved in the solution. To demonstrate the process of crystallization from solution, leave a bowl of seawater out in the sun. As water evaporates from the bowl, the salt concentration in the remaining water increases. There will be less and less water, but the total amount of salt remains the same and, therefore, the concentration of salt in the water increases. As the water begins to dry up, white salt crystals form. The same white precipitate will form on your skin as you dry out after a swim in the ocean. Salt deposits, known as **evaporites**, are chemical sediments that form naturally by this process when desert lakes, shallow bays, or seas dry up (Fig. 4.18).

The occurrence of evaporite sediments tells us that the climate at the time of their deposition must have been dry and probably hot to cause evaporation. The discovery of sequences of evaporites nearly 2000-m thick from drill cores in the floor of the Mediterranean Sea was interpreted as evidence that the entire Mediterranean dried up some 6 million years ago. Both the famous Bonneville Salt Flats and the Great Salt Lake in Utah are the evaporative residue of a huge, ancient lake that had formed during the ice ages (Pleistocene). At its peak capacity, the lake covered approximately half of western Utah and the water was sufficiently fresh that it supported such fish as trout. Upon evaporation, the dissolved salt in this enormous lake precipitated to form the salt on the Bonneville flats and in Great Salt Lake.

FIGURE 4.18 Evaporites (salts of sodium, calcium, magnesium, and potassium) deposited after multiple evaporations of a desert lake (Atacama Desert, Chile).

Biochemical Sediments Some minerals crystallize as a result of organic processes. Many marine organisms—corals, shellfish, and tropical algae—secrete a protective external skeleton, or exoskeleton, of mineral material. Accumulation of these skeletons on the sea floor after the organisms die produces biochemical sediment. The material may then be broken up, transported, and deposited in a fashion similar to that affecting the pebbles or sand grains that form clastic rocks. For this reason, most biochemical rocks can also be referred to as **bioclastic**.

We can easily see the fossil corals and shells responsible for producing **limestones**, such as those shown in Figure 4.19. Many biochemical sedimentary rocks form in the deep ocean from the remains of microorganisms—organisms of microscopic size that secrete external shells. The most prolific sediment-producing microorganisms are *foraminifera* and *coccoliths*, which secrete calcium carbonate (calcite), and *diatoms* and *radiolaria*, which secrete silica (Fig. 4.20a). Chalk, a particular type of limestone that forms a brilliant white rock formation (Fig. 4.20b), is

FIGURE 4.19 Limestone formed by the accumulation of the calcium carbonate shells and skeletons of marine organisms. It has a clastic texture because it is formed from fragments; it is termed bioclastic because the fragments are derived from organic material.

made almost entirely of microfossils. The famous White Cliffs of Dover, England, are made of chalk.

Coal is formed from the accumulation and compaction of dead plant material. It consists of nearly pure carbon from carbon compounds that made up the plants. Note that coal does not satisfy the descriptions of types of sediment or sedimentary rock that we have encountered so far. The plant material that accumulates to form coal is truly organic rather than the inorganic mineral precipitate of an organism. It is arguable, therefore, whether coal should be classified as a biochemical rock—or even as a rock at all. Some argue that it is technically a metamorphic rock because it requires the action of heat and pressure to form. We will return to coal later (∞ see Chapter 16).

Chemical sediments are precipitated directly from solution. Biochemical sediments involve organic activity such as the generation and subsequent accumulation of shell and coral fragments.

Sedimentary Structures

Sediments and sedimentary rocks commonly contain clues concerning the conditions of their formation. These clues may include the distribution of grains or the disposition of fine-scale layering and are referred to as *sedimentary structures*.

One of the most common types of sedimentary structure is **cross-bedding** (Fig. 4.21a). Cross-bedding is formed as grains are deposited by a moving current, such as wind or water. The sandy bottom of a riverbed often contains ripples, or regularly spaced ridges of sand, that run roughly perpendicular to the direction of the current. In cross section, ripples are formed from very finely bedded layers that are inclined downstream. This is a reflection of the way the ripples actually form; sand grains are rolled up one side of the ripple by the current, and when they reach the crest, they simply fall down the steep side (Fig. 4.21a). Cross-bedding occurs on larger scales too, reflecting the migration of channels and sandbars.

(a)

(b)

FIGURE 4.20 (a) Shells or skeletons of microorganisms—foraminifera and radiolaria. This photograph was taken with an electron microscope; its field of view is 1 mm. (b) Cliffs of chalk, a type of biochemical limestone formed largely by the accumulation of microscopic fossils (Bats Head, Dorset, England).

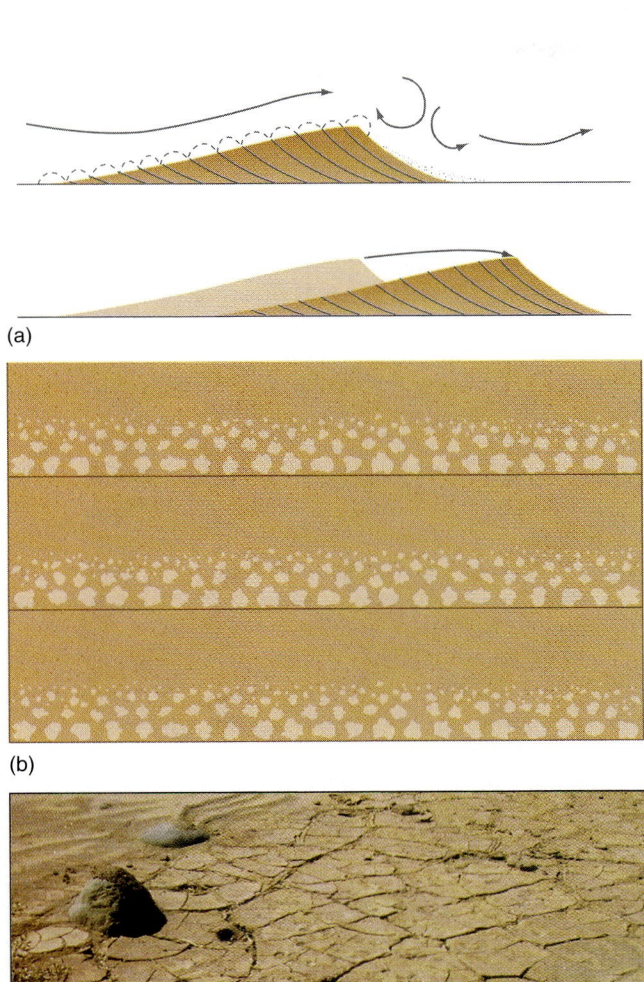

FIGURE 4.21 Different types of sedimentary structures. (a) *Cross-bedding*: A rippled surface is formed as sediment is carried by a moving current. Sand grains are rolled up one side of the ripple and fall down the other, causing the ripple to migrate in the direction of the current and form fine-scale, inclined layering. A cross-bedded sandstone is shown on the right, which was formed by flow from several different directions. (b) *Graded bedding*: Each graded layer is made of grains that become gradually finer upward through the layer. A graded sandstone is shown on the right. (c) *Mud cracks*: Dried out mud that has shrunk and cracked may later be preserved beneath the layer of sand.

Cross-bedding is not just restricted to riverbeds; sand dunes in a desert are like giant ripples and they also have internal cross-bedding. The presence of cross-bedding in a sediment or sedimentary rock tells us that it was deposited from a current. Furthermore, the orientation of the fine layers tells us the direction in which the current was moving. This is of great value in trying to interpret the conditions of formation of ancient rocks, such as current direction and strength.

Normally, rivers and winds flow with rather constant currents. In contrast, some transport conditions can be episodic. For example, during storms, loose sediment in shallow water can be disturbed and can slide down steep slopes as submarine flows. In such cases, the sediment will gradually settle out after the flow stops—the heaviest and largest grains first, followed by progressively smaller ones. This gives rise to **graded bedding**, where, in a single layer, the clast size becomes progressively smaller upward through the layer (Fig. 4.21b).

Mudstones may also contain structures that tell us about conditions of formation. When wet mud dries out, it shrinks and cracks (Fig. 4.21c). You may have seen mud cracks in dried reservoirs and lakes during a drought or even in puddles after a storm. These crack structures can be preserved by later burial. Again, they are important when we find them in ancient rocks because they tell us that the mud was originally deposited in shallow water and the climate was arid to allow the water to dry out.

Organisms can also leave distinct features in sediments, such as burrows or tracks and trails. Again, if these features (referred to as *trace fossils*) are preserved in sedimentary rocks, they may enable us to infer what conditions were like at the time when the sediment was deposited.

In addition to their use as indicators of depositional environment, sedimentary structures can help us determine the orientation and order in which a sequence of rocks was originally deposited. This information becomes critically important when layers of rocks are strongly deformed, because deformation can be sufficiently intense that sequences of sedimentary rock are completely overturned, and the original disposition of the rocks is ambiguous.

Lithification and Diagenesis

Sediments are unconsolidated material, yet sedimentary rocks are hard and rigid. What happens to transform the loose sediment into sedimentary rock? Two processes are important for this transformation: lithification and diagenesis.

Lithification (literally, the formation of rock; from the Greek word *lithos*, "rock") involves the compaction of sediment by burial and the subsequent expulsion of water or air from spaces between the grains. Through compaction, the *porosity* (the amount of space, or pores, among grains in sediment) is greatly reduced and the volume of the sediment decreases accordingly (Fig. 4.22). Even though the grains are much closer together, however, they are still loose. For a rock to form, either compaction must continue until the grains are deformed or partly dissolved into an interlocking arrangement, or the grains need to be joined together by a process called **cementation**.

During compaction, the sediment grains are in contact with one another over a very small area. If the grains are spheres, the contacts are points. Because of the small area of contact between grains, the weight of the overlying column of sediment exerts great pressure on this contact area; an analogy might be a 100-kg man standing with his full weight on a 1-cm square peg. The pressure exerted on the bottom of that peg is 100 kg per square cm. If the peg is tapered to a 0.1 cm square, the pressure would be 10,000 kg per square cm—enough pressure to force the peg easily into a wooden floor.

Similarly, the weight of overlying sediment on a point-to-point contact between two grains is sufficient to cause one grain to distort or indent the other. By this process, multiplied many times, sediment grains can be forced into a tight, interlocking arrangement. At the same time, the high pressures at grain contacts where two mineral grains are squeezed together can cause the mineral to dissolve locally; in other words, the mineral is taken into solution at the grain contacts and may be deposited in nearby pore spaces, causing the grains to be cemented together (Fig. 4.22).

Cement can consist of any mineral deposited from fluids into the pores between grains. The most common

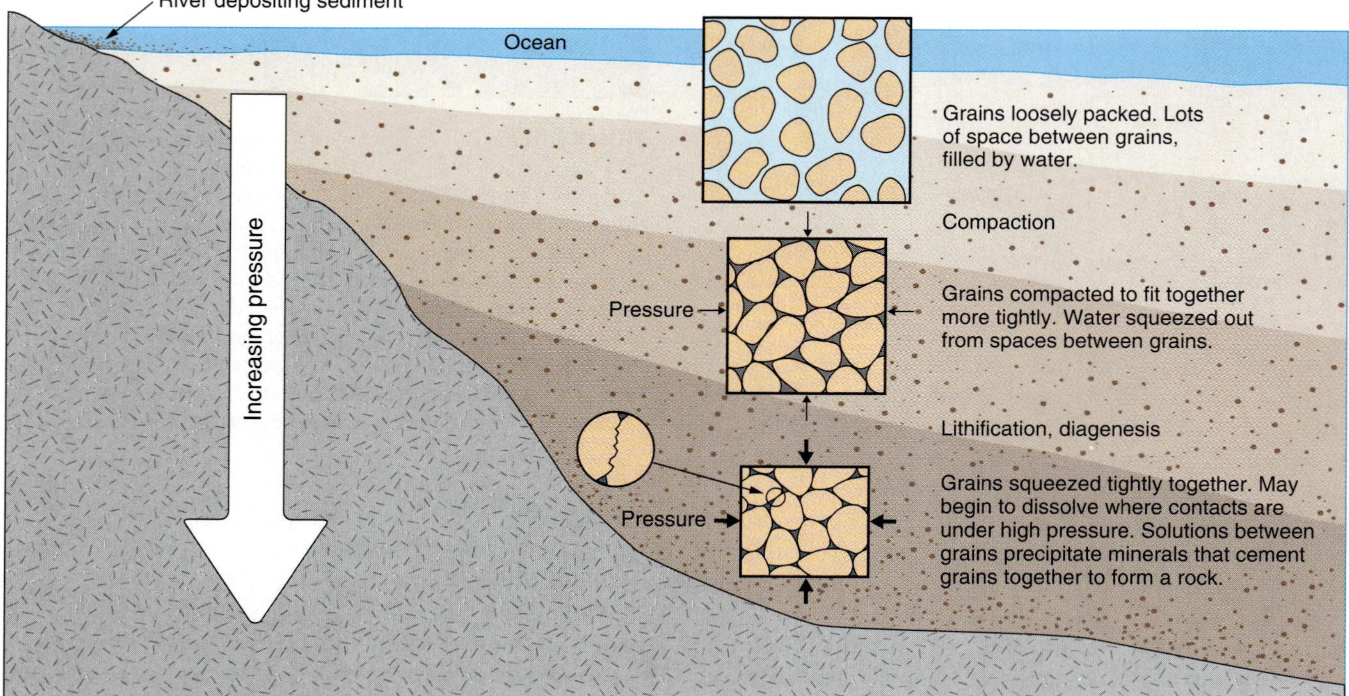

FIGURE 4.22 Schematic illustration of the process of compaction, lithification, and diagenesis that takes place with increasing depth in a thick pile of sediments in a sedimentary basin.

cements are silica and calcite, but other minerals are not uncommon, such as gypsum, anhydrite, or even pyrite. The fluids within the sediment pore spaces may have been present originally in the sediment, or may have been inherited from a different source, such as groundwater. As rocks are buried more deeply, the fluids react with constituent minerals forming solutions or brines. Such brines may be important in transporting metals that are later deposited as economically important ores. In some instances, the pore fluid is of organic origin and may ultimately form oil or gas (∞ see Chapter 16).

The term **diagenesis** (from the Greek *dia-*, for "change" and *genesis*, for "origin") is a collective term for the changes occurring in the minerals that make up a sediment in response to increasing pressures and temperatures and the effects of fluids due to burial. In fact, diagenesis might be considered as low-temperature metamorphism because it represents an alteration of minerals that were in equilibrium at Earth's surface in response to conditions of increasing temperature and pressure during burial. Formation of the mineral dolomite ($CaMg(CO_3)_2$) is among the more common diagenetic changes. It occurs by reaction of magnesium-bearing solutions with calcite ($CaCO_3$) in limestones. Flint and chert, both microcrystalline silica rocks, are also formed by diagenesis.

Thus, the processes of metamorphism and diagenesis merge into one another. Most geologists would agree to this assertion, but there is no strictly defined upper limit to diagenesis or lower limit to metamorphism. An arbitrary definition of 100° to 150°C as the boundary between the two processes would probably meet agreement, and pressure is generally considered to be of far less importance in diagenesis than is temperature. With increasing depth as material is buried, mineral solubility increases so that reactions may contribute dissolved material to the circulating fluids, which may, in turn, assist in cementing the sediment. Lithification and diagenesis can occur over a very wide range of depths depending on specific conditions, starting very close to the surface.

The processes of lithification and diagenesis involve compaction and cementation of loose sediment into hard rock.

Metamorphic Rocks

Metamorphism is the process by which rocks are subjected to heat, pressure, and reaction with chemical solutions and thereby transformed into metamorphic rocks. The transformation can involve changes in mineralogy, texture, fabric, and even chemical composition. A common example is the transformation of a sedimentary mudstone into the metamorphic rock *slate*.

Metamorphic rocks are the result of alteration of preexisting rocks (igneous, sedimentary, or metamorphic) that had been formed previously under different conditions of pressure and temperature. The distinction between igneous and sedimentary rocks is clear; sedimentary rocks form from sediments at low temperatures at or near Earth's surface, whereas igneous rocks form at high temperatures commonly well below the surface. The distinction between metamorphic rocks and either of the other two rock types is less clear because there is a wide range of pressures and temperatures in Earth's crust. Sedimentary and igneous rocks form at opposite ends of the temperature range, whereas metamorphic rocks form between these two extremes (Fig. 4.23). The point at which a sedimentary rock becomes a metamorphic rock is rather arbitrary. Loose sediment at the surface (such as sand) may become a hard sedimentary rock (sandstone) after it has been buried to depths of only a few hundred meters. Is this metamorphism? Conventionally, no; we would say that the loose sediment has undergone the sedimentary processes of lithification and diagenesis, which occur at low pressures and low temperatures.

Sedimentary rocks can lithify simply because of progressive burial under more and more sediment. Some sedimentary basins contain thicknesses of sediment in excess of 10 km. But there is a limit to burial simply as a result of sediment deposition. If detritus is deposited in a basin, the basin will eventually fill up with sediment, allowing no further deposition. Pressures and temperatures, even at the bottom of the basin, may not be high enough for significant metamorphism. But we know that metamorphic rocks have formed from sedimentary rocks originally near Earth's surface that were subsequently subjected to very high temperatures and pressures; in extreme cases, the rock may have even melted to produce magma.

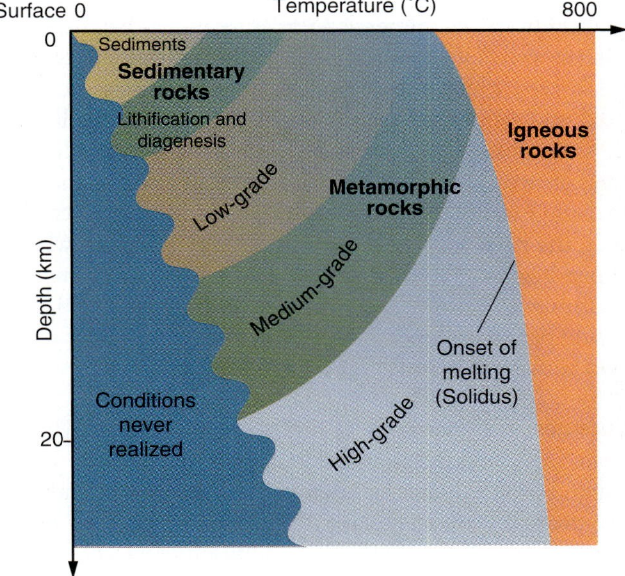

FIGURE 4.23 A pressure-temperature diagram, indicating conditions under which metamorphic rocks may form. As stated in the text, increasing depth corresponds with increasing pressure.

If such high pressures and temperatures cannot be generated simply by burial during sedimentary deposition, then how are they attained? The answer again lies in the process of plate tectonics. Plates moving against each other or sliding past each other generate huge forces, which, in turn, cause rocks to deform by bending or breaking. The mechanisms and results of such deformation are discussed later (∞ see Chapter 7). Rocks can be buried when one plate collides with another. Huge masses of rock can be pushed on top of one another and high pressures and temperatures generated as a result.

As we saw earlier, temperatures and pressures within Earth increase with increasing depth, so rocks at depth in the crust will be hot and under high pressure. We have also seen how magmas are generated at plate margins, such as subduction zones. Rocks can be subjected to high temperatures in such regions of volcanism or pluton intrusion because extra heat is supplied by the magmas themselves.

Metamorphic Grade

We refer to the intensity, or degree, of metamorphism as the **metamorphic grade.** As pressures and temperatures increase with burial over time, the metamorphic grade increases. For example, this would happen if a rock were buried deeper and deeper in Earth's crust over time.

To illustrate the concept of metamorphic grade, consider a layer of mud deposited in a lake or ocean (Fig. 4.24). As it becomes buried beneath subsequent layers of sediment, the mud is compacted and ultimately lithifies into mudstone. If the rock is buried more deeply and pressure increases, it is metamorphosed progressively to higher grades. It is metamorphosed first to a *slate*. During this process, increased pressure and temperature squeeze the rock into a hard, flaky stone, and the process of recrystallization from clay minerals to oriented micas begins, but is not yet well developed. Subsequently, the slate will become a *schist*, in which most minerals are completely recrystallized and reoriented into near-perfect parallelism. It then metamorphoses to a *gneiss*, in which many new minerals have grown. The gneiss bears little resemblance to the original mudstone. As metamorphic grade increases still further, the rock will begin to melt. The rock formed when a gneiss begins to melt is called a *migmatite*. If the melting continues, the entire rock will melt and a magma will form, giving rise to an igneous rock. There exists, then, a gradation between metamorphic rocks and igneous rocks as well.

Changes in Mineralogy The type of mineral that will form in a rock of a given bulk chemical composition depends not only on the chemical composition but also on the pressures and temperatures at which the rock exists. Because metamorphism is the result of increased temperature, or pressure, or both, upon a rock, it may involve changes in the *mineral assemblage* present; in other words, the different types of minerals and their abundances in a given rock change during metamorphism.

We can, in turn, use the mineral assemblage of a metamorphic rock to gather evidence about the conditions existing when the rock was metamorphosed. For instance, basaltic lava commonly contains the minerals olivine, pyroxene, and plagioclase feldspar. At high pressures (equivalent to approximately 100 km depth within Earth), this mineral assemblage is not stable and is replaced by a mixture of the minerals *garnet* and *clinopyroxene*. The metamorphic rock that results is called *eclogite*. The chemical compositions of eclogite and basalt are very similar; the overall chemical composition of olivine, pyroxene, and plagioclase feldspar mixed together as found in basalt, and the overall chemical composition of garnet and clinopyroxene mixed together as found in eclogite, are equivalent.

The chemical composition of the rock has not changed significantly as a result of metamorphism, even though its mineralogy has changed. The crystal structures of the mineral components have been altered into a more compact arrangement. Minerals of increasing density tend to form from a given composition as a rock is subjected to increasing pressure, as is the case when a rock is being buried. This is the response to pressure that would be predicted from theoretical considerations (thermodynamics); denser minerals occupy a smaller volume for a given amount of material. Thus the minerals that we find in a given metamorphic rock provide information about the pressures and temperatures to which the rock has been subjected (Focus 4.4). The principles at work here are the same as those described earlier (∞ see Chapter 3) when we discussed graphite and diamond, the polymorphs of carbon.

The different types of minerals and their abundances in a given rock will change during metamorphism. For a given chemical composition, the metamorphic minerals that form in a rock are a function of temperature-pressure conditions.

Minerals with water in their chemical structures lose that water as pressures and temperatures increase. They are converted to *anhydrous* (water-free) equivalents. For instance, amphiboles, which contain water or (OH^-) ions in their crystal structure, are converted to pyroxenes, which are anhydrous. Thus, low-grade metamorphic rocks contain many minerals such as micas and amphiboles that are hydrous, and high-grade rocks contain anhydrous minerals such as quartz, feldspar, and pyroxene. But what happens to the water that is lost from the minerals as metamorphic reactions progress? These fluids, which are an important part of the metamorphic process, generally move upward through the crust, reacting with the rocks through which they pass, forming new minerals. Such hot fluids dissolve material at depth and then redeposit it near the surface. This is how some of our economically important metal deposits are formed, as we will see in Chapter 16.

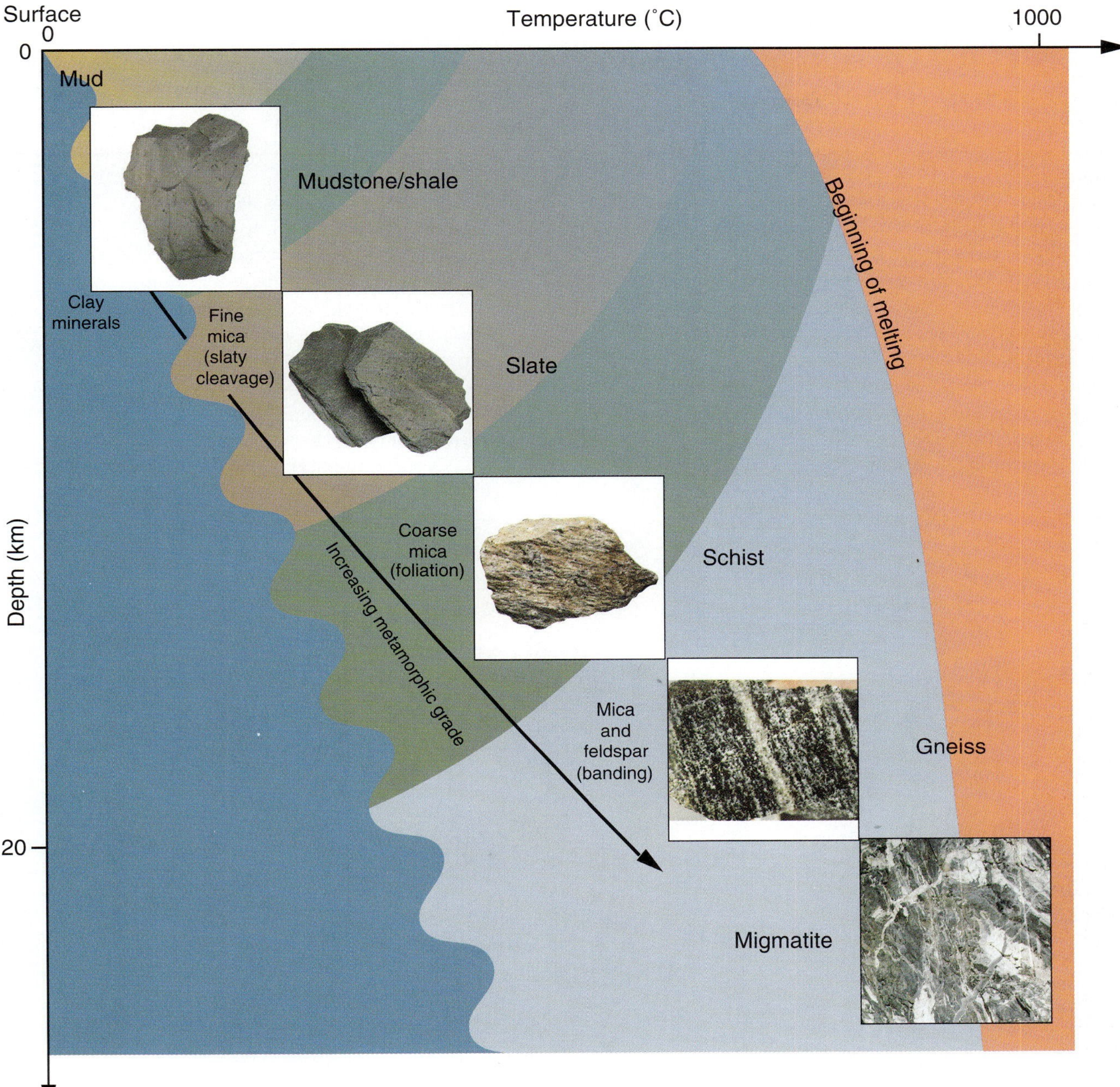

FIGURE 4.24 Effects of metamorphism on a mudstone. With increasing temperature and pressure, mudstone is metamorphosed into a slate, a schist, and then a gneiss. Finally, it is so hot that it begins to melt and forms a migmatite.

Changes in Texture and Fabric The terms **texture** and **fabric** refer to the organization of the minerals in a rock. As we learned in the section on igneous rocks, *texture* is used to describe grain shape, grain size, and the distribution of grain sizes in rocks. *Fabric* describes how mineral grains are arranged relative to one another.

Metamorphism can change the texture and fabric of a rock. Textures may change as new minerals grow and existing minerals disappear by reacting to form new minerals (if they are unstable under the pressures and conditions of metamorphism). Some minerals might grow particularly large, forming **porphyroblasts**. Just as we might describe a *porphyritic* igneous rock based on the type of phenocryst mineral (for instance, olivine basalt), we can describe *porphyroblastic* metamorphic rocks according to the porphyroblast mineral (for instance, garnet schist).

Some metamorphic rocks, such as hornfels and marble, have an interlocking fabric of mineral grains with no preferred orientation and have a granular texture. Such rocks generally form in response to high temperatures (hornfels) or are formed from rocks with a simple mineralogy (calcium carbonate and calcium-magnesium car-

DETERMINING METAMORPHIC PRESSURES AND TEMPERATURES

To understand how minerals can be used to infer metamorphic conditions, let us consider the compound aluminosilicate (Al_2SiO_5), which forms three important polymorphs—each stable over a particular range of temperatures and pressures. An aluminosilicate mineral will form in metamorphic rocks containing sufficient alumina, such as metamorphosed mudstones (schists and gneisses). As in the example of water discussed in Focus 3.4, the stability of the three mineral polymorphs of Al_2SiO_5 can be expressed in terms of a phase diagram (Fig. 1). In this diagram, the three lines are phase boundaries separating the fields of stability of the three polymorphs andalusite, kyanite, and sillimanite. These three polymorphs of Al_2SiO_5 can be distinguished by their different crystal structures. If we find a rock containing the mineral sillimanite, we know that the rock formed somewhere in the range of temperatures and pressures over which sillimanite is stable. We know from Figure 1 that the minimum temperature at which the rock could have formed is about 500°C. A rock containing kyanite, however, must have formed at relatively high pressures, and the occurrence of andalusite indicates lower pressures. In this way, we can use the presence of certain minerals as an indicator of the pressures and temperatures experienced by the rocks that contain them.

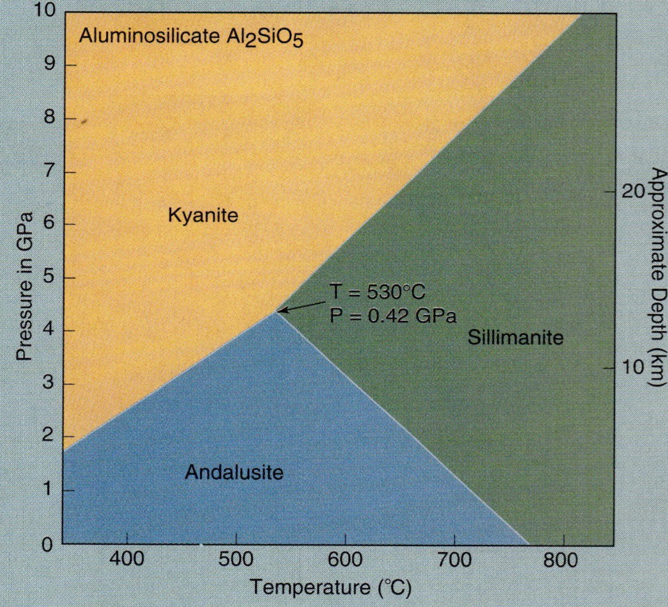

FIGURE 1 A pressure versus temperature diagram—a phase diagram—showing the stability ranges of polymorphs of aluminosilicate (Al_2SiO_5). The diagram is divided into three fields in which kyanite, andalusite, and sillimanite, respectively, are stable. The lines between these fields delineate the stability limits of these minerals. Pressure is given in gigapascals (GPa) where 1 Pa = 1 Newton/m^2.

bonates, in the case of marble). Many metamorphic rocks contain a wide range of minerals and form at high pressures where tectonic forces are active. These metamorphic rocks are commonly characterized by a distinctive fabric in which mineral grains exhibit a preferred orientation that sets them apart from most of their igneous or sedimentary counterparts. They develop a **foliation** (from the Latin *folium*, for "leaf"), which is a texture that resembles a stack of sheets or leaves perpendicular to the direction of maximum force (Fig. 4.25). This feature is characteristic of the common metamorphic rocks slate, schist, and gneiss (Fig. 4.24).

In some rocks, as in slates, the foliation is so closely spaced and the tendency to break into very flat, parallel sheets is so perfect that it is referred to as a rock **cleavage**—just like the cleavage in the *mineral* mica (see Table 3.1, Fig. 7). In fact, a close connection exists between the two since it is the parallel alignment of platy minerals such as mica that gives rise to cleavage or foliation in the rock. The very fine cleavage makes slate a very useful rock to us—it splits into very flat slabs and can be used for roofing tiles, chalkboards, and as a stable, level base for fine billiard tables. In schists, the platy minerals (micas) are a little more coarsely crystallized, and the foliation is called **schistosity**.

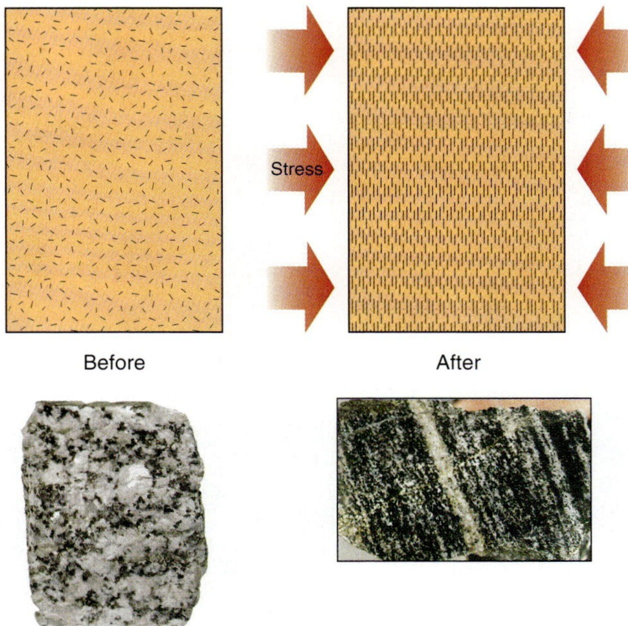

FIGURE 4.25 The generation of foliation by the orientation of minerals in a direction that is perpendicular to the force (stress) direction.

You can test the development of foliation in response to stress by means of a simple experiment. With a ball of modeling clay, knead in some flakes of biotite (or any other platy material). Break open the ball and you will see that the flakes have a random orientation. Now flatten the clay with a rolling pin or pound it, and you will notice that the flakes have become aligned parallel to the flat surface and perpendicular to the maximum force—that is, the force exerted on the clay by either the rolling pin or your hand.

An analogous process occurs in nature when rocks are buried under pressure and squeezed by tectonic forces. In rocks, though, in addition to physical rotation of some mineral grains, the pressure orientation may cause crystals to dissolve and grow in a preferred direction perpendicular to the maximum pressure. Slates and schists form in this way. With increasing metamorphic grade, both recrystallization and development of a more coarsely grained texture lead to a banding of minerals into light feldspar-rich bands and dark amphibole- or biotite-rich bands to form a rock called a gneiss (Fig. 4.25).

The action of tectonic forces during metamorphism may cause the mineral grains in a rock to become oriented, producing a foliation.

Types of Metamorphism

It is clear from the depth (pressure)-temperature diagram of Figure 4.23 that metamorphic rocks can form over a wide range of pressures and temperatures in Earth's crust. The type of metamorphic rock that forms is a function of both the original rock type (protolith; Fig. 4.26), the pressure and temperature to which it was subjected, and the nature of fluids that pass through it during metamorphism. We recognize three main types of metamorphism: *regional*, *high-pressure*, and *contact metamorphism*, which correspond to different processes within the crust (Fig. 4.27). The classification is based on the metamorphic *gradient*, or the amount of temperature change for a given increase in pressure. (Be careful not to confuse metamorphic gradient with metamorphic grade.)

The pressure-temperature gradient of a metamorphic rock is represented by the slope of an arrow in Figure 4.27. For example, the uppermost arrow in Figure 4.27 tends toward high temperatures without experiencing much increase in pressure. Such a gradient is characteristic of contact metamorphism, in which high temperatures result from contact with hot magma. A fourth type of metamorphism—*cataclastic metamorphism*—is associated with deformation along fractures in the crust and only affects a limited volume of rock.

Regional Metamorphism Most metamorphic rocks form by regional metamorphism, which is so named because it affects large tracts of crust. The pressures and temperatures characterizing regional metamorphism occupy the middle part of the pressure-temperature diagram in Figure 4.27. Regional metamorphism generally results from mountain-building processes, which, in turn, are caused by collisions between tectonic plates. Such collisions give rise to considerable deformation and compress or shorten the crust, thus thickening it. Many rocks that originally had been at or near Earth's surface become deeply buried as material is thrust over them in a collision, and they are metamorphosed as a result of the greatly increased temperature and pressure. Where such regional metamorphic rocks are exposed at the surface today as a result of subsequent uplift and erosion, they form extensive outcrops. These outcrops are elongated parallel to the line of plate collision and show systematic variations in metamorphic grade, which increase toward the center of the mountain belt.

As we have learned, the conditions of metamorphism generally affect the texture of the metamorphosed rock. Deformation and the directed pressure that causes preferential orientation of minerals in metamorphic rocks are reflections of regional plate-tectonic forces. Mudstones and shales contain clays that change to micas during metamorphism; these platy minerals become aligned and produce a metamorphic foliation. Of course, foliation can only be produced if such minerals are present or can be formed from the protolith. If the starting rock consists of pure-calcite limestone or a pure-quartz sandstone, for instance, then no platy minerals are present, and the metamorphic rocks that form from them would be marble and quartzite, respectively (see Fig. 4.26).

PROTOLITH	METAMORPHIC ROCK
Sandstone	**Orthoquartzite** Quartz is very unreactive and contains only the elements silicon and oxygen; therefore, metamorphism of a pure quartz sandstone will simply compress the grains into an interlocking fabric. If the sandstone contains other minerals, then some metamorphic minerals, such as garnet, may grow. 
Mudstone	**Slate, schist, gneiss** With increasing pressure and temperature, mica minerals may grow and define a distinct cleavage or foliation (see Fig. 4.25). At high pressures and temperatures, minerals such as feldspars and garnets may grow. 
Limestone	**Marble** Limestone is mainly calcium carbonate. It does not contain the necessary elements to grow micas and therefore will not develop a significant foliation during metamorphism. If the limestone contains some sand or mud, then calcium-rich garnets and pyroxenes may grow. 
Basalt	**Amphibolite, eclogite** Basalts contain a lot of calcium, aluminum, and magnesium. As a result, metamorphism will cause amphiboles and plagioclase feldspars to grow, forming amphibolite. These minerals commonly align perpendicular to the direction of maximum stress so that amphibolites are foliated. At very high pressure, an amphibole reacts to become pyroxene and magnesium-rich garnets grow to form a dense metamorphic rock called eclogite. 

FIGURE 4.26 Common types of metamorphic rocks and the starting rocks from which they are formed.

High-Pressure Metamorphism Certain metamorphic rocks form at high pressures but at lower temperatures than rocks typically produced by regional metamorphism. Some geologists consider these rocks to be a special kind of regional metamorphic rock as there is clearly complete gradation in the pressure-temperature conditions among the three metamorphic gradients illustrated in Figure 4.27. The tectonic settings of such high-pressure/low-temperature gradients where high-pressure metamorphism occurs are subduction zones (Fig. 4.27). High-pressure (low-temperature) metamorphic rocks form in the subducted plate.

At subduction zones, high pressures are caused by the subducting plate sinking into the mantle. Temperatures are kept relatively low because the subducted plate has been at the surface; the upper part of the plate in particular has been cooled by interaction with seawater. The subducting plate has a strong cooling effect as it sinks into the asthenosphere, where it forms a cool zone surrounded by hotter asthenospheric material. Thus, the subducted slab reaches high pressures more quickly than it heats up to high temperatures.

Sediments carried down into the subduction zone may be metamorphosed to **blueschists.** The bluish color of this rock results from a blue amphibole, which crystallizes under high-pressure conditions at depths of 60 to 100 km. The basalts of the subducted plate metamorphose to eclogite (see Fig. 4.26). As we saw earlier, eclogite has the same chemical composition as basalt, but its mineral components (garnet and clinopyroxene) are denser than the minerals found in basalt (plagioclase, olivine, and pyroxene). Eclogite is therefore more stable at higher pressures.

Contact Metamorphism Some metamorphic rocks form as a result of high temperatures without accompanying high pressures. Such conditions occur near hot igneous intrusions in the crust. Rocks that are in contact with these intrusions are said to undergo contact metamorphism, which is also known as *thermal metamorphism.*

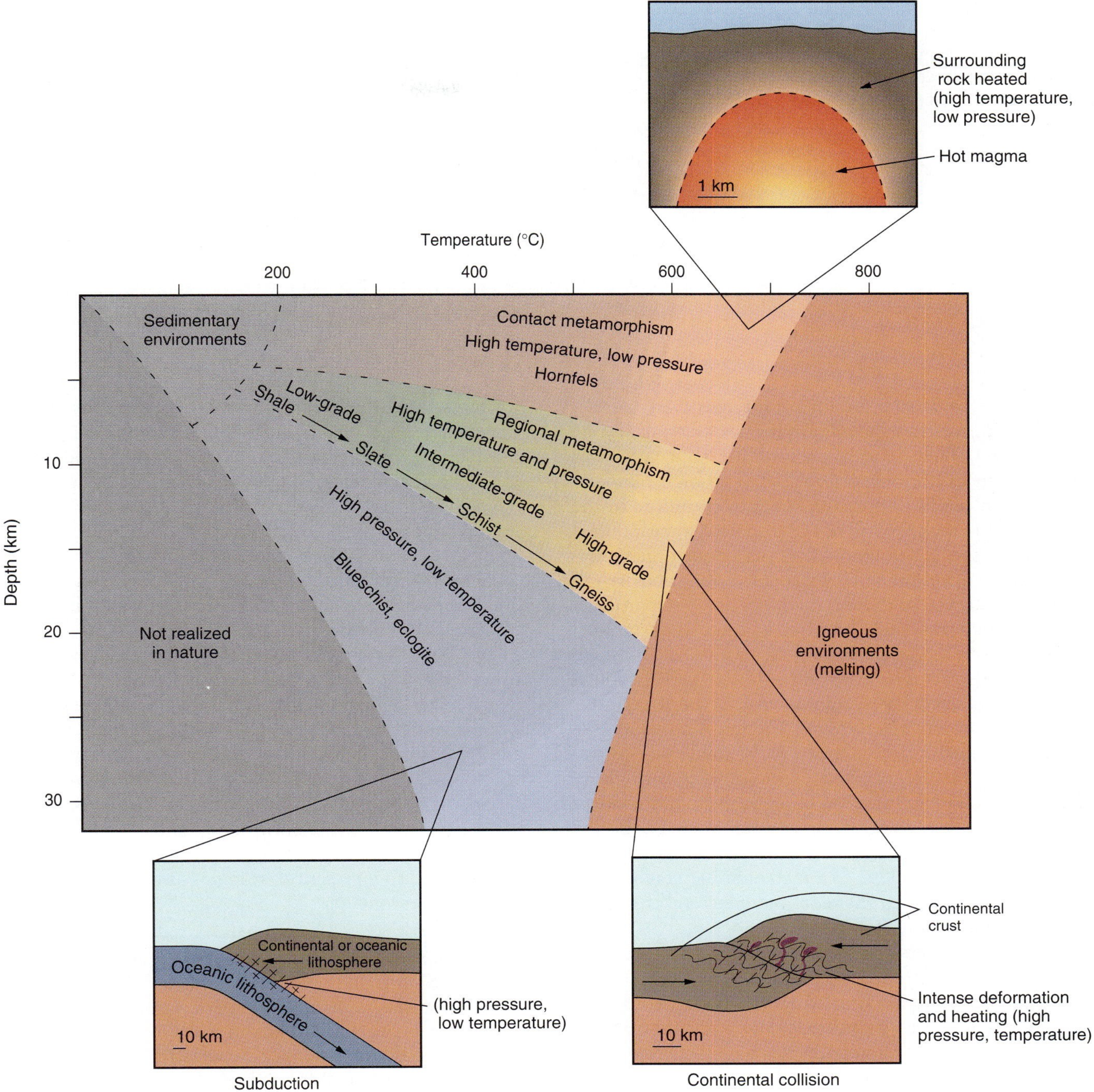

FIGURE 4.27 The main types of metamorphism, corresponding to different pressure-temperature conditions. *Contact metamorphism* is caused by high temperatures, typically by heating adjacent to a magma body. *High-pressure metamorphism* is commonly encountered at subduction zones, where the sinking lithosphere is relatively cool but encounters progressively higher pressures as it descends into the mantle. The most common type of metamorphism—*regional metamorphism*—occurs at high pressures and temperatures, as might be generated when two continents collide.

Metamorphic grade, or the degree of the metamorphism, decreases with distance from the contact, and the extent of metamorphism depends largely on the size of the intrusion. A small dike chills rapidly as it intrudes, and has little effect on the surrounding wall rock. But a pluton made up of hundreds of cubic kilometers of magma is a significant heat source. It cools slowly and transfers a good deal of heat to the surrounding rock (Fig. 4.28).

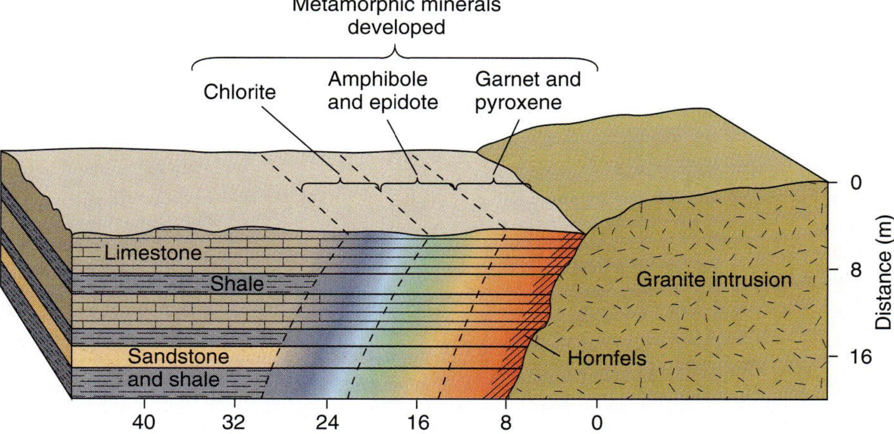

FIGURE 4.28 Metamorphic effects adjacent to a hot intrusion: contact metamorphism. Metamorphic minerals are developed depending on the protolith composition.

As pressure is not an important agent in contact metamorphism, these metamorphic rocks typically are not foliated. They tend to recrystallize to a very hard rock known as hornfels (Fig. 4.28). The high temperatures may cause large porphyroblasts to crystallize in the metamorphic rock, giving rise to a spotted appearance.

Clearly, the controlling factor in contact metamorphism is the intrusion and subsequent cooling and crystallization of magma, so there is no direct link to a particular tectonic environment. Nevertheless, there is an indirect link inasmuch as the generation of magmas is controlled by the plate-tectonic environment. Extensive contact metamorphic effects are common around the large intrusive bodies at convergent plate margins. The volume of magma intruded into the crust can be immense—as was the case for the Sierra Nevada batholith shown in Figure 4.6. The result of an intrusion of successive plutons over a geologically short span of time is an increase in the amount of heat in the crust, which increases the geothermal gradient (Fig. 4.12). The metamorphic grade of the rocks that are intruded therefore increases and the metamorphic effect will be regional.

An important effect of the heat input that occurs as large plutons are intruded is the generation of *hydrothermal* systems. Such systems occur when the hot igneous body acts as a heat source that drives fluid circulation in the crust. Hydrothermal circulation is a highly effective heat transfer mechanism because water moves readily through small openings in rocks and has a high specific heat; it can contain a great deal of heat per unit volume. The migration of water is undoubtedly a major factor in the extent of contact metamorphism. Hydrothermal circulation is an extremely important mechanism for transporting and depositing metals of economic importance, both around mid-ocean ridges and around plutons intruded into the crust. Heated water from deep in the crust can also be exploited as a valuable source of geothermal energy.

Cataclastic Metamorphism Intense metamorphism occurs within and adjacent to fault zones, shear zones, and fractures along which huge masses of rock are thrust during compression. This cataclastic metamorphism involves mechanical processes such as physical crushing and grinding and occasionally even melting at the base of the thrust faults. The metamorphic rock *mylonite* forms in this way. All mylonites have one thing in common: Their deformation occurred at pressures equivalent to depths of approximately 15 km or more. Mylonites commonly show broken and strung-out crystals of feldspar. There may be some recrystallization, and it is quartz that is generally recrystallized. The rocks are pervasively and intensely deformed at every scale, from millimeters to many meters.

There is general agreement that mylonites have formed rather deep in the crust and probably represent the "root zone" of large faults. Some mylonites have been interpreted as having formed in the deep extension of large strike-slip faults, such as the San Andreas; others appear to have formed horizontally and probably represent great faults rising out of the lower crust. Still others are believed to represent the "fault" surface at the top of a subduction zone.

Regional metamorphism is generally associated with plate collisions and mountain building. High-pressure metamorphism is associated with subduction zones. Contact metamorphism occurs in rocks that are in contact with igneous intrusions. Cataclastic metamorphism involves physical crushing and grinding along faults.

The Rock Cycle

It is clear from our discussion thus far that a wide range of rock types make up Earth's crust. Geologists must be able to organize their knowledge of these rocks and use names to refer to, or classify, the different kinds of rock. The names are a shorthand designed to convey as much information about the rock as possible, based on observations. These observations, which include mineralogy, texture, and fabric, provide important clues about the history and origin of the rock (Focus 4.5).

We can now relate the materials that make up Earth to the dynamic processes of Earth. The relationship shown in Figure 4.29 is commonly referred to as the **rock cycle**. We classify the major processes of this cycle according to the two important energy sources that shape Earth: internal energy and external energy.

Internal energy is identified as the heat contained within Earth. This heat sometimes escapes dramatically in the form of volcanoes, but it is more usually released during the constant cooling of Earth's core and mantle, which, in turn, drives convection currents in those regions. The *external energy* reaching Earth's surface, in the form of solar radiation, is responsible for driving weather systems in Earth's atmosphere. In turn, these weather systems cause weathering, erosion, and transportation of sediments.

Internal energy (heat) drives plate tectonics and produces magmas in certain tectonic environments. Molten magma moves up into the crust, where it cools into intrusive igneous rocks or reaches the surface to form extrusive rocks. Igneous rock produced at divergent plate margins forms the oceanic lithosphere, which is returned to the mantle by subduction. Igneous rock of continental regions may be buried deeply at convergent plate margins and metamorphosed or may be uplifted and eroded to form sediment and sedimentary rock.

External energy drives Earth's hydrologic system (including rain, rivers, and glaciers), which acts to wear down Earth's surface by weathering and erosion. These processes produce sediment by eroding uplifted regions and ultimately deposit this sediment in ocean basins, where it may eventually lithify into sedimentary rocks.

On one hand, then, the forces of plate tectonics transfer material between the mantle and crust and change the shape of the surface by deforming the lithosphere. On the other hand, the forces of weathering and erosion constantly shape and mold the surface of the solid Earth. The rock cycle is nothing more than a conceptual aid in visualizing interconnections between geologic processes. For example, weathering, erosion, and deposition of sediments may be followed by burial to great depths to form metamorphic rocks. A portion of the metamorphic rocks may be uplifted into mountain ranges and a portion may be melted to form igneous rocks. The uplifted metamorphic and igneous rocks or the mountains are once again subject to weathering and erosion as part of the sedimentary process. Thus, the cycle continues.

The rock cycle describes the interconnections between geologic processes. Internal forces build up mountains; external forces wear them down.

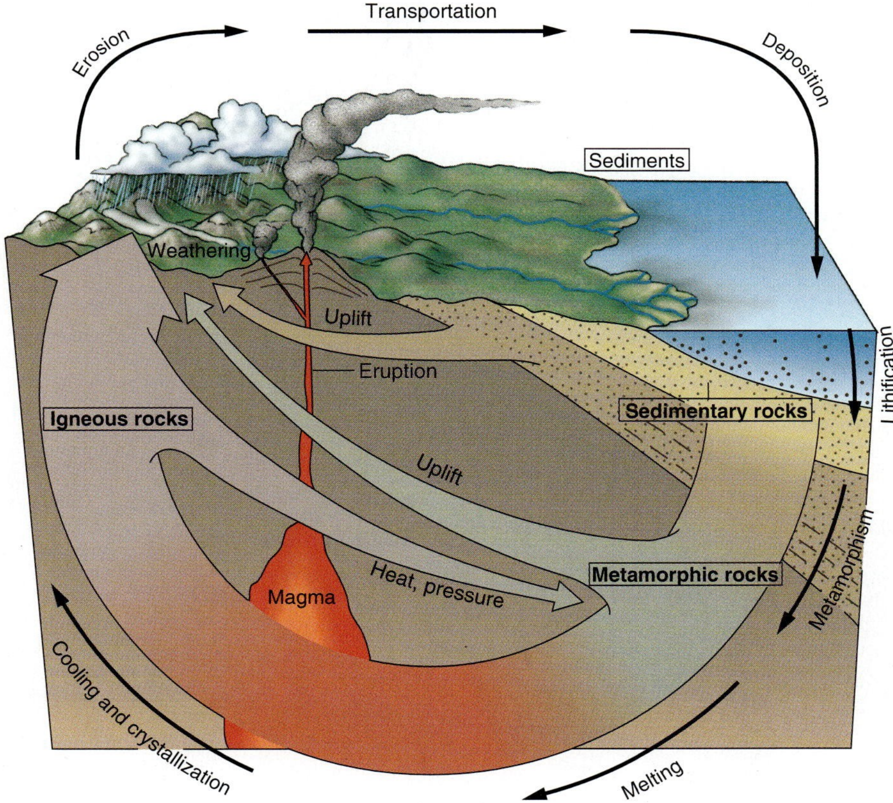

FIGURE 4.29 The *rock cycle* is a useful way to visualize the relationship between different rock types. The different rock types (igneous, sedimentary, and metamorphic) and materials that form rock (sediment and magma) are shown in boxes. The processes that relate them to one another are shown as labeled arrows.

Focus On 4.5 GETTING INFORMATION FROM ROCKS: OBSERVATION AND INTERPRETATION

Now that you are familiar with the main rock types that make up Earth's crust, we can see how each rock type can tell us something about how the rocks formed. To do this, we must understand some elementary principles of magmatism, metamorphism, and sedimentation, which we just described in the text. But most importantly, we must make good observations. Observation is one of the key elements of geology. We will consider three simple examples to illustrate this concept by discussing observations at three different scales.

1. *The outcrop scale:* This enables us to place a rock in a regional context. From this occurrence "in the field," we can consider a rock's relation to the land surface and to other rocks in the area.
2. *The hand-sample scale:* This enables us to see any texture or fabric, the sizes of individual mineral grains, and other features of the minerals that make up the rock.
3. *The microscopic scale:* This allows us to examine the texture on a grain-to-grain basis and to identify with certainty the minerals that make up the rock. Microscopic observation usually involves making a *thin section*, an ultrathin slice of the rock, which can be examined under the microscope (see Focus 4.1).

Note that we could get a lot more information than these simple observations provide if we were to analyze the chemical composition of the rock and its minerals. But direct observation is the first step to interpreting the history of a given rock. The examples given next indicate the types of interpretations geologists can make from observations at each of the three scales.

BASALT—AN IGNEOUS ROCK

OUTCROP

- Observation: The rock occurs in the form of a "frozen river" associated with a volcano (Fig. 1a).
- Interpretation: It is a lava flow and was erupted as a hot liquid from a volcano. It flowed downhill, cooled, and eventually solidified.

HAND SAMPLE

- Observation: The rock is dark (mafic) and largely glassy, although it contains some big (>1 mm) green crystals and bubble holes (Fig. 1b).
- Interpretation: The bubble holes suggest that gas was escaping from it as it solidified, which confirms that the rock was originally a molten liquid. The glassy texture means that it chilled quickly. The large crystals (phenocrysts) are difficult to produce in a rock that cooled quickly, so these

(a)

(b)

(c)
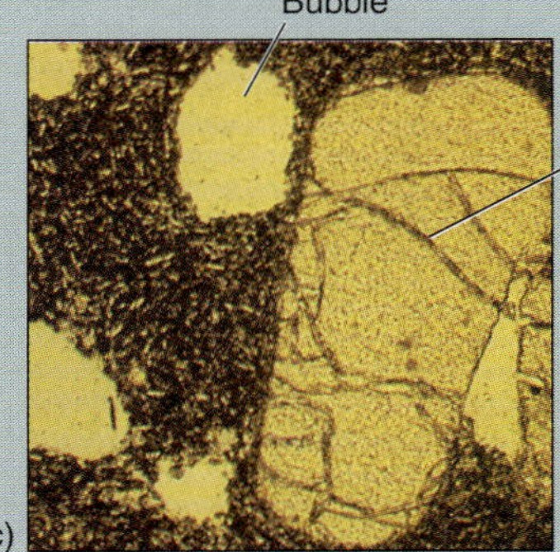

FIGURE 1

crystals must have grown during a period of slower cooling, presumably below the surface. They must therefore have already been in the lava as it erupted. The green color of the crystals suggests they are of the mineral olivine.

Microscopic scale

- Observation: The phenocrysts stand out, surrounded by glass, and can be identified as olivine. Some tiny, needlelike crystals can be seen in the glass (Fig. 1c).
- Interpretation: The needlelike crystal shapes indicate rapid cooling, as you might expect when a molten rock at 1200°C erupts at the surface. The presence of forsteric olivine (Mg_2SiO_4; Focus 3.5) suggests that the lava was rich in magnesium and is therefore probably a basalt.

SANDSTONE—A SEDIMENTARY ROCK

Outcrop

- Observation: The rock outcrops in flat beds but with fine-scale, inclined layering (Fig. 2a).
- Interpretation: The rock is cross-bedded, which indicates that the sediment that now forms the rock was deposited from a current (flowing water or wind).

Hand sample

- Observation: The rock is light-colored and homogeneous, made up of transparent mineral grains of similar size and type (Fig. 2b). The mineral grains are hard enough to scratch a pocket knife.
- Interpretation: This rock would be called well sorted and mature. Thus, it has probably been subjected to a lot of sedimentary processing, probably by rivers or desert winds. The hardness of the minerals and their transparency suggest the mineral is quartz.

Microscopic scale

- Observation: The grains are well rounded and composed of a transparent mineral with no cleavage. A second mineral fills in some of the gaps between the grains (Fig. 2c).
- Interpretation: The mineral grains are quartz. The abundance of quartz and the rounding of the grains confirm that it is a mature sediment. The mineral that fills in the gaps was deposited after the sediment, during diagenesis and lithification (it was probably precipitated by fluid that moved though the pore space), and it is the cement that holds the grains together and makes it a rock rather than loose sediment.

(a) 

(b) 

(c) 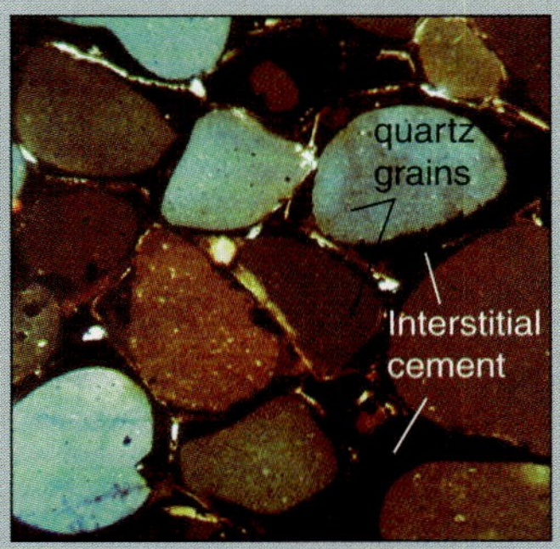

FIGURE 2

(a)

(b)

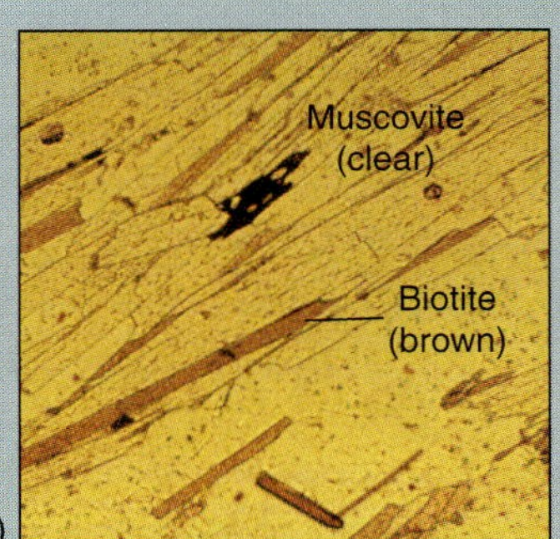

(c)

FIGURE 3

SCHIST—A METAMORPHIC ROCK

Outcrop

- Observation: The rock is highly folded and fractured (Fig. 3a).
- Interpretation: The rock has been deformed, so it must have been subjected to rather high pressures and temperatures; it is metamorphosed. Such deformation is generally encountered only where there are concentrations of tectonic forces, such as occurs when two continents collide to form a mountain belt.

Hand sample

- Observation: The rock is strongly foliated. This results from the alignment of platy micas (Fig. 3b).
- Interpretation: The mica grains grew and became oriented in response to strong squeezing during metamorphism.

Microscopic Scale

- Observation: Even under the microscope, the strong alignment of mica is clear. The micas are colorless and have very well-developed crystal shapes (Fig. 3c).
- Interpretation: The dominant mica can be identified as the mineral muscovite, distinguished from biotite, which is dark brown.

SUMMARY

- Igneous rocks may form either by eruption of lava at the surface to form extrusive rocks or by solidification of magma below ground to form intrusive rocks.
- The type of mineral that crystallizes from a magma depends on the chemical composition of the magma. Whether minerals crystallize at all depends on the cooling rate.
- A wide range of igneous rock compositions is generated by various processes of differentiation such as partial melting, fractional crystallization, and contamination.
- Intrusions are generally named according to their shape, which is controlled by the volume of magma and by the way in which the magma intrudes the crust.
- The shapes of volcanoes are determined in part by the viscosity of the lava they produce. The viscosity, in turn, is controlled by chemical composition, temperature, and water content.
- The geothermal gradient is the rate of temperature change with depth in Earth.
- Melting can be achieved by decompression or by lowering the melting point of the mantle by adding water. These

conditions are encountered at divergent and convergent plate margins, respectively.

- As the maturity of a sediment increases, average grain size decreases, the grains become more rounded and better sorted, and only the most resistant mineral types, such as quartz, survive.
- Chemical sediments are precipitated directly from solution. Biochemical sediments involve organic activity such as the generation and subsequent accumulation of shell and coral fragments.
- The processes of lithification and diagenesis involve compaction and cementation of loose sediment into hard rock.
- The different types of minerals and their abundances in a given rock will change during metamorphism. For a given chemical composition, the metamorphic minerals that form in a rock are a function of temperature-pressure conditions.
- The action of tectonic forces during metamorphism may cause the mineral grains in a rock to become oriented, producing a foliation.
- Regional metamorphism is generally associated with plate collisions and mountain building. High-pressure metamorphism is associated with subduction zones. Contact metamorphism occurs in rocks that are in contact with igneous intrusions. Cataclastic metamorphism involves physical crushing and grinding along faults.
- The rock cycle describes the interconnections between geologic processes. Internal forces build up mountains; external forces wear them down.

KEY TERMS

igneous rock, 72
sedimentary rock, 72
metamorphic rock, 72
magma, 72
lava, 72
extrusive, 72
intrusive, 72
phaneritic, 72
aphanitic, 72
porphyritic, 72
phenocryst, 72
differentiation, 74
solidus, 74
liquidus, 74
partial melting, 76
magma chamber, 76
fractional crystallization, 77
solid solution, 77
batholith, 77
pluton, 77
dike, 79
sill, 79
diapir, 80
assimilation, 80
shield volcano, 81
composite volcano, 81
viscosity, 81
cinder core, 81
pumice, 81
pyroclastic flow, 82
geothermal gradient, 85
strata, 87
clast, 88
sorting, 90
maturity, 91
solubility, 91
evaporite, 91
bioclastic, 91
limestone, 91
cross-bedding, 92
graded bedding, 93
lithification, 94
cementation, 94
diagenesis, 95
metamorphism, 95
metamorphic grade, 96
texture, 97
fabric, 97
porphyroblast, 97
foliation, 98
cleavage, 98
schistosity, 98
blueschist, 100
rock cycle, 103

QUESTIONS FOR REVIEW AND FURTHER THOUGHT

1. What are the two main ways in which Earth's mantle melts?
2. Why are there volcanoes in Washington and Oregon today, but none in Kansas?
3. If you lived within 10 km of a volcano that was liable to erupt at any time, which type would you hope it was, and why?
4. Use Figure 4.2 to estimate the types of minerals and their percentages that you would expect to find in:

   (a) a granodiorite; (b) a basalt; (c) rhyolite
5. Describe the kind of sediment that you might find in a mountain stream in terms of the types of clasts, their sizes, and their shapes. If you followed the stream all the way to the ocean, how will the sediment change?
6. A sandstone composed entirely of well-rounded quartz grains would be called mature. If the original source rock for the sediment was a granite, what processes must have acted on the granite to produce a pure quartz sand?
7. We know that hard detrital sedimentary rocks must have been originally deposited as loose sediment. What mechanisms can you suggest that would convert a loose sediment into a hard rock?
8. (a) If a mudstone was subjected to *regional* metamorphism (high pressures and temperatures), what sort of metamorphic rock might form? (b) If the same mudstone was subjected to *contact* metamorphism (high temperatures but low pressures) what sort of metamorphic rock might form? (c) What would be the difference in texture between the metamorphic rocks formed in (a) and (b), respectively?
9. What are the major differences among contact metamorphism, regional metamorphism, and cataclastic metamorphism? Can you imagine how contact metamorphism might become a type of regional metamorphism? Devise a scenario that could produce eclogite by metamorphosis and return it to Earth's surface.

5 EARTH PROCESSES: PHYSICAL PRINCIPLES

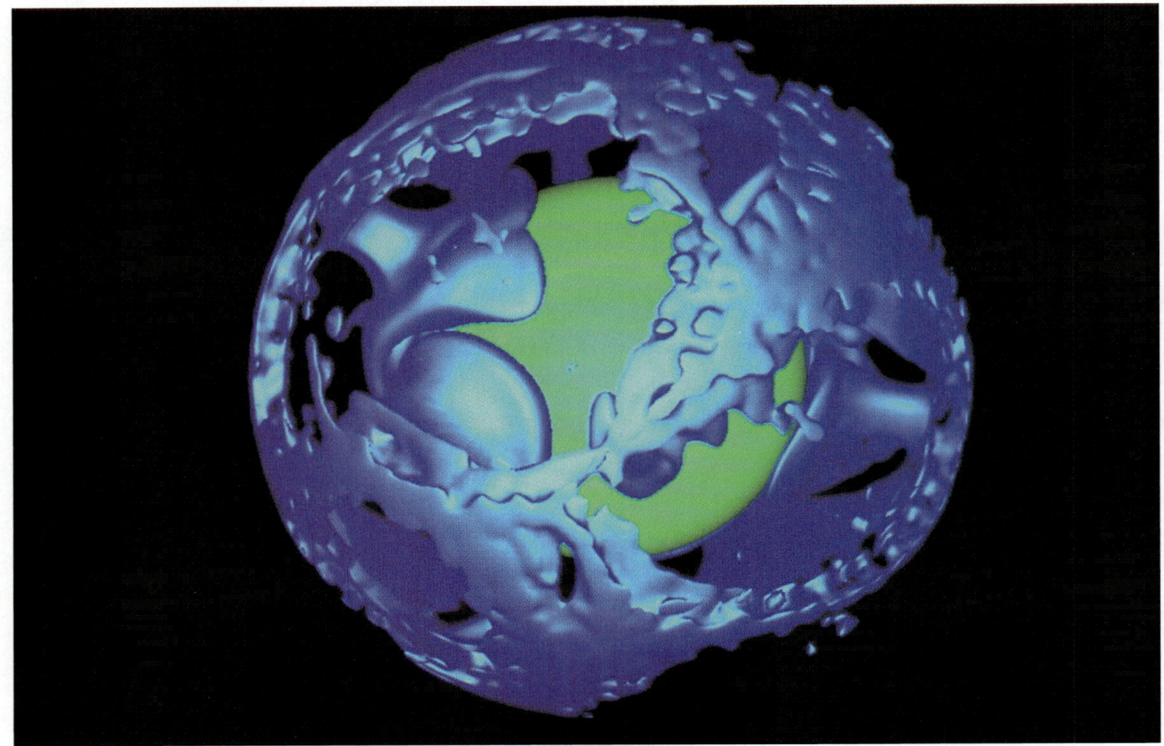

A computer model of three-dimensional convection of the mantle.

OVERVIEW

- Rock can elastically deform by a change in volume or by a change in shape.
- Velocities of seismic waves are controlled by the properties of the rocks through which they travel. These waves tell us about Earth's interior.
- The gravity on Earth can be approximated by assuming that Earth is a homogeneous sphere. Because of Earth's spin, however, a gravity correction is required.
- Because temperature in Earth increases with depth, the mantle tends to flow if subjected to a load for a long period of time. Mountain ranges can be thought of as floating in the mantle.
- Earth's magnetic field is generated by electrical currents in the liquid outer core.
- The geomagnetic field at Earth's surface is similar to the field of a bar magnet or a loop of electric current.
- The polarity of Earth's magnetic field has reversed many times through geologic history.

INTRODUCTION

The driving engine of plate tectonics is the flow of heat within Earth. To understand this engine, we must explore the interior of the planet. Although Earth's average radius is 6371 km, the deepest drill hole currently penetrates only 15 km into the planet. How do we probe below this shallow depth? Although Earth is 4.55 billion years old, historical reports—including the oldest records of earthquakes—span only the last 3000 years. Thus, we are barely able to scratch the surface of the planet's geologic history.

Most of our knowledge of Earth's interior stems from analyses of seismic waves and Earth's magnetic and gravity fields. In this chapter, we investigate the physical basis of the methods used to probe Earth's depths and to infer its history. Within the last hundred years, geologists and physicists have joined forces to study Earth; the resulting discipline is called *geophysics*. Here, we examine some fundamental principles of physics that have been crucial to our understanding of Earth, and we will discover how much they have revealed about our planet.

SEISMIC WAVES

A pebble dropped into a pond generates waves that emanate as circles outward from the point of impact. We can clearly see these waves, but we also frequently experience waves that are invisible. When we speak, we generate invisible waves in the air. As air is exhaled, our vocal cords vibrate and create a disturbance in the air, which travels to the eardrum of the listener. If we could examine this disturbance, we would find that pressure changes have been generated. The air in a given area becomes compressed; the air molecules are more densely packed together. The compressed air then expands and presses on surrounding molecules, which, in turn, push on their surrounding air molecules, and the disturbance travels progressively outward like a water wave in a pond. These are sound waves in air.

Seismic waves in rock are similar to sound waves in air. Seismic waves are generated by an earthquake, an explosion, or some similar disturbance. These waves travel through Earth, and we use them to probe its interior in a way similar to our use of X-rays to examine the internal anatomy of an organism. Study of seismic waves involves understanding the stress changes in Earth and how these waves propagate.

Waves, including seismic waves, can be characterized by certain distinguishing features. The *wavelength* of any wave is the distance between its peaks or any other repeated feature, and the *period* is the time required to complete an oscillation of a single wave. The *amplitude* is the height of the oscillation (Fig. 5.1). Amplitude of sound waves corresponds to loudness. The speed of a traveling wave is determined by dividing the wavelength by the period. The speed of seismic waves depends on two material properties: **density** and **elasticity**. We will examine these two properties in the following sections.

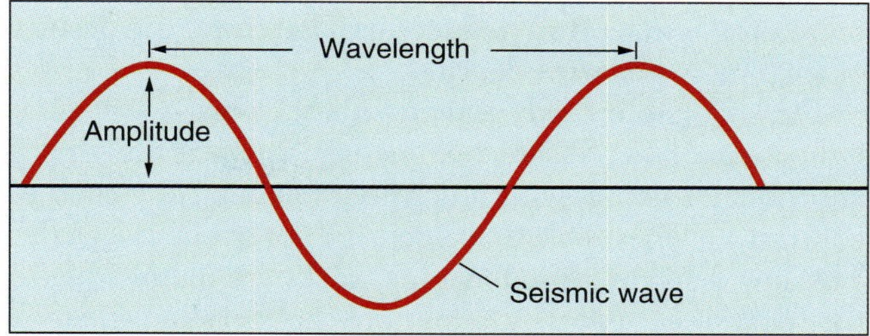

FIGURE 5.1 The height of a seismic wave is called the *amplitude*. The *wavelength* can be measured as the distance between peaks (or troughs, or any other corresponding successive points in a wave). Frequency is the number of oscillations per second.

Density

At Earth's surface, a rock is denser than water, but it is less dense than aluminum. Deep within Earth, however, rock densities become even greater than the density of aluminum as measured at Earth's surface because of the great pressures in the interior (Fig. 5.2). How do we know this? If a rock is subjected to increasing pressure, its volume decreases while its mass remains the same. Therefore, its density increases—since the density of a material is defined as the ratio of its mass to its volume (mass per unit volume; see Focus 2.3).

If you dive to the bottom of a swimming pool, you feel pressure on your ears because your body has to support the weight of the overlying water. You would feel this pressure to the greatest extent on your eardrum because this is the most pressure-sensitive part of your body that is in contact with the water load. In the same way, rocks at depth within Earth are compressed by the weight of the rock above them. In the mantle, pressure increases the density of mantle material from about 3.3 g/cm^3 in the upper mantle to about 5.5 g/cm^3 at the core–mantle boundary (see Fig. 5.2). The density then jumps to 10 g/cm^3 in the outer core at the transition between the silicate rocks of the mantle and the iron of the core. Finally, the density increases to approximately 13 g/cm^3 in the inner core at Earth's center (Fig. 5.2).

How do temperature and pressure interact to affect the density of a rock within Earth? When a rock is heated, it expands. Because its mass remains the same, its density must decrease. We learned earlier (∞ see Chapter 4) that temperatures increase with increasing depth within Earth. If rocks within Earth were subjected only to increasing temperatures, we would expect their densities to decrease with depth. Rocks are also subject to pressure changes though, and pressure has the opposite effect of temperature. Increasing pressure increases the density of a rock within Earth to a greater extent than increasing temperature reduces it. Therefore, the density of rock increases with depth.

Density within Earth depends on rock type as well as on temperature and pressure. Sedimentary rocks at the surface are the least dense, with a density of about 2.0 g/cm^3. In contrast, igneous rocks have densities of about 2.7 g/cm^3. The low densities of sedimentary rocks are mainly due to the rocks' pore spaces—the minute gaps between the rock grains that are filled with gas, water, or other fluids.

Olivine is thought to be the dominant mineral in the upper mantle, extending to a depth of 410 km. At that depth, the mineral structure is unstable and it changes to one that is more compact, a polymorph called the *spinel structure* (∞ see our example of polymorphs in Chapter 2).

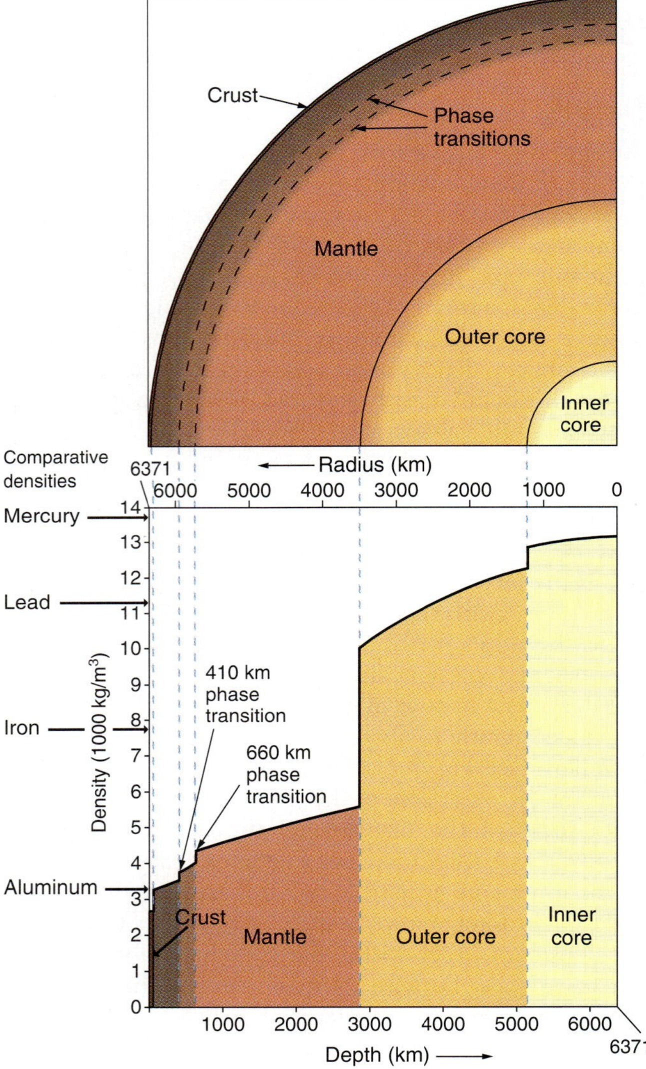

FIGURE 5.2 Density as a function of depth within Earth. Sudden increases in the density of Earth's interior occur at the base of the crust, at the mantle-phase transitions, at the mantle–core boundary, and at the inner core boundary. Densities of some common materials (at Earth's surface) are shown for comparison.

Further changes in mineral structure occur at discrete depths, such as 660 km, within the mantle. These changes, which were inferred to exist by observing the behavior of seismic waves, are referred to as **phase transitions**.

Elasticity

A rubber band will stretch if you pull it; it will return to its original shape once you release it. This behavior is called elasticity. If you pull on plasticine, it will not return to its original shape. This behavior is referred to as **inelasticity**, or **plasticity**.

Perhaps surprisingly, rocks are elastic. If you were to pull on a rock, it would extend. Of course, the amount it extends is so small that you could not see it with your eyes; you would need special instruments to measure the extension. If you were to release the rock, it would return to its original shape. Both the rubber band and the rock are elastic, but the rubber band will extend more than the rock if the same amount of force is applied to both. We need to define a material property that indicates how much elastic response a material exhibits in response to a given force.

Stress is the amount of force per unit area that is applied to an object. If the force is a pull, the stress is a **tension**; if the force is a squeeze, the stress is called a **compression**. Stress can change both the shape and size of an object. If you were to squeeze a sponge on all sides, it would spring back elastically when you released the stress. If you were to twist each end and then release the sponge, it would again regain its original shape. These two actions represent two fundamentally different ways of stressing an elastic material. The first is compression, *which changes an object's volume*, and the second is shear, *which changes an object's shape* (Fig. 5.3). **Strain** is a measure of the change in volume or shape produced by stress. For compression, strain is the change in an object's volume divided by its original volume. For shear, we use angle changes to describe the change in shape. Specifically, shear strain is given as half the angular amount of change in the right angles after the stress is applied.

The response to either compression or shear can vary within the same object because the bonds between the atoms are affected in diverse ways by different applied stresses. If the object is compressed, all the bonds act somewhat like internal springs and are shortened until they balance the applied external stress. If the object is sheared, the bonds are subjected to a sideward motion that is similar to bending springs. Note, however, that the bonds in liquids have no resistance to bending. A sheared liquid remains sheared; it does not return to its original shape. Its response to a shear stress is inelastic. A liquid, however, like a solid, can be compressed elastically.

We are now ready to quantify the observation that some materials are more difficult to strain than others. The ratio of the amount of stress to the strain induced by that stress is called the **elastic modulus**. A modulus is a quantity that expresses the response of a material to an

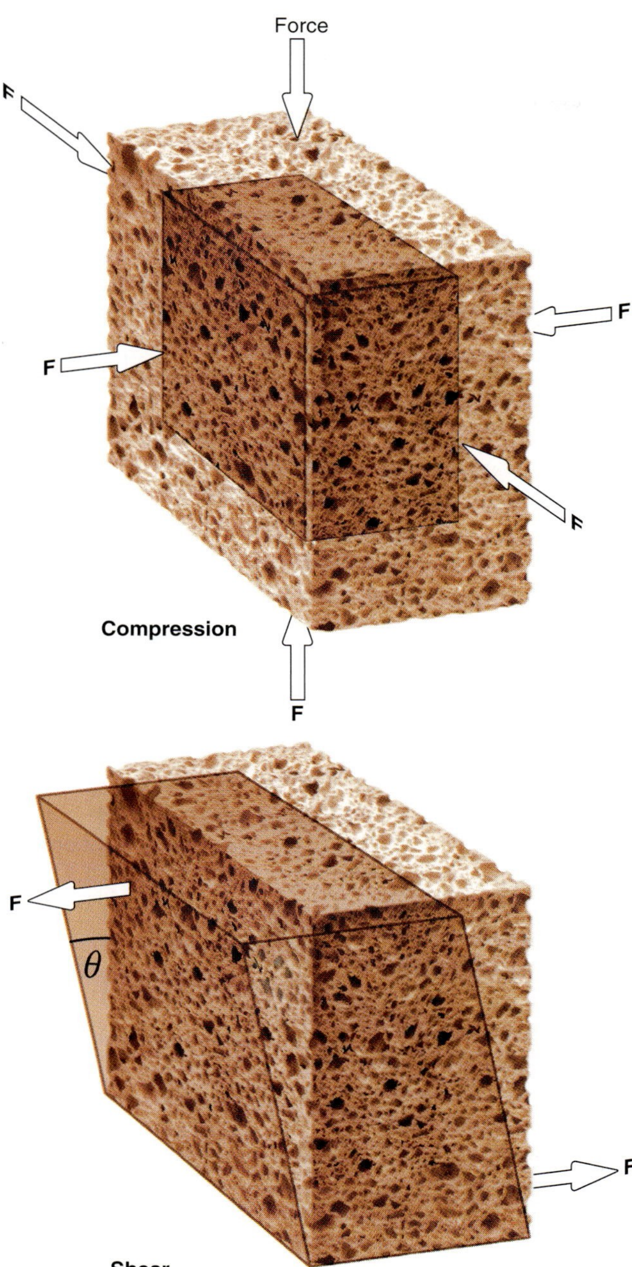

FIGURE 5.3 Two different ways of straining a rock (illustrated with a sponge): compression and shear. *F* represents the applied force. The shear strain is $\theta/2$, which is half the angle by which right angles change after shearing.

external stimulus. For example, the elastic modulus of a sponge is a comparison between the amount of stress applied to the sponge and the resulting amount of change in the sponge's shape resulting from that force. The modulus describing shear is called the *shear modulus*, while that describing compression is called the *bulk modulus*.

The elastic modulus of a rock is a material property that depends on the rock's composition and fabric; this, in turn, depends on the types of minerals contained in the rock and the way those minerals are arranged relative to each other. The elastic modulus can be measured by applying a known stress (force per unit area) to a rock and measuring how much the rock changes shape.

As with density, the elastic modulus increases with pressure but decreases with temperature. Within Earth, the pressure effect dominates; thus, the elastic moduli increase with depth. The outer core is liquid, however, and here the shear modulus drops to zero.

Seismic waves are used to probe the density and elastic moduli of Earth's interior. Elastic moduli give the amount of stress needed to cause a given strain. Density and elastic moduli increase with depth in Earth.

Seismic P and S Waves

Earthquake waves are monitored using a **seismometer**, an instrument that measures ground motion. A record of this motion is called a **seismogram**. Seismograms are used both to locate earthquakes and to measure their magnitudes.

Should you ever experience the waves from a nearby earthquake, you may first feel a relatively low-amplitude shaking, which alerts your senses but generally does not cause much damage. This first wave is called the **P wave** (P for "primary"). It takes the form of a burst of oscillation, as shown on the seismogram in Figure 5.4. Some seconds to minutes after arrival of the P wave—depending on your distance from the earthquake source—shaking of a higher amplitude can be felt. The **S wave** (S for "secondary") has arrived (Fig. 5.4). It is also a burst of oscillation and usually has a greater amplitude than the P wave that emanated from the same source. Both P and S waves travel directly from their source through the body of Earth, and they are therefore called **body waves**.

Almost immediately after the S waves arrive, high-amplitude shaking occurs as the straining builds to a crescendo. You may seek shelter in a doorway and hope the construction can withstand the strain. This shaking signals the arrival of **surface waves**. Surface waves do not build up immediately above an earthquake, but require the body (P and S) waves to reach the surface at a large angle before they become established. Surface waves can have very large amplitudes because Earth's surface is free to move. Therefore, less energy is required to generate a large surface wave than to generate a body wave within Earth where the rock is constrained. Because of their larger amplitudes, surface-wave disturbances cause more damage than does any other type of seismic wave. They are distinctive in that they have much lower frequencies than do the body waves, so that structures will sway to and fro for several seconds to several minutes.

The P and S waves generated by an earthquake are used to locate the source of the disturbance and to determine their cause (∞ see Chapter 7). An earthquake applies sudden shears and compressions to its surroundings, causing the two types of waves to progressively propagate outward.

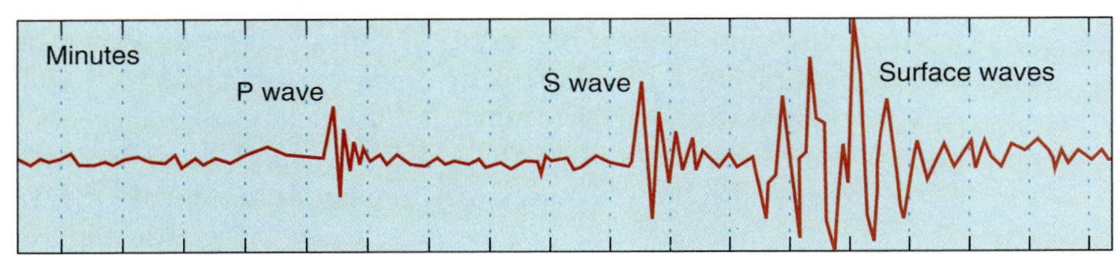

FIGURE 5.4 A seismogram is a record of ground motion detected by a seismometer. Here, the seismogram shows the arrival times of primary (P), secondary (S), and surface waves from a seismic event.

In contrast, an underground explosion is a seismic wave source that generates a sudden compression in a confined zone; this compression then propagates outward as a P wave only. Analyses of such P waves are used to monitor nuclear explosions, and enforce test-ban treaties.

The difference between P and S waves can be understood by carrying out a simple experiment. Place a long spring, such as a Slinky™, on a horizontal surface with the other end attached to a wall. Quickly push the spring toward the wall and then return your hand to its original position (Fig. 5.5a). A P wave (or compression wave) will propagate along the spring. The motion of the springs is directed along the coils in the same direction that the wave travels. Alternatively, if you flick the spring sideways (Fig. 5.5b), an S wave (or shear wave) travels along the spring. In this case, the motion of the springs is at right angles to the direction of wave propagation.

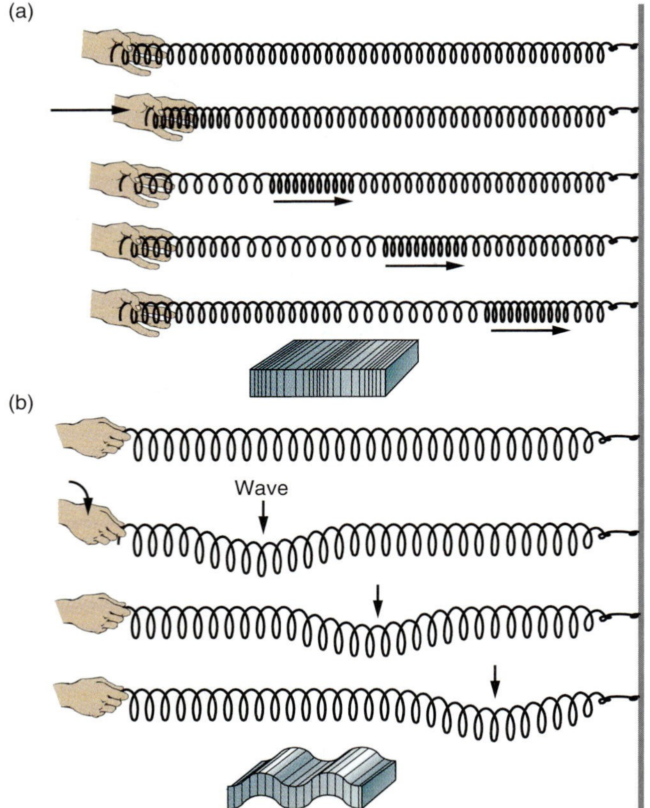

FIGURE 5.5 Types of seismic waves and associated motions. (a) P waves oscillate in the direction of wave travel; (b) S waves oscillate at right angles to the direction of wave travel.

The velocity of seismic waves within Earth varies with the elastic modulus and the density of the medium through which they travel (see Focus 5.1). The velocity varies according to the equation

$$\text{Velocity} = \sqrt{\frac{\text{elastic modulus}}{\text{density}}}$$

Because both P and S waves travel through the body of Earth, the density of the material through which they both travel is the same. The P waves are a combination of volume and shape changes; S waves comprise just shape changes. For most rocks, the modulus for P waves is about three times that for S waves. Using the seismic-wave velocity equation, the velocity of P waves is about 1.7 times that of S waves. This explains why P waves arrive at a given location first (see Fig. 5.4). Further, the velocity of surface waves is about 0.9 times the velocity of S waves, which explains why the destructive energy of surface waves arrives at a given point shortly after the arrival of S waves.

In simple experiments, we can determine the seismic-wave velocities through rocks within Earth by measuring the density and modulus of laboratory samples. By subjecting them to high pressures and temperatures, we can estimate the changes in densities and moduli that would occur at conditions inferred to exist deep within Earth. Scientists can compare calculated velocities for different rock types with measured velocities and infer the variation of rock type, density, and modulus within Earth.

An earthquake generates two types of body waves: P waves arrive at a given location first, followed by S waves. Both P and S waves combine near the seismic source to generate surface waves, which travel along Earth's surface, causing the most damage.

Velocity Variations Within Earth

Travel-Time Curves

We can use velocity measurements of seismic waves to estimate the properties of materials at great depths within Earth. For example, velocities within Earth's core are consistent with the core being composed largely of iron. The velocity of a homogeneous, or uniform, medium can be obtained from the familiar formula: velocity = distance/time.

To measure the velocity of a seismic wave originating from an explosion, we simply measure the elapsed time between the explosion and the arrival of the wave at a monitoring station; this time is then divided into the distance traveled from the explosion to the station. For example, if the distance from an explosion to a seismometer is 12 km, and it took two seconds from the time of the explosion for the waves to register on the seismometer, the velocity of the seismic wave would be 12/2 = 6 km/sec. This is a straightforward calculation for waves that travel through a single material in a straight line between the source and the receiver.

We have already noted that a pebble dropped into a pond produces water waves that spread out in ever-increasing circles; this is because the velocity of each water wave is constant. The waves travel in straight lines radially from the source of the disturbance. In such cases, it is easy to determine the velocity. When the wave velocity is not constant—for example, within a heterogeneous material such as Earth—the direction of the waves is not straight, but curved.

Refraction An underground explosion of dynamite radiates waves in all directions (Fig. 5.6). Chemical energy in the dynamite is converted into heat and the kinetic energy of wave motion in the surrounding elastic material. The outermost surface of expanding energy is referred to as the *wavefront*. This is analogous to the first wave that propagates out from a pebble dropped into a pond. The wavefront marks the first arrival of the seismic pulse, which takes the form of several oscillations generated by the explosion and consequent reverberations of the surrounding material. If a portion of the wavefront encounters a different material as it travels, that portion of the wave will start to travel at a different velocity. This causes the wavefront to become distorted.

We can trace the direction of movement of a wavefront by drawing a line perpendicular to the wavefront at

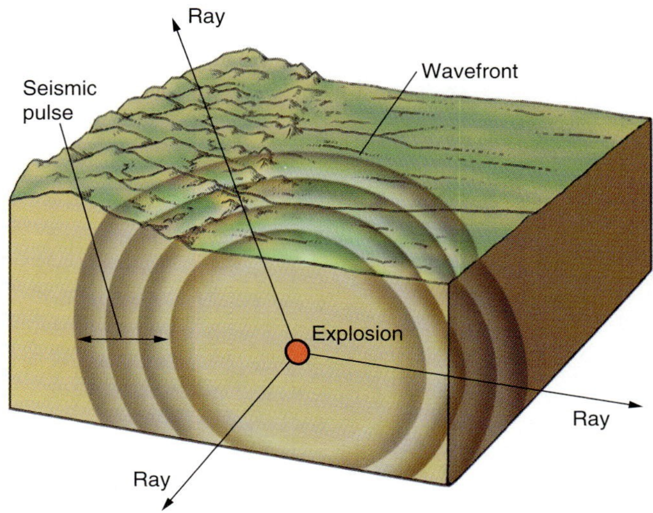

FIGURE 5.6 The wavefront from an explosion moves out in all directions like an expanding balloon. The wavefront traces the outside envelope of the expanding energy. The ray traces the direction of travel. The seismic pulse is the burst of energy that arrives immediately after the wavefront.

each point of the wave. Each of these lines is called a *ray*. The rays drawn for a wave traveling at a constant velocity through a uniform medium are always straight lines between the wave source and the receiver. Appropriately, such rays are called **direct rays**.

When the wave front encounters a region that causes it to change velocity, the rays bend. This bending is called **refraction**. Within Earth, the direct rays travel in a line between the source and the receiver—where a seismic pulse is detected. Rays that dive into Earth encounter material that causes them to travel at higher velocities, which bends them back up to the surface (Fig. 5.7).

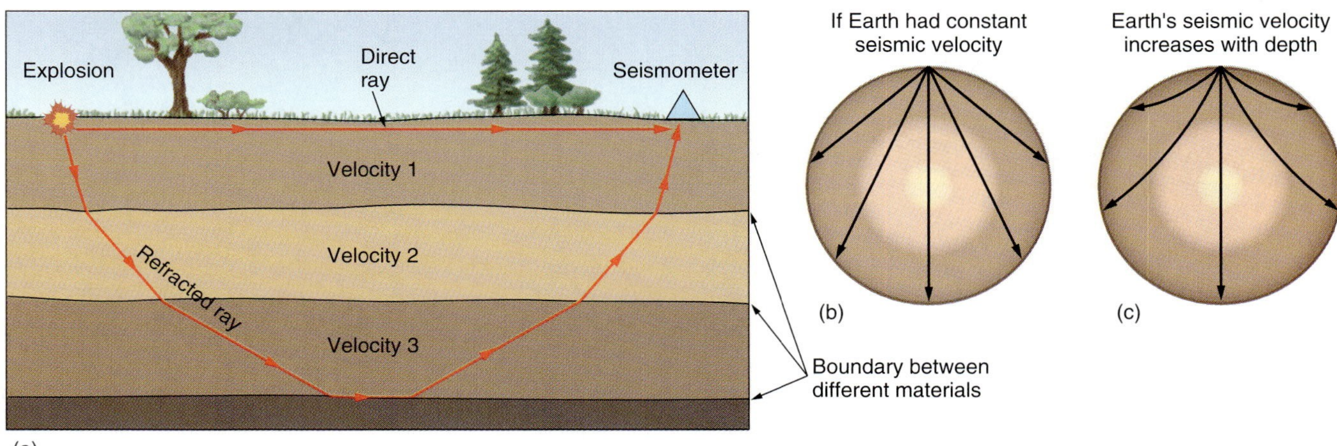

FIGURE 5.7 Seismic waves travel at different velocities through different materials. These waves refract, or bend, upon encountering the boundary between two different materials. They also bend if velocity varies continuously. (a) A refracted ray traveling through a layered region in which the velocity of each layer is higher than the one above it. (b) If Earth had constant velocity, seismic rays would be straight lines. (c) The velocity of a seismic wave actually increases with increasing depth in Earth. As a result, the seismic waves refract (bend)—except for a ray that travels vertically downward, which passes right through undeflected.

Focus On 5.1 THE VELOCITIES OF SEISMIC WAVES

The velocity of P waves depends on both the bulk modulus and the shear modulus. This is because P waves involve both compression of the medium, as described by the bulk modulus, and shearing, as described by the shear modulus. For volume changes, the bulk modulus, K, equals the stress divided by the change in volume/original volume. For shape changes, the shear modulus, μ, equals the stress divided by half of the angular change in right angles. Stress is measured in Pascals (Pa), which is the Standard International unit of stress or pressure equivalent to 1 Newton/square meter, and strain (volume or shape changes) are unit-less ratios. By measuring these quantities in the laboratory, we can calculate seismic-wave velocities for conditions deep in Earth. The P-wave velocity is given as

$$V_p = \sqrt{\frac{K + \frac{4\mu}{3}}{\rho}}$$

where ρ is the density of the medium through which the P wave is traveling.

The velocity of S waves depends on the shear modulus and the density according to

$$V_s = \sqrt{\frac{\mu}{\rho}}$$

From the formula, you can see that if the medium is liquid, which implies $\mu = 0$, then

$$V_p = \sqrt{\frac{K}{\rho}} \text{ and } V_s = 0$$

(P waves travel more slowly than they do in a solid, and S waves cannot travel through a liquid.)

Next, consider crustal rocks, which have a density of about 2670 kg/m^3 (or 2.67 g/cm^3). For such rocks, the bulk modulus $K = 6.3 \times 10^{10}$ Pa; and the shear modulus $\mu = 3.7 \times 10^{10}$ Pa. If we substitute these values into the velocity equation, we find:

$$V_p = \sqrt{\frac{(0.63 + \frac{4 \times 0.37}{3}) \times 10^{11}}{2670}} = 6.5 \times 10^3 \text{ m/sec} = 6.5 \text{ km/sec; and}$$

$$V_s = \sqrt{\frac{0.37 \times 10^{11}}{2760}} = 3.7 \text{ km/sec.}$$

These values are typical of crustal rocks such as granites.

You may be familiar with a common example of the refraction of light rays: A straight stick with one end immersed in water appears to be bent. Light waves travel at different velocities in air and in water. Therefore, rays traveling upward from the submerged part of the stick bend at the water–air interface, whereas those from the protruding part of the stick travel directly to the observer. This gives the illusion that the stick is bent at the water–air interface.

Within Earth, in addition to the direct-wave pulse, refracted seismic rays generate another pulse at the monitoring receiver. The refracted pulse can arrive before or after the direct wave, depending on the wave velocities through the materials encountered and the distances traveled. For example, when an explosion is detonated at Earth's surface, the first seismic pulse may not be the direct one. Surprisingly, a pulse that dives into Earth (Fig. 5.7) and travels at higher velocities might arrive first. It is similar to different traffic routes across a town. One driver might take the most direct route and travel side streets to reach a given destination. A second driver might choose a longer route that involves driving faster on a freeway. This second person—traveling the longer route—may arrive first as a result of the time saved on the freeway.

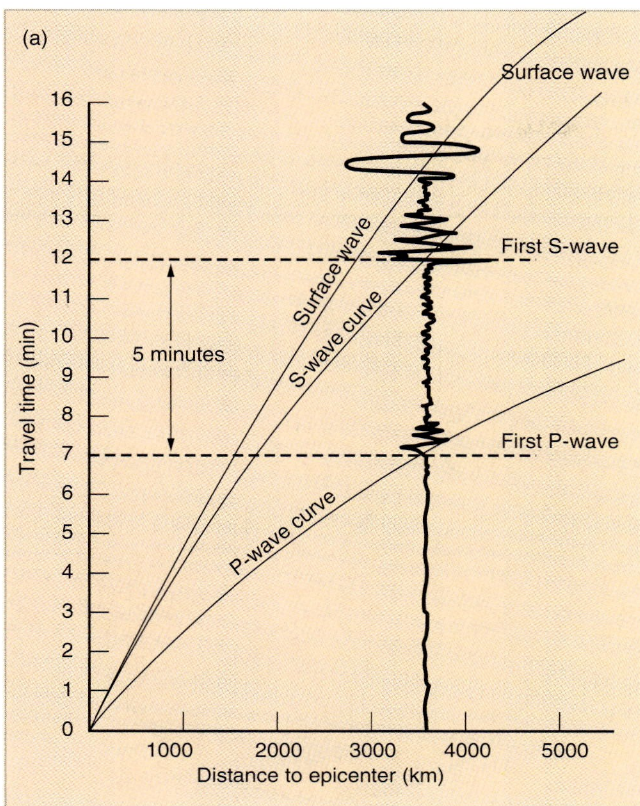

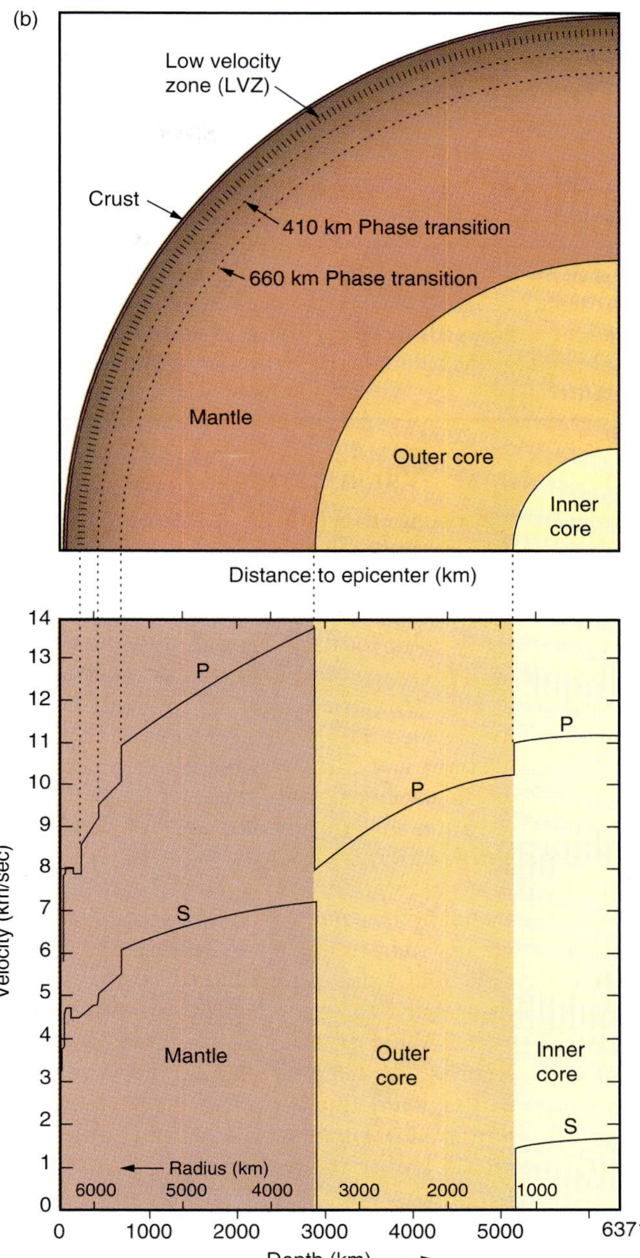

FIGURE 5.8 (a) A curve of the travel times of seismic waves as a function of distance to the epicenter of an earthquake. An example of a seismogram recorded on a seismometer that was located 3700 km from this earthquake is overlaid onto the graph. By timing the arrivals of the P, S, and surface waves at this station and at many others, the travel-time curve is constructed. (b) Both P-wave and S-wave velocity variations as a function of depth are determined from travel-time curves such as that shown in (a). Note that the S-wave velocity is zero in the outer core, which indicates that it is liquid; in contrast, S-wave velocity does have a value in the inner core, suggesting that the inner core is solid. Compare this with the density profile in Figure 5.2.

By successively timing seismic pulses over greater and greater distances from a given source, a **travel-time curve** is constructed (Fig. 5.8a). The shape of a travel-time curve can be used to determine wave-velocity variation as a function of depth (Fig. 5.8b). If this plot is a straight line, it indicates that the wave velocity is constant with depth. A curve indicates that the seismic-wave velocity changes with depth. In practice, both P and S travel-time curves are analyzed, to obtain P and S velocities, respectively.

Reflection If a seismic wave encounters an abrupt change in the medium through which it is traveling, such as at the interface between sediments and bedrock or between the solid mantle and the liquid core, some wave energy is reflected; the remainder may pass through the interface and may be refracted. The path of the reflected energy—the **reflection**—is described by a reflected ray (Fig. 5.9). The angle at which the wave reflects off the surface equals the angle at which the wave initially strikes that surface; the angle of reflection equals the angle of incidence. Reflected waves are distinguished from refracted waves by the shapes of their travel-time curves (Fig. 5.9c). By recognizing reflected pulses in seismic signals, seismologists are able to identify reflecting surfaces that separate different compositional layers within Earth, such as the core, as shown in Figure 5.9b.

We have seen that both refracted and reflected waves can be used to explore Earth's interior. Seismic pulses generated by an earthquake may arrive at a receiving station by various paths, which take different amounts of time. These include the direct path, refracted paths, and reflected paths. Some important paths in Earth include reflections and refractions from the base of the crust, reflections from the surface, and reflections from layers within the mantle (at the 410-km and 660-km depths of the phase transitions), and reflection from the outer liquid core (see Fig. 5.8). Each of these paths has its own travel-time curve.

Energy from a wave source spreads out in all directions. A direct ray traces energy that travels directly from the source; a refracted ray traces energy that follows a curved path; and a reflected ray represents energy that bounces off an interface.

Travel Times Within Earth

Early in the twentieth century, seismologists Harold Jeffreys and Keith Bullen compiled travel times for seismic waves traveling to monitoring stations around the globe. They found that travel time was almost independent of the geographic location of either the source or the receiver, depending only on the angular distance around Earth from the earthquake source to the station. This measurement is called the *epicentral angle* (Fig. 5.9a), and it is the angle formed at Earth's center by the arc between the source of the earthquake and the seismic station.

The observation that travel times are independent of the geographic location of either the origin of the seismic wave or the seismic station was a remarkable demonstration of the lateral homogeneity of our planet. Even though seismic waves may take as much as 20 minutes to reach a seismic station from a distant earthquake, the variability in travel times around the globe for different earthquake-station pairs, separated by the same distance, is usually only a few seconds. The main variation in the wave velocity within Earth is a function of depth. The P-wave velocity increases from a few kilometers per second in the upper crust to nearly 14 km/sec in the deep mantle.

Travel-time curves express the time it takes for a seismic wave to travel from its site of origin to a receiving station as a function of the angle these two sites make at Earth's center. These curves are used to find the velocity of seismic waves within Earth and to determine the variation of composition with depth.

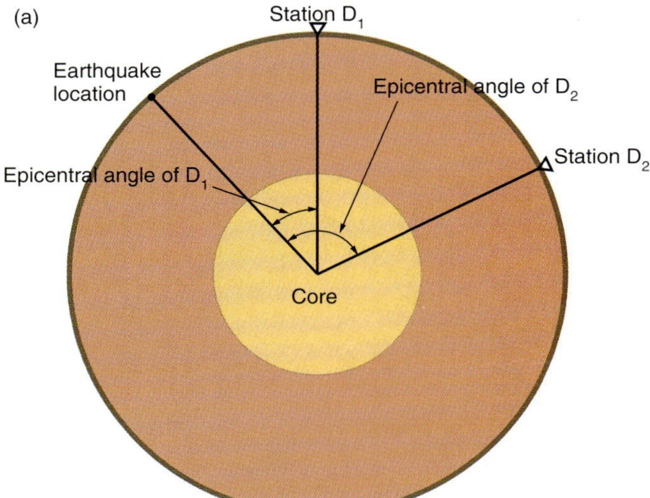

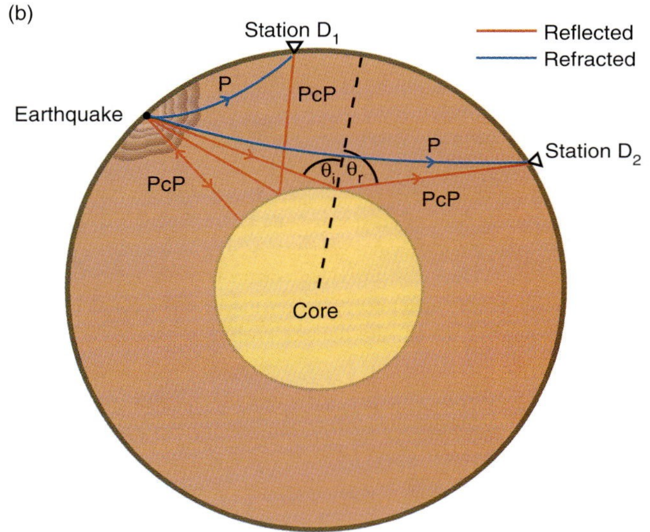

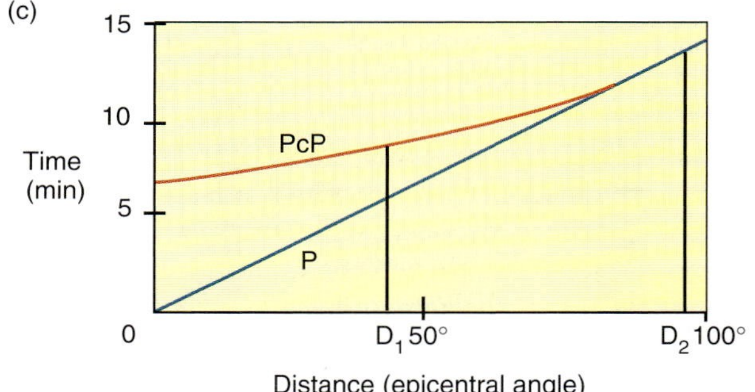

FIGURE 5.9 (a) The epicentral angle is the angle made at Earth's center by the arc on the surface that joins a seismic detecting station and the source of an earthquake or other seismic event. (b) Seismic waves reflect when they encounter a layer of material with contrasting mechanical properties. The reflected wave from the core is called *PcP*. The angle of incidence (θ_i) equals the angle of reflection (θ_r) . The direct P wave is refracted, or bent. (c) Travel-time curves for P waves within Earth showing refracted and reflected waves. A seismometer placed at the seismic source detects a P wave that travels from the source directly down into Earth, reflects off the core boundary, and retraces its path to arrive back at the source eight minutes later. At a distance D_1 from the source, a refracted P wave takes 7.7 minutes to arrive at the receiver, but a reflected P wave *(PcP)* travels a longer distance and arrives 2.0 minutes later. As the angle of reflection off the core boundary becomes very large, the paths traveled by refracted and reflected waves become virtually the same, as do their travel times.

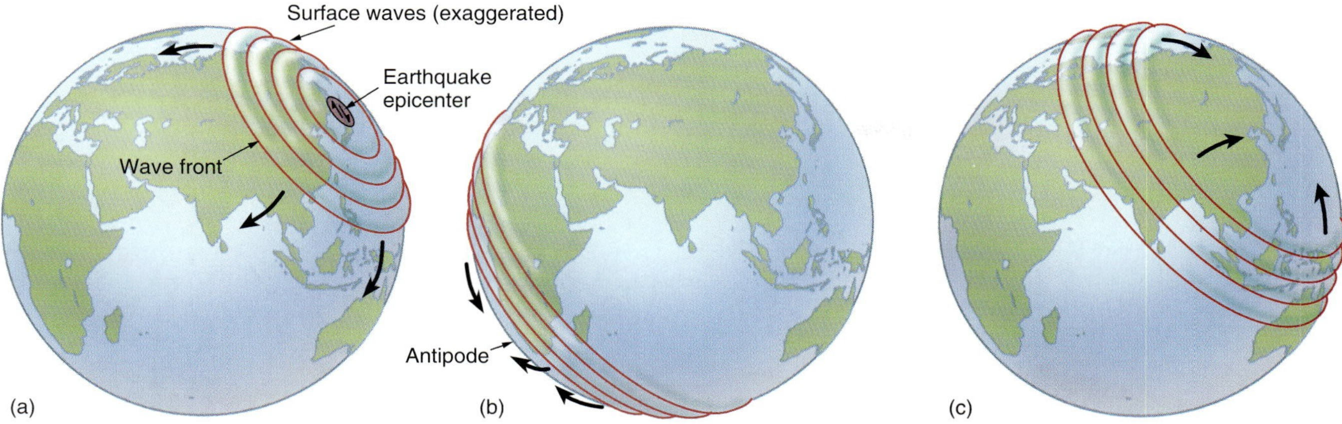

FIGURE 5.10 Surface waves from an earthquake spread out across Earth's surface like ripples in a pond. (a) Surface waves from the earthquake radiate out from the epicenter. (b) Surface waves reach the antipodal point from the epicenter, where they cross over one another. (c) After crossing at the antipode, surface waves complete the circumnavigation, the first of many such cycles. These waves circumnavigate the globe many times, eventually building up an interference pattern with each other that is referred to as a *normal mode* of Earth. Many different interference patterns, or normal modes, are possible.

Normal Modes

We have learned that the velocity of seismic waves traveling through Earth is determined from their travel times. We noted earlier that velocity is equal to the square root of the elastic modulus divided by the density. Therefore, if we want to distinguish the variation in density from the variation in the elastic modulus, we must have additional information. Normal modes provide this information.

Consider taking a sping, and attaching a mass to it, such as a tennis ball, and setting the system into motion. Then, time the oscillations of the ball while it is suspended from the spring. If we were to replace the tennis ball with a baseball (with the same volume), we would observe that the increased mass—and therefore density—of the baseball results in a lower frequency of oscillations. This example shows that the frequency of wave oscillation can be used to distinguish between two oscillating systems that have different densities but that share the same elastic modulus. This mode of oscillation is called a **normal mode**. It is the natural resonance of the spring-ball system where the natural frequency provides us with information about density. Normal modes exist for all elastic bodies, including Earth, but how do we get Earth to resonate?

Earth resonates after a large earthquake. Surface waves travel out from the earthquake and circumnavigate the globe numerous times (Fig. 5.10). Body waves pass through Earth and reflect from its surface and internal boundaries. Because waves radiate in all directions, surface waves circling the globe in one direction interfere with waves traveling in the opposite direction. Reflected body waves interfere with incident body waves. Soon, a vibrating pattern is established similar to the oscillatory pattern of a ringing bell or a plucked guitar string.

When Earth vibrates, it changes shape. The longest period of Earth's oscillation is called the "football mode"; it has a period of 54 minutes (Fig. 5.11a). Earth initially deforms into a football shape; 27 minutes later, it changes into a flattened sphere. It then changes back into a football after another 27 minutes. This cycle repeats many times.

Another simple vibration pattern exhibited by Earth is the radial mode, or "balloon mode," which causes the planet to swell every 20 minutes like an inflated balloon and subsequently to contract (Fig. 5.11b). This oscillation can be observed for months after the occurrence of an earthquake but it eventually dies out. Many different vibration patterns—other than the football and balloon modes—are also generated, including one in which the two hemispheres twist in opposite directions.

Of course, the amount of deformation is minuscule. A point on Earth's surface might rise 10 cm above its original position for a large earthquake. We do not sense the motion without special instruments, because it is so gentle and everything that the eye can see is doing the same thing, so there is no stationary frame of reference.

Each vibration pattern is referred to as a normal mode of Earth. A *characteristic frequency* is associated with the oscillation of each normal mode. Over a thousand different normal modes have been identified. Just as with our tennis ball and baseball analogy, the frequency of oscillation depends on the mass of the region oscillating and on the elastic modulus through fairly involved mathematical relationships. These mathematical relationships have been solved to find the variation of velocity and density within Earth (see Figs. 5.2, 5.8).

Velocity Variations in the Crust

Earth's crust is made up largely of rock that has differentiated from the mantle by partial melting of mantle rock. Basalt is the most common type of rock in oceanic crust, whereas the continental crust has an average composi-

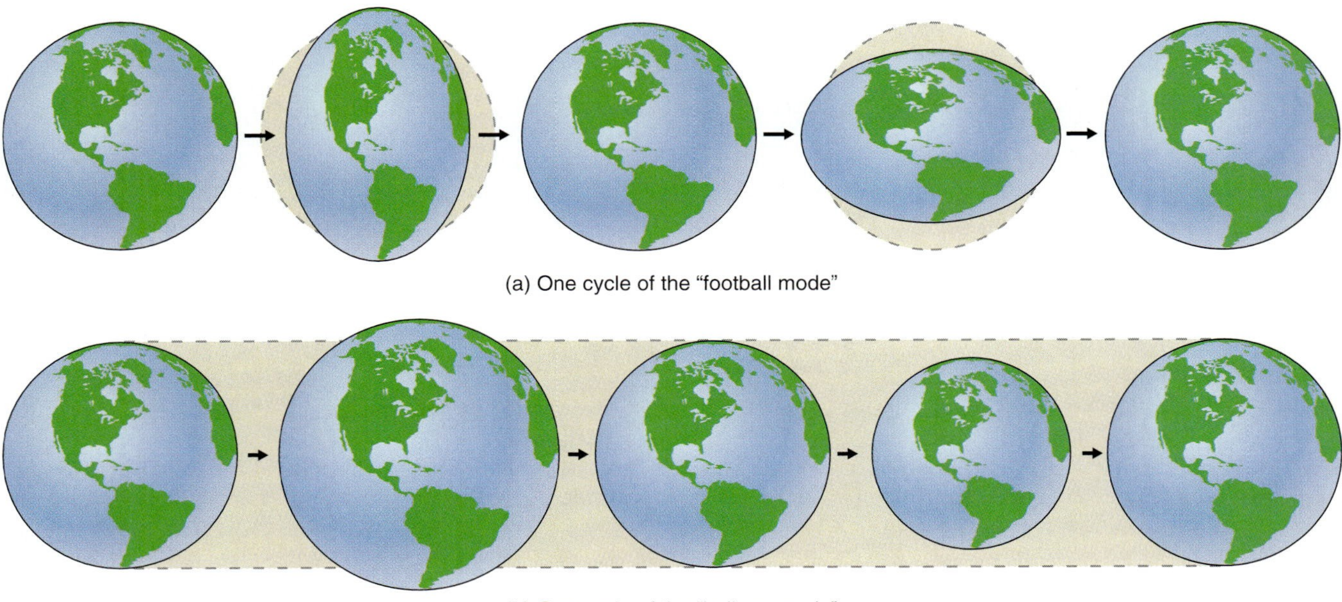

FIGURE 5.11 Two examples (greatly exaggerated) of possible normal modes of Earth. (a) The "football" mode is generated by very large earthquakes. It has a period of 54 minutes. (b) The "balloon" mode has a period of 20 minutes and oscillates for months after a large earthquake.

tion of andesite (∞ see Chapters 3,4). Although both rock types are ultimately derived from the mantle, they are lighter than mantle rock, and both P waves and S waves travel at lower velocities in these crustal rocks than they do in the underlying mantle rock.

The contrast between wave velocities through the crust and the mantle has been the key property in allowing seismologists to map variations in crustal thickness across the globe. The ocean crust has an average thickness of about 7 km, but it can reach a thickness of 24 km in elevated regions of the ocean floor, such as near Iceland. On the continents, crustal thickness averages about 30 km but may reach values as high as 70 km in continental mountain belts, such as the Himalayas and the Andes.

The lower boundary of the crust is known as the Mohorovičić discontinuity, or the **Moho**, discovered in 1909 by Andrija Mohorovičić. The crust has an average P-wave velocity of 6.5 km/sec; the mantle just beneath the crust has an average velocity of 8.0 km/sec. This sudden contrast in velocities between the two layers causes seismic waves to reflect off the Moho. It also causes some waves that reach the Moho to refract, or bend, so severely that they travel along the Moho and subsequently refract back into the crust and return to the surface (Fig. 5.12). The thickness of the crust in a given region is determined using travel-time curves of waves that reflect off, or refract from, the Moho.

Velocity Variations in the Mantle

The Low-Velocity Zone We have remarked that seismic-wave velocities generally increase with increasing depth. At depths greater than 100 km in the oceans and about 150 km on the continents, however, this trend reverses for an interval of 100 or 200 km (see Fig. 5.8). The depth interval at which seismic waves decrease in velocity is known as the **low-velocity zone.** It is at these depths that the temperature approaches the melting point of mantle rock. Low wave velocities may be due to partial melting of the rock in this region; alternatively, high temperatures in the region where the geotherm lies close to the solidus (Fig. 4.12) may be sufficient to decrease the velocity of seismic waves. The low-velocity zone is thought to be a zone of decoupling between the motion of the lithospheric plates and the deeper mantle; thus, it corresponds to the asthenosphere.

Mantle: Phase Transitions Beneath the low-velocity zone, seismic-wave velocity in the mantle increases steadily (see Fig. 5.8b). This increase, however, is interrupted by sudden jumps in wave velocity that are detected at the two major mantle discontinuities; these velocity jumps are detected at depths of 410 km and 660 km (see Fig. 5.8b) and are associated with phase changes (or polymorphic changes). (∞ see Chapter 3).

Seismic Velocity Variations in the Core

Below the low-velocity zone, P-wave velocities in the mantle increase steadily, which causes P waves to refract upward. However, across the core–mantle boundary, the P-wave velocity drops by 40 percent. This contrast causes the waves to refract downward upon crossing the boundary. The result is a shadow zone in which no direct P waves are received (Fig. 5.13). Small amplitude (diffracted) P waves are, however, detected in the shadow zone.

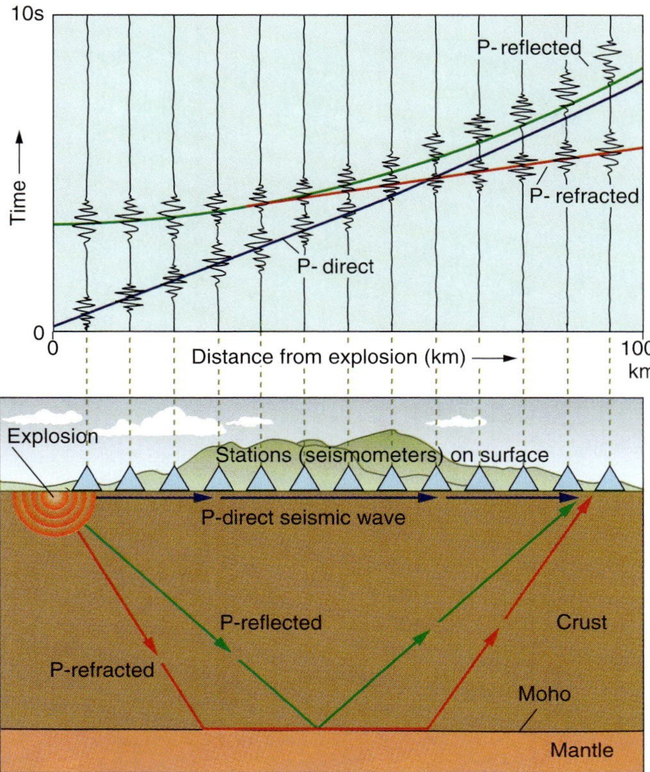

FIGURE 5.12 Three important pathways for seismic waves in the crust. A direct wave travels directly between the seismic source and the receiver; a reflected wave bounces off the Moho, which is the boundary between the crust and the mantle; a refracted wave is bent at the Moho to travel along this interface, and this wave then bends again to reenter the crust, where it travels to the receiver and may actually arrive before the one that traveled through the crust. Seismogram in which each wave type is identified is shown at top.

An even larger shadow zone is observed for S waves, which do not pass through the liquid outer core. This observation provides evidence that the outer core is liquid and therefore has a shear modulus of zero. The S waves reflect strongly off this boundary. Measurements of Earth's normal modes have indicated that shear waves are present in the inner core, which have been generated by P waves passing from the outer core.

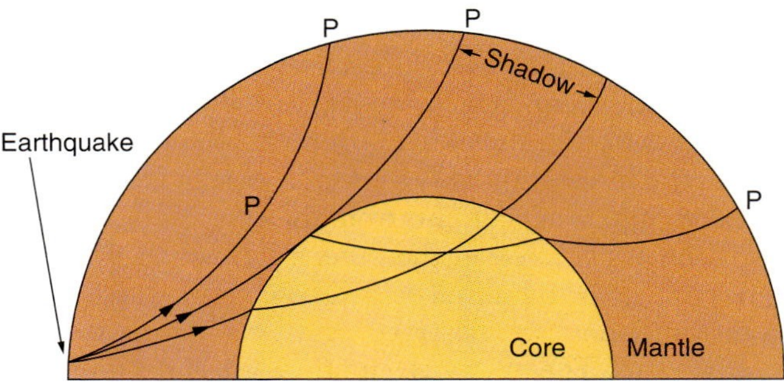

FIGURE 5.13 The P-wave shadow zone is bounded by P waves that graze the core and a P wave refracted by the core, as shown. The P wave that enters the core at an angle slightly greater than grazing incidence is refracted (downward) by the core's low velocity, which gives rise to the shadow.

Seismic-wave velocities increase with depth suddenly at the crust–mantle transition and then decrease at the base of the plates. The absence of S waves in the outer core and their inferred existence in the inner core from normal modes suggest that the outer core is liquid and the inner core is solid.

Seismic Tomography

In the field of medicine, a computer can generate a three-dimensional image of the internal structure of a human head. This is accomplished by transmitting information to the computer from X-ray detectors placed around the head. Seismic tomography performs essentially the same task for geology—generating a three-dimensional image of Earth's interior. This technique uses earthquakes or explosions as transmitters and seismometers as detectors (Fig. 5.14a). Rather than obtaining a one-dimensional velocity profile through Earth as shown in Figure 5.8b, we can build a three-dimensional seismic velocity image of the interior of Earth.

Below the crust at any given depth, Earth is fairly homogeneous. In fact, we can compare it with an onion in which each layer has an average seismic velocity, and these velocities differ from layer to layer. Seismic tomography searches for the smaller, more subtle variations within each layer, called *velocity anomalies*—localized regions in Earth having velocities different from the average values at those depths. In the uppermost layer—the crust—there is a major departure from the "onion Earth" model, which is based on layers of equal thickness. Recall that the ocean crust is only 7 km thick, whereas the continental crust is 30 km thick. Other anomalies occur where oceanic plates subduct, the cold material generates high-velocity anomalies. Beneath mid-ocean spreading ridges, hot upwelling mantle generates low-velocity anomalies relative to cooler surroundings.

To illustrate seismic tomography, imagine there is a velocity heterogeneity in the mantle, caused, for example, by a mass of subducted lithosphere. Seismic waves passing through this high-velocity region will be detected sooner than will waves that do not pass through it. Arrival times at monitoring seismometers will depend on the location of the high-velocity region relative to both the source of the waves and the receiving stations.

After travel times from earthquakes around the globe are measured at all the stations worldwide, a pattern starts to emerge. Anomalous arrivals are identified as ones with travel times significantly different from those expected from the average Earth model. The anomalous pattern is found to change position depending on the direction of the incident seismic waves (Fig. 5.14b). This effect is similar to the moving shadow of an object when it is illumi-

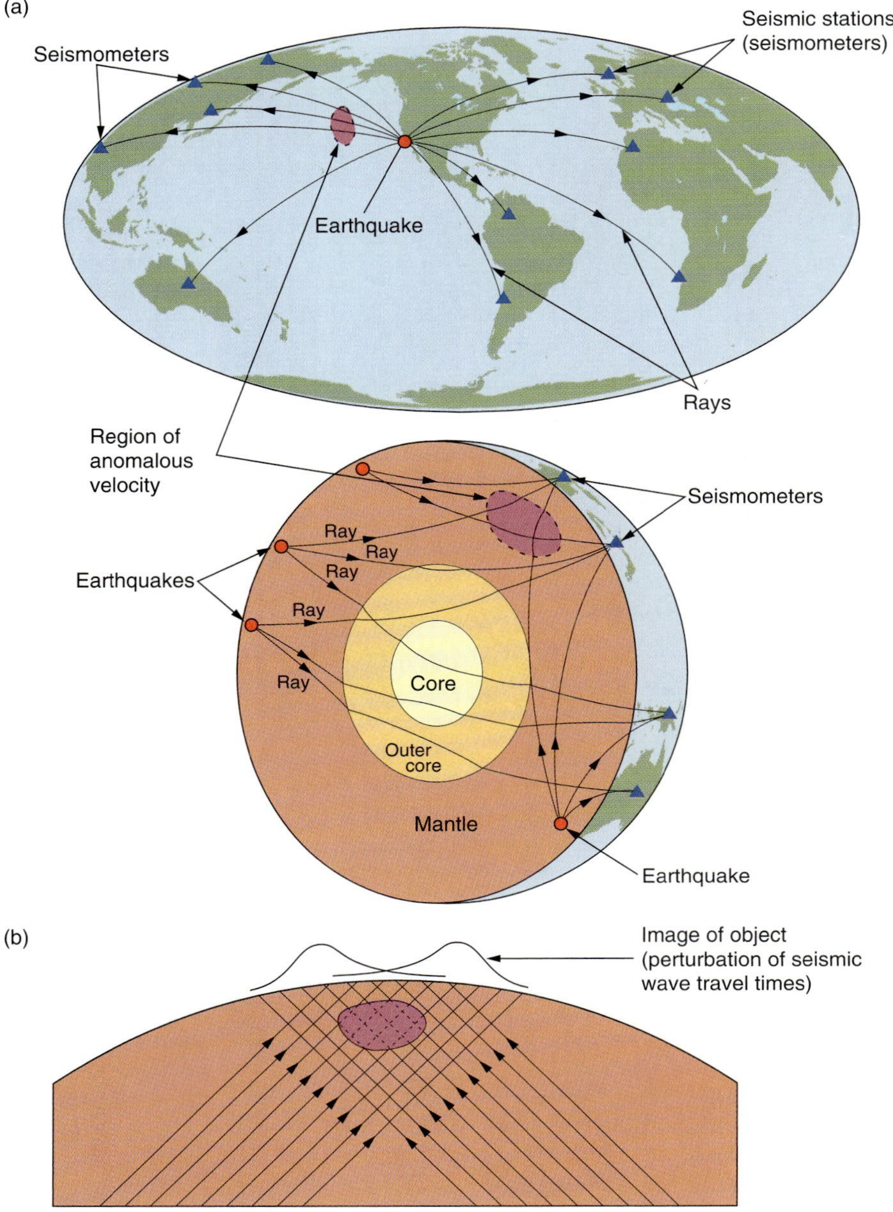

FIGURE 5.14 (a) Seismic tomography constructs images of Earth's interior using seismometers stationed around the globe to detect the seismic waves from earthquakes. Behavior patterns of seismic waves as they travel to seismometers are used to detect a heterogeneity in material at depth within Earth. (b) Like a shadow, the apparent location of the image depends on the angle of the incoming rays and the actual position of the object. After enough signals have been measured from different earthquakes, the position and shape of the heterogeneity can be obtained.

nated from different directions. Computer programs then determine the location of the velocity anomoly.

In the crust, low-velocity anomalies are associated with heat and magma beneath volcanoes (Fig. 5.15). Low-velocity anomalies are also associated with crust that has been intruded by hot mantle rocks, such as in continental rift zones. Seismic tomography experiments have been carried out on a small scale between boreholes to map oil reservoirs.

In the mantle, low-velocity anomalies are associated with both continental rift zones (Fig. 5.16) and mid-ocean spreading ridges (∞ see Chapter 8) and upwelling convection currents (Fig. 5.17). High-velocity anomalies are associated with present and ancient subduction zones (Fig. 5.17). Much of the variation in seismic-wave velocity occurs in the upper 660 km of the mantle. Below this depth, the mantle variation is relatively gradual. At the core–mantle boundary, however, a 200-km thick zone shows remarkable lateral heterogeneity, which is probably due to the interaction of the lower mantle, with heat provided by upwelling currents from the core. Little velocity variation is observed in the outer core, which is consistent with its being a well-stirred convecting fluid. Inner core variations are observed with maximum velocity north–south. The observed variations are consistent with its being a crystalline solid.

The behavior of seismic waves clearly provides useful information on the density and material properties of Earth's interior. The gravitational field of Earth can also be used to infer internal density differences.

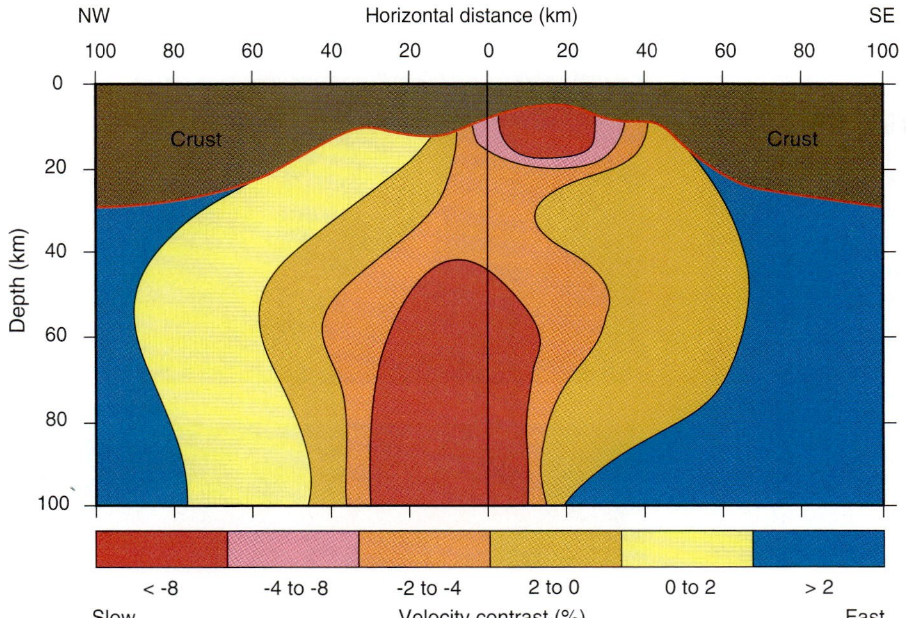

FIGURE 5.15 Seismic tomography has revealed a region in the mantle beneath a hot spot in Wyoming's Yellowstone National Park through which seismic waves travel at low velocities. This area is thought to be associated with a mantle plume that generated the volcanism.

GRAVITY AND ISOSTASY

Gravity

How are large-scale geologic features related to gravity? Why does continental crust rise above sea level, whereas oceanic crust is submerged? Why do some plateaus sit several kilometers above sea level when the mean elevation of the surrounding land is less than a kilometer? We can answer these questions if we first study some basics of Earth's gravity.

Any two masses exert a force of gravitational attraction on each other. The magnitude of the force is proportional to the product of the masses; it is inversely proportional to the square of the distance between them. The force of attraction between your body and the mass of Earth keeps you on Earth's surface. But did you realize that gravity "grounds" us more effectively in some places than in others? Precise measurements of variations in this force on Earth impart information about the distribution of mass beneath us.

The force of gravity at a point on Earth's surface is measured by determining the extent to which a weight stretches a spring at that point. For example, when you stand on a scale, the scale measures the amount of gravitational force exerted on your body by Earth: your weight.

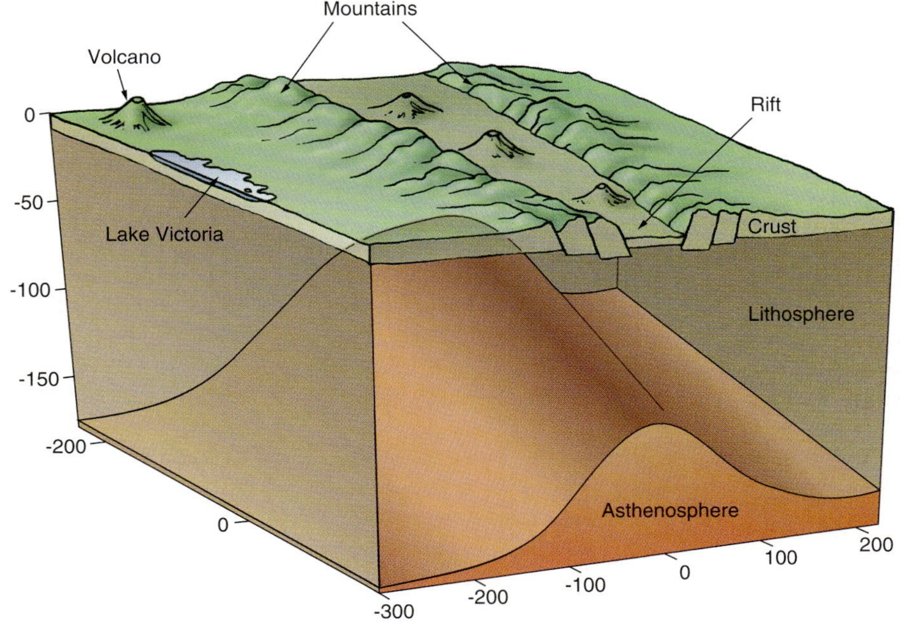

FIGURE 5.16 Seismic tomography in East Africa has produced images of the processes associated with continental breakup. Beneath the Red Sea and the East African Rift system, low seismic-wave velocities are found in the upper mantle, where melting takes place and basalt flows to the surface to create new ocean floor.

If you were to move that scale to the North Pole and the equator, you would find that, even though your mass (your "amount of matter") remains the same, you would weigh about 0.5 percent *more* at the North Pole than at the equator. If you were to take the scale to the top of Mt. Everest, you would weigh 0.1 percent less than you do at sea level. Thus a 200-pound person residing at the equator weighs 201 pounds at the pole; a 200-pound person living at the base of Everest would weigh 199.8 pounds at the top, excluding weight loss from the exertion. Instruments called **gravimeters** measure the force of gravity precisely and are based on the same principle as the scale.

The force of gravity is defined as force = mass × g, where g represents the acceleration due to gravity—the rate at which an object speeds up as it falls. Acceleration is a measure of the change in velocity per unit time. Velocity has units of meters per second (m/sec). The units used to describe the rate at which velocity changes are meters per second per second, abbreviated m/sec^2. On average, g is 9.8 m/sec^2, but it has small variations that are of interest to scientists because they tell us about density variations beneath the surface (Fig. 5.18).

A gravimeter uses a standard mass to measure g (gravity). We quote our weights in kilograms, which is not strictly correct; mass is measured in kilograms but weight is a force and depends on g. Because g is nearly the same value at every point on Earth's surface, we do not usually bother to calculate the product of mass × g to obtain our actual weights. On the Moon or Mars, however, g is considerably less, because they have less mass than Earth, and thus you would weigh much less there than on Earth (1/6 on the Moon; 1/2.6 on Mars).

Variations in gravity are expressed in units of milligals (mgal). One gal, named after Galileo, equals 0.01 m/sec^2, which is equivalent to 1 cm/sec^2. Because g is equal to 980 gals (9.8 m/sec^2), 1 milligal is about one-millionth of g.

A gravity survey across a sedimentary basin might find a gravity variation corresponding to a decrease in g of about 10 milligals because the sediments have lower density than the surrounding rock. Thickness variations in the outer layers of Earth cause gravity variations because the density of one layer is replaced by the density of the other. For every kilometer of variation in crustal thickness, the gravity change is about 20 milligals. Lithospheric thickness changes of over 100 km can cause variations as large as 200 milligals.

Gravity is the force of attraction between any two masses. Gravity variations give information on density variations beneath Earth's surface.

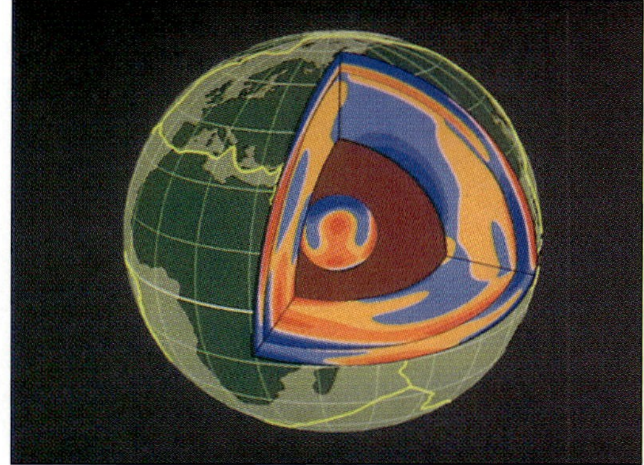

(a)

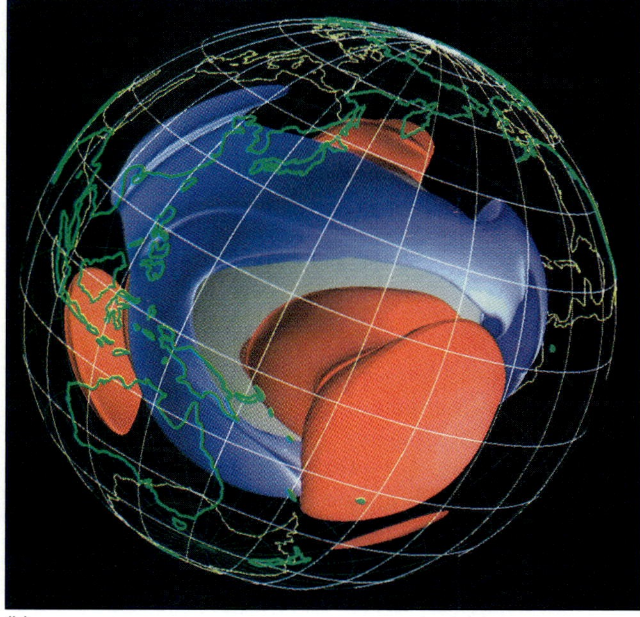

(b)

FIGURE 5.17 (a) Global seismic tomography. Red colors indicate slower velocities and hotter mantle; blue colors, faster velocities. Note that the outer core has a homogeneous velocity consistent with its being a homogeneous fluid stirred thoroughly by convection. The inner core is a heterogeneous crystalline solid. Low velocities lie beneath the oceanic ridge as well as deep beneath South Africa. Low velocities are also seen in the upper mantle beneath the East African Rift system. (b) Tomographic model of the mantle.The red represents slow regions (0.35%); blue, fast (+0.35%). A ring of fast velocities surrounding the Pacific is though to be due to subducted cold material. In the central Pacific, a plumelike low velocity volume extending from the core–mantle boundary is seen. A second plume under South Africa is visible on the left. Subduction beneath Asia appears to have cooled the mantle, as revealed by high velocities.

Free-Air Gravity

A gravity map provides contours indicating the strength of gravity as a function of position. These maps differ depending on the nature of the corrections made to gravity readings taken at survey points. Several types of correction can be made: latitude corrections, elevation corrections, and Bouguer corrections (pronounced boo-gay).

The latitude correction takes Earth's spin and flattened shape into account, which causes the strength of gravity to increase in either hemisphere as one approaches a pole. The elevation correction is made to estimate gravity values at a constant height above Earth's surface. Because the force of gravity falls as 1/(distance between two objects)^2, an object elevated above Earth's surface will experience less gravity than an object at the surface. Gravity measurements made at a range of heights, such as on hills and in valleys (Fig. 5.18b), would be very difficult to interpret because many of the differences are due to elevation effects. Gravity maps that have been corrected to show values at a constant elevation are known as **free-air gravity** maps (Fig. 5.19). By convention, the constant elevation displayed in these maps is taken to be sea level, and mathematical calculations correct measurements made on land to appropriate values at sea level. Gravity measurements made aboard a ship are automatically free-air gravity readings because they are actually made at sea level.

Bouguer Gravity

The **Bouguer gravity map** corrects for topography and density variation at Earth's surface; it shows gravity measurements at a uniform elevation as if all mountains had been scraped off Earth's surface and all valleys were filled with material of the same density as the surroundings. A Bouguer gravity map allows us to investigate mass variations at depth. Any remaining gravity variations—anomalies—on this type of map are either positive, caused by mass concentration underground (see Fig. 5.18c), or negative, caused by mass deficit underground.

The Bouguer gravity map of North America is shown in Figure 5.20a, with a coast-to-coast profile depicted in Figure 5.20b. Crossing the continent from west to east, the gravity steadily decreases by about 250 mgal to a continent-wide low just north of the Colorado–New Mexico border before it increases again to the east. This negative anomaly is caused by a combination of thick crust and thinning of the lithosphere. Where lithosphere has been replaced with hot asthenosphere, the lower density of the hot asthenosphere results in a mass deficit compared with that of the surrounding lithosphere.

Gravity measured at a constant height is known as free-air gravity. Bouguer gravity shows gravity measurements as if all mountains were scraped off and all valleys were filled with material of the surrounding density.

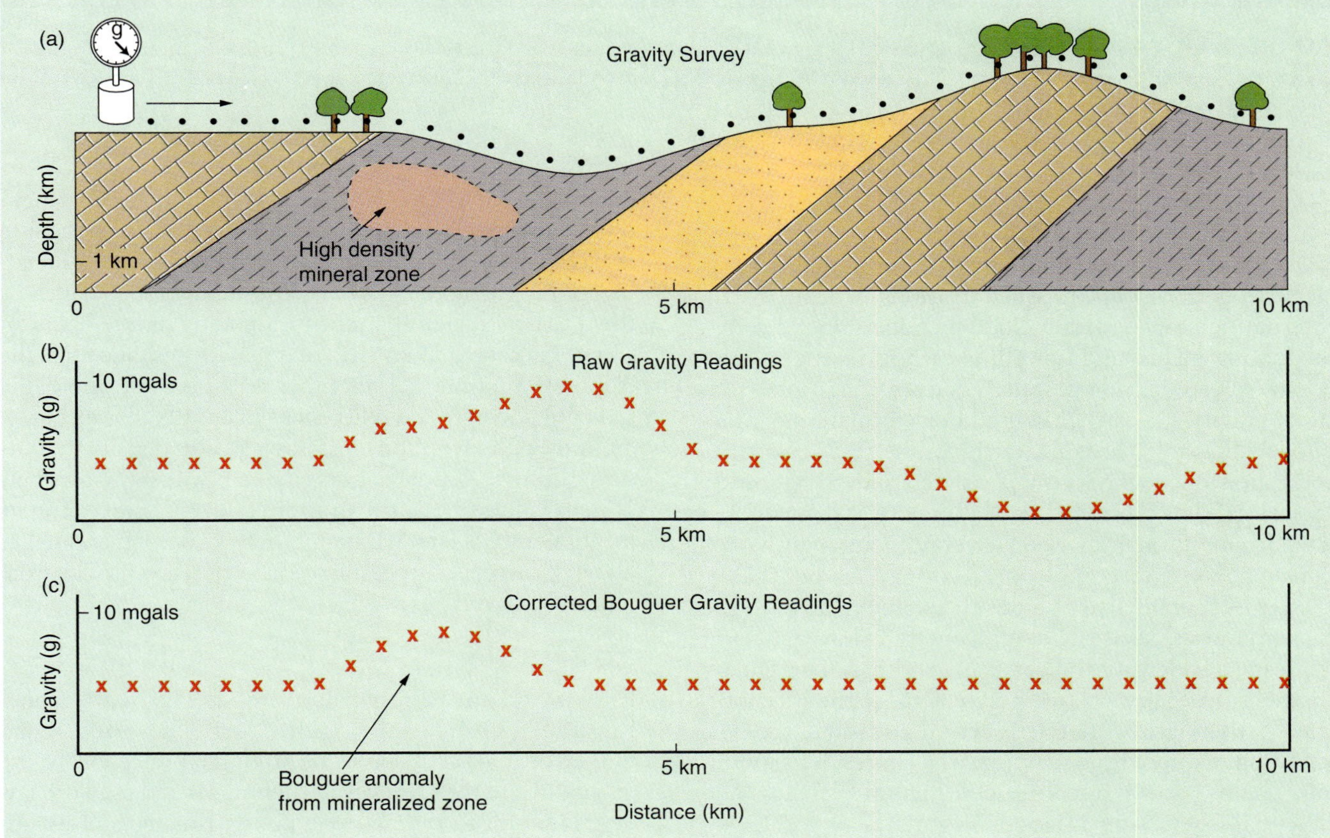

FIGURE 5.18 Changes in Earth's gravity due to topographic effects and density variations. (a) Taking a gravity survey, gravity values are read on the portable gravimeter, and elevations measured at each station (dots); (b) raw-gravity readings; (c) corrected gravity for topography (Bouguer gravity values) isolates the gravity anomaly caused by a buried high-density object such as a metal ore deposit.

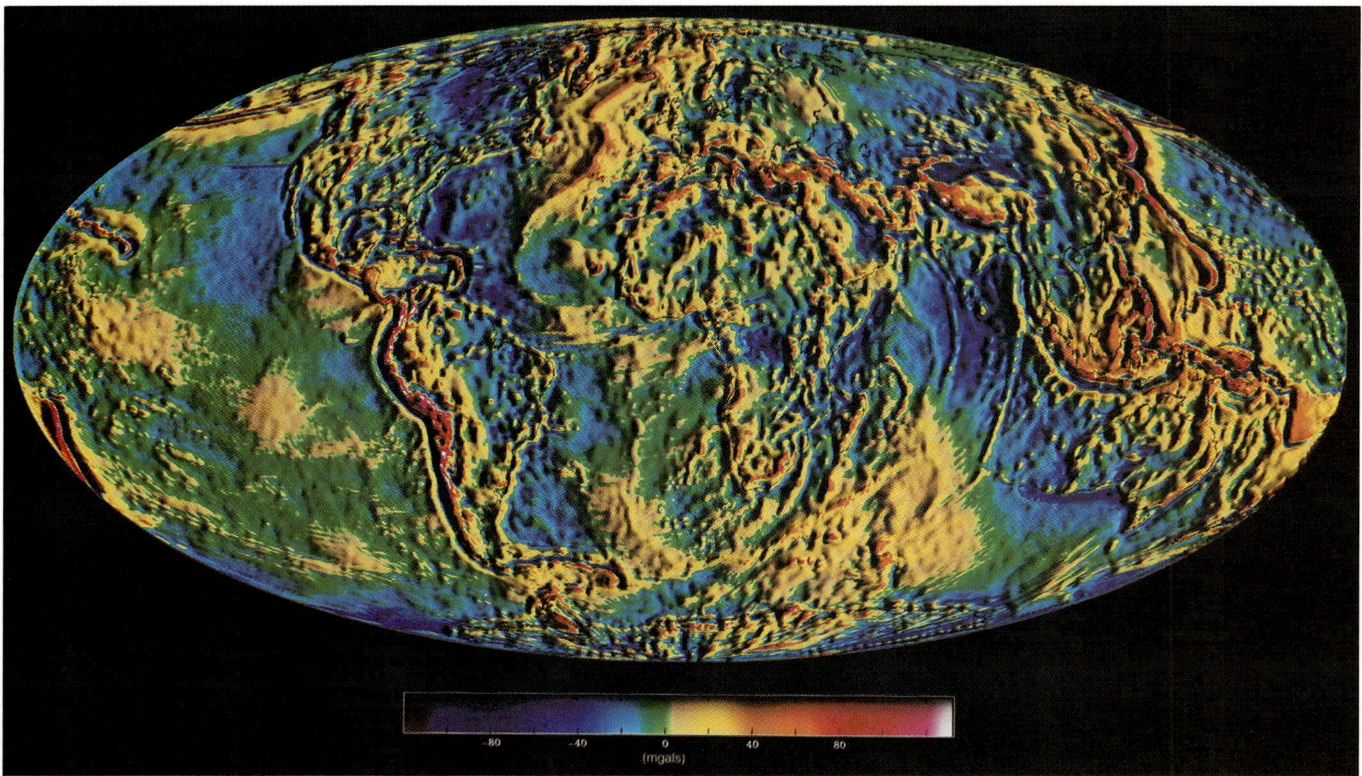

FIGURE 5.19 A global free-air gravity map. High free-air gravity anomalies (red) occur over regions of high elevation due to the excess mass in such regions. These anomalies are not as large as expected, and suggest that many mountainous regions are underlain by low-density "roots."

Gravity Anomalies

Variations in the force of gravity detected on Earth's surface depend on mass concentrations within Earth. Internal heterogeneities include metallic mineral zones, subducting plates, and upwelling of hot mantle at mid-ocean spreading ridges. Small-density variations that affect gravity can also be caused by deformational structures such as folds or faults in the crust. Such variations contoured on a gravity map are referred to as *gravity anomalies*: They are called free-air gravity anomalies on a free-air gravity map or Bouguer gravity anomalies on a Bouguer gravity map.

A gravity anomaly has a positive value relative to the background values if the heterogeneity below this point has a greater density than its surroundings. The gravity anomaly is negative above hot regions of the asthenosphere, where the material is less dense than the surrounding mantle. Topographic variations such as mountain chains, ocean trenches, continental plateaus, and valleys also generate variations in gravity.

We might expect a free-air gravity map to show a large positive anomaly above a mountain range (Fig. 5.19); however, the anomaly is often quite small. This can be explained if there exists a deficit of mass under the mountain range that compensated for the excess mass of the mountain. We will see in the next section that such deficits do indeed exist. Mountains and plateaus can be thought of as blocks of crust floating in the mantle. Like icebergs floating in the sea, mountains have low-density roots that cause them to float higher than the surrounding surface of Earth.

Bouguer anomalies are of great importance in the exploration of natural resources because these anomalies indicate underground differences in density. For example, an iron-ore body in the crust has a greater density than its surroundings. A Bouguer gravity map will show contours of high values above the iron-ore body, thus revealing its location and its lateral extent. Analysis of such anomalies allows prospectors to estimate the potential yield from mining such an area.

Isostasy

Early in the nineteenth century, the English Astronomer Royal, Sir George Airy, carried out The Great Land Survey of India. His surveyors attempted to calculate the gravity due to the Himalayan Mountains and used the law of gravitation to estimate the pull of this mass of mountains on a plumb bob (a suspended weight) (Fig. 5.21). The surveyors were greatly surprised to find that the measurements gave significantly lower numbers (seventy times less) than expected from applying the law to the protruding mass of the mountains. Airy surmised that the expla-

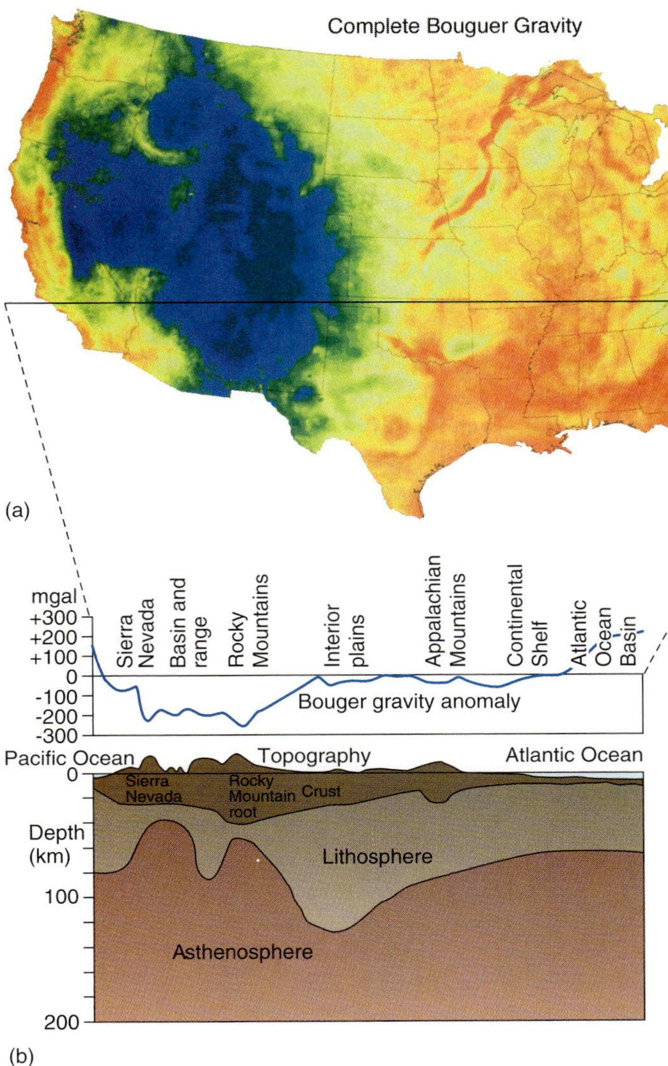

FIGURE 5.20 (a) The Bouguer gravity anomaly map for the United States of America. (b) A transcontinental profile of the Bouguer gravity from the Pacific Ocean to the Atlantic Ocean. The negative values over the mountainous regions in the west are due to thick crust and hot, low-density mantle (asthenosphere) associated with recent tectonic activity; compare this with the cooler cratonic mantle (lithosphere) in the east. The elongated anomaly running northwest–southeast in the north central United States is the trace of the mid-continent rift (Chapter 8).

nation must be that the mountains were underlain by a low-density root protruding into the high-density mantle. This is one of the earliest formulations of the concept of **isostasy**: The great mass of a mountain that extends above Earth's surface is compensated by a low-density mass in the mountain's underlying crustal root.

We can understand the concept of isotasy by considering a simple model. Imagine several blocks of wood of different thicknesses floating in a tank of water (Fig. 5.22a). These blocks float because wood is less dense than water, and the thicker blocks protrude higher above the surface of the water than do the thinner blocks. At the same time, however, the thicker blocks also extend deeper into the water than do the smaller blocks; they are more deeply submerged to compensate for their larger size. We could think of the submerged part of a block as the root of that floating block.

Now, consider the vertical column of rock beneath a mountain. The weight of the rock applies a load to Earth at the base of the column. If such a load is applied for a sufficiently long time, it will cause the mantle below it (and in some cases the lower crust) to flow until a balance exists between the crustal load and flotation forces of the mantle. The amount of crust that extends into the mantle is called the *crustal root*. The crust has a lower density than the mantle, and the crust is therefore analogous to the lower-density wood floating in water; here, crustal blocks can be thought of as floating in mantle (Focus 5.2).

In the upper mantle, temperatures become sufficiently high for the rock to behave like a fluid over the geologic time scale. The depth at which this occurs is known as the *depth of compensation*. Material above this depth acts like a solid floating in the fluid mantle. In fluids at a given depth, the pressures are isostatic or equal. The pressure is given by the weight of the overlying rock; thus, *the principle of isostasy states that the weight (solid and fluid) per unit area is the same in any column of material above the depth of compensation*. Note, if g (gravity) is assumed to be constant, the pressure is proportional to the *mass* per unit area in the vertical column.

A floating block of wood or a mountain range is said to be in **isostatic equilibrium**: that it is neither sinking nor rising. Now imagine placing a block of wood in honey. Initially, the block will not be in isostatic equilibrium until sufficient time has passed for it to sink to the

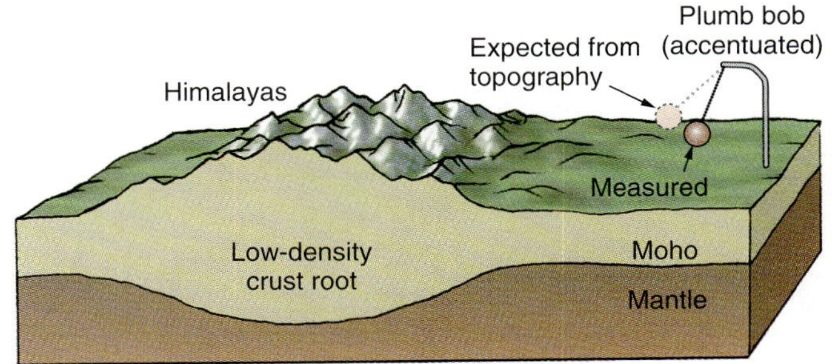

FIGURE 5.21 A cross section depicting the low-density crustal root that supports the elevation of the Himalayan Mountains. The gravitational attraction, measured by the deflection of a plumb bob (greatly exaggerated), was found to be 70 times less than calculated for the mountains alone, suggesting they must have a root.

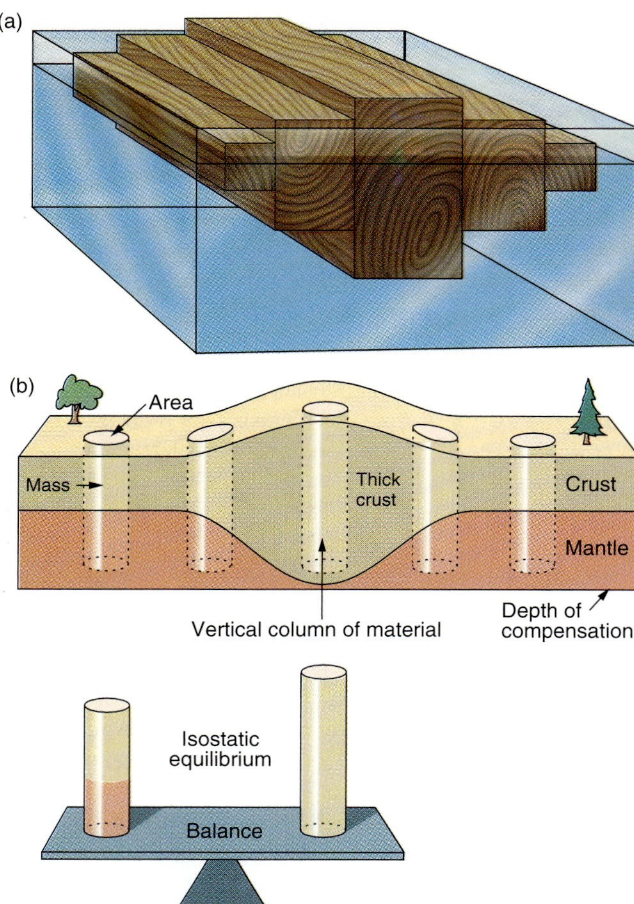

FIGURE 5.22 (a) The principle of Airy isostasy. The larger blocks of wood floating in the tank of water have a proportionately larger submerged "root," which supports the portion extending above the water's surface. The force of buoyancy depends on how much of the block is submerged. (b) The depth of compensation is the point within Earth where flow occurs to ensure that the pressure is the same, or isostatic, everywhere. The weight per unit area in each vertical column of material above this depth must be the same.

level of isostatic equilibrium. Likewise, a mountain may be either rising or falling, depending on whether its current elevation is below or above the elevation needed for it to be in isostatic equilibrium.

Another way to attain isostasy was first described by J.H. Pratt in 1854. Recall the floating-blocks-of-wood experiment, but now imagine the blocks to be composed of different woods with different densities. An extremely light balsa wood block would float the highest in the water, followed by a heavier block made from pine; the heaviest block—made from, say, teak—would not extend far above the waterline at all. The blocks vary in height such that their bases align horizontally under the water (Fig. 5.23).

From the principle of isostasy, the mass of each wood block must be the same. This is why the blocks made of lighter material are longer and float higher in the water. Pratt isostasy achieves isostatic equilibrium by a *lateral variation in density and a constant root thickness*; whereas Airy isostasy proposes *a lateral variation in root thickness and a constant density*. A useful mnemonic to remember the essential differences between the root structures of Pratt and Airy isostasy is "Pratt's is flat and Airy's varies."

Within Earth, where Pratt isostasy operates, the different levels of surface elevation are determined by a lateral variation in density. Such a compensation mechanism is thought to occur at mid-ocean spreading ridges, where lateral density changes are due to heat concentration in the upper mantle beneath the ridge axes. The underlying mantle is hot because it is rising from deep within Earth to fill the gap caused by the separating lithospheric plates. Rock expands when it is heated, which reduces its density. The hot rock under the ridge has a lower density than does the cooler rock on either side. As a result of isostatic compensation, the ridges have higher elevations than the surrounding ocean floor.

In Focus 5.2, we calculate the height of continents above sea level based on the principle of isostasy and crustal thicknesses determined from seismology. Continental crust has been measured to rise approximately 0.84 km above sea level, whereas the oceanic crust lies 3.80 km below sea level. Not all oceanic crust, however, lies beneath sea level. For example, the crust in Iceland—found to be 24 km thick, which is similar to the thickness of the continents—emerges above the surface of the ocean owing to its buoyancy. The Iceland hot spot has caused greater production of crustal rock than elsewhere along the Mid-Atlantic Ridge, where heating is not as concentrated. The lower density of the thick Icelandic crust and hot mantle supports the elevated topography.

The mass of a mountain above sea level is compensated by the mountain's underlying crustal root. A mountain is in isostatic equilibrium if it is neither rising nor sinking. The depth at which this isostatic adjustment occurs is called the depth of compensation.

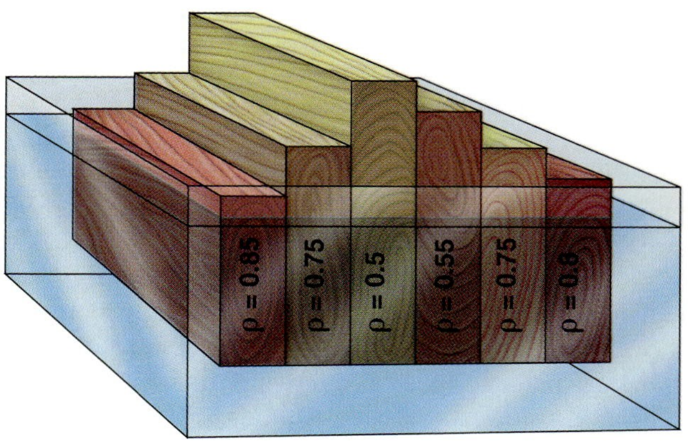

FIGURE 5.23 An example of Pratt isostasy in which blocks of wood of different densities float in water. Note that they are all submerged to the same depth.

Plate Flexure Not all topographic masses are compensated in the column of material directly beneath them. If the load is not of sufficient mass—a hill rather than an entire mountain range, for example—the stresses do not extend sufficiently far into the planet to regions where rock is hot enough to flow. The material is held up by the elastic forces of the upper region of the lithospheric plate.

It is generally thought that bodies more than several hundred kilometers wide are isostatically compensated because their stresses reach down several hundred kilometers to the point where the asthenosphere flows. At intermediate sizes (tens of kilometers), some bodies such as seamounts are supported by broad bending or flexure of the upper elastic part of the lithospheric plates (∞ see Chapter 7). The broad bending is isostatically compensated and is called *regional compensation*; that is, in contrast with Airy isostasy, the support is spread over a much larger region than the size of the load.

Geomagnetism

Like gravity, Earth's magnetic field depends on Earth's internal properties and so has provided a wealth of information on the interior of the planet; it has also been an aid in navigation.

The first compass was made of a lodestone—the mineral magnetite, which is one of the naturally occurring ores of iron—suspended on a string. In 1600, William Gilbert, physician to Queen Elizabeth I, explained how the compass works. It had been noticed that two compasses brought close to each other tend to align with one another so that north poles attract south poles. Gilbert proposed that the whole Earth is a big magnet, the field of which acts on the magnet of the compass needle to align it in a north-south direction (Fig. 5.24a).

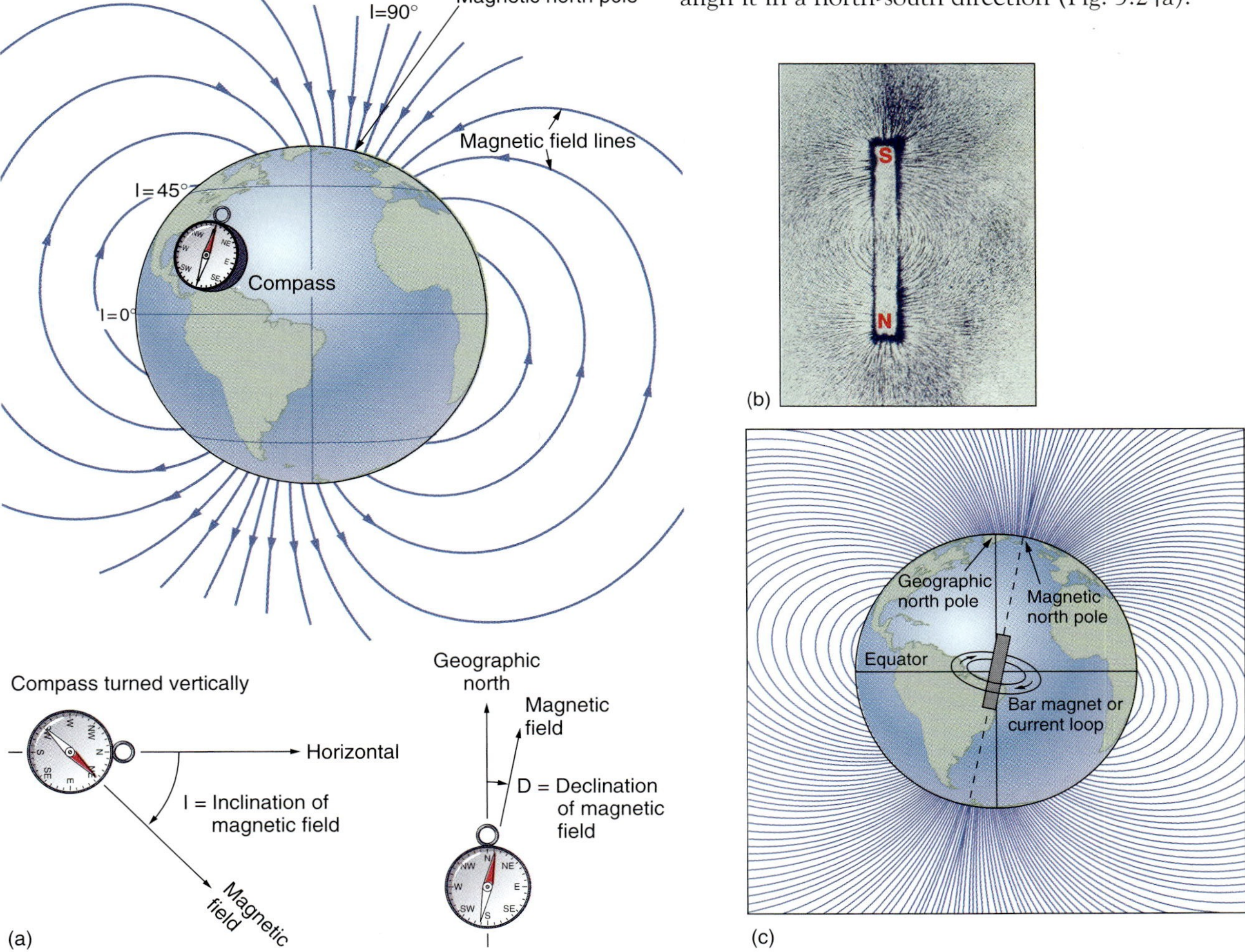

FIGURE 5.24 (a) A compass is a suspended magnet. Its north pole aligns with Earth's magnetic field and indicates the direction of magnetic north. Turned on its side, the compass direction measures the inclination, I, or dip of Earth's magnetic field. Note I = 0° at the equator and 90° at the poles. The angle between the magnetic field of Earth and true north is the declination. The magnetic declination is provided on maps so that a compass can be used to obtain the direction of geographic north (true north). (b) The magnetic field generated by a bar magnet is like that of a dipole. Lines of magnetic force are illustrated here by the alignment of iron fillings around a bar magnet. (c) Earth's current magnetic field is similar to that generated by a bar magnet or a loop of electric current placed near Earth's center and inclined 11° from the planet's axis of rotation.

Focus On 5.2 ISOSTASY: CONTROLS ON SURFACE ELEVATION

Why does continental crust rise above sea level, whereas oceanic crust is submerged? Why do some plateaus sit several kilometers above sea level, whereas the mean elevation of land is little less than a kilometer? Basic mathematics helps to answer these two questions.

To reach isostatic equilibrium, the weight per unit area in any vertical column of material above the depth of compensation must be the same. Recall that the density of an object is equal to the object's mass per unit volume. Now, if both the density and the height of a column of material are known, we can calculate the mass per unit area of this column by multiplying these two values. Suppose that the density varies down the length of the column. The column can then be divided into individual sections of constant density. The total mass per unit area in this column then becomes the sum of the products of individual densities and heights for each section. The weight per unit area is found by multiplying these products by g (assumed constant over this depth range).

Why Does Continental Crust Rise Above Sea Level?

The oceanic crust and the continental crust are in isostatic equilibrium. But, because the continental crust is several times thicker than the oceanic crust—like a larger block of wood in our example of the floating wood blocks—it floats higher in the mantle than does the oceanic crust.

We can determine the mean elevation above sea level of the continents from the principle of isostasy. Consider the situation shown in Figure 1. The mean depth of the oceans is taken as 3.8 km. The mean thickness of the oceanic crust is 7 km. We take the effective depth of compensation (c)—the depth at which both continents and oceans are in isostatic equilibrium—to be 30 km below sea level. The actual depth of compensation is deeper—in the upper mantle—but we can use the Moho as the effective depth of compensation because the mantle densities between the Moho and the real depth of compensation do not vary significantly. If we let x equal the mean elevation above sea level of the continents, then the principle of isostasy requires that a $30 + x$ -km column of continental crust must have the same weight per unit area as a 3.8-km column of water plus a 7-km column of oceanic crust plus a 19.4-km column of oceanic mantle. Now we can determine the value of x :

Let ρ_{crust} = density of crust; ρ_{water} = density of water; ρ_{mantle} = density of the mantle; h = thickness of the oceanic crust; w = thickness of the water column; and m = thickness of the mantle column. We take $\rho_{crust} = 2.85\ g/cm^3$ and $\rho_{mantle} = 3.3\ g/cm^3$. (For simplicity, we consider oceanic and continental crust to be of similar densities.) ρ_{water} is $1\ g/cm^3$. Isostatic equilibrium dictates that the weight per unit-area in each column of material must be the same.

For the continents:

$$\text{weight/area} = (c + x)\ \rho_{crust}\ g = (30 + x)\rho_{crust}\ g$$

For the oceans:

$$\text{weight/area} = (w\rho_{water} + h\rho_{crust} + m\rho_{mantle})\ g$$

For isostatic equilibrium, both expressions—for continent and for ocean—must be the same; this allows us to solve for x:

Through history, mariners have used the compass to navigate while at sea. As data comparing the direction of magnetic north with the direction of geographic north accumulated, it soon became apparent that the two were not identical. Geographic north is the point where Earth's spin axis intersects the Northern Hemisphere (the North Pole). The compass, though, points to **magnetic north**. Nonetheless, Earth's magnetic field could still be

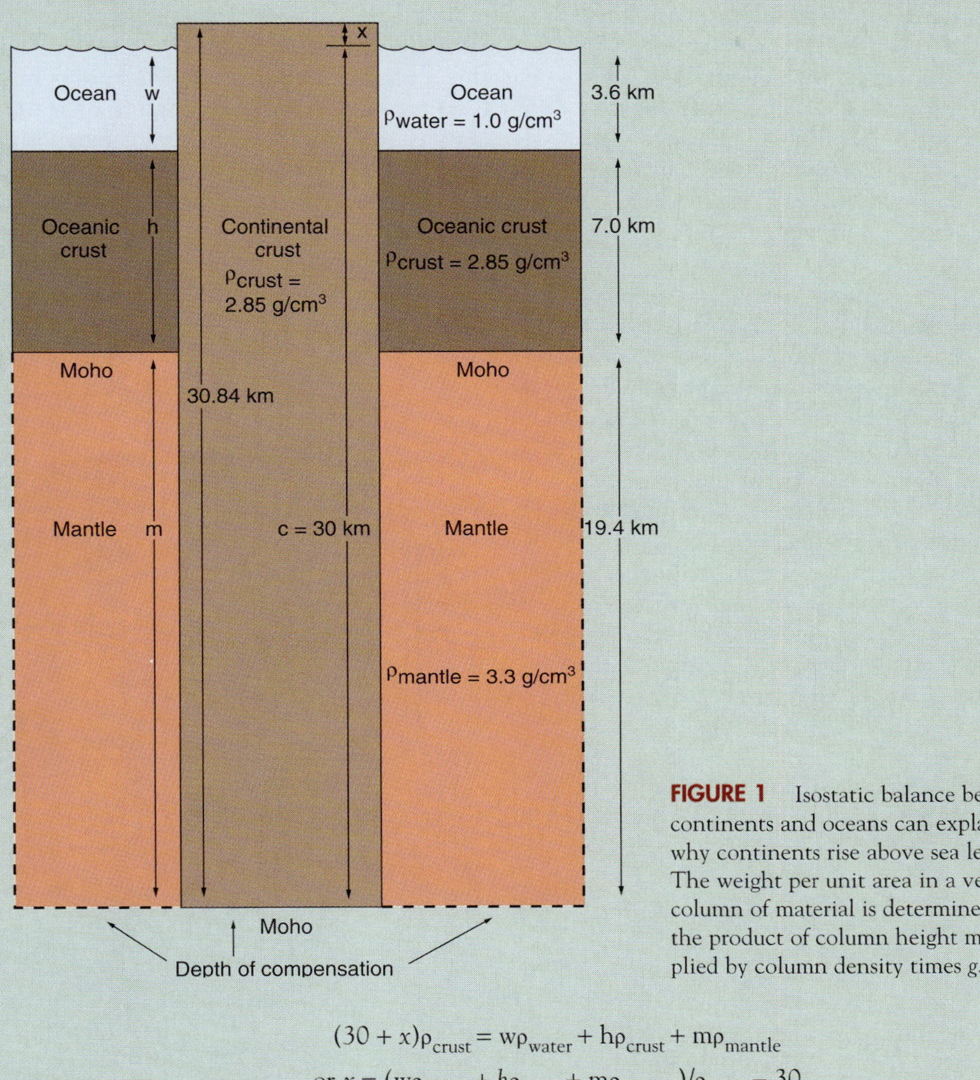

FIGURE 1 Isostatic balance between continents and oceans can explain why continents rise above sea level. The weight per unit area in a vertical column of material is determined from the product of column height multiplied by column density times g.

$$(30 + x)\rho_{crust} = w\rho_{water} + h\rho_{crust} + m\rho_{mantle}$$

$$\text{or } x = (w\rho_{water} + h\rho_{crust} + m\rho_{mantle})/\rho_{crust} - 30$$

If we now substitute the numerical values into this equation, we find that $x = 0.796$ km, or 796 m, above sea level. This value is in reasonable agreement with the average continental elevation value of 840 m above sea level, given that approximate values for thickness and densities were used here. Thus, the fact that oceanic crust and continental crust are in isostatic equilibrium can explain why continents rise above sea level. The effect is slightly accentuated when we take into account the slightly greater density of oceanic crust compared with that of continental crust. We will apply this principle to the Tibetan Plateau in Chapter 11.

explained by imagining a bar magnet placed near its center but inclined from Earth's axis of rotation (Fig. 5.24b and c). The angular difference between the compass direction of magnetic north and the geographic north–south direction is known as the magnetic **declination** (Fig. 5.24a). The angular difference between horizontal and a compass needle turned on its side so that it can rotate in a vertical plane is known as the **inclination** (Fig. 5.24a).

A significant puzzle confronted the early analysts of Earth's magnetism. It was discovered that if lodestone is heated to temperatures greater than about 570°C, it becomes nonmagnetic. This critical temperature, known as the **Curie point**, is a property of all magnetic materials. Both molten rock and extremely hot rock have no magnetism. Rocks at temperatures below the Curie point are magnetic, but they are extremely weak magnets. Temperatures within Earth increase with depth from the surface at a rate of about 25°/km. Therefore, rocks located below about 23 km from the surface should have no magnetism. How, then, could a magnetic field be generated from the center of Earth where temperatures reach thousands of degrees?

The Geodynamo

One way to create a magnetic field—other than using a magnet—is to generate an electric current. A huge electric current of about a billion amperes would be required to generate Earth's magnetic field. Electric power stations generate electricity by using a dynamo, which is a device that converts mechanical energy to electrical energy (also called an *electrical generator*). The dynamo is based on the principle that a moving electrical conductor in a magnetic field generates a current. Earth is thought to have a natural dynamo in its outer core, which we call the **geodynamo**. The outer core is made of liquid iron, which is a good electrical conductor. This molten iron is presumed to be moving because heat flows through it as Earth cools. We have already learned how heat flowing through a fluid will cause that fluid to convect.

Thus, we have a necessary component for a natural dynamo: a moving conductor. A moving conductor on its own, however, is not sufficient. To generate a current, there must be a magnetic field through which the current moves. With a suitable geometry, such a system can be self-reinforcing because the current generates a magnetic field. The motion can generate the current. This then becomes a "chicken-and-egg" argument. Which came first: the magnetic field or the dynamo current? If there were no magnetic field, then fluid motions in the conductor would not generate currents. Such a scenario, however, is highly unlikely. The Sun generates a small interplanetary magnetic field. Thus, fluid motions in the core would begin generating currents. Those currents, in turn, could generate magnetic fields that reinforce the original field. The enhanced magnetic field would then generate larger currents. The field and currents would grow until internal resistance of the system caused them to level off. This is the principle of the self-excited dynamo.

The origin of Earth's magnetic field is believed to be a self-excited dynamo in the outer core, where a moving electrical conductor in a magnetic field induces currents that reinforce the magnetic field.

A simple disc dynamo illustrates the principle of self-exciting dynamos (Fig. 5.25a). A spinning copper disc in a magnetic field generates a current that flows from its center to its rim, then flows through the coil of wire responsible for the magnetic field (Fig. 5.25b). As long as the system spins, the current will flow and generate the magnetic field. Although this example illustrates the basic principle of a geodynamo, the relationship between currents and magnetic fields in the core is undoubtedly much more complicated. One important feature of such dynamos is that they can generate magnetic fields in either direction—either up or down in the diagram. The direction depends on the direction of the background magnetic field at the time of initiation.

A complete explanation of Earth's geodynamo has not yet been achieved. We do not know the geometry of the

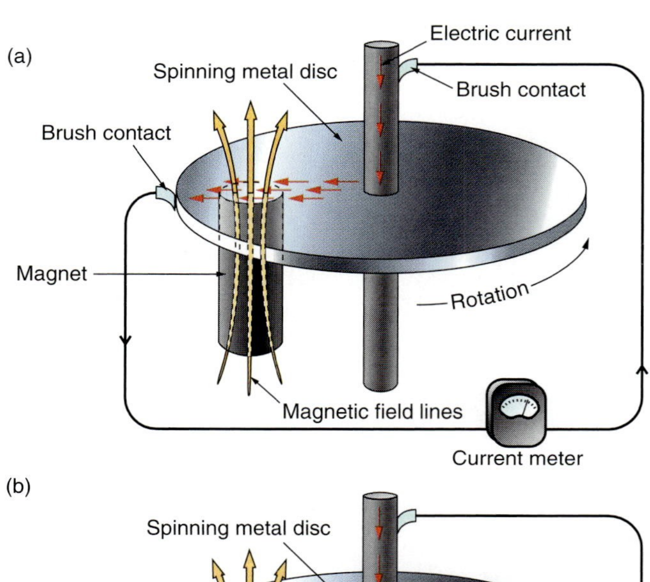

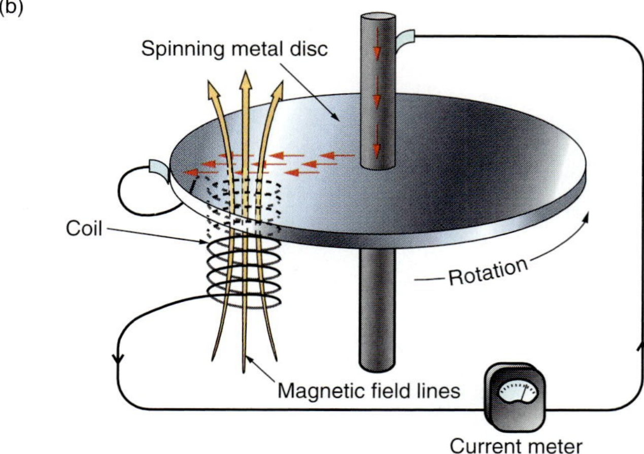

FIGURE 5.25 (a) The principle of the geodynamo. An electrical conductor moving in a magnetic field generates an electric current. The magnetic field is generated by the magnet on the left. (b) An electric current can be used to generate a magnetic field. The magnet is replaced by a coil through which current flows. Earth's geodynamo is thought to be generated in this fashion. Although there are no magnets in the core, the complicated motions of the conducting material in the core are thought to set up currents that sustain the dynamo action.

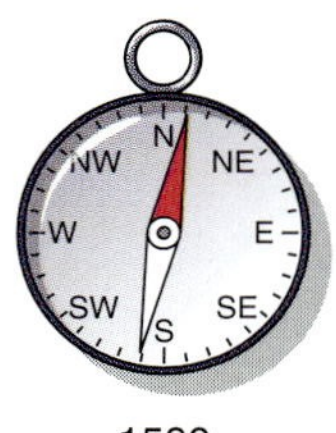

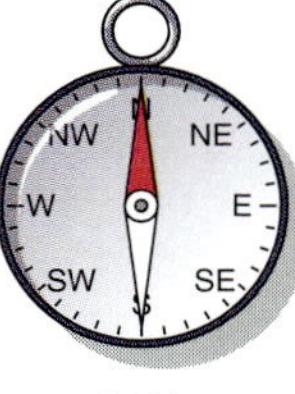

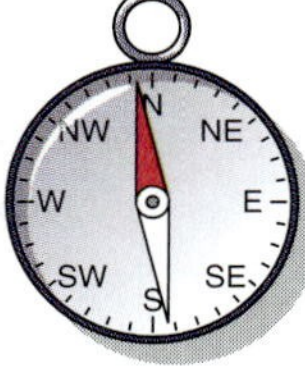

FIGURE 5.26 Change in the declination of a compass needle at the geomagnetic observatory in London (1580–1996).

fluid motions in the core, which, judging from the variations of the magnetic field at the surface, must be quite complicated. Helical motions are a sufficient condition for generation of a dynamo, but detailed information about these motions is needed. Solving the combined problem of magnetic fields and current flow along with core motions is very difficult.

Most of the magnetic field measured at Earth's surface is generated in the core. In some areas, however, a significant component of the magnetic field is caused by the magnetism in near-surface rocks. Such magnetic anomalies are useful in the exploration of magnetic minerals and can also provide information about events of the past.

A smaller component of Earth's magnetic field that is detected at the surface comes from space. Earth is the only terrestrial planet to have a strong magnetic field (see Focus 5.3); Mercury may have a small one. The giant planets have huge magnetic fields generated by dynamos over a thousand times stronger than that of Earth; the Sun's magnetic field is more than a million times stronger (Focus 5.3).

Changes in Earth's Magnetic Field

The location of the Earth's magnetic north pole changes with time. Early mariners were well aware of the changes in the direction of magnetic north as indicated by a compass. Over the years, magnetic declinations were plotted on maps. By 1860, it became apparent that the declinations were changing with time. The contour of zero declination, which was in the mid-Atlantic in 1883, had drifted to South America by the early 1900s. Further, when magnetometers—devices measuring the strength of the magnetic field—came into use in the 1900s, it was discovered that the magnetic-field strength also varied over time.

Figure 5.26 shows the change in declination of a compass needle in London between 1580 and 1996. The total change is as much as 35°. A captain sailing from London for Iceland with out-of-date charts might end up in Greenland.

Today, the magnetic field is measured worldwide, and changes are noted in the International Geomagnetic Reference Field, or IGRF, consisting of maps compiled every five years (Fig. 5.27). Recent magnetic-field measurements made in space by the satellite Magsat were used to obtain global coverage of the geomagnetic field. The IGRF maps have shown that about 70 percent of the geomagnetic field can be represented as a simple dipole, which is similar to the magnetic field from a bar magnet (see Fig. 5.24b).

What of the remaining 30 percent of the geomagnetic field that is not accounted for by the simple dipole model (Fig. 5.28)? Features in contour maps of this nondipole component drift across the globe at rates of tens of kilometers annually, with a tendency for most to drift westward. This variation is thought to be due to the westward drift of eddies in the flow of the outer core, giving us an estimate of core motions, which at tens of kilo-

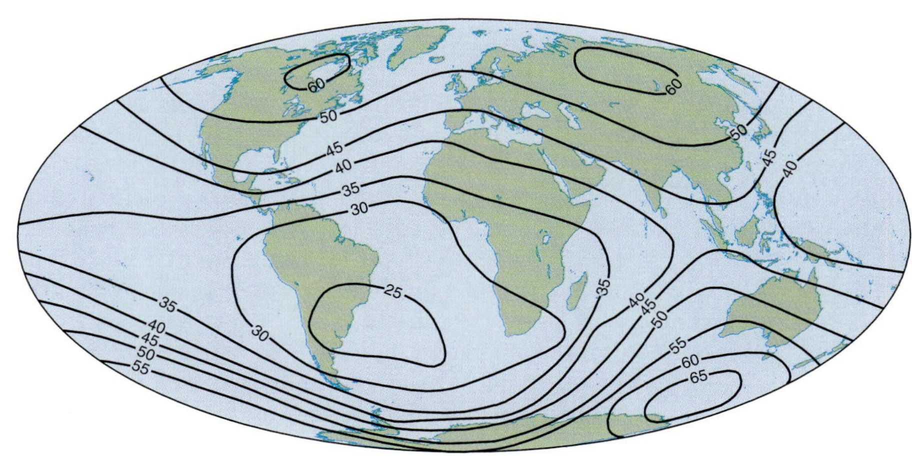

FIGURE 5.27 Contours indicating Earth's geomagnetic field are shown on the International Geomagnetic Reference Field (IGRF). Field values are contoured in five thousands of nanoteslas (nT) and range from 25,000 nT to 65,000 nT.

MAGNETIC FIELDS IN SPACE

Why is Earth the only terrestrial planet with a magnetic field? The two planetary aspects thought to be important for a planetary dynamo are planetary spin and a large, liquid metallic core. Mars may not have a dynamo because it does not possess a sufficiently large core or perhaps the core is solid; however, there is some controversy about this. Meteorites have been found in Antarctica that many believe to have come from Mars, ejected into an Earth-crossing orbit possibly as a result of an impact on Mars. These meteorites are thought to be related to Mars because of geochemical similarities with Martian rocks sampled during the Viking Mission in the late 1970s. Geochemists argue that if these are indeed representative of Martian rocks, Mars may have a large sulfur content (about 16 percent) in its core in addition to iron. This would lower the melting point of an iron sulfide (FeS) core relative to iron (Fe). Thus, the core of Mars may still be molten. Seismic experiments to discover the nature of this core have been proposed for new space missions to Mars in the future.

Venus is compositionally very similar to Earth and would be expected to have a geomagnetic field. Remote sampling of surface rocks by the Soviet Venera missions indicates a basaltic composition similar to Earth's, which suggests an iron core. If planetary spin is necessary to generate a geomagnetic field, however, this may account for the lack of a geomagnetic field because Venus has a very slow (backward) spin.

Mercury has no significant magnetic field and also rotates slowly. Its density indicates that its core extends across three-quarters of the planet's radius. Our Moon has no magnetic field; however, Moon rocks recovered from the Apollo missions indicate that it had one in the past (3 billion to 4 billion years ago). The Moon may have had a small liquid iron core that subsequently froze.

Planetary dynamos operate in the giant planets Jupiter, Saturn, Uranus, and Neptune, and in the Sun (Fig. 1). Recent theories of planetary dynamos postulate that the deep fluid parts of the giant

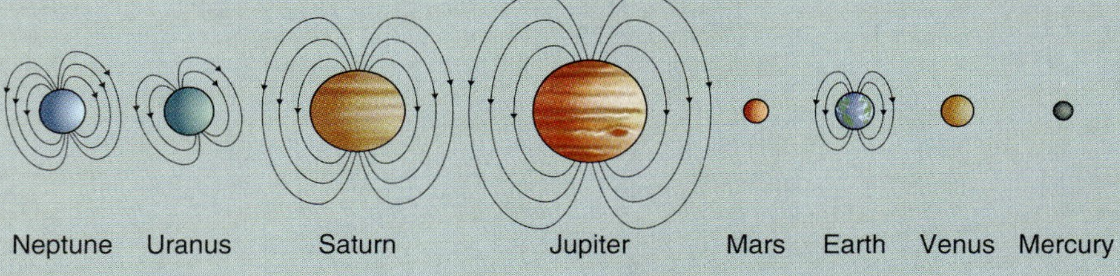

FIGURE 1 Magnetic fields for the planets in the Solar System.

meters each year are a million times faster than the centimeters-per-year rate of plate motions.

The change in Earth's magnetic field with time is referred to as *secular variation*. Secular variations in which the magnetic north pole deviates from true north by as much as 24° occur on the historical time scale (see Fig. 5.26). On the geologic time scale there is evidence for even more dramatic variations in Earth's field. It is now known that the magnetic field actually flips, or reverses, in direction; the north magnetic pole becomes the south magnetic pole and vice versa. The direction of the present geomagnetic field is defined as *normal polarity*. Geologic epochs during which the field had the opposite direction are known as times of *reversed polarity*.

Paleomagnetism

To reconstruct the behavior of the geomagnetic field prior to the times of mariners and nautical charts, some type of natural recorder is necessary. Almost all rocks contain some magnetic material. Igneous rocks contain small amounts (up to a few percent) of the magnetic mineral magnetite (Fe_3O_4). Magnetism in a small grain of magnetic mineral is like a compass needle. When the rock is molten (for instance, as it erupts from a volcano) or hot (soon after crystallizing), its temperature is above the Curie point. The magnetism is then so thermally agitated that it points in all directions; this leaves the rock, as a whole, nonmagnetic.

planets are sufficiently conductive to produce dynamo action. Of particular note are the magnetic fields of Uranus and Neptune as determined by the recent Voyager Mission. Uranus's spin axis lies on its side relative to the plane of its orbit. For the other planets, the spin axis is perpendicular to the plane of orbit. The magnetic field of Uranus is off-centered and makes a large angle to the spin axis (Fig. 2). Similarly, Neptune's field is off-centered at a large angle to the spin axis. These planets may have a large, solid inner core and their magnetic fields may be created by complicated fluid motions in a liquid outer shell. Crystallization of the solid inner core from the liquid outer core is thought to be an effective way of producing a dynamo. Even without fully understanding the details of the mechanisms, we see that dynamos are common in the planets of our Solar System.

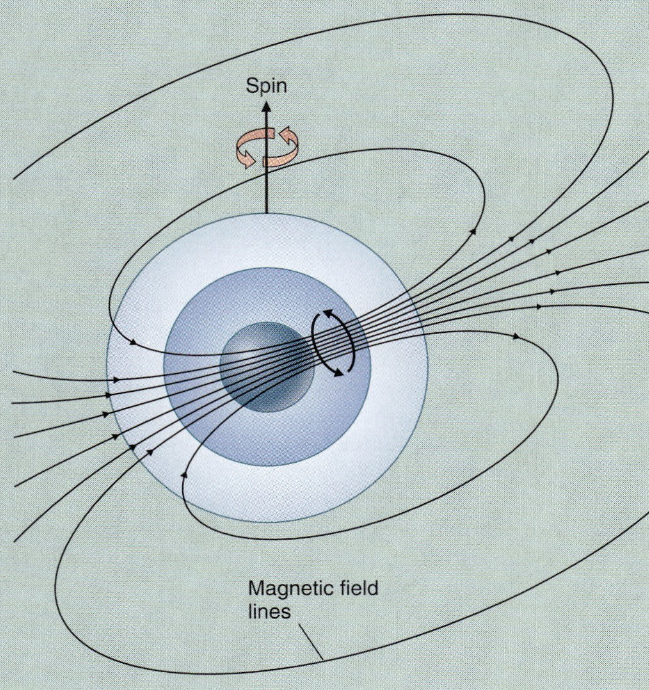

FIGURE 2 An off-centered planetary dynamo in the fluid layer of a giant planet can explain the lack of alignment of the magnetic field with the spin axis of the planet.

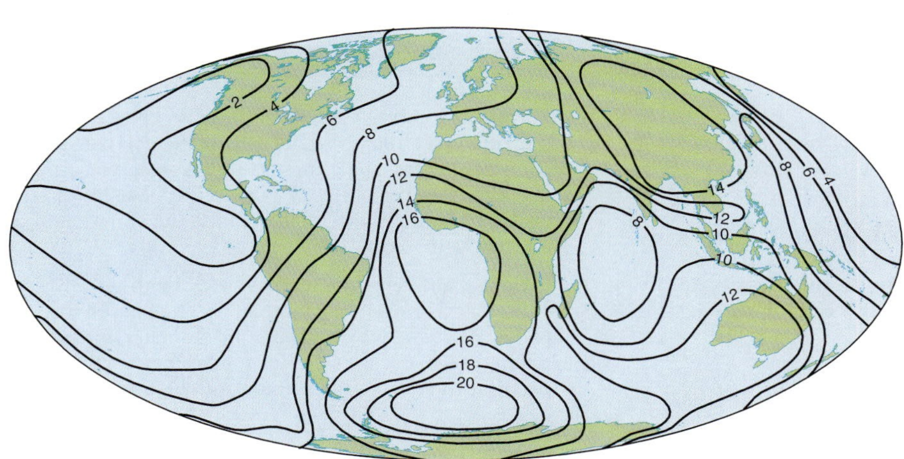

FIGURE 5.28 The geomagnetic field (IGRF) in which the field from a dipole is removed. Compare with Figure 5.27. The contours of such magnetic-field strength are found to move in time. Though they drift across the globe in all directions, there is a general tendency to drift westward, which is thought to reflect the drift of current eddies in the outer core.

Just below the Curie point, however, a mineral grain becomes magnetic. At first, the magnetism acts like an excited compass needle, swinging wildly in all directions within the grain owing to the energy from the heat (Fig. 5.29a). Note that the grain remains stationary. It is the magnetism within it that moves. As the mineral grain cools, however, the magnetism settles down. At a temperature that is about 100°C below the Curie point—called the **blocking temperature**—the magnetism becomes aligned with the direction of the geomagnetic field (Fig. 5.29b). The mineral grain is then cool enough that any further motion of the magnetism is blocked. The direction of Earth's magnetic field at the time that the rock reached the blocking temperature is recorded in the magnetism of the grains.

Provided that the rock is not reheated to a point above the blocking temperature, the magnetism remains unchanged throughout geologic time. In effect, a permanent record of Earth's magnetic field at the time the rock cooled and solidified is locked into its magnetic grains. The study of such records is called **paleomagnetism**.

On Earth's surface, the inclination of the magnetic field direction varies as a function of latitude (see Fig. 5.24). At the north magnetic pole, the magnetic field points downward. At the magnetic equator, however, it is horizontal and points north. At other magnetic latitudes, intermediate angles are found. Over geologic time, continents have drifted relative to the magnetic poles. We can use the magnetic signature recorded in rocks to measure the inclinations and declinations of the geomagnetic field at the time they were formed and thereby calculate their location with respect to the magnetic poles—paleomagnetism—to chart these movements (∞ see Chapter 6). Thus paleomagnetism has provided clinching evidence for the theory of plate tectonics.

The direction of Earth's magnetic field at the time that a rock reached the blocking temperature is recorded in the magnetism of the mineral grains in the rock.

(a)

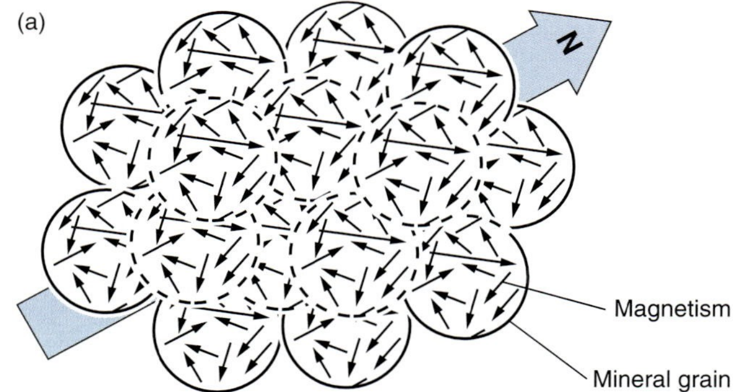

(b)

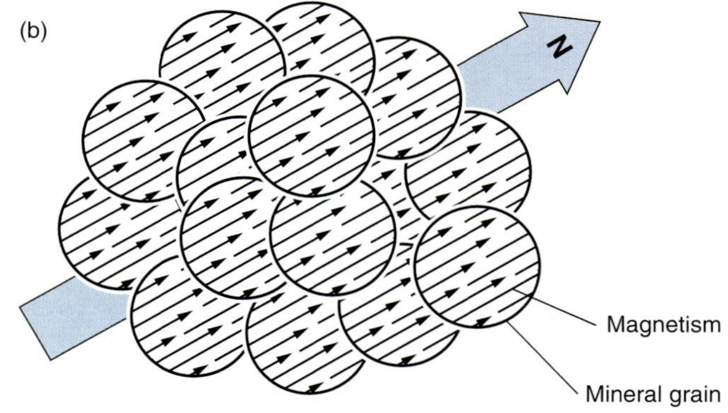

FIGURE 5.29 (a) Above the temperature called the Curie point, the magnetism of mineral grains is oriented in random directions. (b) As these grains cool to a temperature about 100°C below the Curie point—the blocking temperature—the magnetism in the grains aligns in Earth's magnetic field analogous to the alignment of a compass in Earth's magnetic field. Further movement, or shifts in alignment, of the magnetism in mineral grains is blocked by continued cooling. These mineral grains therefore provide a permanent record of the direction of Earth's magnetic field at the time of mineral crystallization.

SUMMARY

- Seismic waves are used to probe the density and elastic moduli of Earth's interior. Elastic moduli give the amount of stress needed to cause a given strain. Density and elastic moduli increase with depth in Earth.
- An earthquake generates two types of body waves: P waves arrive at a given location first, followed by S waves. Both P and S waves combine near the seismic source to generate surface waves, which travel along Earth's surface, causing the most damage.
- Energy from a wave source spreads out in all directions. A direct ray traces energy that travels directly from the source; a refracted ray traces energy that follows a curved path; and a reflected ray represents energy that bounces off an interface.
- Travel-time curves express the time it takes for a seismic wave to travel from its site of origin to a receiving station as a function of the angle these two sites make at Earth's center. These curves are used to find the velocity of seismic waves within Earth and to determine the variation of composition with depth.
- Seismic-wave velocities increase with depth suddenly at the crust–mantle transition and then decrease at the base of the plates. The absence of S waves in the outer core and their inferred existence in the inner core from normal modes suggest that the outer core is liquid and the inner core is solid.
- Gravity is the force of attraction between any two masses. Gravity variations give information on density variations beneath Earth's surface.

- Gravity measured at a constant height is known as *free-air gravity*. *Bouguer gravity* shows gravity measurements as if all mountains were scraped off and all valleys were filled with material of the same density as the surrounding rock.
- The mass of a mountain above sea level is compensated by the mountain's underlying crustal root. A mountain is in isostatic equilibrium if it is neither rising nor sinking. The depth at which this isostatic adjustment occurs is called the *depth of compensation*.
- The origin of Earth's magnetic field is believed to be a self-excited dynamo in the outer core, where a moving electrical conductor in a magnetic field induces currents that reinforce the magnetic field.
- The direction of Earth's magnetic field at the time that a rock reached its blocking temperature is recorded in the magnetism of the mineral grains in the rock.

KEY TERMS

density, 109
elasticity, 109
phase transitions, 110
inelasticity, 110
plasticity, 110
stress, 110
tension, 110
compression, 110
strain, 110
elastic modulus, 110
seismometer, 111
seismogram, 111
P wave, 111
S wave, 111
body wave, 111
surface wave, 111
direct ray, 113
refraction, 113
travel-time curve, 115
reflection, 115
normal mode, 117
Moho, 118
low-velocity zone, 118
gravimeter, 122
free-air gravity maps, 123
Bouguer gravity maps, 123
isostasy, 125
isostatic equilibrium, 125
magnetic north, 128
declination, 129
inclination, 129
Curie point, 130
geodynamo, 130
blocking temperature, 134
paleomagnetism, 134

QUESTIONS FOR REVIEW AND FURTHER THOUGHT

1. Why do P waves travel faster than S waves?
2. What volume would 1 kg of Earth occupy (a) near the surface; (b) at the base of the mantle; (c) just inside the outer core; and (d) at the center of the Earth? [Hint: Refer to Fig. 5.2.]
3. Density increases with increasing depth in Earth. From the formula for seismic velocity, explain why velocity also increases with depth even though density is in the denominator.
4. From Figures 5.2 and 5.8, estimate the average density and P-wave velocity of the outer core and that of the mantle below the 660-km phase transition. The P-wave velocity drops at the transition from the mantle into the core yet the density increases. Suggest why this occurs.
5. A wave traveling downward from the surface of the ocean at 1.5 km/sec takes six seconds to reflect off the ocean floor. How deep is the ocean at that site?
6. (a) When we consider isostasy, what do we mean when we refer to the depth of compensation? (b) If a mountain range is in isostatic equilibrium, which gravity is larger: the free-air or Bouguer? Explain why.
7. What are the maximum and minimum values of Earth's magnetic field and where do they occur? [Hint: Refer to Fig. 5.27.]
8. How would you tell the variation in latitude of a continent that has drifted in geologic time by using rock samples taken from lava flows of different ages?
9. What is the blocking temperature for the magnetism in rocks? Is it higher or lower than the Curie temperature?
10. Why does Earth have a magnetic field and the other terrestrial planets do not?

6 Plate Tectonics

Active plate tectonics—rifting and volcanism—characterize the Afar region of northeast Africa, where the African and Arabian plates move away from each other.

Overview

- Most earthquakes, volcanoes, and rock deformation occur at plate margins. In general, plate interiors are tectonically quiet.
- The theory of continental drift assumes that continents move through or across ocean basins. Plate tectonics suggests that continents are parts of plates that move.
- If a continental margin coincides with a plate boundary, it is referred to as an *active continental margin*. Otherwise, it is a *passive margin*.
- The margins of plates can be divergent, convergent, or transform. The geometry of plate margins changes with time.
- Changes in Earth's magnetic field direction are recorded in igneous rocks as they cool.
- Mantle plumes arise within the mantle and leave a trail of volcanoes on a lithospheric plate if it moves over the plume.

Introduction

Within the time scale of human experience, the shapes and positions of Earth's continents seem fixed—although we are able to observe some minor modifications directly. Erosion by surf action changes the rugged coastlines of places such as Maine, Oregon, and California; sedimentation modifies the outlines of the Carolina coast, Cape Cod, and the Texas gulf coast.

On a time scale of millions of years, however, the continental masses are seen to be quite mobile. Through Earth's history, continents have moved relative to one another—they have collided and have been torn apart again several times. This can be explained by plate tectonics, the grand unifying theme for much of the geologic sciences. The world map shown in Figure 6.1a delineates the boundaries of the lithospheric plates. Figure 6.1b clearly illustrates that the locations of volcanoes and earthquakes are not randomly distributed. In fact, both are concentrated in narrow belts that serve to identify the lithospheric plate boundaries. Most volcanoes and earthquakes are associated in time and space with plate boundaries; thus, plate boundary processes must in some way cause them. In contrast, plate interiors are associated with far fewer earthquakes and volcanoes.

Plate boundaries can be identified on Earth's surface by the locations of volcanic and earthquake activity. In general, plate interiors are geologically inactive.

Continental Drift: The Beginnings of the Idea

For many years, the approximately parallel outlines of the African and South American continents situated on opposite sides of the Atlantic Ocean have attracted attention. When the shorelines of the continents are traced and overlaid, there is quite a good fit between some continents—almost like the pieces of a jigsaw puzzle. This interlocking fit works well with the continents as they are represented on a globe, which lacks the distortions introduced by most map projections. Although Sir Francis Bacon is commonly credited with publishing the first allusion to continental drift in 1620, it appears that he actually pointed out the similarity of the *shapes* of some continents, suggesting that they could have been cut by the same mold. In fact, as early as 1596, the Dutch cartographer Abraham Ortelius suggested that the continents were once joined and subsequently separated.

Possible causes of the parallelism in continental outlines were proposed in the early twentieth century by F.B. Taylor and Alfred Wegener. In 1908 and 1910, Taylor proposed that the Moon had been captured by Earth between 60 million and 150 million years ago. He thought the Moon was initially much closer to Earth than it is today. Further, he proposed that the gravitational pull (tidal friction) was sufficiently strong to cause the continents to move. Taylor's idea was immediately pounced upon by Sir Harold Jeffries, an eminent physicist of the time, who pointed out that if tidal forces were as strong as Taylor suggested, then the same tidal forces would have stopped Earth's rotation.

At about the same time, Alfred Wegener was also engaged in work on this topic. He was not a geologist by training but rather a climatologist and an arctic explorer. Nevertheless, the congruence of the coastlines on either side of the Atlantic continually aroused his curiosity, and in 1910, he proposed the idea of **continental drift**. His biographers speculate that his experiences in the arctic—watching sea ice continually split, drift apart, and collide with neighboring ice floes—undoubtedly contributed to his ideas about continental movement. However the thought came to him, Wegener initially rejected the concept of drifting continents for lack of a conceivable mechanism to cause the drift.

One year later, he began to read accounts of unusual similarities in plant and animal fossils between parts of South America and Africa. World War I kept him from research for several years, but he was able to continue his studies on an extended leave from the military in 1915. His research resulted in the publication of his monumental work *Die Entstehung der Kontinent und Ozeane (The Origin of the Continents and Oceans)* that same year. The third edition of his book was finally translated into English in 1924. Although widely read, Wegener's book was largely rejected by the established scientific community because the process of continental drift seemed improbable. Figure 6.2 is a reproduction of Wegener's continental fit. It is enlightening to compare his reconstruction with those of modern earth science. Wegener's notion was not seriously entertained by the established scientific community because of the very problem that he had earlier acknowledged—no one could conceive of the phenomenal driving forces required to move such large blocks as the continents.

Earth scientists in Europe and North America found Wegener's theory unsatisfactory because they disliked the idea of continents "plowing" through the ocean basin floors. However, geologists working in South Africa and Brazil supported the theory. The first definitive evidence for the separation of continents had been found in these regions; therefore, these scientists did not dismiss Wegener's ideas but used them to guide further work. In 1937, Alexander Du Toit, a South African geologist, published *Our Wandering Continents*, which he dedicated to Wegener. He did not agree with Wegener on all matters, but he was in complete agreement with Wegener in his opposition to the strongly *fixist* philosophy of the scientific establishment.

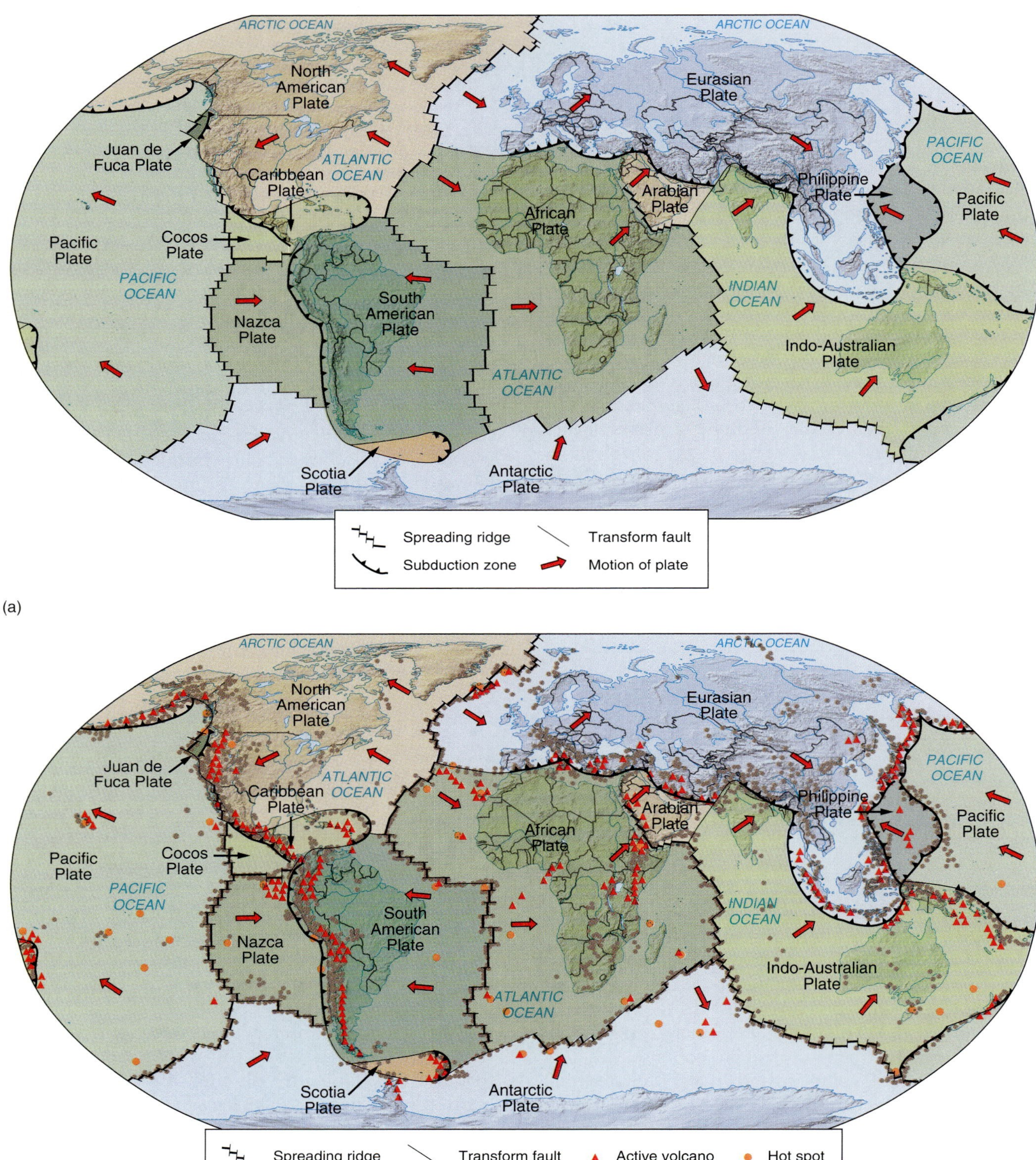

FIGURE 6.1 (a) A world map of the lithosphere plates. The boundaries are defined by subduction zones, by spreading ridges, or by transform faults. (b) The same map showing hot spots, active volcanoes, and areas of abundant seismic activity. (Submarine volcanoes are not shown here.)

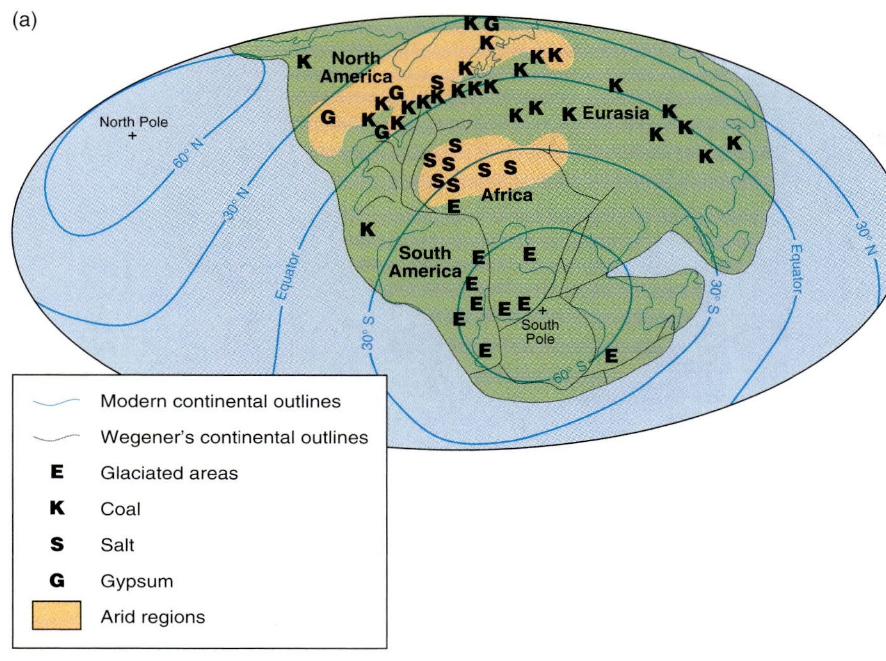

FIGURE 6.2 (a) Wegener's reconstruction of continents 300 million years ago. He based his reconstruction on the distribution of glacial features (implying polar regions), the presence of coal (tropics), and the presence of salt, gypsum, or sand dune deposits (desert). (b) Bullard's fit of South America and Africa. The outlines correspond to the 914m contour. Shaded areas represent gaps (light red) or overlaps (gray) in the fit.

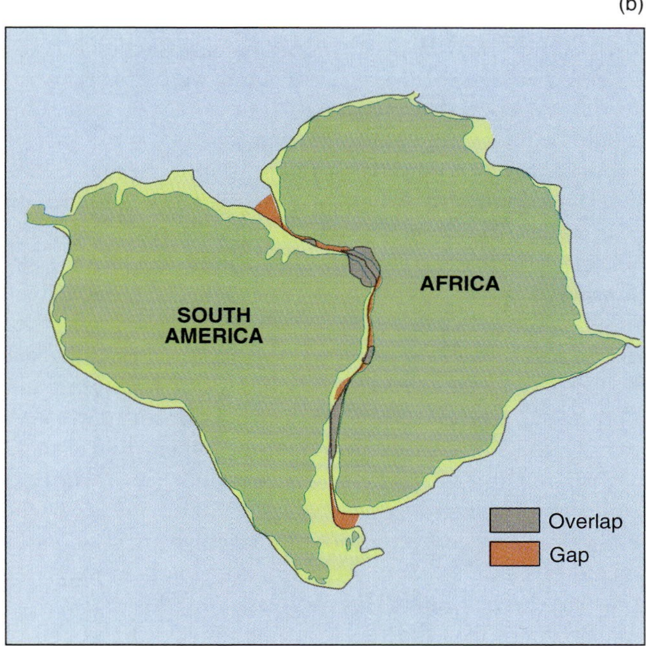

Wegener's Theory

The essence of Wegener's idea was sound, and it represented *a valid scientific hypothesis based on observations*. He recognized that the observed paleontologic correlations between South America and Africa meant that these two continents must have been joined at some time in the past. A set of very peculiar, but identical, fossil plant remains (a seed fern known as *Glossopteris*) and small reptiles (*Mesosaurus*) were found to occur in nearly identical sedimentary rock types of precisely the same age in several widely separated locations (Fig. 6.3): near the eastern coast of South America, in Africa, in India, and in Antarctica (some of these locations were found after Wegener did his work).

Wegener attempted to reconstruct the continents using the distribution of these fossils and geologic evidence from rocks that were approximately 300 million to 250 million years old. He concluded that a single great continent had encompassed most of the continental regions of Earth today, which he named **Pangaea** (Greek, meaning "all land"). He also concluded that the plants and reptiles lived in a relatively small and confined area on that continent (Fig. 6.3). Millions of years after these plants and animals became extinct, this supercontinent *rifted* apart. With a few minor differences Wegener's continental reconstruction for that time period is remarkably similar to modern reconstructions made using much more data.

Alfred Wegener proposed that continents were once joined together as a supercontinent that he called *Pangaea*. This supercontinent subsequently broke up and drifted apart, a process known as continental drift.

Problems with Continental Drift

The theory of continental drift involved two critical assumptions: The shapes of the continents—as defined by their coastlines—are permanent, and the continents are unattached and move independently through or over the ocean basins.

Wegener's first assumption is in error because continental outlines are not fixed. We can observe changes in coastal form as a result of sedimentation or erosion. The coastline is arbitrary and is defined by sea level, which is

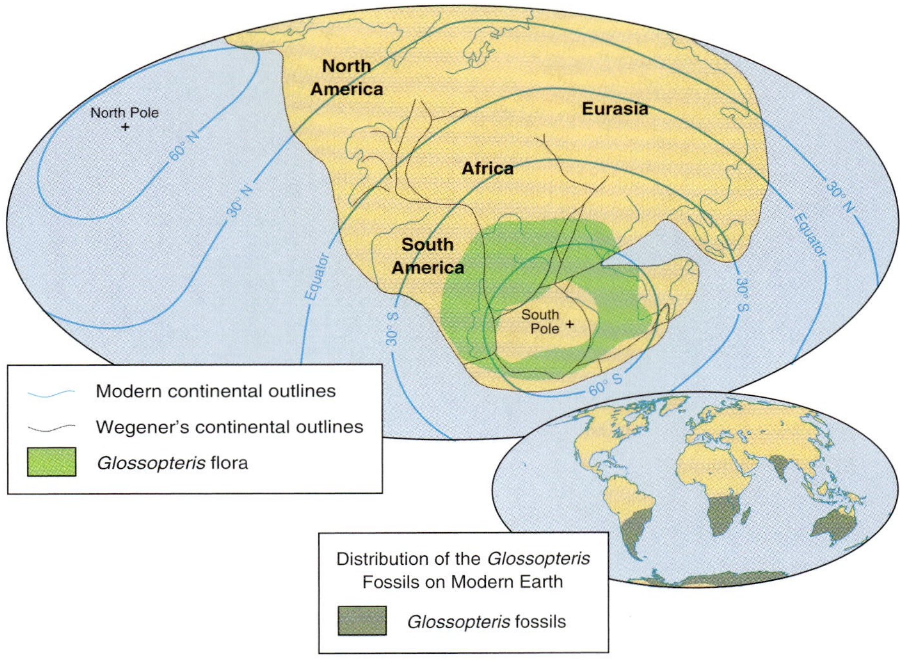

FIGURE 6.3 Wegener's reconstruction showing the distribution of the *Glossopteris* plant fossils. The small inset shows the distribution of these fossils in the modern continental configuration. Wegener's reconstruction accurately accounts for the distribution of these plant fossils.

known to fluctuate depending on a number of variables, including the amount of polar ice on Earth. Therefore, the shape of the continental shoreline has little or no relevance to the geophysical definition of a continent.

In the 1950s, Edward C. Bullard, an English geophysicist, realized that shorelines are transient features and are thus a poor outline upon which to attempt a fit between continents. His idea was to draw maps on which the edges of the continents were defined from a more geophysical perspective rather than a geographic one, and thus he tried several definitions for the edges of the continents. The different boundaries that he used ranged from the edge of the continental shelf (200 m water depth contour) to the base of the continental slope, which is where the flat plain of the deep ocean meets the edge of the continent. Bullard believed that the ocean floor–continental slope boundary corresponds most closely to the real continental margin. Indeed, Bullard's computer simulations for the best fit determined that the least number of overlaps or gaps in the fit exist when continent outlines are compared at a water depth of 914 m (Fig. 6.2b).

Wegener's second assumption—that the continents are not attached and move independently—is also incorrect, because the continents are actually an integral part of lithospheric plates. Wegener proposed that continents moved westward *through* or *over* the basement rocks of the ocean basins. Frictional drag created by this movement was presumed to cause a "wrinkling" into mountains along the leading margins of the continents. He believed that the mountain belt that extends along the entire length of the west coasts of North and South America represented a dramatic demonstration of his frictional-wrinkle theory, and that the eastward-directed "tails" at the southern tips of South America and Greenland represented evidence that the continents were drifting westward. Wegener also postulated that detached fragments would be left behind at the trailing edge of continents. As proof of this, he cited the strings of islands along the eastern shores of North America. Wegener faced criticism on this issue and was never able to rebut convincingly the objection that weaker granitic continents could not plow through stronger basaltic oceanic basement rocks.

The theory of continental drift made no effort to explain the locations of either volcanoes or earthquakes, except where the earthquakes were located along mountain fronts. In the early 1900s, earthquake locations were mainly defined by direct observation—where strong shaking was felt and where damage occurred. During Wegener's time, mid-oceanic ridges were largely unknown, as was the existence of a nearly continuous string of submarine earthquake epicenters along these ridges—simply because no one was there to make observations.

Wegener's idea that the continents plowed through or rode over the ocean basins could be considered a predecessor to the idea of plate subduction. However, there is a major difference. Embedded within the theory of continental drift is the fundamental premise that there must be a major zone of slippage that either surrounds each continent or lies beneath each continent where it either pushes through or slides over the basement rocks of the ocean floor. We know from modern marine geophysical research that no such fault zones surround any of the continental masses; further, continents do not possess any near-horizontal shear zones at a depth of 4 to 5 km, where shearing, or sliding, would occur were the continents riding over the ocean basins. To be sure, there are mechanical discontinuities within the continental crust, but these are at greater depths. A good example is the Mohorovičić discontinuity (∞ see Chapter 5), which is the boundary between the crust and mantle. It occurs at depths between 30 and 70 km beneath the continents, but there is no evidence that slip commonly occurs along this discontinuity.

In the 1940s the British geologist Arthur Holmes suggested that continental drift was related to mantle convection. He proposed that the ocean floor spread apart at the top of a mantle convection cell, carrying the continents with it. This idea, published in his book *Principles of Physical Geology*, was remarkably similar to our present understanding.

Geosynclines and Mountain Building

Prior to the emergence of "The New Global Tectonics," as plate tectonics was known as in the early 1960s, other explanations accounted for some of Earth's geologic features, such as mountain ranges. Most mountain belts are relatively long and narrow features. They were shown to be composed of thick sequences of deformed sedimentary rock. These observations were commonly explained by a "geosynclinal theory": It was thought that elongate belts of deep subsidence and thick sedimentation existed, which were called **geosynclines**. The thick geosynclinal strata were folded and metamorphosed and then uplifted and eroded to form the mountain belts that we see today.

In the early to mid-twentieth century, field observations from many mountain ranges had been accumulated and compared. One key observation was that there appeared to have been two parallel geosynclines side by side: one in which the sediments were very thick and strongly deformed (*eugeosyncline*), and a second belt that exhibited a thinner sequence of sediments that were largely undeformed (*miogeosyncline*). The nature of the basin fills in these two geosynclinal types were different. The miogeosynclinal succession contains mature and undeformed sandstones and limestones, with plentiful fossils. Eugeosynclinal rocks were characterized by immature sediments and submarine volcanic rocks. These sediments appeared to have been deposited rapidly in very deep marine water.

Miogeosynclines were interpreted to be shallow marine settings, and these sediments resemble modern deposits from that environment (∞ see Chapter 13). But the eugeosynclinal deposits presented a difficult problem. The shallow marine fossils found in some of the eugeosynclinal sediments indicate deposition close to continental shelves, but such environments in the modern world commonly lack active volcanism—which is implied in the eugeosynclinal environment by means of volcanic deposits. Moreover, no one was able to relate the strong deformation of these sequences to any known mechanism or environment.

This is where the field of geology stood in the late 1950s when marine geophysicists began to analyze information from the world's ocean floors. Information included magnetic measurements, determination of the topography of the ocean floors, and actual sampling of the rocks beneath the oceans. To their astonishment, researchers found features not apparent on continents, and initial hypotheses engendered excitement among geologists and geophysicists. We now know that geosynclines were an attempt to explain valid geologic observations using an inadequate model. It was recognized that miogeosynclinal sediments represent *passive continental margin* sedimentation—the kinds of sequences that we find today around the edges of continents away from plate margins. Eugeosynclinal deposits are composites of deep marine sediments, volcanics, and material derived from the continents that accumulate at *convergent plate margins* ("active" continental margins).

As we have seen, volcanism is a distinguishing characteristic of convergent plate margins, so the occurrence of volcanic deposits is to be expected. In the plate–tectonic model, the concept of a eugeosyncline is effectively replaced by the oceanic trench that marks convergent plate boundaries. The eugeosynclinal deposits are not a continuous depositional sequence, but a tectonic stacking of material scraped off the oceanic crust at a subduction zone. This, in part, explains the juxtaposition of shallow marine, deep marine, and volcanic deposits, and the deformation that typifies eugeosynclinal sequences.

But what of the juxtaposition of miogeosynclinal and eugeosynclinal sequences that had been observed in old mountain belts? Again, this can be explained by appealing to plate tectonics. Although they represent two very different environments (passive and active continental margins, respectively), they have been brought together in mountain-building continental collisions, as ocean basins that originally separated them eventually closed because of continued subduction.

Plate Tectonics: A Unifying Theory

The Ocean Bottom and Sea-Floor Spreading

The first clues leading to plate tectonics came from research in the ocean basins. Until the middle of the twentieth century, little was known about the ocean floor. With the development of submarine warfare during World War II came a pressing need for accurate and detailed maps of the ocean bottom, and extensive sonar surveys were carried out. The principle of sonar works very much like seismic reflection, described in Chapter 5. A sound burst is directed from a ship to the ocean bottom, and then recorded as it bounces back, or echoes. Knowing the speed of sound in water, the time taken for the signal to travel to the sea floor and back can be used to determine the depth of the sea floor at that point.

The American geophysicist Harry Hess proposed the concept of **sea-floor spreading**, which was based primarily on the topography of the ocean floor (more correctly referred to as bathymetry). The ocean-floor bathymetry that was determined formed an unexpected and remarkable picture. Huge submarine mountain ranges—the ocean ridges—were discovered. The distribution of these ridges (see Figs. 6.1 and 8.1) and the symmetry of the ocean floor led Hess to suggest that the sea floor is created at these ridges and spreads outward from them. Confirmation of this rather speculative idea later came from paleomagnetism and measurements of the magnetic field of the ocean floor.

Paleomagnetism and Polar Wandering

We saw in Chapter 5 that the geomagnetic field resembles that of a dipole and that rocks record the vector direction of the magnetic field at the time they were formed and cooled through the blocking temperature. By measuring the declination and inclination of the magnetism blocked in a rock, the position of magnetic north at the time the rock formed can be determined by comparison with the dipole field pattern (see Fig. 5.29).

Research on paleomagnetism in rocks surged after the discovery of evidence that the magnetic poles appeared to wander through geologic time. Geophysicists measured paleomagnetic data from rocks collected from a variety of locations, including Europe, North America, Australia, South America, Africa, and Antarctica in an effort to identify the mechanism that may have caused the magnetic poles of Earth to alter their position with time. **Apparent polar wander curves** were used to plot the positions of the magnetic north pole at different geologic times and were generated from several continents. The polar wander curves determined from rocks 70 million to 400 million years old from North America and from Europe are remarkably similar in shape, but the data from North America always plot west of the data from Europe (Fig. 6.4a). The variation in the implied location of the north pole through time is far greater than the variations that have been recorded historically (∞ see Chapter 5). Furthermore, at a given time, there appear to be *two* north poles—one for Europe and one for North America. Pole positions determined from older rocks on the two continents plot farther apart than do polar locations determined from younger rocks. These puzzling observations are explained when we realize that it is the *continents* that have moved relative to the pole, not the pole itself.

If the continents have moved relative to each other, then continents located on different plates will have different polar wander curves. In the case of North America and Europe, they have moved apart from each other as a result of divergent margin activity along the Mid-Atlantic Ridge. If the plate motion is reversed, then the curves are nearly identical (Fig. 6.4b). But if we did not know what the original configuration was, we could determine it from the apparent pole positions by moving the continents so that, at any given time, they share a common north pole. Therefore, we can use paleomagnetic data to determine relative plate motions through time.

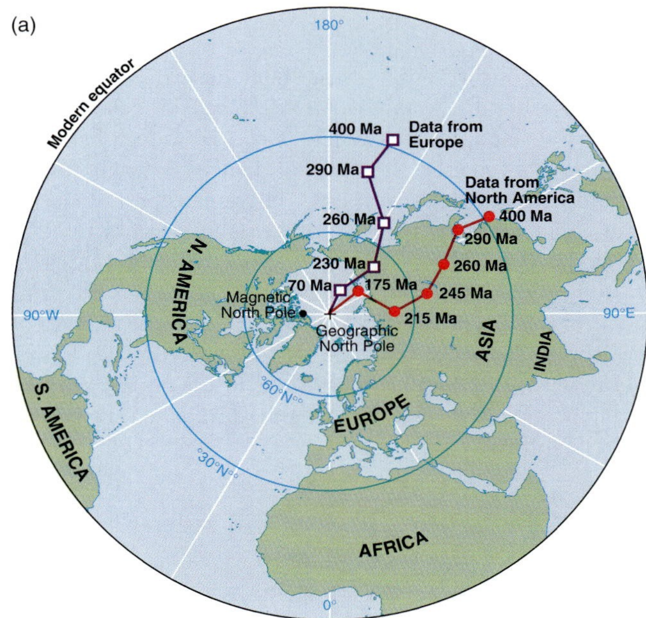

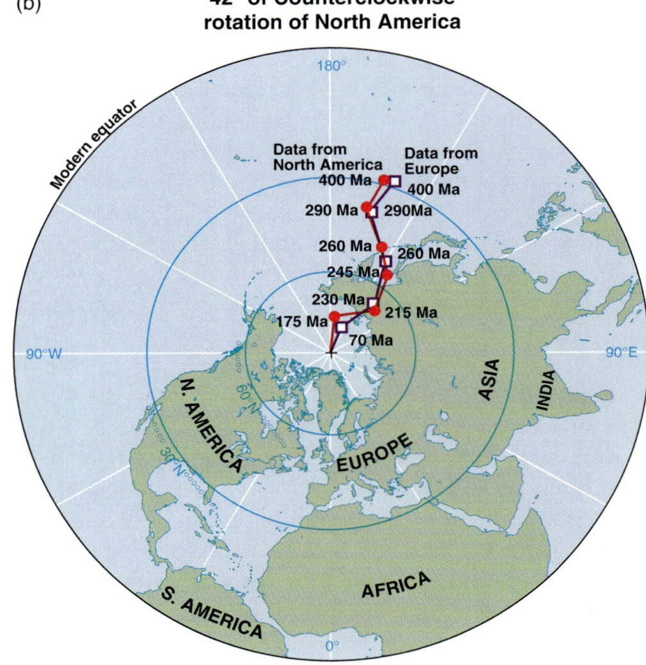

FIGURE 6.4 Apparent polar wander curves for Europe and North America shown on a polar projection of the world. (a) Apparent polar wander curves shown on a map of the modern world; some data points are omitted for clarity. (b) Calculated pole positions if plate motion is reversed so that continents occupy their relative positions at the times the rocks formed. Note the poles now nearly coincide, providing evidence that North America and Europe have moved relative to each other.

Paleomagnetism and Sea-Floor Spreading

In the late 1950s, two young earth scientists at the University of California, Berkeley—a geologist named Brent Dalrymple and geophysicist Allen Cox—began a scientific collaboration that was to have important consequences. Cox was interested in both the apparent wandering of the magnetic pole through geologic time and some intriguing initial evidence that Earth's magnetic field had, in fact, changed polarity several times in the geologic past. Dalrymple was working on his doctoral thesis using isotope dating (∞ see Chapter 1) to measure the decay of the naturally occurring radioactive isotope of potassium (^{40}K), a fraction of which decays to the inert gas argon (^{40}Ar). Potassium is an abundant element in common rock-forming minerals and so is very useful for determining the ages of rocks—particularly volcanic rocks. The collaboration of these two scientists ultimately led them to the Hawaiian Islands, where they were able to sample a very thick sequence of basalt flows.

Dalrymple dated these flows using the potassium-argon (K-Ar) technique, while Cox collected carefully oriented samples to measure the strength and orientation of Earth's magnetic field as it was at the time when the lava cooled below the blocking temperature. As we discovered (∞ see Chapter 5), both the strength and the direction of Earth's magnetic field are preserved in erupted molten rocks that become magnetic when they cool (Fig. 6.5).

Using these samples and others from different locations, Cox and Dalrymple were able to document reversals of Earth's magnetic field and the age of each reversal. Subsequent work by these two investigators has shown that many magnetic reversals have occurred irregularly in the geologic record and that the reversals were not restricted to Hawaii but occurred simultaneously worldwide. Over the past 5 million years, Earth's magnetic field has reversed, on the average, once every 200,000 years.

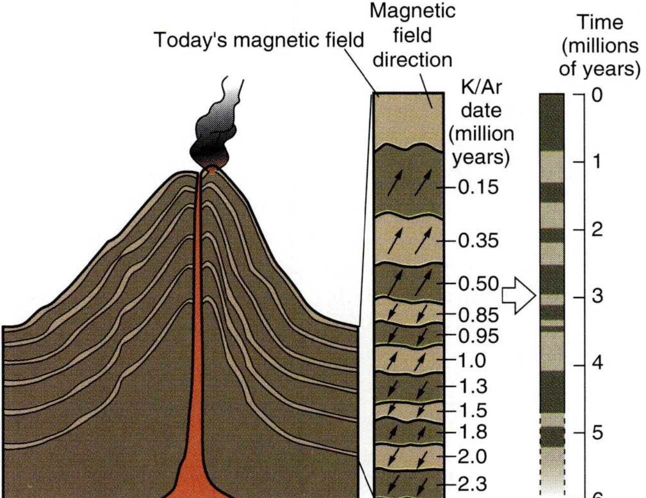

FIGURE 6.5 Layers of lava from a volcano cool below the blocking temperature and thus also record the direction of Earth's magnetic field at the time of cooling.

Reversals of the magnetic field have occurred many times and are simultaneous worldwide events.

Paleomagnetic research proceeded vigorously through the early 1960s, and a time sequence of reversals in Earth's magnetic field was established. In 1960, scientists from Scripps Institute of Oceanography in California used sensitive instruments (magnetometers) that determine Earth's magnetic field to demonstrate that the ocean floor off the west coast of North America shows a symmetric, striped pattern of varying magnetic intensity.

Across mid-ocean ridges, such as the Mid-Atlantic Ridge or the East Pacific Rise, a pattern of alternating high and low magnetic intensity bands was also observed. Moreover, these magnetic stripes stretch for long distances parallel to the ridges and are symmetrical on opposite sides of the ridge axis (Fig. 6.6). By 1963, it was generally known that similar, consistent magnetic patterns on the ocean floor existed in many places, but no one knew why.

In 1963, Fred Vine and Drummond Matthews of Cambridge University published a new idea that resulted from combining the results of the ocean magnetic surveys with those of Cox and Dalrymple. Vine and Matthews suggested a link between the magnetic stripes and sea-floor spreading. They proposed that magma rises to the surface at ocean ridges, forming new sea floor, which subsequently moves away from either side of the ridge to make room for more magma. As the magma cools, the magnetism in the minerals in the basalt becomes oriented into alignment with Earth's magnetic field. During times of normal polarity, the magnetic orientation preserved in newly crystallized minerals of the ocean floor basalts is oriented more or less parallel to modern north. Minerals that formed during times of reversed polarity, however, bear an alignment that is oriented toward the present-day south. Thus, when magnetic measurements are made across an oceanic ridge, basalts emplaced during times of normal polarity are recorded as a higher intensity field. The reason for the higher than normal intensity is that the field from the magnetic minerals in the basalt combines with that from Earth's present-day magnetic field to produce one that is stronger than normal. Basalts emplaced during times of reversed polarity are recorded as weaker-intensity magnetic bands because the magnetic field from the rocks in effect subtracts from Earth's present-day magnetic field.

Beginning in 1963, Harry Hess directed a major magnetic survey and sample-collecting expedition along the Mid-Atlantic Ridge south of Iceland. This cruise served as a major test of the Vine and Matthews hypothesis. Results of this survey, published in 1967, conclusively demonstrated that magnetic banding is symmetrical on the two sides of the Mid-Atlantic Ridge (Fig. 6.6a).

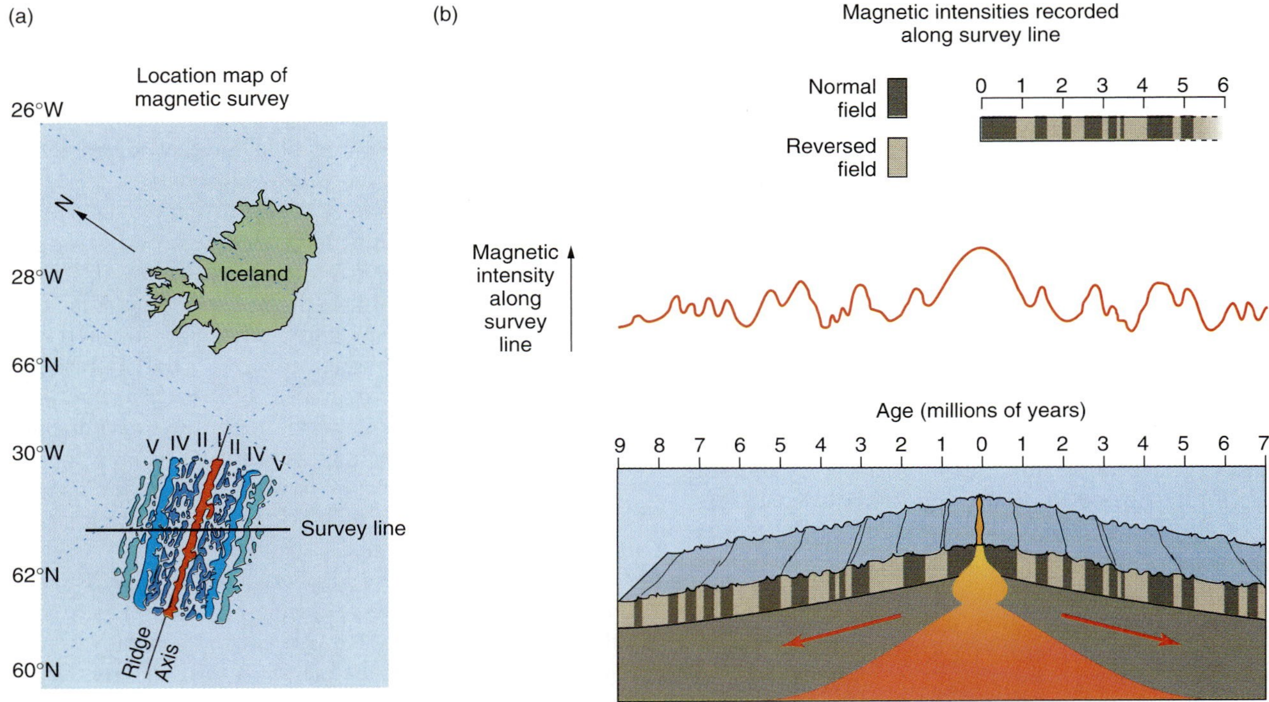

FIGURE 6.6 Magnetic reversals recorded in rocks of the oceanic crust. (a) Location of a magnetic survey across the Reykjanes Ridge—part of the Mid-Atlantic Ridge—near Iceland. Dark-shaded bands represent crust with a normal polarity magnetic field; light blue shading is reversed. (b) Magnetic-field intensity variations along the survey line. High and low intensity fields correspond to normal and reversed polarity, respectively, as summarized in the strip at the bottom of the figure. Ages are determined by correlation with sequences of lavas on land (Fig. 6.5), which were dated isotopically using the K-Ar method.

Moreover, the sequence of polarity reversals recorded in the basalts is identical to the magnetic reversal stratigraphy earlier documented by Cox and Dalrymple as well as other researchers (Fig. 6.6b).

Hess's interpretation that the sea floor is spreading outward from mid-oceanic ridges due to magmatic activity at the ridges was substantiated in this survey. Together with fellow marine geologist Robert Dietz, he popularized the concept of sea-floor spreading. Thus, the sea floor at oceanic ridges is geologically young, becoming progressively older as it moves away from the ridge (Fig. 6.7).

A symmetrical magnetic pattern is recorded in the ocean-floor rocks across spreading ridges. The alternating high and low magnetic-intensity bands mirrored on opposite sides of a ridge are created by basaltic magma rising into a mid-ocean ridge during times of normal and reversed polarity, respectively.

The Rock Record: Fossils, Climate, and Continents

The paleomagnetic record not only records relative plate motions, but also motions with respect to the geographic poles. On average, over geologic time, the magnetic poles of Earth have been found to correspond to the geographic poles. Thus the polar wander curves of Figure 6.4 show that four hundred million years ago Europe and North America were much further south than they are at present. The climates then would have been different from those of today.

There are striations in 400-million-year-old rocks in the middle of the Sahara Desert that could only have been made by glaciers. In Brazil, glacial deposits that are 250 million to 300 million years old contain boulders of a rock type unknown elsewhere in South America but common in southwest Africa. Across Europe, central North America, and the northern tip of Africa, clear evidence shows that tropical rain forests existed 300 million years ago, which were replaced by deserts 50 million years later. These are striking indicators of continental climatic changes, but not necessarily of global climate changes. We know that the climate varies in the modern world, with its polar ice caps and tropical rain forests, and it seems reasonable to believe that similar climate variations existed during the geologic past. In fact, one of the fundamental principles in geologic reasoning is the idea that *processes that functioned in the geologic past are similar to those we observe on Earth today*. This is James Hutton's "principle of uniformitarianism," which we encountered in Chapter 1.

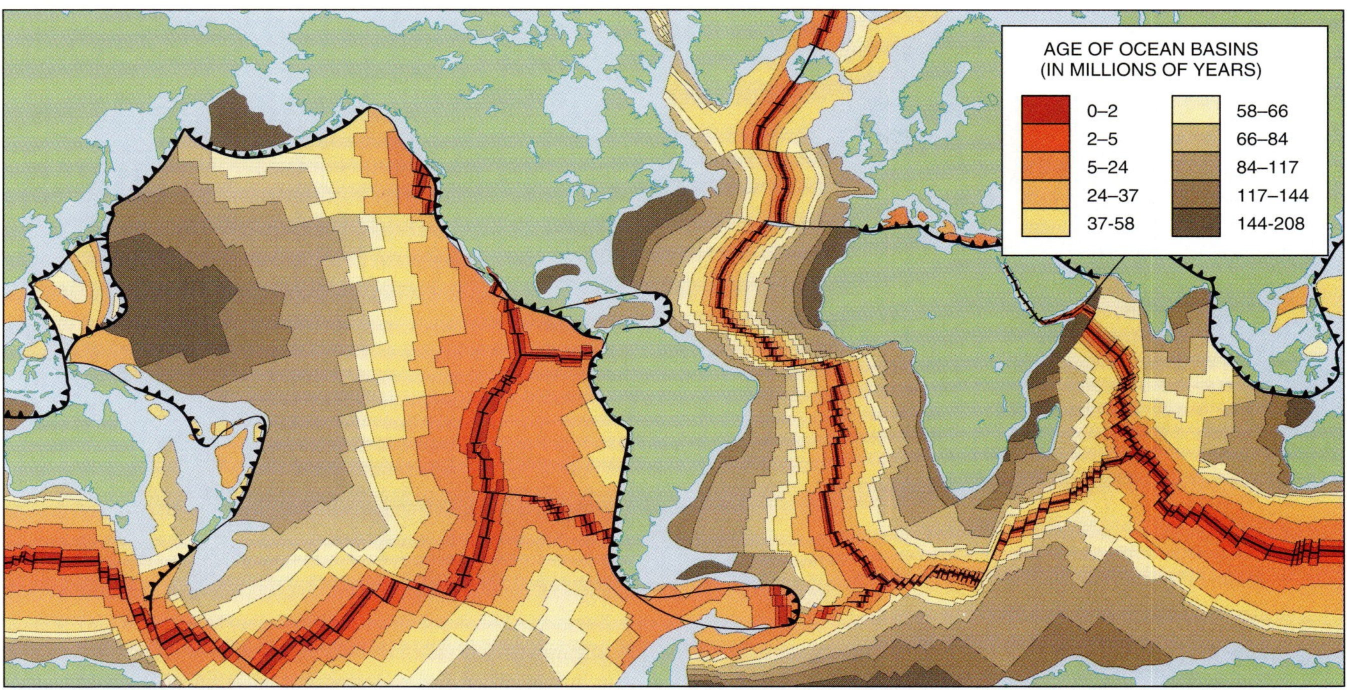

FIGURE 6.7 Global map of the age of the ocean floor determined from patterns of magnetic-field intensity.

Climate Dynamics

Abundant evidence shows that Earth's climate has fluctuated substantially—from temperatures that were much warmer than those of the modern climate to much colder temperatures. We know there have been periods when Earth lacked polar ice caps and there have been times of major glacial advances. Even in the absence of polar ice caps, however, the polar regions will always be colder than equatorial areas because the angle of sunlight incidence is much lower at the poles (Fig. 6.8). At high inclinations—when the sun is directly overhead—the maximum intensities of light and heat reach Earth; at low inclinations, the minimum heat and light intensities reach Earth. In Florida or Hawaii, a person can get a tan in December as well as in June because these regions of low latitude have a high angle of sunlight incidence even in the winter (see Fig. 6.8). The climate is warm and moderate, with little temperature variation through the year.

In contrast, polar regions above the Arctic Circle (66.5°N) experience long hours of daylight in summer and long hours of darkness in winter. But even during the arctic summer, the sun's angle is so low that temperatures rarely reach as high as the lowest temperatures in Hawaii. Thus, the most important control on Earth's climate belts is the relative angle of incidence of sunlight. We are therefore fairly certain that Earth's climatic belts throughout geologic time have remained more or less as they are today.

Atmospheric circulation on modern Earth is divided into a series of cells, in which circulation is nearly parallel to Earth's longitude lines (Fig. 6.9). Near the equator, air rises in a narrow region. Warm air can hold more moisture than cold air; there are many everyday illustrations of this phenomenon. Water beads form on a cold glass of water in a warm room because the moisture in the warm air condenses onto the cold glass. Air cools as it rises near the equator and the moisture picked up during its transport over warm ocean water is released, forming a climatic belt of high rainfall. The resulting cool, dry air circulates to higher latitudes, where it descends in a broad belt at about 30° latitude (north and south). The descending air creates a zone of high pressure that is, in fact, the downward flow of air in the same circulation cell that had its upward flow at the equator. As the dry air flows downward, it is strongly heated by compression, much as a tire heats up as it is pumped full of air. This belt of hot, dry air explains the distribution of many of the world's deserts around 30° latitude. The Sahara Desert of North Africa, for instance, stretches from about 25° to 35° and the desert of western North America is centered at approximately 30° of latitude.

Polar circulation cells are much weaker than the equatorial cells. Air rising slowly at approximately 60° north and south latitudes creates a secondary belt of high rainfall for the same reason that the primary rainfall maximum is created at the equator. Using the Northern Hemisphere as an example, the rising air of these circulation cells splits and part of the air flows north and part flows south; the southward flow intersects and joins the descending dry air at 30°N derived from the tropics, and the northward flow ultimately descends at the North

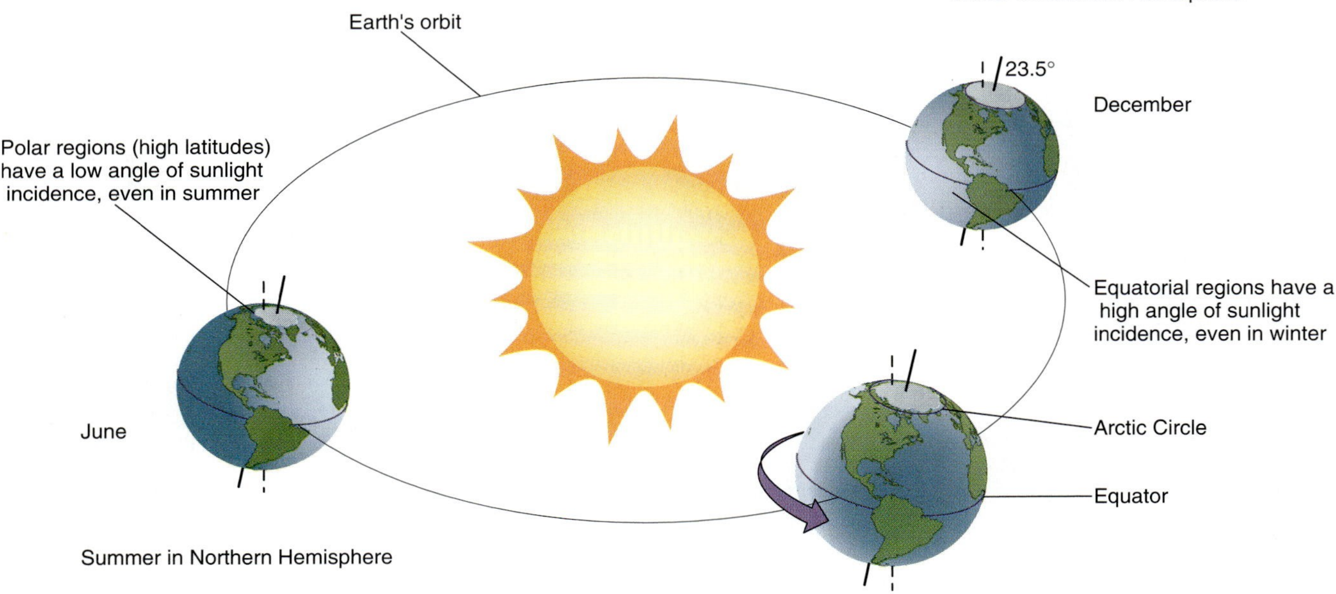

FIGURE 6.8 The tilt of Earth's axis of rotation causes climatic variations with latitude; varying amounts of sunlight—and thus heat—strike Earth at different stages in its orbit. When the Northern Hemisphere is tilted away from the Sun, it is winter, and the angle of light incidence on the North Pole is so low that there are nearly 24 hours of darkness above the Arctic Circle. Note that tropical and subtropical areas, such as Florida and Hawaii, are at sufficiently low latitudes that they receive abundant solar intensity even during winter. Six months later, the Northern Hemisphere is tilted toward the Sun, bringing the summer season.

Pole. The descending dry air in the polar regions is also heated by compression but, because of the low angle of the sun at high latitudes, the average temperature of the polar regions remains much cooler than that at 30° north and south. Polar areas are much like deserts; humidity and precipitation are very low.

Although *global* climate patterns are largely controlled by the angle of incidence of sunlight, *local* climate is strongly affected by topography, wind directions, and other localized conditions. The temperature of a landmass influences the outcome of an onshore flow of moist air from the ocean. If the land surface is colder than the air, fog or rain will result. And, if the onshore flow of air encounters a mountain range, the air flow is forced upward, is cooled, and rainfall results (Fig. 6.10).

As the now dry air flows on beyond the mountains, it descends on the other side of the mountain range, is heated by compression and becomes warm, dry air. The effect is a *rain shadow* on the lee side of the mountains. Along the western coast of the United States, air flows west to east. Rain and snow fall in the western foothills and in the high mountains of the Sierra Nevada range, resulting in a rain shadow east of the mountains. The Mojave Desert and the western Great Basin are examples of this phenomenon. In Brazil, moisture-laden air flows east to west across the Brazilian rain forests. As it flows upward and over the high Andes Mountains, the air loses all of its moisture; the Atacama Desert in Chile is in the rain shadow west of the Andes, and is one of the driest places on Earth (Fig. 6.11). The extraordinarily arid conditions of the Atacama Desert reflect not only the rain-shadow effect but also its location in the desert climate belt.

Global climate is fundamentally controlled by sunlight's angle of incidence. Local climates are strongly influenced by topography, wind direction, and proximity to the ocean.

The History of Earth's Climate

Any explanation of the occurrence of glaciations and tropical rain forests as observed in the rock record must be consistent with our knowledge of climatic controls in the modern world. Geologists have concluded that the explanation for climatic changes interpreted from the rock record through time is more often associated with the positions of the continents relative to the stable climatic belts on Earth, rather than to global climatic changes.

The distribution of Earth's climate belts is reflected in the populations of plants and animals across the globe. Many organisms are restricted to specific conditions—for instance, corals favor warm, shallow subequatorial seas, amphibians require humid, warm conditions, and penguins are adapted to cold polar conditions. As shown in the satellite image of Earth in Figure 6.12, vegetation over the surface of the continents is likewise systematically distributed, with dense forest and jungle in equatorial and temperate regions, and sparse vegetation in the dry desert regions of the tropics and the poles.

This general observation becomes a vital clue in the interpretation of geologic history. At any specific site,

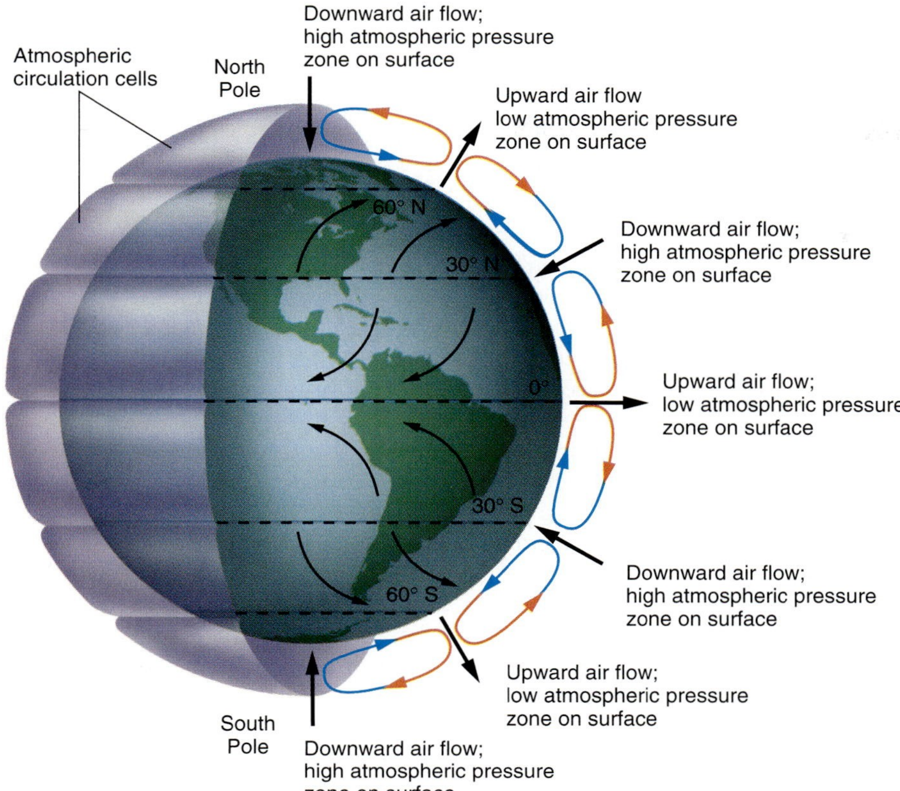

FIGURE 6.9 Atmospheric circulation cells, showing the upward air flow at 0° and 60° latitudes and the downward air flow at 30° and 90° latitude.

fossils of plants and animals from a given time, together with the types of rocks deposited at that time, can be used to infer the climate that existed then, and accordingly can be compared with the fixed distribution of climate belts to see how the site has moved since that time. By comparing paleoclimates with present-day climates and the paleomagnetic field with the current magnetic field, we can, in principle, locate the paleopositions of continents. It is worth noting, though, that while paleoclimatic and paleomagnetic evidence in surface rocks give scientists an idea of relative north–south motion (changes in latitude), east–west motions (changes in longitude) are difficult to decipher.

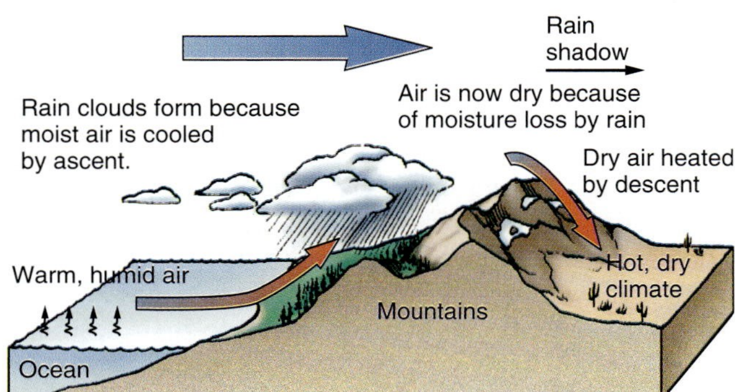

FIGURE 6.10 If the onshore flow of air encounters a mountain range, the air flow is forced upward and rainfall results from the cooling as the air rises. As the air flows on beyond the mountains, "robbed" of its moisture, it descends on the other side of the mountain range where it is heated by compression and becomes warm, dry air. The effect is a rain shadow on the "lee" side of the mountains.

Pangaea: Reconstruction of the Lost Continent

We next explore the development of the ideas that geologists offered to explain the climatic variations in Earth's history. The following discussion examines the pieces of the geologic puzzle that scientists worked with in the mid-1960s. It almost takes the form of a mystery story to illustrate how the science of geology progressed from the idea stage to the proposal of a hypothesis, to testing and modifying that hypothesis (or rejecting it, if necessary), to eventually drawing conclusions about the validity of the hypothesis. These conclusions, of course, are still subject to change.

A reexamination of Wegener's idea for a supercontinent provides a plausible explanation for the climatic variations in times past. Figure 6.13 illustrates a continental reconstruction in which the continents are positioned by paleomagnetic latitudes measured in rocks of different geologic ages collected from different continental areas (compare with Fig. 6.1).

Clues as to the past climatic conditions are contained in ancient rocks and fossils. Sand dune deposits and evaporites were most likely formed in arid desert envi-

FIGURE 6.11 A false-color satellite image of a portion of central South America. North is at the top. To the right of the photo, the green areas are the rain forests and grasslands to the east of the Andes Mountains. The brown zone corresponds to the Andes; the black areas are Lake Titicaca and Lake Poopó; the pale blue areas are dry lakes. West of the Andes is the Atacama Desert, near the bend in the coastline. Note the clouds are heavily concentrated east of the mountains, and Atacama Desert is in the rain shadow.

ronments. Rocks containing tropical plant fossils were likely deposited in hot, humid swamps such as exist in equatorial regions today. Glacial deposits and the effects of glacial erosion were probably formed in polar regions, where the most extensive glaciers are found today.

In the reconstruction of Figure 6.13, the glaciated areas group together, as do the tropical forests. Thus, the glaciated areas of South America and Africa fit together to provide a reasonable explanation for the South American location of exotic rocks that probably originated in southwest Africa. Evidence supports the proposal that South America and Africa were glaciated before Madagascar and India, which, in turn, were glaciated prior to Australia and Antarctica. If the large Pangaean continent slowly drifted across the South Pole, the glaciations associated with a southern polar ice cap would appear to migrate across the continent with time. Thus, the known timing of these old glacial episodes in the Southern Hemisphere is conveniently explained.

A somewhat similar explanation works for the occurrence of tropical rain forests, but in this case, it is the locations of the equator, the tropics, and the adjacent desert belts that we must consider. On modern Earth, the transition from polar regions into temperate zones is gradual, marked by continuously increasing average temperatures (∞ see Chapter 12). In contrast to the gradual transition from arctic to temperate conditions, at the northern and southern edges of the tropical climatic zones, the change is sudden—from hot, moist conditions near the equator to hot, dry deserts at about 25° to 35° of latitude.

Therefore, the same slow northward drift, 300 million years ago, of the great continent that caused the shifting locus of polar glaciation also caused the sudden transition from tropical rain forests to hot, dry deserts (see Fig. 6.13). Evidence for this change in conditions is preserved in the rock record of central North America and Europe.

Global changes in Earth's climate have occurred in the geologic past. The slow movement of continents across Earth's stable climatic belts has resulted in apparent climatic successions through time at a given location.

Flood Basalts and Continental Rifting More than 100 years ago, scientists noticed that similar rock types of similar ages existed on the edges of continents that are now separated by wide ocean basins. One such group of rocks is the vast provinces of lava flows erupted over relatively short periods of time (a few million years) known as **flood basalts**. Located mainly at the edges of the present-day continents, the oceanward sides of some of these provinces are faulted downward into the ocean basins as though they have been pulled apart (see Fig. 11.21). When the continents are rearranged into their positions of 150 million years ago, the flood basalt provinces are continuous. For example, the Etendeka flood basalts of southwest Africa and the Paraná flood basalts of southeast South America are both approximately 130 million years old. In the Paraná basin of Brazil, the flood basalts rest upon 250-million-year-old desert deposits, and the Etendeka flood basalts of southwest Africa lie on desert deposits of a similar age.

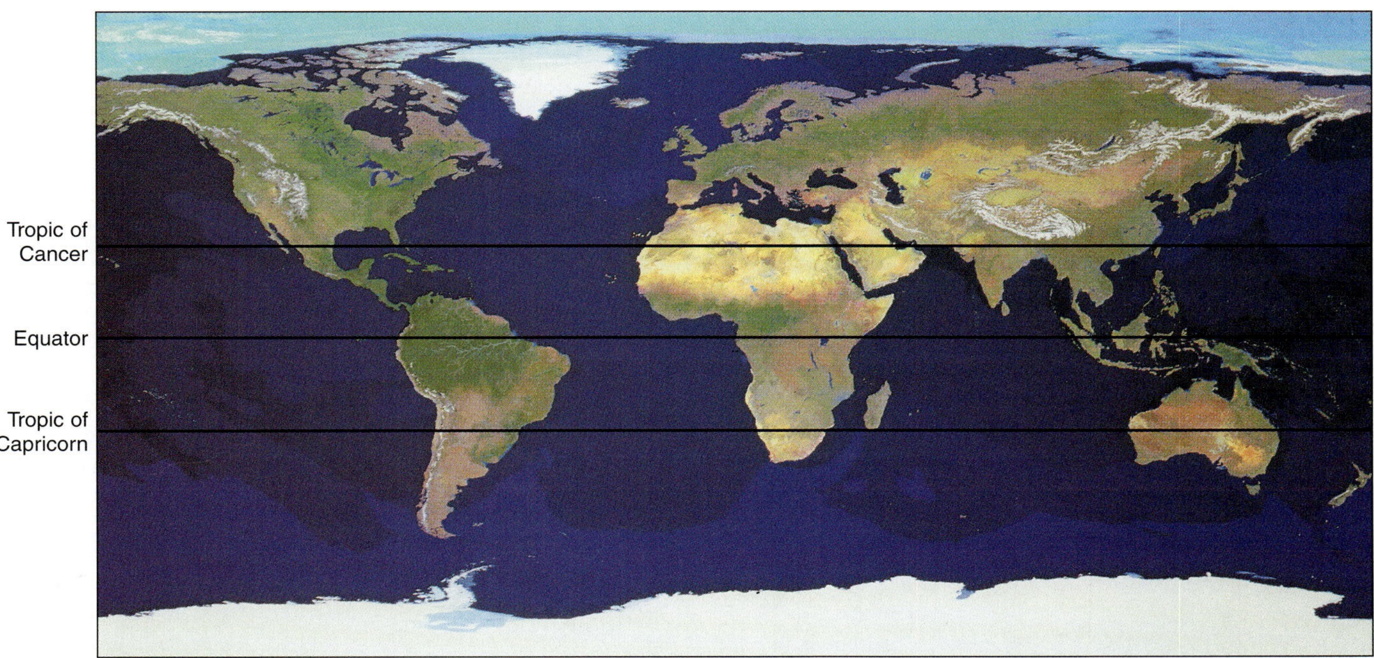

FIGURE 6.12 A satellite view of Earth, showing climate belts.

Why is rifting associated with flood basalt provinces, and why are the flood basalts almost invariably affected by the rifting? The flood basalt provinces were not simply split apart coincidentally; they appear to be an integral part of the rifting process. The ages of the flood basalts in any given province correspond closely with the age of continental rifting there. For instance, the Paraná basalts of Brazil and the Etendeka basalts of southern Africa were erupted at the same time that the southern Atlantic began to open—approximately 130 million years ago (cf. Fig. 6.14a and b). Flood basalts were also erupted 180 million years ago in Antarctica and southeast Africa as a result of the separation of those two continents. The northern Atlantic began to open much later—approximately 60 million years ago; therefore, the North Atlantic basalt provinces of Greenland, western

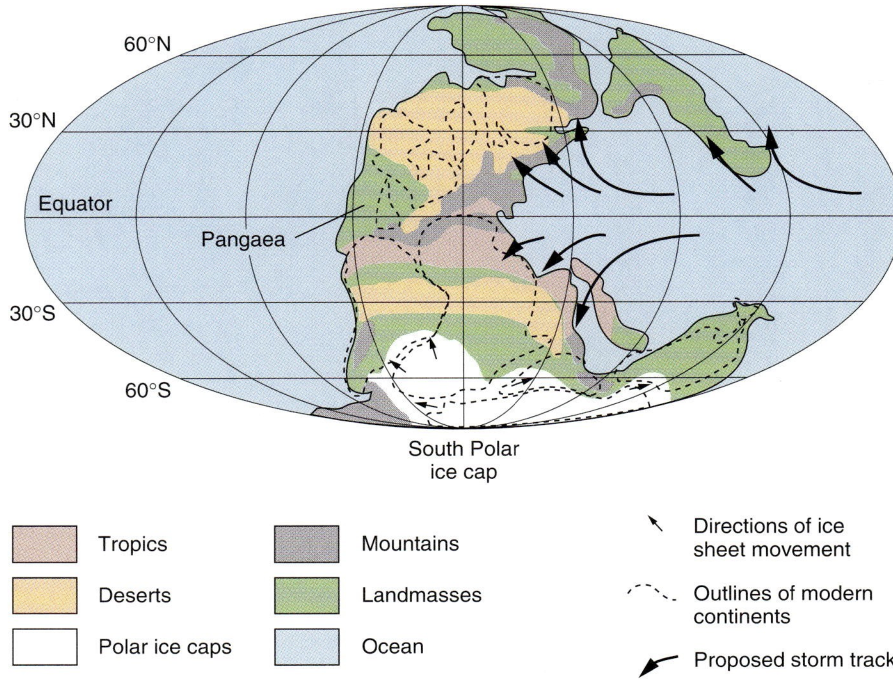

FIGURE 6.13 A reconstruction of the continents and paleoclimates of approximately 255 million years ago. The glaciated areas group together, with ice movement directed outward from the South Pole. A band of tropical and subtropical environments is interpreted from sedimentary rocks to exist between the latitudes of 20° north and south. Note that most of the Pangaean landmass was located in the Southern Hemisphere; the central part of present-day North America is within the tropical climatic belt.

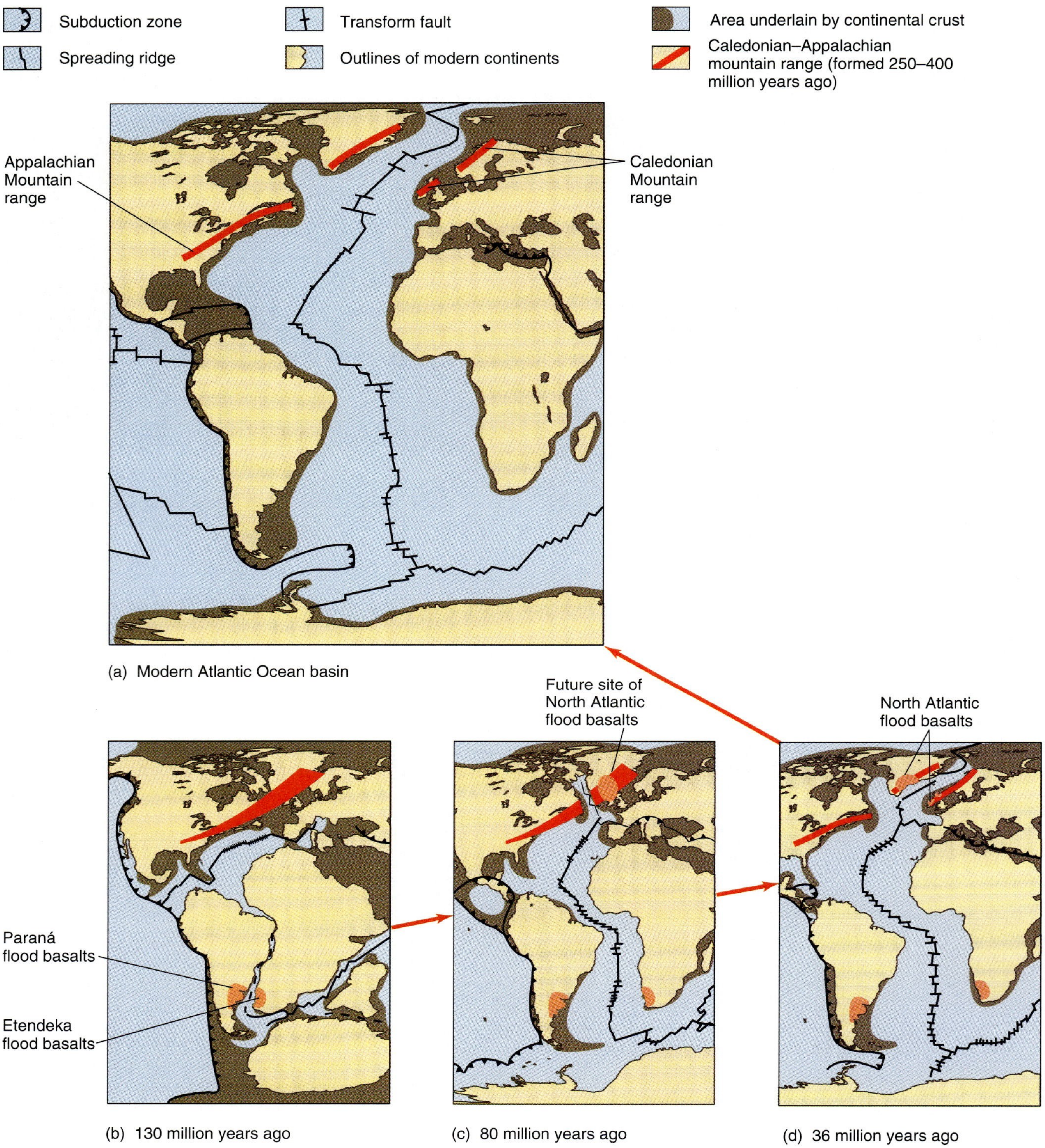

(a) Modern Atlantic Ocean basin

(b) 130 million years ago

(c) 80 million years ago

(d) 36 million years ago

FIGURE 6.14 The opening of the Atlantic Ocean basin. (a) The modern Atlantic Ocean and surrounding continents. (b) A continental reconstruction of 130 million years ago. Compare with the configuration of Pangaea at 255 million years ago shown in Fig. 6.13. Africa and South America have moved east relative to North America. The South Atlantic has begun to open although opening of the Central Atlantic is well underway, and the Paraná and Etendeka flood basalts have already erupted. The north Atlantic has not started to open. (c) The Atlantic as it looked 80 million years ago. The ocean basin separating Africa, South America, and North America has developed and widened. The rift between northern Europe and North America (Greenland) has yet to develop. The position for the rifting of 60 million years ago is indicated by the future location of the North Atlantic flood basalt province. (d) By 36 million years ago, the north Atlantic has opened and separated the North Atlantic flood basalt province.

Scotland, and northern Ireland are correspondingly much younger (Fig. 6.14c and d).

We will delve into a more detailed discussion of the causes for the breakup of large continental masses later (∞ see Chapters 8 and 11). Briefly, we now conclude that *mantle plumes*—localized zones of upwelling hot mantle—may have contributed to the processes that caused continental breakup. These plumes give rise to *hot spots*, a term emphasizing the localized character of the high heat flow. A mantle plume that rises beneath a continent causes the continent to bulge upward, creating an uplift. In response, the parts of the continent on opposite sides of the bulge tend to "slide" or be pushed away from the uplift. As well as exerting an outward force, the plumes represent zones of weakness in the lithosphere which, rather like perforations in a block of stamps, may have channeled the break to intersect their locations. Whether plumes actively break continents or just weaken them is a subject of ongoing study.

Many of the larger plumes are associated with flood basalts that were erupted at their point of inception. Flood basalts are believed to be a result of the melting of the upper mantle in the zones of excessively high heat flow and, therefore, represent the initial stages of continental rifting. For this reason, then, flood basalt provinces have a special place in the plate-tectonic puzzle.

Mantle plumes rising beneath a continent may melt the upper mantle, producing huge volumes of basalt and weakening the continent where it splits apart. Therefore, the initial stages of continental breakup are often characterized by flood basalts.

Plates and Plate Boundaries

The term "plate" defines a coherent unit of lithosphere that is bounded by some combination of spreading ridges, subduction zones, or transform faults. The crucial test of coherence is that a plate displays no internal relative motion; it moves as a unit relative to its neighboring plates. If we focus on one plate, each of the neighboring plates is likely to be moving in a somewhat different direction. Some plates may be converging; others may be diverging. Still other plates may be moving parallel to the boundary between them along transform faults. The boundaries between plates can be divergent, convergent, or transform margins, depending on whether the plates move away, toward, or past each other (∞ see Chapter 1).

It is important to realize that the boundaries of plates do not necessarily correspond to the boundaries of continents. In fact, this is the major distinction between plate tectonics and continental drift. The latter hypothesis suggested that the continents themselves move across the globe as discrete units. We now know that continents are simply a low-density crustal composition that floats isostatically higher in the mantle, and that they are an integral part of the lithospheric plates. Crust (continental and oceanic) and mantle are distinguished on the basis of composition, whereas lithosphere and asthenosphere are distinguished on the basis of mechanical properties. Where lithospheric boundaries do correspond to the edges of continents, such as along the western edge of South America, the continental margin is referred to as **active**. In contrast, a **passive continental margin** occurs where the edge of the continent—which corresponds roughly to the juncture between continental and oceanic crust—is not located at a plate boundary.

The lithosphere is broken into 8 major plates and several smaller ones—microplates—that are in contact with one another (see Fig. 6.1a). The lithospheric plates "slide" over a zone of weakness that has been identified as a region of low seismic wave velocity called the *low velocity zone* (Focus 6.1). On Earth's surface, the plates are always in contact with each other; when one plate moves, it causes other plates to move.

Plates typically move at rates of 0.1 to 10 cm per year—rates too slow for us to notice. We know from the rock record, however, that plates have moved huge distances over geologic time. The Atlantic Ocean, for instance, did not exist 250 million years ago, and therefore must have opened over a period of many millions of years. How fast is the ocean basin now opening? These movements can be measured today using Global Positioning System (∞ see Chapter 7). The rate at which New York and London are moving apart can be quite accurately determined by taking measurements over the course of a few years with the use of satellites, which transmit radio signals to ground stations located in both cities. Because New York and London are located on different plates (North American and European, respectively) on either side of the Mid-Atlantic Ridge plate margin, the rate obtained is a direct measure of the spreading rate along the Mid-Atlantic Ridge. The measurements show that the northern Atlantic Ocean is opening at a rate of about 2 cm annually.

Each type of plate margin exhibits distinctive patterns and types of earthquake activity, volcanic activity, and plate thickness. The characteristics of plate margins are discussed in later chapters. The distribution of plate margins is an integrated system, though, and we must consider what happens when plate margins come together, or where one type of plate margin changes into another.

Triple Junctions

If three plates meet, their boundaries must converge at a single point; this is called a **triple junction**. If the plates at a triple junction can continue to move without changing the geometry—the relative positions—of the three plates, then the triple junction is termed *stable*. If not, it is termed *unstable*. An example of a stable configuration is a triple junction at a spreading center where three plates are in contact at mutual angles of approximately 120° (Fig. 6.15a).

Focus On 6.1 THE NATURE OF THE LOW VELOCITY ZONE

The main characteristic of the low-velocity zone (LVZ) is a rapid drop of about 6 percent in the velocity of seismic waves, with a more pronounced decrease for S waves than for P waves (∞ see Chapter 5 for a discussion of P and S waves). In addition, a significant attenuation of seismic waves occurs within the low-velocity zone—greater than that in other regions of the mantle.

At the base of the lithosphere on the Pacific Plate, seismic velocities drop dramatically (approximately 10 percent), and the depth of the low-velocity zone coincides with the depth at which the mantle temperature is closest to its melting point (solidus) (see Figs. 4.12 and 5.9). For this reason, the LVZ under the Pacific Plate could be attributed to the presence of a small percentage of melt.

Under the continental part of the North American Plate, the seismic velocities at the LVZ drop by only about 5 percent or less, and the depth does not correlate with the expected depth for peridotite melting.

Recent work on the behavior of minerals at high pressures suggests that melt may not be a necessary ingredient to diminish the shear strengths of minerals and therefore of rocks. Most crystals have imperfections called *dislocations* in their lattices, which greatly lower the shear strength of a mineral. Experimental data suggest that deformation by crystal dislocation occurs at great pressures and, therefore, will occur at great depths within Earth. Such effects could lower the velocity without requiring the presence of melt.

The origin and character of the low-velocity zone is the subject of current research. It is clear, however, there must be more than a single explanation for the LVZ; further, apparently different mechanisms may be at work under the continental lithosphere and under the oceanic lithosphere.

In contrast, a triple junction where three transform faults come together is always unstable (Fig. 6.15b). The stability of triple junctions is limited by the relative motions among the plates. Stability should not be interpreted to mean that the junctions may not move; in fact, triple junctions can be quite mobile. It is the *configuration* between the plates that is termed stable or unstable.

Geologists have developed a simple shorthand to discuss triple junctions: A spreading ridge is defined as R (R for **R**idge), a subduction zone (trench) as T (T for **T**rench), and a transform fault as F (F for transform **F**ault). There are many possible combinations of these elements in triple junctions—FFF, RRR, RTT, and so on—but there are only a few that are always stable or always unstable. The RRR geometry is always stable. The absolute position of the RRR triple junction may migrate, but as long as each one of the three ridges continues to spread, the triple junction geometry will remain constant (Fig. 6.15).

A triple junction is the point at which three plates meet. Stability of the triple junction is controlled by the relative motions among the plates. A triple junction can be defined as stable if the relative motions of the plates are able to continue without changing the geometry of the junction.

A Plate-Tectonic Link to Earthquakes and Volcanoes

The rim of the Pacific Ocean basin has long been called the "Ring of Fire" because of its abundant volcanoes. With equal justification, it could be called the "Ring of Shaking" because this same area is also the locus of many great earthquakes, which are felt over vast areas. The rim of the Pacific basin is different from the "Atlantic Rim" in a very important way: the Pacific basin is surrounded by plate margins. The rim consists of a series of subduction zones and transform faults from the southern tip of South America northward around Alaska and the Aleutian Islands southward past Kamchatka and Japan, the Philippines, and across past the Tonga Island chain northeast of Australia (Fig. 6.1a). For that reason, the populations living on the edge of continents bounding the Pacific coast are situated directly on the plate margins, and may potentially be threatened by earthquakes and volcanoes. In contrast, the Atlantic basin is largely bounded by passive continental margins, which tend to be tectonically quiet.

Major earthquakes are exceedingly frightening events, even though we now know that they are a natural part of the working processes of Earth. A great deal of folklore has developed around the association between earth-

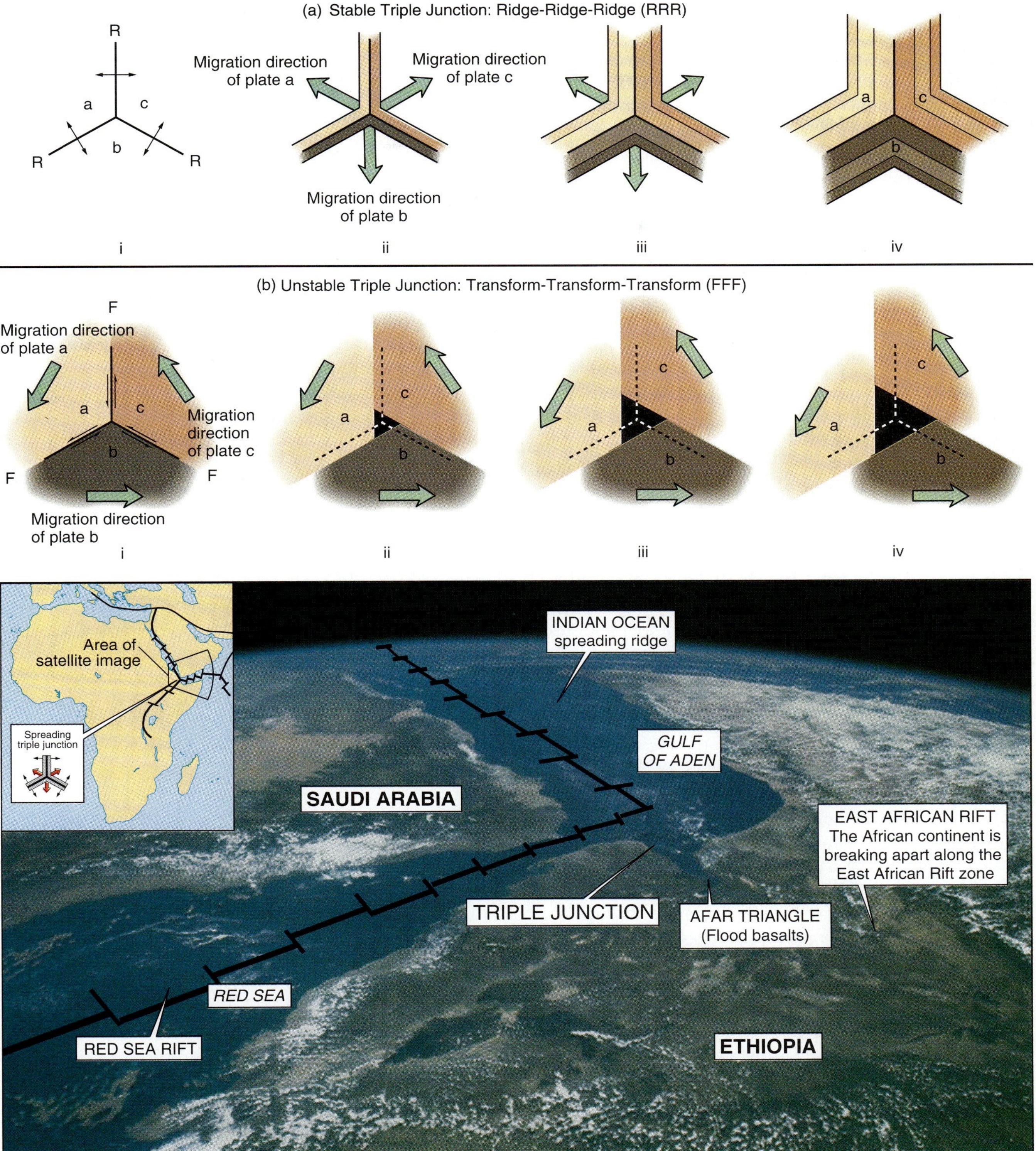

FIGURE 6.15 Triple Junctions. (a) A stable triple junction among three plates (a, b, and c). The ridge-ridge-ridge triple junction (RRR) is always a stable geometry. Stage i: Initiation of the RRR triple junction. Arrows indicate spreading. Stages ii–iv: Spreading continues as the junction maintains a constant geometry. (b) A transform-transform-transform triple junction (FFF) is always unstable. Stage i: Initiation of the FFF triple junction. Arrows indicate the direction of movement for plates a, b, and c. Stages ii–iv: With equivalent movement along the three transforms, a void would open in the lithosphere (indicated in black), which is impossible; thus, this geometry is unstable. (c) A triple junction comprising three active rifts at the intersection of the Red Sea, the Gulf of Aden, and the East African Rift, which runs across Ethiopia and south along eastern Africa (see inset). If this rifting continues, both the Red Sea and the East African Rift will eventually open as oceans, with East Africa drifting away as a separate continental fragment.

quakes and the seacoast. The ancient Greeks attributed earthquakes to the effects of subterranean winds blowing through cave systems. Early Japanese folktales described Earth resting on a great fish. When the fish would periodically move, its movement caused earthquakes. Other coastal countries had similar tales in which Earth was resting on a whale or some other sea animal. One idea proposed that Earth was largely hollow and contained a great internal ocean. When gigantic storms swept across this internal sea, the resulting waves caused earthquakes.

Our modern understanding of earthquakes is less colorful, but more accurate. We know that faulting causes earthquakes, and that volcanic activity and earthquakes are commonly closely associated. However, devastating earthquakes can also occur long distances from volcanic activity, although these are usually still associated with plate margins. For example, the 1906 San Francisco earthquake was on the San Andreas transform fault and affiliated structures. Earthquakes in the Los Angeles area of 1971 and 1994 were caused by stresses associated with the San Andreas fault, even though the actual faults that moved were not exactly along the line of the San Andreas.

Prior to our present understanding of plate motions, subduction, and volcanism, it was observed that andesitic volcanoes are usually confined to the margins of the oceans. Even before the concept of plate tectonics was developed, geologists observed that explosive volcanoes tend to occur in narrow zones such as the line of volcanoes—the Ring of Fire—surrounding the Pacific basin. In fact, it was suggested during the mid-1700s that the distribution of andesite volcanoes around the edges of ocean basins may be due to the involvement of seawater gaining access to Earth's interior along zones of weakness between landmasses and ocean basins. We now know that magmas are formed as the result either of mantle decompression or of lowering the melting point of the mantle through the addition of fluids. These processes directly relate to how plates interact at their margins.

Earthquakes and volcanoes are concentrated along plate boundaries and are the consequence of processes occurring along these boundaries. However, there are instances of earthquakes and volcanism occurring well away from plate margins. Some of Earth's largest currently active volcanoes, such as on the Hawaiian Islands, are, in fact, located near the centers of plates. Such magmatism is the result of mantle plume activity. The mantle plumes give rise to hot spots, which, in turn, may form hot spot trails.

Hot Spot Trails

Hot spot trails provide a very convincing piece of evidence that illustrates plate migration over time. As we discussed in Chapter 2, Earth's mantle convects in order to lose heat. Mantle plumes are an important component of this convection system: They are concentrated regions of hot (and therefore lower density) upwelling material, which are independent of plate boundaries, and therefore most easily distinguished at the interiors of plates (∞ see Chapter 11).

One characteristic of these plumes is that the hot, ascending material from the deep mantle will melt when it reaches the shallow mantle; this produces basalts (∞ see Chapter 4). The basalt magma rises rapidly to the surface to erupt at volcanoes. If we examine the distribution of the volcanoes formed by a hot spot, we notice that the volcanoes are commonly arranged in a linear chain in which the volcanoes become progressively older along the chain in a direction away from the currently active volcano (Fig. 6.16). This observation is curious and generally defied explanation until it was put into a plate-tectonic context.

Mantle plumes arise deep in the mantle—some may come from the core–mantle boundary—and the position of these plumes is believed to be relatively stationary over long periods of time (possibly hundreds of millions of years). In contrast, the lithospheric plates on Earth's surface move relative to each other and relative to the underlying mantle. Therefore, the plates must move over the plumes, and the volcanism produced from a plume may generate a hot spot trail. However, many plumes are short-lived and do not develop hot spot trails.

Figure 6.1b shows the distribution of hot spots around the world. In the middle of the huge Pacific Plate are a number of hot spot trails—chains of volcanoes that form islands or are submerged as seamounts (Fig. 6.16). The Hawaiian Island–Emperor Seamount Chain is perhaps the best-known example. The volcanoes become older in a northwesterly direction: The volcanoes in the northwestern part of the chain are inactive, deeply eroded, and mainly submerged. By using radioactive dating techniques (∞ see Chapter 1), such as the potassium-argon technique, we can date the volcanic activity. This information is used to calculate the velocity of the Pacific Plate.

The volcano on the big island of Hawaii is currently active and therefore has zero age. Midway Island, 2432 km to the northwest, is 27.2 million years old. Therefore, the Pacific Plate moved 2432 km in 27.2 million years, which translates to a plate velocity of 8.8 cm/year. Note that velocity is a vector quantity and not simply a speed because it has both a *rate* of movement and a *direction* of movement (to the northwest).

Take note of another important feature shown on the map in Figure 6.16. There is a bend in the hot spot trail for the Hawaiian Islands as the islands pass into the submerged Emperor seamounts. A similar bend occurs in the hot spot trails of the Marshall and Ellice Island chains in the South Pacific. Rocks from volcanoes located near these bends have been determined to have ages of about 43 million years. This indicates that the Pacific Plate *changed its direction of movement* approximately 43 million years ago; its migration altered from a north–northwestern direction (the orientation of the oldest parts of the hot spot trail) to the present northwesterly direction.

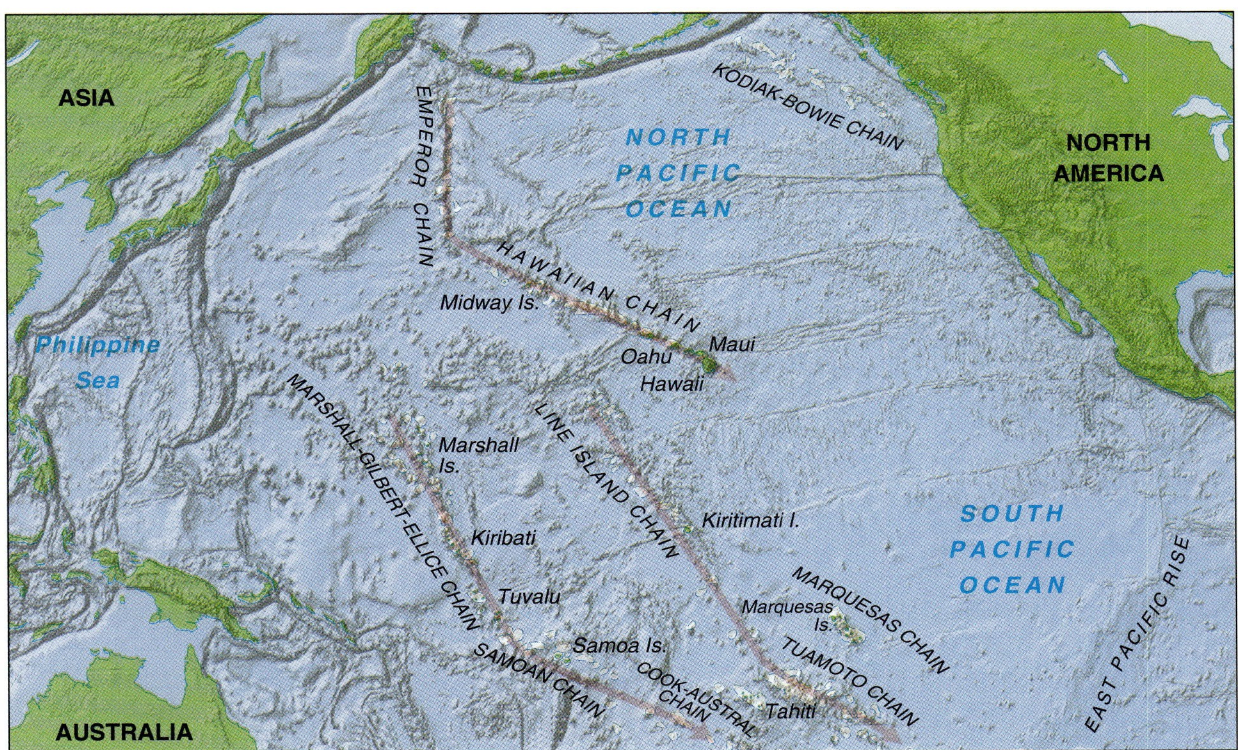

FIGURE 6.16 Several volcanic island chains on the Pacific Plate show a characteristic track in which the active volcano is at the southeastern end of the chain and the islands become progressively older toward the northwestern end of the chain. Each chain contains a bend, and the age of each island at that point is 43 million years in all measured cases. A volcanic chain is formed by movement of a plate over a hot spot, which melts successive holes through the crust to create a string of volcanoes, becoming progressively younger in the direction of the arrows. The Pacific Plate was moving in a north–northwest direction prior to 43 million years ago; subsequent to that time, it has been moving in a more westerly direction.

These observations are the only direct evidence for the way plates have moved in the past. The age of the bend in the Pacific Plate correlates with the age of some tectonic events on the western edge of North America, including complete disappearance of a plate subducted beneath Alaska called the Kula plate—Indian for "all gone;" further, it corresponds to the time that the Indian subcontinent collided with Asia. This is important because it suggests that changes in plate direction are in response to plate interactions elsewhere on the planet.

A hot spot trail is formed as a plate moves over a mantle plume. These trails can be used to determine absolute plate motions.

Topography of the Lithosphere

One of the fundamental features of ocean-basin morphology is that the oceanic plate gradually deepens and thickens with distance away from the mid-ocean ridges. This is understandable because, at greater distances from the ridges, the lithospheric plate has had a longer time to cool. The lithosphere—which is *defined* as being made of strong material—therefore thickens because cool material is stronger than hot material. From the principle of isostasy (∞ see Chapter 5), we saw that the plate can be thought of as floating on the asthenosphere. As oceanic lithosphere cools, it contracts and becomes more dense and so floats at a lower level.

Continental lithosphere, although also relatively cool, has an upper surface that generally rises above sea level. Although continental crust is slightly less dense than oceanic crust, this observation alone cannot explain why the average elevation of continental lithosphere is about 840 m, whereas oceanic lithosphere lies about 3800 m below sea level (Fig. 6.17). Remembering that lithosphere is made of both crust and the uppermost part of the mantle, and that the crust is some 20 percent less dense than the mantle, we can understand the higher elevation of continents based on the relative *proportions* of crust and mantle in the lithosphere (see Focus 5.2). Continental plates have thick crust (30 km) above the lithospheric mantle. The combination gives overall lower lithospheric densities for continental plates, which account for their floating at higher elevations. In contrast, the oceanic plate is made up of a thin (7 km) crust and therefore has a lower proportion of crust to mantle and a higher overall density; thus, it lies beneath the oceans.

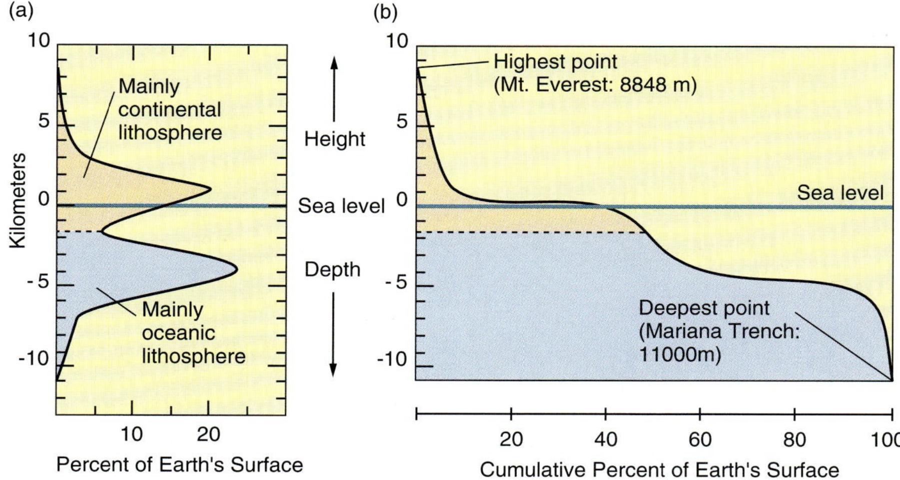

FIGURE 6.17 Distribution of elevations for the surface of Earth. (a) Percentage of Earth's surface corresponding to a given elevation above or below sea level. There is a *bimodal* distribution (two "peaks") with average continents at about 840 m above sea level, and average oceans 3800 m below. (b) The same information shown as the *cumulative* percentage of Earth' surface that lies above a given elevation. This can be thought of as a hypothetical cross section from the highest point to the lowest point on Earth's surface.

Submarine mountains built on the oceanic crust rise every bit as high above their respective bases as do continental mountains. The volcano Mauna Loa on the island of Hawaii has a greater elevation difference from its base to summit than has Mt. Everest, but, because of the greater density of the oceanic lithosphere, Mauna Loa cannot ride as high topographically as its continental counterparts.

What Drives Plate Motions?

You now have a picture of the configuration of lithospheric plates and how they move relative to each other. The following chapters will examine these issues in more detail and describe the processes and features associated with specific plate-tectonic environments. At this point, though, it is worth asking: What are the mechanisms driving plate tectonics?

Why plates move is a subject of ongoing research. One of the difficulties is that details of convective motions in the mantle and how they interact with the plates are not yet completely known. Nonetheless, considerable progress has been made and we will describe current thinking on this topic.

The forces that drive most plates are in balance; that is, plates are neither accelerating nor decelerating. Should two continents collide, the plates that are involved will decelerate, which occurred when India collided with Asia, but, in general, plates travel at constant velocity. Plates are the cool upper layer of the global convection system. The motion of the plates is the return flow of cool material that sinks because of gravity. The gravity driving force is then balanced by viscous drag of the asthenosphere.

The effects of gravity drive have been separated into three categories: slab pull, ridge push, and slab suction.

Slab pull arises from old oceanic lithosphere (a **slab**) sinking into the mantle at subduction zones and literally pulling the remainder of the plate in behind it (Fig. 6.18). This force is thought to be the most important of the forces acting on plates, but it is opposed by an equally large force of viscous resistance of the mantle.

Ridge push is a consequence of the structures of oceanic ridges. The lithosphere of ocean ridges thickens steadily from a value near zero at the ridge crest to about 100 km on either side (Fig. 6.18). The fluid asthenosphere is therefore uplifted at the ridge axis and drops slowly with distance in a manner similar to the topography. Pressure in the uplifted asthenosphere exerts an outward force on the overlying lithosphere. The sum total of this force across the full topographic structure is referred to as ridge push. A similar mechanism is important on continents in regions of elevated topography where continental rifting occurs. Ridge push is also counteracted by viscous drag in the asthenosphere.

Slab suction is the result of interaction between two plates at a convergent margin. Some slabs not only descend along their length but also sink vertically. This is referred to as *slab rollback*. The slab acts like a paddle sweeping back through the mantle, which has the effect of creating a suction on the overriding lithosphere (Fig. 6.18). As a result, subduction zones can migrate opposite to the direction of plate movement. The suction force pulling on the overriding lithosphere can move it toward the trench or even rift it apart.

Plates are thus subjected to ridge push, slab pull, or slab suction, which are opposed by viscous drag in the asthenosphere. Which of these dominates in a given situation depends on multiple factors such as height of the ridge, angle of subduction, temperature of the mantle, and age of the plate.

Certainly plate tectonics would not exist without slab pull, which is responsible for the return flow of mantle convection. The importance of slab pull is illustrated through observations on several different continents of the clear relationships between the plate migration rate and the proportional amount of plate margin associated with subduction zones. Rapid plate movement occurs

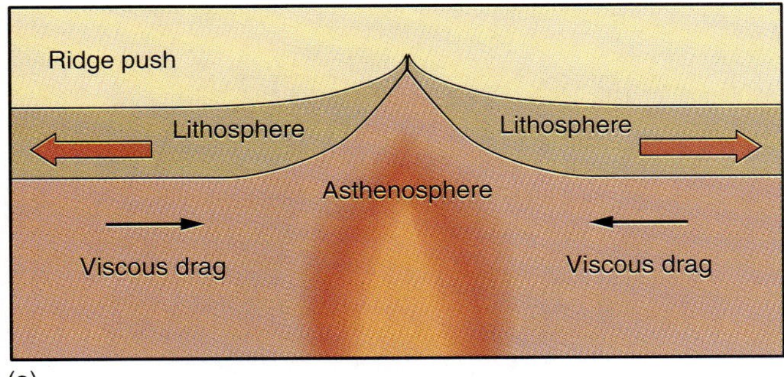

(a)

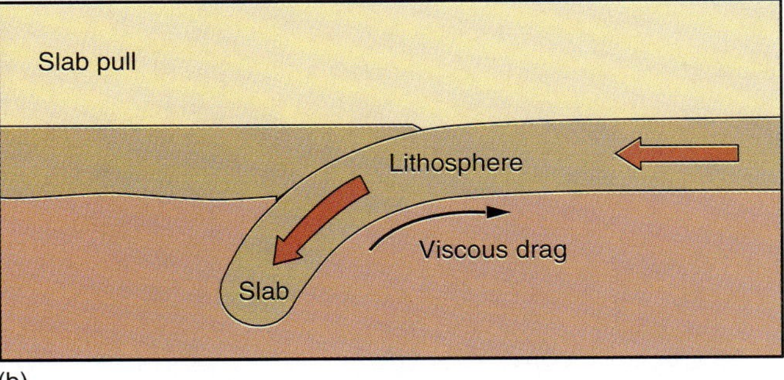

(b)

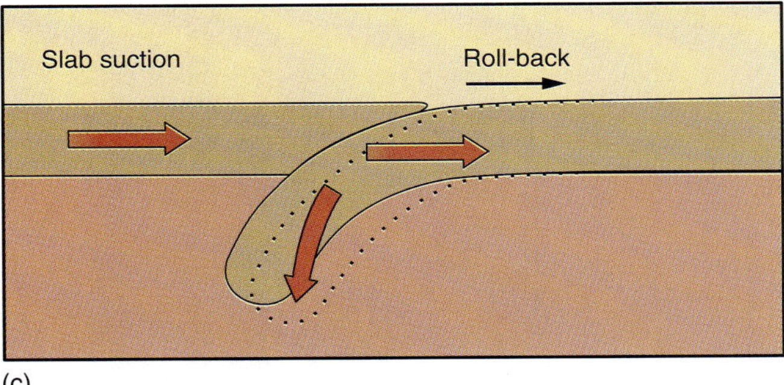

(c)

FIGURE 6.18 Gravity forces acting on plates. (a) Ridge push: pressure exerted on the lithosphere from uplifted asthenosphere. (b) Slab pull: force exerted by sinking old oceanic lithosphere. (c) Slab suction: rollback of subducted lithosphere.

where a subduction zone along one edge of the plate involves more than 20 percent of the plate edge.

As the downgoing slab sinks into the asthenosphere, several related phenomena occur. The cold slab is enclosed in hot asthenosphere and effectively cools off a portion of the asthenosphere. At the same time, the slab itself is being heated by the surrounding hot material. Because the slab is thick and rocks are good insulators, the slab remains cool for a relatively long period of time, extending the time of negative buoyancy necessary for continued sinking into the asthenosphere. As the slab sinks more deeply into the asthenosphere, transformations to denser mineral phases occur in response to the increased pressure at greater depths, which may emphasize the effective negative buoyancy. One example occurs at the 410-m transition—the phase transition to the spinel structure (Chapter 5). However, at the 660-km transition, the base of the upper mantle, the phase transition exerts a resistive force on the slab. Scientists currently debate whether slabs push directly through this transition or are sufficiently impeded to remain in the upper mantle. One set of calculations suggests that they pile up above the transition until the weight becomes so large that an avalanche of slab material descends into the lower mantle.

As the Atlantic Ocean opened, subduction off the west coast of North and South America has migrated or rolled back westward. The Atlantic ridge has exerted a push on the two continents. A new area of research in the earth sciences is the identification of areas of extension or compression in large areas of the crust (∞ see Chapter 7). Recent research has shown that the North American continent west of the Rocky Mountains is under extensional stress, and east of the Rockies the continent is under compressive stress. Eastern South America is also under compression. These observations may indicate that ridge push has been important in opening of the Atlantic, while slab suction and rollback have affected stresses in the west and allowed the continents to migrate westward.

Plate movement may have multiple causes, including slab pull, ridge push, and slab suction.

Plate Motions Through Geologic Time

To reconstruct the geography of ancient continents, many different types of evidence are used. Geologic evidence—similar rock types in identical sequences from different localities, or specific events such as glaciations—can be used to link one continental area with another. Paleomagnetism is an important tool used in continental reconstructions. The paleomagnetic pole for rocks of a given age can be determined; rocks of similar age that show the same pole position regardless of their present location and orientation were very likely part of the same continent at one time.

Using such methods, it has been possible to reconstruct plate movements, continental collisions, and continental breakups over the geologic past. We are hampered in attempts at continental reconstructions much before about 500 million years ago, however, because the rocks are either poorly exposed or so affected by more recent events that earlier events are difficult to decipher. But from about 450 million years ago forward, reconstructions can be made with some confidence; the more recent the reconstructions are in geologic time, the more accurate they are likely to be. The most important information that we strive to obtain is the direction *and* rate of plate migration—the plate velocity vector (Focus 6.2).

Focus On 6.2 VECTOR NOTATION FOR PLATE MOTION

We must use vectors to understand plate motions because both the geometry and the movement of plates are dependent on the sum of their motions (vectors). A vector has both a direction and a magnitude. The vector notation for an aircraft flying north at 600 km/hr is an arrow pointed north, with the length of the arrow properly scaled to represent a speed of 600 km/hr (Fig. 1). The resulting vector represents the velocity, which has both magnitude (speed) and direction; it is depicted by a bold small letter label—in this case, we will use **a**.

Imagine that the aircraft is trying to fly due north, but there is a wind blowing west at 200 km/hr, represented by vector **w** (Fig. 2). Airplanes are carried along with the wind, so the path of the aircraft is now deflected toward the northwest—even though the pilot's compass heading is true north. The path of the airplane relative to the ground (**g**) is the vector sum of the airplane velocity and the wind velocity (**g** = **a** + **w**; Fig. 2).

The movement of plates is also described as a velocity vector, having both a speed and a direction. However, the speed and direction must be specified with respect to a reference frame. For the aircraft, we used the ground. For plates, *it is the ground that is moving!* One choice is to use an adjacent plate and give the motion of a plate as if the adjacent plate were stationary. Alternatively, we can use the mantle beneath the plates, which the hot spots indicate is a stationary (or nearly so) reference frame. Consider the subduction of the Pacific plate at 7.5 cm/year in the vicinity of New Zealand, which is directed approximately west-southwest. The subduction vector **s** is the motion of the Pacific plate relative to the Indo-Australian plate. Relative to the mantle the Pacific plate moves at 8 cm/year northwest, given by the vector **P**, whereas the Indo-Australian plate moves at 5 cm/year close to due north given by **I** (Fig. 3). Relative to an observer on the Indo-Australian plate the Pacific plate has a velocity (of subduction) given by

$$\mathbf{s} = \mathbf{P} - \mathbf{I}$$

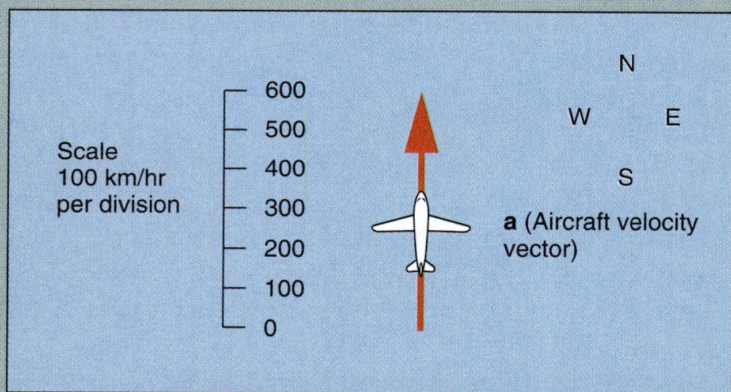

FIGURE 1

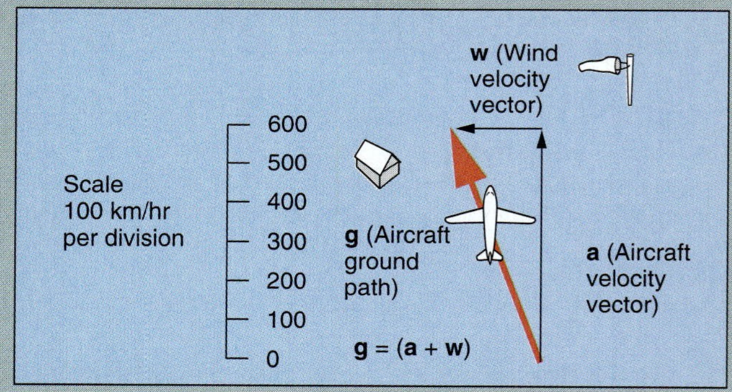

FIGURE 2

Uniformitarianism and Plate Tectonics

Uniformitarianism asserts that all the geologic processes that we can observe today, such as volcanic activity, plutonism, metamorphism, and faulting, occurred in the distant geologic past as well—and for the same reasons and by the same mechanisms. This is the fundamental guiding principle we use in deciphering the ancient geologic past. Geologic processes functioned in the past as they do now, with the exception of those physical processes that might be strongly influenced by biological activity.

The continents of 450 million to 500 million years ago had a different form, outline, and makeup than do those of today, but the landmasses were continents much like those of the present day—with central plateaus, mountains, lowlands, coastlines, rivers, and lakes. There were winds, storms, and ocean currents just like we find in our modern environment, although perhaps with dif-

that is the sum of the Pacific plate vector and the negative of the Indo-Australian plate vectors. This can be summarized in a general law that, for two objects A and B, the velocity of A relative to B is given by the vector A added to the negative vector of B.

One final note: We have assumed that plates are moving everywhere at the same velocity. However, large plates move on a sphere and thus some parts will move faster relative to the stationary mantle than will others. Consider a large plate rotating along lines of latitude. Portions at lower latitudes will travel faster than those at higher latitudes. A full description of plate motions involves rotation vectors in spherical geometry. This explains the apparently puzzling observation that the velocity vectors within a given large plate are not necessarily parallel on a flat map such as Figure 6.1. The linear vectors treated here, however, are adequate to treat limited regions of plates in relative motion.

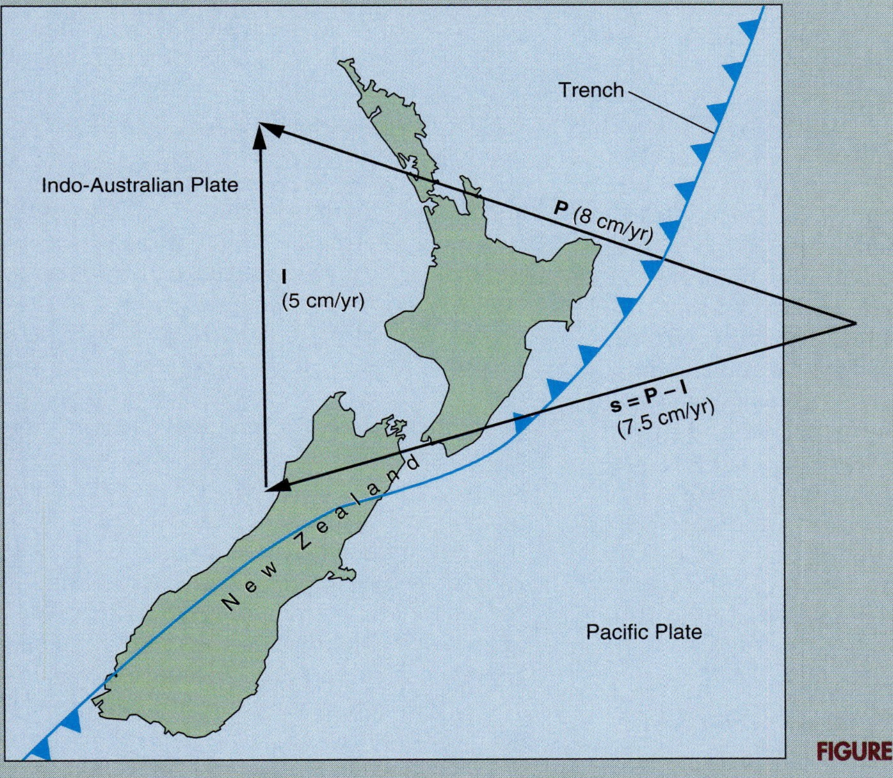

FIGURE 3

ferent directions and geometries. There were some major differences: Nothing was living on the land surface. At that time, all life was confined to the sea and perhaps also to some large lakes or rivers. As a consequence, weathering of minerals on exposure to the surface and the processes of sediment deposition were probably somewhat different from those we observe today.

The atmosphere may have had a very different composition than that of today; the oxygen or carbon dioxide content may have been significantly higher or lower than at present. However, weathering did occur and sediments were still deposited in streams, in lakes, and in the oceans. It is in the resulting rocks that we can recover the record of what happened and when.

How far back can we extend the geologic records based on plate-tectonic processes? We know that the mantle convected more vigorously early in Earth history, when it was hotter (∞ see Chapter 2), and it is possible that these conditions were unable to support plate tectonics as we know it. Consequently, it is worth modifying our uniformitarian view to accommodate the possibility of very different processes—or at least rates of processes—in the earlier stages of Earth evolution. Over the past 500 million years, representing less than one eighth of Earth history, we can, however, reconstruct plate movements reasonably well based on our current knowledge.

Continental masses of the distant geologic past were modified by processes similar to those of the modern world—erosion and sedimentation, volcanic activity, plutonism, metamorphism, and deformation.

If we return to consider Alfred Wegener's contribution, we find that, in many respects, he was not far wrong—although sometimes his mechanisms were in error or oversimplified. He understood that the present distribution of certain fossils is related to their having lived together on a single continent that is now in many fragments. Wegener's continental reconstruction for 300 million years ago (see Fig. 6.2) is reasonably accurate. He saw clearly that the Himalayas were the result of a continental collision, and this realization prompted further inquiry. His idea that mountain chains are wrinkles in the continental crust as a result of interaction with the sea floor is true in a sense, but his mechanism was wrong. Wegener did not know about subduction zones and spreading sea floors, but he did see the larger picture—that the entire Earth works together in a unified fashion. He observed volcanism, but could not have understood the interactions between the subducted slab and an overlying asthenosphere. It would also be difficult to explain the juxtaposition of eugeosynclinal and miogeosynclinal sequences in mountain belts without recourse to ocean basin closure, which brings together passive and active margin environments. The step forward in understanding Earth, then, was from Wegener's ideas of continental drift to plate tectonics.

Living with Geology: Can Plate Tectonics Cause Ice Ages?

An illustration of the complex interrelationships between climate and plate tectonics comes with recent suggestions that the ice ages of the past 2 million years may have been a consequence of the uplift of the Himalayan Mountains and the Tibetan Plateau, which is believed to be the result of the collision between the continents of Asia and India along a convergent plate margin. The uplifted region of the Tibetan Plateau is so immense that it affects the atmospheric circulation pattern. Heating of the high plateau by sunlight warms the air, which rises and causes an atmospheric low, sucking in air from surrounding regions. Much of the air so drawn in is laden with moisture from the warm Indian Ocean, which precipitates as the annual monsoon deluges over India and the southern Himalayas before ever reaching the plateau. But the effects are more far-reaching than simply disturbing the regional climate.

In Chapter 16, we will learn about the global carbon cycle. You have probably already heard of the "greenhouse effect." It refers to global warming that occurs as a consequence of increased carbon dioxide (CO_2) levels in the atmosphere. The current concern is that the carbon dioxide we are producing from burning fossil fuels is causing a greenhouse effect. The fundamental principle behind the effect is that carbon dioxide gas in the atmosphere traps infrared radiation (heat), making the Earth's surface warmer. Ignoring the recent contributions of humans, the carbon dioxide budget is controlled by:

1. Weathering and erosion, which takes carbon dioxide from the atmosphere and combines it with elements derived from weathered rock to form carbonates, which are then deposited in the oceans as rocks such as limestone.
2. Volcanoes, which return carbon dioxide to the atmosphere.

The uplift of the Himalayas and Tibetan Plateau, and the monsoon rains, led to greatly increased rates of weathering, which removed carbon dioxide from the atmosphere. As the mountains rise, and the valleys are carved deeper by erosion (Fig. 6.19), the topography becomes more extreme and the surface area of erodable rock increases, and carbon dioxide stripping from the atmosphere becomes even more efficient. The lower levels of carbon dioxide causes global cooling—sometimes known as an "icehouse" effect to complement the "greenhouse" effect. The exact causes of the ice ages are still not well understood (∞ see Chapter 14). There appears to be a strong control from orbital dynamics—how far Earth is from the Sun, which way the hemispheres are tilted from the Sun, to name but two. However, it appears that the average global temperature needs to drop to some critical threshold for these external effects to "kick in."

Once global cooling had taken effect in response to uplift from the India-Asia collision, the system may have been self-reinforcing. Cooler climates lead to higher erosion, particularly when the climate is sufficiently cold to promote widespread glaciation. This, in turn, removes huge masses of rock from the mountains, and, just as a cargo ship rises in the water as its cargo is off-loaded, the mountains rise in isostatic adjustment (∞ see Chapter 5). One final interesting speculation is that the dramatic change in climate in the last few million years that resulted in the ice ages may have optimized the environmental conditions necessary for the evolution of humans.

(a)

(b)

FIGURE 6.19 Uplift of the Himalayas. (a) Himalayan Mountains photographed from the space shuttle. The monsoons deposit snow and rain on the Himalayas, the runoff from which erodes the rock. (b) Rapid erosion is evident from the deep valleys cut through the mountains. Weathering of the exposed rock strips carbon dioxide from the atmosphere, causing a reduction in the greenhouse effect.

SUMMARY

- Plate boundaries can be identified on Earth's surface by the locations of volcanic and earthquake activity. In general, plate interiors are geologically inactive.
- Alfred Wegener proposed that continents were once joined together as a supercontinent that he called Pangaea. This supercontinent subsequently broke up and drifted apart, a process known as continental drift.
- Reversals of the magnetic field have occurred many times and are simultaneous worldwide events.
- A symmetrical pattern of magnetic intensity is recorded in the ocean floor rocks across spreading ridges. These alternating high and low magnetic intensity bands mirrored on opposite sides of a ridge are created by basaltic magma rising into a mid-ocean ridge during times of normal and reversed polarity, respectively.
- Global climate is fundamentally controlled by the sunlight's angle of incidence. Local climates are strongly influenced by topography, wind direction, and proximity to the ocean.
- Global changes in Earth's climate have occurred in the geologic past. The slow movement of continents across Earth's stable climatic belts has resulted in apparent climatic successions through time at a given location.
- Mantle plumes that rise beneath a continent may melt the upper mantle, producing huge volumes of basalt and weakening the continent where it splits apart. Therefore, the initial stages of continental breakup are often characterized by flood basalts.
- A triple junction is the point at which three plates meet. Stability of the triple junction is controlled by the relative motions between the plates. A triple junction can be defined as stable if the relative motions of the plates are able to continue without changing the geometry of the junction.
- A hot spot trail is formed as a plate moves over a mantle plume. These trails can be used to determine absolute plate motions.
- Plate movement may have multiple causes, including slab pull, ridge push, and slab suction.
- Continental masses of the distant geologic past were modified by processes similar to those of the modern world—erosion and sedimentation, volcanic activity, plutonism, metamorphism, and deformation.

KEY TERMS

continental drift, 137
Pangaea, 139
geosyncline, 141
sea-floor spreading, 142
apparent polar wander curves, 142
flood basalts, 148
active continental margin, 151
passive continental margin, 151
triple junction, 151
slab, 156

QUESTIONS FOR REVIEW AND FURTHER THOUGHT

1. How does the modern concept of plate tectonics differ from Wegener's ideas of continental drift? What are the fundamentally important assumptions of the continental drift hypothesis that are the most erroneous?
2. Why are active plate margins characterized by volcanic activity or earthquakes or both? Why are plate interiors usually geologically quiet areas?
3. Explain what polar wandering represents in the context of modern plate tectonics.
4. What is the explanation of parallel magnetic anomalies (magnetic stripes) in oceanic crust?
5. How can you explain a geologic sequence representing glaciation, temperate forests, tropical rain forests, and dry, hot deserts succeeding one another at a given location over a relatively short geologic time span?
6. Explain the fundamental principles of sea-floor spreading.
7. How are flood basalts related to the breakup of continents? What do flood basalts suggest about the mechanism of continental fragmentation?
8. What is the importance of a triple junction in plate tectonics? Why are some triple junctions stable and others unstable?

7 Deformation, Earthquakes, and Formation of Geologic Structures

The San Andreas fault cutting through California shows up clearly on this aerial photograph.

Overview

- Many geologic structures are reflections of the stresses affecting the crust. Compressional stresses form folds and thrust faults. Extensional stresses cause normal faults and rift valleys.
- Earthquakes occur when opposite sides of a fault slip relative to each other.
- At a plate margin, the two plates are held together by friction across faults. An earthquake occurs when the rocks move to catch up with the plate.
- Over long periods of time, rates of movement across a plate margin can be estimated from the offset of geologic features of known age. Over short periods, rates of movement can be measured directly.
- Inflation of a volcano is caused by pressure buildup in its magma chamber, which is followed by eruption or by intrusion of magma to form sills or dikes. The volcano then deflates through the loss of magma.

Introduction

You have probably seen examples of Earth's deformation in roadcuts where layers of rock are contorted into folded structures. In some cases, faults can be observed as linear fractures that offset layers. The most spectacular manifestations of Earth's deformation are mountain ranges such as the Himalayas. Most of Earth's deformation is ultimately driven by convection in the mantle, which involves flow of rock. Deep in the crust, rock may deform by folding and flow. Near the surface, the rock is brittle and deforms by breaking, which produces earthquakes, releasing enormous amounts of energy in a few seconds or minutes. This energy often has accumulated over hundreds of years. In fact, earthquakes can be more devastating than the largest atomic weapons. They are an inescapable natural phenomenon generated by the driving forces behind plate tectonics. The ground also deforms before a volcanic eruption. Measurements of volcanic deformation can be used for prediction. In this chapter, we will examine structures and events caused by deformation of Earth. We will discover that there is an earthquake cycle; earthquakes occur in the same regions time and time again. There is also a volcanic cycle in which volcanoes erupt after periods of dormancy.

Not all ground deformation is associated with violent events such as earthquakes or volcanic eruptions. The style of deformation depends on how much and how fast the ground is deformed and on the mechanical properties of the region being deformed. If you hit a rock with a hammer, it fractures. If some force is applied to the rock slowly, it does not fracture, but flows. For example, some old limestone benches have sagged in the middle, deforming into a gentle fold under the pull of gravity. Geologic structures show many examples of deformation caused by either fracture or flow.

Crustal Deformation

Most crustal deformation occurs at plate boundaries, where the forces are concentrated; however, stress can be transmitted through the rigid lithosphere to plate interiors. The most intense deformation occurs during continent–continent collisions. An analogous situation might arise during a head-on collision between two cars. In this case, most of the crumpling and breakage occurs at the front of each vehicle, with the amount of damage decreasing toward the rear. In the same way, deformation is concentrated at plate edges where collisions occur while the interiors remain intact.

Increases in pressure and temperature within Earth may produce metamorphic rocks by the process of *metamorphism*, introduced in Chapter 4. This process naturally accompanies deformation. The high pressures and temperatures associated with continental collisions may transform the types of minerals present in a given rock. At the same time, the forces at work may change the fabric of the rock. In particular, a strong alignment of platelike mineral grains—schistosity—is created by preferential alignment of mineral grains perpendicular to the direction of maximum stress.

Stress and Strain

We should examine the concepts of stress and strain to understand how rocks in Earth's crust will respond to plate-tectonic forces. In Chapter 5, we saw that stress is the amount of force applied to an object per unit area. Strain may be expressed in terms of the percent change in an object's length due to an applied stress (Fig. 7.1a). Either the compression or the extension of a given material can be expressed in terms of strain. The relationship of stress to strain gives us information about the mechan-

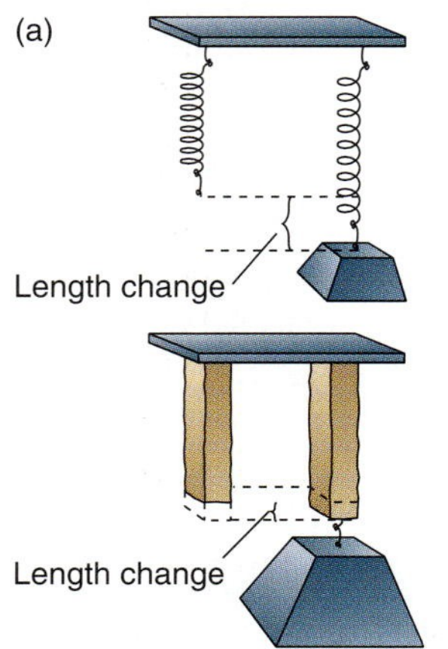

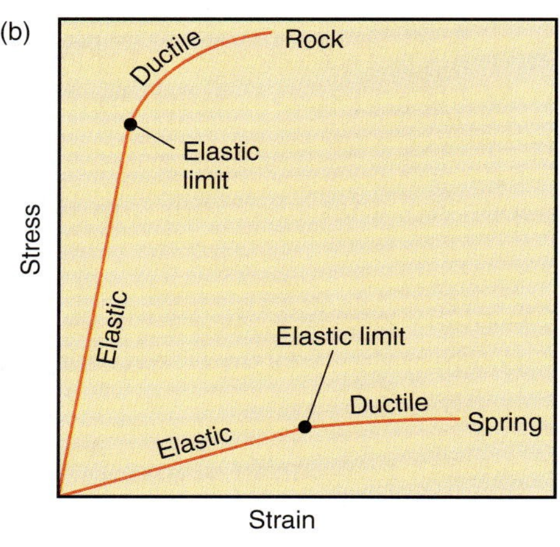

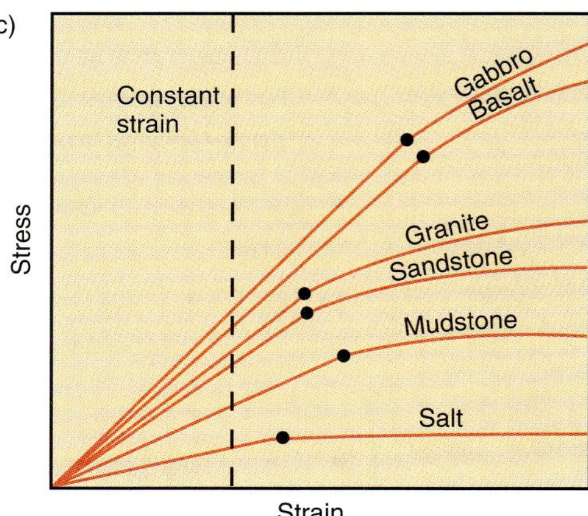

FIGURE 7.1 (a) A weight extends a spring. The strain is the proportional change in length. The stress is the force per unit area. The same experiment performed on a column of rock causes significantly less extension, but the principle is the same. (b) Stress versus strain variation. Elastic (straight line from origin to dot) to ductile or inelastic behavior (curved line) occurs when stress is greater than the elastic limit. The slope of the elastic part is modulus = stress/strain. (c) Stress-strain curves for different rocks. To generate a fixed strain as depicted by the vertical line requires different amounts of stress (vertical axis), with gabbro requiring the most.

ical properties of a material. For example, for the same applied stress, a rock will undergo much less strain than will a metal spring (Fig. 7.1a).

Imagine a metal spring hanging from a beam to which you successively add small weights, gradually stretching its length. The stress in the spring is proportional to the amount of weight you have applied; the strain is the fractional stretch experienced by the spring. If you were to plot the stress of the spring relative to its strain, you would see a clear linear relationship between stress and strain (Fig. 7.1b). The behavior exhibited by the spring is called *elastic* behavior; when the amount of stress is divided by the strain, the result is always the same. The ratio of stress to strain (the slope of the elastic portion of the plotted line shown in Fig. 7.1b) gives a measure of the *strength* of the material. This quantity is known as the *elastic modulus*, which is a term that was introduced in Chapter 5. According to this definition, you can see from Figure 7.1b that the rock is much stronger than the spring. Geologic materials (rocks) have varying strengths and, accordingly, behave differently under a given stress (Fig. 7.1c).

If you now remove the weights, the spring will return to its original length represented by the origin of the graph. You cannot, however, add weights indefinitely to the spring. There will come a point, termed the **elastic limit**, where the spring has stretched too much and will not resume its original shape after removal of the weights—it becomes permanently deformed.

Once the elastic limit is exceeded, the behavior of the spring is said to be **ductile** and the amount of stretching produced by a given weight is no longer expressed by a simple linear relationship (Fig. 7.1b). In other words, the spring undergoes irreversible deformation when it is stretched beyond its elastic limit. In some cases, the spring fractures; this break is **brittle deformation**.

In studies of mechanical properties of materials, nonelastic deformation is called **plastic deformation**; ductile deformation is restricted to describing nonelastic extension. Thus plastic is a more general term. However, the term ductile has commonly replaced the term plastic to describe all types of nonelastic strain, including contraction, extension, and shear.

Some materials are barely elastic at all; instead, they are ductile and will not return to their original shape upon removal of an applied stress. Try hanging weights on a piece of chewing gum. After a certain point, it will stretch indefinitely (or flow) until it finally breaks. If you stretch chewing gum and release it, it will not return to its original shape; it is a ductile material.

The type of geologic structure that forms in response to deformation will depend in large part on whether the rocks deform as elastic, brittle, or ductile materials. If stress is applied above the elastic limit, the material will either flow or suddenly fracture. The fractures that are formed within Earth in this manner are called **faults**. If the rocks behave in a ductile manner, they may thin when extended and they will tend to undulate and bend when compressed. Compression of ductile rock layers results in the formation of **folds** rather than actual breakage, or fracture, of the rock.

Strain in rocks can be elastic, ductile, or brittle. Rocks exhibiting elastic strain recover their original form upon release of the applied stress. Rocks undergoing ductile strain exhibit a permanent deformation. If a material breaks, the associated deformation is brittle.

How Rocks Respond to Deformation

Two main factors determine how a rock will respond to deformation over a given time period. The first is the nature of the rock itself: its mineralogy and texture. The second is the nature of the external conditions present during deformation—in particular, pressure, temperature, and effects of fluid trapped in the rock.

The Effects of Rock Type Minerals comprising a rock differ in certain properties such as their strength. Quartz, for instance, is both a hard and a strong mineral. A rock made largely of pure quartz (such as a quartzite) will therefore tend to be strong and brittle. It requires high stresses to deform it elastically, and it will break as these stresses become sufficiently large. At the other extreme, minerals such as halite (common salt) are very weak and will readily flow when subjected to relatively low stresses. Beds of salt are notoriously weak and commonly flow underground, deforming the surrounding sediments. Beds of schist also tend to be weak because of their fine-scale, layered structure. If a succession of rocks containing weak beds are subjected to a shear stress across the beds, fracture commonly occurs on the weakest bed. As a result, huge masses of rock may slide along extensive, nearly horizontal faults known as **décollements**, after the French word for "ungluing."

Igneous rocks are generally strong. They are formed from fairly strong minerals, such as feldspar, quartz, pyroxene, and olivine, and they crystallize in a strong interlocking texture. Coarse-grained (intrusive) rocks tend to be stronger than their fine-grained (extrusive) equivalents—probably because the interfaces between grains are areas of weakness and a fine-grained rock has more grain boundaries for a given volume.

Sedimentary rocks differ greatly in strength; a quartz-rich sandstone may be quite strong, whereas a mudstone containing mostly weak clay minerals will be weak. Clay minerals are weak because they are made up of weakly bound silicate layers that readily slip relative to each other. A rock is only as strong as its weakest component, which in many cases will be the grain boundaries or the cement holding the grains together. In a single outcrop, it is common to see different rock types that have been

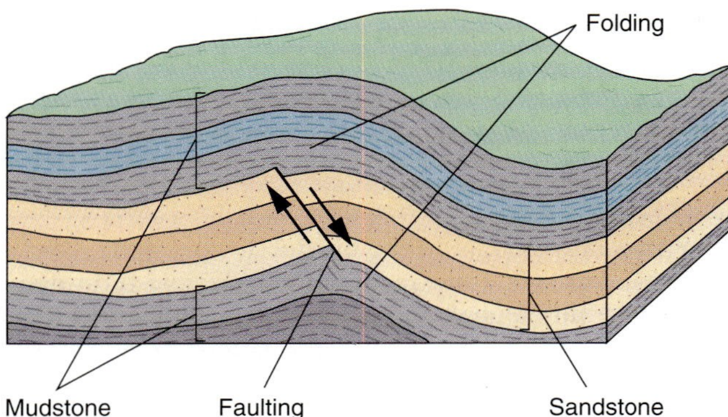

FIGURE 7.2 Combined brittle (fault) and ductile (fold) deformation. The layers of mudstone have folded ductily but a fault has occurred in the sandstone layers. Whether faulting or folding occurs depends on the rock type, temperature, pressure, and fluid content, and also on the rate of straining.

subjected to the same stress but have responded in different ways (Fig. 7.2).

Such large variations in rock strength clearly have a bearing on how we can use geologic materials. If we want to construct robust, sturdy buildings, we need to use strong rock types; hence, granites, limestones, and sandstones are preferred building stones.

The Effects of Pressure, Temperature, and Fluids Any rock subjected to higher pressures and temperatures will change its properties and become progressively more ductile. The limit of this trend occurs when a rock eventually melts and becomes a fluid. As a consequence, more intense deformation involving higher pressures and temperatures will tend to be characterized by ductile behavior, with evidence of folding and flow. Clearly, as temperature and pressure increase with increasing depth, the behavior of rocks during deformation will differ according to their depth within Earth. At shallow levels in the crust, we expect brittle behavior such as faulting and fracturing, together with some folding but little metamorphism. Deeper in the crust, however, we expect high-grade metamorphism as well as folding and flow of some rock types. This means that, at some level in the crust (about 15 km depth), brittle structures such as faults will pass down into ductile structures. This change in behavior is termed the **brittle-ductile transition**. It is a loose definition because folding also occurs above this transition and brittle deformation in the form of earthquakes can occur below it—depending on material properties and strain rate.

Quartz, which is very common in crustal rocks, is very weak in the presence of a small amount of water, compared with dry quartz. This is due to a chemical reaction between quartz and water, called *hydrolytic weakening*. At higher temperatures, this reaction is more effective and quartz becomes weaker. At a depth of about 15 km, the combination of hydrolytic weakening and increased temperature causes quartz to become ductile. Because many crustal rocks contain quartz, this controls the location of the base of the brittle part of the crust. In more mafic mantle rocks, hydrolytic weakening is less important because the water content is very low, but it is thought there may be another brittle-ductile transition deeper in the mantle determined by temperature effects.

Whether a rock deforms in a brittle or ductile manner depends not only on its composition, mineralogy, and texture but also on such physical conditions as pressure, temperature, and fluid content.

Geologic Structures

Joints

Most rocks, whether igneous, metamorphic, or sedimentary, are systematically jointed; that is, there are fractures, cracks, and other mechanically weak surfaces throughout them. Both **joints** and faults are fractures or discontinuities in a rock—the difference is that joints tend to be more closely spaced and regular in their occurrence and there is no relative sliding of the rock on either side of a joint.

Fractures can originate for a variety of reasons, but joints resulting from the release of confining stresses are probably the most common and the simplest in origin. Although these joints may occur in many different types of rock, we can use the example of a granitic intrusion to illustrate the process. At depth, the granite is subjected to the great confining pressure exerted by the overlying column of rock. Subsequently, as the granite is uplifted and ultimately exposed at Earth's surface, the confining stresses on the rock have been reduced to nearly zero. Because of this relaxation of confining pressure, the rock expands and cracks into great sheets. Two terms are commonly used to describe such features. If the joints are planar, parallel, and very closely spaced, the fracturing process is called **sheeting**. If the joints are curved, concentric, parallel, and with uniform spacing, the fracturing process is termed **exfoliation** (Fig. 7.3a). The concentric pattern of joints may resemble layers of onion peel and is sometimes referred to as *onion skin weathering*. Exfoliation joints are approximately parallel to topography, and exfoliation occurs in exposures of most rock types. In hard rocks, the joints ultimately control topography (Fig. 7.3a), but they are much less obvious in softer or more easily weathered rocks.

Joints can also occur in igneous rocks as a result of cooling. As the rock crystallizes and cools from the molten state, it contracts, setting up stresses in the rock that cause cracks to form. These cracks or joints are often very regularly spaced and form geometric patterns. Cooling of basalt lava flows can form columnar jointing

(a)

(b)

(c)

FIGURE 7.3 Examples of different joint types. (a) Exfoliation sheets form the characteristic domes of Yosemite Valley, California. (b) Regular hexagonal columns formed by cooling joints in basalt lava (Giant's Causeway, N. Ireland). (c) Sandstone joints formed when rocks under pressure due to burial are exposed at Earth's surface (Checkerboard Mesa, Zion National Park, Utah).

in which the rock cracks into columns with a hexagonal cross section (Fig. 7.3b). These columns stand upright, perpendicular to the cooling surfaces.

Igneous rocks are not alone in showing systematic jointing; massive rocks of all types are jointed. Massive sandstones—for example, those that form the spectacular cliffs in the area of Zion National Park, Utah—are strongly jointed. Checkerboard Mesa is a famous example; it shows three sets of joints that are nearly perpendicular to one another (Fig. 7.3c). The joints on Checkerboard Mesa are expansion cracks formed in the same fashion as the expansion joints in granites. The sandstone had been buried and strongly compacted; on exposure at Earth's surface, the surrounding stresses on the rock were reduced and the sandstone expanded slightly, cracking as it expanded.

Faults

Faults are fractures in rocks across which there has been some relative movement, or displacement. They most frequently form above the brittle-ductile transition. We can classify faults by the movement that has occurred relative to the *fault surface*. In most cases, the fault surface is approximately flat, and can then be described as a **fault plane**.

The two blocks on either side of an inclined fault are called the **hanging wall** and the **footwall** (Fig. 7.4a). The hanging wall lies above the fault plane; the footwall lies below. These two terms were coined during the early days of mining. Many of the old mines were located along fault zones because mineralization tends to occur along faults. In a mine tunnel intersecting or driven along a fault, the miners stood on the footwall, while the hanging wall of the fault was overhead. These terms, then, are used to describe inclined faults; they do not apply if the fault plane is vertical.

The amount of relative displacement across a fault plane is the **throw** (Fig. 7.4). The **strike** of a fault is the geographic bearing of the line formed by the intersection of the fault with the horizontal plane. The **dip** of a fault is the angle between the fault plane and the horizontal measured perpendicular to the strike (Fig. 7.4a).

Three main types of faults can be recognized: normal faults, reverse faults, and strike-slip faults. Consider the block models in Figure 7.4, with fault planes as indicated. If the faulted block is pulled at each end—and thus put under tension—it will extend and the hanging wall block will drop down relative to the footwall. This is known as a **normal fault** (Fig. 7.4b) and is common in regions of extension, such as rifts and spreading ridges. As a result of extension, the distance between any two points on either side of the fault plane will increase, as indicated. A rift valley bounded by normal faults is known as a **graben**.

In contrast, if the block is pushed at each end and is put under compression, the footwall drops relative to the hanging wall, forming a **reverse fault** (Fig. 7.4c). Reverse faults are common in regions of compression such as collisional mountain belts. Even more common at convergent plate boundaries is a special type of reverse fault called a **thrust fault** (Fig. 7.4d). Thrust faults are low-angle (small dip angle) faults in which the fault plane is often parallel to the layering or bedding in the rock because these are the planes of maximum weakness. Occasionally, the thrust fault may cut across the layers in short reverse-fault steps or **ramps** (Fig. 7.4d). Thrust faults may transport huge rock masses over large distances (tens to hundreds of kilometers) across the underlying rock.

Another type of fault occurs when the block is sheared horizontally (Fig. 7.4e); movement on either side of the fault plane is entirely horizontal. In this fault type, called a **strike-slip fault**, the two blocks of rock simply slide past

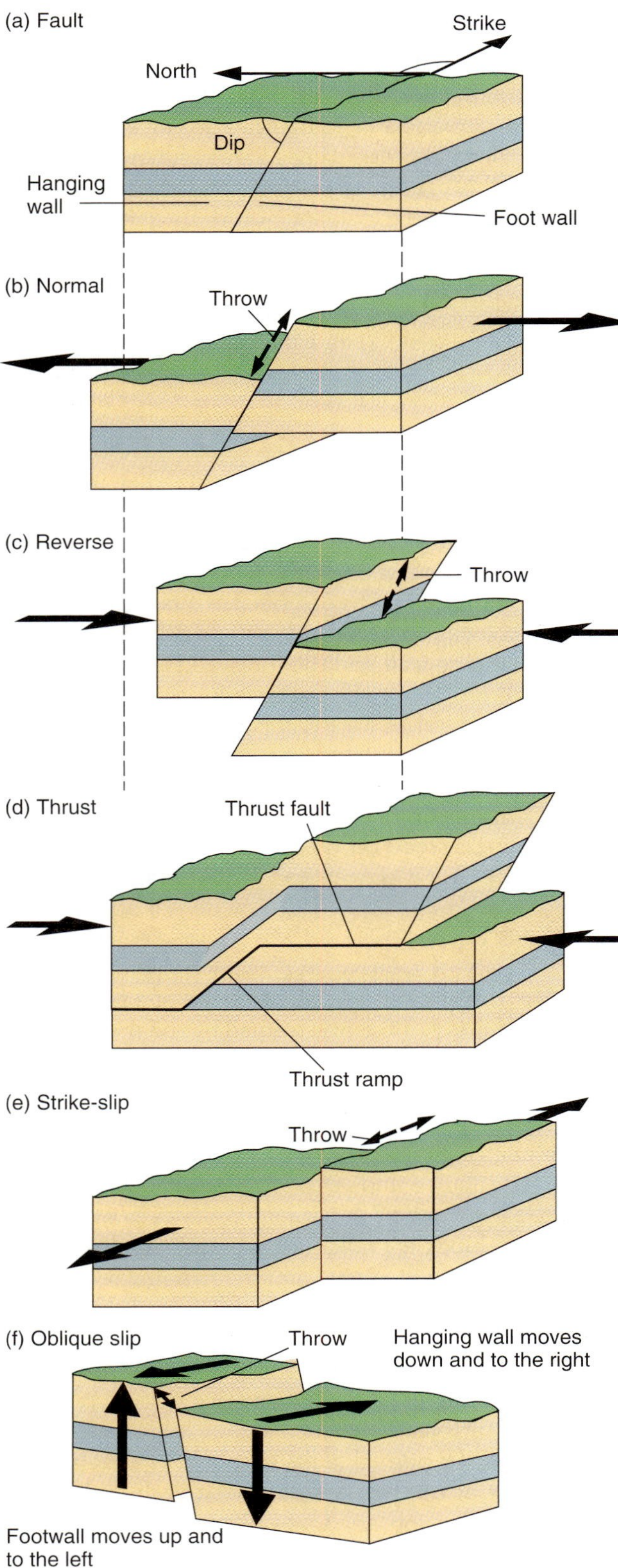

FIGURE 7.4 Characteristics of the main types of faults. (a) Prior to faulting, showing the dip and strike of a fault plane. The fault plane separates rocks below it in the footwall from those above it in the hanging wall. (b) A normal fault and, on the right, a photograph of a normal fault. (c) A reverse fault with a photograph of an example on the right. (d) Deformation into thrust (near-horizontal) and ramp (inclined) faults. On the right, the Lewis thrust fault (Glacier National Park, Montana). (e) A strike-slip fault. On the right, San Andreas strike-slip fault showing an offset stream (Carizzo Plain, California). (f) Oblique slip.

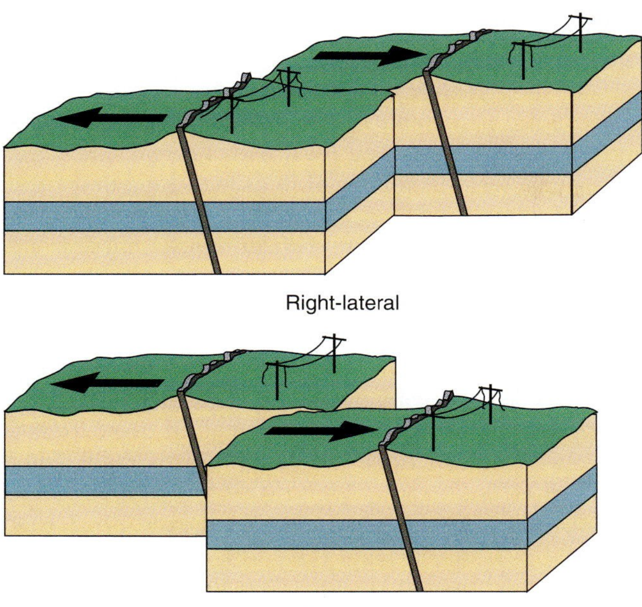

FIGURE 7.5 Left-lateral and right-lateral faults.

each other with no net compression or extension. The slip is in the direction of the strike, in contrast with normal and reverse faults, which may be termed *dip slip* because slip occurs parallel to the dip of the fault plane. If the motion of the opposite, or far, side of a strike slip fault is to your right, it is called a **right-lateral fault**; this holds true when the fault is viewed from either side. If the motion is to the left, it is a **left-lateral fault** (Fig. 7.5). Strike-slip faults are the most common style of faulting at transform margins.

In many faults, the slip is not pure strike-slip, pure normal, or pure reverse. The faulting can be combinations of strike-slip and normal or strike-slip and reverse. In such cases, a slip making an angle between 0° and 90° with the strike of a fault is called **oblique slip** (Fig. 7.4f).

Over geologic time, the total movement taking place on a fault plane is not usually continuous, nor does it all happen at once. The blocks of rock move in a series of sharp jerks, each one representing the sudden release of accumulated stress in an earthquake. When we see a fault crosscutting a series of rocks and causing a significant displacement of the rocks on either side, this probably represents the cumulative effect of many small movements. In some cases, we can tell when the fault has slipped by determining the ages of the sediments that have been displaced across the fault.

The topographic break, or cliff, where a fault plane of a normal or reverse fault cuts the surface is called a **fault scarp**. Strike-slip faults also have fault scarps where they intersect the surface, but the disturbance is often more subtle. If a strike-slip fault cuts a hill, motion on the fault displaces parts of the hill on either side. The location of the fault can be recognized as a break in the topography. Scarps often erode rapidly, leaving little evidence of the fault at the surface.

One indication of the existence of a fault can be a mismatch in the rock types on either side of the line where the fault intersects the surface (the *fault trace*). In the fault zone, repeated movements may crush the rock to form a *fault breccia*, or grind up the rock to a fine powder called *fault gouge*. The gouge zone may be impermeable to water and can dam the water table on the uphill side of the fault. A line of foliage fed by the trapped groundwater can be an excellent indicator of a fault trace. Commonly, scratches or striations (*slickensides*) are found on the fault plane itself, and are formed as one block slides past the other. These scratches enable geologists to determine the relative sense of the most recent movement across the fault. If the slickensides on a dipping fault plane are horizontal, it is a strike-slip fault; if they are up and down the fault plane, it is a dip-slip fault and may be either normal or reverse.

At depth, normal or reverse faults may simply die out in regions where the rock flows. Some natural faults also curve into a near-horizontal orientation, called a **listric fault** (Fig. 7.6a). Faults that extend quite deeply into Earth will eventually encounter sufficiently high pressures and temperatures that the rocks no longer behave in a brittle fashion but are ductile and deform by flow (Fig. 7.6b).

Folds

In contrast to the brittle failure characterized by faulting, rocks can also deform into broad or tight undulations called folds (Fig. 7.7). Folds can be smoothly curving with rounded sides or sharply angular; they can be upright or inclined, symmetrical or asymmetrical, or brit-

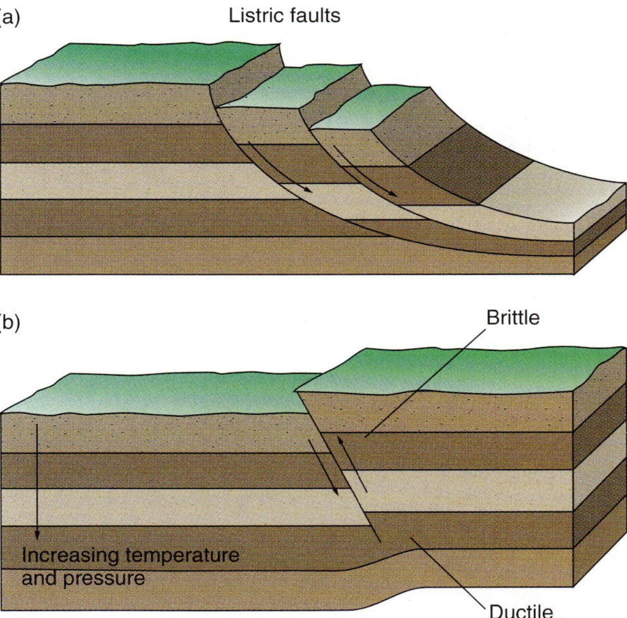

FIGURE 7.6 (a) Listric faults curve and flatten out with depth. (b) In the upper crust, failure can be brittle and can generate faulting. At depth, temperatures and pressures increase, and failure becomes ductile and generates folding and distributed shear.

tle or ductile. These different forms tell us much about the history of forces that acted on the rocks and about the properties of the rocks.

The two main types of folds are anticline and syncline (Fig. 7.7b). An arched fold is called **anticline**; a down-warped fold is a **syncline**. If we consider the eroded surface of a folded region, an anticline will have the oldest rocks in the middle, whereas a syncline will have the youngest rocks in the middle (Fig. 7.7c). A fold that is neither up- or down-warped and appears more as steplike flexure is called a *monocline*.

The general anatomy of a fold is shown in Figure 7.7d. The region of curvature of the fold is called the **hinge,** and the line of maximum curvature is the **axis.** Those parts of the structure that are not curved are called the **limbs.** Folds may differ in the relative extent of hinges and limbs—from rounded folds with no real limbs and large hinge regions to angular folds with long limbs and a sharp, abrupt hinge.

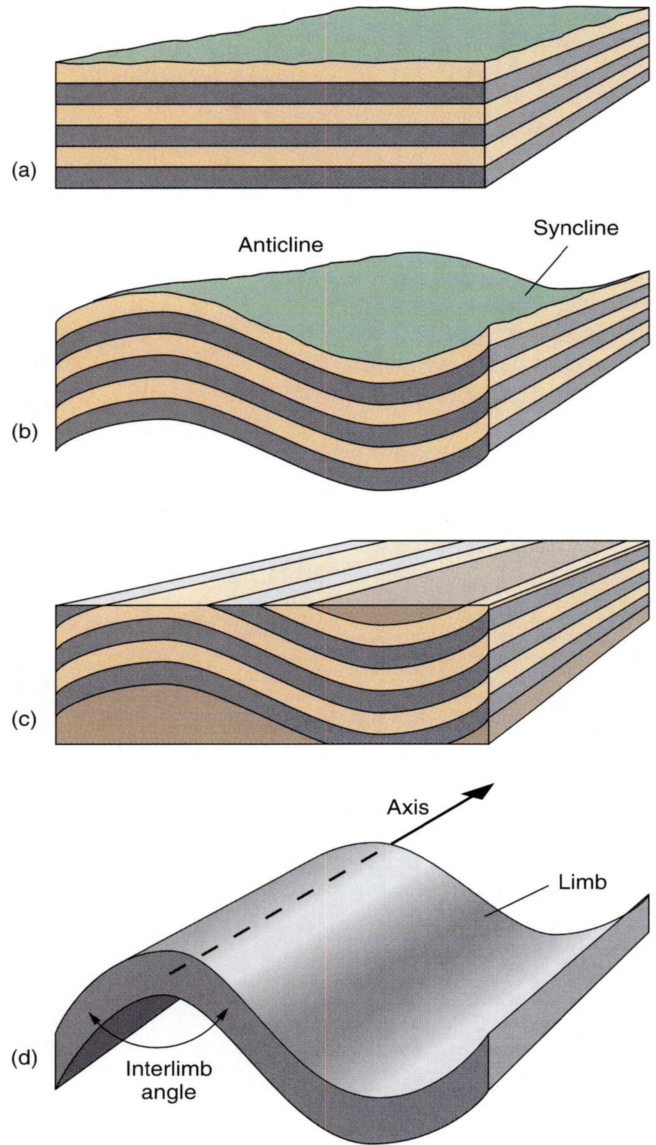

FIGURE 7.7 (a) Flat-lying layers of rock before folding. (b) Anticline and syncline. (c) Eroded fold structure. (d) Fold axis, limb, and interlimb angle.

These characteristics are seen in the sketches of fold profiles (a view of the fold in a section perpendicular to the axis) shown in Figure 7.8. The angle between the limbs (the interlimb angle) will vary according to the intensity of the fold deformation. This gives rise to a range of fold types varying with increasing compression from a very wide fold with an *open* angle at one extreme, to a closely folded *isoclinal* pattern with parallel limbs (Fig. 7.8b).

Folds can be upright, asymmetric, or overturned, that is turned over on one side (Fig. 7.8c). If a fold is completely turned over, it is called *recumbent*. The fold axis can be thought of as a line in space along which a layer of rock has been folded. Many layers of rock exist in a given fold, and we can define an axial plane as the plane that contains all the fold axes. The axis of a given fold may be horizontal or it may be inclined to the horizontal, or *plunging* (Fig. 7.8d, e). The plunge of a fold is given by the dip of the axis down from horizontal, and the strike is given by the bearing of its projection on the surface—the axial trace. If folding is associated with uplift due to intrusion from below—for instance, a rising diapir of buoyant magma or salt—the structure may not have a true axis but, instead, will form a dome.

Just like fault scarps, folds are readily eroded. We can tell that the rocks have been folded by looking at the outcrop pattern, which is the pattern that the layers of rock make on the ground surface. In general, older layers of rock are found with increasing depth. If we examine a horizontal section through a nonplunging anticline (Fig. 7.8d), we find that the layers of rock form a series of parallel symmetrical stripes, with the oldest located in the middle. For a plunging anticline (Fig. 7.8e), the outcrop pattern forms a series of symmetrical **V** shapes in which the oldest rock layer is also located in the middle. The **V** shapes are closed *in the direction that the fold is plunging*. For a plunging syncline, the youngest rocks are in the center and the **V** is *open* in the direction that the fold plunges.

During folding, the rocks may undergo brittle deformation, ductile deformation, or some combination of the two. If you bend a thick stack of paper—say 5 cm thick—into a curved shape, you will observe that there is slippage between the individual sheets of paper (Fig. 7.9a). A flat stack of paper lying on a table has vertical edges, but the edges of the curved stack of paper flare outward or inward depending on whether the center of the curve moved downward or upward. This type of folding occurs in rocks, and is called *cylindrical* or *concentric folding*. In thick sequences of sedimentary rocks that are concentrically folded, there must have been slippage along the bedding planes as folding occurred.

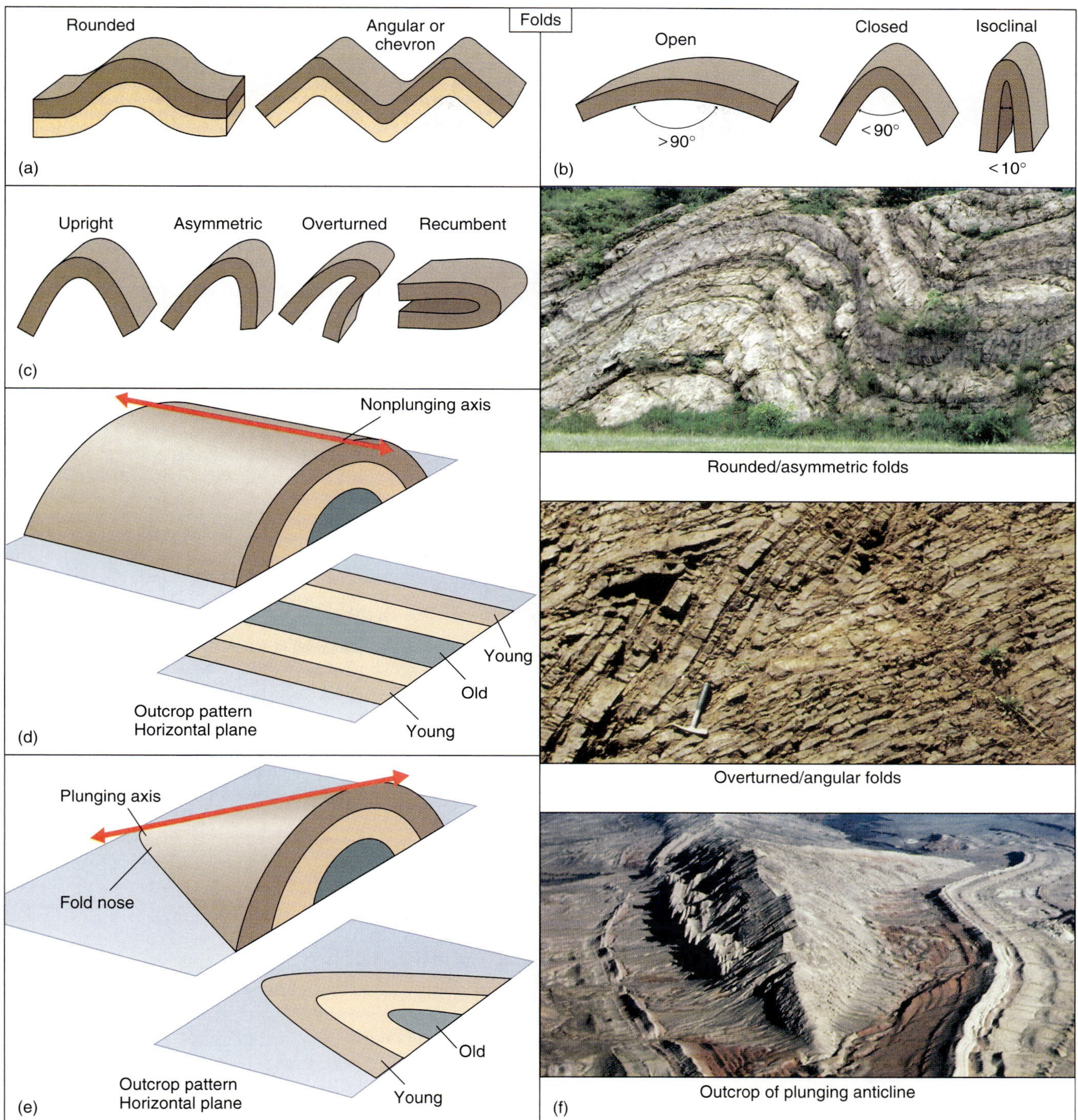

FIGURE 7.8 (a) Fold characteristics. (b) Fold profiles. (c) Types of folds. (d) Fold with nonplunging axis and outcrop pattern. (e) Fold with plunging axis and outcrop pattern (compare with Fig 7.7c). (f) Geologic outcrops illustrating different types of folding. Top to bottom: Anticline and syncline (Newfoundland, New Jersey). Overturned angular folds (Marin headlands, California). A plunging anticline (Sheep Mountain, Wyoming).

In metamorphic and some igneous rocks, other mechanisms for folding are apparent. In slates with closely spaced cleavage, slight shearing along the cleavage surfaces causes a type of folding called *shear folding*. The layers of rock are not truly bent; rather, the beds appear to be folded because of pervasive and closely spaced shear surfaces (Fig. 7.9b). Rocks that have been shear-folded have behaved in a brittle manner. With deeper burial, this type of folding may start to exhibit ductile behavior and give rise to *flow folding* (Fig. 7.9c). In this style of folding, material flows in

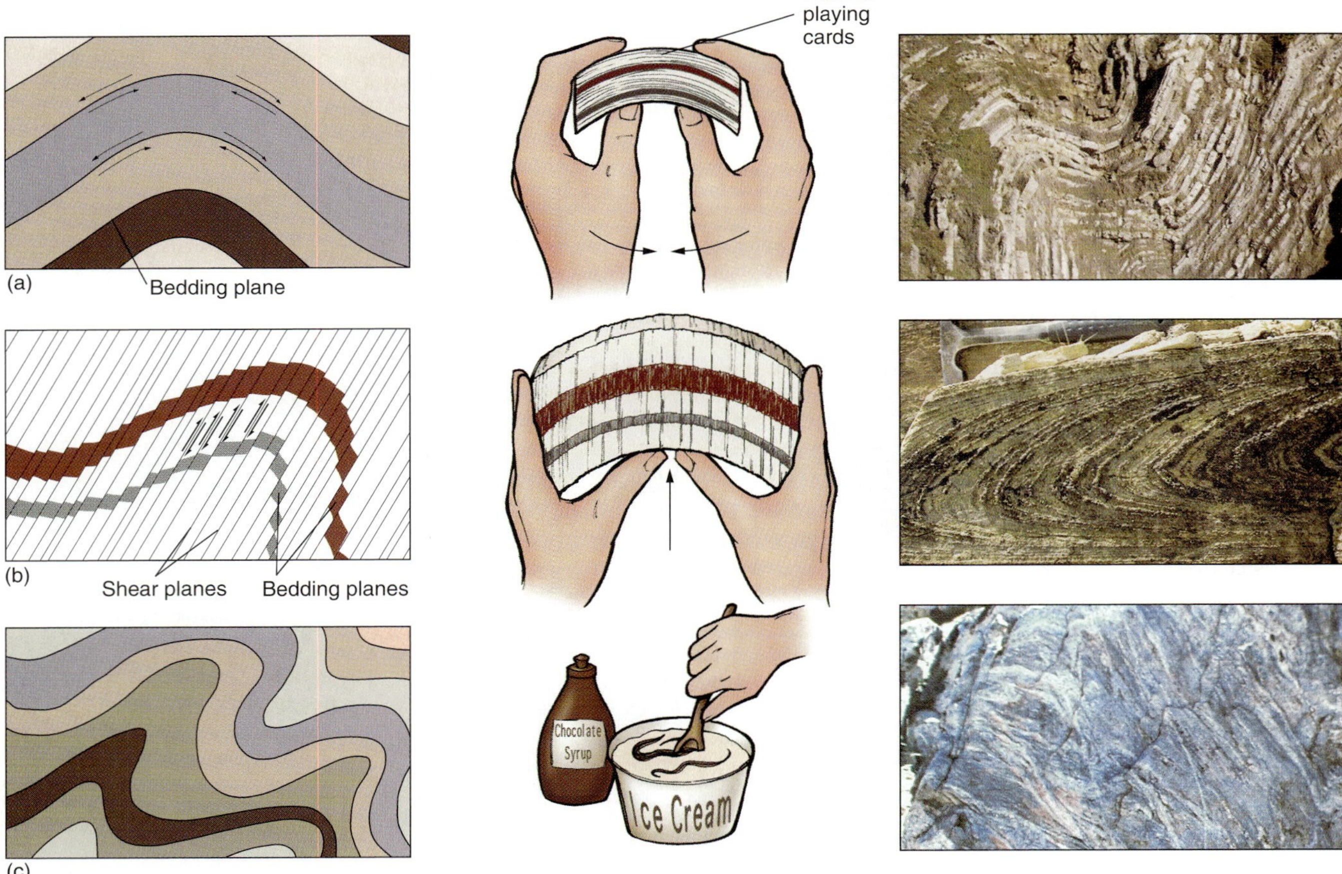

FIGURE 7.9 Three different mechanisms for forming folds: (a) Concentric or cylindrical folding. Bending of layers is accommodated by slip along bedding planes. As an analog, take a stack of cards or a sheaf of paper held horizontally, with a thick pen line marked *along* them, and bend it. (b) Shear folding. Slip along closely spaced shear fractures or cleavage that crosscuts layers such as bedding effectively bends the layers on a large scale. To reproduce this, take a stack of cards held vertically, with a thick pen line marked *across* them, and press on the edge of the pack. The cards act like finely spaced cleavage surfaces. (c) Flow folding. The rocks behave plastically, and the original geometry (e.g., bed thicknesses) is lost. Stirring chocolate syrup into vanilla ice cream will produce similar features.

a ductile (plastic) manner from the limbs of the fold into the hinges. Under conditions of higher temperature, true ductile folding occurs. This style of folding is pervasive, in that the same type of folding is seen under the microscope as can be observed in the outcrop.

Using the patterns generated by folding and erosion, geologists can reconstruct the strata before folding and model the stresses that caused the deformation. Folding of sedimentary rocks is common because they are weak and slippage between layers permits flow. Crystalline rock is more likely to fracture than fold, but under high temperatures and pressures it will also fold.

Joints are closely and regularly spaced fractures in a rock. Faults are brittle fractures in rocks across which relative movement displaces the crust on either side. Folding is a compressional deformation that can be both brittle and ductile.

Earthquakes

Earthquakes result from brittle fracture and involve rapid movements of fault blocks on either side of a fault plane. They occur mainly at plate margins (see Fig. 1.6). The point within Earth at which an earthquake begins is called the **focus** or **hypocenter**; the point on Earth's surface directly above the hypocenter is called the **epicenter** (Fig. 7.10), which is the point that is usually plotted on maps indicating earthquake locations. Hypocenters at convergent plate margins extend down to depths as great as 700 km. At divergent and transform margins, hypocenters are much shallower and rarely exceed 15 km. To understand this difference, we will explore the nature of earthquakes in this section.

If you bend a stick far enough, it will eventually snap (Fig. 7.11). This is analogous to what happens during an earthquake. As your hands bend the stick, they act like Earth's tectonic plates, which move under the forces of

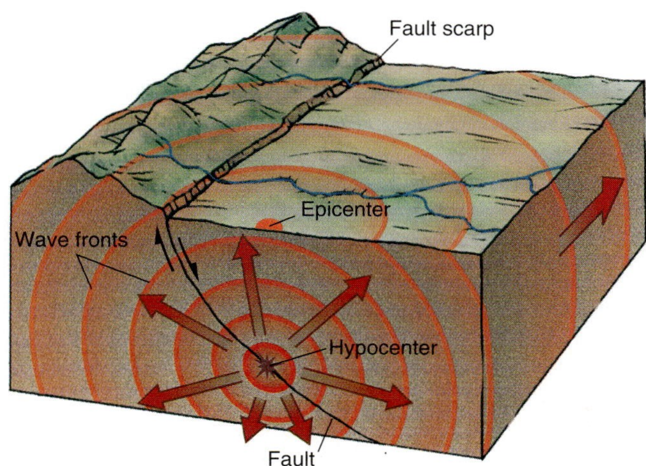

FIGURE 7.10 Hypocenter of an earthquake, epicenter, and fault scarp. The fault rupture begins at the hypocenter, radiating seismic waves. As the earthquake develops, the rupture grows to cover a large region of the fault.

plate tectonics. Within Earth, the upper crust holds together across a fault zone because of frictional resistance in the rocks. The upper crust that remains together across the fault zone must eventually slip; in the interim, however, it bends like the stick, storing elastic energy. Eventually the frictional force can no longer hold the crust together and it breaks suddenly—just like the stick. The snap you hear when the stick breaks is caused by sound waves generated by the rapid motion of the ends of the stick in the air. Similarly, the damage in the vicinity of an earthquake is due to the seismic waves that radiate through Earth from the rapid motion of the crustal rocks (See Chapter 5).

Consider two plates moving past each other (Fig. 7.11). Imagine that the plate margin is a transform margin running in a north–south direction. The plate on the western side of the margin moves to the north relative to the plate on the east, which is stationary. Typically, the earthquake hypocenters at transform margins only extend down to about 15 km, although the plates are about 100 km thick. At depths greater than 15 km, however, the rock is so hot that it is ductile and it flows. The deep rocks of one plate slide past those of the other as if the margin were lubricated. In the upper 15 km, however, temperatures are much lower, and frictional forces develop across the margin, preventing such smooth movements. The crust of the northward-moving plate pulls that of the stationary plate so that it is bent to the north in the vicinity of the margin. In contrast, the crust of the moving plate is bent to the south of where it would otherwise be had it kept up with the motion of the remainder of the plate. The result, as shown in Figure 7.11, demonstrates that a previously straight fence across a fault will take on a bend. The bend indicates that elastic strain has been stored in the rock below.

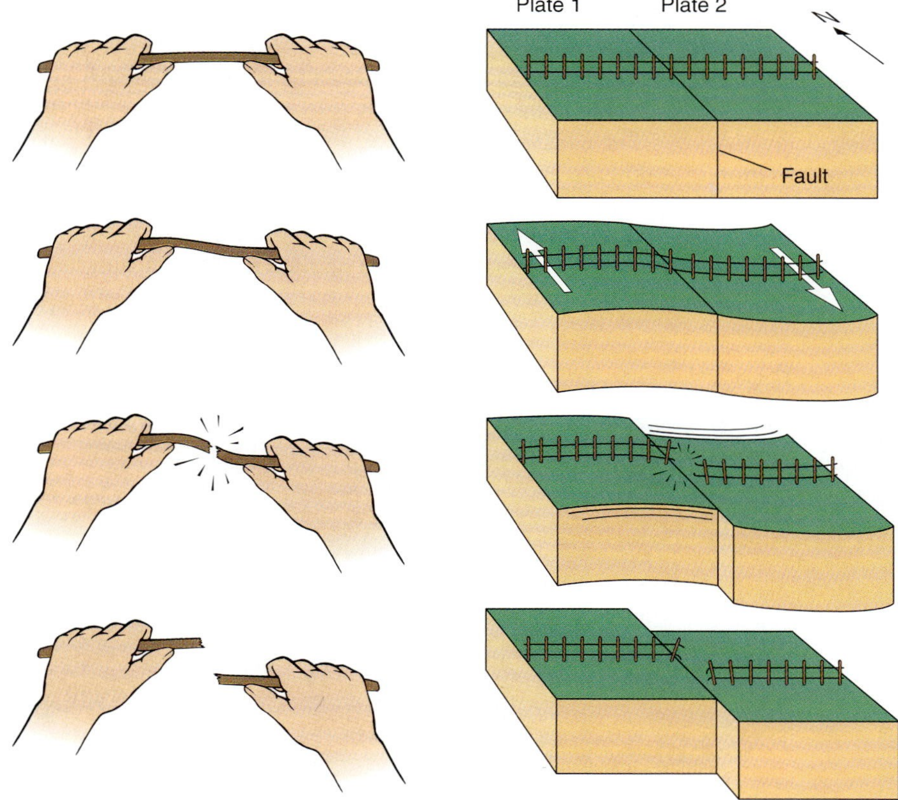

FIGURE 7.11 The bent stick breaks upon reaching the elastic limit. On the right-hand side, the crustal blocks break along the fault surface. A fence built across a strike-slip fault becomes deformed into a bent shape as the blocks of the crust are elastically sheared with time. The elastic blocks rebound to their unstressed state. The fence, though broken at the fault line, rebounds to two straight segments dislocated by the amount of slip on the fault.

Eventually, however, the margin of the moving plate does catch up with the rest of this plate, releasing the stored strain. The friction is overcome and the plate boundary rips along a fault causing an earthquake. The fault may have been frictionally locked for hundreds of years. This sequence of crust sticking together and suddenly slipping apart is appropriately called *stick-slip motion*. For the moving plate, crust near the margin has been dragged behind the remainder of the plate. It catches up in the earthquake, and a length of crust sometimes up to 1000 km can move as much as 10 m in just one or two minutes. This process is referred to as **elastic rebound**. In a strike-slip region such as the San Andreas fault in California, the amount of crust that suddenly moves extends to a depth of at least 15 km, and extends out on either side of the fault to a similar distance.

During an earthquake, the rapid accelerations of the ground near the fault cause damage to buildings. This damage can extend to distances of up to hundreds of kilometers because seismic waves generated by the sudden movement carry damaging energy away from the fault zone.

Earthquakes occur at depths of up to 700 km in subducting slabs at convergent margins. The cold brittle material of the oceanic lithosphere is transported to great depths so fast that it does not have time to heat up and become ductile. As the slab sinks under the overriding plate and into the mantle, forces from the surrounding material oppose the motion and exert a stress on its interior, causing earthquakes.

There is an increase in the number of earthquakes in the region between the 410- and 660-km-phase transitions. Recent theories attribute this increase to incomplete transition of olivine to the spinel structure (∞ see Chapter 5). Cooler temperatures of the slab delay this transition to depths greater than 410 km where it normally occurs. The transition is energetically favored because the olivine transforms to a denser structure. Volume changes caused by the phase transition set up stresses that, in conjunction with the stresses due to the moving slab, cause fracturing and the increased number of earthquakes. Beyond a maximum depth of 700 km, sufficient time has elapsed since the plate left the surface for it to be heated all the way through. The material is no longer brittle and earthquakes no longer occur.

In the vicinity of divergent margins, most large earthquakes occur on transform faults that separate offset ridges through strike-slip motion (Fig. 7.12). Along the divergent margin itself, some earthquakes are generated by normal faults that run parallel to the rift. The depth of earthquakes that do occur near ridges increases with distance from the ridge axis, which is consistent with cooling of the plate and consequent thickening of the brittle layer away from the axis.

Intraplate earthquakes are those that occur away from plate margins. These earthquakes happen infrequently compared with the activity observed at the margins. Over long geologic time scales, plates break apart and reassemble to form new plates. The forces moving the plates give rise to variable stresses within the plates. For example, forces in the North American Plate west of the Rocky Mountains give rise to east–west directed extensions; however, stresses to the east are east–west directed compressions. These stresses are among those that give rise to intraplate earthquakes.

Large intraplate earthquakes occurred in the Mississippi Valley in 1811 and 1812 near New Madrid, Missouri (see Focus 11.2) and in Charleston, South Carolina, in 1886. Australia sits in the midst of a plate, yet it has experienced several large earthquakes that have caused little damage owing to Australia's sparse population. In 1989, however, the town of Newcastle, Australia, was badly damaged by a moderate earthquake.

Intraplate earthquakes also occur in the ocean floor. The earthquakes that occur under Hawaii to depths of 60 km are an example. These are some of the deepest earthquakes other than those located at convergent margins. It is thought that deep beneath Hawaii, stresses associated with rising magma are applied rapidly, causing brittle behavior in a region that is normally ductile. Convective

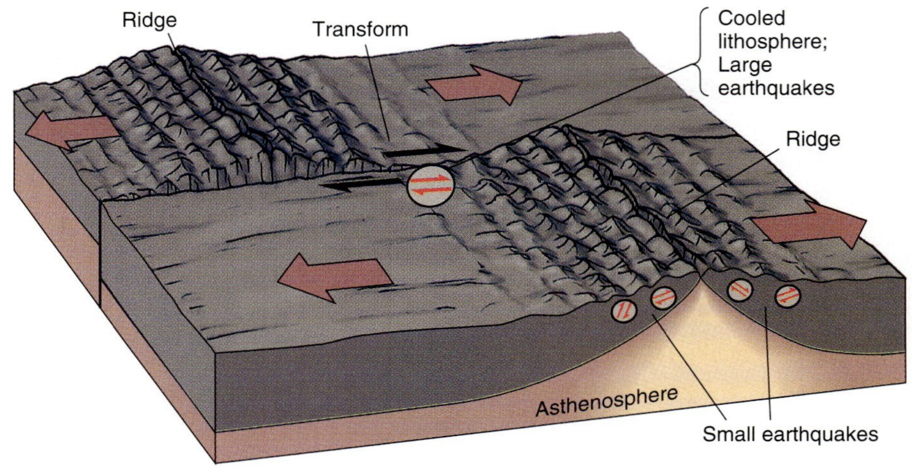

FIGURE 7.12 At a divergent margin, the largest earthquakes occur at transform faults where the lithosphere has cooled and rocks are brittle and in relative motion. Small earthquakes are common at the ridges where the rocks are too hot and weak to generate large earthquakes.

motions in the mantle, such as those that generate hot spots, that form continental rifts, and that ultimately break up plates, are probably responsible for the stresses generating some intraplate earthquakes.

Earthquakes result from rapid movements of fault blocks on either side of a fault plane when the stress overcomes friction on the fault. At transform and divergent margins, earthquakes do not occur at depths greater than about 15 km—the transition from brittle to ductile failure. Deep earthquakes occur at convergent margins to depths of 700 km.

Measuring Earthquakes

We saw in Chapter 5 that the seismic waves generated by an earthquake travel outward as expanding shells of energy, which are similar to water waves generated by dropping a pebble into a pond. Earthquake waves are measured using instruments called seismometers (∞ see Chapter 5). The simplest seismometer consists of a mass suspended on a spring attached to a support structure. When the seismic waves reach this instrument, the movement of the ground causes the mass to appear to bob up and down. Actually, the suspended mass of an ideal seismometer should remain still due to its inertia, while the ground bobs up and down due to the earthquake. A pen attached to the mass will then trace wiggly lines on a moving sheet of paper, indicating the extent of ground movement (Fig. 7.13a). Such tracings are called seismograms (Fig 7.13b).

Figure 7.13c shows an ancient Chinese version of a seismometer from the second century A.D. It consists of an elaborately decorated urn displaying eight carved dragons on its outer circumference. The urn is surrounded by eight metal toads positioned below each dragon. Bronze balls held in the dragons' mouths drop into the mouths of the toads stationed below when even slightly disturbed by the vibrations from an earthquake. The dragons aligned with the direction of the earthquake would release their balls first. The noise made upon striking the toad's mouth sounded a warning, and the dragon with an empty mouth indicated the direction of the seismic event.

Modern seismometers located right in earthquake zones are called *strong-motion seismometers*. If you have ever experienced an earthquake and observed furniture crashing around and lights swinging, you will appreciate that the spring and mass system of a strong-motion instrument need not be sensitive. In fact, you could build one yourself. A swinging pendulum with a pencil attached that writes on a moving sheet of paper would suffice. A seismometer that detects waves from an earthquake originating on the other side of the planet, however, must be very sensitive. Seismic waves detected at large distances from an earthquake epicenter are called *teleseisms*. They have given us information not only about features of the earthquakes but also about the internal properties of Earth (∞ see Chapter 5).

Earthquake Locations

Suppose we measure the time of arrival of the P wave and that of the S wave. We saw in Chapter 5 that P waves travel at about 1.7 times the speed of S waves. Thus the difference in P and S arrival times increases with distance from an earthquake. Because P-wave and S-wave speeds are known in a region, we can use the S–P arrival time interval to determine the distance from a seismic station to the earthquake. We then draw a circle on a map—centered on that station—with a radius equal to this distance. The earthquake could have occurred anywhere on the circle because every point would have the same S–P interval. However, if circles are drawn for three separate stations, the circles will intersect at one point—the point corresponding to the epicenter of the earthquake (Fig. 7.14). This technique, known as *triangulation*, is the same process described for the accurate location of points on Earth's surface by the Global Positioning System. In practice, S and P waves are measured by a seismic array that can have more than 100 stations. Other arrivals are also measured, including surface waves or body waves that travel in the mantle and reflect off or travel through the core. A location of the earthquake is then found that best matches the various travel times.

Earthquake Sizes

The **magnitude** of an earthquake is a measure of its size. This measure is classified using the **Richter magnitude scale**, which is named after its inventor, Charles Richter. The Richter magnitude is proportional to the logarithm of the amplitude of the largest waves in a seismogram, which are measured on a standard seismometer at a standard 100-km distance from the epicenter (Fig. 7.13b). Distance corrections are applied for seismic stations not located at the standard distance. The Richter magnitude scale was developed for earthquakes in California. Other magnitude scales developed from this concept are now in practice worldwide. The two most commonly used are the surface and body wave magnitude scales based on amplitudes of body and surface waves. Typically, reported magnitudes vary between 0 and about 9. Earthquake magnitudes can even be given as negative numbers for very tiny earthquakes, because a logarithm is used and the logarithm of a number between 0 and 1 is negative.

Possibly the largest instrumentally measured earthquake occurred in Assam, India, in 1897; it had a magnitude of 8.7. Because magnitude is a logarithmic scale rather than a linear one, an increase of one on the magnitude scale means that the amplitude is 10 times larger; the amount of energy released is 30 times greater. An earthquake that measures 7 on the Richter scale causes ground shaking that is 10 times greater than an earthquake with a magnitude of 6. Seismic waves generated by the largest exploded nuclear weapons are similar to waves from a magnitude 6.5 earthquake.

Damage to buildings and effects on humans in populated areas can also be used to estimate the intensity of an

(a)

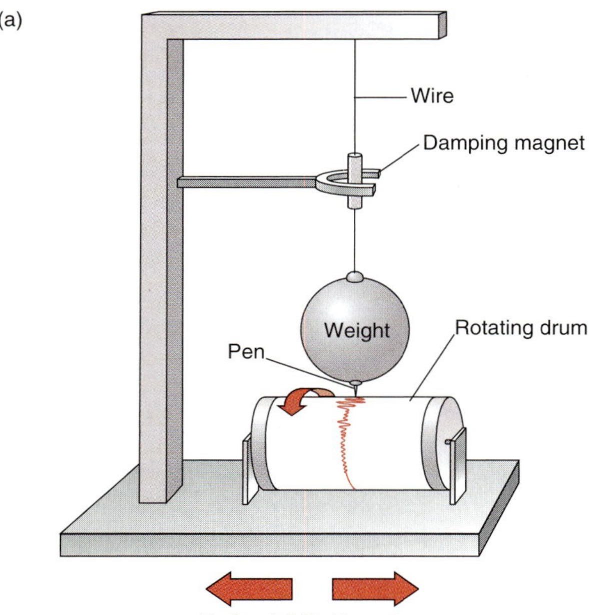

(b)

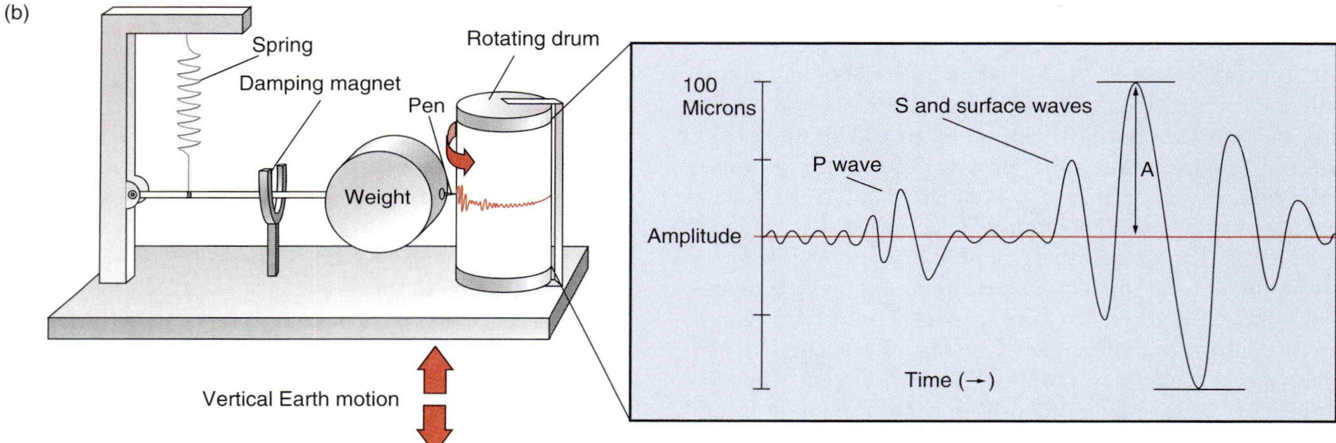

(c)

FIGURE 7.13 (a) A simple seismometer. A large suspended inertial weight remains fixed in space while Earth vibrates about it. The paper drum anchored to the ground records the vibrations as a stationary pen affixed to the weight traces the path of the relative motion. This instrument is used to measure horizontal vibrations. The seismograph on the right uses an electrical signal generated from the moving weight to drive the pen. (b) This instrument measures vertical vibrations. The seismogram on the right shows P, S, and surface waves from an earthquake. The magnitude is usually determined by measuring the maximum amplitude of the surface wave. The magnitude is proportional to the logarithm of the amplitude (A), with corrections to a standard distance and instrument. (c) A Chinese dragon seismometer. The dragons mounted on the outside of the pot hold bronze balls in their mouths. An internal hair-trigger mechanism opens the mouths of the dragons that are aligned with the incoming seismic waves. The clanging noise of the balls as they drop sounds the earthquake warning, while the direction to the earthquake epicenter is ascertained by noting which balls have dropped.

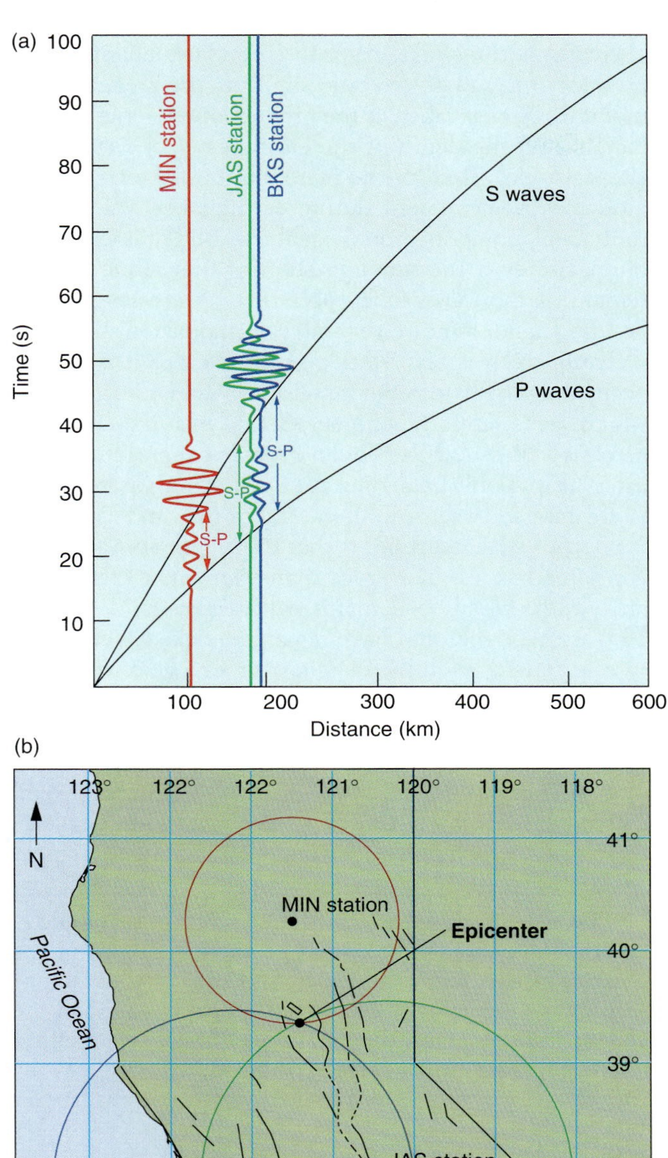

FIGURE 7.14 (a) Locating earthquakes: seismograms from an earthquake are recorded at three seismic stations: Mineral, Jamestown, and Berkeley. The time difference between arrivals of S waves and P waves (S–P) is read off seismograms from an earthquake, which can be used to obtain the distance from each station to the earthquake. (b) A circle of radius scaled to the distance of each station from the earthquake locates the epicenter of the earthquake where the three circles intersect.

earthquake. Long before the introduction of the Richter scale, an intensity scale was developed that is now called the **Modified Mercalli Intensity Scale** (∞ see Chapter 15). Intensity measurements are given in Roman numerals ranging from I to XII. Intensity XII corresponds to nearly total destruction; VI, to being felt by everyone, with furniture overturned; an earthquake that is not felt by anyone has an intensity measure of I.

Small earthquakes occur much more frequently than do large ones. If you examine a broken stick or a broken rock, you will see that the fractured region is very rough and contains many small fractures, a fewer number of intermediate fractures, and only a few large fractures that cover most of the area. If you were to listen to the stick with a sensitive seismograph just before it broke, you would have heard popping and cracking as small fractures broke before the large, main snap. The same concept holds true within Earth at a plate margin. Earthquakes with low magnitudes—of about magnitude 3—happen every day. Magnitude 5 earthquakes happen at a given plate margin about once a year, and really big earthquakes, having magnitudes greater than 8, occur—fortunately—only about once every 200 years.

The graph of the logarithm of the number of earthquakes above a given magnitude against the magnitude of each earthquake is a straight line called the *Gutenberg-Richter relation*, after the two seismologists who first presented such plots (Fig. 7.15). This plot shows that there are 10 times more earthquakes of magnitude 5 than there are of magnitude 6. The relationship holds not only for the global catalog of earthquakes but also holds regionally, as in an active region at a plate margin.

Most of the plate-margin readjustment, or catch-up, occurs during large-magnitude earthquakes. Small earthquakes contribute significantly less to this readjustment and simply rearrange the stresses as they continue to build up for the large event. You can work out how often a large earthquake is expected at a plate margin if you divide the distance the fault moves by the plate velocity. This will tell you how far the crust must suddenly move to catch up with the rest of the plate. The 1906 San Francisco earthquake was caused by about 6 m of sudden movement along the fault. The motion of the Pacific Plate relative to the North American Plate is estimated to be 5 cm/yr (0.05 m/yr). Therefore, the fault movement released 120 years (6/0.05 = 120) of accumulated plate motion during that event. Perhaps another such earthquake will occur next century when once again the plates have moved 6 m while the plate margins remain stuck together.

Ruptured faults of very large earthquakes—those above magnitude 8—measure about 1000 km long. Analysis of the displacement pattern and seismic waves caused by an earthquake shows that the faults typically break through the seismically active zone—that is, the brittle upper crust. On average, the slip is 0.03 percent of the smaller dimension of the fault; for example, 0.03 percent of 15 km,

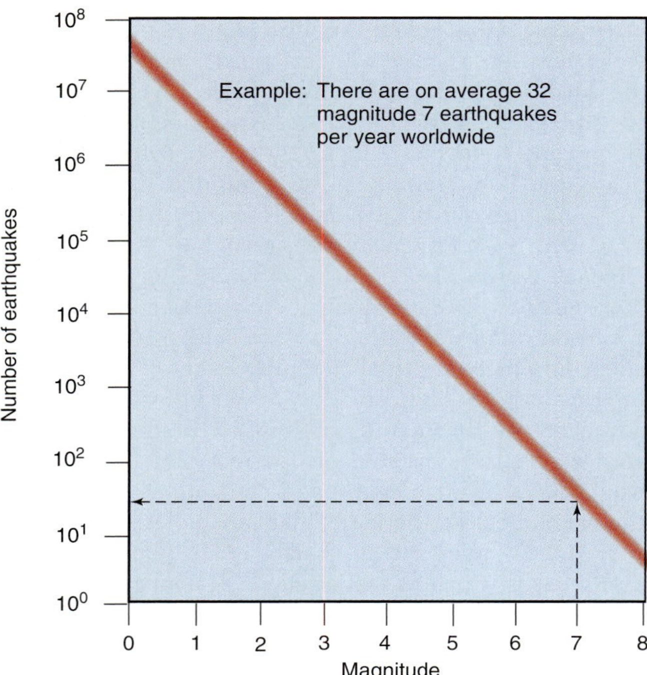

FIGURE 7.15 A plot of the number of earthquakes worldwide of a given magnitude (plus or minus 0.5 magnitude units) as a function of that magnitude. The resulting straight-line relation is known as the Gutenberg-Richter magnitude-frequency plot.

which is about 5 m. As we decrease by one unit of magnitude, the average length and recurrence interval decrease by a factor of 10. For example, the Loma Prieta earthquake, which struck just south of San Francisco in 1989, had a magnitude of 7.1 with a fault length of about 50 km. The San Francisco earthquake of 1906 ruptured 400 km of the fault and is estimated to have had a magnitude of 8.2.

Although magnitude has been extremely useful for comparing the sizes of different earthquakes, a new measure called **moment** is now regarded as a much more accurate measure of the strength of large earthquakes. The moment of an earthquake is defined as the product of the slip, the fault area, and elastic modulus of the rock.

$$\text{Moment} = \text{slip} \times \text{area} \times \text{elastic modulus}$$

You can see that the product of slip and area is a natural measure of the size of the disturbance. Multiplication by the elastic modulus includes the strength of the rock. The moment can be estimated for a given earthquake from the seismic waves it generates. It can also be determined directly by measuring the amount of slip if the fault breaks Earth's surface. The fault area is estimated from the length of the fault and the distribution of aftershocks with depth. The elastic modulus, discussed in Chapter 5, is calculated from seismologic models of Earth.

We can test whether all the relative motion at a plate margin is accounted for by earthquakes. One can compare the total moment release per year by earthquakes with the total required at a plate margin to keep up with the overall plate motions. The total for the plate margin is the product of the plate-tectonic slip rate, the area, and the modulus. Area is taken as the product of the length of the margin and the depth of the earthquake zone (typically about 15 km). If all of the plate motion is being accommodated by movement during earthquakes, the sum of individual moments should equal the total moment. Very often, however, the sum falls short of that required. The remaining motion is taken up either by stresses that are building up for a future very large earthquake or by creep.

Fault creep is the relative motion across a fault that occurs too slowly to generate seismic waves. Creepmeters, which are essentially automated tape measures installed across a fault, show that creep events of slip occur at odd intervals and only on certain faults. The San Andreas fault is creeping between Hollister, California, and Parkfield, California. The net result is that this segment of the fault is regarded as less dangerous than segments to the north and south, which had major earthquakes in 1906 and 1857, respectively, and have remained locked ever since.

Slip can occur on the margin over a range of velocities from 0 for a locked fault up to the slip velocity in earthquakes. Earthquake ruptures grow rapidly at a rate of about 3 km/sec, whereas the slip velocity, which is the relative motion of points on either side of a fault, occurs at several meters per second, a fast catch up compared with the plates themselves, which move at an approximate rate of 5 cm/yr.

The magnitude measures the size of an earthquake. Another measure, the moment of an earthquake, is the product of slip, area, and elastic modulus. If the sum of earthquake moments over a time interval is less than the plate-margin moment, the deficit may be taken up either in creep or by a future large earthquake.

Foreshocks and Aftershocks

Earthquakes cluster in time. By definition, the largest event in a cluster is called the *main shock*; those that precede it are *foreshocks* and those that follow are *aftershocks*; for the largest events, the time duration of the clustered events can last several years.

Because earthquake foreshocks do not always occur, they are not reliable earthquake predictors. The only major earthquake to be predicted in modern times was the 1975 Haicheng quake in China (see Focus 7.1). This prediction was largely based on the increasing number of foreshocks that occurred over the month prior to the event.

In southern California, estimates from historical earthquake data indicate that when a magnitude 6 event occurs, there is a 6 percent chance that it is a foreshock of a larger earthquake that will occur within the following six days.

Aftershocks, in contrast, occur after most earthquakes. The number per unit time decreases rapidly with time after the event. A rule of thumb holds that the largest aftershock of a sequence is usually one unit of magnitude less than that of the main shock. The number of aftershocks per unit time is inversely proportional to the time since the main earthquake; that is, to 1/time. This is known as Omori's law—named after its discoverer, a Japanese seismologist. If we express time in days, this law becomes the number of aftershocks on the first day divided by time. For example, on the first day there might be 100 aftershocks, on the tenth day (100/time=10) there would be 10, and on the twentieth (100/time=5) only five aftershocks would occur. For moderate earthquakes, such as those of magnitude 6, the aftershocks continue for several weeks. For great earthquakes, aftershocks can continue for years. They are thought to result from the movement of crustal material, which redistributes stress after the main event.

The range of earthquake intervals is so scattered that it appears random. Many earthquakes have aftershocks but few have significant foreshocks that could be used for prediction.

Earthquake Motions

The initial seismic waves from an earthquake exhibit a specific pattern related to the fault motion. To understand why let us consider a right-lateral strike-slip fault running in a north to south direction. When an earthquake occurs, the rock on the west side moves northward while that on the east side moves southward. We can divide the fault and its surroundings into four quadrants centered on the epicenter: NW, NE, SE, and SW (Fig. 7.16). The rock in the NW and SE quadrants will be compressed due to the fault motion, while the rock in the other two quadrants will be extended.

The very first motions of the seismic waves radiating out of these quadrants will have opposite polarities (Fig. 7.16). In the compressed or compressional quadrants, the motion is radially outward; in contrast, the motion is radially inward in the extended or dilational quadrants. This radial motion forms P waves (∞ see Chapter 5).

At right angles to the fault, no radial motion is detected at all because these directions represent the transition

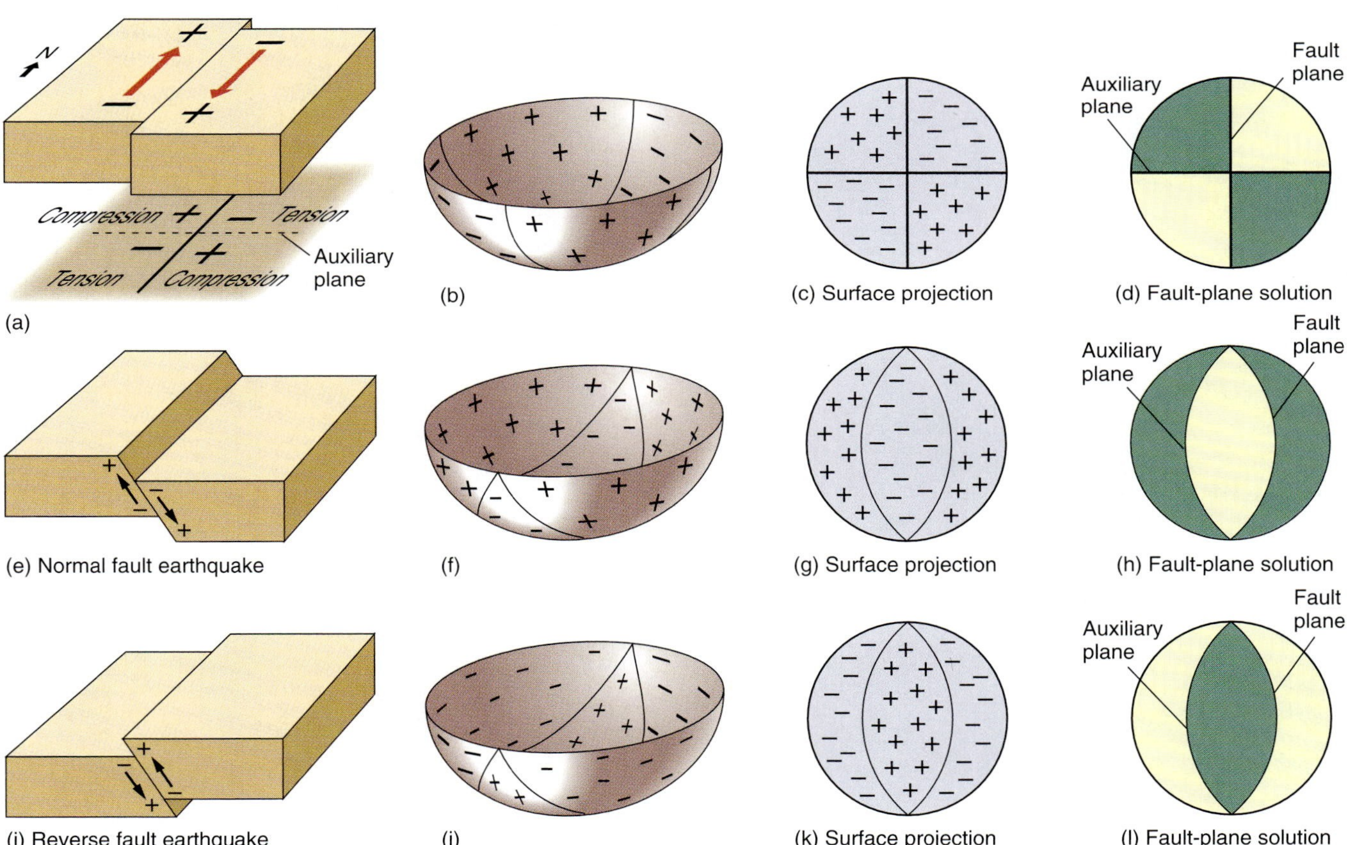

FIGURE 7.16 (a) Fault motion: When the fault moves in the direction of the arrow, the material suffers compressions denoted by the (+) sign and extensions denoted by the (–) sign. Because of the symmetry of the pattern, other information is needed to distinguish between the fault plane (solid line) and the auxiliary plane (dashed line). (b) Seismometers detect positive (+) or negative (–) ground motions. An imaginary hemispherical bowl is placed around the fault and the direction of motion of its surface denoted by a + (outward) or – (inward). (c) Projection of the motions onto the Earth's surface. A plot of these motions yields a fault-plane solution. (d) Geologists shade the (+) regions, leaving the (–) regions blank. The shaded diagram is a fault-plane solution. (e–h) same as (a–d), but for a normal fault earthquake. (i-l) same as (a-d) but for a reverse fault earthquake.

Focus On 7.1 PREDICTION OF THE HAICHENG EARTHQUAKE OF 1975

At 7:36 on the evening of February 4, 1975, a magnitude 7.3 earthquake occurred near the town of Haicheng, in the People's Republic of China (Fig. 1). That very afternoon, Chinese authorities had issued an earthquake warning that had resulted in a massive evacuation of inhabitants from their dwellings. Although this was not the only earthquake that had been predicted in China, there had also been some false alarms and failures to predict other earthquakes. Teams of Western scientists traveled to China to discover how it had been predicted, since prediction of earthquakes in the West had been, and still is, elusive.

The most important observation that prompted the prediction was the foreshock buildup. The first foreshock occurred three days before the earthquake, on February 1. A day later, several foreshocks occurred. The following day, February 3, the number and intensity of foreshocks increased further until just before noon on February 4; the number of foreshocks per hour rose to a peak of over 60, and then decreased over the next 8 hours. This was followed by a few hours of relative calm just prior to the major event (Fig. 2a).

Because Chinese scientists had observed this pattern before—building to a peak, calm, and then an earthquake—they issued the warning. Other phenomena were observed over the same period: anomalous animal behavior, unusual taste and quality of water in communal wells, pre-earthquake lights (sometimes referred to as "earthquake lightning"), and unusual ground fog. These, however, were variable in reliability and thus do not provide a scientific basis for prediction. One animal reaction of note was the appearance of a snake in the snow—snakes normally remain underground in winter.

Long before the earthquake, unusual changes in various measurements of geophysical phenomena had alerted scientists, so that when the activity started increasing rapidly, they were prepared to issue a public warning. Since 1971, the number of earthquakes per month had been steadily increasing (Fig. 2b).

FIGURE 1 The region of the Haicheng earthquake. The red dot indicates the epicenter along the Shihpengyu fault.

Significant ground uplift had been measured using precise surveying techniques. Electrical signals in the ground were detected, and tilting of the ground surface showed anomalous trends. Water wells were found to undergo changes in level, and even Earth's magnetic field underwent unusual local variation.

When the warning was issued, people moved out of their homes into tents and temporary huts even though temperatures were below freezing. Ninety percent of the homes collapsed and certainly would have killed many of the inhabitants. It has been estimated that over 100,000 lives were saved due to that successful prediction.

This seeming breakthrough in earthquake prediction, however, was delivered an alarming blow the following year when the Tangshan earthquake occurred in China with absolutely no foreshock or any other kind of warning. It has been estimated that this earthquake was responsible for about a quarter of a million deaths.

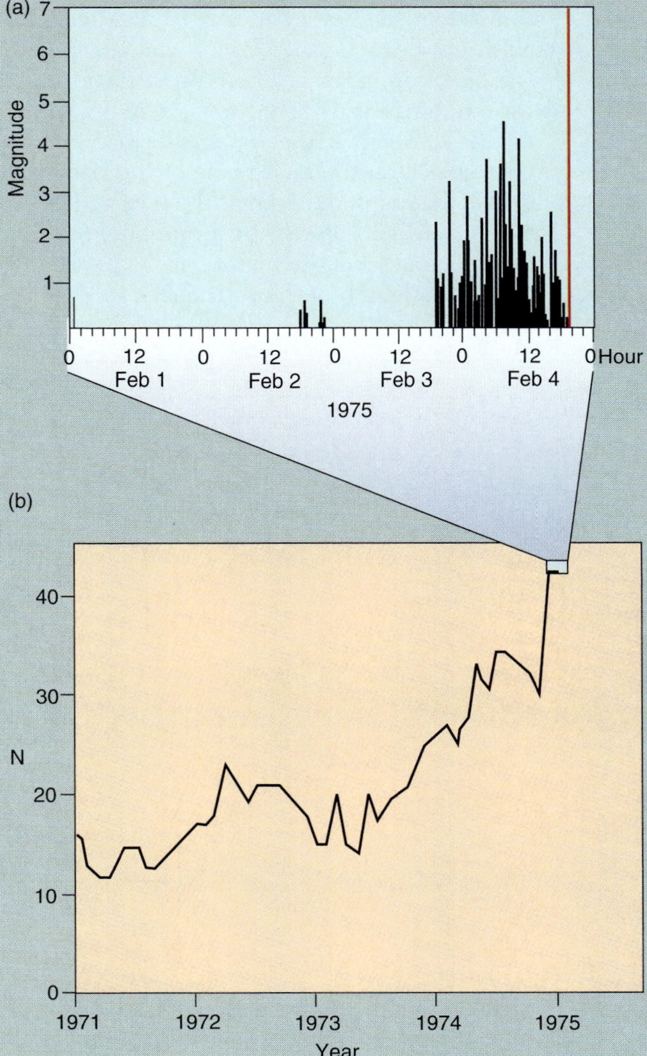

FIGURE 2 (a) Foreshocks for the four days prior to the Haicheng earthquake. (b) Monthly numbers of earthquakes for four years before the Haicheng earthquake.

from compression to tension. Tangential motions, however, are detected. These are S or shear waves, so named because they are the secondary waves (after the primary P waves) recorded on a seismogram (∞ see Chapter 5).

If a sufficient number of seismometers surrounds the fault to detect and record the radial or P-wave motion, the results can be used to determine the orientation of the fault. This is accomplished by plotting the first motions on an imaginary hemisphere surrounding the fault. In practice, projection of the hemisphere onto a sheet of paper is used, with pluses (+) plotted for compressions and minuses (–) for extensions (Fig. 7.16c). Two surfaces at right angles are then defined, separating the positives from the negatives. One of these is the true fault plane; the other is known as the *auxiliary plane* (Fig. 7.16d).

Such patterns, known as **fault-plane solutions**, can also be made for thrust faults and normal faults. The first motions are rotated on the hemisphere relative to their positions for a strike-slip earthquake. When projected onto a plane, they take the forms shown in Figure 7.16. Imagine looking into a bowl placed beneath the earthquake. Positive and negative motions plotted on the bowl separate the compressional and tensional motions. The boundaries between them give the fault plane and the auxiliary plane. If (–) motions lie mainly along the bottom and (+) motions lie along the sides, the fault is normal (Fig. 7.16f). Alternatively, if (+) motions lie along the bottom and (–) motions lie along the sides, the fault is reverse (Fig. 7.16j). In Focus 7.2, we consider the 1989 Loma Prieta earthquake in California, which had a combination strike-slip and reverse-fault focal mechanism.

No easy way exists to distinguish between the true fault plane and the auxiliary plane because the wave-radiation pattern of the quadrants is symmetrical. Other information must be used. If there is a surface break, then it is easy to tell. Other distinguishing clues include the direction of plate motion, or if the earthquake is followed by many aftershocks, the aftershock distribution in space will generally follow the trend of the fault. After the very first wave arrives at the seismometer, later seismic wave arrivals depicted on the seismogram contain information about the direction of rupture. For example, the waves pile up in the direction of rupture and become peaked as the fault extends along with the propagating waves. In contrast, the waves in the opposite direction become drawn out as they must travel farther from the receding rupture front (Fig. 7.17). This effect is analogous to the Doppler effect discussed in Focus 2.1.

Modern seismology uses the full waveform to distinguish not only the fault from the auxiliary plane but also the direction in which the rupture traveled, how large it was on different parts of the fault, and when it occurred. Figure 7.18 shows the rupture history inferred to have taken place on the fault plane of the Loma Prieta earthquake. The rupture began at the hypocenter and expanded outward and upward at a rate of about 3 km/sec. At this speed, it took the rupture about 13 seconds to travel the 40-km length of the earthquake.

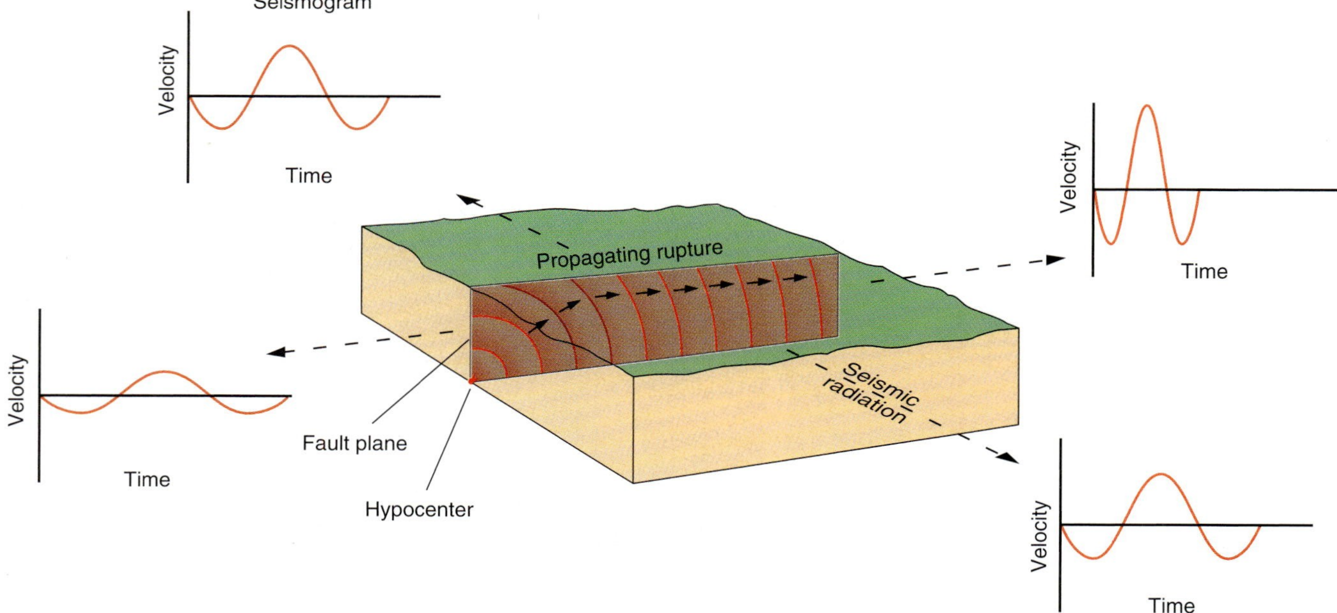

FIGURE 7.17 Seismograms take on different shapes at different directions for an earthquake rupture that propagates in one direction. In the direction of propagation, the energy builds to give a peaked shape. In the opposite direction, the energy spreads out because it must travel farther from the retreating rupture along the fault. In other directions, the shapes are intermediate between these two. This pattern can be used to distinguish the fault plane from the auxiliary plane.

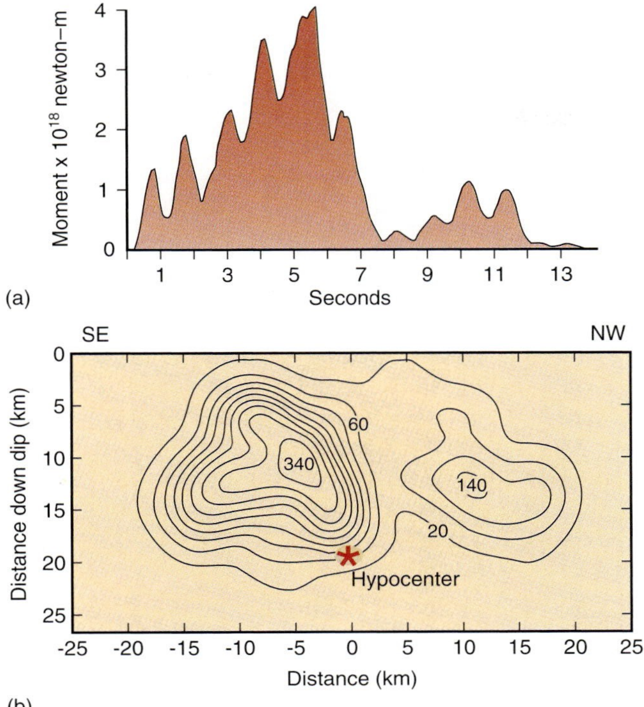

FIGURE 7.18 Slip on the fault plane as a function of time for the Loma Prieta earthquake in California in 1989. Part (a) shows the time history of the moment (slip × area × modulus). The bottom diagram (b) shows the final distribution of slip after faulting ceased. Contours are in centimeters. The star indicates the hypocenter, where slipping originated. The fault grew upward and outward.

The fault-plane solution gives two planes at right angles: the fault and auxiliary planes that separate the seismic-wave radiation into compressional and extensional quadrants. Distinguishing the fault plane from the auxiliary plane requires extra information such as distribution of aftershocks, waveforms, or surface breaks.

Plate Motions and Earthquakes

Measurement of Past Plate Motions

Earthquakes do not recur at regular and predictable intervals; in fact, only the long-term average time between earthquakes appears to be calculable. As we have seen, this interval is expressed by the average displacement of great earthquakes divided by the rate of plate motion:

Time interval = displacement of earthquake/plate velocity

For example, if the average displacement is 10 m and the rate of plate motion is 5 cm/yr, the time interval between earthquakes is expected to be 200 years. Individual intervals, however, exhibit too much variability for the time of a given event to be predicted at all. Perhaps this occurs because the crust is very heterogeneous. How and when the rocks fail is a complicated process that makes the record of individual events appear to be more random rather than a regular, periodic one.

Times of ancient earthquakes are estimated by measuring the offsets of sedimentary strata in fast-filling sedimentary basins. Scientists have developed this method for the San Andreas fault at a location north of Los Angeles (Fig. 7.19), and they have determined that over the last 1500 years major earthquakes occurred on this fault at irregular intervals, averaging about every 136 years. Because we have historic records only for the last 200 years, this finding is of great interest—especially for the inhabitants of Los Angeles, where the last major earthquake occurred in 1857. This earthquake had a magnitude of about 8 compared with recent events such as the 1994 Northridge earthquake, with a magnitude of 6.7, which, though severe, was too small to have had a significant role in catching up plate movement. Without this more extended record, it would be difficult to estimate when the next earthquake might occur; given the short historic record, however, it appears that a large event could take place quite soon. The record also shows that the time between earthquakes has been highly variable. Thus, predicting when the next event will occur is extremely uncertain.

Geologic evidence of fault movement includes offset stream beds across a fault or displaced rocks that originally spanned a fault. On the San Andreas fault, some rock sequences have been displaced by as much as 250 km in the 5 million years of its existence, corresponding to movement at an average rate of 5.0 cm/yr.

Another way to estimate plate motions is to use plate-separation velocities deduced from the magnetic stripes

FIGURE 7.19 Excavation at Pallet Creek, California, reveals times of ancient earthquakes on the San Andreas fault. Professor Kerry Sieh of the California Institute of Technology is shown pointing to offset sedimentary layers that are geologic markers of prehistoric earthquakes.

Focus On 7.2 THE LOMA PRIETA EARTHQUAKE

The San Andreas fault is the primary boundary between the Pacific and North American plates. It is a right-lateral transform, or strike-slip, fault, and it transforms spreading motion in the Gulf of California into subduction of the Juan de Fuca Plate, which it meets at Cape Mendocino. In southern California, the fault takes a bend to the west (Fig. 1), called the "big bend."

Two major earthquakes have occurred on the San Andreas fault in recorded history. In 1857, the Fort Tejon earthquake ruptured a section that was 300 km in length, in the region of the big bend. The 1906 San Francisco earthquake ruptured the fault farther north, along a 400-km segment. Both fault segments have remained locked ever since. In the region of Parkfield, the fault is creeping, or continuously slipping. Orange trees that cross the fault, originally planted in rows, show offsets on either side of the fault. The fault can be traced through the town of Hollister, where it dislocated sidewalks and white lines in the middle of the road. In one case, the fault passes through a winery causing a remarkable distortion of the building.

In 1989, a World Series baseball game in San Francisco's Candlestick Park was brought to an abrupt halt when the magnitude 7.1 Loma Prieta earthquake struck near the San Andreas fault in California; its epicenter was located 9 miles northeast of Santa Cruz and 60 miles south-southeast of San Francisco. The Loma Prieta earthquake occurred at the junction of the 1906 locked section to the north and the slipping section to the south. In 1988, the Working Group on California Earthquake Probabilities (a team of scientists convened by the United States Geological Survey) recognized that displacement in the 1906 earthquake was significantly less in the Santa Cruz Mountain section of the fault compared with the displacement in sections farther north. Because this zone had remained locked since 1906, it still had some plate motion to catch up. The team's 1988 report had allotted this segment a 30 percent probability of a magnitude 6.5 or larger event between the years 1988 and 2018 (Fig. 2).

The other hint that an earthquake might be imminent was the observed increase in the number of large events that occurred in that part of California prior to the earthquake. Using a computer program that recognizes patterns in earthquake data, Russian scientists had issued a warning for northern California in 1988. The warning came in this form: Owing to the increased seismic activity, among other criteria, for the next five- to seven-year period, the northern part of the state and northern Nevada was in a "Time of Increased Probability" (TIP), for an earthquake of magnitude equal to or greater than 7.

Both of these forecasts are too limited in resolution to be useful on a practical level but do serve to alert authorities to take necessary precautions. The former was precise in space but spread out in time, whereas the latter was more precise in time but spread out in space.

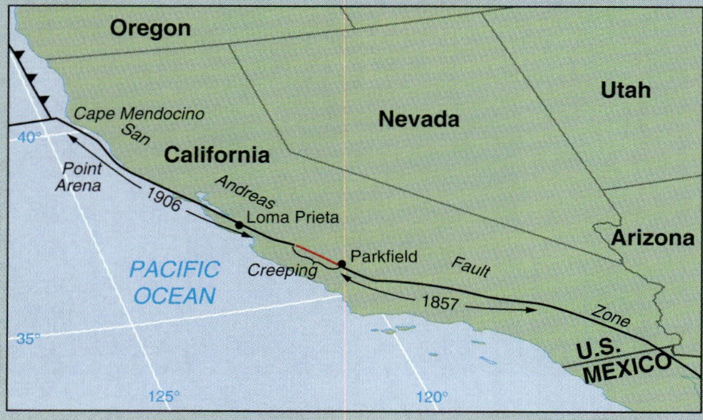

FIGURE 1 The San Andreas fault system showing the northern and southern locked sections, which ruptured last in 1906 and 1857, respectively. They are separated by the central creeping section.

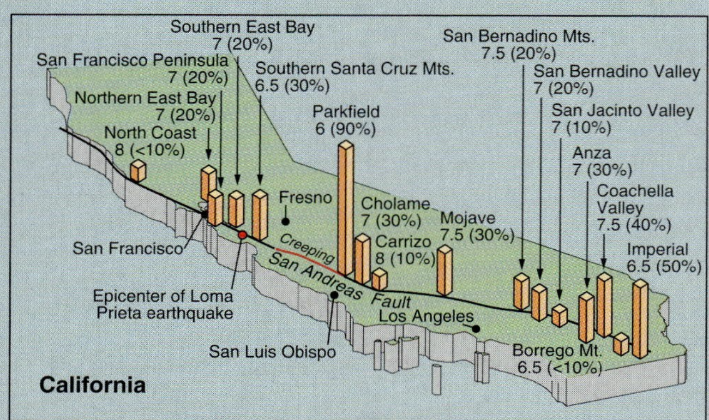

FIGURE 2 Earthquake probabilities above a given magnitude along the San Andreas fault for a 30-year period, calculated by the U.S. Geological Survey's Working Group on Earthquake Probabilities.

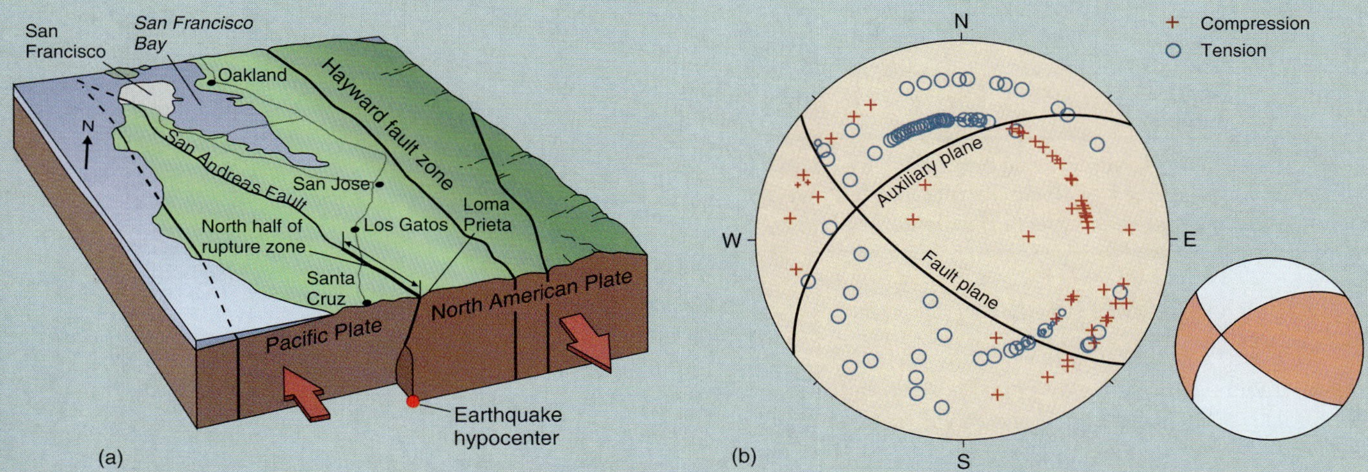

FIGURE 3 (a) Northern half of the rupture zone of the Loma Prieta earthquake showing the section of the San Andreas fault that broke. The rupture began at the hypocenter 18 km deep and spread upward to a depth of 6 km and outward for a distance of about 15 km. The slip on the fault was oblique, with 8 m in the horizontal direction and 4 m in the vertical direction. (b) The fault-plane solution for the Loma Prieta, California, earthquake. Circles are extensions; (+) represent compressions. The solid curve running NW–SE is the fault plane; the other curve represents the auxiliary plane.

The fault rupture of the Loma Prieta earthquake began at a hypocentral depth of 18 km (Fig. 3a). The Pacific block moved 2.3 m (7.4 feet) relative to the North American block, 1.9 m (6.2 feet) horizontally to the northwest, and 1.3 m (4.2 feet) upward over the North American Plate. Thus, the Pacific block was the hanging wall and the North American Plate was the footwall. This earthquake was a combination strike-slip and thrust event. Fault-plane solutions (Fig. 3b) showed that the fault plane dipped at about 70° to the southwest. Analysis of ground displacements revealed that the fault plane did not break the surface but stopped 6 km below it. A plot of aftershocks made it easy to distinguish the fault and auxiliary planes (Fig. 4).

It was fortunate that this event occurred in the remote Santa Cruz Mountains and that the death toll was no greater than 62. If a great earthquake were to occur closer to a major population center, the death toll might rise to thousands and damage costs might be in the tens of billions rather than the $6 billion of damage incurred during Loma Prieta.

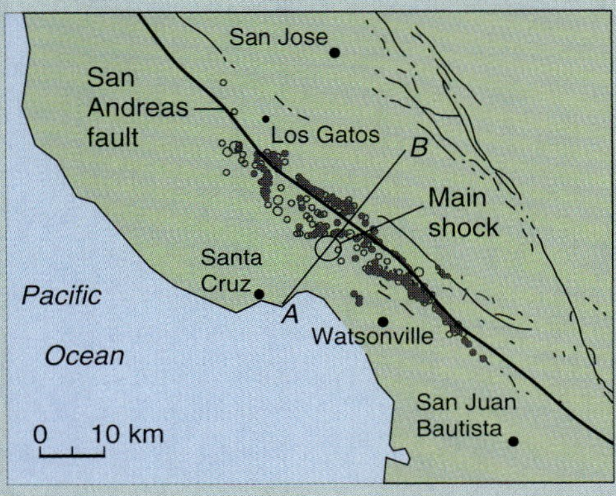

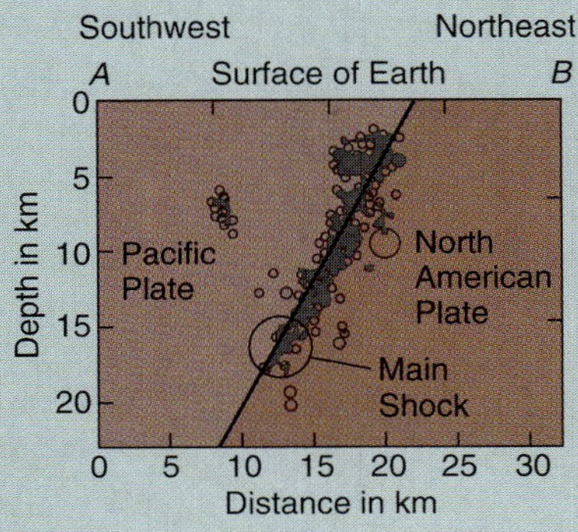

FIGURE 4 Aftershocks of the Loma Prieta earthquake clearly allow the NW—SE orientation of the fault plane to be distinguished from the auxiliary plane.

generated at mid-ocean ridges (∞ see Chapter 2). These are combined with plate velocities determined from hotspot trails such as the Hawaiian-Emperor seamounts. Scientists have recently gathered all this information together and used a geometrical model to calculate motions of all the plates. The latest version (called the NUVEL model for **N**orthwestern **U**niversity **Vel**ocity) is based on geologic evidence of the last 10 million years. It predicts plate motions that agree with current measurements remarkably well.

Long-term geologic slip rates across faults are estimated by measuring displaced geologic features of known age, such as stream beds, rock types, or sedimentary layers.

Geodesy and Measurement of Present-Day Plate Motions

Measurements of present-day plate motions fall into two categories: land-based or space-based measurements. Land-based measurements use *geodimeters* and *optical levels* to determine the *geodesy* of Earth, which is a measurement of Earth and its geography. Geodimeters measure horizontal distances by timing laser pulses between the instrument and a reflector. A geodimeter can measure the distance between two points across a fault (Fig. 7.20).

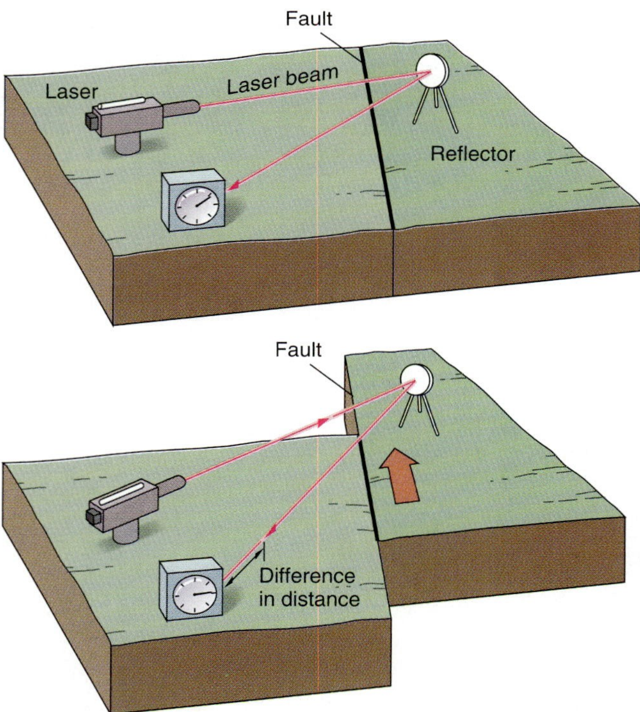

FIGURE 7.20 Geodimeters measure change in distance across a fault using a laser beam. The time that the laser takes to travel to the reflector and back is measured with great precision using a geodimeter. Any fault movement changes the distance, and hence the travel time of the laser pulse.

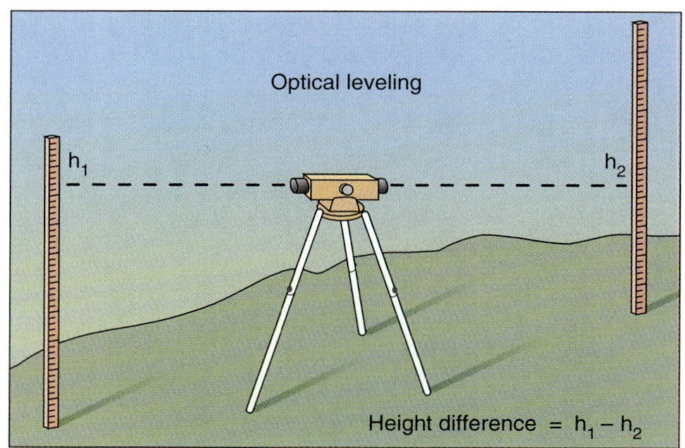

FIGURE 7.21 Optical leveling measures change in height by sighting on two measuring rods and subtracting the two readings.

Optical leveling uses sea level as a reference and two vertical measuring rods to obtain relative elevations from sea level to *benchmarks*, which are permanent steel pins driven into the ground at places where the elevation above sea level has been accurately measured on a regular basis. The technique of optical leveling uses a telescope to read the scales on the two vertical rods (Fig. 7.21). The telescope is carefully adjusted using bubble levels so that it is horizontal. One reading is then subtracted from the other to obtain the difference in height. By leapfrogging the telescope and the backward rod, the process is repeated from sea level to the benchmark. The sum of all the height differences gives the total elevation of the benchmark.

Space-based methods used to measure plate motions fall into two categories: Very Long Baseline Interferometry (VLBI), which relies on signals from deep space, and the Global Positioning Satellite system (GPS), which uses signals from satellites orbiting Earth. The VLBI antennae (Fig. 7.22) detect radio signals from quasars (*quasi*-stell*ar* radio sources), which are thought to be the most distant objects known in the universe. If one antenna is farther away from the quasar than the other, the same signal arrives a little later. We know the speed of light; therefore, if we time the delay in the signal arrival, we can accurately determine the distance. If we repeat the measurement at a later time in the day when Earth has rotated relative to the position of the quasar, we obtain the difference in distance from a different angle. For example, when the quasar is on the horizon, we measure the difference in horizontal position. Six hours later, with the quasar overhead, we measure the difference in elevation. When the VLBI antennae are on different plates, measurement of the plate separation as a function of time gives the relative plate motion.

The Global Positioning Satellite system (Fig. 7.23) uses a series of 24 satellites to perform the same function. In recent years, the two methods have become so accurate that plate movements of just a few millimeters can be detected. The NUVEL model predicts that plates move

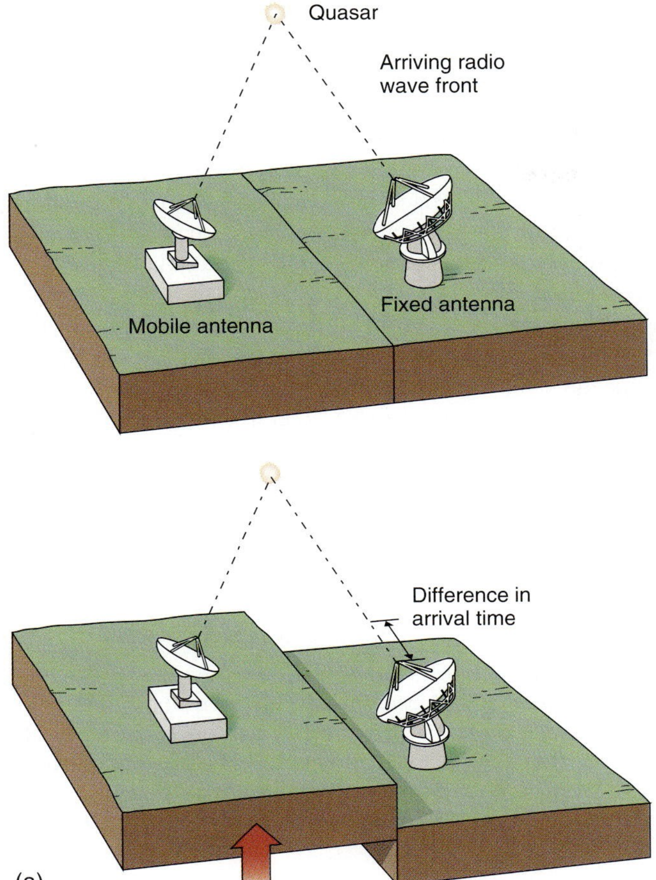

FIGURE 7.22 Very Long Baseline Interferometry (VLBI) uses signals from quasars to measure relative positions between giant radio antennae (a) such as the one pictured here (b).

many centimeters annually. A triumph of the space and geodesy methods has been to demonstrate that they indeed detect this motion as it occurs today, with the measured magnitudes and directions of plate movement consistent with those inferred from geologic observation.

Geodetic results show that, in areas away from plate margins, the plates travel in their separate directions unaffected by the havoc they cause in earthquakes at the margins. For example, across the San Andreas fault, the VLBI site on the California coast at Vandenburg Air Force Base (on the Pacific Plate) is traveling at 40 mm/yr relative to the site at Goldstone, California (on the North American Plate). This rate is similar to the 50 mm/yr predicted from the NUVEL model and also matches estimates from offset streams (the 10 mm/yr discrepancy is the subject of ongoing research and may occur either farther inland or offshore or both). Across

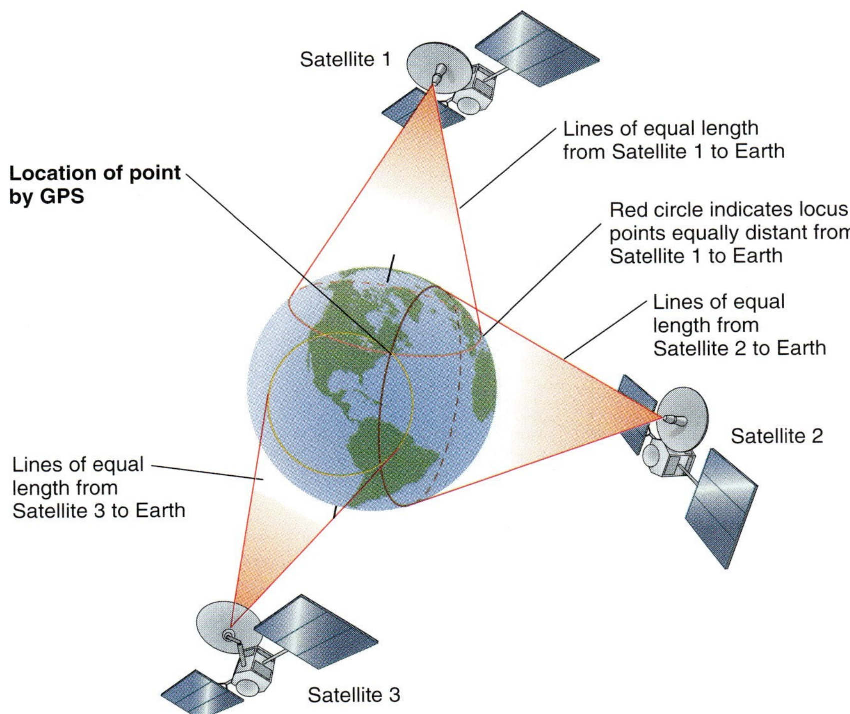

FIGURE 7.23 The accurate location of a single geographic point on Earth can be determined by measuring the distances from three or more satellites—the same technique of triangulation used to determine earthquake locations (Fig. 7.14).

FIGURE 7.24 The earthquake cycle consists of stress buildup about a fault, resulting in an earthquake with immediate stress reduction, healing of the fault, followed by stress buildup again.

the margin, geodetic measurements show that relative motion steadily decreases as the San Andreas fault is approached. The crust there is bent back from where it would be if had it kept up with the remainder of the plate (see Fig. 7.11). After an earthquake, measurements show that the crust catches up. The fault zone then heals and friction takes hold again, setting the stage for a repeat of the **earthquake cycle**, which can be summarized as plate motion, causing an increase in strain across a locked fault, strain release during an earthquake, and healing of the fault (Fig. 7.24). Healing occurs after an earthquake when the rocks on either side of the fault stop slipping and adhere, reestablishing the friction.

Strain Measurement: Strainmeters and Tiltmeters

A major step toward accurate earthquake prediction involves measurement of the strain accumulation in a region. *Strainmeters* are automatic devices that accurately measure changes in the distance between two points (Fig. 7.25a). Some versions use laser beams in evacuated tubes; others use electromechanical transducers to measure the extension of a wire or a quartz rod connected to the points in question. Devices installed at depths of 200 m in boreholes achieve maximum sensitivity because they are isolated from near-surface noise associated with rainfall or temperature changes. Incredibly small motions that are less than the size of an atom can be detected by these sensitive instruments.

Tiltmeters (Fig. 7.25b) measure the rotation of the ground relative to vertical. They are ideal for measuring volcano deformation, which has a fairly simple and predictable geometry with large detectable tilts. In earthquake zones, the tilts are much smaller. Reports of anomalous tilt and strain occurring before earthquakes indicated that such measurements may be used for earthquake prediction. As these instruments have improved and become more stable, however, such reports are decreasing in frequency, indicating that earlier reports may have been due to spurious causes and that the real changes are extremely small.

Measurements of present-day plate motions show that plates move on the order of a few centimeters each year. The earthquake cycle, a consequence of plate motion, consists of an increase in strain across a locked fault, strain release during an earthquake, and healing of the fault.

INTRAPLATE DEFORMATION

Satellite images of the topography of continental plates show domed regions and basins. The doming is generally associated with heating of mantle and crustal rocks. The heating causes them to expand and become lighter than

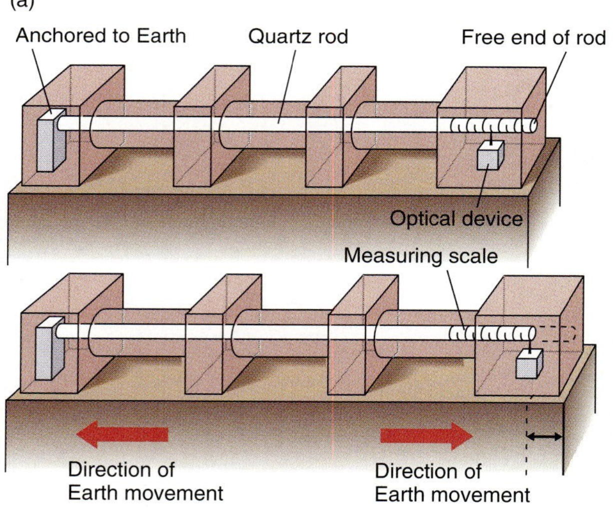

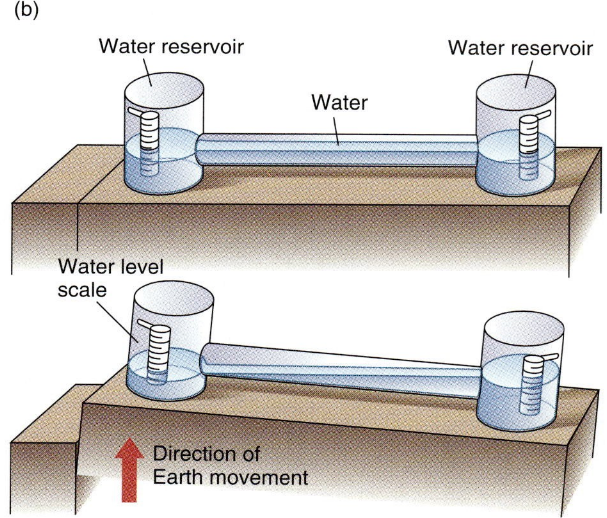

FIGURE 7.25 (a) A strainmeter. A long quartz rod is attached to Earth at one end and is free to move at the other. At the free end an optical device is used to make precise measurements of the end of the rod relative to the surroundings. Because the length of the rod remains fixed, the optical device is measuring the difference in motions between rock at one end and that at the other. The strain is this difference divided by the length of the rod. (b) A water tube tiltmeter registers change in the slope of the land. A tube connects two reservoirs separated by several hundred meters. When the land tilts, the water flows downhill from one reservoir to the other; by precisely measuring the depth of water in each reservoir, the tilt can be measured. Such instruments can detect tilts as small as 1/10,000,000 of one degree.

the surrounding rock. The surface bulges up in response to such buoyancy forces. In extreme cases, the rock is so hot it is molten, and the forces are so large that fractures occur. Magma erupts onto the surface at volcanoes. Some examples include hot-spot volcanoes such as Nyiragongo in Zaire, or Yellowstone in Wyoming.

In contrast, when a hot region cools over geologic time, it contracts and the surface deformation is prolonged subsidence. Continental margins subside as the hot material that was formed at the rift loses its heat. In addition, lithosphere that moves away from a hot spot will cool and subside. Intraplate vertical motions (uplift and subsidence) are called *epirogenic*. These motions are driven by growth and decay of convective currents in the mantle.

Elastic Flexure of the Lithosphere

If a concentrated mass, such as a seamount, is placed on the lithosphere, it will depress, or flex, the upper lithosphere elastically (∞ see Chapter 5). After a very long time (millions of years), the material below the brittle-ductile transition zone will have had sufficient time to flow. The flexure of the upper layer increases with time as the lower layer flows to accommodate the stress applied from above. On a short time scale, the lower part of the lithosphere acts as a solid. For example, seismic shear waves, for which shear stresses build up and decay in seconds, do not pass through a fluid such as water or the outer core but do pass through the lower layer of the lithosphere (∞ see Chapter 6). For stresses that are applied for a long time, the lower layer can be regarded as a fluid. Over short geologic periods, such as thousands of years, the effective elastic layer in the oceans is equivalent to the thickness of the lithosphere, about 100 km. For longer periods, however (millions of years), the thickness of the effective elastic layer decreases to about 20 km.

Figure 7.26 shows cross-section of a seamount (a submerged volcano) in the northeastern Atlantic Ocean, based on downward flexure, or bending, of the lithosphere. It was estimated that a total flexure of several kilometers spread out over several hundred kilometers is required to explain these data. The effective elastic thickness is estimated to be about 19 km in this case.

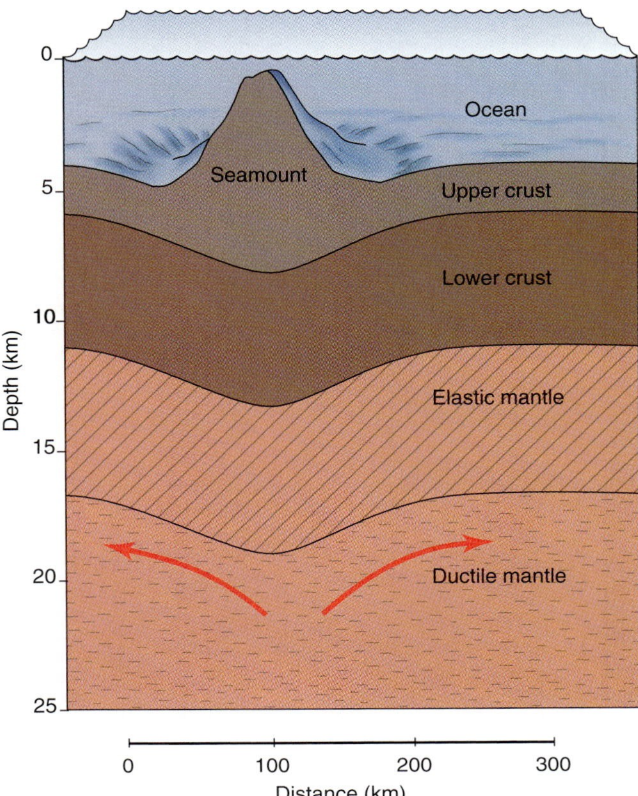

FIGURE 7.26 Plate flexure from the load of a seamount in the northeastern Atlantic Ocean. The weight of the seamount has caused the crust to bend downward. The elastic part of the upper mantle also bends. The plastic part of the mantle flows to accommodate the deformation.

The upper layer of the lithosphere flexes elastically when subjected to a load such as a seamount. Over long stretches of time, the lower layer flows to accommodate the load.

Volcano Deformation Cycle

Volcanic earthquakes nearly always precede a volcanic eruption. As magma forces its way through the crust, it generates stress in the upper brittle region. Sudden release of this stress results in earthquakes, which can be used to predict impending volcanic activity. Seismometers on the island of Hawaii are connected into an automated warning system, day and night. If the seismic activity increases above a certain threshold, the warning system automatically telephones scientists at the Hawaiian Volcano Observatory to alert them of an impending volcanic eruption. Earthquakes also preceded the violent eruptions of Mount St. Helens in Washington (see Focus 9.3). They were such reliable predictors that scientists were successfully evacuated from the crater before each of the eruptions that took place after the major event of May 18, 1980.

Another class of seismic activity detected on volcanoes is a sustained oscillatory motion usually at a dominant, fixed frequency. This vibration is called *harmonic tremor*. Models of harmonic tremor have been used to explain it as the result of magma reverberating in conduits as the magma forces its way to Earth's surface. If harmonic tremor is detected at a volcano, there is a good chance the tremor will be followed by an eruption.

A magma chamber can be thought of as a holding tank several kilometers deep where magma accumulates prior to the eruption of a volcano. Magma is thought to feed into the base of the chamber through a narrow conduit from sources in the mantle. Before a volcanic eruption, the volcano inflates as the magma chamber fills with pressurized magma. The ground surface can rise as much as a meter. The buoyancy of the underlying magma column

Focus On 7.3 VOLCANIC DEFORMATION IN HAWAII

Deformation of a volcano is a key measurement for predicting its eruption. On Kilauea volcano, Hawaii, staff members of the Hawaiian Volcano Observatory use geodimeters and optical levels to measure the deformation of the volcano about every six months. In vaults in the side of the volcano, tiltmeters continuously record the tilting of Earth's surface as the volcano swells or inflates owing to the pressures of upwelling magma (Fig. 1).

Careful analyses of these data have revealed that the surface deformation is caused by a nearly spherical magma chamber at about 2.6 km depth beneath the summit of Kilauea. When magma is injected into this chamber, the surface domes upward. The geodetic surveys demonstrate that the dome has a radius of about 5 km and can rise as much as a meter, although you would not see it by eye.

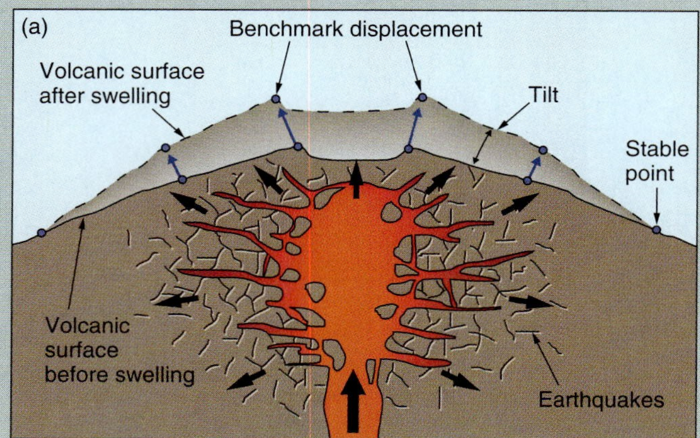

FIGURE 1 (a) Diagram illustrating the causes of volcano deformation. Magma moving into the magma chamber beneath the volcano builds up pressure. The increased pressure stresses rocks surrounding the chamber, leading to cracking and the generation of small earthquakes. The intrusion of magma also inflates the ground beneath the volcano, causing the ground surface to rise and stretch. This deformation can be measured by recording the change in distance between two fixed points (benchmarks) or by measuring changes in the slope (tilt) of the ground surface. (b) A record of tilt near the summit of Kilauea volcano, Hawaii, over a 30-year period. A six-month period during 1986 is expanded to show correspondence among the measured tilt, the occurrence of earthquakes, and the timing of volcanic eruptions.

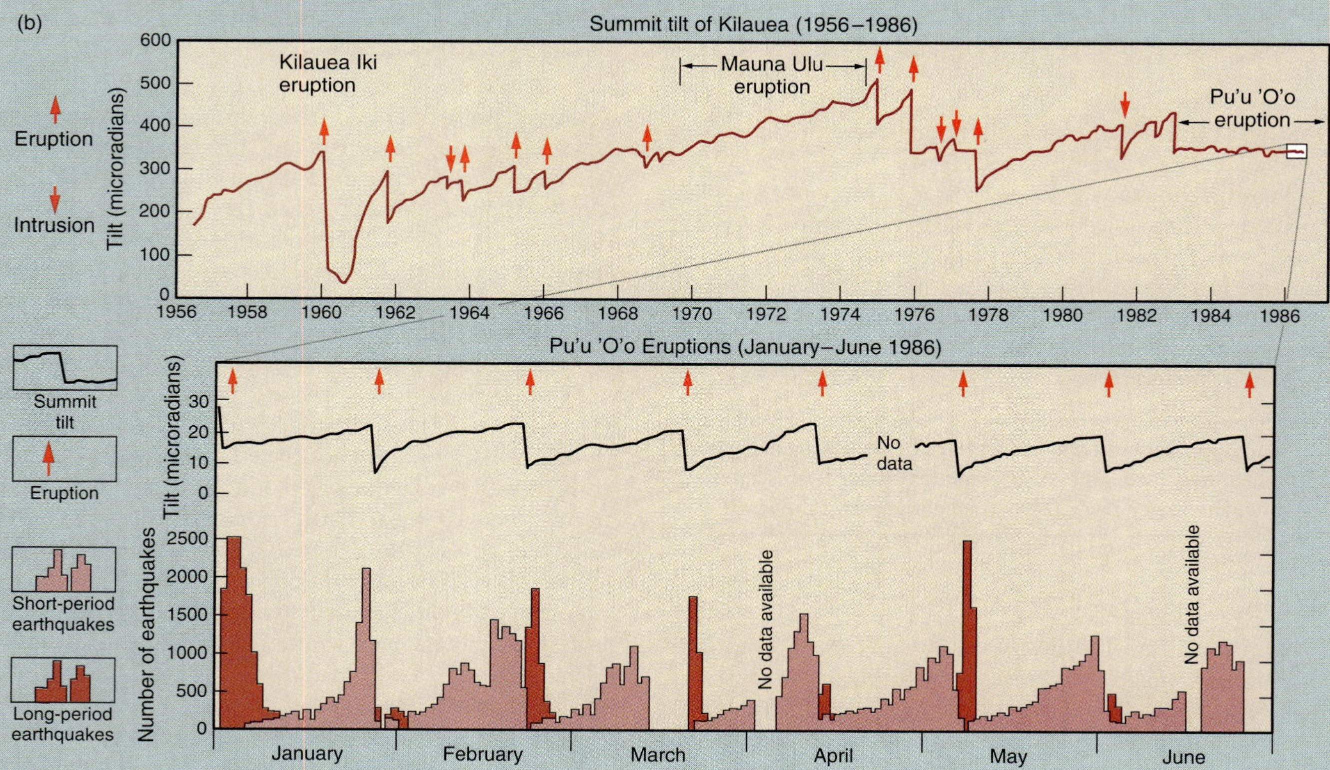

Inflation of Kilauea may continue for about six months—only to be terminated by an eruption. Magma occasionally flows out at the summit; at other times, it extrudes somewhere down the volcano slope as a flank eruption in one of the rift zones. Deflation may take less than a day, and generally progresses more rapidly than does inflation. After the eruption, the cycle repeats.

The driving mechanism for Kilauea eruptions is thought to be related to the hot spot beneath Hawaii, which provides heat to the magma column, keeping it molten and lighter than the surrounding cooler, solid rock. Earthquakes extending down to depths greater than 60 km indicate the magma comes from the asthenosphere. By forcing its way through the lithosphere, it generates earthquakes. Because magma is relatively low density, this 60-km column tries to rise and protrude above the surface, rather like a vertically submerged stick will rise out of water once it is let go. On its way up, however, magma ascent is impeded by the brittle crust, and the magma collects in the magma chamber. When pressures in the magma chamber have built up sufficiently, the crust breaks. A summit eruption occurs if the top is breached. If the sides are breached, the magma may generate flank eruptions.

Figure 2 shows a geologic cross section of exposed dikes of Koolau volcano on Oahu Island in the Hawaiian chain. The rock is composed of more than 30 percent intrusions; the remainder consists of layered extrusive lava flows. Koolau is an older volcano, and erosion has exposed its interior. By analogy, we would expect the interior of the currently active Kilauea volcano to consist of many dikes and sills. Thus a large fraction of the growth of Kilauea volcano occurs through intrusion.

Intrusion of dikes is especially prevalent along Kilauea's southwestern and eastern flanks. Forceful injection of the dikes causes the volcano to be rifted apart. After many such episodes, the stresses rifting the volcano become so great that eventually the entire side of the volcano moves toward the sea in a giant earthquake.

The magnitude 7.1 Kalapanna earthquake of 1975 was related to such a failure: The fault-plane solution showed that the rupture occurred along a horizontal plane near where the volcanic pile rests on the sea floor. The water-filled sediments on the sea floor may have provided a lubricated zone on which the fault blocks slid. Recent ocean-bottom surveys reveal that huge volumes of the Hawaiian Islands have, on occasion, slid into the sea. Destruction caused by such landslides and accompanying tsunamis (∞ see Chapter 15) must have been truly catastrophic.

The pathways for magma flow beneath Kilauea are complex (Fig. 3). The deformation at the summit, however, is clearly explained by a series of volcano deformation cycles—inflation of the central chamber, dike intrusion, or eruption accompanied by dike intrusion. This process is punctuated by occasional great earthquakes when the volcano adjusts itself to the stresses of intrusion.

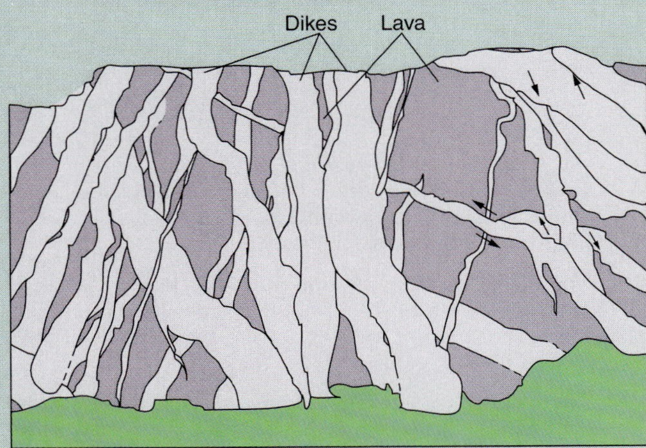

FIGURE 2 Sketch of dikes and inclined sheets in Koolau volcano, Hawaii.

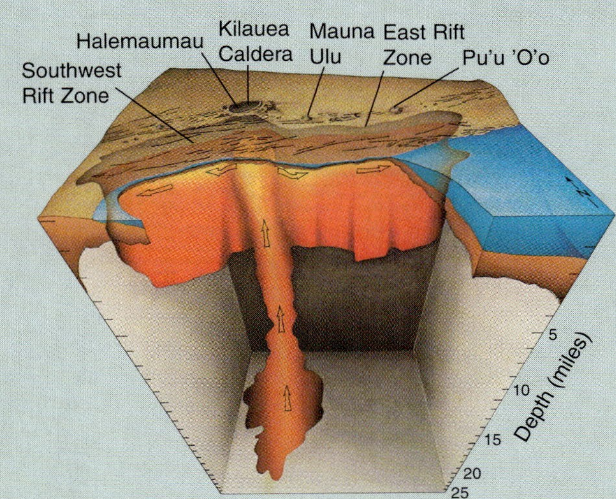

FIGURE 3 Internal structure of Kilauea volcano.

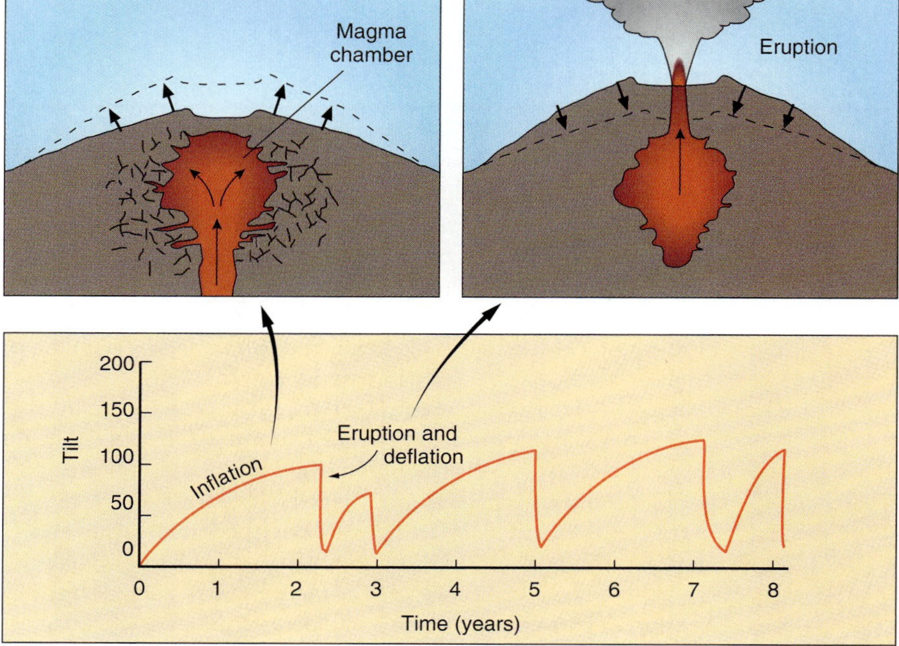

FIGURE 7.27 The volcano cycle consists of inflation of the volcano, building up stress in the surrounding rock, eruption when the elastic limit of the edifice is reached, and deflation, which reduces the stress. The fractures through which the magma flowed heal, and the process repeats.

forces more and more magma into the chamber until the walls can no longer withstand the pressure and they break. If the magma breaks upward, a summit eruption might occur. If it breaks laterally, a crack opens, usually a vertical one, and the magma travels along it, forming a dike. The dike may break through to the surface along the flank of the volcano. In some cases, the crack is horizontal and forms a sill, which eventually freezes in the surrounding rock. Many small earthquakes are detected and recorded during the formation of sills and dikes. Eroded volcanoes may expose dikes that radiate out from the central chamber for many kilometers (∞ see Chapter 4).

At the time of eruption or intrusion into sills or dikes, the pressure in the magma chamber drops as magma is expelled. The volcano deflates. Once the magma cools, the cracks in the chamber walls heal, and the **volcano deformation cycle** (Fig. 7.27) of inflation, eruption, and deflation repeats (Focus 7.3).

The volcano deformation cycle starts with inflation of a magma chamber by a supply of new magma. It is followed by deflation. The magma leaves the chamber either to feed into dikes or sills or to erupt at the surface. Movement of magma underground generates small earthquakes.

SUMMARY

- Strain in rocks can be elastic, ductile, or brittle. Rocks exhibiting elastic strain recover their original form upon release of the applied stress. Rocks undergoing ductile strain exhibit a permanent deformation. If a material breaks, the associated deformation is brittle.
- Whether a rock deforms in a brittle or ductile manner depends not only on its composition, mineralogy, and texture but also on such physical conditions as pressure, temperature, and fluid content.
- Joints are closely and regularly spaced fractures in a rock. Faults are brittle fractures in rocks across which relative movement displaces the crust on either side. Folding is a compressional ductile deformation that can be both brittle and ductile.
- Earthquakes result from rapid movements of fault blocks on either side of a fault plane when the stress overcomes friction on the fault. At transform and divergent margins, earthquakes do not occur at depths greater than about 15 km—the transition from brittle to ductile failure. Deep earthquakes occur at convergent margins to depths of 700 km.
- *Magnitude* measures the size of an earthquake. Another measure, the *moment* of an earthquake, is the product of slip, area, and elastic modulus. If the sum of earthquake moments over a time interval is less than the plate-margin moment, the deficit may either be taken up in *creep* or by a future large earthquake.
- The range of earthquake intervals is so scattered that it appears random. Many earthquakes have aftershocks, but few have significant foreshocks that could be used for prediction.
- The fault-plane solution gives two planes at right angles: the fault and auxiliary planes that separate the seismic-wave radiation into compressional and extensional quadrants. Distinguishing the fault plane from the auxiliary plane requires extra information such as distribution of aftershocks, waveforms, or surface breaks.

- Long-term geologic slip rates across faults are estimated by measuring displaced geologic features of known age such as stream beds, rock types, or sedimentary layers.
- Measurements of present-day plate motions show that plates move on the order of a few centimeters each year. The earthquake cycle, a consequence of plate motion, consists of an increase in strain across a locked fault, strain release during an earthquake, and healing of the fault.
- The upper layer of the lithosphere flexes elastically when subjected to a load such as a seamount. Over long stretches of time, the lower layer flows to accommodate the load.
- The volcano deformation cycle starts with inflation of a magma chamber by a supply of new magma. It is followed by deflation. The magma leaves the chamber either to feed into dikes or sills or to erupt at the surface. Movement of magma underground generates small earthquakes.

KEY TERMS

elastic limit, 165
ductile, 165
brittle deformation, 165
plastic deformation, 165
faults, 165
folds, 165
décollement, 165
brittle-ductile transition, 166
joints, 166
sheeting, 166
exfoliation, 166
fault plane, 167
hanging wall, 167
footwall, 167
throw, 167
strike, 167
dip, 167
normal fault, 167
graben, 167
reverse fault, 167
thrust fault, 168
ramp, 168
strike-slip fault, 168
right-lateral fault, 169
left-lateral fault, 169
oblique slip, 169
fault scarp, 169
listric fault, 169
anticline, 170
syncline, 170
hinge, 170
axis, 170
limb, 170
focus, 172
hypocenter, 172
epicenter, 172
elastic rebound, 174
magnitude, 175
Richter magnitude scale, 175
Modified Mercalli Intensity Scale, 177
moment, 178
fault creep, 178
fault-plane solutions, 182
earthquake cycle, 188
volcano deformation cycle, 192

QUESTIONS FOR REVIEW AND FURTHER THOUGHT

1. What geologic features would you associate with brittle failure of the crust and what features with ductile failure?
2. What change in mechanical properties determines the brittle-ductile transition in the crust?
3. Suppose you were to drill vertically downward into Earth and you intersected an inclined fault plane. In what order would you intersect the hanging wall and footwall?
4. Describe strike-slip, thrust, and normal faults and give examples of tectonic situations in which each is expected to occur.
5. A plunging syncline is eroded. Sketch the exposed rock strata and mark the strata expected to be oldest and youngest.
6. Why does the damage for earthquakes extend hundreds of kilometers away when the major permanent displacement may be confined to only 20 km on either side of a fault zone?
7. How much larger are the amplitudes of waves from a magnitude 8 earthquake than those from a magnitude 6 earthquake?
8. Use the Gutenberg-Richter relation (Fig. 7.15) to determine the ratio of the number of magnitude 8 earthquakes worldwide to the number of magnitude 6.5 earthquakes.
9. A magnitude 8 earthquake has a slip of 10 m on a fault 300 km long and 20 km deep. Assume the elastic modulus is 1 and calculate the moment.
10. How do scientists discriminate between the fault and auxiliary planes?
11. (a) Suppose that the typical displacement of great earthquakes on a plate boundary is 10 m. If the relative plate velocity is 0.1 m/yr, how often are such earthquakes expected if great earthquakes account for all the relative motion? (b) If great earthquakes account for only 60 percent of the motion, what processes might account for the remainder of the motion?
12. Compare the earthquake cycle and the volcano deformation cycle.

8 Divergent Plate Margins and the Ocean Floor

The mid-ocean ridge system runs through the world's ocean floors like the seam on a giant baseball, as can readily be appreciated from this relief map.

Overview

- A divergent plate margin forms where a plate splits and the two pieces separate. On continents, the valley floor that forms as the two pieces separate is called a continental rift; on the ocean floor, it is an ocean rift.
- Recently formed major rifts generally lie on top of broad ridges supported by low-density mantle material. Most oceanic rifts probably start as continental rifts. As a continental plate separates, a new ocean basin is formed.
- New oceanic plates are formed at oceanic rifts; they then cool, thicken, and subside as they move away, until they are eventually consumed at subduction zones.
- The topography on either side of continental rifts decreases in elevation. The uplift is probably supported by upwelling asthenosphere, which is hottest beneath the rift, and becomes cooler with increasing distance from the rift.
- Convection causes the continental surface to bulge. When the surface fractures and these fractures link up, continental rifts form.

Introduction

Rifts: Windows to the Mantle

Hot rock from the mantle rises to the surface at oceanic rifts, where an array of fascinating structures, minerals, and life-forms are found. Underwater mountain ridges run across all the ocean floors, and the longest valleys in the world are located at the crest of these ridges (Fig. 8.1). At ocean ridges, hydrothermal vents on the sea floor have recently been discovered that discharge superheated, mineral-bearing water. The discharge looks like black smoke pouring out of chimneys, and previously unknown organisms flourish in their vicinity. These vents are the sites of both ocean rifting and the creation of new plates.

On land, continents rift apart to form giant fault-bounded valleys (grabens), which become loaded with sediments and can extend to a depth of 6 km (Fig. 8.2a).

At continental and oceanic rifts, the lithosphere splits apart to form a divergent plate margin (Fig. 8.2b and c), while mantle material flows into the resulting gap. We will examine how continental and oceanic rifts are surface manifestations of Earth's internal motions.

Continental Breakup and Plate Migration

The breakup of Pangaea (∞ see Chapter 6) provides evidence that oceanic rifts between the separating continents started as continental rifts. This transformation is readily demonstrated in the Atlantic Ocean because the oceanic lithosphere is attached to continents that were once joined together (∞ see Chapter 6; Fig. 6.14). The origin of the Pacific Ocean, however, has been masked by subsequent plate motions. The Pacific Plate is almost completely surrounded by subduction zones, and it is not connected to continents as one would expect if it had begun as a continental rift. Perhaps the oceanic part of the plate was detached from the surrounding continents and began subducting, but more research is required to unravel this history.

Other than at the mid-ocean ridges themselves, oceanic lithosphere appears to be very strong and able to resist being rifted apart; continental lithosphere does not seem to be as strong. The continents are made up of accreted blocks of andesitic composition crust of different ages, and the resulting continental crust is weaker than the basaltic crust of the ocean floor. This mechanism supposes that the crust is important in determining the strength of the lithosphere; however, the mantle portion of the lithosphere, which makes up 50 to 75 percent of the lithosphere, must also be very important.

An alternative view holds that continental and oceanic lithospheres both have similar strengths and that their ability to move as a whole is what determines whether or not they break up. Plates that contain a continent move more slowly than do purely oceanic plates. Oceanic plates lie above weaker asthenosphere, which allows them to slip readily without breaking up. Plates containing continents

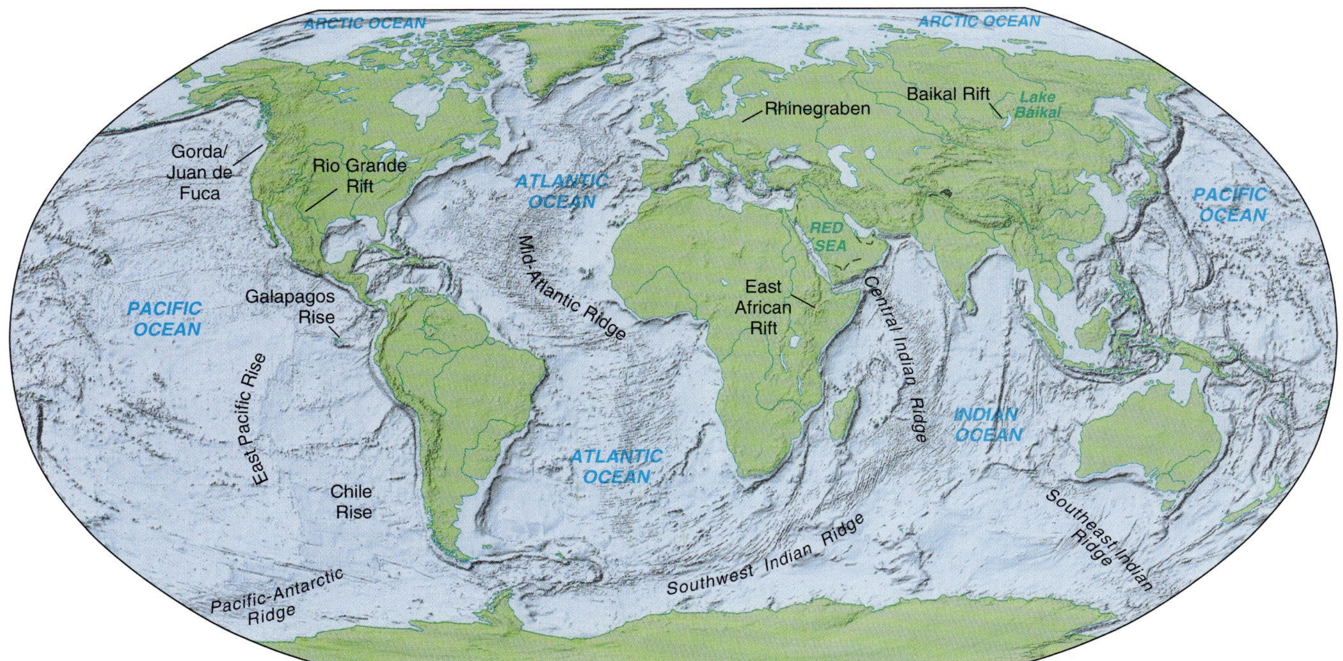

FIGURE 8.1 Topography of Earth's solid surface marking the locations of major oceanic rifts as well as four major Cenozoic continental rifts: Rio Grande, Lake Baikal, East African, and Rhinegraben.

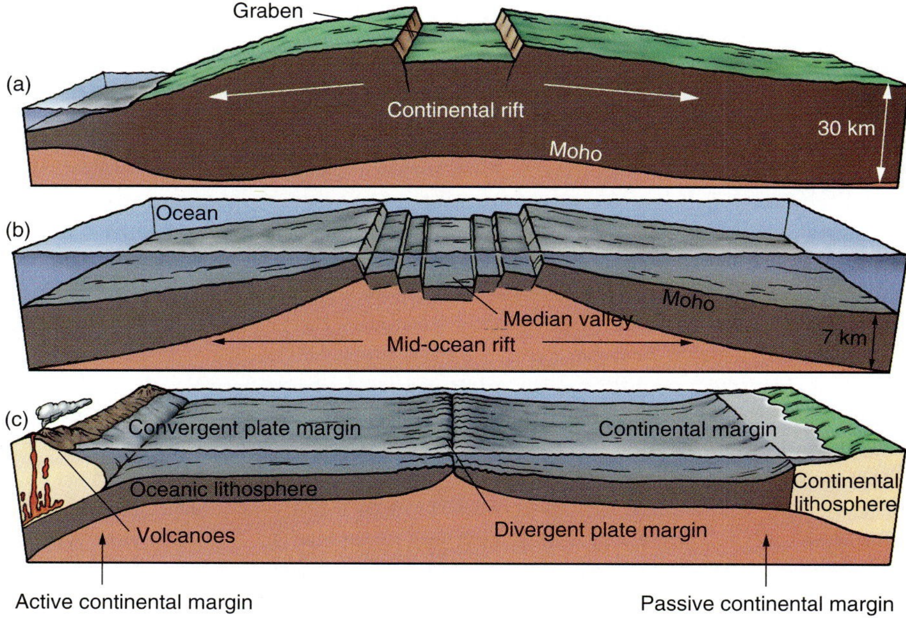

FIGURE 8.2 Rifts in their plate-tectonic settings: (a) A continental rift; (b) a mid-ocean rift; (c) margins associated with rifting.

may be more firmly attached to the underlying mantle by a stiffer asthenosphere, especially beneath cratons. When subjected to a stress, the younger, more mobile regions of the continents overlying weak asthenosphere will tend to rift apart. The full explanation may be a combination of both processes, and is a subject of much study.

Once formed, oceanic lithosphere tends not to rift, whereas continental lithosphere does. Continental rifts become oceanic rifts when the rifting progresses to the extent that separating continental lithosphere is replaced by oceanic lithosphere as the mantle flows in from below.

Oceans Subduct and Continents Rift

Why Pangaea formed and subsequently dispersed is not known with certainty. It has been suggested that these events were due to changes in the circulation pattern within Earth's mantle. At the time of Pangaea, 300 million years ago, three-quarters of the globe was covered by a vast ocean. If massive subduction of oceanic plates brought Pangaea together—as India has recently been brought together with Asia—then mid-ocean ridges must have been a dominant feature on the floor of Pangaea's ocean. As the continents locked into place, buoyancy of the continental lithospheres resisted subduction, which then ceased. New subduction zones formed in the oldest oceanic lithosphere; subsequently, subduction in the ocean became an important plate force, which, along with plumes, began the job of tearing the supercontinent apart. When a plate is pulled by oceanic lithosphere sinking into a subduction zone, it will be stressed and will break, or rift, along the weakest part of the plate. If continental lithosphere forms part of the plate, the rift will form there.

From the reconstruction of the plates of Pangaea, scientists have concluded that most oceanic rifts actually started as continental rifts. We can think of this behavior as "oceans subduct, continents rift." Oceanic rifts provide a line of weakness about which new ocean floor forms, but their origins ultimately trace back to the splitting of a continent. Once formed, the new oceanic plate tends not to rift, but sinks back into Earth at subduction zones. Oceanic rifts do propagate into new areas of oceanic lithosphere, so the ridge system is an ever-changing pattern, but older ocean floor appears to resist rifting. Continental evidence older than 200 million years indicates that ancient rifting and continental collisions have taken place, but the record requires more work to reconstruct the movements of the continents through time.

Oceans subduct, but continents are too buoyant to subduct; instead, they rift apart.

If we assume that both subduction in the Pangaean ocean (named Panthalassa) and plumes beneath the continent caused continental breakup—a process repeated many times in the past—what might happen were the same mechanisms operative today? Can these processes explain modern continental rifts—in particular, their similarity to mid-ocean ridges?

Rifting at mid-ocean ridges is one of the better-understood processes of plate tectonics. We will examine this process before trying to explain rifting on the continents. We have found that if a continent cracks apart, an oceanic rift may eventually form. The two fragments of the continent migrate across the globe, trailing newly formed ocean floor in their wake and leaving a ridge behind—a memento of their former union (Fig. 8.3).

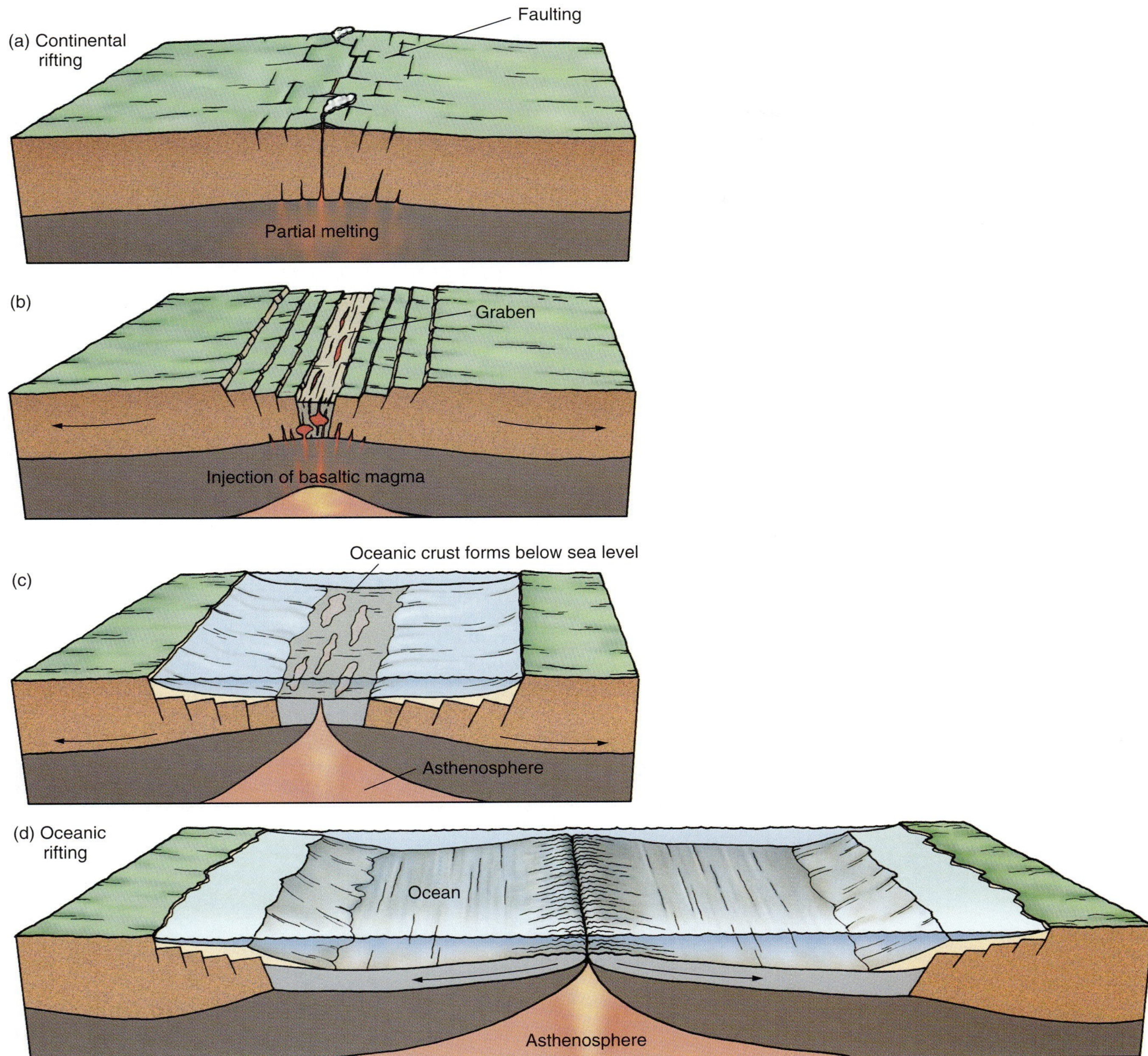

FIGURE 8.3 The transition from continental rifting to oceanic rifting. Continental crust above sea level breaks by normal faulting to form a graben. As rifting progresses, oceanic crust forms below sea level.

At mid-ocean rifts, the plates separate at a rate of a few centimeters per year, whereas the separation at continental rifts is about 10 times slower. Seen in profile, however, the topography across each rift type is similar. In Figure 8.4a and b, we compare the topography of continental rifts and slow-spreading oceanic rifts. As well as having a central graben, or valley, both types of rifts lie in linear zones of regional topographic uplift extending many hundreds of kilometers to either side of a central rift zone. The topographical similarities in these two rift types, however, arise from entirely different processes. The continental process may eventually transform into the oceanic one as the continental rifting becomes so extreme that a new ocean forms.

Topographic profiles across slow-spreading continental rifts resemble profiles across oceanic rifts even though continental-plate separations occur at a rate that is 10 times slower than that of oceanic rifting.

Oceanic Rifting

Mid-Ocean Ridges

If the water were drained from the oceans and all sediments were removed from the continents, the mid-ocean rift system and the continental rifts would form some of the most spectacular topography on the face of the globe (see Fig. 8.1). Most rifts lie along the crests of broad ridges. In the oceans, such ridges are referred to as mid-ocean ridges because, ideally, they run along the middle of an ocean basin. Such positioning is indeed true for the Atlantic Ocean ridge; in the Pacific Ocean, the ridge, known as the East Pacific Rise, is displaced to the east.

The main topographic feature of a mid-ocean ridge is the regional uplift. Depending on the spreading rate, a central rift graben may be superimposed on this broad topography. The regional uplift rises over the ocean floor to reach a maximum at the edges of the median valley. Oceanic ridges spread at different rates: Fast-spreading oceanic ridges, such as the East Pacific Rise, open at rates of up to 20 cm/yr; slow-spreading ridges open about 10 times slower, or 1 to 2 cm/yr. At fast-spreading ridges, the central graben is virtually absent, whereas it is about 10 km wide at slow-spreading ridges, such as the Mid-Atlantic Ridge (Fig. 8.4b and c). In contrast, we will discover in the next section that the central grabens at continental rifts are many tens of kilometers wide.

Oceanic Crust, Lithosphere, and Asthenosphere

Seismic refraction surveys (∞ see Chapter 5) of the ocean floor reveal that, on average, the ocean crust is 7 km thick. The ocean crust is thought to be made mainly of basalt and gabbro. Deep-sea drilling of the crust has shown that it can generally be divided into four distinct layers (Fig. 8.5), although the layered structure is not so well-developed at very slow spreading ridges. The first layer corresponds to up to several hundred meters of sediments; the second layer is formed from about a kilometer of pillow basalts; the third layer is a sequence of basalt dikes; and the fourth layer, rarely penetrated by drilling, is gabbro.

In various parts of the world, ophiolites (∞ see Chapter 3) have been thrust up (obducted) onto the continents. These slices of ancient sea floor provide geologists with rare opportunities to view a cross section of oceanic lithosphere to much greater depths than have been obtained by drilling. Ophiolites confirm that gabbro is found beneath the dikes and it, in turn, overlies peridotite. Gabbro is a rock of similar composition to basalt, but is coarser-grained due to its slower cooling (∞ see Chapter 4). At mid-ocean ridges, gabbro is the rock formed by the settling of crystals (mainly plagioclase, pyroxene, and olivine) at the base and sides of a basalt-filled magma chamber (see Fig. 4.5). The transition to peridotite below

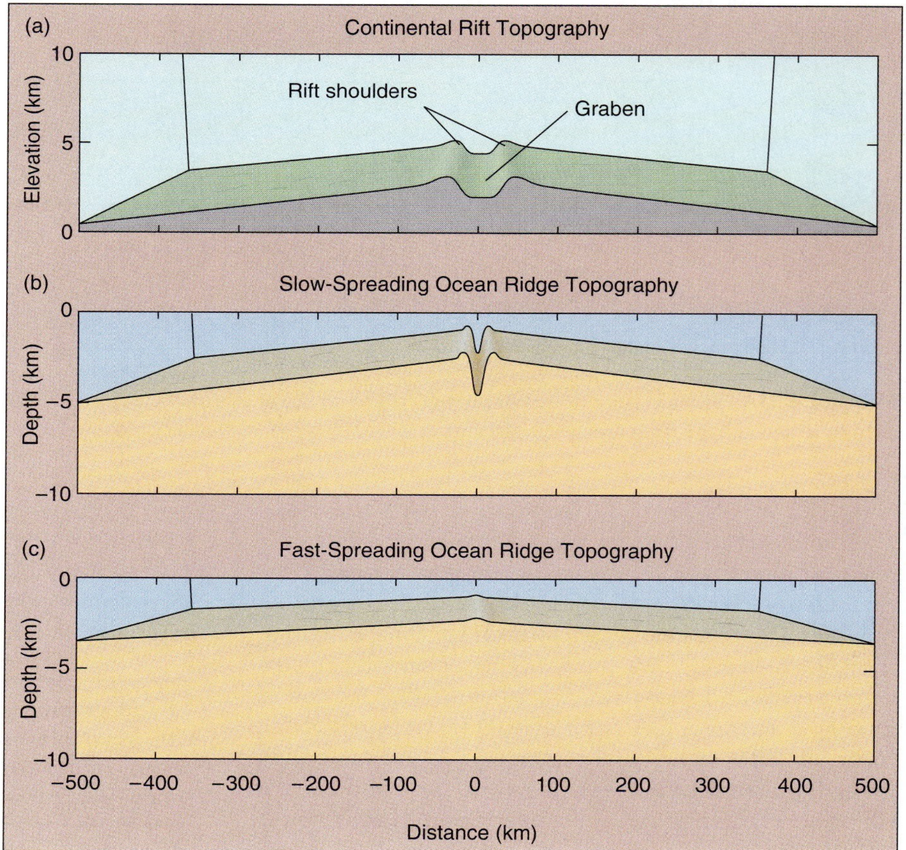

FIGURE 8.4 Comparison of the topographic profiles across continental rifts as well as fast- (or slow-) spreading oceanic rifts. The curves are based on averaged profiles from the East African Rift, the Mid-Atlantic, and the Pacific ridges, respectively. Similarities between the structure of continental rift topography and slow-spreading ocean ridge topography are explained by mantle and plate motions.

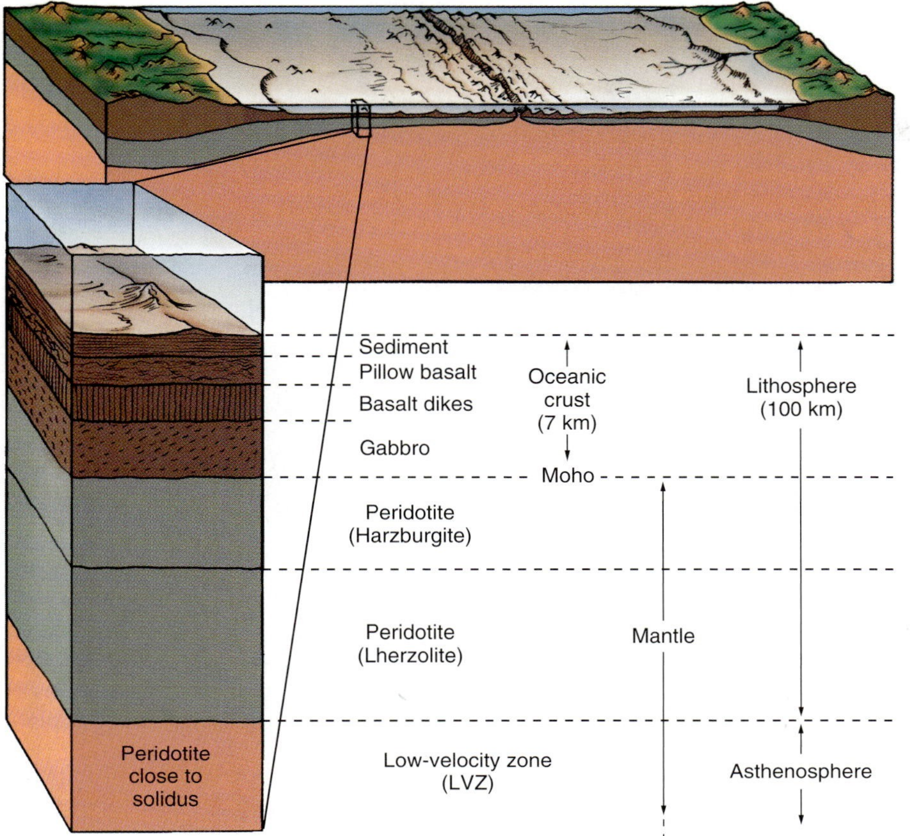

FIGURE 8.5 The composition of oceanic lithosphere that was formed from the mid-ocean ridge is shown in the upper diagram. The lower diagram (not to scale) illustrates rocks found in ophiolite complexes. Resemblance to the results from deep-sea drilling strengthens the belief that ophiolites are oceanic lithosphere that has been obducted onto continental margins at the time of subduction.

gabbro is thought to represent the Moho: the transition from crust to mantle (Figs. 8.5 and 8.6).

The mantle immediately beneath the oceanic crust is still part of the oceanic plate, but how can we tell how deep this lithosphere reaches?

Recall that the lithosphere is the strong outer layer of Earth lying above the weaker asthenosphere. Strong materials generally have higher seismic-wave velocities than do weak ones. This is thought to be true for the rocks making up the lithosphere compared with those of the asthenosphere. Thus, the velocity of seismic waves can be used to find the transition from high velocities in the lithosphere to low velocities in the asthenosphere. On this basis, it has been determined that the average thickness of the oceanic lithosphere is 100 km, but it thins toward the ridges.

The most important constraint on the depth to the base of the lithosphere has come from studying the surface waves of earthquakes. Seismic surface waves consist of components having a range of frequencies and wavelengths; the longer the wavelength, the deeper the wave action penetrates into Earth. High-frequency, short-wavelength waves sample the upper regions of the planet—the crust, for example. Longer wavelengths penetrate the upper mantle beneath the crust, which is found to transmit waves at a higher velocity than does the crust. For even longer wavelengths, however, a change in this trend occurs. Instead of traveling faster, the waves move slower, indicating that the velocity decreases at greater depth. For still longer wavelengths, the speed picks up again and continues to increase thereafter. This region immediately below the lithosphere where velocity decreases with depth is the low-velocity zone or LVZ, introduced earlier (∞ see Chapter 5).

When seismic waves of different frequencies travel at different speeds, the result is *dispersion*. As a result, the seismogram from an earthquake changes shape with time traveled. Seismologists have used measurements of such dispersion to define the lithosphere as the crust and that part of the upper mantle lying above the low-velocity zone. Thus, it is thought that the LVZ is the region of decoupling between the moving plate and the underlying mantle.

Oceanic lithosphere is approximately 100 km thick and exhibits high seismic-wave velocity. A downward section through the lithosphere reveals 7 km of oceanic crust, comprising sediments, pillow lavas, basalt dikes, and gabbro, overlying peridotite mantle.

Anatomy of a Mid-Ocean Ridge

At divergent margins, the upward movement of mantle causes decompression, which has the effect of lowering the melting point. The geotherm here lies above the mantle solidus, causing melting to produce liquid basalt

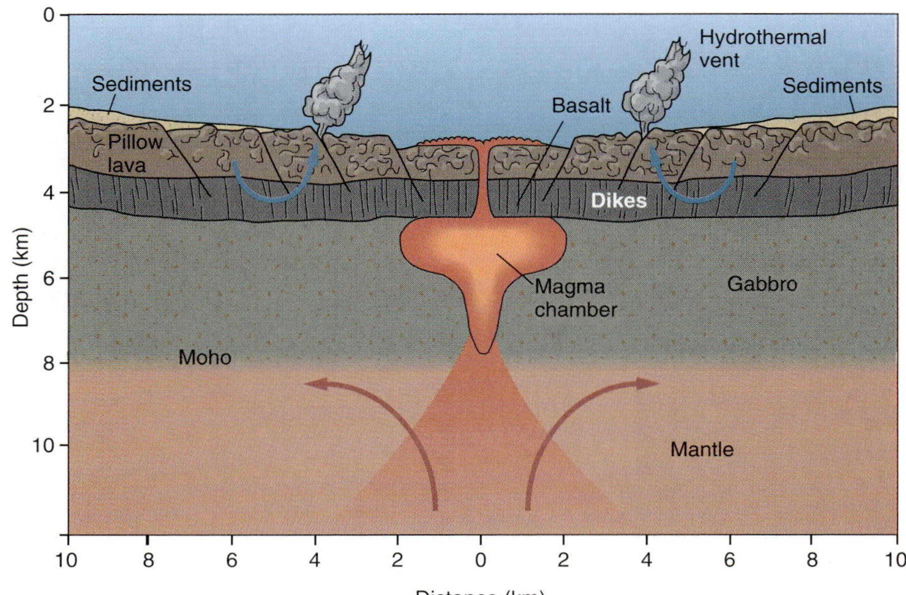

FIGURE 8.6 Cross section of the axial region of a mid-ocean ridge. Magma collects in an axial magma chamber. Gabbro crystallizes on the walls. Basaltic dikes inject into the crust to form a sheeted dike complex. Magma erupts at the surface to form pillow lavas. Stress from magma intrusion generates normal faults and the median valley. Sediments slowly accumulate on older ocean floor. Water heated by circulating into the hot crust generates hydrothermal vents.

(∞ see Chapter 4). The volume of basalts added along ridges to form new oceanic crust has been estimated to be about half a cubic kilometer for every kilometer of ridge length per thousand years.

The ultramafic peridotite of the mantle partially melts (about 5 to 20 percent) and forms molten basalt, leaving behind solid *harzburgite*, a type of peridotite (Fig. 8.5). The molten basalt forces its way up and out through channels in the solid peridotite and then moves rapidly into the gap created by the parting plate. Basalt accumulates in a magma chamber at the base of the newly forming crust. As the plates separate, the chamber widens. At the same time, magma freezes to the walls of the chamber and crystals settle to its floor, keeping the chamber size, on average, constant over time. Magma is also injected into cracks above the chamber where it cools and solidifies to form dike structures between the chamber and the surface (layer 3 of Fig. 8.5). These dikes record the direction of Earth's magnetic field and give the ocean floor its characteristic magnetic stripes, each corresponding to the polarity of the geomagnetic field at the time of their formation. New dikes intrude older dikes or intrude between older dikes until the upper few kilometers of the ocean floor become a **sheeted dike complex** (Fig. 8.7).

When magmas from the dikes erupt onto the surface and are quenched by seawater, submarine pillow basalts form (layer 2 of Fig. 8.5). Rapid cooling in seawater gives them their pillowlike shape. The thickness of sediments that form layer 1 increases with distance from the ridge because the ocean crust gets systematically older, allowing more time for sediments to accumulate.

Pillow lavas, dikes, and chamber walls cool and solidify and move away from the ridge axis, with the remainder of the lithosphere. Cooling in the upper dike and pillow basalt regions is rapid (hours to years), aided by circulation of oceanic water (Fig. 8.8). At depth, the hot, newly formed lithosphere cools more slowly (millions of years) as heat is conducted to the surface and the plate moves away from the ridge.

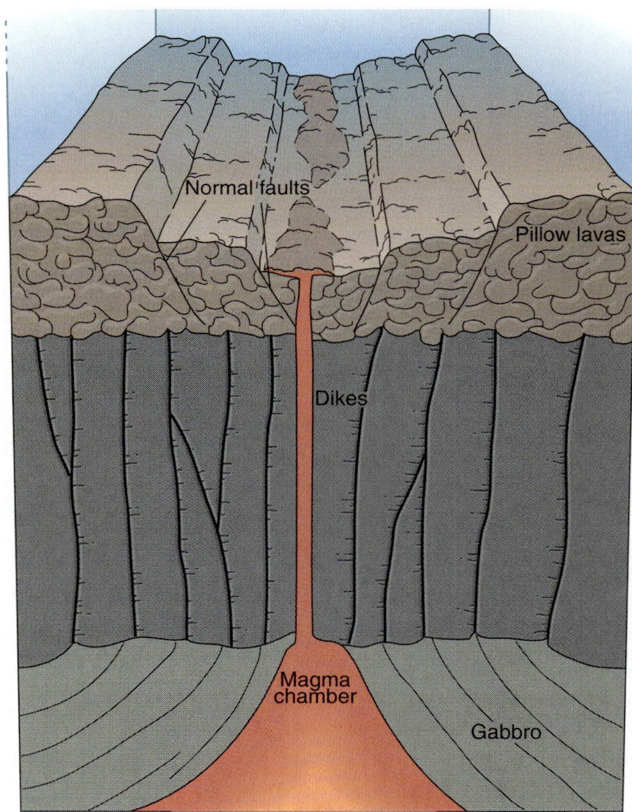

FIGURE 8.7 Formation of a sheeted dike sequence. Each new dike wedges in between the older dikes, pushing them aside as the sea floor spreads. Stresses from the intruding dike generate normal faulting in pillow lavas.

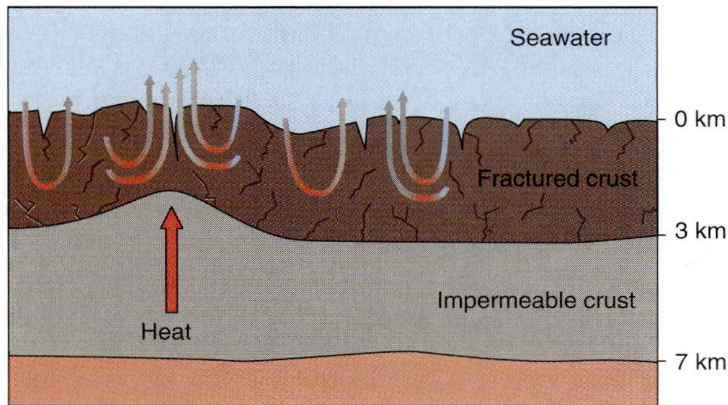

FIGURE 8.8 Percolation of water into the fractured basaltic oceanic crust cools the crust and reacts with the basalt to form minerals.

Interaction of Mid-Ocean Ridge Basalts and Seawater

Several consequences result from the contact between hot basalt and seawater at mid-ocean ridges. Not only are reactions between the basalt and the seawater fundamental to changing the composition of both the basalt and the seawater itself, but the heating of seawater is the major mechanism for dissipation of Earth's internal heat. Sediments and fractured rocks near the ridge crest are permeable; water flows through them via cracks and pores. This permeability allows seawater sufficient penetration into the crust (Fig. 8.8) that it approaches the magma chamber, and is heated.

Geologists have drilled several boreholes near the crests of active rises. In most cases, the crust was found to be quite cold to depths of 1 to 2 km. Such observations led these researchers to conclude that large hydrothermal circulation cells penetrate to these depths, and that hydrothermal circulation represents a very important planetary cooling mechanism. Recently discovered deep-ocean hot springs, called **black smokers**, are spectacular surface manifestations of this circulation (Fig. 8.9).

Dives to the Galapagos Ridge on the East Pacific Rise by the submersible *Alvin* in 1977 revealed this remarkable hydrothermal activity. Scientists observed towering mineral deposits on the ocean floor, discharging sulfides in water that had been heated and mineralized underground by magmatic rocks. This discharge of mineral-laden water looks like black smoke pouring out of chimneys sitting upright on the sea floor; hence, they were named black smokers. Temperatures as high as 450°C have been measured, although they are more commonly about 350°C.

One fascinating outcome of these investigations was the discovery of previously unknown marine organisms flourishing around the hot springs (Fig. 8.10). Near the outer edge, organisms nicknamed "dandelions," galatheid crabs, and "spaghetti worms" were found, while tube worms, crabs, and mussels were located close to the vents. Remarkably, these new species live 2.4 km below the surface of the ocean—completely beyond the penetration of sunlight; thus, these organisms cannot depend on the process of photosynthesis for survival. They constitute the first known communities of higher animals not directly dependent on sunlight. Instead, they rely on bacteria, which, in turn, derive their energy from the geothermal energy of the hot oceanic rocks by a process called *chemosynthesis*. In this process, as in photosynthesis, inorganic carbon dioxide is converted to organic matter. The energy source is chemical, however, not solar.

FIGURE 8.9 Black smokers are hydrothermal vents that liberate mineral-laden water from mid-ocean ridges.

This discovery has added a whole new perspective to theories on the existence and sustenance of life in extreme environments, and it raises many interesting questions. If life can exist in an environment so different from Earth's surface, what unusual forms might it take on, or within, other planets of our Solar System?

Hydrothermal vents on the sea floor, called *black smokers*, are important heat exchangers between the interior and the surface of Earth. New life-forms that utilize chemosynthesis have been discovered at such vents.

Interactions between seawater and hot rock at mid-ocean ridges trigger extensive chemical changes. Downwelling water in the ocean crust reacts with the basalt of the ocean floor. Some chemical components are leached out of the basalts and replaced by components from seawater, with the most commonly leached elements being silica, iron, sulfur, manganese, copper, calcium, and zinc. The basalt incorporates magnesium and sodium, which is dissolved in seawater as a result of weathering and erosion. Chlorine combines with sodium remaining in the seawater to form dissolved salt. Thus, both the basalt and seawater are profoundly changed by this interaction.

(a)

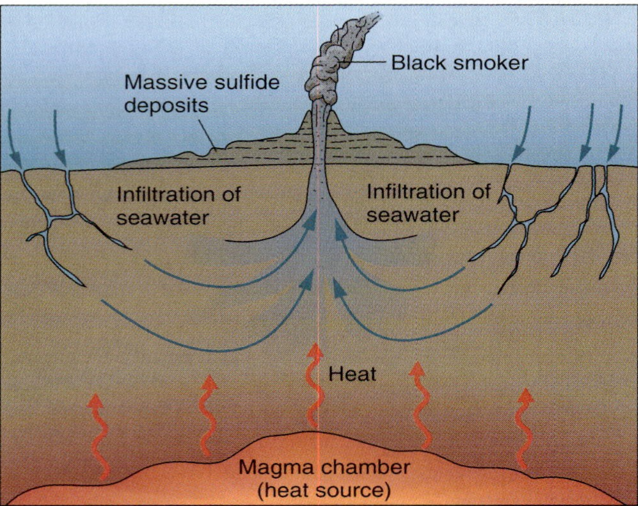

(b)

(c)

As the water rises through smokers, iron, copper, zinc, and other metals quickly precipitate as sulfides on the sea-floor surface and along channels or pipes that develop in the rock, which carry the hot, mineral-laden water upward through the basalt. Rich mineral deposits of iron sulfide (pyrite), calcium sulfate (anhydrite), or copper sulfides (chalcopyrite, bornite) are found adjacent to and within the smokers (∞ see Chapter 16). Given the quantity of water cycled through these hydrothermal systems, it is possible that the basalt–seawater interaction at mid-ocean ridges represents a primary mechanism contributing dissolved salts to the oceans.

From such interactions, the primary mineral components within the basalts are altered to new mineral assemblages (∞ see Chapter 4). Plagioclase feldspars are metamorphosed from a high-calcium content to a high-sodium one, which results from the loss of calcium and the addition of sodium during the water–rock hydrothermal reaction. The rock formed by this process is called *spilite*. Formation of spilites had been a puzzle to geologists for years, and would have remained so without the observations of sea-floor interactions between basalt and seawater.

Reactions between seawater and basalt rising at the spreading ridges change the compositions of both basalt and seawater. Basalt contributes dissolved salts to the seawater and metals to the sea floor; seawater alters the basalt by addition of sodium and magnesium.

Ocean-Ridge Uplift

Figure 8.4 shows that at fast-spreading ridges, the regional uplift decreases slowly from its maximum elevation at the ridge over a distance of a thousand or more kilometers from the ridge axis. At slow-spreading ridges, the ocean floor drops down more rapidly within a few hundred kilometers of the ridge. We can explain these features by modeling the cooling of a newly formed plate as it moves away from the ridge axis.

Below a depth of about 5 km from the surface, pressures within Earth are so great that pore spaces in the

FIGURE 8.10 (a) The deep submersible—named *Alvin*—operated by the Woods Hole Oceanographic Institute to probe the sea floor. (b) Schematic of a black smoker. Arrows indicate the flow of seawater, which is heated by hot rock and rises, to be replaced by cool water that flows downward. The upwelling hot water dissolves minerals that precipitate out when the water is discharged from the chimneys. These minerals contribute to the chimney structure and also give the appearance of smoke issuing from it. (c) Tube worms and organisms unlike any others are found in the extreme environment of mid-ocean ridges, where sunlight is nonexistent. These deep-sea organisms rely on nourishment and energy from the thermal processes in the hydrothermal zone.

rock (holes and channels between rock grains) are closed off. This means that oceanic waters are unable to circulate or convect below this depth. Deeper than 5 km in the oceanic crust, then, the plate cools by thermal conduction. Conduction (∞ see Chapter 2) is a mode of heat transport in which strongly vibrating molecules in a hot region act on their cooler neighbors to increase their vibration, and the disturbance progressively diffuses outward into cooler regions. Conduction involves no actual transport of material. Compared with convection, it is a very slow process.

Near the ridges, where new plate has been formed, temperatures are close to the melting point of rock. As the plate moves, this heat starts working its way slowly to the surface, where it is eventually removed by hydrothermal circulation. The farther the plate moves from the ridge axis, the longer the time available for it to cool. This cooling relationship explains the different uplift profiles for fast-spreading and slow-spreading ridges. At the ridge crest, the hot material is uplifted because heat causes it to expand, making it lighter, and thereby causing it to float higher than the cooler surrounding mantle as required by the principle of isostasy (∞ see Chapter 5). As the plate moves away, its heat is lost to the ocean. Using a mathematical model of heat conduction, it has been calculated that the amount of heat lost is proportional to the square root of time. Thus, slow-spreading oceans have narrower ridges than do fast-spreading oceans because, after a given cooling time, the crust at slow-spreading ridges has not traveled as far from the ridge.

Because plate cooling is proportional to the square root of time, so too is the subsidence of the cooling sea floor. This gives rise to the remarkable result that all ocean-ridge profiles look the same when compared on one graph *if time is used as the x-axis rather than distance* (Fig. 8.11). It illustrates an incredible uniformity in the mid-ocean ridge process. It does not depend on the rate at which spreading occurs; the Pacific Ocean spreads at a rate of 20 cm/yr, and the Mid-Atlantic Ridge at a rate of 2 cm/yr, yet the topographies of their ridge systems are identical when plotting depth against age (see Focus 8.1).

The conduction model does not work for ages greater than about 100 million years. At these ages, the conductive cooling has reached the base of the plate where convection (∞ see Chapter 5) in the asthenosphere

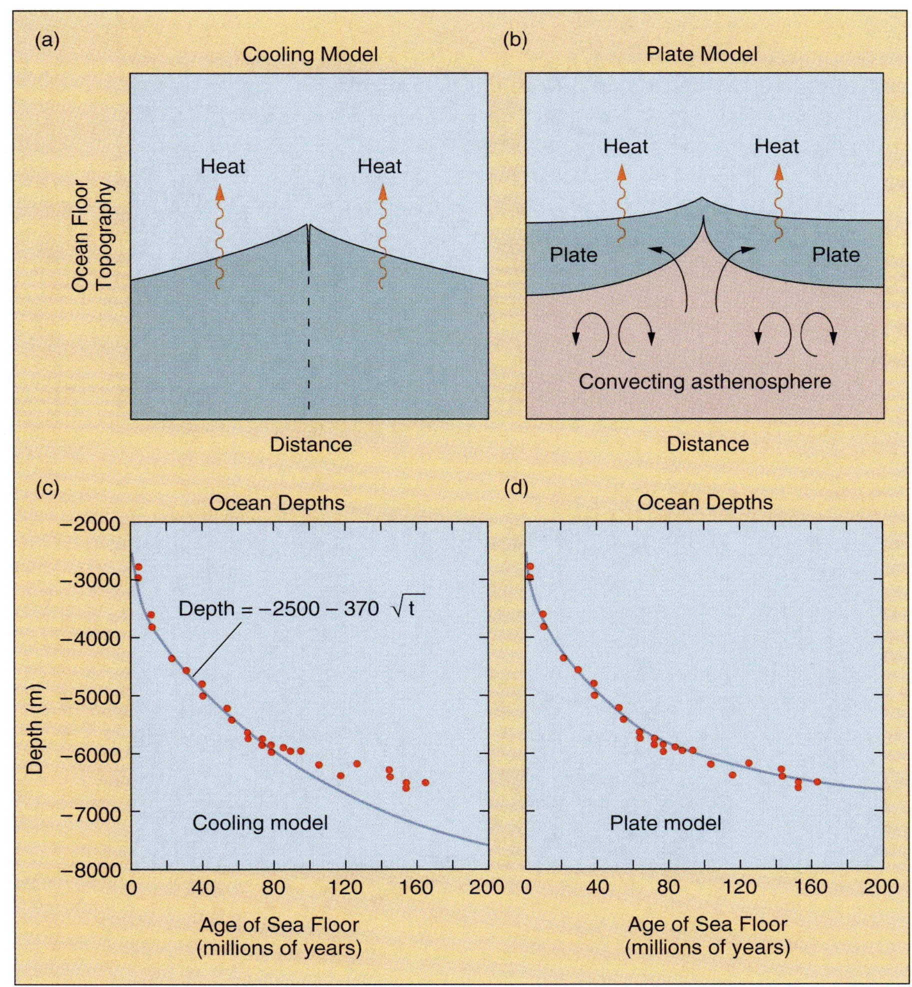

FIGURE 8.11 The topographic profile across a mid-ocean ridge for about 100 million years varies as a function of the square root of age (a,c) as predicted by the thermal conduction model. The fact that profiles from all the ocean ridges lie on the same curve even though spreading rates differ by a factor of more than 10 is evidence that a common process of plate formation and conductive cooling with spreading operates in each case. The depth of the ocean floor is given by: Depth = $-2500 - 370\sqrt{t}$, where depth is in meters and t is time (in millions of years). For ages greater than 100 million years, convection at the base of the plate counteracts cooling by conduction (b) and a plate model is used to describe the topography (d).

Focus On 8.1 THE EAST PACIFIC RISE AND MID-ATLANTIC RIDGE

The East Pacific Rise and the Mid-Atlantic Ridge are two oceanic rifts with spreading rates that differ by a factor of 10. Yet the topography, gravity, and heat flow that are measured across each can all be understood using the theory of plate tectonics.

The East Pacific Rise runs from the Gulf of California southward, where it joins the Pacific Antarctic Ridge. It then loops south of New Zealand and Australia to form the Indian Ocean Ridge (see Fig. 8.1). The East Pacific Rise opens at variable rates, of up to 20 cm/yr, which makes it the fastest-known spreading ocean ridge. It generates plate material that is subducted at the northern and western edges of the Pacific, and to the east beneath Central and South America. Figure 1a and b shows the topography and the Bouguer and free-air gravity anomalies for the East Pacific Rise and those for the Mid-Atlantic Ridge. From this figure, it is clear that the East Pacific Rise is much broader than the Mid-Atlantic Ridge. In addition, the Mid-Atlantic Ridge has a pronounced central graben that is absent from the East Pacific Rise. Both of these features are a result of the difference in spreading rate. Because it spreads so slowly, the plate close to the Mid-Atlantic Ridge has had more time to cool and has settled lower than a plate at a similar distance from the East Pacific Rise. Further, the lower rate of opening at the ridge results in less heat rising from the mantle. It is cooler and the mantle flows less readily, which results in a delay of mantle inflow into the gap generated by the parting plates. The net result is the central graben.

Even though the topography of the ocean floor has a relief of some 3 km, free-air gravity (∞ see Chapter 5) profiles across it remain almost flat. This implies that the mass in vertical columns on and off the ridge is the same—that is, the condition for isostatic equilibrium. When the gravity is corrected for the additional heavy material associated with the rise in the sea floor, the result is the Bouguer anomaly. Bouguer anomalies in each case are large and negative over the ridges, indicating that low-density material lies beneath the ridges. The ridges are therefore in isostatic equilibrium. The rise in the ocean floor must be compensated by lighter material below it so that the net result is the observed zero free-air gravitational anomaly. As we have already seen, the lighter material is produced by higher temperatures under the rise, which cause the rock to expand, thereby reducing its density.

One way to test this model is to measure the amount of heat flow from the ocean floor into the oceans. It is expected that the heat flow will be highest at the rise and will drop off systematically with distance from the rise—as does the topography. Heat flow is measured by dropping probes into the sediments on the ocean floor. The heat flow and topography for the Mid-Atlantic Ridge and East Pacific Rise are plotted together in Figure 2a and b. The results indicate a progressive decrease in heat flow with age (distance) from the ridge and confirm this model of mid-ocean ridge formation.

Also plotted for comparison is the average continental heat flow (60 mW/m^2). The maximum heat flow at the mid-ocean ridges is five times this value. Instead of plotting Figure 2a and b as a function of distance, they are plotted as a function of time. The result is that the plots from the two oceans lie on top of each other even though the actual distances from the two ridges at which the measurements were made differ by a factor of 10. Again, in the topography we see the departure from the cooling model, but in both cases, this departure can be explained by the plate model.

Finally, the result is that all the plots lie on top of each other. This remarkable result also works for other oceanic ridges. It confirms the proposal that the same physical process can explain the topography at all the ridges and the occurrence of sea-floor spreading. It is a key point for the unifying theory of plate tectonics.

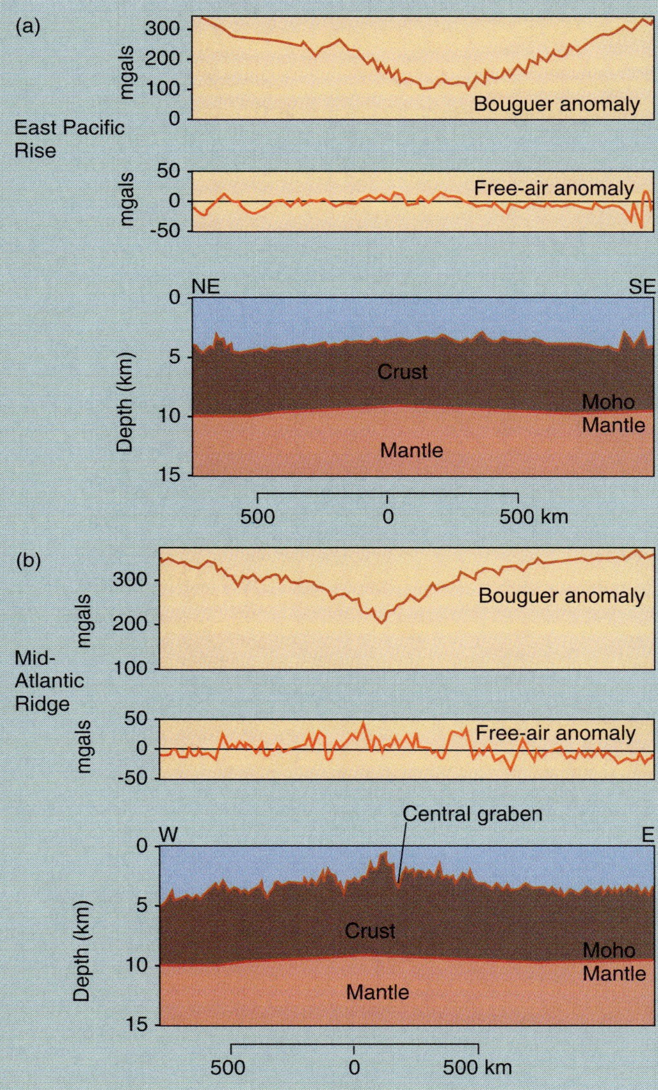

FIGURE 1 (a) Bouguer and free-air gravity anomalies across the East Pacific Rise compared with topography. Note the free-air anomaly is nearly zero, indicating that the topography is in isostatic equilibrium. (b) Bouguer and free-air anomalies across the Mid-Atlantic Ridge. Note the central graben on the Mid-Atlantic Ridge that is not present on the East Pacific Rise. The topography of the Mid-Atlantic Ridge decreases much more with distance away from the ridge, as predicted by the cooling model for a slow-spreading ridge.

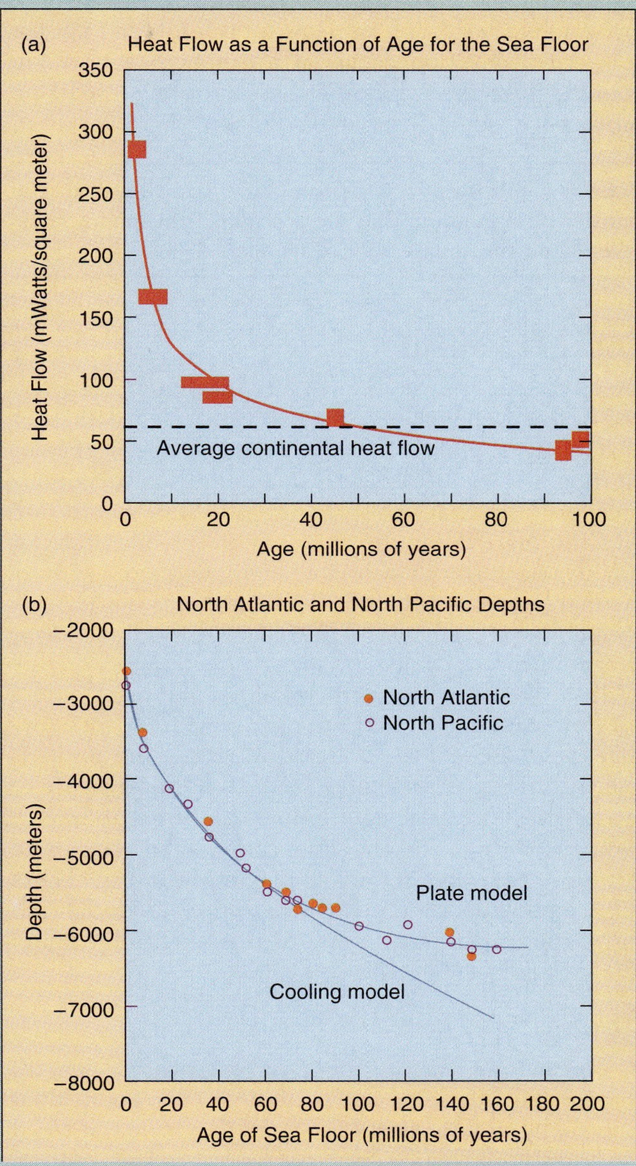

FIGURE 2 (a) A curve describing heat flow for both the Mid-Atlantic Ridge and the East Pacific Rise plotted as a function of time. Boxes show range in observed values. For a comparison, the average heat flow on the continents is plotted. (b) Comparison of topography for the Mid-Atlantic Ridge and the East Pacific Rise plotted as a function of time. The plate model curve better fits the data for ages greater than 100 million years.

transports heat. Conductive cooling is then no longer effective because the heat conducted out is rapidly replaced by heat supplied through convection. The square root of time relationship derived from the conduction model breaks down. The deepening of the ocean levels off. At these distances, a model of a plate above convecting asthenosphere is needed to explain the topography (Fig. 8.11). For slow spreading—at 1 cm per year—this transition occurs at a distance of 1000 km from the ridge axis, whereas for fast spreading—at 10 cm per year—it would occur at 10,000 km.

Cooler lithosphere is denser and floats lower in the mantle. The amount that the oceanic lithosphere cools depends on the square root of its age.

Central Grabens

We have seen that slow-spreading ridges are characterized by a central graben or median valley (Fig. 8.4b), but such a central graben is nearly absent along fast-spreading ridges (Fig. 8.4c). The explanation for this difference has been that fast-spreading ridges are hotter and thus the material at the axis is more mobile and can quickly rise to the equilibrium height associated with its thermal buoyancy, which is the height of isostatic equilibrium (∞ see Chapter 5). At slow-spreading ridges, temperatures are cooler and magma freezes more readily in dikes. Thus, as the plate separates, the mantle does not flow smoothly in to fill the gap. The forces of buoyancy drive dikes into cool, brittle crust. The result is that the lithosphere solidifies before it has time to rise to the height of isostatic equilibrium, causing it to have a permanent valley at the ridge crest. The stresses of intrusion generate normal faults in the crust. These faults mark steps along which the crust moves upward out of the graben until it attains the height of isostatic equilibrium (see Fig. 8.6).

Propagating Rifts

A propagating rift is one that extends its length. Propagating rifts have left patterns in the ocean floor topography and the magnetic record that have been used to characterize them. On fast-spreading ridges, such as the East Pacific Rise, ridge offsets by transform faults can be transitory. Cases have been observed where the rifts in each segment extend or propagate until they overlap (Fig. 8.12). Then they impose stresses upon each other that cause them to curve in on themselves until one links with the other (Fig. 8.13). The end of the linked rift is then cut off and abandoned to drift away from the spreading axis as new ocean floor forms.

The direction of sea-floor spreading undergoes small changes over time, which are related to changes in the forces acting on the plates. The rifts are then no longer oriented perpendicular to the direction of plate motion. Growing evidence indicates that new spreading centers form that are oriented perpendicular to the new direction of motion. They grow at the expense of the old ridge system, which continues until completely replaced. A new segment of spreading ocean ridge will generate a new set of magnetic stripes (regions of positive and negative polarity; Chapter 5) at an angle to the old set. If this segment also increases in length as it replaces the old direction of spreading, the new stripes will form an arrow shape (Fig. 8.14). The net result is two sets of magnetic lineations: those corresponding to the old direction of spreading and those associated with the new (Fig. 8.15). The arrow point indicates the direction of propagation of the new rift tip, which leaves a wake of ocean floor behind it—rather like the wake of a moving ship—with its magnetic lineations in the new direction.

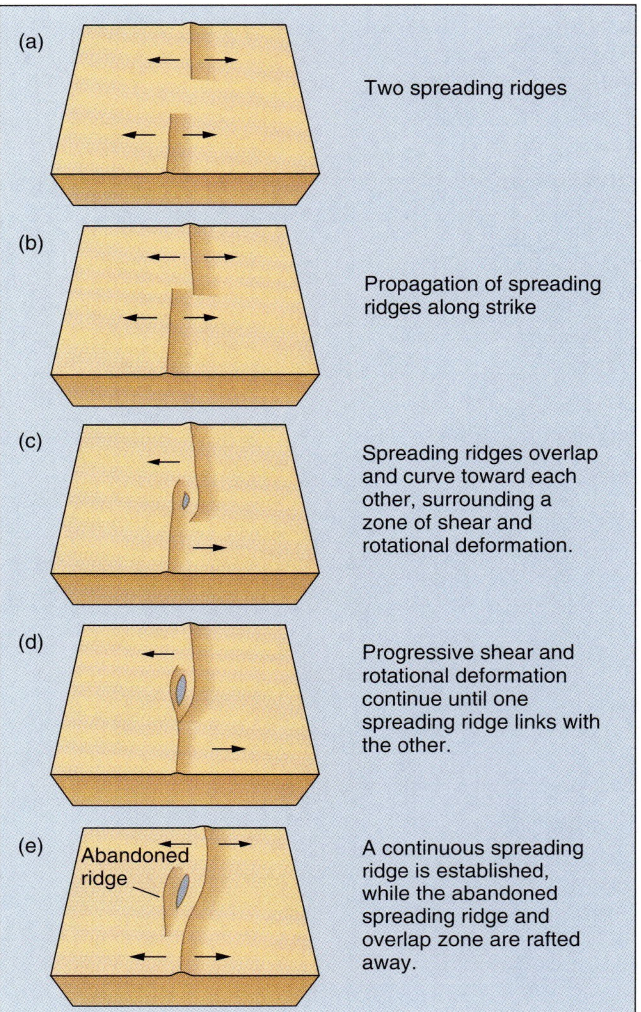

FIGURE 8.12 Rift segments on fast-spreading ridges propagate past each other. Stresses from interaction cause them to curl in on one another until one is cut off. The segment that is cut off then becomes a failed rift.

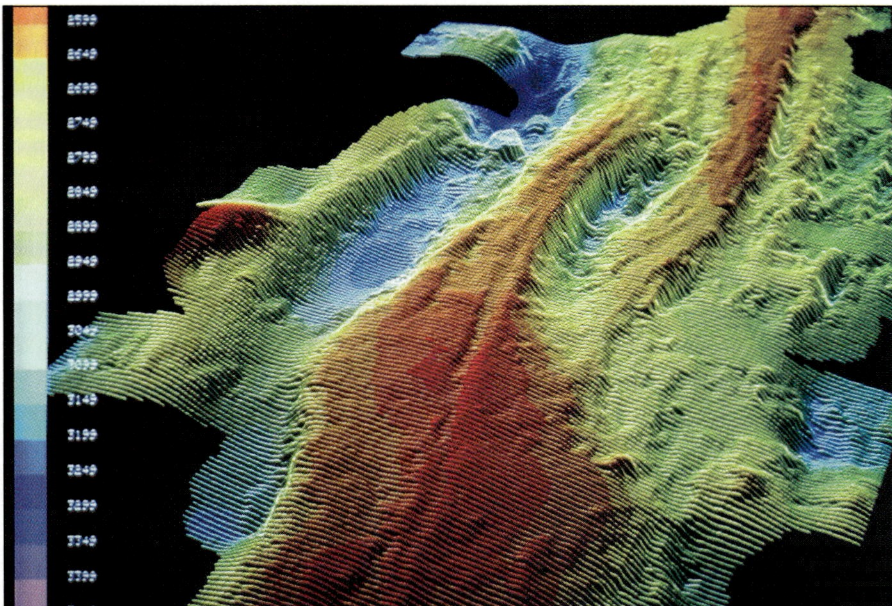

FIGURE 8.13 Sonar image of overlapping spreading centers showing the curved features sketched in Fig 8.12. Color-shaded scale is labeled in meters.

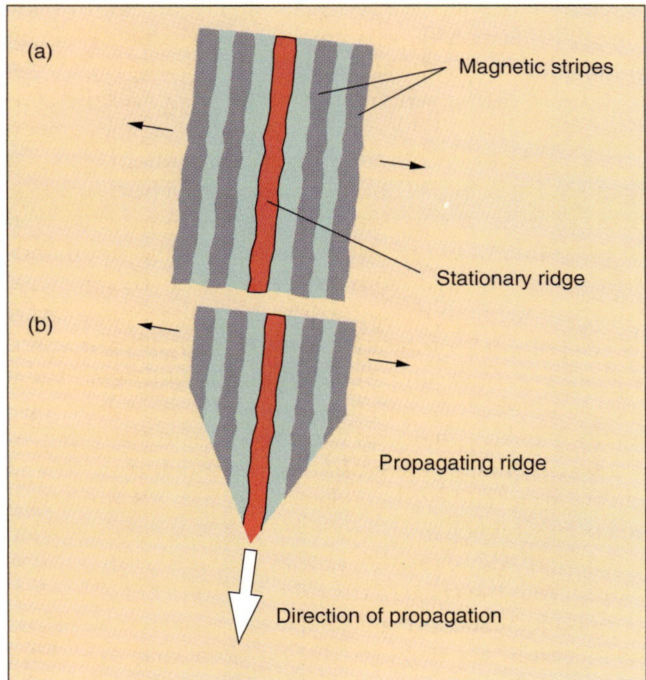

FIGURE 8.14 (a) Magnetic stripes from a spreading center. The light stripes correspond to reversed magnetic polarity; shaded stripes to normal polarity. (b) Magnetic stripes from a propagating spreading center develop a characteristic arrow shape. The tip of the arrow is the current position of the spreading center. The truncations of the shaded stripes show locations of the tip at the times of magnetic-field reversals.

Examination of the geometry of magnetic lineations in the sea floor shows that rifts propagate and alter direction in response to plate-direction changes.

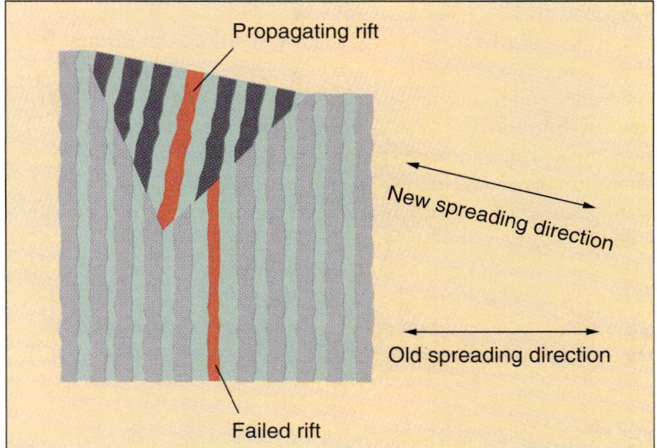

FIGURE 8.15 An old spreading center is being replaced by a newly propagating one that has a slightly different angle of spreading. The result is a pattern of magnetic stripes in the form of an arrow. The tip of the arrow shows the direction of propagation of the new rift.

The Driving Mechanism for Sea-Floor Spreading

The observation that mid-ocean rifts propagate, intersect, and either re-form or are superseded by a new set indicates that forces concentrated at the ridge axis are not the primary driving mechanism for plate motions. Nor are rifts likely to be markers of upwelling limbs of convection cells in the mantle since they migrate so readily in response to external influences such as overlapping rifts or changes in plate direction.

As mentioned earlier (∞ see Chapter 6), the asthenosphere at mid-ocean ridges exerts an outward push on the

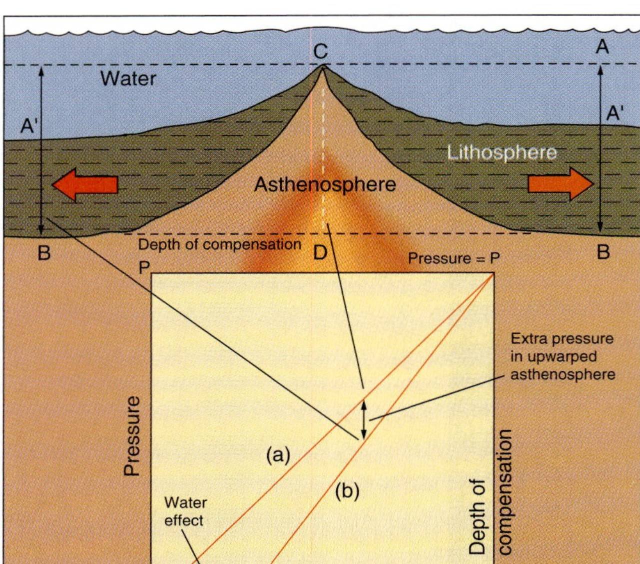

FIGURE 8.16 The principles of ridge push. The pressure variation with depth is plotted for two profiles as illustrated in the figure: from the ridge crest to the depth of compensation (curve a), and from the same depth in the ocean through old oceanic lithosphere to the depth of compensation (curve b). The rate of pressure increase is proportional to density. The lower curve initally has a small slope because of the low density of water. The upper curve has a slope determined by the thermally expanded asthenosphere, which is less dense than the lithosphere. Both meet at the depth of compensation where pressures must be equal. The difference between these curves gives the amount of extra pressure the asthenosphere exerts on the lithosphere. The sum total is expressed as ridge push.

lithosphere, called *ridge push*. As the lithosphere moves aside, the asthenosphere cools onto the underside of the lithosphere, creating new lithosphere. The lithosphere subsides, owing to cooling, the net result being that the lithospheric profile remains constant in time. Upwelling asthenosphere at the ridge axis replaces that lost to the lithosphere.

We can understand ridge push by examining the variation of pressure with depth (Fig. 8.16). The depth of isostatic compensation is the depth at which pressures are equal because the mantle behaves like a fluid (∞ see Chapter 5). In old ocean lithosphere, this occurs at the lithosphere–asthenosphere transition at about 100 km below the sea floor. At the ridge crest, which sits an average of 3 km higher than old ocean floor, this depth of compensation is 103 km below the sea floor. In Figure 8.16, we plot pressure as a function of depth for these two cases. Below the ridge crest, pressure increases linearly, as given in Figure 8.16, curve (a). If we then plot pressure as a function of depth from the same level for the old ocean lithosphere, we obtain the second lower curve (b) in Figure 8.16. We can see from these curves that, owing to the higher elevation of the ridge, the pressure in the asthenosphere at every depth above the depth of compensation is greater than that in the old lithosphere. It is the summation of these extra pressures over the region of ridge uplift that gives rise to ridge push.

Continental Rifting

Structure of Continental Rifts

Continental rifts form in regions of extensional stress. Recall (∞ see Chapter 7) that extensional stress generates normal faults. One of the main features of continental rift topography is the central graben, which can be some 50 km wide. It is flanked by steep boundaries made up of a series of normal faults with fault scarps that step down toward the central trough. The size of a fault scarp is a measure of the *throw* (∞ see Chapter 7) or displacement on the fault. One side of the graben frequently exhibits greater throw than the other, resulting in what is referred to as an asymmetrical rift (Fig. 8.17a). In some cases, the graben floor is covered by hundreds of small normal faults called *grid faults*.

Prior to rifting, sedimentary rocks may cover the continental surface. The surface subsides as the graben forms. In Kenya, limestones over 600 million years old are found at the base of rift escarpments. These old sedimentary rocks are, in turn, covered by sediments derived from erosion of the rift walls, volcanoes, and rim mountains. Volcanic eruptions pour lava into the grabens, which, along with the sediments, form alternating layers of volcanic and sedimentary rock. As the graben deepens and widens, the sedimentary infill itself undergoes faulting. Seismologic experiments reveal that graben sediments may be as much as 6 km thick. The main sources of these sediments are erosion of the shoulder uplift and fault scarps. The rift-boundary faults initially form with steep dips (about 60°), but their scarps are soon eroded to much shallower angles, so that the fault scarps are displaced from the trace of the original fault (Fig. 8.17b, c).

Seismic reflection images (∞ see Chapter 5) of the faults within the graben sediment show that they differ from the rift-boundary faults, which are thought to be straight and to dip at about 60°. In some cases, faults in sediments appear to flatten out at depth, forming a curved rather than straight surface. Such listric faults (∞ see Chapter 7) can be responsible for large rotations of sedimentary blocks as extension takes place (Fig. 8.18). Basins formed by such listric faults are filled, so that the sediment thickness increases toward the fault, and are referred to as **half-grabens** (Fig. 8.18).

Seismic refraction studies (∞ see Chapter 5) also reveal that crustal thickness on either side of a continental rift has high values—about 35 to 50 km. In the rift zone, the thickness may decrease to about 25 km,

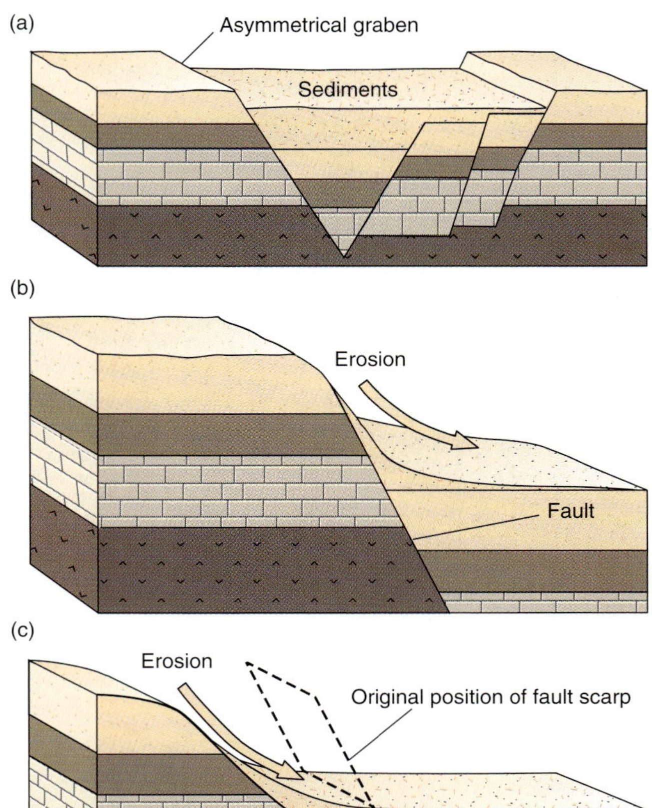

FIGURE 8.17 (a) A continental rift basin in which one of the rift-bounding faults has a greater throw than faults on the other side. The result is an asymmetrical graben. (b) Rift formation through normal faulting displaces rock types. The fault scarp erodes into the graben (one side only of the rift is shown). (c) Further erosion leaves the rift escarpment displaced from the original position of the fault scarp.

and the seismic-wave velocity of the upper mantle also decreases from 8.0 km/sec outside to 7.6 km/sec beneath the rift (Fig. 8.19), indicating that the asthenosphere reaches right to the base of the crust.

The shoulders, or edges, of continental rift grabens are at a higher elevation than is the surrounding plateau. This is referred to as **shoulder uplift** and is attributed to two sources: The first is doming of the plateau—possibly prior to rifting—that continues during the rifting process. It is an isostatic adjustment to balance relatively low densities in the hot upwelling mantle. The second is the result of volcanic eruptions along the rift margins that have built up large volcanoes.

In some cases, rifting and doming occur in regions of the continent that have undergone plateau uplift over hundreds to thousands of kilometers. As we shall see later in the chapter, the driving force of continental rifting must explain three phenomena: plateau uplift on the broad scale, shoulder uplift, and graben formation in the rift zone.

Continental rifting is nearly always accompanied by volcanism, and most volcanoes lie close to the faults associated with this rifting, although there are some spectacular exceptions, such as Kilimanjaro in Tanzania, which is over 100 km from the East African Rift margin. Continental rift volcanism spans the whole range of volcanic types, from flood basalts (∞ see Chapters 6 and 11) to explosive eruptions that deposit tuffs (∞ see Chapter 4).

Magma is not thought to rise to the surface along rift-bounding faults, which, even though they permit relative slip, are thought to remain tightly pressed together. Instead, magma appears to rise along mainly vertical fractures within the graben, forming dikes. Pressure from the magma and stresses associated with rifting force the rock apart. Frequently, in such fissure eruptions, one region along the fissure begins to dominate the outpouring of magma and a *central volcano* forms (Fig. 8.19).

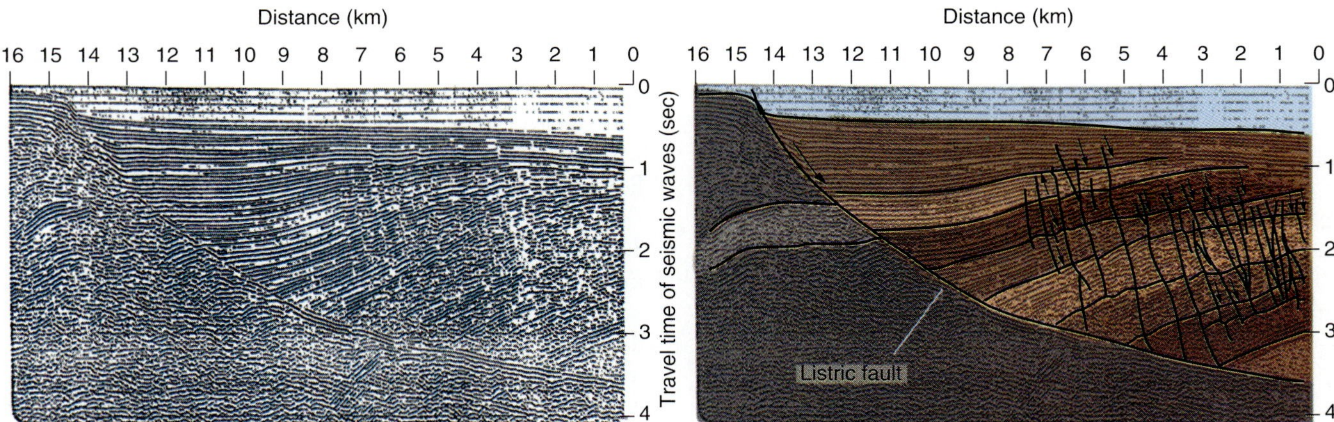

FIGURE 8.18 Listric faults in a seismic section from a sedimentary basin. The left-hand figure shows seismic reflections from the basin. The curved feature on the left side is interpreted to be a listric fault. The right-hand provides an interpretation of the seismic section, where bold lines indicate faults, and the structure is described as a half-graben.

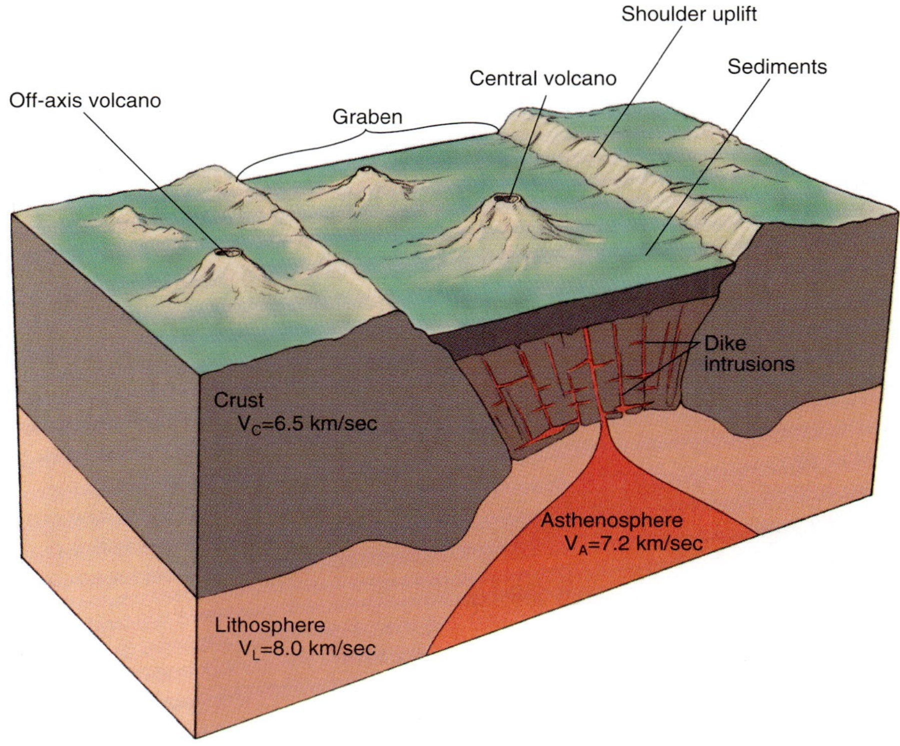

FIGURE 8.19 A schematic cross section showing the main features of a continental rift. High-velocity mantle material invades the base of the crust, whereas low-velocity material invades the lower lithosphere. This indicates that mantle motions have eroded the base of the lithosphere, so that the asthenosphere upwarps into it. Seismic velocities have been determined from seismic tomography, surface-wave studies, and refraction and reflection studies.

Gravity

Figure 8.20 shows the topography and Bouguer gravity variation across the world's four major young (the last 30 million years) continental rift zones: the Rio Grande Rift, East African (Kenyan) Rift, the Baikal Rift, and the Rhinegraben (see Fig. 8.1). Bouguer gravity anomalies (∞ see Chapter 5) across rifts such as the Rhinegraben are quite small—less than 25 mgals. Bouguer gravity profiles across some continental rifts can give rise to the large anomalies. The Rio Grande and East African rifts exhibit regional negative troughs of about −300 mgals. Like the mid-ocean ridges (Fig. 8.21), this low gravity is associated with low density at depth, which compensates the regional uplift of the topography. This compensation is attributed to thermal effects associated with upwarp of the asthenosphere into the lithosphere. The hot, partially molten asthenosphere has a lower density (by 50 kg/m^3) than the cooler surrounding lithosphere, which it has replaced. Being lighter, it gives rise to a negative gravity anomaly and also exerts a buoyancy force on the crust, accounting for the uplift.

At the deepest part of the East African gravity trough, a small rise occurs in the gravity field (see Fig. 8.20). Bearing in mind that regardless of temperature, mantle rocks have higher densities than do crustal rocks, it can be seen that this rise in gravity can be attributed to mantle rocks intruding into the continental crust. It marks the beginning of the transition from continental to oceanic crust, which, when complete, will form a new ocean basin. This phenomenon is less developed in the Rio Grande Rift, indicating that the East African Rift is closer to making the transition from continental to oceanic rift.

Although topography and gravity of continental rifts bear a striking similarity to those of mid-ocean ridges, heat flow is quite different. Within the rift valleys themselves, heat flow is generally high. On either side of a rift, however, heat flow is not significantly greater than that measured in stable continental regions. This can be explained in the case of rifting where the lithosphere is extended through plate interactions. Stresses generating the grabens need not be associated with any heating mechanism. It is surprising, though, in the cases of rifting thought to be due to localized convection in the mantle where you might expect the surrounding plateau uplift to have high heat flow. Keep in mind, however, that these rifts and their plateau uplifts were formed within the last 30 million years. If active convective processes in the mantle have given rise to mantle thermal anomalies, it is unlikely that the associated heat could have been conducted to the surface to show up in heat-flow measurements in such a short geologic time. Heat travels at such a slow rate by conduction within rock—moving only 30 km in 30 million years—that a thermal anomaly would only become apparent in regions where actual upward motion of hot material brings heat near the surface—for example, in the rift zones themselves.

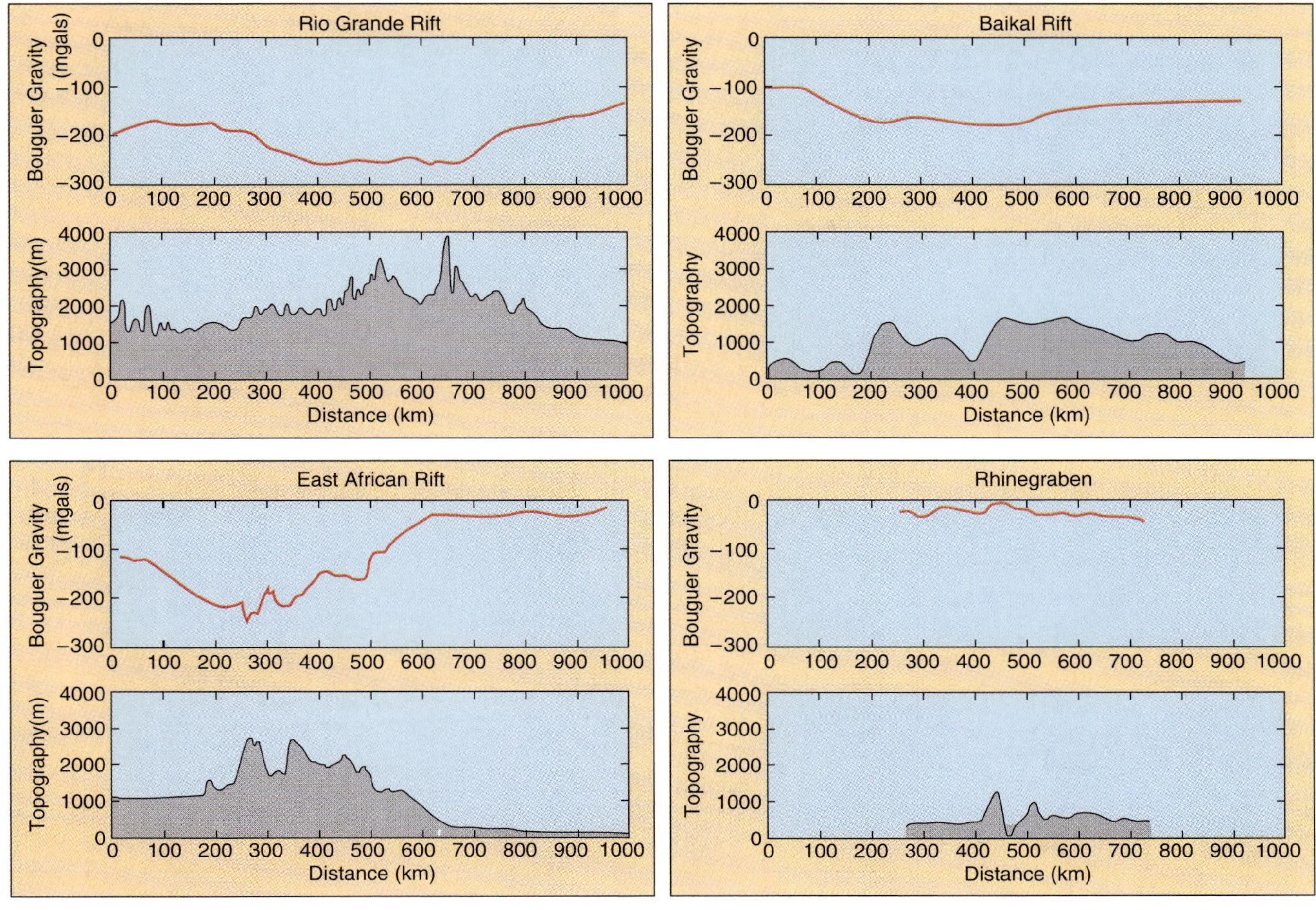

FIGURE 8.20 Bouguer gravity anomaly and topography across the world's major continental rift zones. Because Bouguer gravity corrects for topography, negative anomalies indicate that low densities must underlie the broad regions of topographic uplift. For the Rio Grande, Baikal, and East African rifts, this low density is thought to be due to hot mantle under the rifts. The Rhinegraben has a very small gravity anomaly, which implies that the mantle is at normal temperature. The small positive anomaly along the Bouguer trough for the East African Rift is thought to be due to mantle intrusion at the base of the crust.

Continental rifts are expressed by a central graben bounded by normal faults and containing faulted sediments. Rift volcanism occurs at fissures and central volcanoes along the rift.

Failed Rifts

When Pangaea broke apart, a number of continental rifts formed that did not ultimately become ocean basins. It has been proposed that the supercontinent Pangaea formed an insulating blanket that trapped the heat in the mantle. Hot convective plumes of upper mantle rock, or hot spots, developed and caused the supercontinent to dome upward in various places. When the doming became sufficiently extreme, cracks in the surface formed. Typically, three cracks radiated out from the dome at about 120° to each other, which would form the RRR or ridge-ridge-ridge triple junction that was introduced earlier (∞ see Chapter 6). Like the mid-ocean ridges, the elevated topography above elevated asthenosphere gives rise to a continental ridge push. As well as being pushed, the lithosphere was pulled apart by the forces of plate tectonics, as mentioned earlier; this probably occurred by subduction at distant plate edges. Then, the continent failed as favorably aligned pairs of cracks from the domal uplifts propagated toward each other, linked up to form continental rifts, and eventually formed divergent margins. The remaining crack or rift became inactive, cooled, subsided, and filled with sediments to become a **failed rift**, or **aulocagen**. The Benue Trough of Nigeria and the mid-continent rift of North America are examples of failed rifts. The mid-continent rift, though barely identifiable at the surface, has a clear signature in the gravity field (see Fig. 5.20).

What conditions cause a continental rift to continue spreading and to become a fully fledged divergent plate

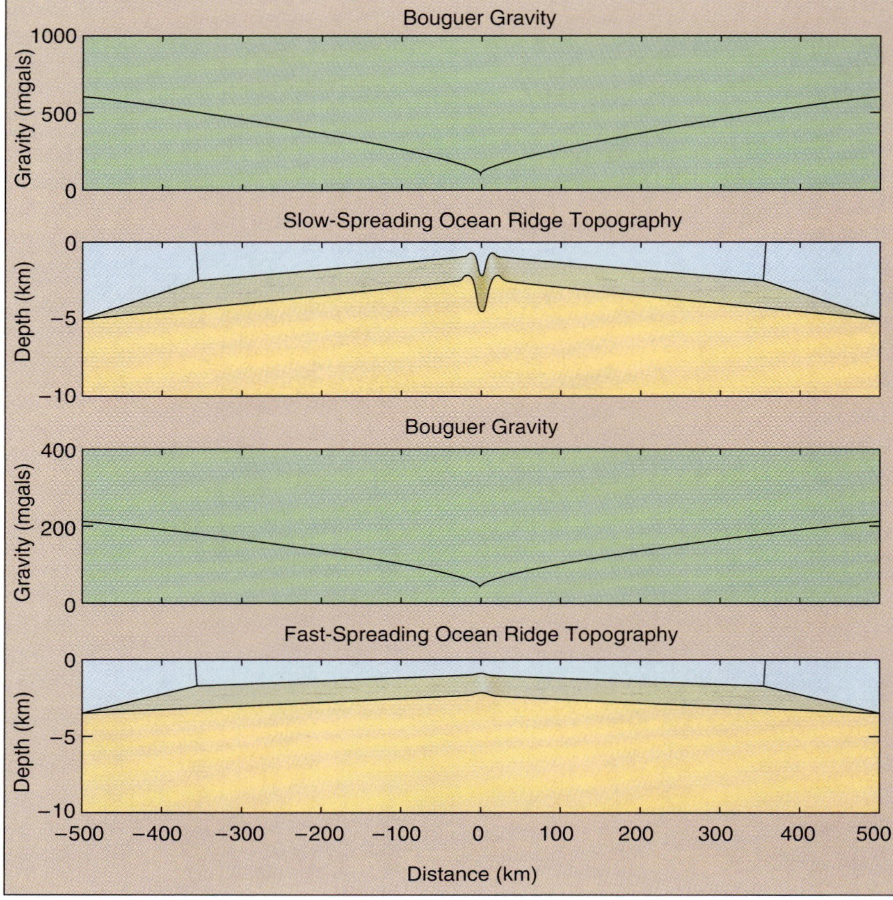

FIGURE 8.21 Bouguer gravity across slow-spreading and fast-spreading mid-ocean ridges. Like the continental rifts, the low Bouguer gravity anomalies indicate low densities in the mantle.

margin or to cease spreading and become a failed rift? The answer is not clear. At the time of breakup of Pangaea, subduction was occurring to the west beneath future North and South Americas. Plumes were generating flood basalts on the continent. Thus the opening of the Atlantic was probably a combination of continental ridge push and slab suction. Perhaps the suction actively pulled on the continent or it might have provided a low resistance (slippery) boundary over which the plate could move, driven by the push. One speculation is that for a continental rift to make the transition to an ocean basin, one of the flanks must be pulled by a subducting slab either through slab suction or through direct pull. For example, the successful rifting of the Red Sea is linked on its eastern flank to subduction of Arabia. In contrast, nearby rifting in East Africa is small in comparison to mid-ocean rifting, but Africa has no direct connections to subducting slabs. Perhaps rifting in Africa will also fail and be barely recognized at the surface, as is the case for the mid-continent rift of North America.

When a rift fails, the region cools. The regional uplift subsides and the central graben fills with sediments. Mafic material intruded into the crust generates positive Bouguer gravity anomalies superimposed on the shallower negative anomalies from the sediments. These gravity anomalies may be the only clue for the existence of an ancient failed rift, after the surface expression of the rift is obscured by erosion and sedimentation.

The Mechanisms for Continental Rifting

Continental rifts can be separated into two types: *passive rifts* and *active rifts*. The Rhinegraben in Germany, Lake Baikal in Siberia, and the Rio Grande Rift of North America have been described as passive rifts. In these cases, remote subduction forces have extended the crust, and associated faulting has resulted in graben formation. It has been suggested that subduction forces that formed the Alps during the collision of Africa and Europe exerted a tensional stress north of the Alps, which has given rise to the Rhinegraben. It has also been suggested that formation of the Baikal Rift is due to stresses associated with the collision of India with Asia and uplift of the Himalayas. Indentation of the Asian Plate by India is thought to have generated far-reaching stresses. The extensional stress is generated rather like bending an elastic beam that extends on the outer side and compresses on the inner side. For the Rhinegraben, the effective beam is the Alps; for Baikal, it is the compressed zone of inner Asia. In western North America, subduction along the edge of the continent has been proposed

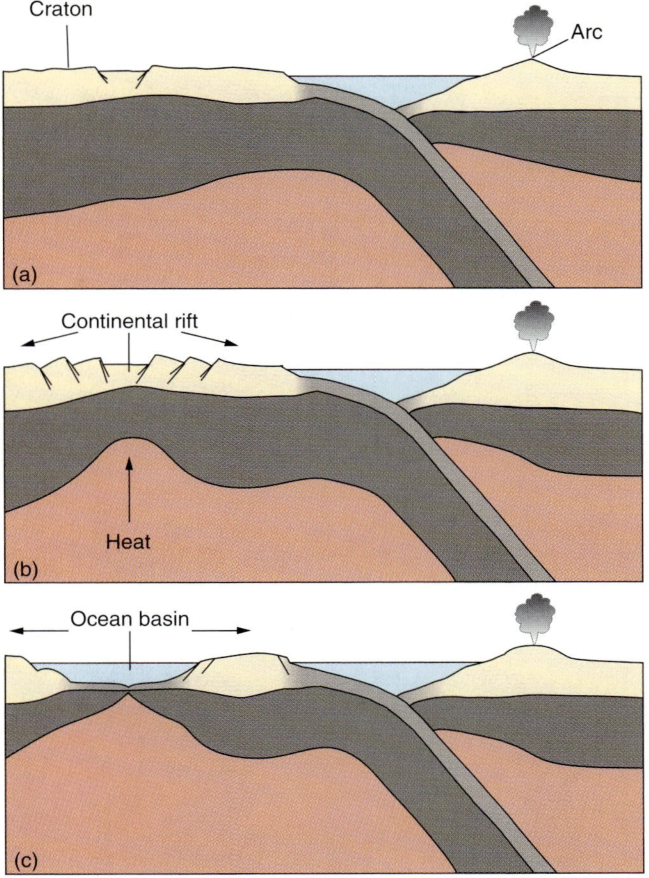

FIGURE 8.22 Passive continental rifting by slab pull. (a) A slab firmly attached to a continent causes normal faulting. (b) A continental rift forms. (c) Continental rifting forms an ocean basin.

to extend the continent lying west of the Rocky Mountains, including the Rio Grande Rift. Other models have been proposed, such as mantle plumes beneath these rifts, but this remains a subject of ongoing research.

Active rifts occur where active motions in the mantle—upwellings of hot material such as mantle plumes—are primarily the cause of rifting. Rifts that form above hot spots are active rifts. The East African Rift (see Fig. 8.1; Focus 8.2) appears to fit into this category, but the distinction between active and passive rifts is difficult to make and is a focus of continuing debate. Some argue that the East African Rift is a passive rift with active convection beneath it in the asthenosphere, causing a weak zone that fails.

Passive rifting occurs above normal mantle and is driven by plate forces. Active rifting is driven by convection in the mantle beneath the rift.

As we have already discussed, a key force of plate tectonics is the force exerted by the sinking of dense oceanic slabs at subduction zones. If the slab is attached to a continent across a passive margin, slab pull could literally pull the continent apart (Fig 8.22). The rifting of Australia from Antartica is thought to have occurred by this mechanism. Slab suction is another mechanism by which continents can be torn apart (Fig. 8.23). This mechanism also gives rise to back-arc basins such as the Sea of Japan and the Tasman Sea (∞ see Chapter 9), and it may have been partially responsible for the breakup of Pangaea.

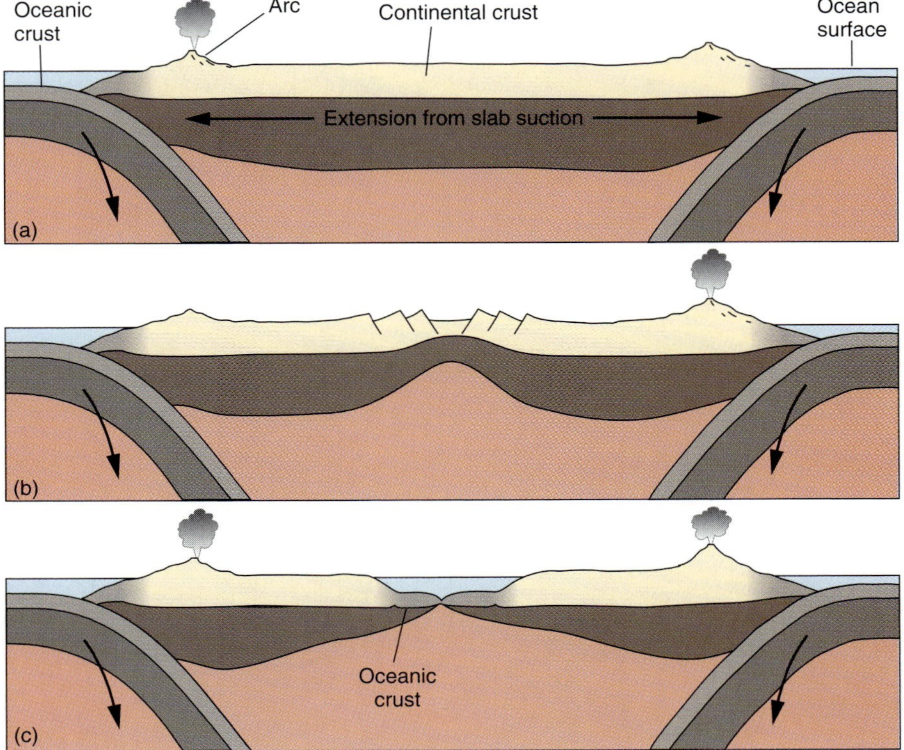

FIGURE 8.23 Passive continental rifting by slab suction. (a) The motion of the ocean slab is downward as well as along its length. As a result, the slab rolls back, exerting a suction on the overlying continent. The continental lithosphere is extended. (b) Rifting occurs and the lithosphere thins. (c) Rifting becomes so extreme that an ocean basin occurs behind the volcanic arc.

Focus On 8.2 CONTINENTAL RIFTING: THE EAST AFRICAN RIFT

The African continent is dominated by a structural pattern of broad basins and swells that rise toward the east, culminating in the eastern plateau uplift (Fig. 1). Plate reconstructions suggest that Africa has remained stationary with respect to the mantle for the last 40 million years. This continental cap on the mantle could have had an insulating effect on mantle heat flow, which caused the mantle to heat up. It is reasonable to conjecture that the basins and swells might be caused by mantle convection, which became more vigorous under Africa because of trapped heat.

Rifting appears to have begun in the Gulf of Aden about 30 million years ago and to have propagated southward, crossing the Ethiopia and Kenya domes (Fig. 2). North of Lake Victoria—the source of the Nile—the rift bifurcates into the eastern and western rift valleys, flanking the lake on either side. The eastern rift is called the *Gregory Rift* after one of the earliest geologists to study it. South of Lake Victoria, the two rifts rejoin in Tanzania. Rifting continues to the south and dies out where it meets the lower Zambezi River.

The eastern arm of the rift is much more volcanically active than is the western. Volcanic cones are found on the rift floor (Fig. 3), but some of the more spectacular volcanoes occur on the rift flanks, such as Mts. Kenya, Elgon, and Kilimanjaro. The rift flanks rise up from sea level to elevations of more than 2 km. This elevation change is partially caused by volcanic rocks erupted onto the surface and partially caused by the forces of rifting, which have uplifted the preexisting basement rock. The elevation causes moist air from the Indian Ocean to precipitate abundant rainfall that, combined with the mineral-rich volcanic soil, makes the highlands extremely fertile.

A number of lakes have formed in the rift graben, including Lakes Turkana, Tanganyika, and Nyasa. The rift walls are spectacular, steeply dipping escarpments thousands of feet high (Fig. 4). Fresh, uneroded fault scarps attest to the recency of rifting. Where the East African Rift meets the sea in the Gulf of Aden and the Red Sea, it forms an RRR triple junction of a continental rift meeting two oceanic rifts (Fig. 2). The Red Sea has formed over the same period of time as the African Rift. Its spreading rate, however, has been significantly greater, and it has made the transition from continental rift to oceanic rift. Near Djibouti, where the African Rift meets the sea, this transition is seen

FIGURE 1 The tectonic basins of Africa with intervening swells, plateaus, and rift valleys.

as rift-bounding faults, which drop the level of the land below sea level (Fig. 5).

Heat flow can be very high within the rift zone itself, especially near active volcanoes. In fact, geothermal energy from hot rocks in the rift provides Kenya with 10 percent of its power. On the flanks, however, the heat flow is equivalent to that of other nonrifted continents.

Gravity profiles across Africa indicate that both the Kenya and Ethiopia domes are regions of the lowest Bouguer anomalies in the entire continent. They indicate that, under the domes, roots of mass

FIGURE 2 The Afro-Arabian Rift system.

FIGURE 3 The central volcano Longonot in the East Africa Rift zone near Nairobi, Kenya.

FIGURE 4 The East African Rift escarpment.

FIGURE 5 Transition from continental to oceanic rifting as the East African Rift meets the Red Sea in Djibouti.

deficiency exist, which isostatically support the uplift. Along the rift axis, a small positive anomaly is thought to be associated with intrusion of high-density mantle into the crust as the rift extends and becomes more like ocean crust.

Seismic exploration of the Gregory Rift has revealed that the crust is thinner (25 km) beneath the rifts than beneath the flanks (35 km), and that the upper mantle velocity is slower (7.6 km/sec compared to 8.0 km/sec for normal mantle). Seismic tomographic experiments have provided images of the mantle intrusion into the crust; these explain the broad gravity anomalies as asthenospheric upwarp into the lithosphere (see Fig. 8.20). By comparing velocity changes and effects on seismic waveforms for different regions of East Africa, scientists estimate that the upwarped asthenosphere is so hot that it contains traces of partial melt. This melt gives rise to volcanism as it escapes through channels in the lithosphere.

The anatomy of the Gregory Rift (Fig. 6) has been compiled on the basis of surface geologic mapping, gravity, and seismic sounding. Isostatic support of the regional uplift comes from the hot mantle, where the asthenosphere intrudes into the lithosphere. Magma chambers form in regions where the fraction of partial melt in the hot mantle becomes so great that it can no longer be contained in the interstices between rock grains, and it flows upward, forming chambers that feed volcanoes. Dikes also propagate in the upper brittle layers of the crust or mantle. Diking will tend to force apart the crust, generating stresses that are possibly the cause of normal faulting at the surface. Dikes may also break through to the surface. The actual shape of the graben, the rift flanks, including the rise to the rift, is a superposition of extension, intrusion, volcanism, and erosion.

What drives East African rifting? Several observations provide a clue as to the source of the rifting, but it is still a subject of debate. Africa is almost entirely surrounded by mid-ocean ridges. Thus, it is unlikely that subduction is pulling the continent apart. The Red Sea and the Horn of Africa are regions of the largest low-velocity tomographic anomalies in the upper mantle. These anomalies are most pronounced beneath the Kenya and Ethiopia domes, and are best explained as being due to higher temperatures associated with an upwelling plume of convection in the mantle. Geologic recon-

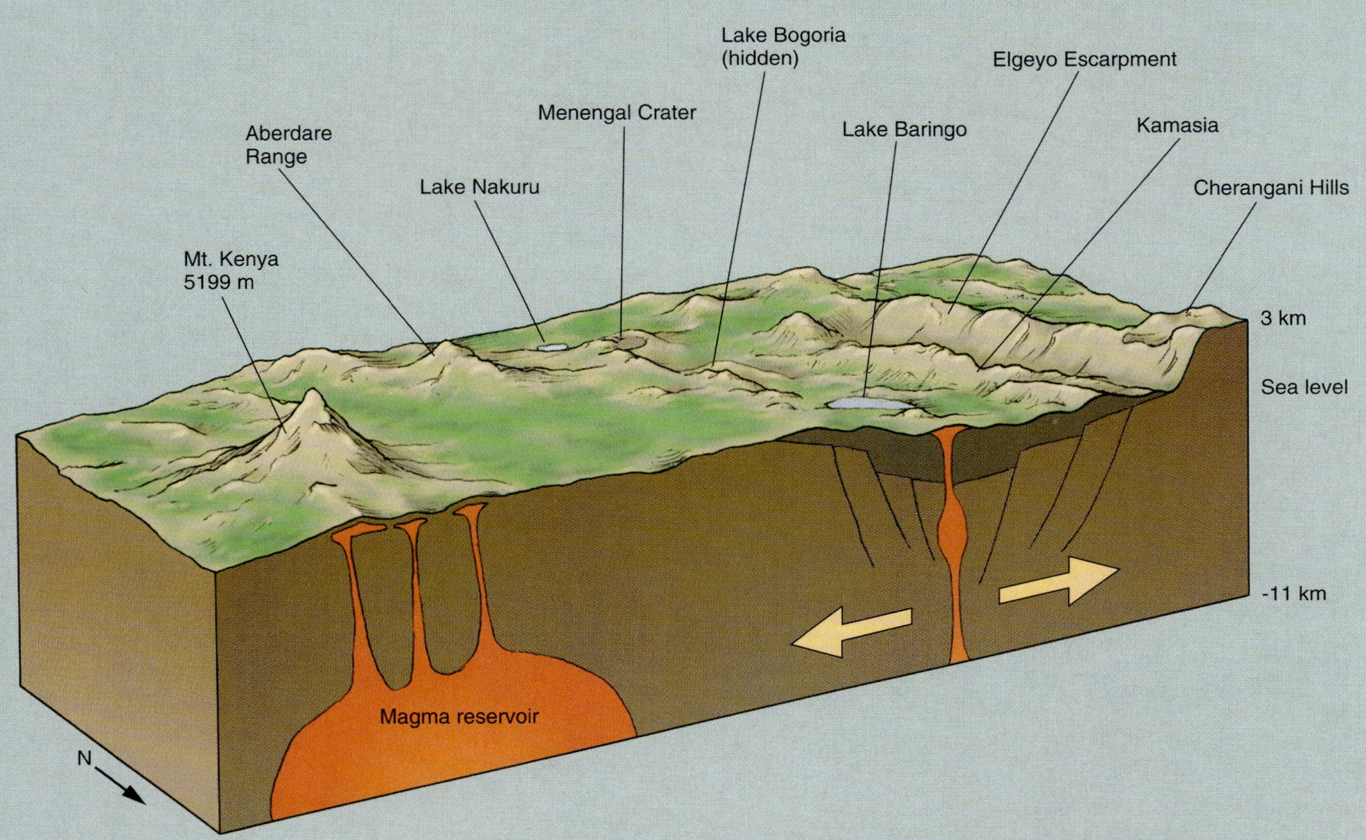

FIGURE 6 A computer-generated view of the internal structure of the East African Rift.

structions of fault slips in the East African Rift system indicate that the separation of the continent is only tens of kilometers. The amount of heat brought up by deep mantle rocks into this small volume is far too small to explain the velocity anomalies, which extend for hundreds of kilometers on either side. Therefore, such a passive mechanism must not apply. Instead, an active mechanism must transport the heat there, which must have begun fairly recently in geologic time since the heat flow on the surface is low, indicating that the increased heat flow has not yet reached the surface.

Rifting began about 30 million years ago—a time when the Red Sea was also opening. One possible model for rifting proposes that, prior to rifting, mantle convection developed an upward flow beneath East Africa and the Red Sea region. In hot-spot regions such as the Kenya and Ethiopia domes, the lithosphere began to dome. Plate forces, such as those associated with subduction beneath the Zagros Mountains of Iran or mantle flow at the base of the plate, may have exerted an extensional stress. Rifting began in the dome areas, and the extra extension imposed on already hot regions caused them to melt more and release voluminous basalts, forming the present-day topography.

In Africa, seismic tomography (∞ see Chapter 5) indicates that high-velocity anomalies exist in the lower part of the crust. These are interpreted as arising from mantle rocks intruding the crust as plate divergence takes place. At greater depth, tomography studies indicate that continental lithosphere has been thinned beneath the plateau uplift, resulting in an upwarping of the asthenosphere that extends over hundreds of kilometers. A similar structure is observed for the deep region beneath the Rio Grande Rift.

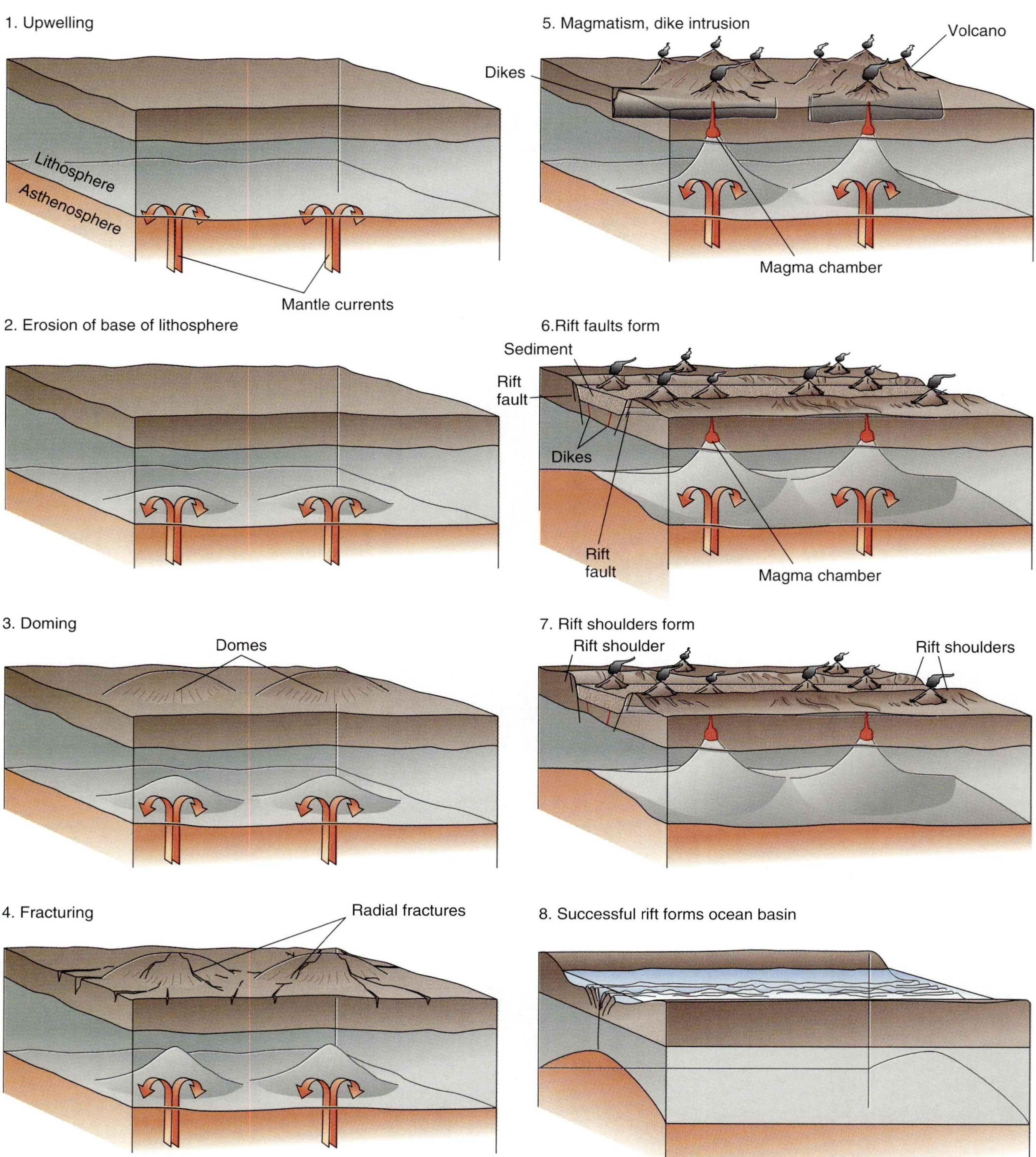

FIGURE 8.24 A model for continental rifting. (1) Excess heat in the mantle causes upwelling convection currents beneath continental lithosphere. (2) The currents erode the base of the lithosphere, replacing it with hot mobile asthenosphere. (3) Doming occurs at the surface, held up by the hot buoyant rock in the eroded base of the lithosphere. (4) Doming becomes so intense that fractures are formed in the lithosphere that radiate out from the domes. (5) Partial melting of the mantle forms magma. Magma reaches the surface to form volcanoes or intrudes into fractures to form dikes. (6) The forces of dike intrusion and buoyant uplift cause rifting at the surface. (7) The wedging effect of the dikes causes the rift shoulders to form. (8) If the plate tectonic forces have the right configuration, the rifting continues to form an ocean basin. Alternatively, rifting may cease and become a failed rift.

Where rifting is intimately connected to upwelling mantle currents, we can envisage the following stages in rift development (Fig. 8.24):

Convective plumes of upwelling material become established beneath the continental lithosphere. This upward motion changes the base of the continental lithosphere to asthenosphere—either by injection of heat, which causes partial melting, or by erosion and replacement by deeper, hotter material, which partially melts as it reaches a region of lower pressure. The resulting thermal anomalies cause the rock in the mantle to be lighter, or more buoyant, which causes the surface to dome upward. The asthenospheric uplift exerts an outward force similar to ridge push at oceanic spreading centers. The elastic limit (∞ see Chapter 7) of the brittle crust is reached and fractures form that radiate out from domes. Magma starts ponding within magma chambers in the upper mantle and intrudes the mantle and crust in the form of dikes and extrudes as volcanoes. Internal magma pressure and intrusion cause rifting to occur, which follows a path of weakness determined by propagation and linking of the domal fracture systems. The forces of intrusion wedge up the rift shoulders. With time, as heat is lost, the mantle currents cease to flow. The mantle cools and contracts and the doming subsides. The rift becomes a failed rift. Alternatively, slab suction or slab pull at a nearby subduction zone allows rifting to continue. In such cases, the continental lithosphere can move aside to be replaced by oceanic lithosphere. The sea flows into the subsided rift, and a new ocean forms. At this point, passive and active rifting become indistinguishable.

This model of continental rifting, while consistent with geophysical and geologic observations, is still undergoing vigorous testing. It does explain, though, why continental topography is similar to ocean topography: The ridgelike topography in both cases is held up by thermally expanded rock in the mantle. In the continental case, however, we appeal to convective motions in the mantle, which transport heat to the region beneath the rift. The lithosphere is relatively stationary and the asthenosphere moves below it. Under the oceans, however, the asthenosphere is relatively stationary, and it is the lithosphere that is moving from its origin at mid-ocean ridges, sliding across the asthenosphere.

Mantle upwelling beneath continental lithosphere causes thinning and doming. Continental rifts form when fractures associated with doming link together.

SUMMARY

- Once formed, oceanic lithosphere tends not to rift, whereas continental lithosphere does. Continental rifts become oceanic rifts when the rifting progresses to the extent that separating continental lithosphere is replaced by oceanic lithosphere as the mantle flows in from below.
- Oceans subduct, but continents are too buoyant to subduct; instead, they rift apart.
- Topographic profiles across slow-spreading continental rifts resemble profiles across oceanic rifts even though continental-plate separations occur at a rate that is 10 times slower than that of oceanic rifting.
- Oceanic lithosphere is approximately 100 km thick and exhibits high seismic-wave velocity. A downward section through the lithosphere reveals 7 km of oceanic crust, comprising sediments, pillow lavas, basalt dikes, and gabbro overlying peridotite mantle.
- Hydrothermal vents on the sea floor, called *black smokers*, are important heat exchangers between the interior and the surface of Earth. New life-forms that utilize chemosynthesis have been discovered at such vents.
- Reactions between seawater and basalt rising at the spreading ridges change the compositions of both basalt and seawater. Basalt contributes dissolved salts to the seawater and metals to the sea floor; seawater alters the basalt by addition of sodium and magnesium.
- Cooler lithosphere is denser and floats lower in the mantle. The amount that the oceanic lithosphere cools depends on the square root of its age.
- Continental rifts are expressed by a central graben bounded by normal faults and containing faulted sediments. Rift volcanism occurs at fissures and central volcanoes along the rift.
- Passive rifting occurs above normal mantle and is driven by plate forces. Active rifting is driven by convection in the mantle beneath the rift.
- Mantle upwelling beneath continental lithosphere causes thinning and doming. Continental rifts form when fractures associated with doming plumes link together.

KEY TERMS

sheeted dike complex, 200
black smoker, 201
half-graben, 208
shoulder uplift, 209
failed rift, 211
aulocagen, 211

QUESTIONS FOR REVIEW AND FURTHER THOUGHT

1. Explain how you would plot the topography for a cross section of the Mid-Atlantic Ridge and one for the East Pacific Rise so that the two curves lie on top of each other.
2. The square root of age dependence of cooling can explain oceanic ridges. Can it be used to explain the uplift associated with continental rifts?
3. Use the global plate-tectonic map (Fig. 1.6) to predict where the next mountains similar to the Himalayas might form in the next 100 million years.
4. Slow-spreading mid-ocean ridges have a median valley, whereas fast-spreading ones do not. Why is this so?
5. The free-air gravity anomaly across mid-ocean ridges has virtually no variations, whereas the Bouguer gravity anomaly has negative values that become most extreme at the ridge itself. What do these two gravity anomalies tell us?
6. At the ocean ridge crest, drill holes in the crust have revealed the crust is cold, yet this is where hot mantle is rising to fill in the gap created by spreading. Explain this observation.
7. What two processes may give rise to the salt in seawater?
8. Describe the difference in gravity, topography, heat flow, and tomography expected for a passive rift and an active rift. Give a sketch of each.

9 Plates That Collide: Convergent Margins

Snow-covered peaks of the Himalayan mountain range—the ultimate result of plate convergence and continental collision.

Overview

- Much of the tectonic activity on Earth's surface occurs at convergent plate margins, including volcanoes, earthquakes, and crustal deformation.
- At convergent margins, oceanic lithosphere sinks back, or subducts, into the mantle, giving rise to volcanic and seismic activity.
- Volcanic arcs, characterized by violent and explosive volcanism, are developed on the upper plate above subduction zones.
- Sediment, including abundant volcanic material, accumulates in a dynamic wedge where two plates converge.
- Most mountain ranges result from the convergence of two plates and the eventual collision of two continents.
- Deformation and magmatism are also associated with the relative movements of many small microplates.

Introduction

Unlike divergent margins, convergent margins are asymmetric, with one lithospheric plate sinking beneath the other. Two possible structures characteristic of convergent plate margins are shown in Figure 9.1: Oceanic lithosphere can subduct beneath oceanic lithosphere at an oceanic convergent margin (Fig. 9.1a). Alternatively, oceanic lithosphere can subduct beneath continental lithosphere at a continental convergent margin (Fig. 9.1b)—which would also be termed an *active* continental margin (Chapter 6). Note that the subducting, or sinking, plate is always oceanic. Continental lithosphere, which contains a thick section of continental crust, far less dense than the mantle, is too buoyant to sink back into the mantle. When two plates with continental crust converge, they collide. Subsequent movement toward each other will cease as the buoyant continents jam against one another, preventing subduction. The results of continent-continent collisions are spectacular mountain ranges such as the Himalayas. We will examine the effects of such collisions later in the chapter.

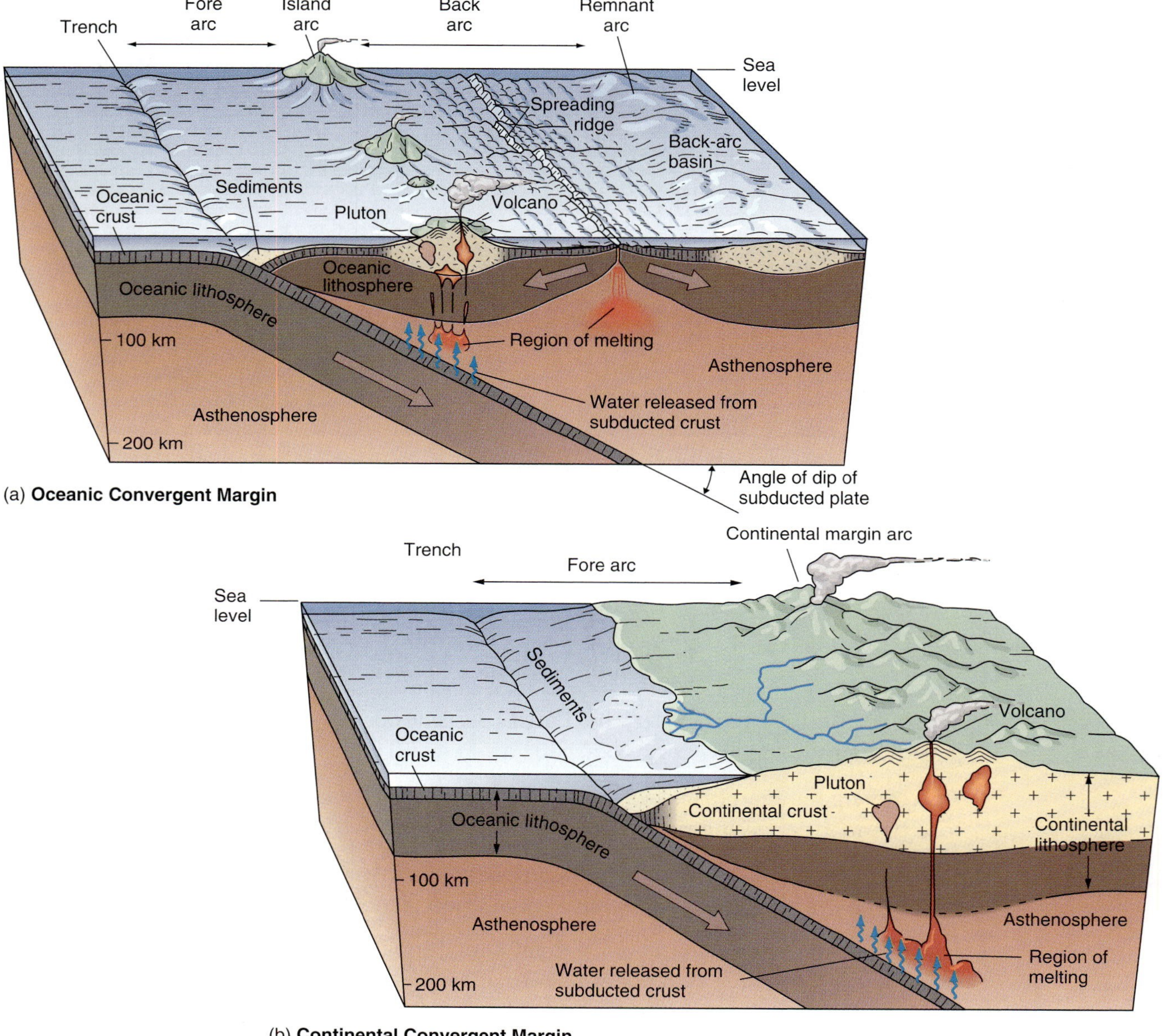

FIGURE 9.1 Structures of convergent plate margins: (a) An oceanic convergent margin with volcanoes emerging above sea level to form an island arc; (b) a continental convergent margin. Note an example of a back-arc basin in (a), with active sea-floor spreading. Back-arc basins can form at any convergent margin, but are not universal. The general subdivision of the arc environment into fore arc, back arc, and volcanic arc is indicated.

Rates of plate convergence vary from 10 cm/yr to less than 1 cm/yr. Subducted slabs dip at a range of angles from about 10° to 90°. The *dip* is the angle between the horizontal and the subducted plate. Dip is controlled by a number of factors, including the rate of plate convergence and the age of the subducting lithosphere. The age of the subducted lithosphere depends upon when the lithosphere was formed at a mid-ocean ridge. Therefore, the age will depend upon the distance of the subduction zone from the ridge and the rate of lithosphere addition at the ridge (∞ see Chapter 8). Old oceanic lithosphere is cold and dense, and will therefore sink back into the mantle. Young oceanic lithosphere is hotter, more buoyant, and less susceptible to subduction than is old oceanic lithosphere. As a result, younger subducted lithosphere also forms shallower slab dips.

The site where the two plates meet on Earth's surface is commonly marked by a topographic depression called a **trench**. Ocean trenches are the deepest features of the sea floor. The Mariana Trench in the western Pacific reaches a depth of nearly 11,000 m below sea level, which is 2000 m deeper than Mount Everest is high!

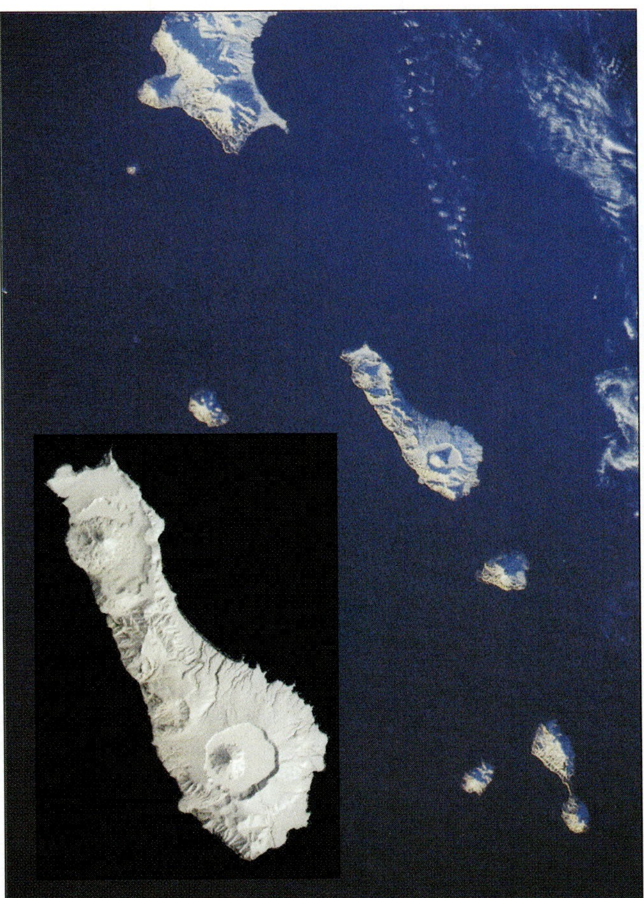

FIGURE 9.2 Satellite photograph of the Kuril volcanic arc. Inset is a satellite view of Onekoten I, clearly showing a volcano in a large crater, or caldera.

The most obvious expression of a subduction zone is a **volcanic arc**, which is developed on the upper plate (Fig. 9.2). Intrusive rocks (plutons) are abundant in the crust below the volcanic arc (Fig. 9.1), so we often use the term **magmatic arc** to describe more accurately such features. At continental margin arcs, the volcanoes commonly form a significant mountain range, reaching heights of over 6000 m in the case of the South American Andes. In contrast, the volcanoes at an oceanic convergent margin are built on oceanic lithosphere, and it is only the peaks of the highest volcanoes that usually emerge above sea level. These peaks will form a chain of volcanic islands known as an *island arc* (Fig. 9.2).

Between the volcanic arc and the trench is a region in which sediments may accumulate, called the **fore-arc** region. Sediments accumulate from both erosion of the active volcanic arc and actual transport of ocean-floor sediments toward the trench on the subducting plate. The area on the side of the volcanic arc away from the trench is known as the **back-arc** region, where, in some cases, a back-arc basin forms (Fig. 9.1a). Such a basin is similar to a mini-ocean basin, where sea-floor spreading takes place on a small scale. The processes occurring here (volcanism, earthquakes, sediment accumulation, and so on) are analogous to those described for mid-ocean ridges and true ocean basins (∞ see Chapter 8).

The idea of lithosphere pulling apart may seem inconsistent with our intuition about convergent margins; after all, if two plates are moving toward one another at a subduction zone, it might seem strange that one of them would be breaking apart only a short distance away—behind the arc. It is important to realize, however, that convergent margins are not necessarily zones of compression. Even though two plates that meet at a convergent margin are moving together in a relative sense, the lithosphere might be under extension or might even be subjected to strike-slip motions. The subduction zone is not a freely moving interface; rather, the subducting slab is coupled to the upper plate through slab suction (∞ see Chapter 6). If the upper plate is not overidding the subducting slab, the downward pull of the subducted slab will pull the upper plate toward the trench and the arc will be under tension (lower panel of Fig. 9.3a). Alternatively, if the dominant force is caused by a rapidly moving plate overriding the slab (top panel of Figure 9.3a), the overall effect will be to generate a compressional stress, and the arc may be subjected to deformation such as folding and thrusting, rather than extension.

Convergent margins occur where one plate sinks (subducts) beneath the other. Subduction is due to dense oceanic lithosphere sinking back into the mantle, largely under its own weight.

The weakest part of the upper plate is the line of the magmatic arc, which we can think of as a series of perforations in the lithosphere. Accordingly, if the upper plate is under tension, it will break along these perforations and rift the arc into two pieces—with the formation of a back-arc basin between the pieces (Fig. 9.3b). Formation of a back-arc basin is probably aided by convection of the asthenospheric mantle, known as *corner flow*, between the subducted and overriding plate (Fig. 9.3b). The piece of volcanic arc left on the trenchward side remains a site of active volcanism, whereas the other volcanic arc fragment becomes an inactive remnant arc on the opposite side of the back-arc basin (Figs. 9.1 and 9.3b).

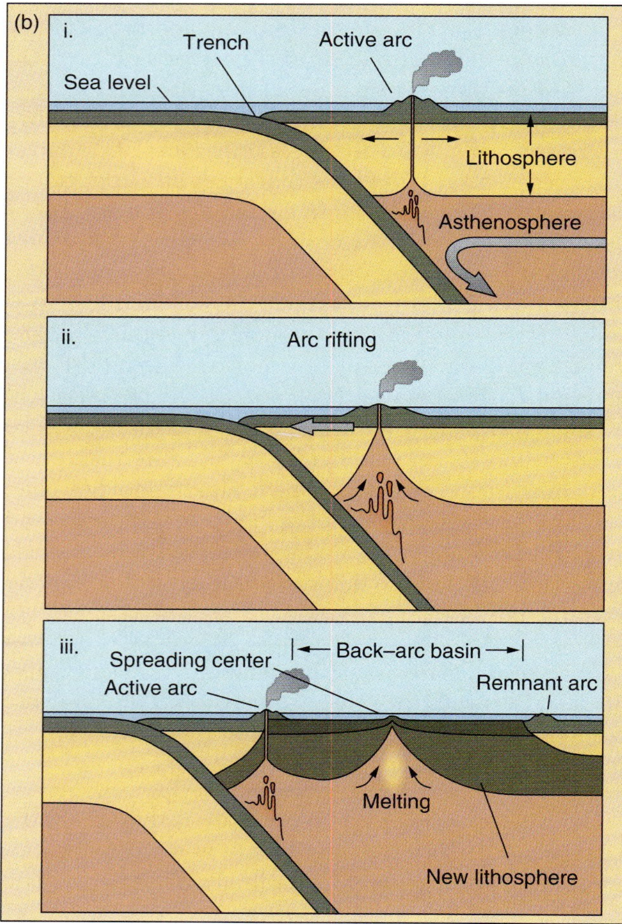

FIGURE 9.3 (a) Examples of compression or tension caused in the upper plate as a result of subduction zone geometry. If movement of the upper plate is dominant, compression will result and may give rise to folding and thrusting. If the downward sinking of the slab is the dominant effect, it will roll back, generating slab suction and causing extension in the upper plate. (b) Schematic illustration of the formation of a back-arc basin. In (i), the upper plate is being "pulled" toward the trench and is therefore under tension. Arrows in asthenosphere indicate probable nature of circulation, known as corner flow. (ii) The upper plate begins to split at its weakest point—the line of magmatism that defines the arc. Hot asthenosphere moves up between the diverging plates and melts, forming new oceanic lithosphere, just as at mid-ocean ridges described in Chapter 8. (iii) A back-arc basin now separates the active arc from a remnant arc—the piece of the arc farthest from the trench in (i)—which has been rifted away. Note that the position of the active arc is fixed relative to the trench because it is a function of the subduction zone geometry. The position of the back-arc spreading center is actually migrating away from the trench because new lithosphere is being added symmetrically about the center, and the back-arc basin is growing wider.

The Action at Convergent Margins

Earthquakes at Convergent Margins

Earthquakes are associated with subduction zones and are a direct result of the stresses generated when one plate moves beneath the other (Focus 9.1). Locations of earthquake epicenters closely correlate with the locations of volcanoes. Nearly all of the seismic hypocenters are located on the upper part of the subducting plate within 300 km of the trench. Analyses of seismic data from earthquakes at subduction zones allow geologists to locate both the hypocenters of the earthquakes and their epicenters. Recall that the hypocenter is the actual location of the seismic event within Earth, and the epicenter is the point on Earth's surface directly above the hypocenter (∞ see Chapter 7).

The inclined array of earthquake hypocenters at a subduction zone defines the **Wadati-Benioff zone**, which dips beneath the volcanic arc (Fig. 9.4). The Wadati-Benioff zone is often considered to be synonymous with the upper surface of the subducted slab. Although this is a reasonable approximation, it is not strictly the case, as shown in Figure 9.4, where it is apparent that, at shallow depths (less than 100 km), earthquake hypocenters are near the upper surface of the descending slab, but at greater depths, the hypocenters are *within* the descending slab. Earthquakes originating at shallow depths are large-

Focus On 9.1 THE (GOOD FRIDAY) ALASKAN EARTHQUAKE

On Good Friday, March 27, 1964, at 5:36 P.M., a magnitude 8.6 earthquake in Alaska affected about 200,000 km^2 of Earth's crust. The northward-moving Pacific Plate sinks into the mantle in the area along the Aleutian Trench, which stretches from southeast of Anchorage to Kamchatka in Siberia. (Fig. 1).

Subduction has formed a trench with depths up to 7600 m. North of the trench are the Alaska Peninsula and the Aleutian Islands. For many years prior to the earthquake, a section of the subducting slab must have been stuck, and this halted the motion beneath Alaska. Most of the Pacific Plate, however, kept moving, and elastic stress developed in a region about 800 km long and 250 km wide. On March 27, the blocks on either side of the plate boundary moved several meters in about six seconds. This generated a huge tsunami (referred to—incorrectly—as a tidal wave) that swept ashore devastating the Alaskan waterfront and carrying ships many kilometers inland. The seismic shaking lasted for several minutes—enough time to cause enormous property damage. Perhaps surprisingly, only nine people died from the effects of the shaking.

One person experienced the earthquake in the basement of a large building. When he noticed daylight outside through cracks opening and closing in the building walls, he feared for his safety and ventured outside. There, he found high-amplitude surface waves causing the conifer trees to oscillate so violently that he was in danger of being swatted by the branches, which pounded the ground from side to side.

A radio announcer's impressions of the earthquake, included these excerpts: "Hey, that's an earthquake for sure! Wheeeee! Boy oh boy. . . . I'm telling you, the whole place just moved like someone had taken it by the nape of the neck. . . . "

One of the most damaging aspects of the shaking was ground liquefaction. The Turnagain Heights area of Anchorage consisted of over 22 m of unconsolidated sediment, forming bluffs above Cook Inlet. Near the base of this bluff was a layer of clay, which became unstable during the earthquake. Vibration caused the clay to liquefy, and the bluffs, together with the houses on them, collapsed in a failure referred to as a *slump* (Fig.2). This extended as much as 300 m back from the former cliff line.

This earthquake was the largest recorded this century in North America. By the time the high-frequency seismic waves reached Anchorage, 130 km from the epicenter, they had been attenuated and longer period waves dominated. Many smaller structures survived the event. Tall structures suffered more damage because they were more susceptible to the longer period waves.

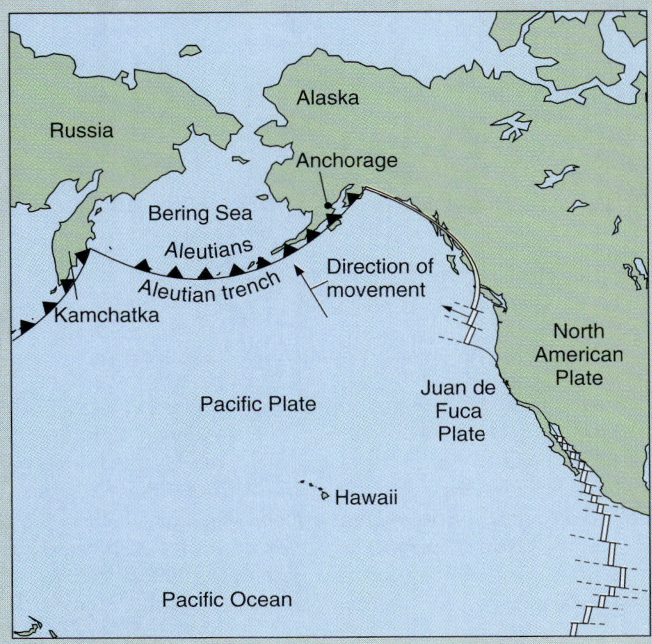

FIGURE 1 Tectonic setting of Alaska.

FIGURE 2 A landslide caused by liquefaction tumbled these houses in Turnagain Heights, Anchorage, Alaska. This damage resulted from shaking during the 1964 earthquake.

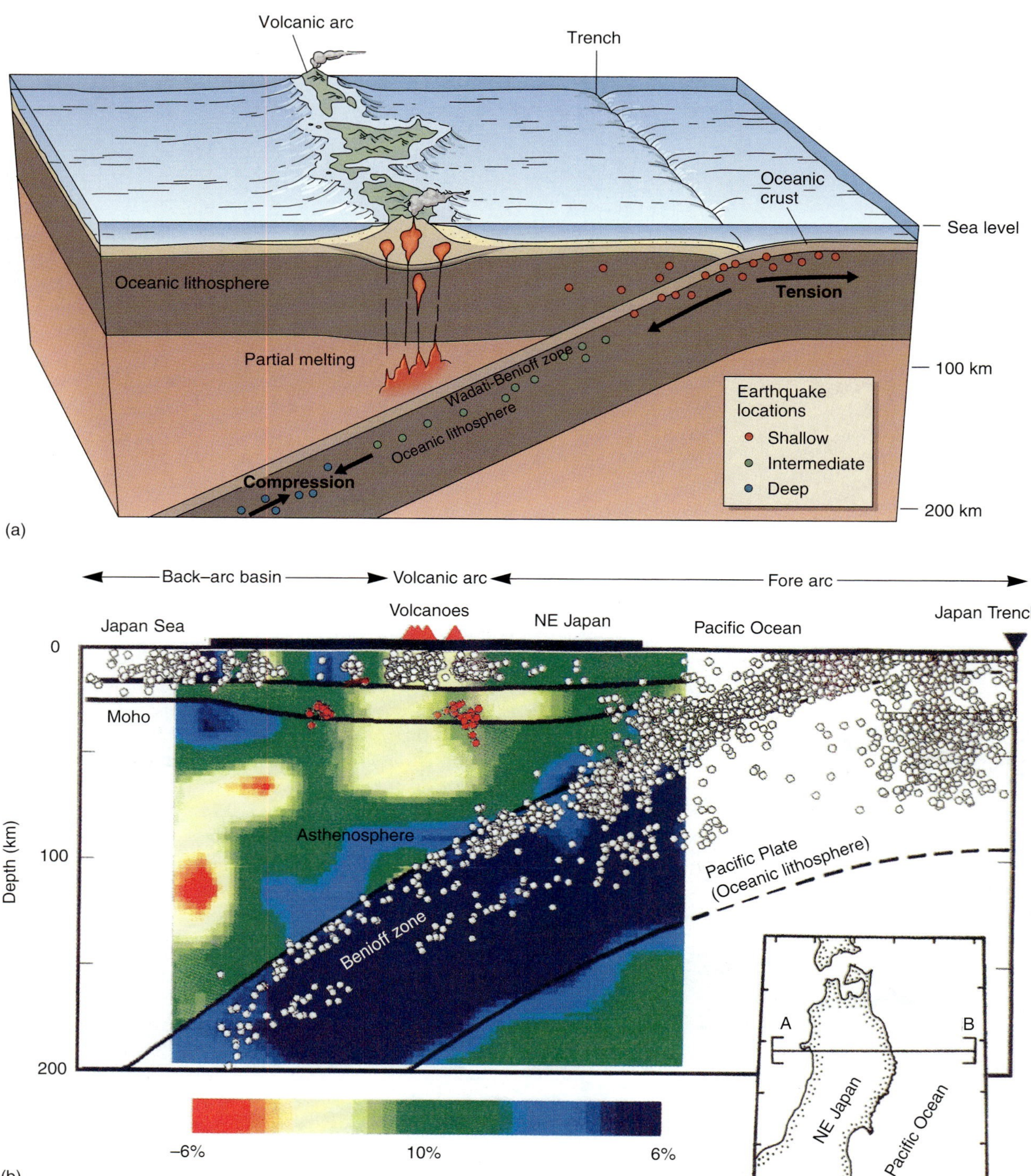

FIGURE 9.4 (a) Distribution of earthquakes in a subduction zone, which is related to zones of tension and compression in the subducted slab. At shallow depths, earthquake hypocenters are located near the top of the subducted lithosphere. At greater depths, earthquakes tend to occur within the subducted slab. Note that some earthquakes also occur in the upper plate, probably due to stresses generated by contact between the two plates. (b) Tomographic image of the subduction zone beneath Japan (see figure inset for location). A color-coded scale defines the percentage of deviation in P-wave velocity from an assumed average value. Blue colors indicate fast seismic velocities (velocities that are up to 6 percent faster than the assumed average), which occur in the slab because it is relatively cool and dense. Yellow and red colors indicate P-wave velocities that are slower than the assumed average, which occur particularly beneath volcanoes. This probably indicates partial melting of the mantle, which would slow the velocity of the seismic waves. Earthquake locations are marked by the open circles.

ly caused by tension on the upper surface of the slab as it bends downward. You can see for yourself how tension due to bending can cause cracking (which will result in earthquakes in the case of rocks) by simply holding a candy bar at both ends and slowly bending it until cracks appear.

At greater depths, earthquake hypocenters fall below the surface of the slab, probably because the upper part of the slab is being heated and behaves in a more ductile manner. Deep earthquakes can have very large magnitudes. They commonly result from compression rather than tension, which is most likely due to the subducted slab meeting resistant material as it pushes deeper into the mantle, perhaps influenced by phase transitions (∞ see Chapter 5). Real subduction zones are more complex than the version shown in Figure 9.4a. Detailed studies of the Japanese convergent margin, for instance, show earthquakes defining a "double" Wadati-Benioff zone, with hypocenters up to 200 km depth both at the surface of the slab and in the slab interior (Fig. 9.4b).

The Wadati-Benioff zone is the array of earthquake hypocenters in a subduction zone; it corresponds approximately to the upper surface of the subducted slab.

Geophysicists also use seismic waves to help determine the overall structure of a subduction zone; the seismic waves may be from any earthquake worldwide—they need only pass through the subduction zone. By recording the velocities of the waves arriving from many directions to many receivers, we can construct a seismic tomography image of the subduction zone (Fig. 9.4b) (∞ see Chapter 5). The colors in Figure 9.4b are used to define deviations of seismic P-wave velocity from an "average" mantle value. The "cold" colors (blues) indicate velocities greater than average, whereas "warm" colors (yellows and reds) are used to distinguish slower velocities. The subducted slab shows up as a region of higher seismic-wave velocities because it is cold and dense, and accordingly stronger than the surrounding mantle. In contrast, seismic waves slow down in the mantle beneath the volcanic arc. This is probably because the mantle is hot here and may include a little partial melt, which would greatly reduce the velocity of seismic waves passing through it.

It is clear from both Figure 9.4b and Chapter 5 that the velocities of seismic waves traveling through the mantle are strongly controlled by temperature. In cold mantle, seismic waves travel at higher velocities than they would through hot mantle. The temperature structure of a subduction zone is shown in Figure 9.5, page 232. You can see that the relatively cold temperatures in the slab correspond well to the faster seismic velocities implied in Figure 9.4. Why is the slab cold? It has been at Earth's surface—which is cold relative to the mantle. As it sinks, it does not heat up quickly enough to reach the same temperature as the surrounding mantle, and thus it forms a cold sheet penetrating the hot mantle. The cool temperatures in the slab not only mean that it will be dense—which is why it sinks—but also that it is unlikely to melt.

The heat flow across the subduction zone reflects the temperature structure (Fig. 9.5). Heat flow is low at the trench where the cold lithospheric plate is subducted and high over the volcanic arc where hot material—magma—is reaching the surface. The Bouguer gravity profile, in turn, shows high values in the fore-arc region because of the dense subducted slab, which lies just below the surface. Recall from Chapter 5 that the strength of the Bouguer gravity field at Earth's surface varies mainly as a function of the density of underlying material. The low Bouguer gravity field in the region of the trench (Fig. 9.5) is a little harder to understand. The main reason for the low gravity here is that the trench is not in isostatic equilibrium (∞ see Chapter 5). The lithosphere here is actually pulled down by the excess weight of the subducting slab and is at a lower topographic level than it really should be if the effects of buoyancy were balanced.

Magmatism at Convergent Margins

Volcanic arcs at convergent margins obey a rather consistent relationship with the geometry of the subduction zone. The upper surface of the subducted slab is typically located at a depth of 100 to150 km beneath the volcanoes (Figs. 9.1, 9.4b, 9.5). The width of the fore arc—the distance between the trench and the volcanic arc—therefore varies according to the dip of the subducted slab. If the slab dips steeply, the arc is close to the trench (50 to100 km); if the dip is shallow, the arc may be 200 to 300 km from the trench. Note that slab dips may vary through time so that the position of the arc on the upper plate might migrate over millions of years, reflecting changes in the dip of the subducting slab.

As explained earlier, we recognize two main types of volcanic arcs: island arcs, in which an oceanic plate is subducted beneath another oceanic plate and the tops of the volcanoes rise above the sea to form islands (Fig. 9.2), and continental arcs, in which an oceanic plate subducts beneath continental lithosphere forming volcanoes along the edge of the continent (see Focus 9.2). The islands of the Lesser Antilles on the eastern edge of the Caribbean Plate are an example of an island arc. The Andean volcanoes of South America and the Cascades volcanoes of North America are continental arcs. In some places, a volcanic arc lies astride both oceanic crust and continental crust. In the case of the Aleutians in Alaska, the western islands are built on oceanic crust, whereas volcanoes of the eastern end of the chain and the Alaskan mainland lie on continental crust.

Mechanism of Magmatism The systematic and almost universal association of volcanism with subduction zones links the process of subduction to magma generation.

Focus On 9.2 THE ANDES AND THE LESSER ANTILLES: A CONTINENTAL ARC AND AN OCEANIC ARC

Now that we have a basic understanding of the structure and magmatism characteristic of convergent margins, we will compare a "typical" continental arc and an oceanic arc.

THE ANDES

The Andes arc is a continental arc running along the western margin of the South American continent (Fig. 1). Here, the oceanic Nazca Plate subducts eastward beneath the South American Plate. The Peru–Chile Trench is almost devoid of sediments in the central part of the arc. The low sediment supply to the trench is probably due to the low erosion rate in the central Andes, which, in turn, is a result of the arid climate there. (The region includes the Atacama Desert, which is the driest desert on Earth.) Farther south, the climate is wetter and the southern Andes are glaciated, so the supply of sediment to the trench is greater. East of the Andes lies a trough of sediments referred to as a *foreland basin*. This has been formed by the weight of the mountains, which have elastically flexed the crust downward (∞ see Chapter 7). Sediment infill has accumulated to a depth of 6 km.

The Andes arc has been active for more than 300 million years, although the exact locations of the arc volcanoes have migrated with time. The peaks of the Andes—particularly in the central

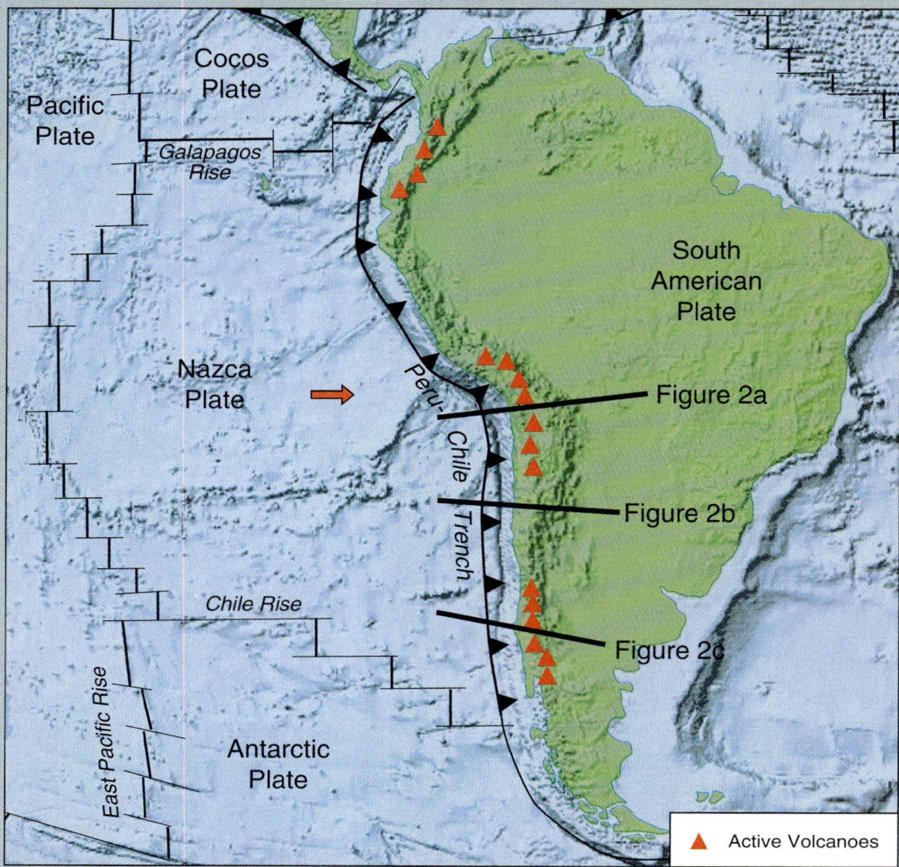

FIGURE 1 Plate-tectonic setting of the Andes arc. Triangles mark currently active volcanic zones.

part—include the highest volcanic summits on Earth, which are over 6000 m high. This extraordinary elevation results from unusually great crustal thicknesses—70 km or more in the central Andes. According to isostatic principles, the thicker the low-density crust, the higher it will stand (like an iceberg floating in water) (∞ see Chapter 5). The bases of many of the volcanoes are as much as 4000 m high; they are built on a vast, elevated plateau called the Altiplano (literally, "high plain").

How can the crust be so thick? Thick crust and high mountain ranges, such as the Himalayas and the Alps, typically occur where two continents have collided. This is not the case with the Andes, which are on the edge of a continent, bordering an ocean. The thick Andean crust could be the result of continuous addition of magma from the mantle, but melts from the mantle are generally basaltic—of high density—and are therefore not likely to form "light" crust.

More likely, the thick crust has been formed by compression and crustal shortening. How can we explain compression at a convergent margin? After all, arcs are often under tension and may split apart to form back-arc basins. It seems that changes in plate motions can lead to compression, which usually occurs when a subducted slab has a shallow or horizontal dip beneath the lithosphere of the opposing plate. The amount of contact between the two plates then increases and they push in opposite directions. The compressive forces eventually result in failure of the continental crust of the upper plate, which is weaker than oceanic lithosphere (Fig. 2). In the central Andes, compression took place some 20 million to 30 million years ago, uplifting the Altiplano. During the period of shallow subduction, magmatism ceased. Since then, the slab dip has increased and magmatism has resumed, building the spectacular volcanoes of the Andes. One model that has been suggested is that as the Atlantic Ocean opened, the ridge push that developed caused the South American plate to override the Nazca Plate more rapidly, causing compression at the margin.

THE LESSER ANTILLES

The Lesser Antilles arc is an oceanic arc developed where the Atlantic lithosphere sinks beneath the Caribbean Plate (Fig. 3). It is one of only two arcs in the Atlantic Ocean (the other being the South Sandwich Islands of the Scotia arc at the southern tip of South America). For the most part, the Atlantic Ocean has passive margins at its edges; in contrast, the Pacific is almost encircled by volcanic arcs, referred to as the "ring of fire."

Oceanic arcs typically have much shorter lifetimes than do continental arcs. The Lesser Antilles arc has been active for about 30 million years. The arc comprises a chain of volcanic islands with a very large accumulation of sediment to the east, which just rises above sea level to form the island of Barbados. West of the arc lies a small back-arc basin called the Grenada basin. It is bounded on its western edge by the largely submerged Aves Ridge, a remnant arc that split away from the Lesser Antilles arc as the Grenada basin began to open by back-arc spreading.

Along the Lesser Antilles arc, the islands consist of volcanic material, most of it pyroclastic. Generally, an individual island is not a single volcano rising above the sea, but rather a series of volcanoes that have been active at various times in the past. The most infamous volcano of the arc is Mount Pelée on the island of Martinique, which erupted in 1902, destroying the town of St. Pierre and killing an estimated 28,000 people. The eruption involved explosions in the summit crater, which destroyed a lava dome and generated pyroclastic flows. The inhabitants of St. Pierre were killed by suffocation or burning by the scorching hot gases associated with the pyroclastic flows. Mechanisms giving rise to the Andes and Lesser Antilles magmatic arcs are really the same: subduction. What is the difference? Why are the Andes volcanoes so high and the Antilles volcanoes just above sea level (in fact, some are still submerged)? The answer lies in the nature of the lithosphere on which the arc is developed. Recall the principles of isostasy (∞ see Chapter 5). The continental lithosphere of the Andes has very thick crust and has an overall low density; therefore, the upper surface is relatively

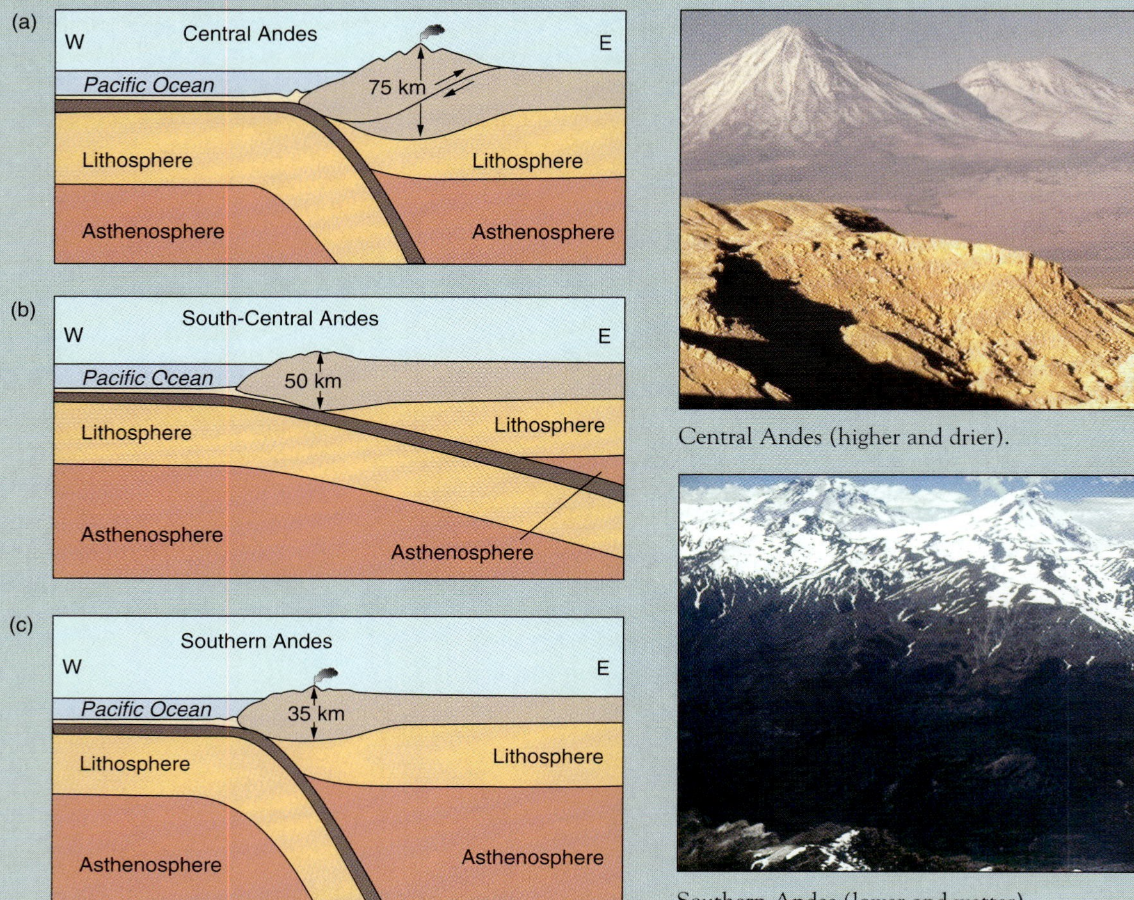

Central Andes (higher and drier).

Southern Andes (lower and wetter).

FIGURE 2 Cross sections of the Andes at different latitudes. Comparison among the three sections shows varying slab dip: about 30° in (a) and (c) and about 10° in (b). Note that there is no active volcanism where the slab dip is shallow (see Fig. 1). This kind of shallow subduction may cause compression in the upper plate, leading to thrusting and crustal thickening, which has happened farther north (a).

high. The oceanic lithosphere that underlies the Lesser Antilles has thin crust and is dense and low lying. As a result, it is submerged beneath the ocean and only the peaks of volcanoes are high enough to rise above sea level to form islands.

Early in the development of the theory of plate tectonics, it was thought that a subducted slab melts as it sinks down into the hot mantle, perhaps as a result of frictional heating. More recent calculations have shown that this explanation is unlikely. In fact, oceanic lithosphere is relatively cold when it sinks at a subduction zone (Fig. 9.5) and will take a long time to heat up to temperatures sufficient for it to start to melt. It is now thought that the mantle beneath the arc melts, rather than the slab. This is due to water being driven off from the subducted oceanic crust, which lowers the melting point of the mantle under the arc, causing it to melt. Chapter 4 describes how melting occurs in this fashion and also provides background on how mantle-derived melts may subsequently differentiate to produce a spectrum of both extrusive and intrusive igneous rocks in the arc.

The nature of the crust also has an effect on the types of volcanic and plutonic rocks characterizing the arc. Where the crust has a low density, such as beneath the Andes, it is more difficult for high-density basalts to reach the surface. Magmas tend to differentiate more until their density becomes sufficiently low to erupt at the surface (∞ see Chapter 4). Andesites, dacites, and rhyolites are common at continental arcs such as the Andes; they are less common—although by no means rare—at oceanic arcs such as the Lesser Antilles.

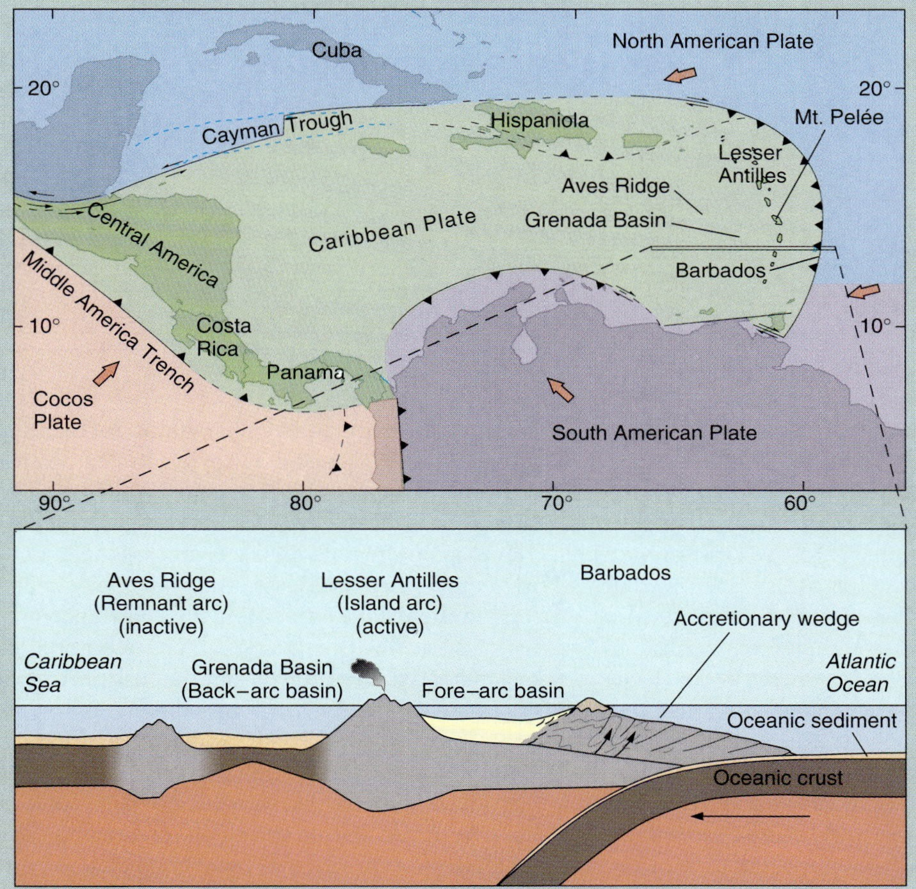

FIGURE 3 Plate-tectonic setting of the Lesser Antilles island arc, with a cross section of the arc.

Where the slab dip is very shallow, such as in parts of the Peru–Chile Trench (Focus 9.2), subduction may not be accompanied by volcanism. The reason for this is unclear, but it is most likely related to the more gradual release of water from the shallow slab into the cool lithosphere of the upper plate, rather than into the deeper, hotter asthenosphere, as is the case for a steeper-dipping slab.

Magmas at arcs are produced from the mantle by adding water from the subducting slab, which lowers the melting point of the overlying mantle and leads to melting.

Characteristics and Compositions of Magmas In tectonic environments other than convergent margins, such as divergent margins and intraplate settings, magma is

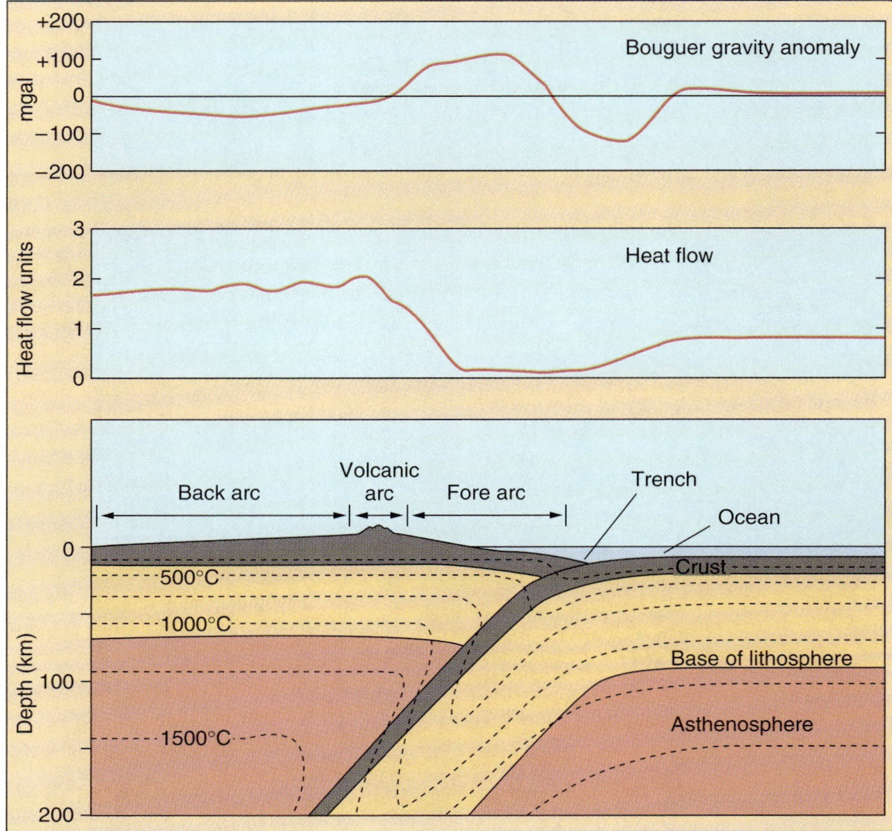

FIGURE 9.5 Schematic cross section through a subduction zone. Temperature structure is indicated with contours of equal temperature. The heat flow and Bouguer gravity profiles are shown in the upper panels.

dominantly basaltic in composition. This probably reflects the dominance of magmas that are derived directly as partial melts of the peridotitic mantle (∞ see Chapter 8).

The compositions of magmas erupted at convergent margin magmatic arcs or intruded into the crust are, on the whole, more differentiated. But studies suggest that the most primitive magmas generated at arcs are also melted from the mantle and are therefore basaltic in composition. Magmas at arcs must therefore be more prone to differentiation than are magmas in other tectonic settings. This is perhaps due to the structure of the arc environment, which impedes the ascent of magma.

Melts accumulate in magma chambers and differentiate until they become less dense and can rise farther. At continental margin arcs in particular, magma cannot rise easily through the thick, low-density continental crust. To rise, magmas must decrease in density and become buoyant relative to the crust. This is achieved by differentiation, in which dense minerals crystallize to leave a more buoyant liquid (see Fig. 4.5). In contrast, magmas can ascend easily at mid-ocean ridges (divergent margins) where the crust is basaltic, thin and relatively dense.

Intrusive Rocks: The Origin of Granites Batholiths are spectacular manifestations of the huge quantities of differentiated rocks generated at magmatic arcs (see Fig. 4.6a). As we already learned (∞ see Chapter 4), they are actually composed of many individual plutons, such as those shown as part of the magmatic arc in Figure 9.1. Although they are loosely termed *granites* in composition, most batholiths, such as those of the Peruvian Andes and California's Sierra Nevada, are generally in the compositional range of diorite to granodiorite (see Figs. 4.2, 4.6). The magmas that formed these batholiths could have been generated in two main ways: by the fractional crystallization of basaltic magma originally derived from the mantle, or by the melting of crustal material such as sedimentary rocks.

In the first case, a basalt derived by melting of the mantle would need to undergo a good deal of crystallization. Crystals of minerals like olivine, pyroxene, and plagioclase must be removed from the basalt, presumably by settling to the bottom of the magma chamber (see Fig. 4.3). More than twice the volume of the granite would have to be removed as gabbro—the rock that would form by accumulating olivine, pyroxene, and plagioclase. We would expect, therefore, to find large bodies of gabbro below granite batholiths. Gabbros are indeed found associated with many granite bodies but generally not in the volumes needed to have generated granites by fractional crystallization of basalt.

In the second case, if granites were formed by partial melting of rocks that already existed in the crust, we might expect to see rocks that represent such melting. In fact, there are metamorphic rocks called *migmatites* that show evidence for partial melting (∞ see Chapter 4).

Also, xenoliths of crustal material may be found in granites, which again indicates that the crustal rocks may melt to form granites. Although granites might form either by basalt fractionation or by crustal melting, in most cases a combination of the two has probably occurred. Basalts rising into the crust will tend to undergo fractional crystallization as they cool. At the same time, the heat provided by the basalts may cause melting of the crustal rocks. The resulting magma, which eventually intrudes the crust as granite, will then be a mixture of differentiated mantle-derived magma and crustal melt.

Granite batholiths may be produced by fractional crystallization of basalts, by melting of continental crustal material, or by a combination of these processes.

Extrusive Rocks: Pyroclastic Deposits Many of the igneous rock types that form at convergent margins, such as andesites and dacites, were described earlier (∞ see Chapter 4) and will not be discussed again here. Convergent margin volcanism, however, commonly occurs as explosive types of eruptions, and we focus on the volcanic products of such eruptions in this chapter.

In terms of the total volume of erupted material, many major eruptions for which we have some geologic record have been associated with subduction and have been pyroclastic in nature (Fig. 9.6a). The recent eruptions of Mount St. Helens in 1980 (Focus 9.3) and Mount Pinatubo in the Philippines in 1991 are typical examples. In the past 200 years, large eruptions have occurred at Tambora and Krakatoa in Indonesia (1815 and 1883, respectively) and at Katmai, Alaska (creating the "Valley of Ten Thousand Smokes," 1912). These eruptions produced up to 30 km^3 of ash and other pyroclastic deposits (Fig. 9.6a) These volumes, however, pale in comparison with the hundreds to thousands of cubic kilometers erupted in places such as the western United States and the Andes in the geologic past. No eruptions of this enormous size have been recorded in human history (Fig. 9.6b, c), so we can only speculate as to the impact of such an event.

Although pyroclastic eruptions can also occur at divergent plate margins or intraplate settings, they are much less frequent in these locations. Magmas at volcanic arcs tend to be rich in dissolved gases, and they tend to be viscous because of their high silica content. As a result, explosive pyroclastic eruptions are common (∞ see Chapter 4).

Pyroclastic eruptions can vary in style and magnitude (Fig. 9.7), and they can produce deposits that are described and classified both in terms of the size and type of fragment (clast) and in terms of the way in which the deposit was formed. Pyroclastic materials range in size from large blocks or **volcanic bombs** (10 cm or larger) that are ejected into the air (Fig. 9.8a) to smaller, pebble-sized fragments to fine-grained ash (less than 2 mm). The dissolved gases responsible for such explosive eruptions form bubbles in the magma as it erupts. The bubbles are preserved in pyroclastic rocks such as pumice (light colored) and cinders (dark colored; Fig. 9.8b) and are called *vesicles*. Pumice contains so many air bubbles that it can float on water. An interesting consequence of this property is that pumice may act as an effective raft to transport living organisms over large bodies of water. Pumice fragments from the 1883 eruption of Krakatoa floated across the Indian Ocean from Indonesia to Africa, carrying a number of insects, small animals, and plants.

Pyroclastic deposits, which are generally termed **tuffs** when lithified, are classified by the way in which they were formed. The two main types are falls and flows. Figure 9.7 shows examples of both being formed. **Pyroclastic fall** deposits form when pyroclastic material is ejected into the air and falls back to the ground. Figure 9.7a shows scoria and ash ejected from a volocano, and falling back around the vent to build a cinder cone. The finest material can reach high into the stratosphere in large eruptions—over 40,000 m—and can be carried long distances in the direction of the prevailing winds. The finest ash can circumnavigate the globe several times, as atmospheric dust. Larger-sized pieces fall back to the ground closer to the volcanic vent. The overall effect is that pyroclastic fall deposits tend to be well sorted. Layers of fall deposits can be spread over a wide area and cover the topography almost like snow (Fig. 9.8c). The main dangers from ashfalls are the possibility of suffocation and burial by the ash and damage to buildings owing to the weight of accumulated ash. Ashfall deposits from ancient eruptions can be traced over large areas. Because they are deposited very quickly, ashfall layers are useful time markers in the geologic record.

Pyroclastic flows commonly form when the material from a rising ash-eruption column suddenly collapses and falls to the ground. The eruption column is supported by gas thrust at the vent (rather like a jet engine) and by the buoyancy of hot gas in the column. Collapse can occur if the exit velocity at the vent is reduced during the eruption, owing to perhaps widening of the vent. Figure 9.7b shows a pyroclastic flow forming down the flank of a volcano while the eruption is still producing a vertical column of ash. The flows follow preexisting valleys as they travel downhill and form poorly sorted deposits (Fig. 9.8d). They are highly fluid and can travel at very high speeds (>100 km per hour), because the particles in the flow are lubricated by an intervening layer of hot gas. The heat and speed of these flows make them particularly hazardous, as was evident from the catastrophic eruption of Mt. Pelée, Martinique, in 1902 (Focus 9.2).

Explosive eruptions produce two main types of pyroclastic deposits: falls and flows. Fall deposits accumulate as erupted material falls back to the ground. Flow deposits form when ejected material accumulates rapidly and flows downhill.

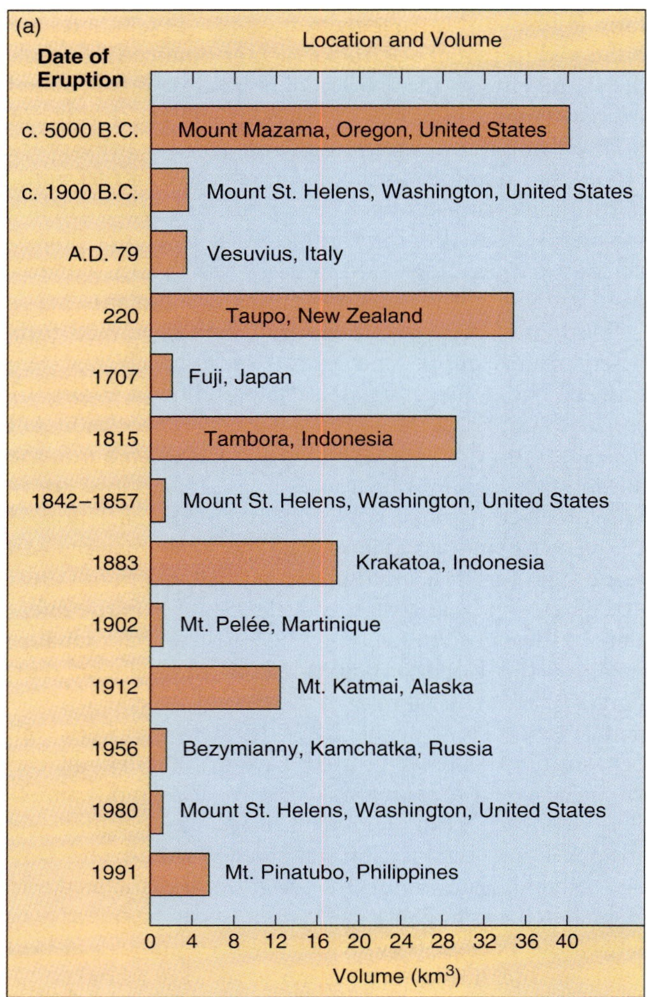

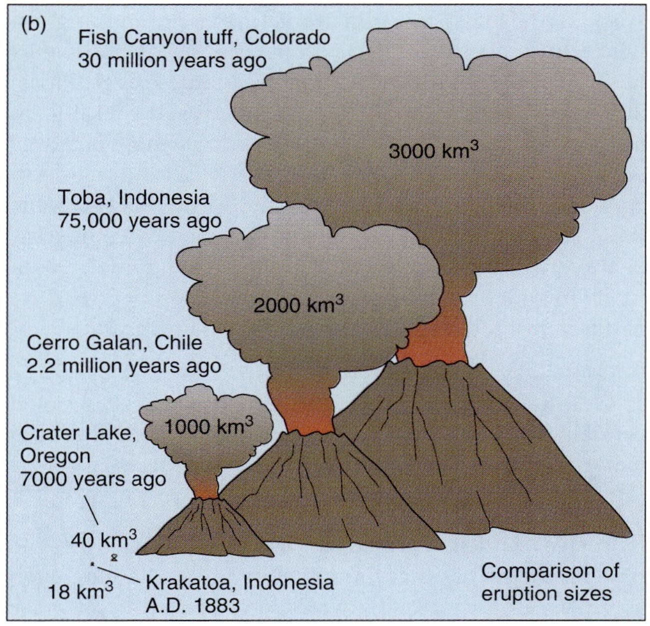

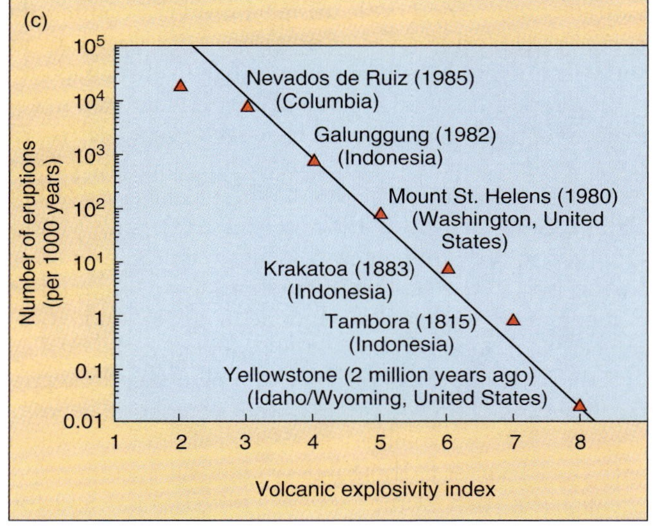

FIGURE 9.6 (a) Approximate magnitudes of some famous volcanic eruptions, which are represented by the volume of material erupted. Although impressive, the 1980 Mount St. Helens eruption is small compared to some eruptions in the past. (b) Relative sizes of some eruptions in the past. The size of the sketched volcano is proportional to the estimated volume of erupted material, which is given in cubic kilometers. Mount Mazama (Crater Lake) and Krakatoa are shown for comparison, with the much smaller sizes of historical eruptions shown in (a). (c) Relationship between the eruption size and the frequency of occurrence. The size of the eruption is given as the "Volcano Explosivity Index" and is roughly proportional to the amount of pyroclastic material erupted. Just as with earthquakes, small volcanic eruptions occur frequently (several per year), whereas the largest events are much more rare.

Many large-volume eruptions, such as those at Crater Lake in Oregon and Lake Taupo in New Zealand, have produced widespread ash-flow tuffs or **ignimbrites**. An ignimbrite is a pyroclastic flow formed of ash and pumice. The amount of pyroclastic material in the largest ignimbrites is too great to be constrained in valleys and spreads out to cover wide areas. The heat and weight of these pumice and ash deposits typically cause them to compact and fuse together to form a hard rock called a **welded tuff**. The effect of welding can be seen where the pumice fragments become squashed into flat plates of dark glass (Fig. 9.8e). The eruption of such a large volume of material removes the support from underneath the volcano, which may collapse in a broad circular depression or **caldera.** The famous and aptly named Crater Lake (Fig. 9.9) formed in this way.

At certain times in the geologic past, continental arcs have produced huge volumes of ignimbrites from large volcanic complexes. A very large magma chamber is required to erupt such large volumes (Fig. 9.6b), and it is possible that the root zones of large caldera complexes are the plutonic and batholithic complexes exposed in eroded continental arcs. Hundreds of thousands of cubic kilometers of ignimbrites found in the western United States

(a)

(b)

FIGURE 9.7 (a) A volcanic eruption producing pyroclastic fall at Tolbachik, Kamchatka (Russia). The deposits from such an eruption may resemble Figure 9.8a and b. (b) Volcanic eruption producing pyroclastic flow at Ngaurohoe, New Zealand, in 1975. This flow reached speeds of up to 100 km per hour. The deposits from such an eruption look like those shown in Figure 9.8d. This flow was accompanied by a large ash plume rising vertically from the volcano, which produced simultaneous ashfall deposits.

(a) (b) (c) (d) (e)

FIGURE 9.8 Types of pyroclastic rock: (a) volcanic bomb (Mojave Desert, California); (b) pumice (light colored) and cinders (dark); (c) pyroclastic fall deposits from a volcanic eruption, showing typical layering (Chile); (d) pyroclastic flow deposits comprising poorly sorted pumice and ash, which have been easily eroded by the stream shown here. (Valley of Ten Thousand Smokes, Alaska); (e) welded tuff, where the black, glassy streaks are flattened pumice fragments.

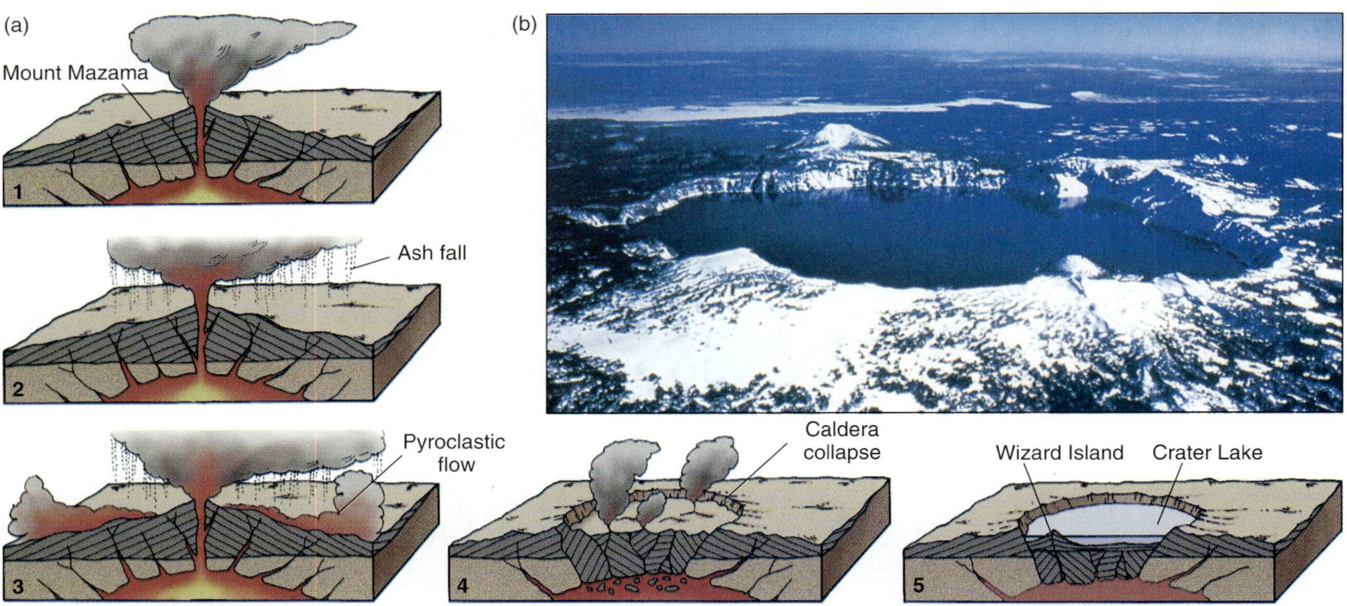

FIGURE 9.9 (a) The successive stages of development of Crater Lake, Oregon. About 7000 years ago, Mount Mazama erupted violently, partially emptying the underlying magma chamber. The summit of the volcano collapsed into the empty space left in the magma chamber, creating a large caldera, which was subsequently filled with water from rain and snow, to form Crater Lake. (b) Crater Lake today. The small island, called Wizard Island, is a cinder cone formed by later eruptions in the caldera.

were produced in eruptions that occurred between 40 million and 20 million years ago. Individual ignimbrite sheets each comprise volumes up to 3000 km^3 in the case of the Fish Canyon Tuff (Fig. 9.6b) and are distributed over thousands of square kilometers. In the last 20 million years, ignimbrites have been produced in the high Andes, and more "normal" volcanic activity has returned to the region only in the last 5 million years. More recently, enormous eruptions at convergent margins include Cerro Galan caldera, Chile (2.2 million years ago) and Toba caldera in Sumatra (75,000 years ago; Fig. 9.6b).

A process known as **sector collapse** has only recently been recognized as a common volcanic phenomenon associated with large stratovolcanoes at volcanic arcs. A sec-

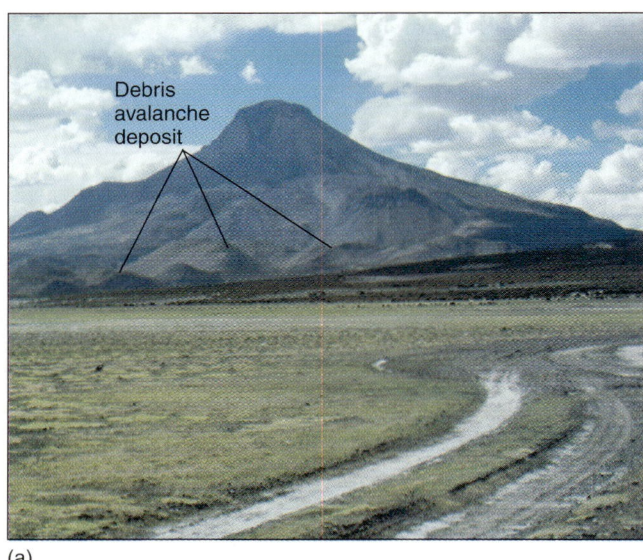

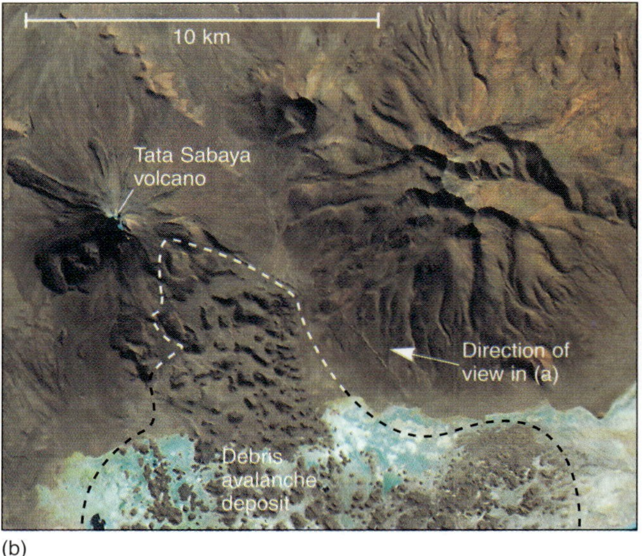

FIGURE 9.10 Debris avalanche deposits form when a volcano partially collapses. (a) A southeastern view of Tata Sabaya volcano in Bolivia. The hummocky material in the foreground is the debris avalanche deposit formed when the southern sector of the volcano collapsed southward (to the left). (b) A satellite image of Tata Sabaya. The southern part of the cone has collapsed as a debris avalanche that flowed southward into the inland lake and can be seen as a blotchy or spotty area. The dashed line outlines the extent of the deposit; the arrow shows the direction of view for the photograph shown in (a). (Scale is in kilometers.)

tor collapse is a giant landslide triggered by gravitational instability or by a volcanic eruption. The extensive deposit formed by a sector collapse, called a *debris avalanche*, represents volcanic rock that previously constituted the volcanic cone but became unstable on the steep slopes and failed in the collapse. Debris avalanche deposits have a characteristic "hummocky" appearance (Fig. 9.10)—each hummock is a huge block of material from the volcano that often has little internal disturbance. The first observed sector collapse occurred at the start of the Mount St. Helens eruption in 1980 (Focus 9.3). Recent satellite observations of thousands of remote volcanoes in the Andes have shown that sector collapse might be a typical stage in the life cycles of these large stratocones. The possibility of such collapse makes debris avalanches one more important element that should be considered in the assessment of volcanic hazards (∞ see Chapter 15).

The common occurrence of debris avalanches at arc volcanoes, and not at volcanoes from divergent margins and intraplate settings, is a reflection of the distinct compositions of magmas at convergent margins. Convergent margin lavas tend to be more differentiated (more silica-rich) and thus more viscous, which, in turn, builds volcanoes with steeper sides that are more prone to gravitational instability and to explosive eruptions.

Frequently, convergent margin volcanoes are highly explosive and dangerous. Huge eruptions of more than 1000 km^3 of ash are known to have occurred at volcanic arcs.

Sedimentation in Arc Environments

There is commonly an abundant supply of sediment at volcanic arcs. The sediments accumulate in the trench, where plate convergence forms a characteristic structure known as an **accretionary wedge** (Fig. 9.11). The structure of an accretionary wedge is not found in any other tectonic environment, and it is a direct reflection of the processes occurring at convergent plate margins.

Types of Sediment

Sediment is supplied to an accretionary wedge from both the volcanic arc and the subducting plate. Sediment derived from the arc is volcanic material accompanied by sediments eroded from other igneous, sedimentary, and metamorphic rocks that may make up the basement of the arc. In particular, the fragmented pyroclastic material that is erupted at arcs erodes much more easily than do solid lavas. Sediment from the arc is transported to the sea by rivers, wind, or glaciers. It may accumulate in the fore-arc region (Figs. 9.1, 9.11) or in the trench. The sediment on the subducting plate is typically fine-grained pelagic (deep sea) muds, deposited on top of the igneous oceanic crust far out to sea. Pelagic muds consist mainly of microscopic plankton skeletons and fine, windblown

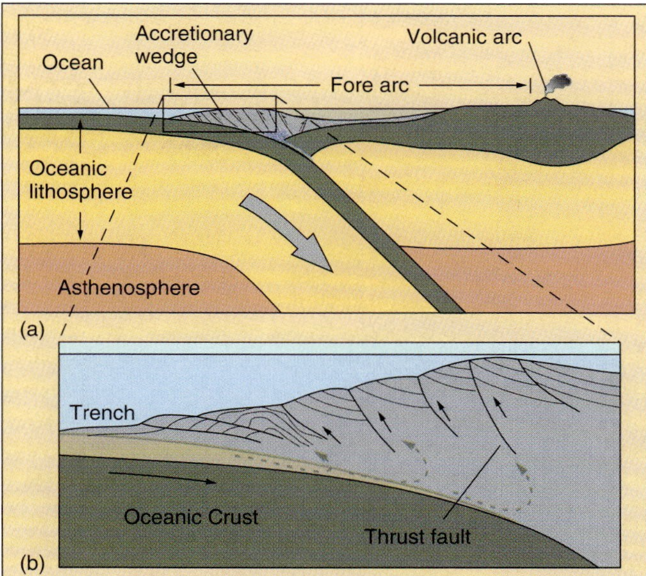

FIGURE 9.11 Structure and setting of an accretionary wedge. (a) The general setting of the wedge is shown relative to a convergent plate margin. (b) An enlargement of the accretionary wedge shows the structure with thrust faults, which become progressively younger toward the bottom of the pile (nearest to the trench). Subducted sediment becomes trapped in the wedge and mixed into the mélange. The relative motion of the sediments is indicated here by the dashed arrows; it is carried down by the incoming plate, then pushed up into the wedge by thrust faults. The mechanics of the wedge are analogous to a bulldozer pushing on a pile of loose sand.

dust, which are carried in conveyor-belt fashion toward the subduction zone. There, they can be scraped off by the upper plate (Fig. 9.11).

We discussed the concept of *maturity* in sediments earlier (∞ see Chapter 4). Sediment maturity will vary with the plate tectonic setting of sedimentation. In the case of convergent margins, sediments are largely immature, comprising sandstones and mudstones containing clay minerals and volcanic rock fragments. These rocks are known as **greywackes**. The immaturity of greywackes is a reflection of two characteristics: the source of the sediment, which is mainly volcanic material that decomposes readily to clay minerals; and the short transport distances from the arc (where most of the detritus is generated and where the highest mountains are located) to the trench (the site of deposition). This brief transport limits the degree of sorting and rounding experienced by the sediment.

The sediment supply at continental arcs may be greater than that at island arcs simply because more land is exposed, allowing for more erosion. Also, a greater portion of the sediment at continental arcs will originate from eroded sedimentary and metamorphic sources. Remember, though, that weathering and erosion are also controlled by geographic factors such as climate (∞ see Chapter 12), which will vary with latitude and may differ between island arcs and continental arcs (e.g., Focus 9.2, Fig. 2).

Focus On 9.3

THE ERUPTION OF MOUNT ST. HELENS

At 8:30 A.M. on May 18, 1980, geologist David Johnston of the United States Geological Survey was stationed on a ridge 10 km from Mount St. Helens volcano in western Washington (Fig. 1). Johnston had just finished using a laser instrument to make some routine measurements of the volcano's shape. Over the previous few weeks, such measurements had shown that a conspicuous bulge was growing on the north side of the volcano, but there were no indications on that particular morning that the volcano was behaving differently. Nothing prepared Dr. Johnston—or indeed the entire Pacific Northwest—for the events that soon followed.

At 8:32 A.M., the bulge on the north side of Mount St. Helens collapsed. An enormous debris avalanche was triggered as the side of the mountain began to move downslope, accompanied by huge volcanic explosions (Fig. 2). Geologist Johnston was one of 60 people killed by the eruption—the first significant volcanic eruption in the contiguous United States since Mount Lassen in northern California (1914–1917). The violent and sudden sequence of events on that morning in May was pieced together later from records and eyewitness accounts.

The debris avalanche at Mount St. Helens was apparently triggered by a magnitude 5.1 earthquake. The "bulge"—until then growing at an astonishing rate of up to 1.5 m per day—was interpreted as deformation due to magma moving into the cone. The removal of such a large mass from the volcano by the sector collapse probably relieved the pressure on the magma that was being injected into the cone. As a result, the magma outgassed rapidly and explosively. The first explosion was a lateral blast

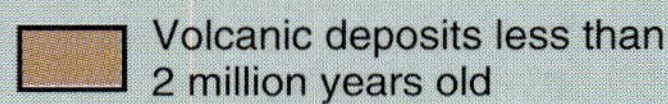

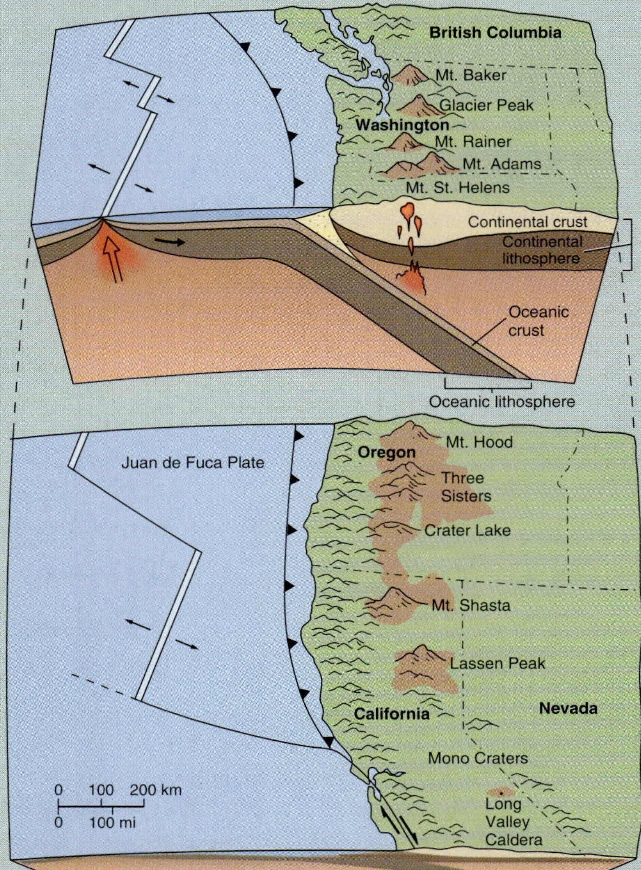

FIGURE 1 Map of the Cascades arc showing Mount St. Helens and other volcanoes relative to the plate-tectonic configuration of the Pacific Northwest

that blew down trees over a distance of about 8 km north of the volcano in the direction of the debris avalanche and probably led to the death of the survey geologist. Seconds later, explosive activity was directed vertically, giving rise to a large eruption column characterized by a plume of ash forced upward by the energy of the blast. This ash plume formed a mushroom-shaped cloud up to 20 km high.

Ash from the eruption was blown downwind (to the east) and eventually settled back to the ground. Thin ash deposits were found as far east as Nebraska and Minnesota. Although the huge eruption column could be seen clearly from the west, little ash fell there. This observation underscores the strong influence that wind direction plays in determining the areas of ash deposition from a volcanic eruption.

In contrast to the ash that blanketed a wide area, a large volume of volcanic material was deposited as flows, which were channeled along preexisting river valleys. Pyroclastic flows were formed either by ash and pumice falling back onto the steep volcanic slopes and then flowing downhill or by pumice

FIGURE 2 Sequence of photographs showing the violent lateral eruption on May 18, 1980, at Mount St. Helens in Washington. (a) The preeruption view of the bulging north flank. These next frames represent only about 30 seconds of real time. (b) A huge earthquake-triggered landslide moving much of the volcano flank. (c) At this time, the landslide unloaded sufficient weight off the side of the mountain that the pressurized magma explodes into a dark billowing ash cloud upward and a lateral blast. (d) The ash clouds continues to build and the lateral blast is beginning to move away from the mountain.

and ash frothing over the sides of the vent. Mudflows, composed of mixtures of volcanic material with water from streams or snowmelt, were much more extensive, traveling over 30 km from the volcano. On the steep upper slopes of the volcano, mudflows traveled at speeds of up to 150 km per hour.

The eruption had subsided by early the following day, but the shape of the volcano had changed dramatically. The once-symmetrical, snow-capped cone that had been mirrored in Spirit Lake was replaced by a gaping crater up to 3 km wide and 700 m deep. The summit of the mountain had been reduced by 400 m. Spirit Lake was choked with volcanic debris and uprooted trees.

At the bottom of the crater, magma now squeezed slowly and gently to the surface and began forming a dome. The new dome has grown sporadically, accompanied by variations in activity, which have included numerous small eruptions. The dome might continue to grow slowly and eventually fill the entire crater. Alternatively, the dome might collapse or explode at some stage, but, for the moment, Mount St. Helens continues to present a threat.

The May 18 eruption was not without some warning. Over the past few hundred years, native Americans and settlers have witnessed occasional eruptions. The most recent period of intense activity prior to the 1980 eruption was in the mid-1800s. Since then, the volcano had been tranquil. Nevertheless, detailed mapping and dating of past volcanic deposits indicated that activity at Mount St. Helens had occurred in repeated cycles. This led geologists of the United States Geological Survey to warn that Mount St. Helens might erupt before the end of this century. In March 1980, the onset of earthquake activity at Mount St. Helens signaled the end of the volcano's dormancy. In the next few days, the number of small-magnitude earthquakes increased.

In April, seismographs began to record "harmonic tremor"; a rhythmic ground shaking that is distinct from the typical jolting movement of most earthquakes. These tremors were interpreted to be the result of subsurface movement of magma (Chapter 7) that was also responsible for the growth of the bulge in the volcano's flanks. On March 27, and continuing sporadically until April 21, Mount St. Helens erupted explosively, producing columns of ash and steam up to 2000 m above the volcano. The ash was composed of material from the existing cone that had been fragmented and pulverized as a result of the steam explosions. Scientists studying the behavior of the volcano recommended to civil authorities that the immediate area be evacuated. These actions prevented the loss of life being greater than it was.

Victims of the eruption died mainly from asphyxiation due to inhalation of hot volcanic ash. Apart from the loss of human life, the effects of the eruption on the region were widespread. Much wildlife was destroyed, although some small burrowing animals and aquatic animals did survive the eruption. The lateral blast and mudflows that followed destroyed more than 200 houses and cabins. Thousands of acres of forest were devastated. Surprisingly, some small trees survived in the blast zone, because they were buried by deep snow that protected them. In areas of thick ash cover, agricultural crops were destroyed, although the ash has ultimately proved beneficial in providing new mineral nutrients to the local soils. The ashfall disrupted air and land transportation systems and water-treatment plants. Mudflows blocked navigation on the Columbia River and endangered the salmon fishery. Ash clogged machinery, blocked oil filters, and short-circuited electrical equipment. The damage and destruction of the 1980 eruption is estimated to have cost some $1.1 billion. The short-term impact on tourism in the region was a considerable loss; over the long term, however, the volcano itself may prove to be a great tourist attraction. In fact, it was declared a national monument in 1982.

The eruption served as a timely reminder that volcanic activity still poses a threat to some areas of the United States. The scientific information gathered at Mount St. Helens has proved an invaluable contribution to worldwide volcanic hazard monitoring. Finally, it is worth remembering that the 1980 Mount St. Helens eruption was extremely small compared with volcanic events that have occurred in the geologic past and are represented today by their deposits (see Fig. 9.6).

The Formation of an Accretionary Wedge

We can think of the formation of an accretionary wedge at a subduction zone in a manner similar to the action of a bulldozer. At the trench, the sediment is scraped off the sinking plate and it piles up there. By analogy, the bulldozer represents the upper plate, and the sediment piles up in front of it.

Continuing subduction pushes additional sediment underneath the accumulating pile. The sedimentary mass is progressively lifted up and increases in volume, forming the characteristic wedge shape (Fig. 9.11). Within the wedge itself, material is carried down short distances and then pushed back upward on thrust faults, as indicated in Figure 9.11. In some cases, however, such sedimentary material can be carried down to substantial depths and by this mechanism be overridden by hot asthenosphere.

Sediments in arc environments tend to accumulate in an accretionary wedge at the trench.

Depending upon a number of factors, including the amount of sediment available, extensive accretionary wedges may develop. At some subduction zones, though, very little sediment accumulates, either because not much is available (continental sediment sources are far away) or because it is subducted beneath the arc rather than scraped off at the trench.

Accretionary Wedges and Subduction Zone Dynamics In some accretionary wedges, so much sediment accumulates that the pile rises above sea level. The Caribbean island of Barbados is the tip of an accretionary wedge formed at the Lesser Antilles subduction zone (see Focus 9.2). Subduction rates are slow—approximately 20 km per million years—and the subducting plate has carried a thick sequence of muddy sediments to the trench. These sediments were originally deposited far from shore and far from the volcanic influence of the Lesser Antilles arc. Because of the slow subduction rate, the trench has been infilled with sediment. The abyssal plain rises along the deformation front of the accretionary wedge (Fig. 3 of Focus 9.2). The sediment pile extends more than 5000 m in thickness at the trench. It appears that approximately 85 percent of the sediment reaching the subduction inlet is scraped off the lower plate by the action of one plate against the other, giving us a rare opportunity to study ocean-floor sediments at the surface, which would otherwise be subducted. The remaining 15 percent travels down into the subduction zone. Within the accretionary wedge, low-angle faults separate a lower, strongly folded and faulted succession of sediments from an upper zone that exhibits less deformation.

If subduction rates are faster, the sedimentation may be considerably different. The accretionary wedge at Kodiak Island in Alaska is different from that in Barbados in every respect, including its rate of subduction, sediment thickness, type of sediment, and amount of sedimentary material subducted. Kodiak is located just east of the Aleutian Islands and immediately south of the Aleutian range, which is the eastern terminus of the volcanic arc. The greater subduction rate, approximately three times that of the Lesser Antilles, has resulted in a clearly delineated trench. Sediments within the trench are primarily sand transported from the nearby arc, and the sequence is approximately 1500 m thick. Calculations suggest that about 60 percent of the sediment carried to the subduction zone is scraped off at the inlet and the remaining 40 percent is subducted.

The mechanics of subduction determine how much sediment ultimately accumulates at the trench, and how much is subducted to greater depths.

Deformation and Metamorphism in the Accretionary Wedge Deformation and squeezing of some of the wet sedimentary material cause some parts of the accretionary wedge to be very fluid-rich and low in density. As a result, buoyant forces cause the intrusive rise of mud diapirs within the wedge. In Barbados, for example, rapid, geologically recent uplift has been attributed to the intrusion of large quantities of diapiric material within and through the accretionary wedge.

Temperatures within subduction zones are anomalously low because of the cooling effect of the subducted oceanic crust (see Fig. 9.5). Given that the slab penetrates hot mantle, the subducted succession will alter to some moderately high-grade metamorphic rocks. But, in the core of the subducted sediment package, protected and insulated from contact with the hot asthenosphere, there is probably a sequence of much less altered, relatively buoyant rocks capable of flowing back out of the subduction zone (counterflow). This counterflow process may result in complex mechanical mixing. Within the portion of the accretionary prism that becomes stacked up at the trench, deformation can be so extreme that the sedimentary material becomes pervasively sheared and folded on every scale. Subduction **mélange** (from the French word for *mixture*) is the name given to strongly deformed and usually jumbled sedimentary materials that have accumulated within the accretionary wedge. The mass of sediment effectively rolls around, caught between the grip of the downgoing slab and the buttress of the overlying plate. You can see by the dashed arrows in Figure 9.11 how sediment can be mixed together by the dynamics of subduction. The resulting mélange consists of tectonically mixed assemblages of lower- and higher-grade metamorphic rocks embedded within a matrix of pervasively sheared, slightly metamorphosed mudstones (Fig. 9.12).

Dynamic tectonic forces at the trench cause deformation and mixing of sediments to form mélanges.

FIGURE 9.12 Typical mélange from western California where blocks of different rock types that vary from pebble-sized to kilometer-sized are mixed together in a chaotic jumble. Here, large, resistant blocks measuring several meters across are surrounded by a softer, sheared, muddy matrix.

Continental Collisions: Plate Suturing

The low density of lithosphere containing continental crust prevents it from being subducted. It is isostatically buoyant and resists subduction—somewhat like a cork that would bob up if you tried to push it below the surface in a bowl of water.

Two important consequences result from this buoyancy. First, unlike oceanic crust, once continental crust has been generated, it is here to stay (although some fraction of the continental crust is undoubtedly eroded, deposited on the ocean floor as sediment, and then subducted). Because oceanic crust continually recycles, its average age is only about 100 million years; the average age of the continental crust is about 2 billion years. Fragments of continental crust have been found that are as old as 4 billion years—80 percent of the age of Earth! Second, when two plates of continental lithosphere converge, they will eventually be crushed together, or sutured. The crust folds and fractures in response to the huge forces involved in the collision, and a mountain belt forms along the line of collision.

Oceanic lithosphere is relatively dense and tends to subduct, whereas continental lithosphere is light and buoyant; the average age of oceanic crust (about 100 million years) is thus much younger than the average age of continental crust (2 billion years).

The Rise and Fall of Mountain Ranges

The growth of mountain belts at convergent plate margins—in response to both magmatic additions at the volcanic arc and the deformation associated with lithospheric plates converging and colliding—is known as **orogenesis**. The elongate tracts of igneous rocks and deformational structures formed at convergent margins are known as *orogens* (Fig. 9.13). The compressional forces involved when two continents collide can give rise to significant vertical transport of rocks. This explains how, in the collision of India and Asia that formed the Himalayas, rocks originally deposited on the ocean floor containing fossils of seashells came to be exposed near the top of Mount Everest, some 9 km above sea level! Careful measurements show that the Himalayas are still being uplifted at a rate of about 1 cm annually.

More than 150 million years ago, India was part of a larger continent that included Africa, Antarctica, Australia, and South America. This supercontinent eventually broke apart along a number of rifts (∞ see Chapter 8), and the plates carrying the continental fragments that we are familiar with today began to move relative to each other.

India traveled northward toward Asia—convergence being made possible by subduction along southern Asia and possibly at a number of island arcs in the intervening ocean (Fig. 9.14). Subduction beneath Asia probably gave rise to a huge continental volcanic arc, much like the Andes today. The northern edge of India was a passive continental margin. As the two continents approached each other, sediments accumulated in the intervening ocean. When all of the oceanic lithosphere had been subducted, the continental lithospheres of India and Asia came into contact. As the ocean finally closed, the sediments, gripped in a vise between the two continents, were squeezed and folded. We can view the collision as being rather like a hard piston ram (India) being pushed into a block of softer, weaker material (Asia), forcing the latter upward and forming the Tibetan Plateau (∞ see Chapter 11; Fig. 11.13). This plateau is one of the highest continuous tracts of land on Earth at an elevation of over 4000 m. Note that the extreme elevation of the Himalayas is a direct reflection of the crustal thickening that occurred in response to continental collision (Focus 9.4).

Some of the Asian crust caught in the collision was squeezed out sideways, displaced to the east and west along strike-slip faults. Many of these faults are still active and give rise to very damaging earthquakes in Asia (∞ see Chapter 11). The crust forced up in the Himalayan range was shortened by folding and faulting to accommodate the convergent movement. Huge slices of this crust were transported tens to hundreds of kilometers southward over the advancing Indian subcontinent.

A spectacular section through the Himalayas is provided by the 3000 m deep Indus River gorge (Fig. 9.15a). The course of the Indus River shows little regard for the

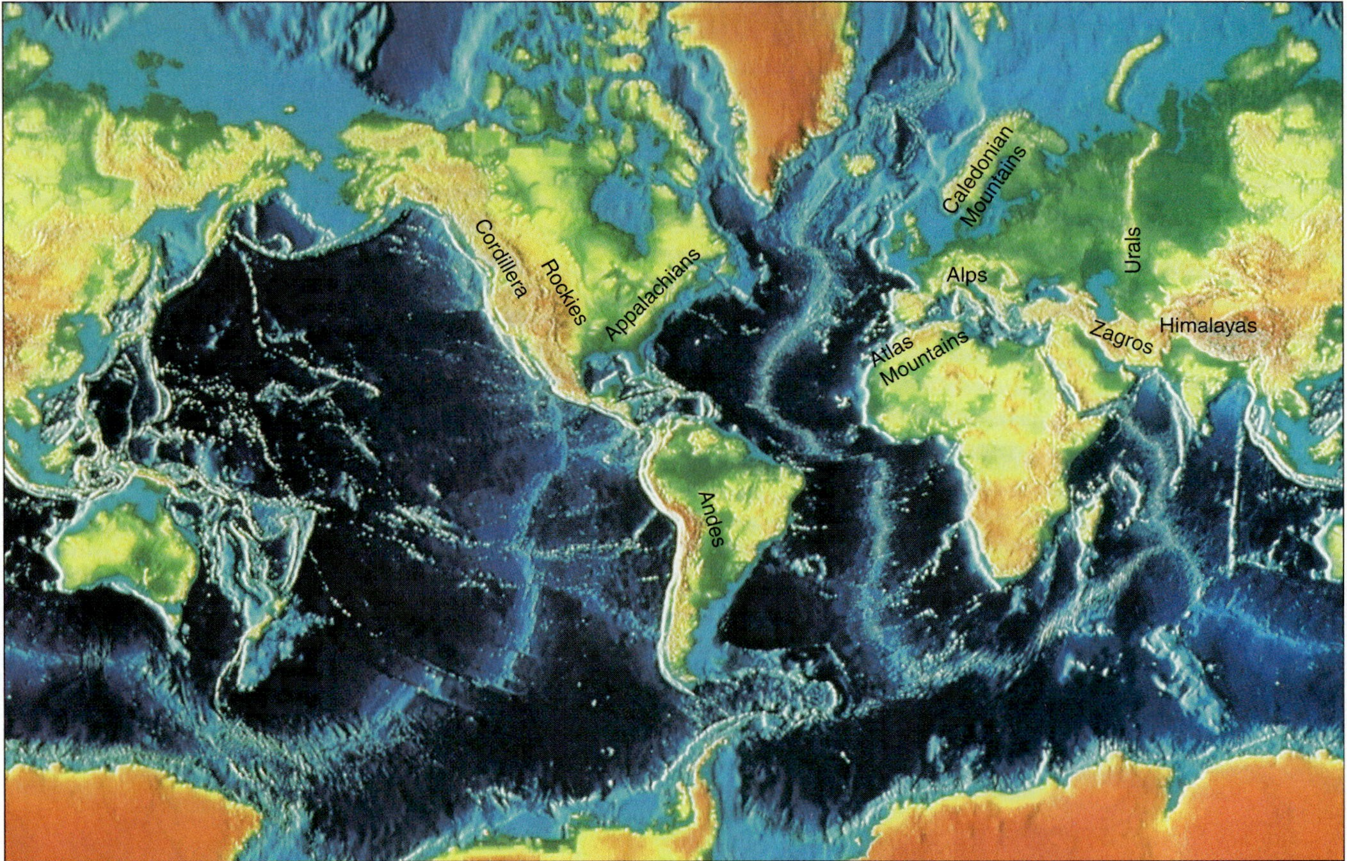

FIGURE 9.13 World topography map emphasizing the distribution of mountain belts (orogens) formed by processes at convergent plate margins. Many of the mountain ranges are still geologically active. The elongate nature of these mountain belts is evidence that they were formed at plate margins.

rock types that it cuts through. Why is the gorge so deep, and why doesn't the river preferentially follow the softer rock types? The Indus drainage system existed *before* the Himalayas were uplifted and is known as an *antecedent stream* (∞ see Chapter 14). At that time, the Indus flowed across relatively low land, and its course was not controlled by the types of rocks buried beneath the sediments over which it flowed. The river has simply been eroding downward as the mountains rose around it.

As mountains are uplifted, they are subjected to faster erosion rates. A constant balancing act occurs between internal, heat-powered forces that drive plate tectonics on Earth and are responsible for the mountain-building processes, and the external, largely solar-driven forces of weathering and erosion that shape the planet's surface and wear down mountain ranges (∞ see Chapters 12 and 13).

During collisions, tectonic forces dominate and the mountains grow taller; erosion does its best to keep up. In fact, erosion may actually increase the rate of uplift through an interesting feedback between the tectonic and hydrologic systems. Uplift leads to more rapid downcutting by rivers and glaciers, which, in turn, removes mass from the region. The removal of mass produces broad isostatic uplift, which accentuates tectonic uplift. The overall effect is to produce extreme topographic relief with high mountain peaks and deep valleys. The huge accumulations of sediments on the ocean floor, known as the Indus and Bengal Fans (Fig. 9.15b) testify to the impressive amount of material that has been eroded from the Himalayan Mountains. Eventually, the lithosphere will adjust to the Indian and Asian continents being jammed; the mountains will settle into isostatic equilibrium as they are worn down (∞ see Chapter 5). The crust will eventually return to normal thickness.

The forces of erosion will then gradually wear down the mountains. Given enough time, they will be reduced to a nearly flat surface. We have only to look at the Appalachian Mountains to see the results of such processes. The Appalachians formed as the result of a collision between the North American continent and the joined continents of Europe–Africa 350 million years ago. The Appalachians were probably nearly as lofty at that time as the Himalayas are today, but erosion has taken its toll, wearing them down in places to gentle rolling hills.

Mountain belts, or orogens, are formed when continents collide. The orogens are then worn down by erosion, and this cycle of mountain building and erosion has been repeated many times during Earth's history.

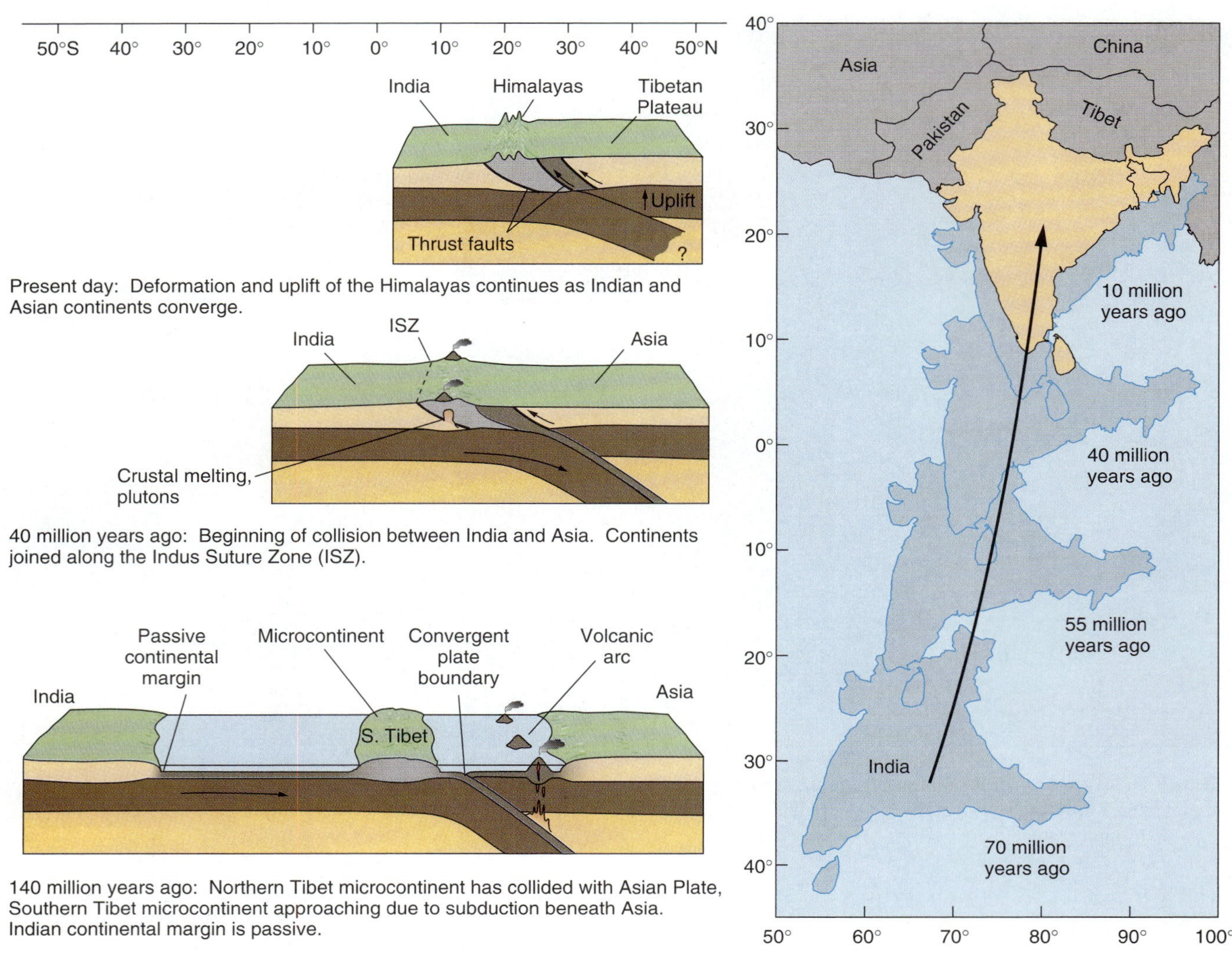

FIGURE 9.14 The possible sequence of events resulting in the collision between India and Asia, which uplifted the Himalayan Mountain range. The map illustrates the northward movement of the Indian continent toward Asia over the last 70 million years. The cross sections show the plate-tectonic configuration at three different times in the past and at the present.

Deformation and Metamorphism

When continental crust is brought together in a collision, most of the deformational structures produced reflect large-scale compression. In the upper layers of the crust, where the rocks are brittle, thrust faulting is very common. Large tracts of rock can be moved tens or even hundreds of kilometers by thrust faults. The sole, or base, of these thrust faults commonly lies in a layer of weaker rock, such as shale, salt, or limestone.

Tight folds are produced where the rocks respond in a more ductile, or plastic, way. Continued compression may overturn these folds. As the fold becomes increasingly stretched, the limbs of these folds may be sheared off completely, giving rise to **nappe** structures, which are a combination of a large fold and a thrust (Fig. 9.16). If you were to approach the center of a collisional mountain belt or travel deep within that belt, you would find rocks that have been subjected to increasing temperatures and intensities of deformation.

Accordingly, the metamorphic grade of the rocks also increases. Typically exposed in the heart of a mountain belt are rocks such as granulites and migmatites, which have been subjected to high pressures and temperatures. Temperatures can become sufficiently high to cause melting of the crust. The magma produced in this way can form large granite plutons, which are common in collisional mountain belts. The chemical composition of such granites differs from the compositions of igneous rocks formed during the episode of subduction at the active continental margin that preceded continental collision. High concentrations of silica and alumina are found in these rocks, reflecting the high silica and alumina concentrations of the continental crust from which they were melted. By using ages that have been determined from radioactive elements in the minerals of the igneous and metamorphic rocks of an orogenic belt (∞ see Chapter 1), we can get a sense of the timing of deformation and meta-

Focus On 9.4 ESTIMATION OF THE CRUSTAL ROOT FOR ISOSTATIC EQUILIBRIUM IN THE HIMALAYAS

Let us estimate the crustal thickness necessary to support the height of the Himalayas. This exercise illustrates an application of the theory of isostasy (∞ see Chapter 5). We will assume the uplift of these mountains is in isostatic equilibrium, compensated by a thickened crustal root extending into the mantle; that is, the Himalayas are "floating" in the mantle.

Consider two columns of crust: one in the Himalayas and the other in the subcontinent (Fig. 1). The crust in the subcontinent is 30 km thick, which is the thickness of normal continental crust. Its average elevation is close to sea level (0 km). Assume the average height of the Himalayas is 6 km above sea level. How far does the Himalayan crust extend below sea level to compensate for this elevation?

Let x = the unknown crustal thickness (the crust and root beneath sea level that supports the Himalayas);

h = the thickness of normal continental crust = 30 km;

H = the thickness of the Himalayan crust = x + 6 km;

m = the thickness of the mantle above the depth of compensation = x − 30km;

$$\rho_{crust} = 2.85 gm/cm^3 = 2850 kg/m^3 = \text{density of the crust}$$

$$\rho_{mantle} = 3.3 gm/cm^3 = 3300 kg/m^3 = \text{density of the mantle}$$

For isostatic balance the weight/area for columns above the depth of compensation must be equal:
For the Himalayas

$$\text{weight/area} = H\rho_{crust}\, g = (x + 6)\, \rho_{crust}\, g$$

For India

$$\text{weight/area} = h\rho_{crust}\, g + m\rho_{mantle}\, g$$

Setting these two equations equal and rearranging terms

$$x = 30 + \frac{6\rho_{crust}}{(\rho_{mantle} - \rho_{crust})}$$

We find that x = 68.0 km and total crust thickness (H) = 74 km.This value corresponds to the seismologically determined depth of the crust, the Moho beneath the Himalayas. The flotation of this root supports the Himalayan uplift, and it is an illustration of Airy isostasy.

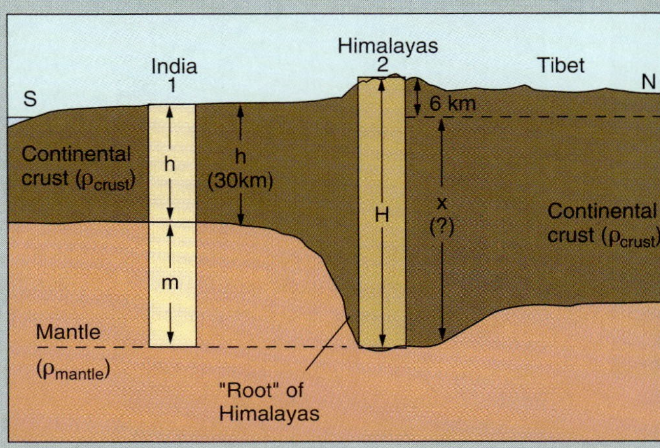

FIGURE 1 Two columns of crust that are in isostatic equilibrium: (1) "Normal" thickness continental crust, and (2) the continental crust of the Himalayan Mountains. The constants shown here are used in the equations.

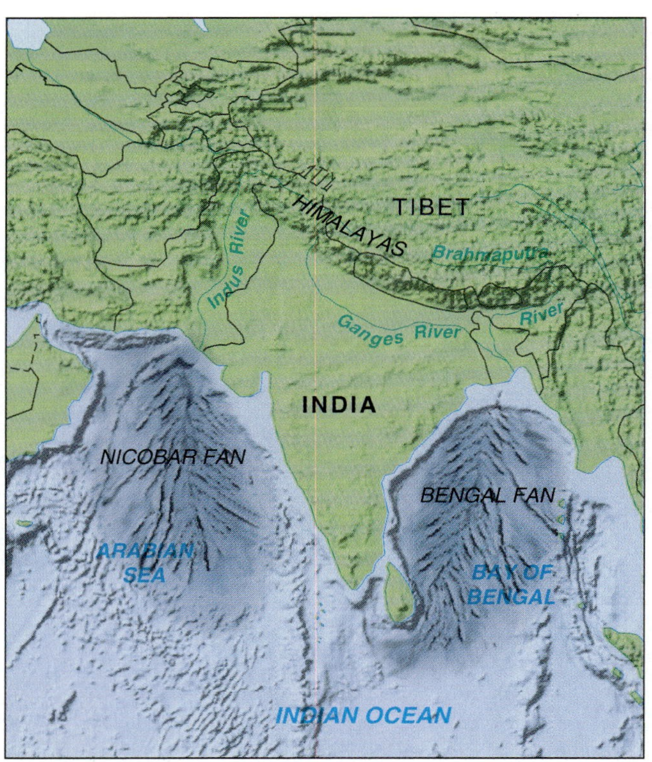

FIGURE 9.15 The Indus River gorge, formed from erosion of the uplifting Himalayan range. The photograph indicates the scale of the feature, while the map shows how the course of the river clearly cuts across the Himalayas.

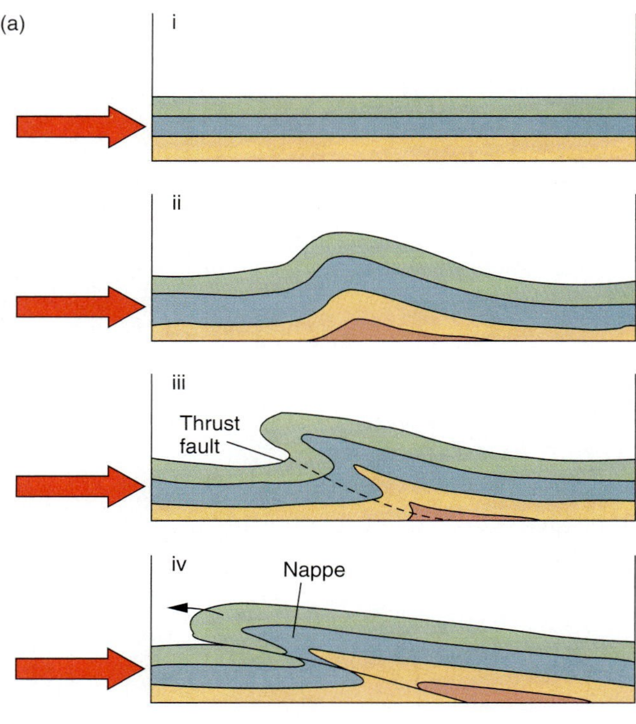

FIGURE 9.16 (a) Cross sections showing the development of a nappe, which is a combination of a very tight fold and a thrust fault. Flat-lying strata are compressed (i), causing folding (ii). Continued compression causes the fold to tighten and overturn. The lower limb is stretched so much that it begins to break, forming a thrust fault (iii). In the last panel (iv), the thrust fault is well developed, and the nappe (the sheet of rock above the thrust fault) has clearly moved from left to right. (b) An example of a nappe: the Morcles limestone Nappe, Matrigny, Switzerland.

morphism and work out the history of a collision in some detail. Large volumes of granitic crustal melts were formed and emplaced in the Himalayas immediately following the most intense period of collision between India and Asia.

When two continents converge and collide, the history of an entire ocean is largely eradicated; most of the ocean floor has been subducted and returned to the mantle. Our only clues to the former existence of an ocean basin are in the deformed ocean sediments and other rocks that were scraped off the subducted slab or trapped in the suture between the plates. Most important among these remnants are slivers of ocean floor—ophiolites (∞ see Chapter 8)—caught up and deformed between the colliding plates. In ancient mountain belts, outcrops of ophiolite indicate the suture zone along which two plates were joined.

Rocks on either side of the suture are typically very different in character and may contain different fossil types. Rocks that might now be only a few kilometers from each other were originally deposited at the opposite sides of an ocean.

Crustal Shortening and Continental Collisions The mechanisms of deformation during the collision of two lithospheric plates are extraordinarily complex. The only clues we are given to decipher the mechanisms come from the rocks and the structures exposed in mountain belts as a result of erosion. We can, however, make a fundamental distinction between the **basement**, the bulk of the crust, which is generally crystalline igneous or metamorphic material, and the **cover**, the younger, more stratified (typically sedimentary or metamorphosed sedimentary rocks) part of the crust. As the basement is not obviously layered but is more homogeneous than the cover, it is difficult to tell whether the basement has been faulted and folded in the same way as the cover. In some cases, most of the deformation is taken up in the cover and there is, in effect, a detachment between the cover and the underlying basement. Because the cover is just a thin skin on the top of the basement, this view of deformation has been dubbed "thin-skinned tectonics." In "thick-skinned tectonics," the basement is deformed along with the cover, so thrust faults and folds would affect both.

During continent-continent collisions, both types of deformation might be encountered. Within the interior of orogens—in the vicinity of the suture between the two plates that have collided—the deformation is typically very intense and the basement is folded and thrust along with the overlying cover (Fig. 9.17). Thin-skinned deformation is commonly found in the outer zones of the mountain belt or the foreland, where the uppermost layers have simply been pushed outward across the basement. Figure 9.17 shows two schematic cross sections through the Appalachian belt. This orogen was formed by crustal collisions 450 million to 250 million years ago. Both sections are based on the geology of the region as mapped at the surface. The difference between them is largely a matter of interpretation—in particular, the shape of the thrust faults below the ground. The upper panel depicts a thick-skinned tectonic interpretation with steep thrust faults that cut through most of the crust. The lower panel is an example of thin-skinned tectonics, with thrust faults that flatten downward and converge on a major décollement, along which the upper thrust slices have been moved hundreds of kilometers.

Below the décollement lies the rifted continental margin that existed prior to the collision. The thrust slices mainly comprise the layered sedimentary rocks (cover) although the thrust slices may include thin slivers of underlying basement. You can see that the orogen is asymmetric (the thrust faults all tend to dip one way), which reflects the asymmetric arrangement of the convergent plate margin that led to the collision. Drilling and seismic reflection surveys across the Appalachian chain have now shown that the thin-skinned interpretation is more realistic in this particular instance.

As a result of continent-continent collisions, the crust is thickened considerably during mountain building. Pressures calculated in some rare metamorphic rocks from the Alps in southern Europe, for instance, indicate they were formed at depths of about 100 km. As these are rocks of crustal composition, we must conclude that the crust beneath the Alps was very thick at some time in the past—in contrast to normal continental crustal thicknesses of some 30 to 40 km. The crust beneath the Himalayas is currently about 70 km thick; a crustal "root" exists beneath these mountains in order to isostatically support them, as outlined in Focus 9.4.

Some increase in crustal thickness in orogenic belts could be achieved by the intrusion of molten material from the mantle; the crust would inflate like a balloon in response to the intrusion of magma. But the majority of thickening probably occurs in response to compression and shortening applied in the horizontal direction. If the mass of crust remains the same before and after collision, then any shortening in the horizontal direction should produce uplift and thickening.

In closing this section, it is worth noting that not all of the structural features associated with continental collisions are the result of compression. One interesting consequence of the vertical thickening caused by collision is that it produces a high-standing mass of rock that may collapse simply under the influence of gravity. It is rather like trying to push loose sand into a steep pile—there comes a point where the sand pile is too steep and too unstable, and sand slips down the sides of the pile. In both the high plateaus to the east of the Andes (Focus 9.2) and to the north of the Himalayas (Figs. 9.14 and 9.15), there are features such as normal faults (∞ see Chapter 7), that indicate local extension and that are interpreted as the result of gravitational collapse.

Continental collision produces intense deformation during mountain building. In particular, the weaker upper layers of the crust may be intensely folded and faulted.

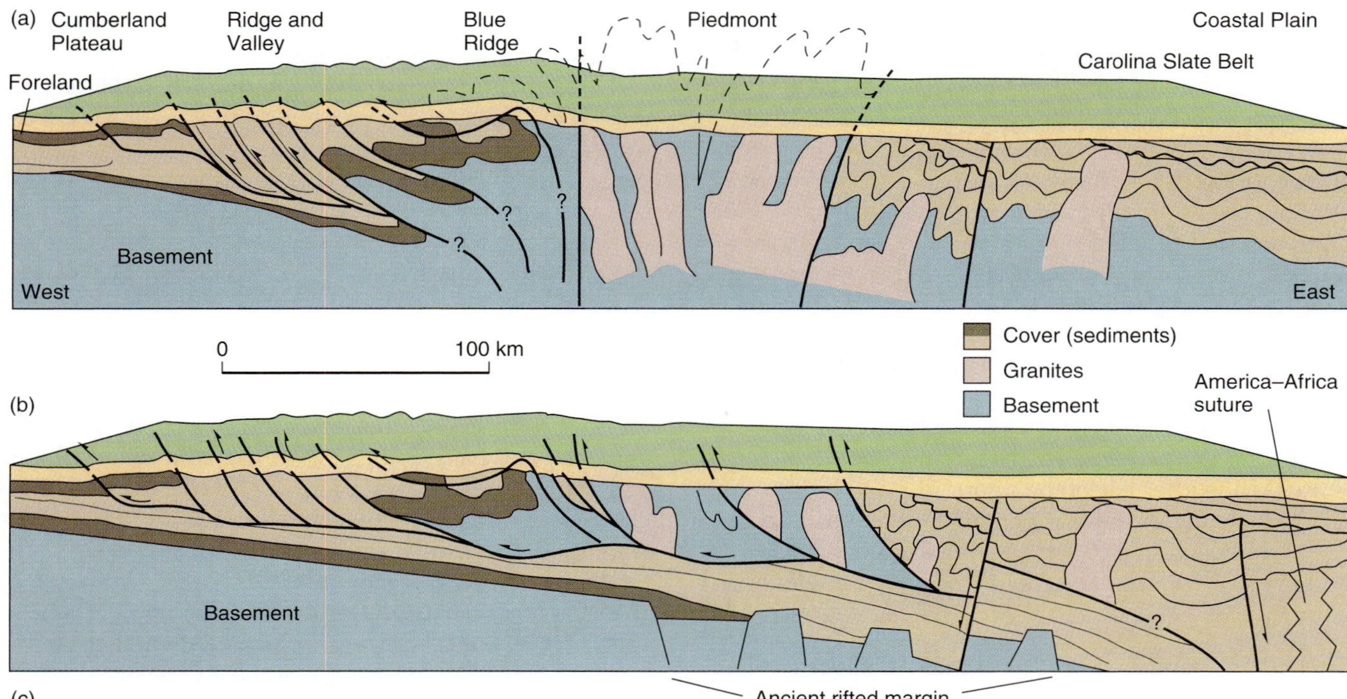

FIGURE 9.17 Two different interpretations of the structure of the Appalachian Mountains in the eastern United States. (a) Thick-skinned tectonic interpretation. (b) Thin-skinned tectonic interpretation. Note that the near-surface geology is quite similar in both (a) and (b). The main difference is in the interpretation of the geologic relationships and structures at depth. Seismic reflection and drilling have now been used to locate the thrust faults below the surface. The data indicate that (b) is a more accurate interpretation. (c) Satellite photograph of the deformation caused by continental collision between North America and Africa more than 250 million years ago in the Appalachians.

Collage Tectonics and Exotic Terranes

Up to now, it may seem there are a relatively small number of rigid plates moving relative to each other, albeit changing shape, size, and location through time due to interactions at their margins. In light of recent geologic studies, however, we should slightly modify this view. Research now indicates that some regions—rather than being large lithospheric plates—actually comprise several relatively small microplates, or "blocks." Moreover, these blocks have experienced independent and different histories prior to being brought together at the locations where they are now found.

Evidence for the existence of these blocks has been compiled from various sources such as paleomagnetism, fossil studies, and the simple mismatch of rock units across the boundaries of these blocks; in fact, many of the same approaches originally led to the concept of plate tectonics. The blocks are too small in size to be considered true plates, so they are referred to as **terranes**.

In some places, such as western North America, it appears that many of these terranes have been sutured or accreted to the continent. In Figure 9.18, many distinct terranes are shown along the western edge of North America, and many more smaller ones have been defined. In each of these terranes, certain packages of rocks and types of fossils can be mapped, but they do not cross the boundary into the adjacent terrane. Many of the terranes were not part of the original North American Plate,

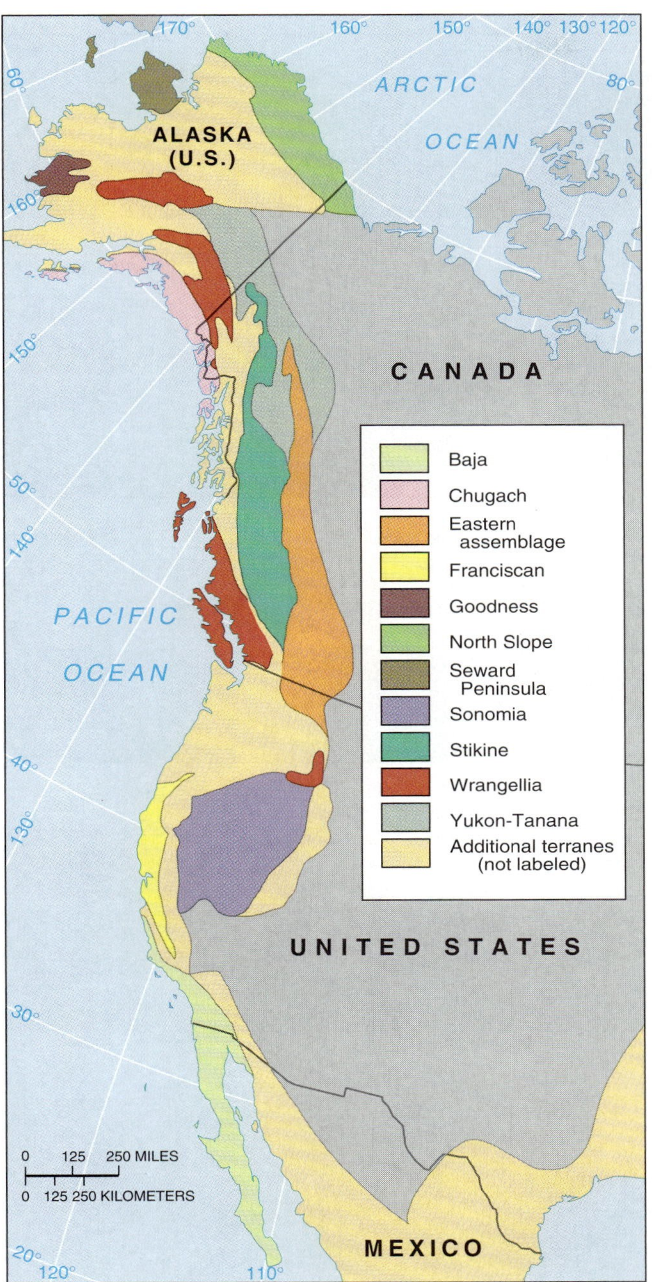

FIGURE 9.18 Map of western North America showing different terranes that have accreted to the plate margin over the last 300 million years. Each terrane (shaded) has distinct geologic characteristics: different rock types, fossil types, and paleomagnetic directions that distinguish it from surrounding terranes.

although they are part of it now. These blocks have been added to the continent at various times in the past and are now known as *exotic terranes* because they have a different history and origin from most of North America. The assembly of several of these terranes over a period of time is occasionally referred to as **collage tectonics**. In effect, the western margin of North America might represent a collage of many different terranes.

How do these terranes assemble and become sutured together? The answer lies in the plate-tectonic processes we have already explored—particularly subduction. Island arcs are difficult to subduct because they constitute a fragment of thicker, lower-density crust. Therefore, as an ocean basin closes, any arcs formed within the basin will eventually collide with a continent. The arc becomes sutured to the continent, and the collision causes deformation—although less deformation than would occur in the collision of two continents.

Figure 9.19 shows how collage tectonics were responsible for shaping the eastern margin of North America. About 600 million years ago, the ancestor of present-day North America (proto-North America) was separated from proto-Africa by the Iapetus Ocean basin, much as America and Africa are presently separated by the Atlantic Ocean. Unlike the present Atlantic Ocean, however, the Iapetus Ocean basin was closing. This movement brought island arcs and small continental fragments toward the eastern margin of the continent, where they collided, causing deformation and uplift (Fig 9.19a–c). Each collision added a new terrane to the edge of the continent. Eventually, a major continent-continent collision occurred (Fig. 9.19d). This collision was the final stage in the uplift of the Appalachian Mountains. When these mountains originally formed, they must have been quite spectacular—and were, perhaps, similar to the present-day Himalayas.

The suturing of these continents was part of the assembly of Pangaea (∞ see Chapter 6). Approximately 200 million years ago, Pangaea began to split apart, and the Atlantic Ocean started to form. The rifting occurred near to the location of the suture between the two continents, where the lithosphere had been weakened as a result of the deformation and magmatism associated with the preceding continental collision. Rifting did not occur precisely at the suture, however, because a remnant of what was originally Africa became stranded as part of present-day North America (Fig 9.19e and f).

Note that the rocks of the island arc forming an accreted terrane may have been formed thousands of kilometers away and thus will record distinct and different paleomagnetic directions. There may also be fossils that are distinct between the arc and the continent, just as countries that are thousands of kilometers apart in the world today often have different fauna and flora. After the arc has collided, there may be a rearrangement of the plate boundary (Fig. 9.19), and the subduction zone might jump to the other side of the now-accreted terrane in order to allow convergence to continue. Keep in mind that only oceanic lithosphere can be subducted.

Another way of accreting exotic terranes is by transform faulting. The role of transform plate margins in plate tectonics is discussed more fully in the next chapter. It is sufficient at this point to say that terranes may be "slid into

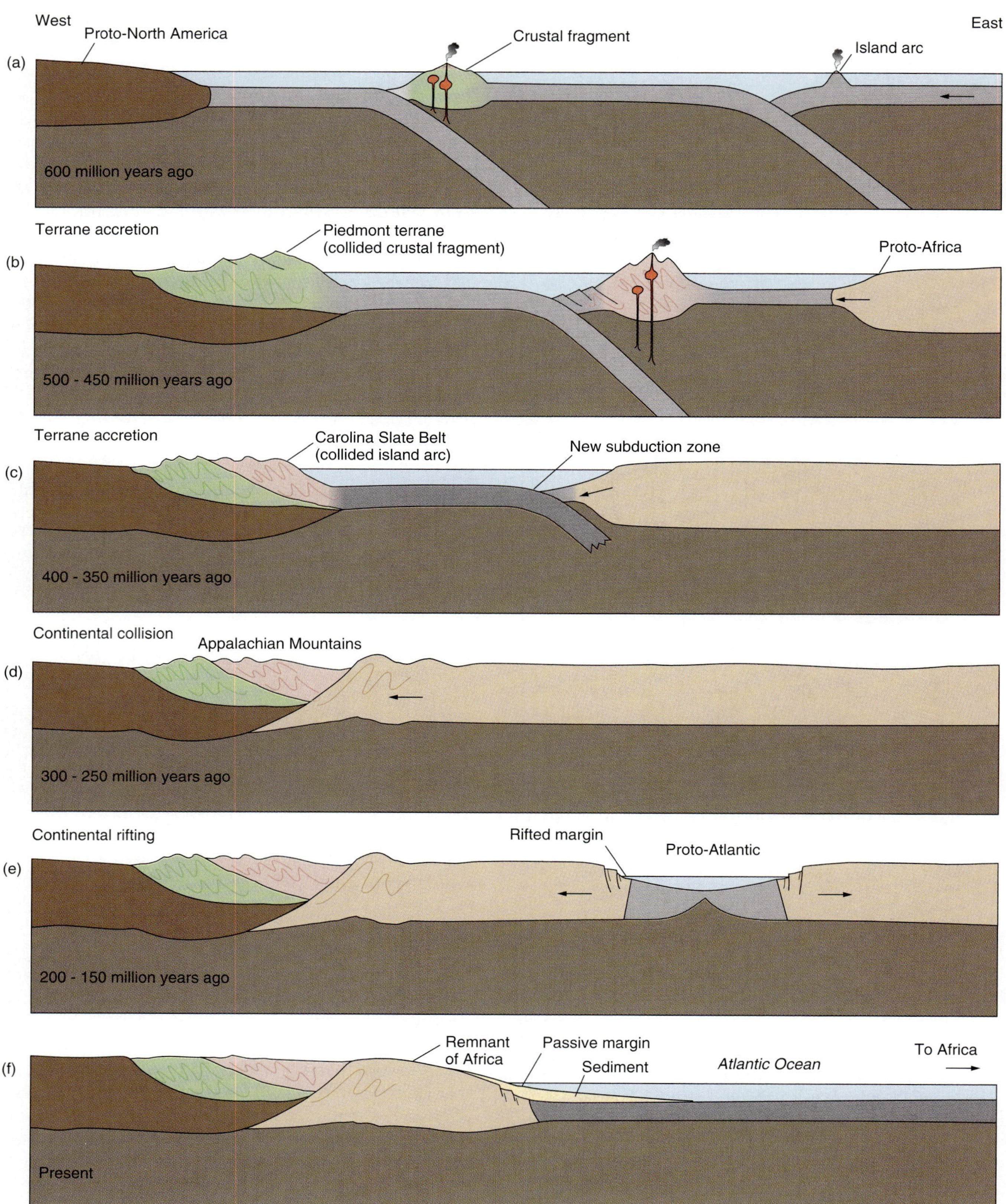

FIGURE 9.19 Example of collage tectonics and effects in eastern North America. Panels show cross sections through time from 600 million years ago (top) to the present (bottom). Much of the Appalachians were built as a result of collisions between proto-North America and crustal fragments (a–b) or island arcs (b–c). A collision between proto-North America and an ancient African continent (proto-Africa) 300 million years ago caused intense deformation and uplift in the final phase of Appalachian Mountain building (d). Since then, Africa has rifted away, leaving a remnant accreted to the eastern margin of the Appalachian Mountains, and the Atlantic Ocean basin has formed by sea-floor spreading (e–f).

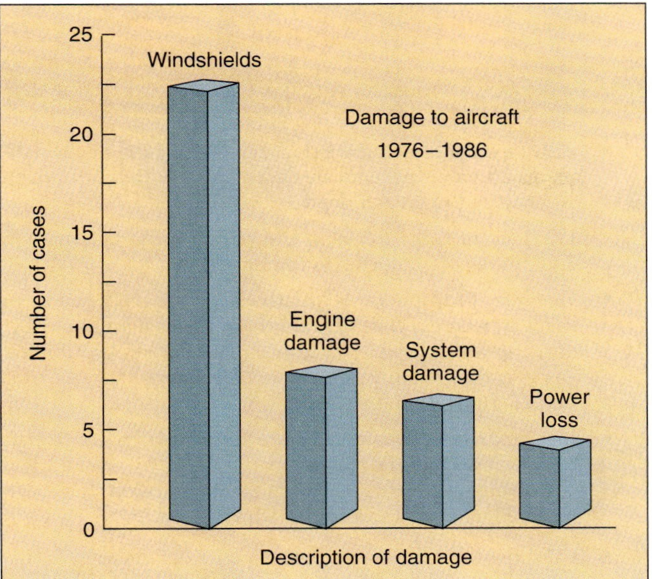

FIGURE 9.20 Encounters between aircraft and volcanic ash clouds over a 10-year period, and the types of damage sustained.

place" along strike-slip faults. In fact, based on paleomagnetic data, many of the terranes that make up western North America may have moved hundreds to thousands of kilometers north or south into their current locations.

Island arcs and continental fragments can become sutured onto continents through plate convergence or strike-slip motions—a process sometimes referred to as collage tectonics.

We have noted that these terranes are now part of the North American Plate, but it is important to remember that the terranes are not equivalent to plates. Many were probably island arcs or small continental fragments that were part of a much larger plate, most of which has been subducted (if it was oceanic lithosphere) or broken away along transform faults during the accretion process.

There are, however, small plates, or microplates, existing today in regions such as the Mediterranean, which exhibit complex responses to the motions of larger plates surrounding them. The oceanic portion of these microplates—which, in some cases, is their entire extent—may eventually be subducted, but any arc material or continental fragments will be accreted as terranes. It should be clear that once a terrane is accreted to a continent, it is difficult to determine the original extent of the lithospheric plate with which it was associated.

The concept of collage tectonics also reminds us that plates are not entirely rigid, especially at their margins. The collage of terranes along the edge of North America, represents continental crust that is crisscrossed with fractures, the suture zones between the terranes. The terranes may jostle about and rotate in reponse to plate-margin processes along the western edge of North America, making plate reconstruction based on paleomagnetism and geologic correlations very complicated.

Living with Geology

Explosive Volcanic Eruptions: A Hazard to Aircraft

The immediate threat of explosive volcanic eruptions typical of convergent plate margins is powerfully illustrated in Focus 9.3. The long-term potential effects on global climates of injecting huge volumes of ash and gas into the stratosphere are explored later (∞ see Chapter 15). When an eruption produces a huge ash plume reaching into the stratosphere, one of the most immediate threats is posed to air traffic (Fig. 9.20). Ash and tiny particles of rock can pit windshields and reduce visibility. More seriously, intake of ash into the hot turbines of jet engines will cause clogging and result in stalling of the aircraft. Also, ash may be heated sufficiently to melt and to coat the turbine blades.

Once a volcanic eruption is underway and recognized, air traffic can, of course, be routed around the hazard. In the early stages of an eruption, however, particularly for remote volcanoes, local authorities might be unaware of the danger. In uninhabited regions such as the Aleutians, aircraft following polar routes may actually provide the first reports of eruptions in progress. The Aleutian Island arc consists of numerous potentially active and explosive volcanoes. Although satellite monitoring can sometimes warn of eruption plumes, notoriously bad weather in the region commonly impedes observations. The Boeing 747 shown in Figure 9.21 flew into the ash plume of Redoubt volcano, Alaska, in 1989. It descended over 4000 m without power, eventually restarting its engines only 1000 m above the mountain peaks over which it was gliding. It landed safely at the Anchorage airport, but it sustained over $80 million of damage.

This example underscores the need to develop and maintain a global volcano-monitoring network, with close communications between scientists and government authorities. Branches of the United States Geological Survey serve this purpose in the United States by working closely with officials of the FAA (Federal Aviation Administration).

FIGURE 9.21 Ash-damaged KLM Boeing 747 at Anchorage airport after encountering the volcanic ash plume from Redoubt volcano, December 14, 1989.

SUMMARY

- Convergent margins occur where one plate sinks (subducts) beneath the other. Subduction is due to dense oceanic lithosphere sinking back into the mantle, largely under its own weight.
- The Wadati-Benioff zone is the array of earthquake hypocenters in a subduction zone; it corresponds approximately to the upper surface of the subducted slab.
- Magmas at arcs are produced from the mantle by adding water from the subducting slab, which lowers the melting point of the overlying mantle and leads to melting.
- Granite batholiths may be produced by fractional crystallization of basalts, by melting of continental crustal material, or by a combination of these processes.
- Explosive eruptions produce two main types of pyroclastic deposits: falls and flows. Fall deposits accumulate as erupted material falls back to the ground. Flow deposits form when ejected material accumulates rapidly and flows downhill.
- Frequently, convergent margin volcanoes are highly explosive and dangerous. Huge eruptions of more than 1000 km^3 of ash are known to have occurred at volcanic arcs.
- Sediments in arc environments tend to accumulate in an accretionary wedge at the trench.
- The mechanics of subduction determine how much sediment ultimately accumulates at the trench, and how much is subducted to greater depths.
- Dynamic tectonic forces at the trench cause deformation and mixing of sediments to form mélanges.
- Oceanic lithosphere is relatively dense and tends to subduct, whereas continental lithosphere is light and buoyant; the average age of oceanic crust (about 100 million years) is thus much younger than the average age of continental crust (2 billion years).
- Mountain belts, or orogens, are formed when continents collide. The orogens are then worn down by erosion, and this cycle of mountain building and erosion has been repeated many times during Earth's history.
- Continental collision produces intense deformation during mountain building. In particular, the weaker upper layers of the crust may be intensely folded and faulted.
- Island arcs and continental fragments can become sutured onto continents through plate convergence or strike-slip motions—a process sometimes referred to as collage tectonics.

KEY TERMS

trench, 223
volcanic arc, 223
magmatic arc, 223
fore arc, 223
back arc, 223
Wadati-Benioff zone, 224
volcanic bomb, 233
tuff, 233
pyroclastic fall, 233
pyroclastic flow, 233
ignimbrite, 234
welded tuff, 234
caldera, 234
sector collapse, 236
accretionary wedge, 237
greywacke, 237
mélange, 241
orogenesis, 242
nappe, 244
basement, 247
cover, 247
terrane, 248
collage tectonics, 249

QUESTIONS FOR REVIEW AND FURTHER THOUGHT

1. What information can we use to show that oceanic lithosphere is subducted at trenches?
2. Suppose two continents are separated by an ocean basin that is 5000 km across. A subduction zone is located along one edge of the ocean (an active continental margin). If the rate of subduction of oceanic lithosphere is 10 cm per year and there is no active spreading within the basin, how long will it take before the continents collide?
3. How do magmas form at convergent plate margins, and how is this different from the way that magmas form at divergent plate margins?
4. If you went to a region where a volcano had recently erupted a lot of pyroclastic material, how would you distinguish between ash*fall* deposits and ash-*flow* deposits, based on the relationships of the deposits to the local topography?
5. The island of Barbados, to the east of the Lesser Antilles island arc (Focus 9.2), is the tip of an accretionary wedge. What types of rock would you expect to find there, and how would they differ from the rock types comprising the islands of the arc?
6. Based on the description of the formation of the Himalayas and on Figure 9.14, how old would you expect the high-grade metamorphic rocks formed as a result of collision to be?
7. What evidence could you use to define the boundary of a terrane?

10 The Conservative Boundary: Transform Plate Margins

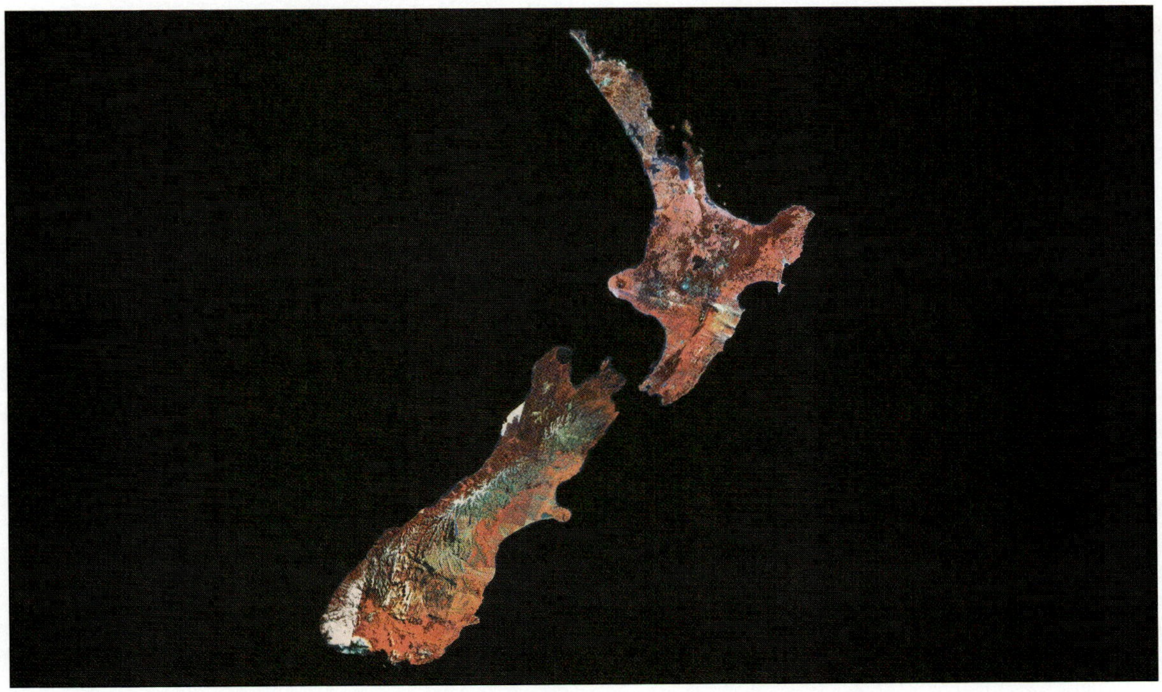

Satellite view of New Zealand. The volcanoes of the North Island form a volcanic arc resulting from subduction from the east. The plate boundary passes southward into a transform fault, which cuts along the length of the South Island. Compression across this fault has caused uplift of the Southern Alps.

Overview

- A transform fault is an offset between two spreading ridges, between two subduction zones, or between a spreading ridge and a subduction zone.
- On oceanic crust, active transforms occur between the axes of spreading ridges or subduction zones.
- On continental crust, transforms occur as long strike-slip faults, such as the Alpine fault of New Zealand.
- Bends in continental transform faults generate zones of compression or extension, depending on the shape of the bend relative to the direction of fault motion.
- Compressional structures form where the geometry causes fault movement to squeeze inward on the bend, creating uplift and becoming a local source of sediment.
- Extensional structures form when the geometry causes fault movement to release outward on the bend; extensional structures most commonly subside, forming local sites for sediment deposition.

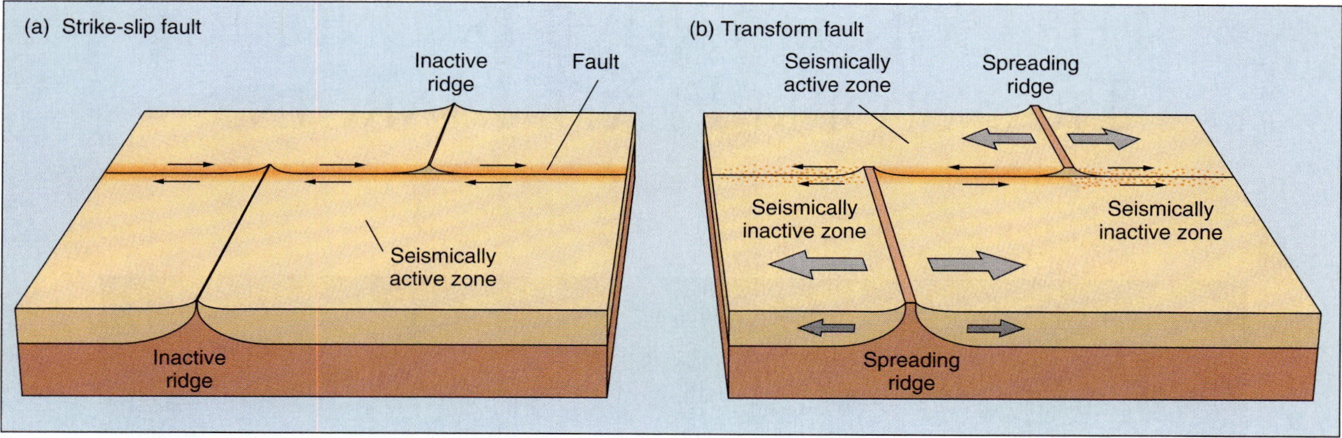

FIGURE 10.1 A comparison of a spreading ridge offset by transform and ordinary strike-slip faults. (a) A strike-slip fault. The seismically active zone goes beyond the ridge axes, and fault movement is right lateral strike-slip along the entire length of the fault. (b) A transform fault. The seismically active zone is confined to the region between the ridge axes, and fault movement between the spreading ridges is opposite to the sense of offset shown by the ridge positions.

Introduction

Most transform faults occur in oceanic crust and, therefore, are concealed beneath the oceans; as a result, their existence has only recently been recognized. For the same reason, the extent of submarine volcanic activity was vastly underestimated in the past, and both the extent and the continuity of the mid-ocean ridges were not fully appreciated. In fact, development of the concepts of plate tectonics coincided with the development of modern marine geology.

Present research—the detailed mapping of submarine topography, studies of the magnetic properties of the oceanic floor, mapping the seismicity of the ocean floor, and first motion studies of submarine earthquakes—has provided strong documentation for the formative ideas of plate tectonics. These studies allowed Canadian geologist J. Tuzo Wilson to identify and describe a new type of fault: *transform faults*. He observed that spreading ridges were offset in one direction by a fault, yet the first motion of earthquakes along that fault indicated an opposite sense of slip. This was indeed a puzzle until 1965 when Wilson was able to interpret this information correctly. He compared offsets of geologic features along scarps associated with transform faults with the near-uniform offsets that would be expected if they were strike-slip faults. He found that the relative movement of transform faults was confined to the zone between the spreading centers. There was no movement across the scarps extending either side (Fig. 10.1). From this comparison, he was able to identify and define transform faulting.

Many transform faults have been documented since Wilson's important finding. Transform faults may offset spreading ridges or subduction zones; transform faults may also lie between a spreading ridge and a subduction zone or between two subduction zones. In his initial paper published on the subject, Wilson drew sketches of the different geometries possible for each type of transform fault (Fig. 10.2). We now understand that transform faults are important tectonic elements of all plate boundaries.

The Nature of Transform Margins

Just as subduction zones characterize convergent margins and spreading ridges characterize divergent margins, the

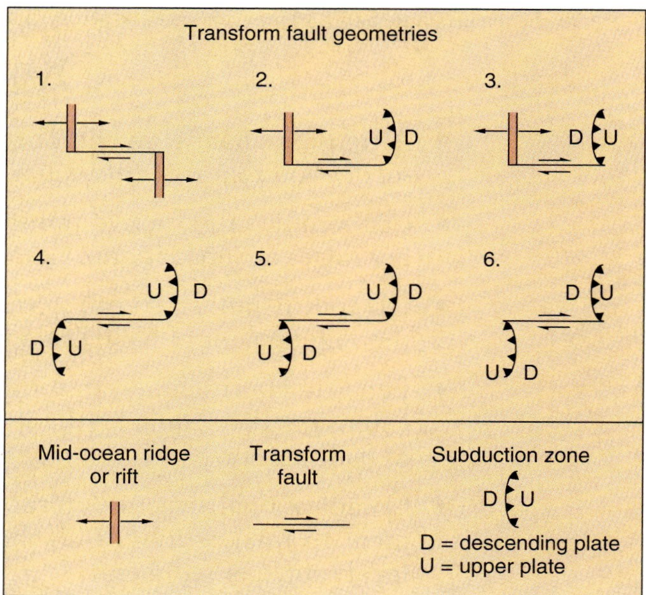

FIGURE 10.2 Six possible geometries for transform faults, which link together convergent or divergent plate margins.

physical feature associated with transform margins are *transform faults*. The name "transform" was coined because it describes the nature of the fault termination. These faults do not continue for long distances, with movement gradually diminishing. Instead, they end abruptly at a plate margin, where the lateral motion of the fault terminates at either an oceanic ridge or a subduction zone (Fig. 10.2). At a spreading ridge, the fault motion is *transformed* into the generation of new lithosphere; at a subduction zone, the fault motion is *transformed* into destruction of old lithosphere. Thus, activity along transform faults is confined to the interval between the axes of spreading ridges or subduction zones. The Alpine fault, New Zealand, links two subduction zones (Focus 10.1). The San Andreas fault, California, links a spreading ridge to a subduction zone/transform boundary at a triple junction.

Major transform faults represent approximately 15 percent of the total length of plate margins worldwide. However, the most common expressions of transform faulting are small offsets along spreading ridges. These small faults were not included in the 15 percent figure, but cumulatively they represent a substantial amount of offset. For example, along the East Pacific Rise there is approximately 4000 km of cumulative offset on the smaller transforms from near the coast of central Mexico (18°N) to the southern tip of South America (55°S) (Fig. 10.3).

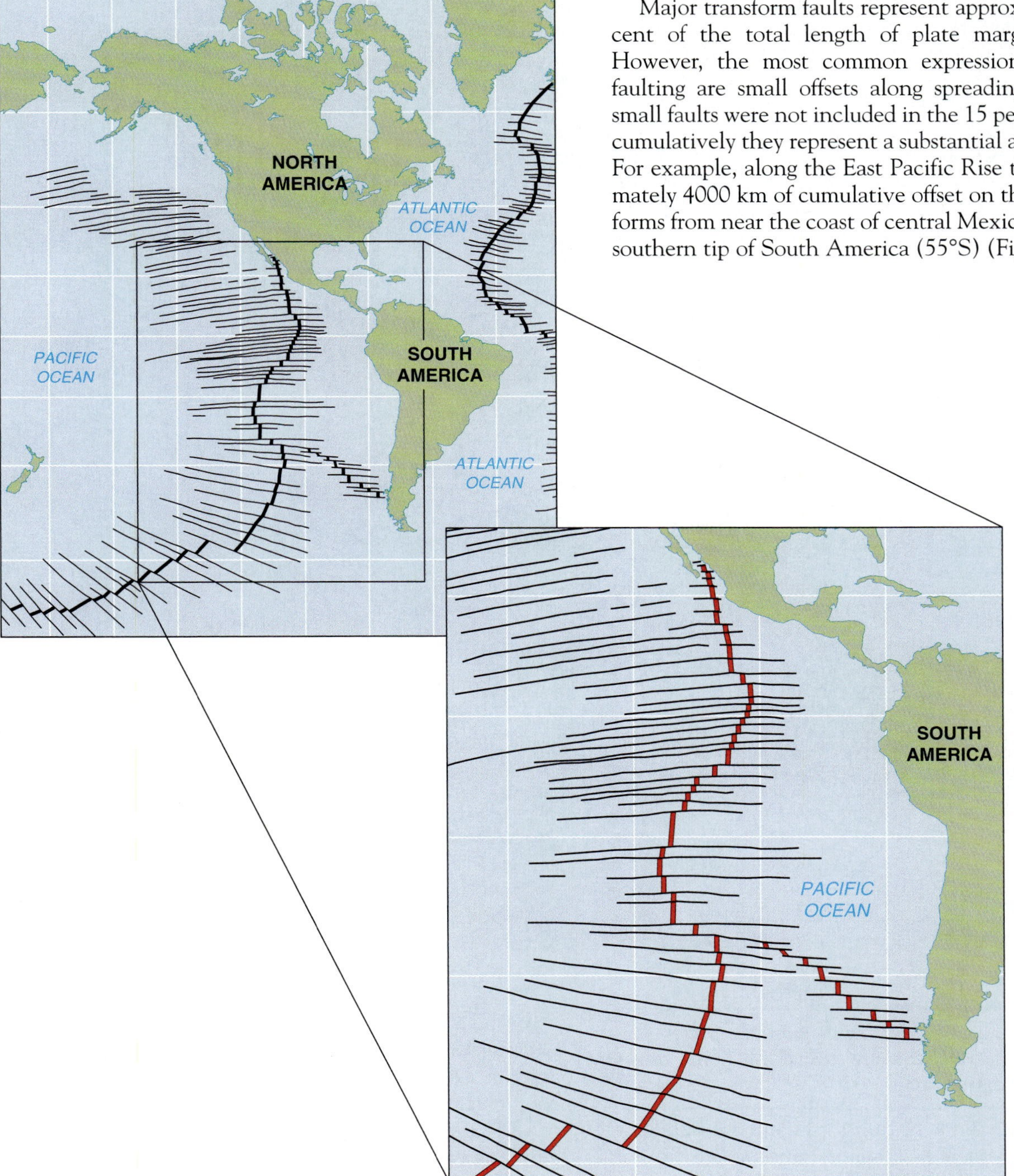

FIGURE 10.3 Spreading ridge segments of the Eastern Pacific offset by transform faults. The spreading ridge segments actively continue to generate new crust, with the relative motion across transform faults confined to the region between ridge segments.

THE ALPINE FAULT, NEW ZEALAND

A stable subduction zone can consume oceanic lithosphere only in a single direction. The development of a spreading ridge in the southern Pacific Ocean, between Australia and Antarctica, created opposing directions of subduction between the Indo-Australian Plate and the Pacific Plate. This caused the junction between these plates to become unstable. To stabilize the geometry, a transform fault evolved between the opposing directions of subduction. This is the mechanism for the Alpine fault as it exists today (Fig. 1a and b). The total right-lateral slip is approximately 480 km. This fault, however, is not strictly a strike-slip fault; it also is an area of active transpression, which is an area that is simultaneously sheared and compressed, and has led to the uplift of the Southern Alps in New Zealand (Fig. 2).

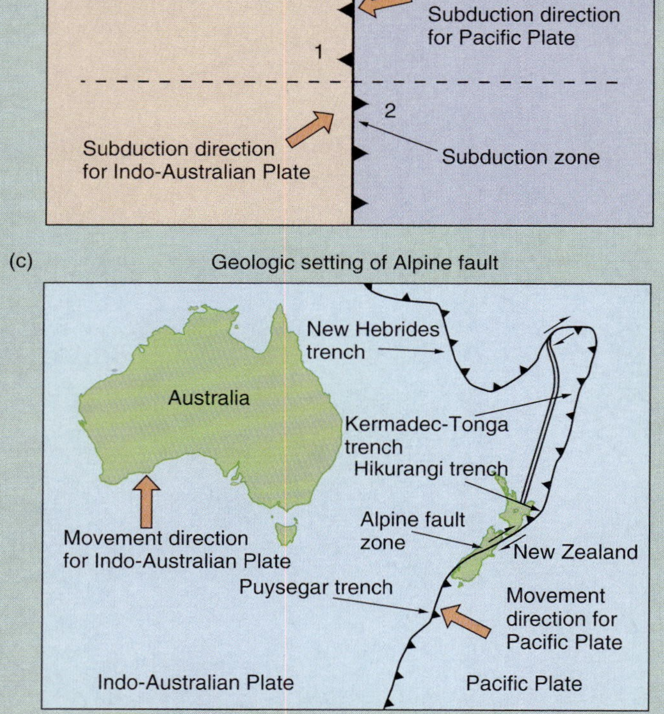

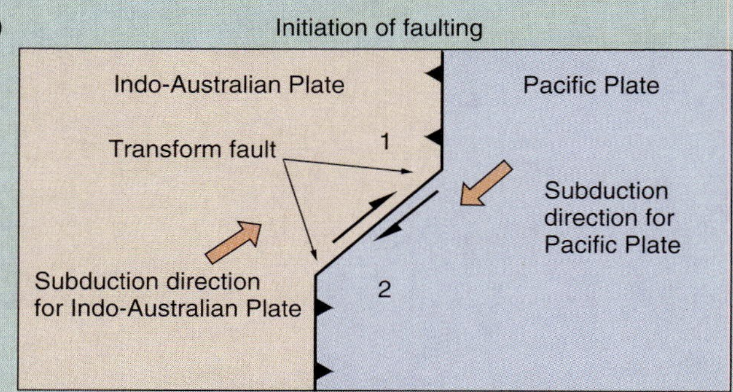

FIGURE 1 Development of the Alpine transform fault in New Zealand. (a) The southern portion of the Pacific Plate is moving southwest, and the Indo-Australian Plate is moving generally northeast to east. (b) A long subduction zone separates the two plates, but the polarity of subduction along the zone must shift from east-directed to west-directed along the zone. (c) Subduction in the New Hebrides and Tonga-Kermadec trenches is directed westward, with the Pacific Plate subducting beneath the Indo-Australian Plate; subduction along the Puysegar Trench is directed toward the east.

In most cases, transform faults laterally offset segments of spreading ridges; a few transform faults offset segments of subduction zones.

Worldwide, there are 14 transform faults of sufficient length to be considered of major importance. Two of these—the San Andreas fault zone of California and the Queen Charlotte fault zone offshore from Canada and Alaska—make up most of the Pacific Plate's northeastern margin (Fig. 10.4).

Just as with convergent and divergent margins (∞ see Chapters 8 and 9), various distinguishing features characterize transform margins. The most distinguishing characteristic is the virtual absence of magmatic activity in the form of volcanoes or intrusions. Transform margins are very active seismically, and earthquake hypocenters are characteristically shallow.

Faulting tends to occur when a portion of Earth's crust is subjected to a shearing stress. If the stresses are horizontal, or nearly so, movement along the fault will be parallel to

The Alpine fault cuts diagonally across the South Island of New Zealand and is continuously exposed for about 450 km, but this is only a portion of its total length (Fig. 1c). This large transform fault forms the eastern margin of the Indo-Australian Plate at its contact with the Pacific Plate.

Migration of the Pacific Plate at these southern latitudes is toward the northwest, and that of the Indo-Australian Plate is northerly; the relative velocities (∞ Focus 6.2) cause subduction along the Tonga-Kermadec subduction zone to be west-directed. For the same reason, subduction in the Puysegar Trench is toward the east. Thus, the *polarity* of the subduction zone north of New Zealand is the opposite of that south of New Zealand.

North of New Zealand, between the northern ends of the Tonga and the New Hebrides trenches, two other large transform faults probably exist, but these have not yet been thoroughly studied (Fig. 1c).

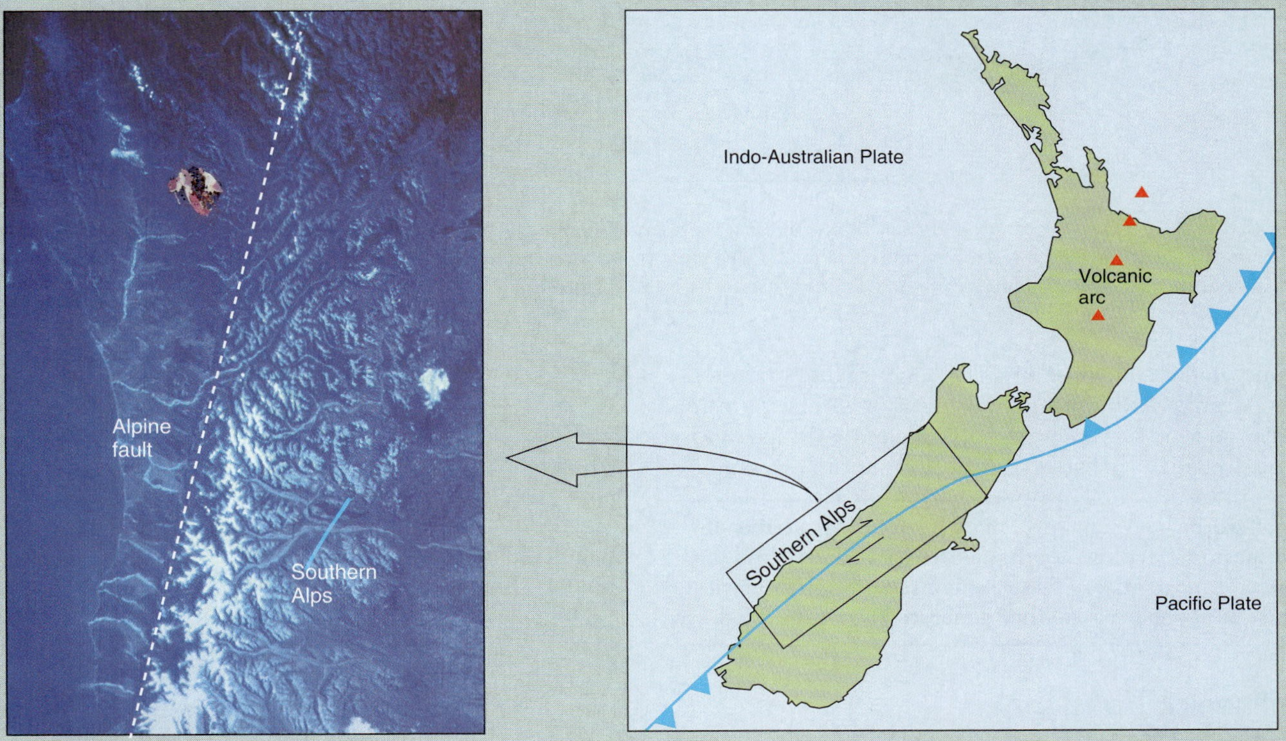

FIGURE 2 Satellite photograph of South Island, New Zealand, showing the trace of the Alpine fault and Southern Alps.

the strike of the fault, and the mechanism of offset is termed strike-slip. This is the case for transform faults. Studies of earthquakes along spreading ridges have shown that much of the seismic activity is along normal faults within the rift, in which the direction of movement is parallel to the dip of the fault (∞ see Chapter 7). However, the energy released along transform faults at ridges by strike-slip earthquakes is much greater than the energy released by normal faulting in the rift center. Maps indicating the frequency of earthquakes (seismicity) show that they are common within the rift and along the transform between ridge segments; earthquakes along the fracture zones outside the axis of the ridge segments, however, are extremely rare (Fig. 10.5).

The relative movement between the two plates along the transform margin is parallel to the fault trace. Thus, transform margins are *conservative margins* in that the plates slide past each other without the loss or generation of new lithosphere. Motion on these faults is more or less pure strike-slip but might not be confined to a single fault line. Instead, the movement might be distributed across

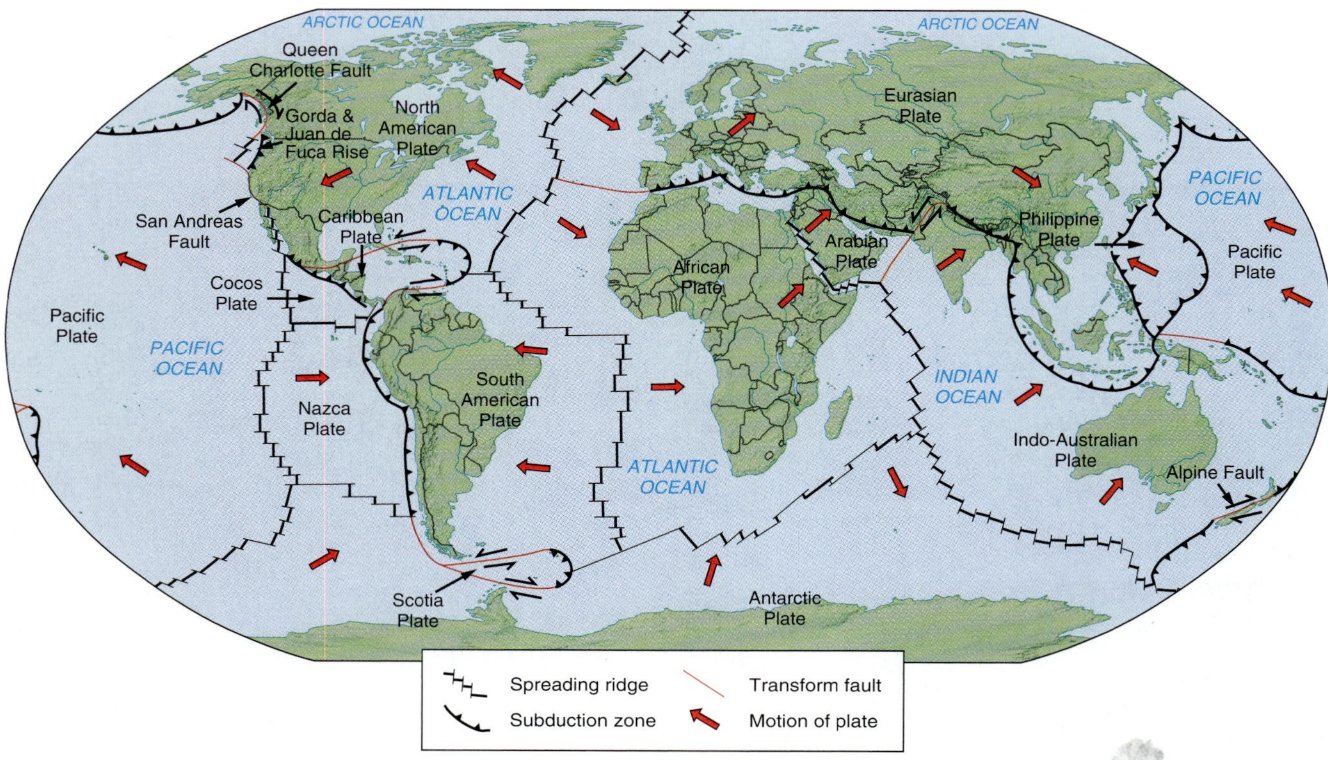

FIGURE 10.4 Global distribution of transform faults that represent plate boundaries. Approximately 15 percent of the total global plate boundary length is accounted for by transform faults. Two great transforms are the Queen Charlotte fault along the western coast of Canada, and the San Andreas fault along the western coast of the United States.

several parallel fault lines or shear zones. This is true for both oceanic and continental transforms, but, until recently, marine surveys were not sufficiently complete to see the similarities between oceanic and continental faults.

Transform margins are characterized by strike-slip fault motion, shallow earthquake hypocenters, and an absence of magmatism. They are conservative boundaries because lithosphere is neither generated nor consumed.

Continental Transforms

We begin by describing some of the major characteristics of continental transform faults simply because these can be directly observed. In contrast, direct observation of oceanic transforms requires deep-diving submarines and scientific ships with sophisticated instruments.

Continental transform faults are large strike-slip faults usually characterized by substantial offset—on the order of tens to hundreds of kilometers. In general, these faults are nearly vertical, and the depth to which they extend is a matter of controversy. To qualify as a plate boundary though, the discontinuity must penetrate completely through the lithosphere. In California (Focus 10.2), earthquake activity along the San Andreas fault extends to depths of approximately 12 to 15 km, and both the distribution and the nature of the earthquakes are consistent with a vertical fault trace. Below 15 km, the crust is ductile (∞ see Chapter 7) and the relative motion of the remainder of the lithosphere occurs by ductile flow.

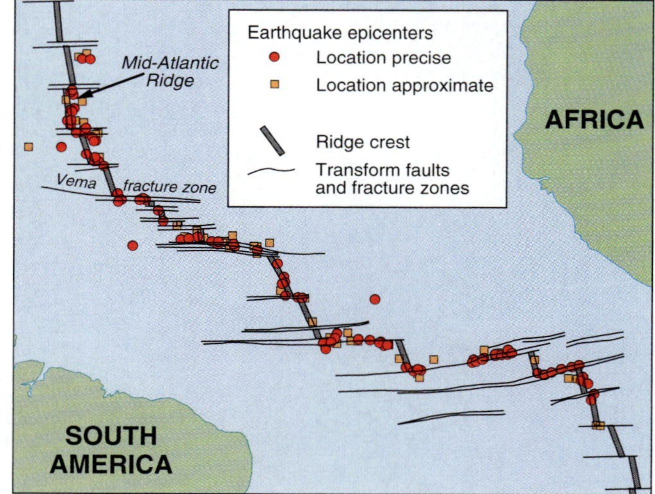

FIGURE 10.5 The occurrence of seismicity along the Mid-Atlantic Ridge and associated transforms. With very rare exceptions, earthquakes tend to occur exclusively along this ridge and along transform faults.

It is inferred from indirect evidence that the fault extends through the lower crust and upper mantle, possibly as a shear zone, but the deep structure of the fault is still the subject of ongoing research. The indirect evidence includes exhumed ductile shear zones that are interpreted to have been the lower crustal extensions of major transform faults. Also, geodetic data of plate motions on either side of a transform fault are explained if a vertical shear zone extends from the brittle region to great depth.

Continental transform faults are large strike-slip faults with up to hundreds of kilometers of offset. These faults are near-surface expression of a transform plate margin.

Traces of recent strike-slip faults on Earth's surface can be seen most clearly where they create distinctive landforms. For example, faulting has a strong influence on stream drainage patterns. The rapid lateral movement of a block on one side of a fault past an adjacent block on the opposite side of the fault causes streams to be progressively offset and sometimes even beheaded; the lower portion of the stream becomes disconnected from the upper portion (Fig. 10.6). The features so formed are termed *shutter ridges* because they shut off stream flow out of a canyon, causing water and sediment to accumulate behind them. Within these ponds, vegetation might be rapidly buried. This buried plant debris can be dated using a radioactive isotope of carbon (^{14}C) if the buried organic matter is less than about 50,000 years old. By dating the sediment in a pond that has been offset, geologists can determine the rate of movement along the fault. The measured offset of a stream channel, in combination with dated material from deposits in the offset channel, provides a direct measure of the average rate of fault movement.

Streams that cross active strike-slip faults are progressively offset by fault movement. Determining the age of materials, such as buried plant matter, in deposits of the offset stream channel provides a measure of the rate of fault movement.

Mechanics of Strike-Slip Faulting

We have discovered that the mechanics of offset for transform faults is strike-slip, in which movement along the fault is parallel or nearly parallel to the strike of the fault. Various physical models have been constructed to understand the mechanics of geologic processes, but few models are as illuminating as those that depict faulting. In these models, clay is spread evenly on a broad, flat base plate. This base plate usually has a seam that divides it into two pieces—or plates—that allow differential movement between them. The clay is given time to "settle"; then, while the clay is still quite wet, small electric motors slowly force the underlying base to move, causing deformation in the overlying clay (Fig. 10.7a).

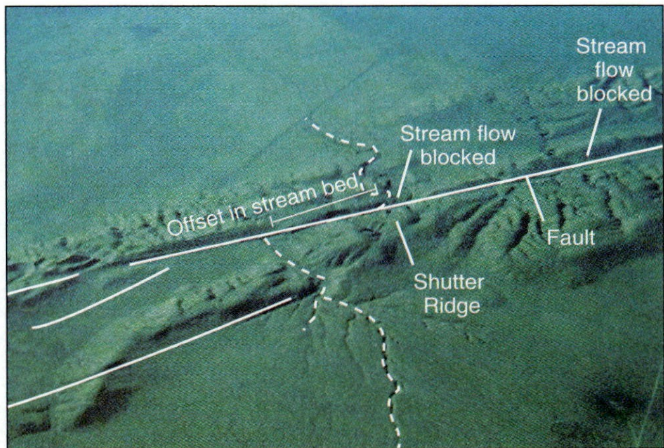

FIGURE 10.6 This oblique aerial perspective indicates how fault movement along the San Andreas fault blocks stream courses, causing streams to flow parallel to the fault.

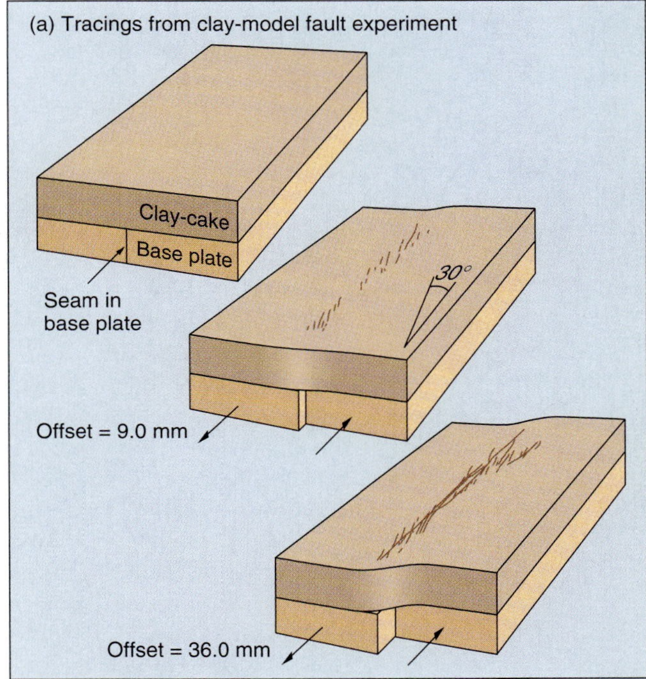

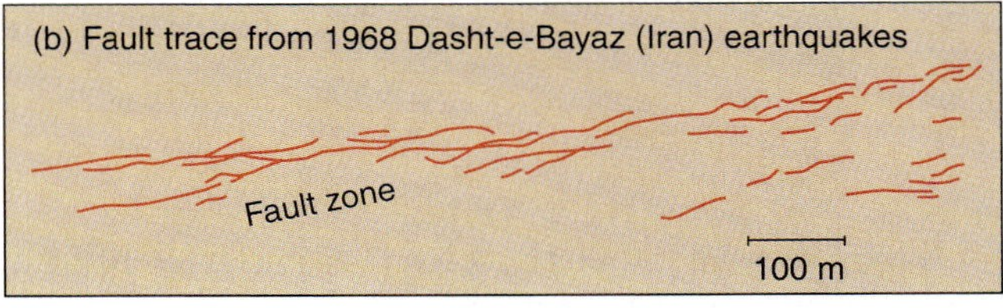

FIGURE 10.7 (a) A clay-model fault experiment. Clay is spread on a flat surface consisting of two base plates. With increasing strike-slip offset, the initial breaks in the clay occur at approximately 30° to the moving seam between the plates. As offset continues, the breaks merge. (b) A fault trace from an earthquake.

Focus On 10.2 THE SAN ANDREAS FAULT, CALIFORNIA

Although it had been generally recognized that a fault of some great length existed in California, its significance was not recognized until the San Andreas fault ruptured over a length of approximately 400 km, causing the San Francisco earthquake of 1906. Following this event, investigation of the fault immediately commenced.

For approximately 40 years after the earthquake, controversy raged as to whether or not the San Andreas fault had major or minor amounts of total offset. Many scientists believed that the fault had just a kilometer or so of offset; others argued for hundreds of kilometers of offset. Most geologists at the time were attempting to correlate a single feature across the fault—for example, a dike, a bed, or some other similar feature. John Crowell, a structural geologist, correlated a whole assemblage of rocks that occurred in outcrops on the two sides of the San Andreas fault, including ancient basement rocks, such as gneisses and anorthosites, as well as younger rocks such as granites and sediments. His correlation of the rock assemblages provided convincing evidence that the San Andreas fault actually had substantial offset—approximately 270 km of offset on the southern portion of the fault. Other geologists correlated a series of volcanic rocks suggesting as much as 365 km of offset on the fault's northern portion. These differing estimates of offset are the subject of continuing research, but the explanation probably lies in the distribution of offset in the southern portion of the San Andreas fault on other parallel faults.

The northern strand of the San Andreas fault is apparently mechanically decoupled from the southern strand by a zone of weakness near the towns of Hollister and Hayward in central California. In these areas, the San Andreas fault is in more or less constant motion, with a creep of 2 to 3 cm annually. This aseismic creep—described in Chapter 7—constantly relieves stress so that no elastic stress can accumulate to be released later in a large earthquake. With such a zone of weakness between the northern and southern portions of the fault, it seems unlikely that the accumulation of elastic stress in the northern strand of the San Andreas fault, which might ultimately lead to an earthquake, can be in any way related to stress accumulation or release in the southern strand. Simply put, the two strands appear to act independently, with little or no influence on each other.

Geologic mapping of the northern terminus of the San Andreas fault posed something of a puzzle until the late 1960s. Marine magnetometer (∞ see Chapter 6) and seismic investigations were able to follow the northern end of the fault offshore of California north to Cape Mendocino, where the fault abruptly turns at nearly right angles into the Mendocino escarpment. Seismicity along the fault drops to zero beyond the intersection of the Mendocino escarpment and the Gorda Rise, a small spreading ridge off the coast of northern California, indicating the end of the San Andreas fault at that point (Fig. 1).

To the untrained eye, the San Andreas fault zone may be difficult to detect. It generally forms a broad linear valley covered with vegetation. Usually, a number of springs lie along the edges of the fault valley and small ponds are scattered along the length of the valley. From a perspective on the ground, these fault zone areas could be most commonly described as gently rolling and rather scenic.

Geologically, the linear valley is created because the rocks within the fault zone are so shattered that there is little resistance to erosion. The low rolling topography is also the result of low erosional resis-

When movement of the plates is purely horizontal and is parallel to the seam in the base plate, strike-slip faults are formed in the clay. A fascinating feature of these experiments is that—contrary to what might be expected—faulting does not begin as a single straight line immediately above the seam in the underlying plates. Instead, the clay breaks into many parallel cracks, each at a slight angle to the seam (Fig. 10.7a). As the shearing motion continues, the cracks in the clay continue to propagate; at the same time, however, these cracks rotate to become parallel with the seam. Eventually, most of the offset in the clay does become concentrated in a single break that lies over the seam in the plates (Fig. 10.7a).

Typically, large faults are not a single trace extending for a long distance; rather, they are usually a series of

tance combined with intense deformation of the bordering rock assemblages. The springs along the edge of the valley are there because the fault represents a barrier to groundwater migration. As groundwater flows toward the fault zone, it cannot percolate through the fine-grained claylike fault gouge. As the water must go somewhere, it simply flows out onto the ground surface along the fault. In places where there is an abundance of water, these springs form small lakes, or ponds, which fill the deepest surface depressions, or *sags*, along the fault zone (Fig. 2). These elongate small lakes, or sag ponds, are a common feature along strike-slip faults and remain one of the easiest features to identify.

An aerial perspective provides a more complete picture of the fault features, the most striking of which is their linearity; the fault zones form nearly straight lines that cut across mountains and valleys with little deviation. Small linear cliffs are common. These are fault scarps from the most recent fault movements and resulting earthquakes.

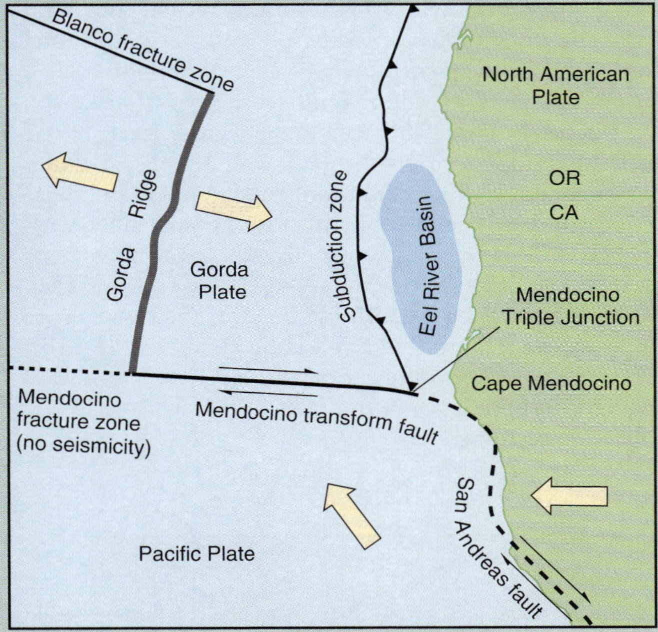

FIGURE 1 The northern end of the San Andreas fault. The fault is underwater just south of Cape Mendocino and bends sharply to intersect with the Mendocino transform fault. The triple junction is formed by the San Andreas fault, the Mendocino transform fault, and the subduction zone.

FIGURE 2 Groups of trees mark sag ponds along the San Andreas fault zone. These ponds are filled with water and are fed by springs on the northern (right) side of the fault zone.

approximately parallel fault strands, slightly offset from one another. These are called **en echelon structures**, and they are usually slightly inclined—sometimes up to about 30°—to the primary direction of fault movement. This feature is clearly shown in the clay-model experiments shown in Fig. 10.7.

Figure 10.7b, a fault map of a recent break along the Dasht-e-Bayaz fault in Iran, illustrates the similarity between the clay-modeling experiments and actual faulting. Note that the ground trace of the Dasht-e-Bayaz fault resembles the clay model after some movement. Moreover, transform faults that offset segments of the Mid-Atlantic Ridge (Fig. 10.5) are not single breaks but multiple fractures with the offset distributed across them. This behavior is also similar to the clay-model experiments.

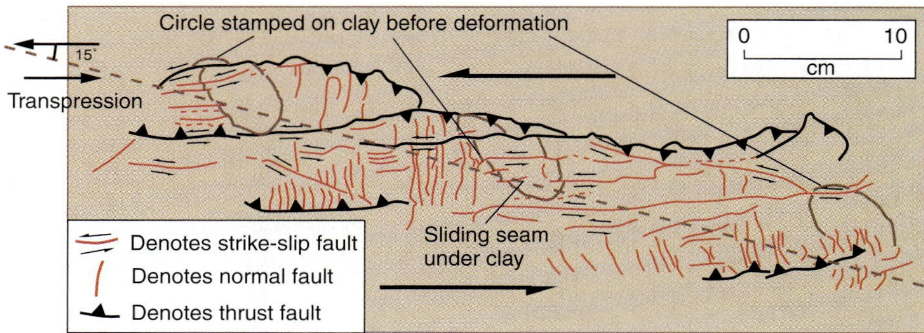

FIGURE 10.8 A clay model of a transpressive, left-lateral strike-slip fault showing sense of offset on faults. Small arrows indicate faults that are primarily strike slip; faults marked by triangles indicate primarily thrust-fault motion.

Other modeling experiments can be equally informative. When movement of the plates on opposite sides of the seam is at an angle to the seam (see Figs. 10.8 and 10.9), the overlying clay can be simultaneously sheared and compressed (**transpression**) or it can be simultaneously sheared and pulled apart (**transtension**). Under transpressive faulting conditions, the stresses compress opposite sides of a strike-slip fault together; this results in uplift near the fault, and in thrust faulting outward away from it. Figure 10.8 depicts a clay model of a transpressive strike-slip fault. In contrast, transtensive faulting, shown in Figure 10.9, causes the opposite sides of a strike-slip fault to pull apart, which results in normal faulting and subsidence.

Along transform faults, the motion of one plate may be oblique to that of the other along its entire length. In such cases, variations in the individual plate velocities can cause the fault boundary to be either transpressive or transtensive. Figure 10.10 gives an example of both a transpressional and a transtensional transform boundary in which the relative rates and directions of plate movement results in the specific boundary indicated.

Fault-Bend Geometries

Strike-slip faults that have experienced substantial offset are seldom straight for long distances; rather, they tend to be slightly sinuous or locally split into several parallel traces that subsequently merge again (see Fig. 10.7b). When a several-kilometer-long section of a strike-slip fault bends or curves beyond a few degrees, parts of the lithosphere adjacent to the fault will be placed under either compression or extension, depending on the sense of curvature relative to the direction of fault movement (Fig. 10.11). If the fault trace curves so that movement along the fault tends to pinch the two sides together (Fig. 10.11a), it is called a **confining bend**. If the curvature is opposite, we refer to it as a **releasing bend** (Fig. 10.11b). The effects of confining or releasing bends are always evident on both sides of the fault, as we will see in the following sections.

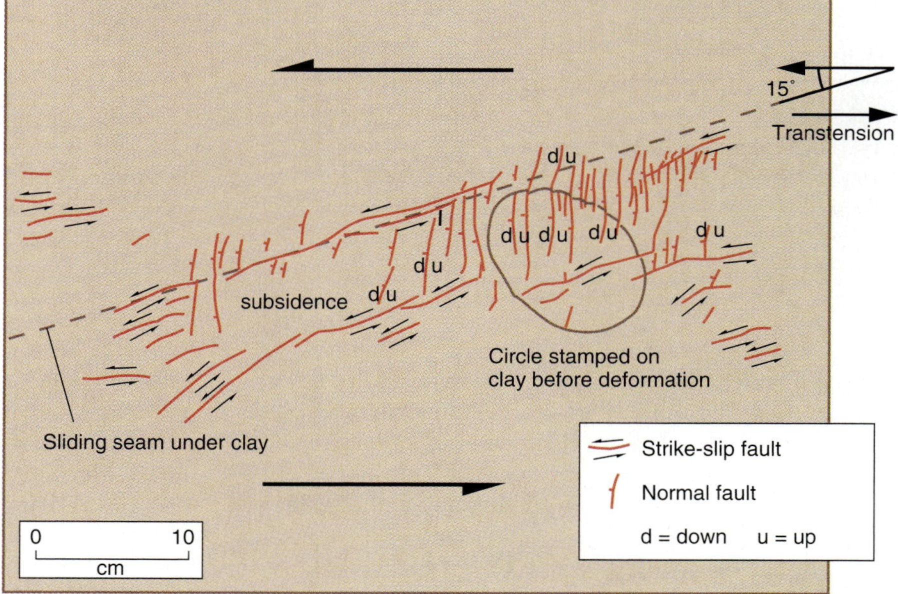

FIGURE 10.9 A clay model of a transtensive, left-lateral strike-slip fault. The deformed oval was a circle at the start of the experiment; the red lines represent faults. Note that most are normal faults in comparison with the transpressive model shown in Fig. 10.8, which produced a number of thrust faults.

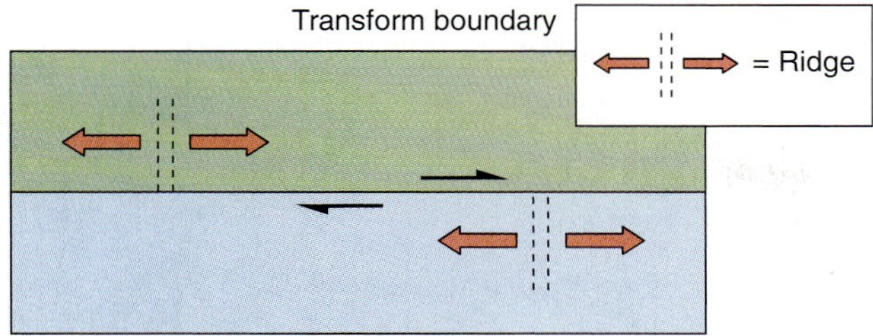

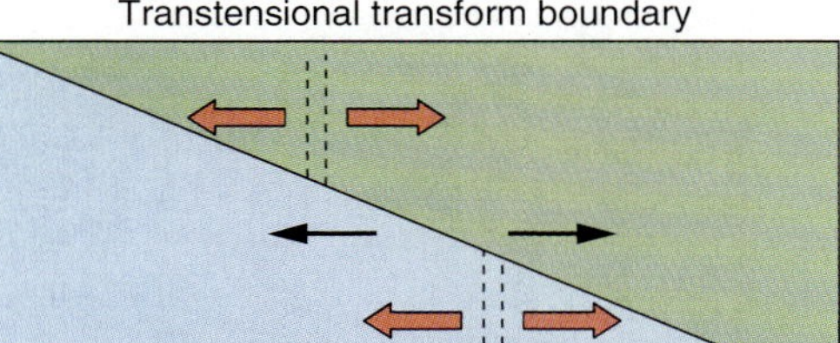

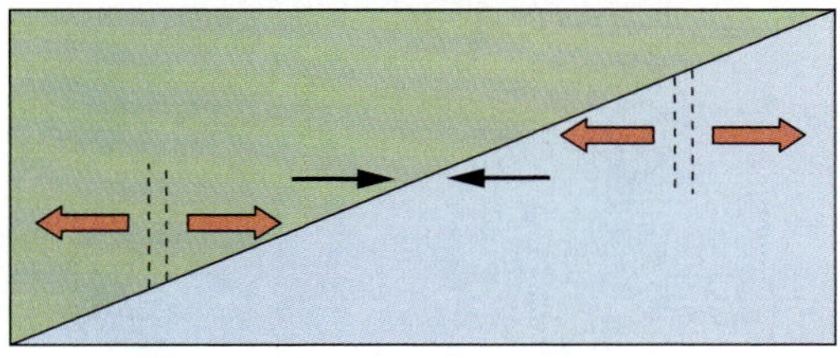

FIGURE 10.10 Examples of transpressional and transtensional plate boundaries.

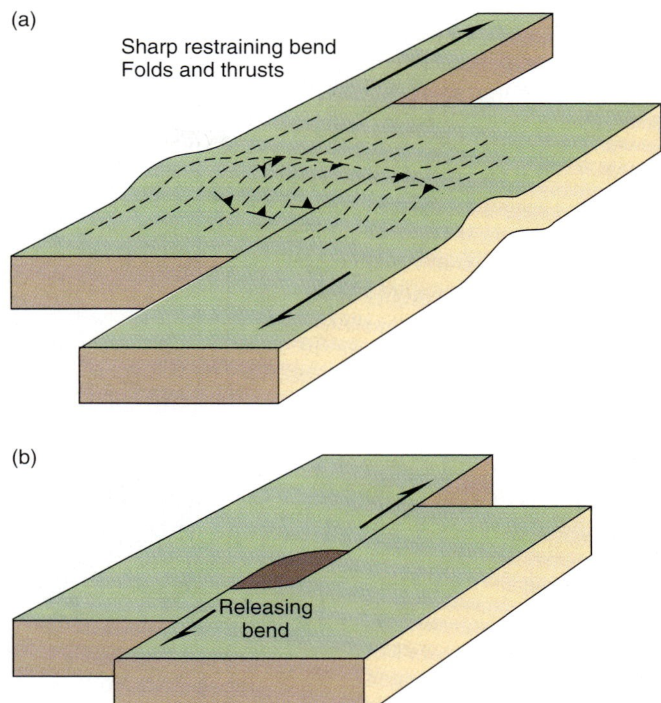

FIGURE 10.11 (a) A confining bend along a strike-slip fault showing the outward thrusting and resultant uplift adjacent to the bend. (b) A releasing bend along a strike-slip fault showing the subsidence between fault strands that results from fault motion.

Large strike-slip faults tend to be slightly sinuous or to split into several traces. If the fault trace curves so that movement along the fault compresses the two sides together, it is a confining bend. If the curvature is opposite, it is a releasing bend.

Confining Bends At a confining bend, the rocks on opposite sides of the fault are progressively forced together; movement along the fault forces material on opposing sides into the same space. As a result of the compression, local areas on both sides of the fault are tectonically uplifted. Thus, one feasible mechanism to accommodate compression is for the rock to heave upward (Fig. 10.11a).

Another mechanism to relieve the stress is the formation of other types of faults associated with the main strike-slip fault. The geometry of these secondary faults is upward and outward away from the compressional zone of the confining bend along thrust faults (Fig. 10.11a; Chapter 7). Depending on the strength of the rocks involved (specifically, their resistance to shear stress), thrust faulting may be concentrated primarily on one side of the fault or may occur on both sides.

Figure 10.12a shows the northern portion of the "big bend" in the San Andreas fault. The bend is a confining bend along this section of the fault, and the effects of this geometry are quite clear. Uplift of the Frazier Mountain area (Fig. 10.12b) is directly related to compression along the northern part of the "big bend." To the north and south of this bend, the fault zone contains many parallel fault scarps, which, side-by-side, represent a fault zone more than a kilometer wide.

Both translational and compressive stresses (transpression) occur in a confining bend of a strike-slip fault. Results of transpression include uplift, folding, and thrust faulting.

Regions on either side of the "big bend" have been subjected to intense compression because of the confining bend, which has caused uplift of this region. How much uplift has occurred? At the apex of the bend, rocks of high metamorphic grade are exposed at the surface (Fig. 10.13). Analyses of specific minerals sensitive to temperature and pressure during crystallization suggest that burial depths may have been as great as 15 to 20 km and that temperatures reached as high as 750° to 800°C. In fact, the metamorphic grade of these rocks is so high that the gneisses of the Frazier Mountain block experienced some melting and recrystallization (Fig. 10.13).

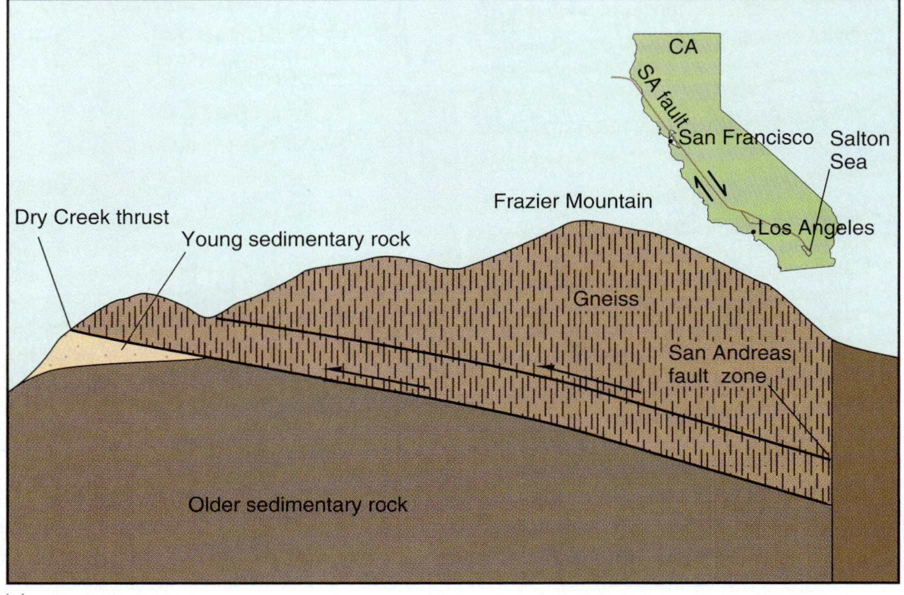

FIGURE 10.12 Characteristics of the northern portion of the confining "big bend" in the San Andreas fault, California. (a) Schematic cross section illustrating the relationship between the several thrust faults upon which Frazier Mountain rests, and the San Andreas fault. (b) An oblique aerial photograph of Frazier Mountain, an uplifted block. The solid bold lines are scarps resulting from the 1857 earthquake.

The observation that rocks that were formed at such great depths are now exposed at Earth's surface in a transpressive bend on the San Andreas fault is evidence for a great amount of uplift. The uplifted terrane of the Frazier Mountain block is a good example of a source area for sediments at a confining bend in a strike-slip fault. Uplift increases the slope of streams in the area and, as a result, also increases the rate of erosion. The uplift rate is geologically rapid; many landslides and rock-falls deposit very coarse-grained sediment in the adjacent basins (Fig. 10.13).

As fault activity decreases, stream erosion and sedi-

FIGURE 10.13 Old, high-grade metamorphic rocks thrust-faulted over young, unaltered sandstones. (a) Dry Creek thrust fault, part of the Frazier Mountain thrust fault system, exposed in wall of stream canyon. The view is toward the north; the fault is inclined northward approximately 30° and crosses the hillside slightly upward toward the right side of the photo. (b) A very high-grade partially melted metamorphic rock (1.7 billion years old) showing several zones of melt accumulation. For scale, the pocketknife is 9 cm long. (c) Young (less than 5 million years old) sedimentary rock, a pebble conglomerate that has been overriden by the Dry Creek thrust. This rock is uncompacted, and can easily be picked apart by hand.

mentation become more important and, ultimately, sediments in the adjacent basins become finer grained. If new uplift occurs, a new flood of coarser sediment is introduced into the basin. Thus, in basins adjacent to faults, the sediments vary from coarse to fine in a somewhat cyclical manner, depending on "cycles" of fault movement.

Over the long expanse of geologic time, sedimentary deposits within and adjacent to confining bends along large continental transform faults have poor **preservation potential**; that is, stream-channel deposits and other small features offset by fault movement are usually not preserved in the geologic record. Such deposits are more commonly eroded as a result of progressive uplift or are tectonically eroded along the confining bend by continued fault movement.

Mechanically analogous to a confining bend, long sections of a continental transform fault may be under transpression. These faults generally show oblique offset, where one side of the fault zone is forced inward, or compressed, against the fault trace. This is a regional effect compared to confining bends along a fault, which are local transpressive environments (Fig. 10.10 and 10.11a). On the scale of lithospheric plates, the compression is much greater than along a confining bend in a fault; for that reason, the effects are larger. The faults are forced upward and outward, bending them into a fanlike geometry called a *flower structure* (Fig. 10.14).

Releasing Bends In contrast to the compression that characterizes a confining bend, releasing bends are areas of relaxation, or extension. Analogous to transpression, the term for translational movement in combination

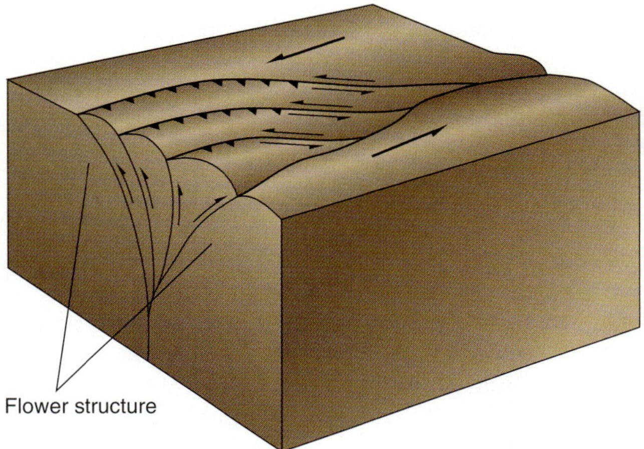

FIGURE 10.14 Diagram of a fault flower structure, which is the characteristic transpressive structure. Compare this feature with the less extensive features shown in Figures 10.8, and 10.11.

with extension is transtension. Thus, releasing bends are transtensive environments; rather than being uplifted, these areas subside (sink) and become sites for sediment deposition. The sedimentary basins in transtensive environments are typically small—usually only a few dozens of square kilometers. Because these basins are bounded on both sides by faults, they will be fragmented by continuous fault movement; if the sedimentary sequences survive at all in the geologic record, they will survive as wedges of sedimentary rock caught between faults.

Releasing bends are closely associated with confining bends. In fact, the sediments eroded from the uplifted transpressive portions of the fault (confining bends) are commonly deposited in nearby subsiding basins in the transtensive portions of the same fault (releasing bends). Because of the association between uplift and subsidence, sediment transport distances tend to be short, occurring approximately along the fault trace from the uplifted regions to the subsided basins. Therefore, both the coarse and fine detritus are not well segregated. Sedimentary rocks deposited in these environments tend to be immature poorly sorted conglomerates and sandstones. Sediments derived from one side of a fault may be simply dumped locally on the opposite side of the fault. This sediment transport and reworking required to segregate coarse and fine sedimentary material does not occur, leaving the sediments poorly sorted.

Fault movement causes rapid local uplift and subsidence, and it also creates abrupt discontinuities among stream sizes, drainage patterns, or topography on the two sides of a fault trace. Stream channels on opposite sides of a fault might not have adequate time to build a drainage basin and a sediment distribution system. Rocks of different hardnesses or soil types might be emplaced on opposite sides of a fault, yielding different river-channel characteristics, plant communities, or soil and slope features.

Releasing bends in the trace of a strike-slip fault experience both translational and extensional stresses (transtension). These bends result in subsidence and, therefore, become small depositional basins. The sediments are commonly provided by the erosion of uplifts at confining bends.

Pull-Apart Basins A "pull-apart basin" is another structure that may be associated with a releasing bend in a large strike-slip or continental transform fault (Focus 10.3). Such a structure is extensional in nature. We can understand how a pull-apart basin forms if we picture a left-lateral strike-slip fault with a right-angle step or jog to the left (Fig. 10.15a). The Dead Sea in Israel is an example of such a structure (Fig. 10.15b). As the fault moves, the two sides of the fault are pulled apart, forming a more or less rectangular, fault-bounded trough. The crust is stretched so that normal faults form across the ends of the offset to accommodate the extension.

Clearly, the formation of a pull-apart basin requires extension of the crust, and continued fault movement may cause sufficient crustal thinning to rupture the crust, giving rise to volcanic activity. Some pull-apart basins contain extrusive volcanic rocks alternately layered with the lower part of the sedimentary succession.

Pull-apart basins are specifically related to large strike-slip faults where the crust has been thinned by substantial fault offset. The crust may be sufficiently thinned that volcanism occurs early in the basin history.

Duplex and Offset Fault Structures At the point where strike-slip faults curve and offset the relative motion, **duplex structures**—a braided pattern of several fault traces—may form instead of a simple bend (Fig. 10.16). They are called duplexes because they duplicate the main fault trace in a series of faults in the bend. Analogous to confining or releasing bends on a curved fault trace, duplexes and offsets can be either compressional or extensional. Compressional duplexes form in a transpressive tectonic environment along a fault trace; extensional duplexes form in a transtensive regime: An example of a compressional duplex on the San Andreas fault is shown in Figure 10.17.

Most of the characteristics of confining and releasing bends also hold true for duplexes. Compressional duplexes form uplifted areas. The bounding fault strands—the outermost faults of the duplex structure—will show oblique offset with a component of outward thrust faulting. Extensional duplexes form small, fault-bounded basins, with the bounding fault strands showing a component of normal slip. As in transpression, compressional duplexes may exhibit substantial uplift, and extensional duplexes will show subsidence. Thus, along and within the duplex structure, basement rocks might be exposed immediately adjacent to thick piles of sediment.

Duplex structures result from the duplication of a fault trace into a braided pattern of several fault traces and may be either compressional or extensional.

Duplex structures tend to be short lived. With continued movement along the fault, the interconnecting fault strands gradually rotate to become parallel to the primary direction of faulting. Thus, duplexes tend to be smeared out to form isolated fragments of rock surrounded by faults (Fig. 10.17). In a compressional duplex setting, the potential for preservation of uplifted sedimentary wedges is considerably less than it is in basins formed extensional duplexes. Of course, larger sedimentary basins are more likely to have segments preserved fairly intact than are smaller basins.

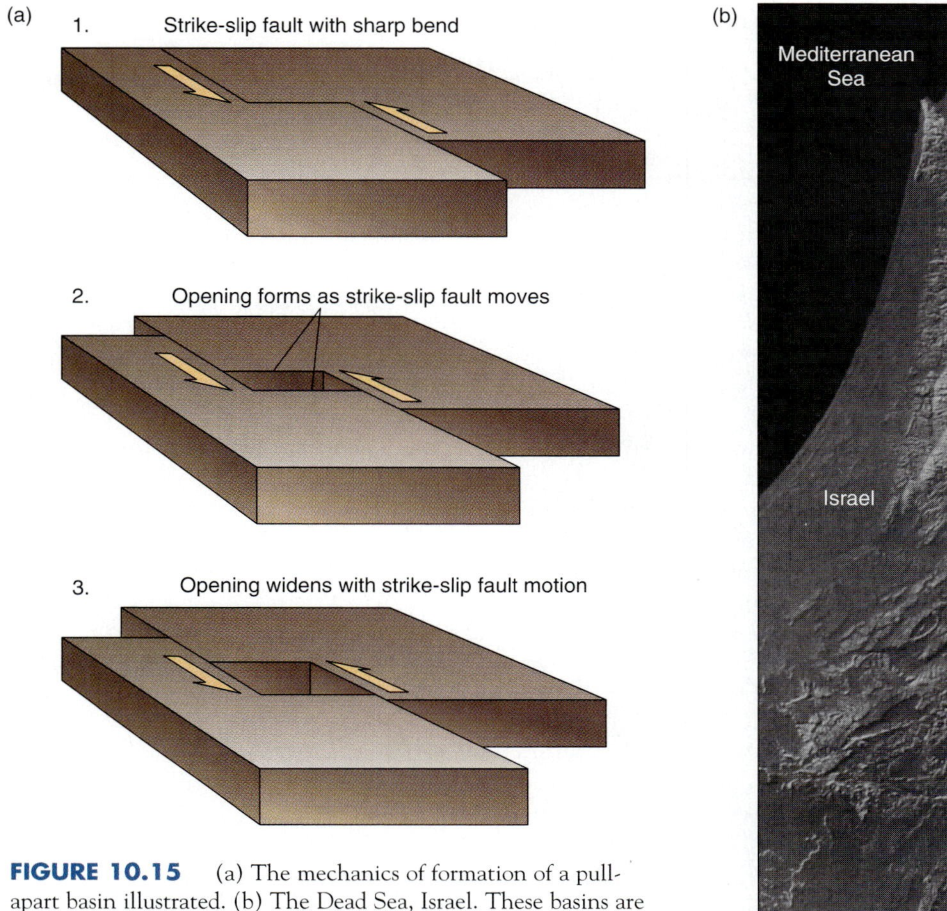

FIGURE 10.15 (a) The mechanics of formation of a pull-apart basin illustrated. (b) The Dead Sea, Israel. These basins are similar to those in a releasing bend; however, pull-apart basins are larger and penetrate the crust more deeply .

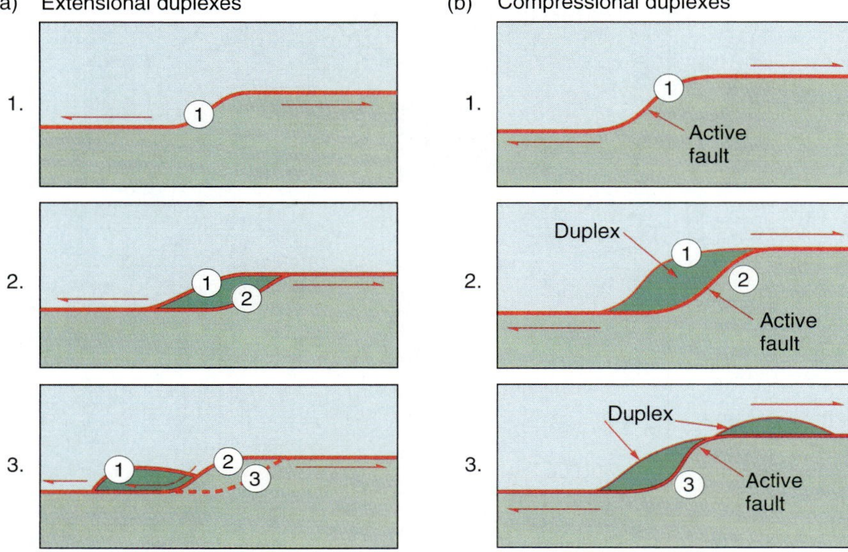

FIGURE 10.16 Progressive development of duplex structures associated with transform faults. Circled numbers indicate order of fault development in duplexes. (a) Extensional duplex formed at a releasing bend in the fault. (b) Compressional duplex formed at a constraining bend in the fault.

Focus On 10.3 THE SALTON TROUGH OF CALIFORNIA: A PULL-APART BASIN

Near the southern end of the "big bend" in the San Andreas fault, the fault trace divides into two major strands: the San Jacinto fault and the Elsinore fault. South of the Salton Sea and into Mexico, the trace of the San Andreas fault is difficult to follow. Geologists working in this area have demonstrated that fault offset, and therefore the interpreted main fault trace, shifts from the San Andreas fault zone east of the Salton Sea to the San Jacinto fault zone on the west side of the Salton Sea. The geometry is that of an extensional offset, providing favorable conditions for subsidence and the creation of a basin (Fig. 1).

Initiation of sedimentation in the basin was marked by the influx of coarse mudflow and debris-flow deposits derived from the margins of the basin. These deposits overlie a slightly older marine sequence (about 3 million years of age) of a finer grain size. In the northern portion of the trough, tectonic activity is illustrated by recurrent earthquakes and uplifted and folded very young sediments. Sedimentation lags behind subsidence. The southern portion of the basin, however, is filled by sediments from the Colorado River delta; here, sedimentation exceeds the rate of subsidence.

Drilling in this area has revealed volcanic rocks (sills and flows) similar in composition to oceanic basalts. Detailed gravity measurements indicate the presence of dense mantle material underlying and intruding the basin floor. High heat-flow measurements are characteristic of the Salton Trough, due to hot saline water circulating within the pores of the sedimentary rocks. Originally, this pore fluid was probably groundwater that reacted with, and was heated by, hot magmas. Sediments at moderately shallow depths are in the processes of metamorphism at high temperatures.

The basin lies in a peculiar tectonic setting; that is, along the extension of the East Pacific Rise in the Gulf of California. What the future holds for preservation of the Salton Sea deposits is unclear. Transform faulting may dissect them, so that they will appear either as isolated blocks of sediment or as metamorphic rocks within the fault zone. Equally likely, continued sea-floor spreading in the Gulf of California may cause it to widen, with the deposits of the Salton Trough partly consumed by magmatism and partly preserved as deposits marginal to an opening ocean basin.

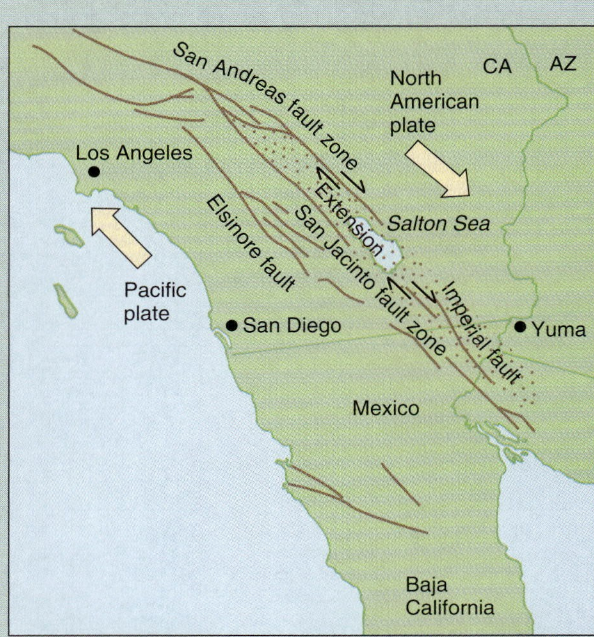

FIGURE 1 Fault setting of the Salton Sea region. Movement along the San Andreas fault diminishes near the southern end of the Salton Sea, and the major transform motion is taken over by the San Jacinto fault zone. The fault displacements create an extensional offset between the San Andreas and the San Jacinto faults. This region (stippled) is now a subsiding basin.

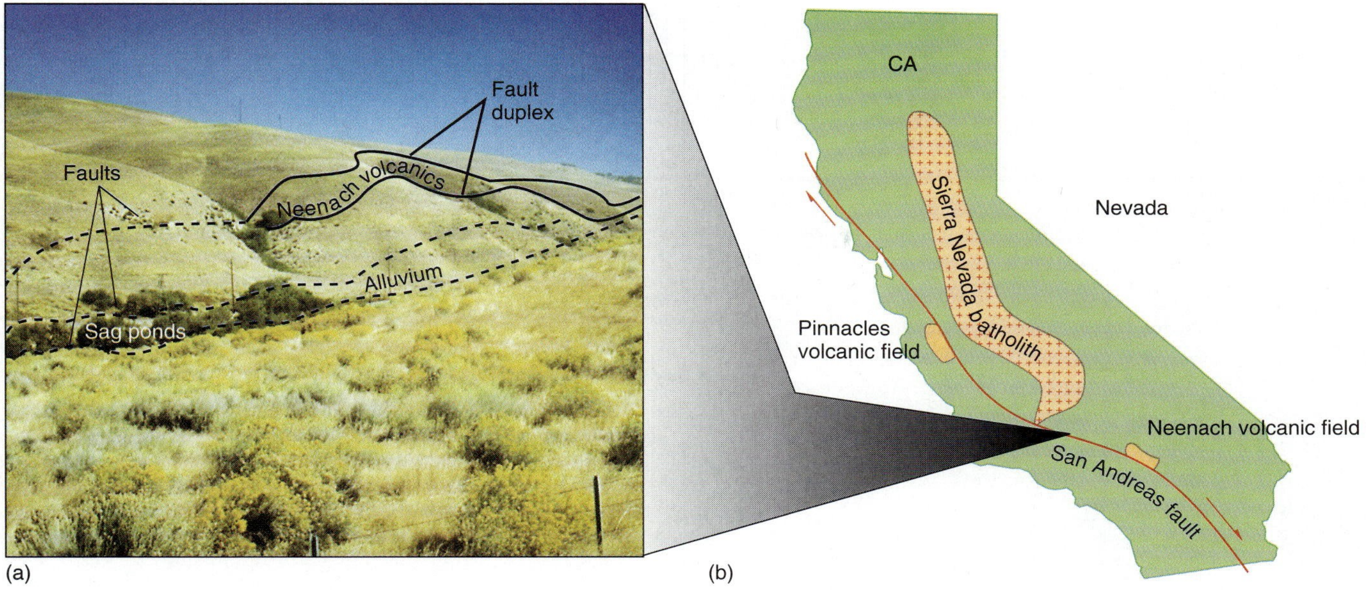

FIGURE 10.17 Example of a duplex formed along the San Andreas fault, California. The photograph (a) shows a fault-bounded duplex of volcanic rocks, which correlate with rocks found in the Neenach Volcanic field farther to the south-east. The map (b) shows the location of this duplex, relative to the Neenach and Pinnacles volcanic fields. These fields were originally one single field of volcanic rocks which has been cut, and the two sides displaced from each other by the right lateral motion of the San Andreas fault. The small duplex shown in (a) represents a sliver of volcanic rocks transported along the San Andreas fault as shown in Fig. 10.16b.

Oceanic Transform Faults

We started this chapter with a consideration of continental transform faults because they are easily observed. Nevertheless, recent oceanographic research has provided a comprehensive picture of the topography (bathymetry), composition, and age of the ocean floor, including the numerous transforms that cross-cut it. Remember that the locations and geometries of transform plate boundaries are continually changing through time as they interact with other plate boundaries.

One major advantage to the undersea examination of spreading ridges and transform faults is that the features are fresh and well exposed, not covered by thick soil or vegetation. Therefore, geophysical studies have clearly revealed the fundamental basement features without confusing information introduced from the surface cover. Numerous research groups have performed visual surveys of oceanic transforms, and many geophysical studies—including seismic profiles and magnetic surveys—have also been conducted across the **transform faults** and fracture zones.

Transform Faults and Fracture Zones

Mid-ocean ridges are segmented, with spreading axes offset from each other along the length of a ridge (Fig. 10.3). These transform faults are usually perpendicular to the ridge segments. Relative motion of the plates generated at successive ridge segments is taken up along transform faults, which terminate in fracture zones. Both transform faults and fracture zones separate lithosphere of different elevations. Across transform faults, the lithosphere on one side moves relative to the other; in contrast, no relative movement occurs across fracture zones.

How do fracture zones form? At the junction of a transform fault and a ridge segment, new ocean floor from the ridge lies next to older, partially subsided ocean floor extruded from the ridge segment at the other end of the transform fault. Cooling of the lithosphere causes the density to increase with age; thus, older oceanic crust has a lower isostatic elevation than does younger crust (∞ see Chapter 8). As spreading continues, this elevation difference propagates out onto the ocean floor. The result is a topographic scar, or fracture zone, extending long distances across the ocean (see Fig. 10.5). The extent of such fracture zones indicates that segmentation is a permanent feature of sea-floor spreading. Fracture zones trace the *relative plate motions* on either side of spreading ridges. For example, the fracture zones in the South Atlantic Ocean trace the motion of South America from Africa.

An empirical relationship has been established between the age of oceanic lithosphere and its elevation. The isostatic equilibrium elevation for oceanic lithosphere diminishes approximately with the square root of its age (∞ see Chapter 8). Thus, the prominent scarps along fracture zones remain, owing to an age difference in the oceanic crust across them; in the case of the Mendocino escarpment, the age difference is approximately 35 million years and the scarp is more than 2 km high. Adjacent to the prominent scarp, the fracture zone is usually marked by

Focus On 10.4 THE BEGINNING OF CALIFORNIA'S TRANSFORM MARGIN

The East Pacific Rise was once a continuous spreading ridge that ran along the entire length of the western Pacific. Geologists have deduced the former location of this rise from magnetic anomalies (see Fig. 10.18) in the oceanic lithosphere. From this work, it is clear that the East Pacific Rise was offset by several large transform faults (Fig. 1a). When this rise impinged upon the North American Plate—approximately 20 million to 30 million years ago—the point of the transform offset touched first (Fig. 1b). Increased impingement of these two lithospheric plates, however, caused the East Pacific Rise to subduct beneath the continental mass; two triple junctions were thus formed: the Mendocino and the Rivera triple junctions. Further subduction of the still-active spreading ridge caused these junctions to migrate away from one another. This migration of the triple junctions is caused by the slight angle between the subducted ridge and the continental margin. These migrating triple junctions moved in opposite directions, creating a propagating transform fault. This transform gave rise to the modern San Andreas fault system (Fig. 1c).

The orientations of transform faults on the small spreading-ridge segments in the Gulf of California are parallel to the San Andreas fault (see Fig. 10.18). The magnetic anomalies offshore from Baja California, however, suggest that the ridge segments and transform orientations in the Gulf of California were not parallel to the ridge orientation prior to subduction. It seems probable that as the ridge approached the continent, the two began to interact because a spreading ridge is the most buoyant of all oceanic crust. For that reason, the ridge tends to resist subduction in somewhat the same way that continents resist subduction. One possible interaction involves distortion of the ridge by drag against the overriding continent. Reconstructions of past locations of dissected and displaced rock types have made it clear that another type of interaction certainly did occur; that is, pieces of the continental mass were slivered off and translated northward by strike-slip faulting. The opening of the Gulf of California occurred 5 million years ago, and this opening corresponds with the origin of the modern San Andreas fault.

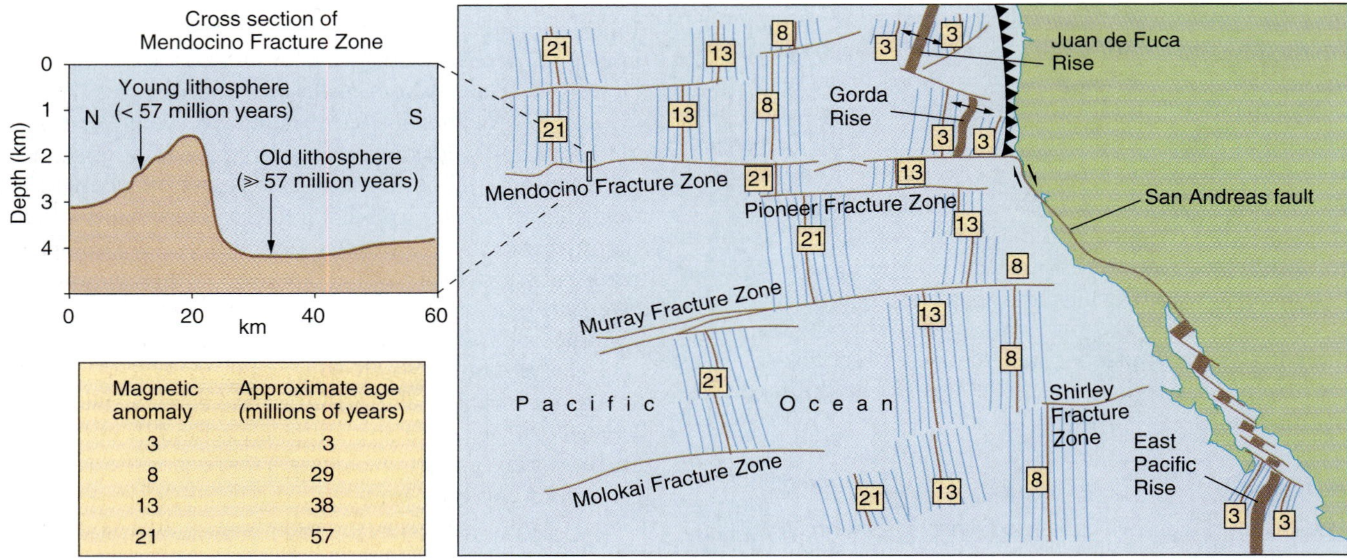

Magnetic anomaly	Approximate age (millions of years)
3	3
8	29
13	38
21	57

FIGURE 10.18 Structure of the northeastern Pacific Ocean floor. The numbered lines represent specific magnetic signals that can be correlated across the entire ocean basin. These magnetic anomalies have been dated (lower chart) and clearly indicate the structure of the ocean lithosphere. Upper chart: Topographic profile across Mendocino fracture zone. This fracture zone is an extension of a transform fault.

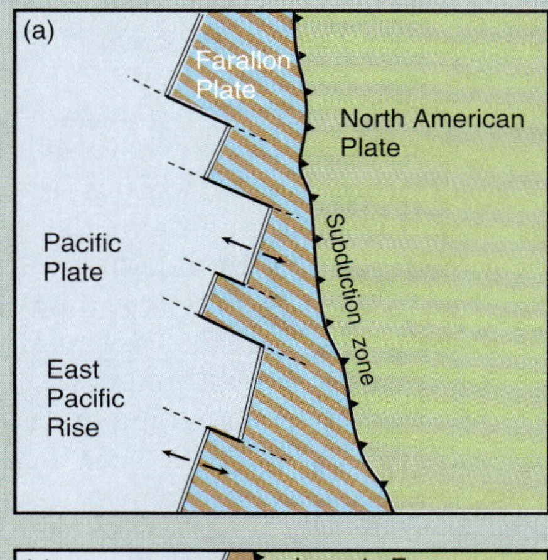

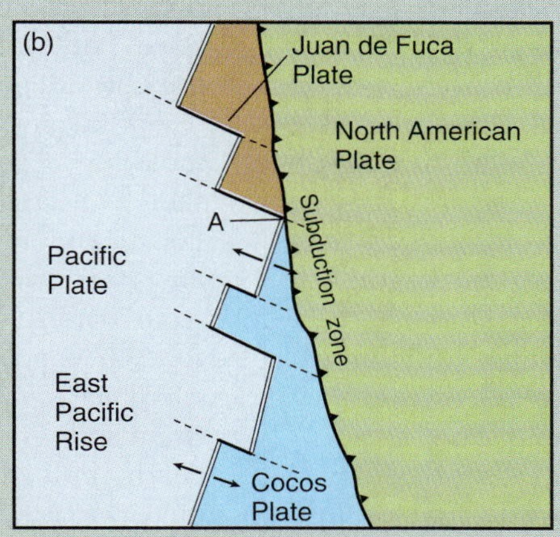

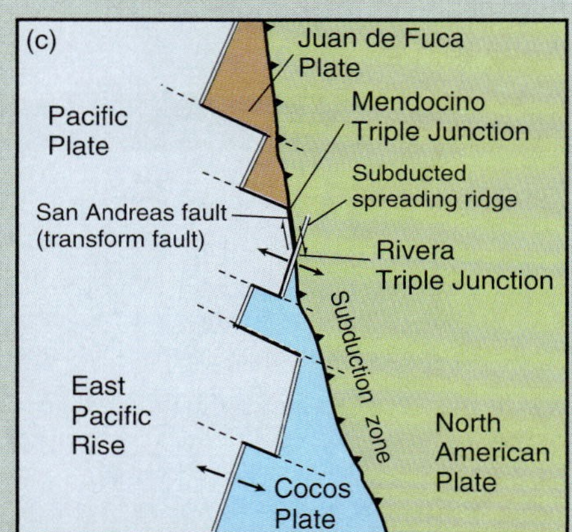

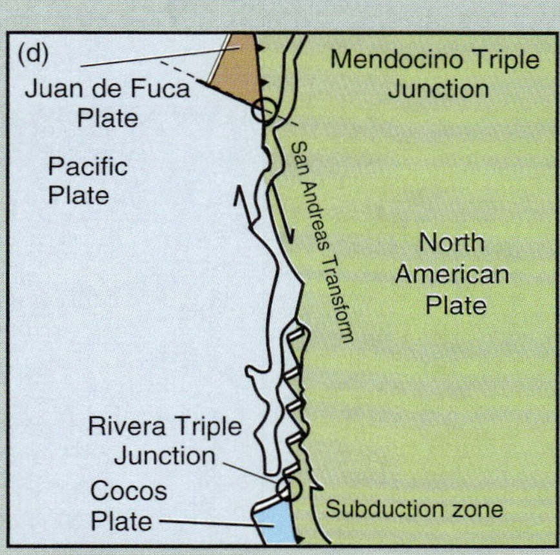

FIGURE 1 The initiation of the San Andreas transform fault. (a) The East Pacific Rise separates the Pacific Plate from the Farallon Plate. Approximately 24 million years ago, these two plates and the North American Plate were converging. (b) Approximately 20 million years ago, the Pacific and North American plates came into contact. The primary contact occurred between the spreading ridge and a transform fault (point A). (c) As the East Pacific Rise and its transform faults are subducted, two triple junctions are formed. The Mendocino triple junction migrates slowly northward, and the Rivera triple junction migrates slowly southward. (d) Present configuration of the San Andreas fault system.

a valley or trough; on the other side is a relatively gentle slope out of the trough away from the zone. Fracture zones are large features; the trough can be as much as 2 km deep and 20 km or more wide (Fig. 10.18).

These fracture zones have been studied in some detail and are quite complex in structure. The fault or fracture is not a simple shear; instead, it may be a series of short, *en echelon* strike-slip faults a few kilometers in length. These smaller features, though dominantly strike-slip in sense, probably open as extensional cracks at the ridge crest, and thus are aligned at an angle of approximately 30° to the transform fault trace, as seen in the clay experiments of Fig. 10.7.

Living with Geology

Can Earthquakes Be Controlled?

Some of the great continental transform faults are near populated centers. The Managua fault in Nicaragua and the San Andreas fault in California are two examples. The potential for widespread damage due to large earthquakes on these faults has led to the proposal that earthquakes might somehow be controlled. In fact, it has already been shown that this is possible on a limited scale. In the late 1960s, scientists conducted experiments in which they initiated earthquakes and then halted them.

It seems that the U.S. Army had a severe problem with the disposal of wastes from chemical manufacturing processes carried out at the Rocky Mountain Arsenal near Denver, Colorado. After much consultation, the ideal solution appeared to be the creation of a disposal well 3670 m deep into the basement gneiss. A well was subsequently drilled. Injection of fluids into the well began on March 8, 1962. At times, fluids were forcibly injected into the well; at other times, gravity flow was used. On April 24, 1962, a series of earthquakes occurred with epicenters within 8 km of the disposal site. Some of the larger earthquakes had a Richter magnitude between

3 and 4, were felt over a wide area, and minor damage was reported. Finally, in 1966, it was established that a clear correlation existed between the volume of fluid pumped into the disposal well and the occurrence of these local quakes.

In 1967, C. B. Raleigh and associates established seismometers at the Rangely Oil Field, near Vernal, Utah. The oil-recovery technique used in this oil field employs forcible injection of water into the oil-bearing zone to displace the oil. In one 10-day period, 40 earthquakes were recorded along a known fault in the oil field. With the cooperation of the Chevron Oil Company, Raleigh and his group varied water injection volumes and pressures and found that they could turn earthquakes on and off by manipulating the fluid pressure. Through these experiments, it might seem that Raleigh and his co-workers had discovered how to control all earthquakes, but the real situation is much more complex.

We know from previous discussions (∞ see Chapter 7) that large faults slowly accumulate elastic stresses until the strength of the rock is exceeded. At that point, the rock yields abruptly in an earthquake. The theory of earthquake control seems reasonable—many small earthquakes instead of one large one. However, two considerations must be taken into account.

First, the number of magnitude 3 earthquakes that would be required to dissipate the stored energy for a potential magnitude 8 earthquake is more than 15 million. The second consideration presents a much more difficult dilemma—one of responsibility.

For instance, suppose that a large fault had been accumulating elastic stress for 150 years. We have only a sketchy picture of how this accumulated stress is stored. For maximum impact in this scenario, imagine that the energy is stored near the fault. If an earthquake control team were to pump water at high pressure into the fault zone, it is possible that they might induce a very large earthquake rather than a series of small ones. Who is financially responsible for the damage and possible deaths brought about by the unexpectedly large earthquake? Until these sociological issues are resolved, earthquake control will probably remain a dream.

SUMMARY

- In most cases, transform faults laterally offset segments of spreading ridges; a few transform faults offset segments of subduction zones.
- Transform margins are characterized by strike-slip fault motion, shallow earthquake hypocenters, and an absence of magmatism. They are conservative boundaries because lithosphere is neither generated nor consumed.
- Continental transform faults are large strike-slip faults with up to hundreds of kilometers of offset. These faults are the near-surface expression of a transform plate margin.
- Streams that cross active strike-slip faults are progressively offset by fault movement. Determining the age of materials, such as buried plant matter, in deposits of the offset stream channel provides a measure of the rate of fault movement.
- Large strike-slip faults tend to be slightly sinuous or to split into several traces. If the fault traces curves so that movement along the fault compresses the two sides together, it is a confining bend. If the curvature is opposite, it is a releasing bend.
- Both translational and compressive stresses (transpression) occur in a confining bend of a strike-slip fault. Results of transpression include uplift, folding, and thrust faulting.
- Releasing bends in the trace of a strike-slip fault experience both translational and extensional stresses (transtension). These bends result in subsidence and, therefore, become small depositional basins. The sediments are commonly provided by the erosion of uplifts at confining bends.
- Pull-apart basins are specifically related to large strike-slip faults where the crust has been thinned by substantial fault offset. The crust may be sufficiently thinned that volcanism occurs early in the basin history.
- Duplex structures result from the duplication of a fault trace into a braided pattern of several fault traces and may be either compressional or extensional.

KEY TERMS

en echelon structure, 261
transpression, 262
transtension, 262
confining bend, 262
releasing bend, 262
preservation potential, 265
duplex structure, 266
fracture zone, 269

QUESTIONS FOR REVIEW AND FURTHER THOUGHT

1. Along spreading ridges, far more earthquake energy is released in transform faulting than is released in normal faulting. Why do you think this is so?
2. If the San Andreas transform plate boundary extends as deep as 100 km, why are the hypocenters of earthquakes along this fault only as deep as 12 to 15 km?
3. Explain how sag ponds or offset stream valleys can be used to determine the *rate* of fault movement.
4. Given that the depth of ocean floor depends on the square root of time since formation, where on an individual fracture zone would you expect the scarp height to be greatest? Explain your reasoning.
4. Give the requirements for fault motion and fault offset for a step in a strike-slip fault to generate a pull-apart basin.
5. Why are slivers of sediment or other types of rock trapped in fault duplexes believed to have only a limited chance for long-term preservation?
6. In a fault basin—particularly in a pull-apart basin—can you predict a distinct order in the sediments? For example, is the age of any of the sediments in the basin related to the direction of fault movement?

11 Plate Interiors

Hawaii, an island in the Pacific Ocean, is built almost entirely from the volcanic products of intraplate magmatism. The Hawaiian Islands are thought to reflect the trace of a mantle plume over which the lithosphere has moved.

Overview

- The interiors of plates are usually tectonically quiet regions.
- The oldest parts of a continent are usually located within plates rather than at plate margins.
- Earthquakes are rare in plate interiors but can be devastating when they do occur.
- Widespread sediment accumulation occurs in shallow seas on continental shelves.
- Volcanoes located away from plate margins are mostly associated with rising mantle plumes.
- Mantle plumes are an integral part of plate tectonics. They can give rise to voluminous flood basalts and are important in rifting continental lithosphere.

Introduction

We have discovered that most geologic activity occurs at plate margins: earthquakes, volcanoes, and the formation of mountains. We might think, then, that the interiors of plates—located well away from their margins—must be very safe places to live. We know, however, that the shapes of plates are continually changing, albeit over geologic time scales that are nearly imperceptible in our lifetimes. The Appalachian Mountains of the eastern United States (Fig. 11.1) are currently located a long distance from a plate margin; the eastern boundary of the North American Plate is the Mid-Atlantic Ridge spreading center located 3000 km to the east of the mountains, while the western margin of this plate is the San Andreas fault system and the Cascades subduction zone, 3000 km to the west. If mountain ranges result from processes at plate margins, as we have claimed to be the case in general, then how do we explain the Appalachians?

We saw in Chapter 9 that this mountain range, in fact, was the site of an ancient plate boundary, where North America collided with Europe some 250 million to 450 million years ago (see Fig. 9.19).

Indeed, remnants of many mountain ranges lie in continental plate interiors. Many of these ranges have been worn down to flat landscapes called **peneplains,** the roots of which are now buried beneath sediments. In contrast, oceanic plate interiors are relatively young and do not bear the scars of ancient collisions. Nevertheless, the interiors of oceanic plates are far from featureless. Most oceanic islands that are not at plate margins are the result of hot-spot volcanic activity. Examination of seafloor topography shows many submarine mountains, or **seamounts**, and huge submarine plateaus exist that stand much higher than the surrounding abyssal plains. These features are the result of within-plate, or intraplate volcanism, which we shall explore in this chapter.

FIGURE 11.1 A satellite photograph of the Appalachian Mountains, a 300-million-year-old eroded mountain belt in eastern North America.

Plate interiors are tectonically quiet today; they are not typically associated with active deformation, earthquakes, or volcanoes. What is now a plate interior, however, could have been a plate margin in the past.

The Structure of Plates

To understand the structure of lithospheric plates (Fig. 11.2), we must consider how they were originally generated. We know that oceanic lithosphere is created along spreading ridges at divergent plate margins, a process discussed earlier (∞ see Chapter 8). The origin of the continental lithosphere, however, is more uncertain. On average, continental lithosphere is much older and more deformed than is oceanic lithosphere. As a result, the history of the continents is more complex and more obscure: Vast tracts of continental lithosphere were generated early in Earth's history, possibly by processes at plate margins or possibly by processes unrelated to plate tectonics. Given the perspective of geologic time, it is possible that some of the processes that generated the continental crust are no longer as important on Earth.

Oceanic Lithosphere

The structure of the oceanic lithosphere is quite well known from both oceanic drilling studies and the study of ophiolites (trapped fragments of oceanic lithosphere caught up in continental collisions [∞ see Chapter 8]).

The oceanic crust is basaltic in composition (Fig. 11.2). The uppermost portion comprises pillow lavas (basalts erupted underwater onto the sea floor) covered by a thin layer of deep-sea sediments. This overlies, and is interbedded with, a sheeted dike complex, which is a series of subvertical dikes intruding the sediments and volcanics. Beneath the sheeted dike complex—and, in fact, the source of it—is intrusive gabbro and mantle peridotite. This layered structure seems to occur rather consistently throughout the ocean basins, and it relates directly to the processes of magmatism and spreading at the plate boundary (∞ see Chapter 8). The mantle part of the oceanic lithosphere is largely composed of peridotite from which basalt melt has been extracted; it is called harzburgite (Chapter 8). It is typically 100 km thick—in contrast with the 7-km average thickness of the crust.

In contrast, even though we know much more about the geology of the *surface* of the continents (after all, it is much

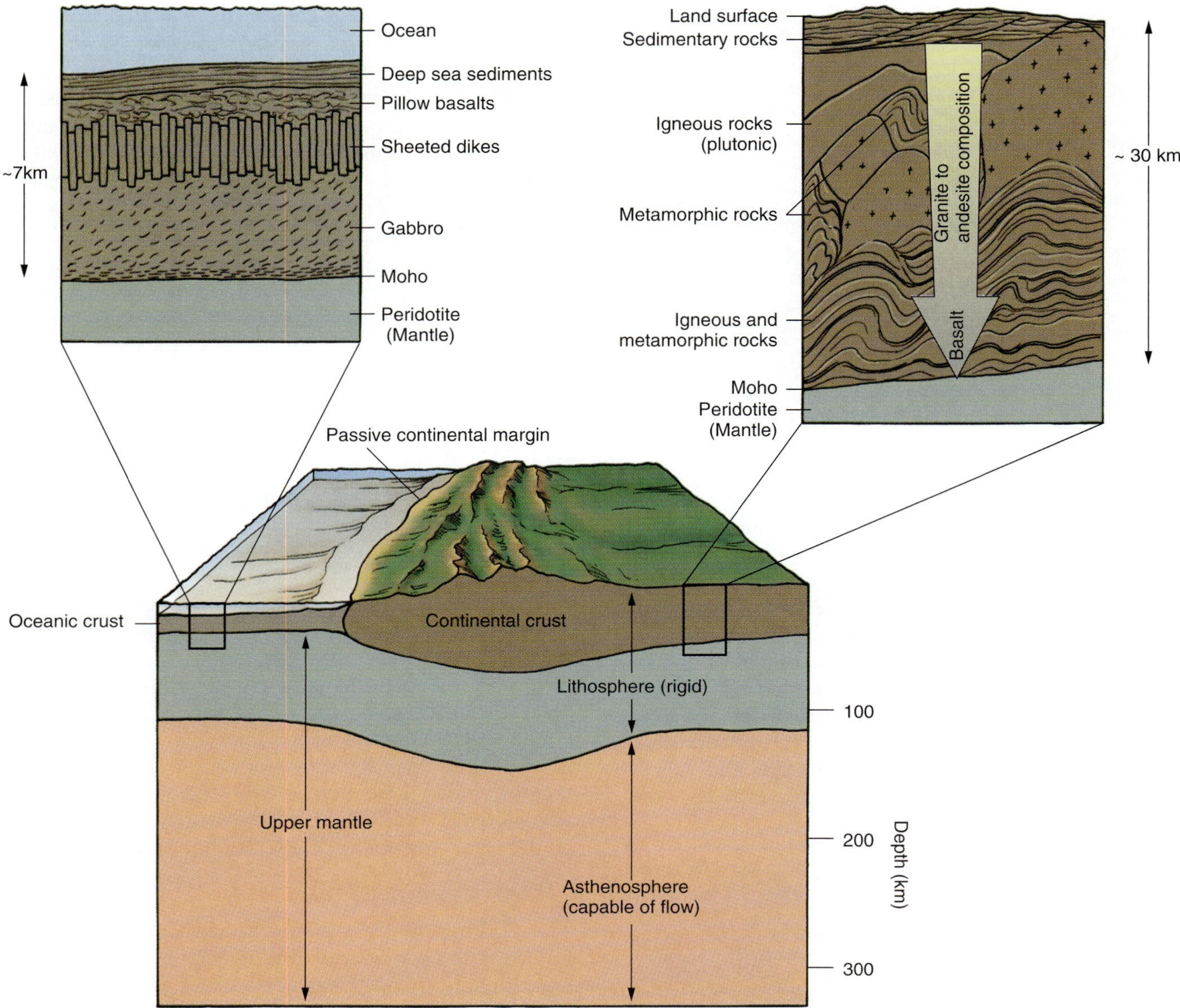

FIGURE 11.2 Typical structures—thickness, rock types, and nature of deformation—of both continental and oceanic lithosphere. The oceanic lithosphere varies little across all the ocean basins and reflects the process of lithosphere generation at mid-ocean spreading ridges. The structure of the continental lithosphere is much more varied, reflecting generation processes less well known than those in the oceans.

easier to study!), we are not yet sure how the continental lithosphere formed. Thus, we will spend some time examining the continents and how they may have been formed.

Continental Lithosphere

The continental lithosphere consists of two major structural components distinguished at the surface: mountain belts, which are generally young, and older regions referred to as **cratons** (Fig. 11.3). The mountain belts are largely the result of recent plate-collision processes discussed in the preceding chapters. As you can see from comparing Figure 11.3 with Figure 6.1, they correspond closely with the locations of currently active plate boundaries. Some older mountain belts such as the Appalachians and the Urals do not coincide with current plate boundaries but are inferred to mark the locations of older margins, which have since become sutured (∞ see Chapter 9).

The composition of the continental crust is much more variable than its predominantly basaltic oceanic counterpart. On average, this crust is made of less dense material (approximately andesite or diorite composition; ∞ see Chapter 4). There is actually a gradation in many regions of the continental crust—from low-density, silica-rich granitic rocks near the surface to silica-poor basaltic rocks in the lower crust.

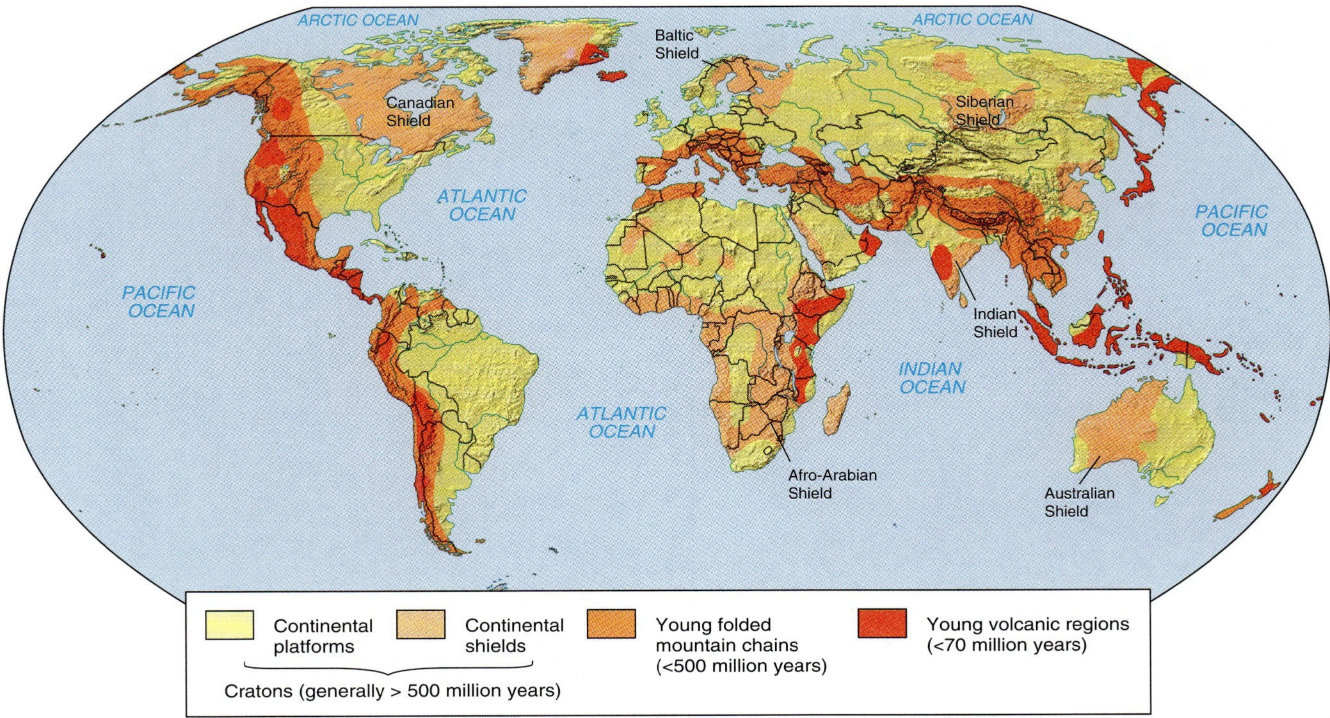

FIGURE 11.3 Distribution of young mountain belts, platforms, and shields. Notice that the location of the young mountains corresponds quite closely to present-day plate boundaries. Some areas defined as platform are actually mountain belts of intermediate age.

In terms of actual rock types, 95 percent of the crustal volume comprises igneous and metamorphic rocks. Sedimentary rocks, however, cover most of the continental surface. The absence of sedimentary rocks deep in the crust is caused by increases in pressures and temperatures, which inevitably transform sedimentary rocks that existed at the surface into metamorphic rocks as they are buried.

Shields and Platforms

Ancient cratons, more than 500 million years old, form most of the area of the continents. Because they have been subjected to weathering and erosion over such a long period of time, cratons generally form vast areas of low relief, lying within a few hundred meters of sea level (both above sea level as land, and below sea level as continental shelves). The low relief may be interrupted in places by intermediate-age mountain belts, many of which formed in the last 500 million years but are no longer tectonically active and are slowly eroding and diminishing in relief. Where extensive tracts of the low-lying, ancient continental crust are exposed at the surface, they are referred to as **shields**, although the terms *shields* and *cratons* are often used interchangeably. More commonly, though, the cratons are covered by layers of sediment; under these circumstances, they are referred to as **platforms** (Fig. 11.4).

One problem geologists encounter when evaluating the origin of the oldest continental fragments is that these fragments are never preserved in a pristine condition. The older crustal fragments have often been subjected to repeated periods of deformation and metamorphism. Each event overprints and masks the effects of previous events, which means that information about the original rock compositions, textures, and structures may be lost. The stylized section in Figure 11.2 is intended to portray a complex history of deformation, metamorphism, and igneous intrusion that may be considered typical of much of the continental crust. In regions such as the Canadian Shield, the oldest crustal terrains not only yield the oldest rocks, but also record the greatest number of metamorphic and deformational events.

Sedimentary processes of erosion and deposition also hamper studies of the oldest continental fragments. Erosion tends to level off continental crust; as a result, fragments of old crust commonly lie at very low elevations. The Canadian Shield, with an area of more than a million square kilometers, typically has elevations very close to sea level. This low-lying land is vulnerable to submergence under shallow seas in which sediments will be deposited. If these submerged regions are subsequently uplifted above sea level—or if sea level drops—they will have been covered by a veneer of sediments and will therefore become platforms. Most of the flat-lying sediments characterizing continental platforms were deposit-

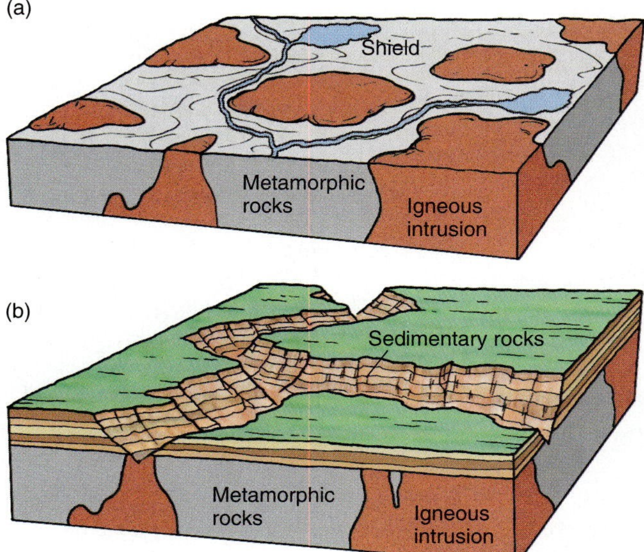

FIGURE 11.4 (a) Typical structure of the continents: exposed, old crystalline rocks (commonly referred to as the *basement*) form a shield. (b) A thin veneer of sedimentary material overlies the crystalline basement to form a platform.

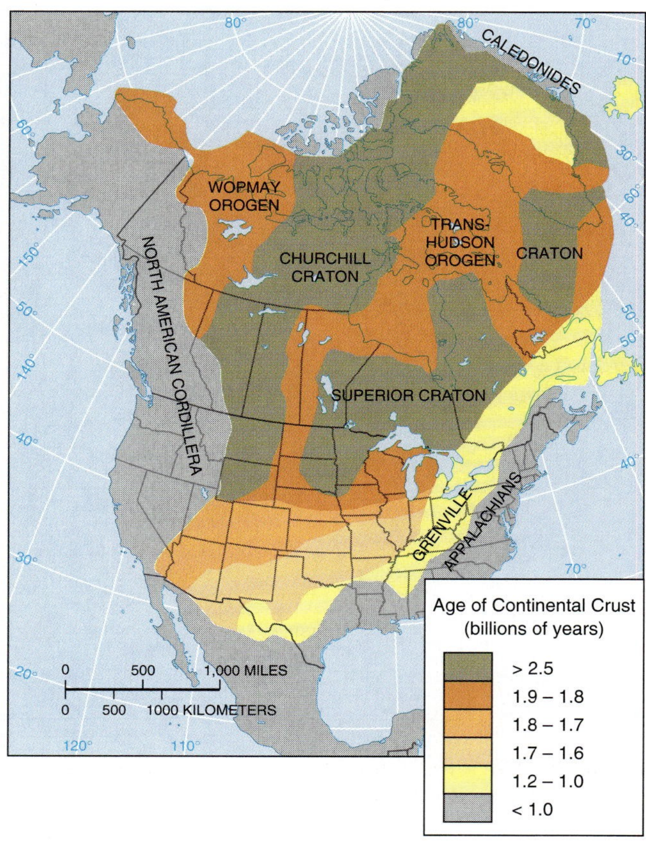

FIGURE 11.5 Distribution of average crustal ages in North America.

ed in such shallow seas on the continental shelves. The agricultural plains of the North American midwest—from Kansas in the United States north to Alberta, Canada—lie on nearly horizontal, sedimentary rocks, but we know from both deep boreholes and seismic investigations that old crystalline basement forms the bulk of this crust below the sediments.

Most of the continental crust comprises older crystalline (igneous and metamorphic) basement covered by a thin veneer of younger sediment.

The Origin of the Continental Lithosphere

Currently, lithospheric material is being added to the continents at magmatic arcs. Volcanism along island arcs or continental margin arcs produces voluminous low-density andesitic to basaltic composition material. Oceanic arcs, which mark ocean-ocean convergent margins, will be carried along with the oceanic lithosphere into the subduction zone. These arcs will eventually be sutured to the edges of continents if the material is sufficiently buoyant to avoid subduction. It follows, then, that the continents grow by **lateral accretion**: Young material produced at arcs is sutured onto existing continental margins. This explains, in part, why we commonly find discrete terrains along active continental margins (∞ see Chapter 9.)

Given this information, we might expect to find older continental crust at the interiors of continents. Indeed, the cores of most continents comprise many old fragments (more than 2.5 billion years old) of continental crust composed of thick, rigid, crystalline material. These old blocks are generally surrounded by progressively younger material extending outward toward the continental edges (Fig. 11.5), although this broad distribution may be complicated by the cross-cutting effects of later continental rifting.

Cratons may be of different ages, but by definition, they must be old (more than 500 million years). In many cases, two or more cratons are separated by belts of younger material. Perhaps the cratons were sutured by continental collisions into the configuration we see today—for example, as the Indian and Asian continents are sutured together today at the Himalayas.

Age patterns and analogy with present-day plate-tectonic processes suggest that much of the continental crust has grown by lateral accretion.

We must still ask, though, how the cratons themselves were formed. We might suggest that they were the products of magmatism above early subduction zones. However, major differences exist between the chemical compositions of old cratons and younger crust, including material currently produced at volcanic arcs. If, as seems reasonable, the chemical composition of the rocks reflects the processes by which they were formed, then cratons cannot have been formed by subduction-zone magmatism as we know it today. Furthermore, although it is tempting to apply the plate-tectonic model universally, no compelling evidence indicates that large-scale creation and subduction of oceanic lithosphere—plate tectonics—operated before about 2 billion years ago. The oldest known fragment of oceanic lithosphere, which escaped subduction and was thrust onto continental crust as an ophiolite in Canada, is about 1 billion years old.

If Earth is continually cooling, then it must have been much hotter earlier in its history. This, in turn, suggests that convection processes within the early Earth would have been more vigorous and the lithospheric plates may have been thinner, hotter, and less likely to be subducted. Thus, it may have been difficult to preserve the integrity of large plates as we observe them today. Indeed, there is good reason to believe that Earth's surface was entirely molten shortly after it formed (∞ see Chapter 2). It may be that the oldest cratons actually represent the earliest preserved crust that collected like a floating skin from this magma ocean.

If the origin of the ancient continental crust is not well known, then the origin of the mantle portion of the continental lithosphere (Fig. 11.2) is even less clear. Beneath the continental crust is a considerable thickness of mantle peridotite, which, in turn, overlies the asthenosphere. The subcontinental mantle lithosphere may have been formed by the same processes as the overlying crust, in which case it is old and has been isolated from the convecting mantle for over 1 billion years.

Alternatively, mantle material may be added to the continental crust from below, and the subcontinental mantle may bear little relationship to the overlying crust. Both mantle xenoliths brought up in basalts from the deep continental lithosphere and basalts believed to have been melted from the subcontinental lithosphere have chemical and isotopic compositions that suggest they may indeed be very old—and very different from the rest of Earth's mantle. Geophysical measurements also indicate that the continental mantle beneath cratons is compositionally different from the convecting mantle. Seismic- wave velocities are faster, suggesting that the continental lithospheric mantle comprises mainly depleted peridotite (like harzburgite) (∞ see Chapter 8). This lithosphere is slightly more buoyant than undepleted peridotite, contributing to the tendency to remain attached to the base of the continental crust. The solidus of harzburgite is also higher, making it more resistant to thermal erosion by the convecting asthenosphere.

Passive Continental Margins: Junctures Between Oceanic and Continental Lithosphere

We have seen that continental and oceanic lithospheres have very different compositions and structures and are formed by different mechanisms. What, then, is the nature of the transition between the two types of lithosphere? The zone of transition from the ocean floor to the continent is known as the *continental margin* (see Focus 11.1). It marks the transition from thin basaltic oceanic crust to thick, dioritic-to-granitic continental crust. We refer to it as a passive continental margin because it is not a plate boundary (active). However, the features characterizing the margin are actually the consequence of the rifting apart of a continent, the subsequent formation of a new divergent plate boundary, and, with the formation of new oceanic lithosphere, the migration of the plate boundary away from the ocean-continent interface (see Fig. 8.3). Thus, continental rift shoulders eventually become passive continental margins if rifting continues and results in the formation of a new ocean basin (∞ see Chapter 8).

Volcanism and earthquakes do not commonly occur at passive continental margins. Thus, these margins differ from margins at which subduction or transform faulting is taking place. The east coasts of North and South America are passive continental margins (see Focus 11.1) and are tectonically quiet, whereas the west coasts are active margins along which subduction or transform faulting occurs.

Passive continental margins are characterized by a shoreline, a continental shelf, a continental slope, and a continental rise (Fig. 11.6, page 282). The *continental shelf* slopes gently away from the shore and extends an average of 65 km out into the ocean; in some places it extends out to a distance of several hundred kilometers. The continental shelf reaches an average depth of 130 m and, in some cases, attains a maximum depth of 650 m. At the edge of the shelf, the sea floor becomes steeper and is called the *continental slope* (Fig. 11.6). The *continental rise* lies at the base of this slope and consists of sediments that have moved down the continental slope to accumulate at the edge of the ocean basin. The rise can stretch out for hundreds of kilometers until it finally becomes the floor of the ocean basin, or the abyssal plain.

Several processes contribute to the evolution of a passive continental margin and the subsidence and submergence of a continent edge to form the current shelf: First, dike intrusion into the continental crust during the rifting phase would leave the margin partially continental and partially oceanic, with the proportion of oceanic crust increasing to 100 percent at the base of the slope. This means that the margin would contain rock denser than average continental material, causing it to settle lower in the mantle. Second, erosion of the continent would lower

Focus On 11.1 THE ATLANTIC CONTINENTAL MARGIN OF NORTH AMERICA

Volcanism was widespread at the time of rupture of many continental margins. The northeast Atlantic margin of North America is associated with profuse igneous activity from Greenland in the west to Scotland on the other side of the ocean (Fig. 1a). The rifting associated with the formation of the Atlantic Ocean was probably caused in part by mantle hot spots, such as the one that now lies beneath Iceland. Volcanic activity during rifting released a vast amount of magma and intruded dike swarms along the margins. A considerable amount of basaltic lava—about 6 km thick—erupted in southeast Greenland about 60 million years ago (the Greenland flood basalts, Fig. 1a). Farther south,

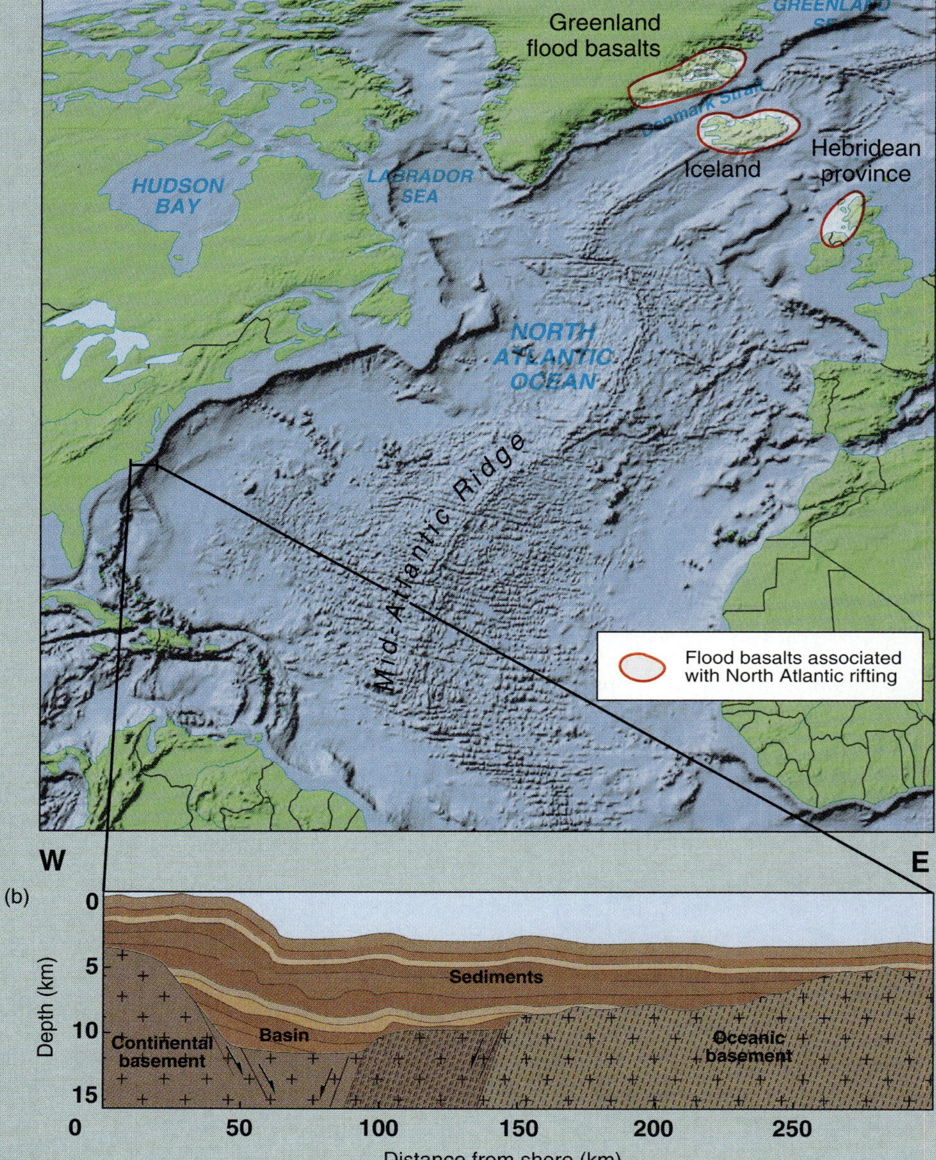

FIGURE 1 (a) Topographic map of the North Atlantic, showing the locations of major basalt provinces associated with rifting of the North Atlantic, the Greenland flood basalts of southeast Greenland, and the Hebridean province of northwest Scotland. One of the mantle plumes that initiated rifting in the region is probably now located beneath Iceland and is responsible for the voluminous magmatism that has built the island more recently. (b) Cross section of the continental margin of eastern North America, based on seismic data. Note the thick section of sedimentary deposits draped over the continental margin.

there appears to be less volcanism associated with continental breakup, although significant volumes of basalt probably exist beneath the thick sediment cover along the continental margin.

In the region of Cape Hatteras (Fig. 1a), the basement at the coast is about 4 km below the surface and drops to a depth of 10 km some 100 km offshore; the basement then rises gradually to 7 km over the next 200 km toward the southeast (Fig. 1b). The basement can be detected because seismic waves reflect strongly from the interface between the basement and the overlying, less dense layers of sediment (∞ see Chapter 5). Although the ocean depth has increased from 0 to 4 km in the 100-km transect away from the coast, the net effect on the sediment is that it thickens from 4 to 6 km, then thins to about 3 km oceanward, so that there is a basin of thick sediments where the oceanic lithosphere meets the continental lithosphere. Continental margin basins like the one shown in Figure 1b are dropped down along normal faults; they were formed during the continental rifting event that split North America away from Africa and Europe.

It has been suggested that sediments in these deep continental margin basins represent a new frontier for oil exploration since they have experienced depths of burial and heat flow sufficient for maturation of hydrocarbons (∞ see Chapter 16). The association of rift basins filled by thick successions of sediments, normal faulting, and voluminous basaltic magmatism, as found along the eastern edge of North America, is typical of passive continental margins.

elevations as sediments were deposited onto the rise; third, the heat beneath the rift would cool as the margin moved away from the upwelling hot mantle that began the process, thereby removing a source of thermal uplift.

A passive continental margin represents what was originally one side of a continental rift. The crust grades from fully continental on land to fully oceanic—that is, from less dense to more dense material.

Sedimentation in Plate Interiors

Sedimentary systems work in such a way that sediment is transported from higher elevations (the continents) to lower ones (the oceans). Ultimately, most sediment is deposited in the oceans surrounding the continents. To help us understand this type of sedimentation, it is important to remember that a significant portion of the continental crust at passive margins is currently covered by seas (Fig. 11.7)—these regions are known as the continental shelves. The amount of the continents covered by seas, or the extent of the continental shelf surrounding the continents, is determined by sea level. We have seen that sea level and therefore the shorelines are not fixed and can vary over time. This can be shown in Figure 11.6; if you place a ruler on the horizontal line that represents the surface of the ocean and move it down slightly, you will see that the shoreline moves to the right and the width of the continental shelf is reduced. Such a decrease in sea level will cause sediments previously deposited near the shoreline to now be exposed, eroded, and moved farther out toward the edge of the shelf. Thus, changes in sea level will control the nature and sites of sediment deposition.

Changes in Sea Level

Over geologic time, sea level actually fluctuates quite rapidly and over several tens of meters. There are three main ways in which the sea level can change: First, the volume of water in the ocean basins can change if a significant amount of water is frozen into ice on land. Second, the average depth of the ocean basins themselves can change because of the differences in ridge volumes, which correspond to changes in spreading rates. Finally, the relative level of the land can be altered locally by tectonics and deformation, such as uplift or subsidence in response to continental collisions (Fig. 11.8).

It has been estimated that the melting of all polar ice on Earth today would result in a rise in sea level of approximately 70 m. This may not sound like a lot; however, if we consider that much of the continental crust has elevations within 100 m of sea level and many of our largest cities are built on shorelines, the consequences of such a rise could be disastrous. Figure 11.9 shows the effects on the North American coastline of both a 100-m rise and a 100-m fall in sea level and the subsequent effects on the amount of continental crust that would be submerged or exposed. You can see that many of the major United States cities, such as Los Angeles, New York, and San Francisco, and much of Florida would actually be flooded by such a rise in sea level.

Over the past 2 million years, there have been at least four major glacial advances, or *ice ages*, and accompanying glacial retreats—although we tend to combine them and refer to the "ice age." In one of the most recent glacial advances, large areas of North America and northern Europe were covered with ice—much more ice than is present on Earth today. These fluctuations in the amount of ice caused sea level to vary considerably (Fig. 11.8a). During an

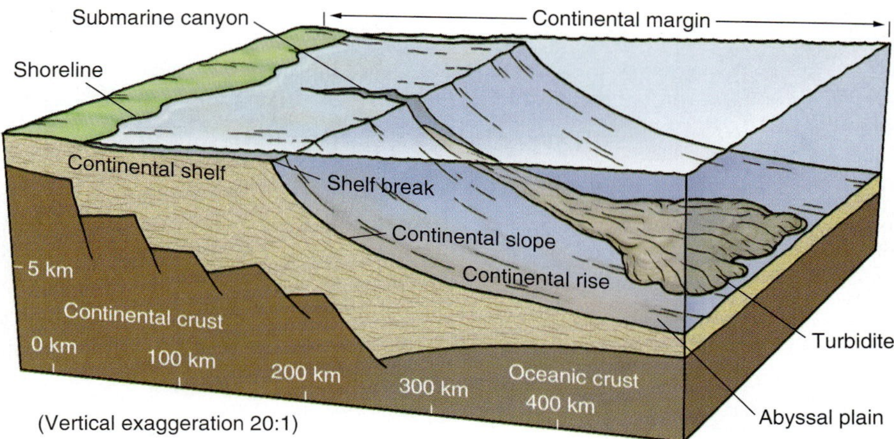

FIGURE 11.6 Cross section of a passive continental margin, indicating the main morphological features.

ice age, a significant amount of water from the oceans is locked up in a huge volume of ice. The total volume of water in the oceans is therefore decreased; as a result, sea level is lower. A corresponding rise in sea level takes place when the ice melts, returning water to the oceans.

The huge weight of ice on the northern continents during the ice ages actually made the continental crust sink several hundreds of meters in order to attain isostatic equilibrium (∞ see Chapter 5). Thus, while global sea level was decreasing owing to water being incorporated into the ice sheets, these regions were actually being submerged because of ice loading. After the ice melted, the crust rose again very slowly by a process called *isostatic rebound*. The relative rise of the land (hundreds of meters) again outweighed the rise in sea level that resulted from melting ice. The net effect, therefore, is of a relative fall in local sea level.

Sea-level fluctuations can also result from variations in the rate of sea-floor spreading. A rapid rate of sea-floor spreading means that the ocean floor will have a younger average age than when the rate of sea-floor spreading is relatively slow. Younger oceanic floor is hotter and its equilibrium isostatic elevation is higher than that of an older, colder oceanic lithosphere (∞ see Chapters 5 and 8). The net result is that the ocean basins are shallower when sea-floor spreading occurs more rapidly (Fig. 11.8b). If we assume an approximately constant volume of ocean water, this water will have to flood more of the continental shelves if the ocean basins are shallower; thus, sea level will rise.

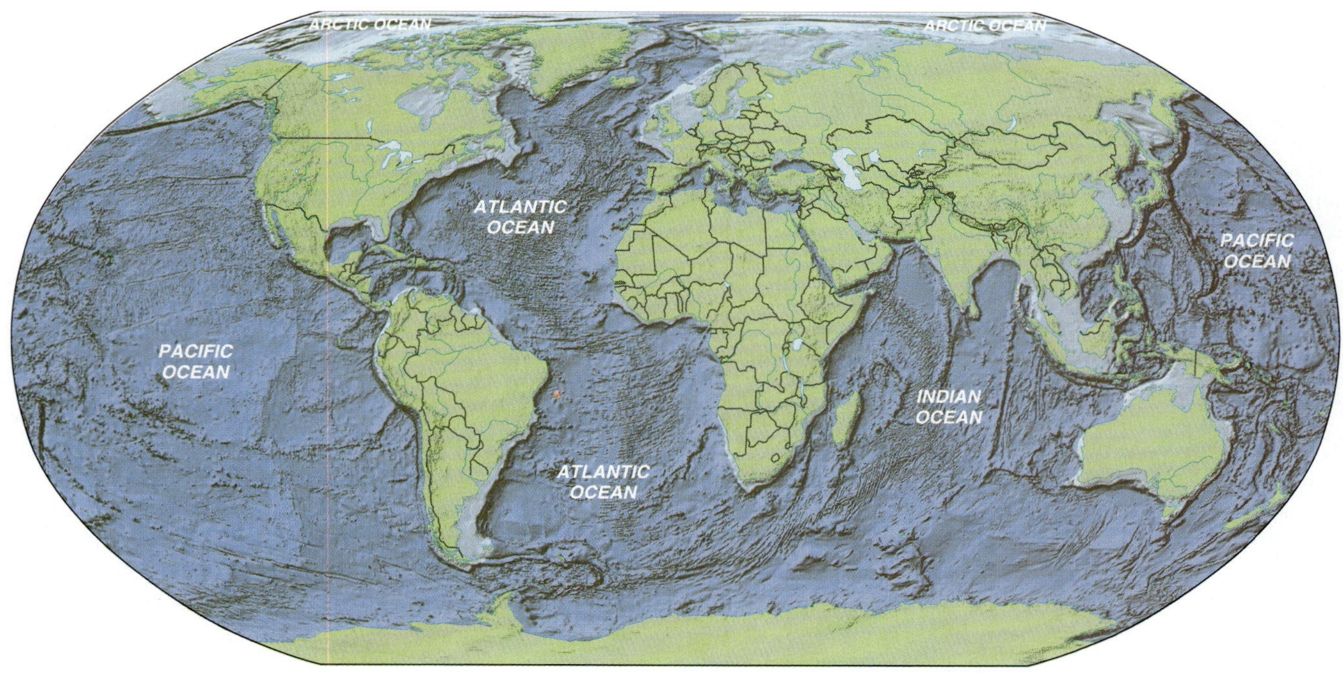

FIGURE 11.7 Distribution of continental shelves (light blue). These shelves lie beneath the sea but still on continental crust and are therefore important regions of sediment deposition.

(a) Varying ice volume (Global and regional)

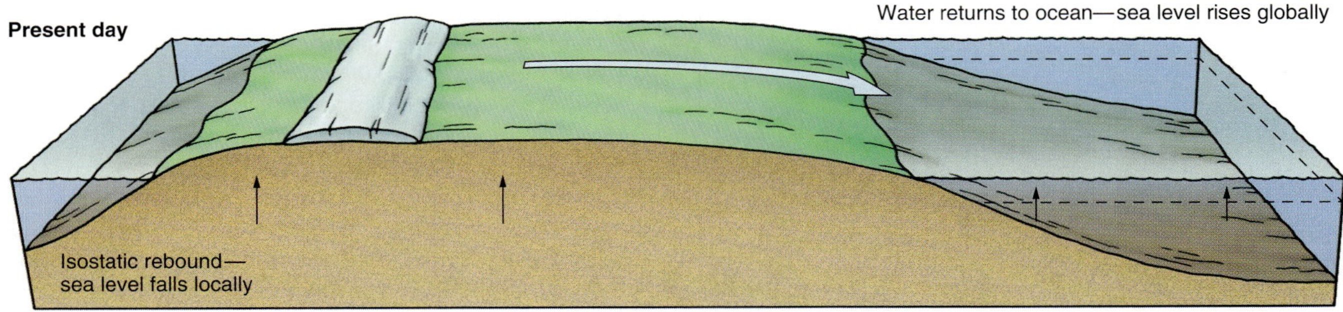

(b) Varying ridge volume (global)

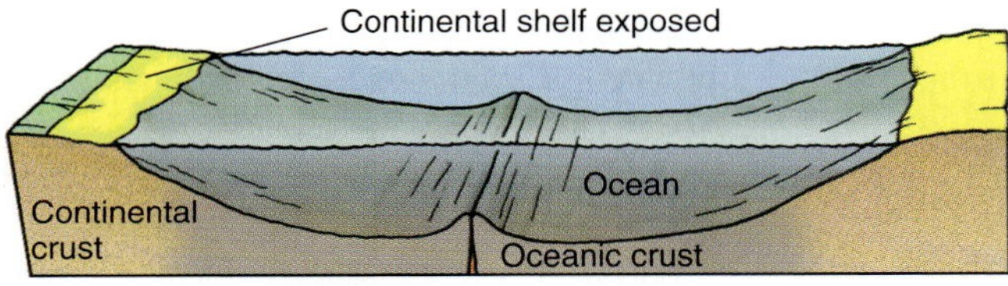

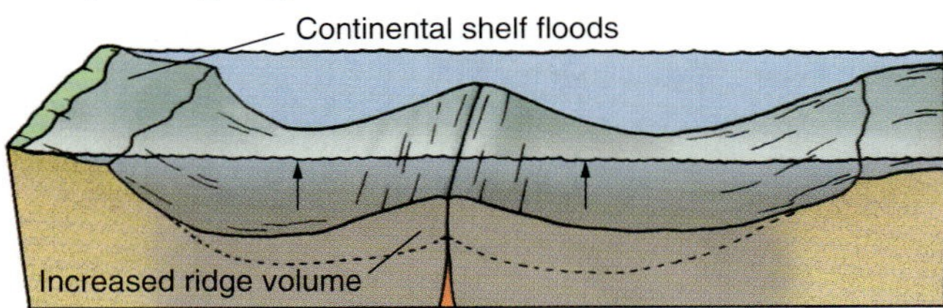

(c) Tectonic effects (regional)

FIGURE 11.8 Mechanisms of sea-level change. (a) Variation in the volume of polar ice; (b) variation in ocean ridge spreading rates; and (c) tectonic deformation. (d) An example of evidence for recent relative changes in sea level: ruins of a Roman marketplace built in the second century B.C. Borings have been made by marine organisms on the lower portions of the columns, indicating that the columns have been submerged since their construction. The columns now stand out of the water and were probably uplifted recently in response to local magmatism, which heats and inflates the crust, causing it to rise.

Local apparent changes in sea level can be seen where tectonic forces have caused the land to be uplifted (Fig. 11.8c). Marine fossils are found on top of Mt. Everest in the Himalayas. This is an extreme example of a relative alteration in sea level as a result of the continental collision between India and Asia. Here sediments that were originally deposited below sea level have been uplifted some 10,000 m over the past 50 million years. In a more general sense, though, relative changes of sea level will occur wherever the continental lithosphere is subjected to stresses that cause it to move vertically. The Roman ruins shown in Figure 11.8d, for example, have been uplifted because the crust has warped upward in the region in response to hot magma intruding below (it is located close to Mt. Vesuvius). The uplift can be estimated confidently because of the clam borings that can be seen on the columns, which were made by marine shellfish. Lest you were to conclude that the Romans originally built their town underwater, we should also point out that the recent uplift was preceded by submergence. Careful studies have indicated that the region has experienced many fluctuations in local sea level as the crust has warped up and down. Because the Roman Forum was only built in the second century B.C., changes in local sea level clearly can occur quite rapidly.

All of these mechanisms undoubtedly have influenced sea level—global and local—through the geologic past. A higher sea level will flood more of the low-lying continental interior. The shoreline might shift many tens or hundreds of kilometers inland when sea level rises a few hundred meters (Fig. 11.9).

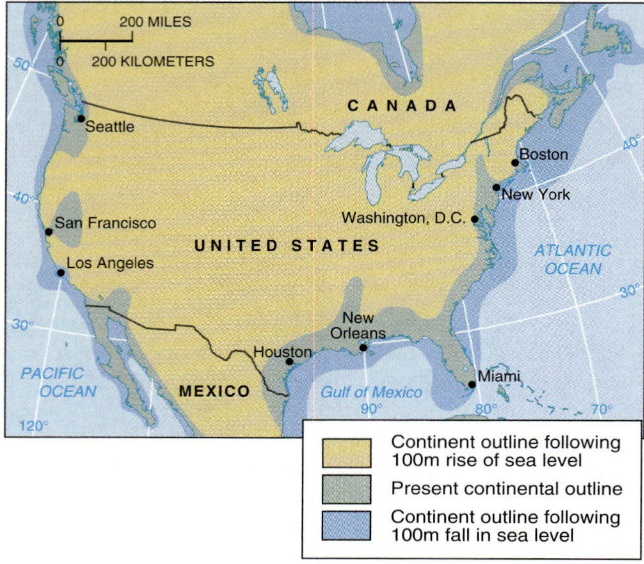

FIGURE 11.9 Illustration of the effects of a 100-m rise or a 100-m drop in sea level on the shoreline of North America. Clearly, the location of the shoreline, and therefore the relative extent of exposed continental landmass and submerged continental shelves, is controlled by sea level.

Much sediment accumulates on continental shelves. The extent of the continental shelves (and therefore the amount of sediment that can accumulate) will depend on sea level, which rises and falls through geologic time.

Sedimentary Successions at Passive Margins

When the sea rises and the shoreline moves inland across the continent, it is called a **transgression**, which means that the sea is encroaching upon, or transgressing onto, the land. The reverse situation—a **regression**—occurs when sea level drops. We can actually tell how the sea level has varied in a particular place just by looking at the record preserved in the sediments (Fig. 11.10).

As explained earlier (∞ see Chapter 4), fine-grained sediments will be carried greater distances than will coarse, heavier sediments. As a result, coarse-grained sandstones are formed near the shore, whereas finer sediment, which will form mudstones, is carried out to sea. Suppose we were to come across a sequence of sedimentary rocks that became finer grained toward the top of the section—as the rocks become younger. This suggests that sea level at that location was rising; the sea had become deeper and the shoreline had moved farther inland at the time the sediments were being deposited.

An important feature of such successions is the unconformity (Fig. 11.10). As we have learned (∞ see Chapter 1), unconformities mark breaks in the depositional record and represent periods of time during which erosion, rather than deposition, occurred. Transgressive sequences will have an unconformity at the base, representing the surface over which the shoreline advanced inland. Regressive sequences after subsequent burial may have an unconformity at the top, where the top of the sedimentary pile was exposed above sea level and eroded.

The study of sedimentary strata (stratigraphy) shows that global transgressions and regressions have occurred many times during Earth's history. Beginning some 600 million years ago, at least six major unconformities are recognized in rock successions worldwide that correspond to these events (Fig. 11.11). At times, vast tracts of North America (and other continents) were submerged beneath the sea, resulting in the deposition of extensive layers of sediment. At other times, sea level fell, exposing the continent and permitting erosion of the land surface. The unconformity-bounded rock successions that represent the times of higher sea level and consequent sedimentation constitute important subdivisions of the sedimentary rock record. This concept, known as *sequence stratigraphy*, is particularly important in the interpretation of seismic reflection data from sedimentary basins (∞ see Chapter 5), a great deal of which now exists as a result of our efforts to search for oil (∞ see Chapter 16).

A large fraction of Earth's life-forms inhabit the shallow seas—this was even more significant earlier in Earth's

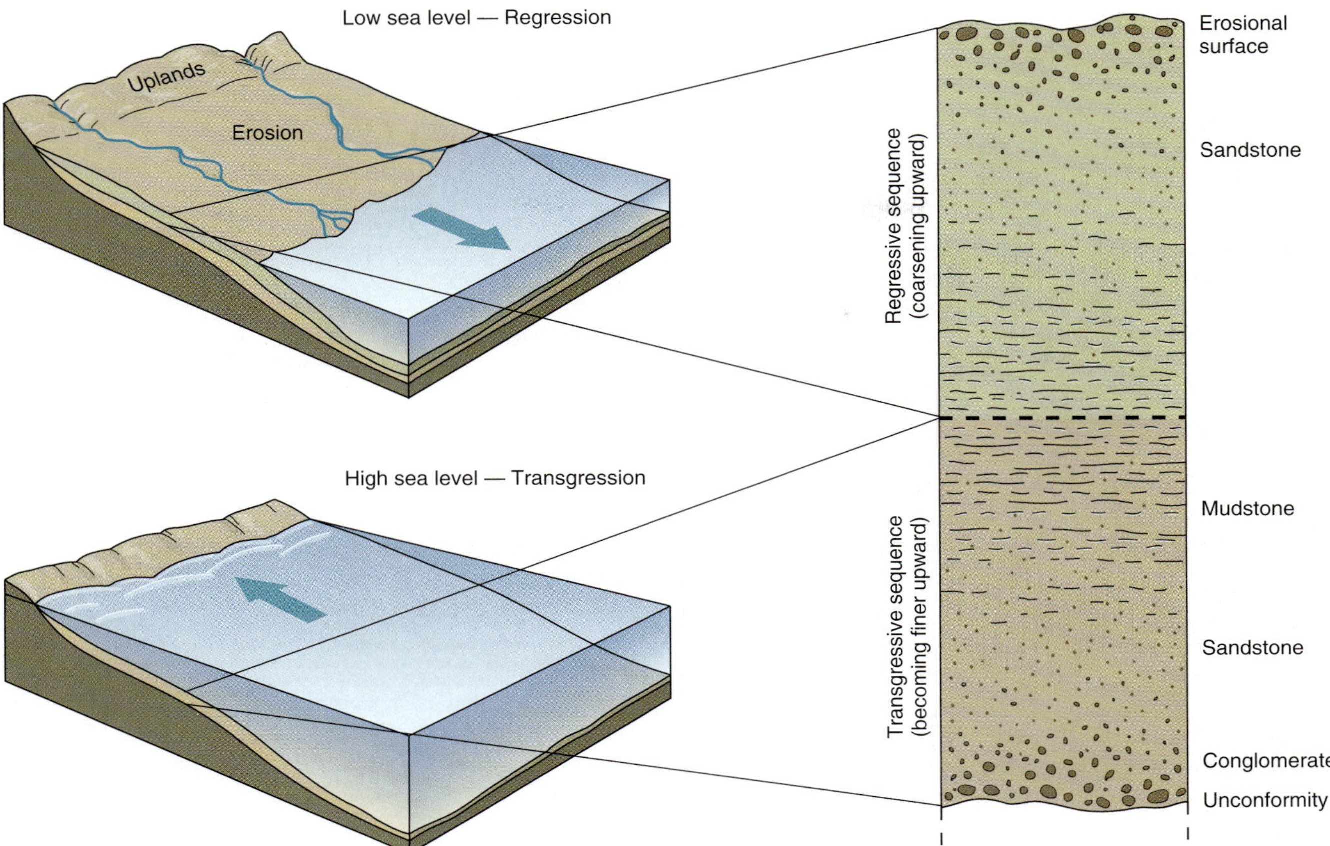

FIGURE 11.10 Effects of a rise or fall in sea level on both the location of the shoreline and the nature of the sediments deposited in the sea. If the sea level rises (transgression) the shoreline will move inland. At a given location, the sedimentary section has an unconformity at the base, which marks the deposition of sediments onto previously exposed land. The sediments become increasingly fine-grained moving upward; this reflects deposition farther from the shoreline. If the sea level then falls (regression), the sediments become coarser-grained, moving upward to the surface, reflecting deposition progressively closer to the shoreline. Regression will eventually lead to the sediments being exposed above sea level, and the top of the succession will be eroded, leaving an unconformity if subsequently buried.

history (more than 400 million years ago) when no life existed on land. Understandably, changes in sea level will therefore have a profound influence on life. Many organisms are adapted to specific conditions of sediment influx, temperature, water depth, and other environmental factors. A change in sea level will affect these factors; if organisms are unable to adapt sufficiently fast, they will die. In the geologic record, global regressions that reduce the shallow sea habitat of the continental shelves are commonly accompanied by extinctions of many groups of marine organisms. In contrast, global transgressions that expand the size of the shallow marine habitat are accompanied by a rapid appearance of new species—a "burst" of evolution. Thus, the evolution of life itself is modulated by geologic processes of sea-level change.

Seismicity and Earthquakes in Plate Interiors

It may come as a surprise that earthquakes can occur far from plate margins. Earthquakes result from a failure of the lithosphere, owing to the buildup of stress. Stress buildup is restricted largely to plate margins, where two plates interact by pushing together, pulling apart, or sliding past one another (∞ see Chapter 5). The stress that builds up at plate margins, however, may be transmitted through a lithospheric plate to its interior (Fig. 11.12). If weak parts exist within the lithospheric plate, they may fracture and give rise to earthquakes. Accordingly, the distribution of earthquakes in plate interiors is not random, but rather corresponds to lineaments of weakness—usually old, preexisting fractures that can be reactivated.

Because the crust of continental interiors is old and strong, it will sustain much more stress before it finally breaks. The resulting earthquake is then the result of a much greater energy release and is of large magnitude. The New Madrid, Missouri, earthquakes of 1811 and 1812 (Fig. 11.12) serve to underscore the reality of an earthquake risk in the central United States (see Focus 11.2). The dangers in this region are particularly severe for three main reasons: First, when this rigid lithosphere fails under stress, the magnitude of the resulting earthquakes is potentially higher than earthquakes resulting from the rupture of weaker lithosphere at plate margins such as in the west-

Focus On 11.2 THE NEW MADRID EARTHQUAKES OF 1811 AND 1812

Despite our perceptions that California and the San Andreas fault represent the greatest earthquake dangers in the United States, the largest known U.S. earthquakes actually occurred in New Madrid, Missouri, in the early 1800s. The occurrence is not entirely isolated—a huge earthquake also struck Charleston, South Carolina, in 1886 (Fig. 1). New Madrid is more than 2000 km from the San Andreas fault and is far from any other current plate boundary. The earthquakes were felt as far away as Canada to the north, New Orleans to the south, and Boston to the east. Fortunately, few lives were lost because the area was sparsely populated at that time.

New Madrid was subjected to a series of three large earthquakes. The first was on December 16, 1811, which was estimated to be of magnitude 8.6; this event was followed by magnitude 8.4 and 8.7 earthquakes on January 23 and February 7, 1812, respectively. Each of these principal shocks was followed by numerous smaller—but damaging—aftershocks. The epicenters of the three earthquakes are not exactly coincident and are aligned in a northeast to southwest direction.

Damage to stone and masonry buildings was incurred more than 250 km away (Fig. 1). Local landsliding and liquefaction of the soft sediments in the Mississippi Valley were major causes of destruction. The old town of New Madrid was destroyed and the course of the Mississippi River was altered—locally the river flow was actually reversed for a short while. Subsidence of blocks of crust was so great in certain areas (up to 5 m) that a number of new lakes were formed.

Studies of micro-earthquakes (numerous small shocks that are too slight to be felt) during the 1970s in the same region showed that the locations of the large earthquakes defined a continuous zone of weakness—probably a major fault. Recent measurements show that strain is building up very slowly. Scientists estimate that the time interval between major earthquakes is likely to be hundreds to thousands of years. This is not certain, however, and researchers continue to keep a watchful eye on any unusual activity that might signal the triggering of another major earthquake in this region.

The New Madrid earthquakes are now known to be the result of a rupture of an ancient fracture in the crystalline basement. The fault was part of a giant rift that began to tear the continent apart about 1 billion years ago, but stopped short of becoming a new plate boundary. Scientists have used stress measurements in boreholes and earthquake focal mechanisms (∞ see Chapter 7) to infer that the entire midwest lithosphere is currently in a state of compression. When stresses are sufficiently high, the lithosphere will fail along the weakest parts, such as old faults. This sort of reactivation is believed to have caused the New Madrid earthquakes.

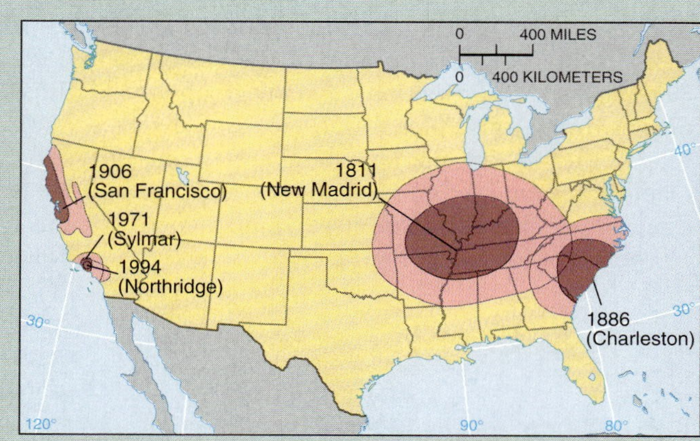

FIGURE 1 Comparison of earthquake intensity of the New Madrid, Missouri, earthquake with other large earthquakes in the United States. The dark areas are Mercalli intensity VIII regions, and lighter shading represents regions VI to VII. Mercalli scale VIII corresponds to the partial collapse of unreinforced buildings and the fall of chimneys, monuments, and VI corresponds to swaying and falling objects, etc. (∞ see Chapter 7, Chapter 15). The large 1886 earthquake in Charleston, South Carolina, is another example of seismicity occurring at a remote distance from a plate boundary.

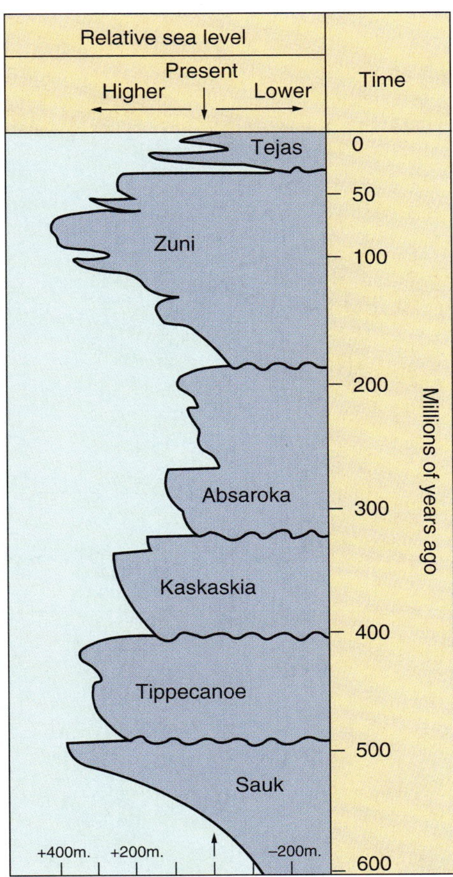

FIGURE 11.11 Variations in global relative sea level as reflected in the sedimentary record. Each of the six unconformity-bounded parts of the column represents a major sedimentary sequence, each of which has been named as shown.

ern United States. Second, large-amplitude seismic waves will travel farther through this old rigid lithosphere than through the weak lithosphere of a plate boundary such as the western United States, where they attenuate more rapidly. Third, the buildings in the region are structurally less robust because, unlike those of California, they have not been built to earthquake code (∞ see Chapter 15).

In 1976, an earthquake of magnitude 7.6 devastated the city of Tangshan in southwest China, resulting in 240,000 deaths. Although Tangshan is not near a plate boundary, the earthquake was not a complete surprise because earthquakes are extremely common throughout Asia (∞ see Chapter 7). In fact, the Chinese built the first earthquake detector (see Fig. 7.13). Possibly the greatest natural disaster ever in terms of human deaths was an earthquake that hit Shensi, China, in 1556, claiming an estimated 830,000 lives. This earthquake also occurred far from an active plate margin.

Most of the earthquakes in China are associated with strike-slip faulting (∞ see Chapter 7). The faults themselves are related to deformation at a plate boundary—specifically, the collision of India and Asia. The Indian Plate is still moving northward, pushing into the Asian Plate. This collision is accommodated in two ways: The crust is squeezed, deformed, and uplifted to form the Himalayan Mountain range and material is also pushed sideways (Fig. 11.13a). It is this lateral displacement of the crust—Southeast Asia being moved out of the way of the colliding Indian Plate—that is responsible for the many earthquakes in this region.

Experiments have been performed to investigate the formation of strike-slip faults in response to continental collision. Specifically, a rigid block (representing India) was pushed into a slab of plasticine (representing Asia); one side of this plasticine was confined in a box while the other side was free (eastern Asia). This experiment represents the collision of India with Asia (∞ see Chapter 9. Consequences of the collision—large blocks of Asian lithosphere being pushed sideways along strike-slip faults—are reproduced well in the simple experiment described here (Fig. 11.13b). This sort of deformation, where plate movements are accommodated on strike-slip faults that cut across plate interiors, is quite common as you can see from the global distribution of earthquakes relative to the distribution of plates (see Fig. 1.6).

Although earthquakes are less likely to occur within plates rather than at plate margins, they can be very powerful when they do occur. They are commonly located at ancient plate-tectonic scars or result from intense crustal collision.

Intraplate Magmatism

Distribution

We can see from plate-tectonic maps that a very strong correlation exists between the locations of volcanoes and plate margins (see Fig. 1.6). But there are also many volcanic centers located far from plate margins (Fig. 11.14, page 292). Most of these volcanoes are believed to be caused by *hot spots*: regions where the underlying mantle is somewhat hotter and melts to produce magma (Focus 11.3). In general, there is no systematic or regular distribution of within-plate, or intraplate, volcanoes. Some occur as isolated small centers in a continental or oceanic plate and exhibit few associated structures other than local doming of the lithosphere owing to the focusing of heat. Most intra-oceanic plate earthquakes are thought to arise from stresses caused by plumes upwarping the oceanic lithosphere. Other volcanoes appear to be related to rifts in the continental lithosphere, such as the volcanoes found along the East African Rift (Fig. 11.14 and Focus 8.2).

On a global scale, hot spots are found in regions where seismic tomography shows that the lower mantle is slow for seismic-wave velocities (∞ see Chapter 5) for example, beneath the position of Pangaea, before breakup, and

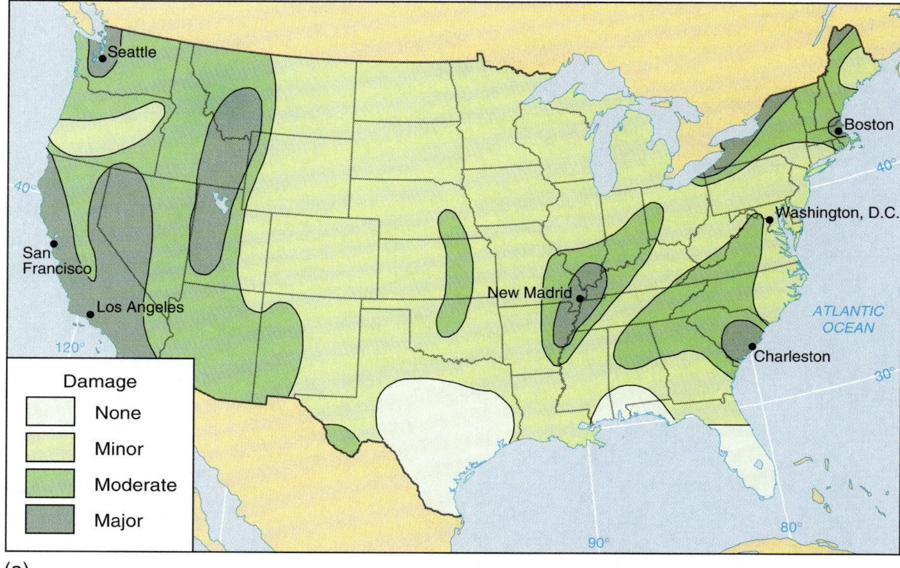

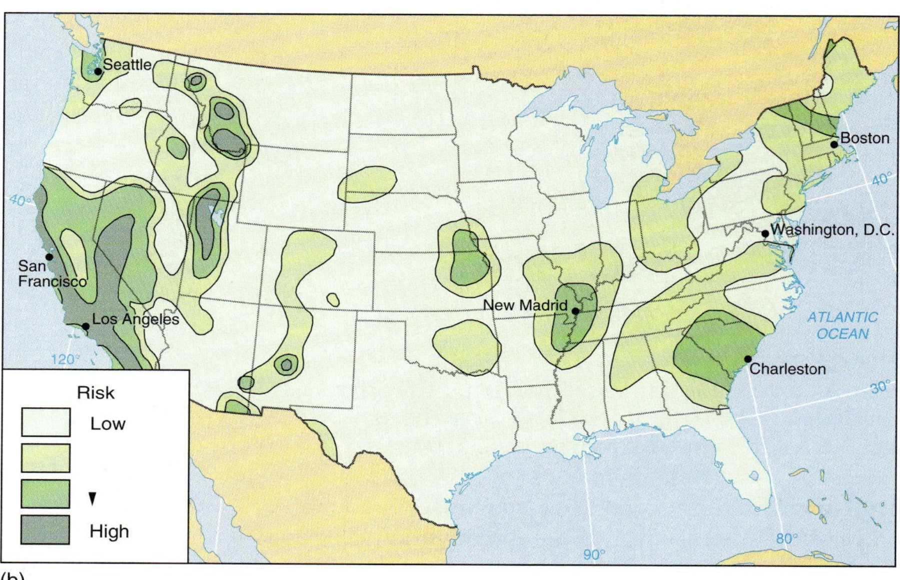

FIGURE 11.12 (a) The potential damage that could result from earthquakes; (b) the risk of occurrence of a large earthquake in the United States.

beneath the Pacific. These are regions where there has been little or no subduction in the last several hundred million years. The low velocities are explained if these regions are hotter than average because they have not been cooled by subducting slabs. In contrast, regions with large amounts of subduction appear fast and have far fewer hot spots. These include the circum-Pacific belt and Asia.

In oceanic regions, islands such as Hawaii, Tahiti, St. Helena, and Reunion are formed as the volcanic products of hot spots that have built up above the sea. The sea floor itself is commonly domed up in these regions; this is caused by the heat from the hot spot. When the volcano becomes inactive—because it has been carried away from the underlying hot spot by the lithospheric plate on which it is built—the addition of volcanic material ceases. Erosion removes material from the island while, at the same time, the entire volcanic edifice subsides as the domed lithosphere relaxes to its original level because of cooling. The volcanic island becomes smaller and smaller, and it eventually sinks beneath the sea to form a seamount.

During his ocean voyage on the *Beagle*, Charles Darwin noted this phenomenon of subsidence. He also observed that the relationship between tropical islands and the coral reefs that form in tropical regions could be used to indicate the relative age of volcanism. Corals must grow at shallow depths in the oceans in order for the organisms to receive sunlight for photosynthesis and will therefore commonly grow as a *fringing reef* in the shallow seas immediately surrounding a volcanic island. When volcanism ceases and the island begins to subside, the corals grow upward to keep pace with the subsidence (and thus remain in shallow water); this results in the formation of a *barrier*

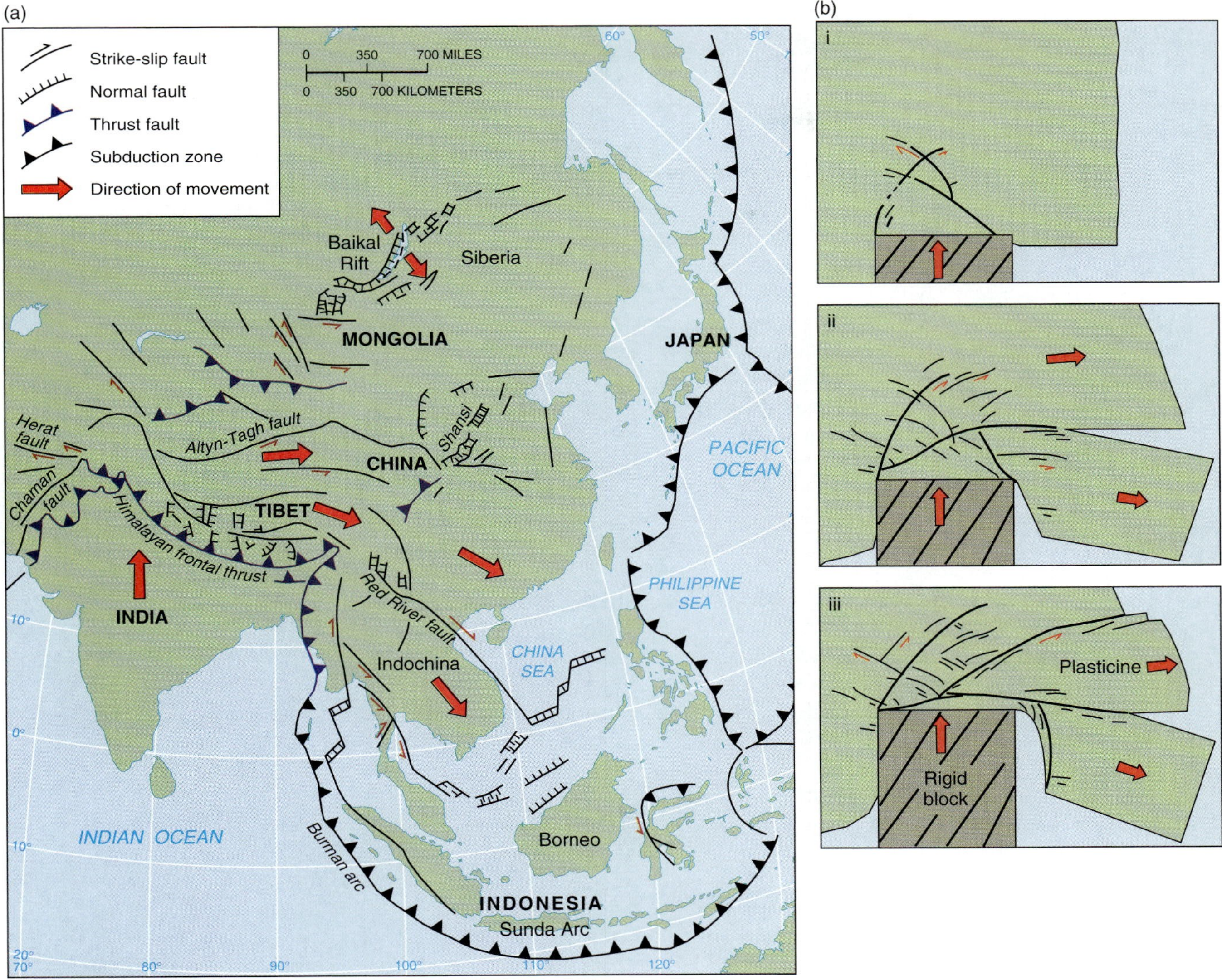

FIGURE 11.13 (a) Schematics of the collision between India and Asia, showing the system of strike-slip faults in Southeast Asia. These are due to the crust being pushed to the side, out of the way of the northward-moving Indian Plate. The subduction zones along the western Pacific allow the lithosphere to move eastward relatively easily. (b) An experiment is used to represent the collision between India and the weaker Asian subcontinent. A rigid block is pushed into a block of plasticine, which progressively deforms through steps i, ii, and iii. Notice the resemblance in fault patterns between the model and those actually observed deforming the continent in part (a).

reef. Eventually, the island subsides completely beneath the sea, but provided the corals can grow rapidly enough, they will still form a ringed reef or *atoll* at the surface just where the original fringing reef had grown. This sequence is illustrated in Figure 11.15, page 293. In time, the entire structure becomes submerged, forming a seamount. If erosion has left the seamount with a planed-off flat top, it can be more strictly referred to as a *guyot*.

Hot Spots, Plumes, and Plate Tectonics

It has recently been shown that hot spots—far from being secondary curios in the plate-tectonics model—are an important element in our understanding of how Earth works.

The entire dynamic Earth system, including plate tectonics, is driven by Earth's losing its internal heat (heat always flows from hot to cold). The majority of Earth's heat loss takes place along mid-ocean spreading ridges. This is not surprising, for mid-ocean ridges are also the location of the most voluminous magmatism at Earth's surface. The lavas that erupt along spreading centers, however, only carry heat up from relatively shallow mantle.

Plumes, in contrast, enable heat to be transferred from deeper regions within Earth, such as the core. It has been said that plate tectonics—the formation of oceanic lithosphere and its subduction back into the mantle—is the mechanism by which the upper mantle loses heat, whereas plumes are the way in which the core loses its heat. If

Focus On 11.3 HAWAII: AN EXAMPLE OF INTRAPLATE MAGMATISM

The Hawaiian Islands are perhaps the most thoroughly studied example of an intraplate volcanic province on Earth. Together with the Emperor Seamounts, the Hawaiian Islands form a chain 6000 km long—stretching northwest from the island of Hawaii to about 700 km northwest of Midway Island and then continuing north-northwest as a submarine chain of seamounts known as the Emperor Seamount Chain. This chain of seamounts is currently being subducted at the Aleutian/Kamchatka Trench.

This island chain is the classic example of a hot spot trail (∞ see Chapter 6). Active volcanism occurs on the islands of Hawaii and Maui, while the volcanic centers become progressively older with increasing distance from Hawaii. The islands become correspondingly smaller with age, as they are eroded and subside, eventually becoming submerged seamounts. This almost-perfect progression of volcanism would lead us to predict that the next focus of volcanic activity will be to the southeast of Hawaii as the big island moves off the underlying hot spot. In fact, an underwater volcano, Loihi, which is active 2000 m beneath the ocean surface, has recently been discovered in this very location.

The mantle plume responsible for the volcanism on Hawaii has domed up the oceanic lithosphere into a shallow "blister" or swell over 1000 km in diameter. The huge extent of this region of uplift on the ocean floor and the longevity of volcanism related to the plume indicate that the plume must be large and must arise from deep within Earth—possibly from the core–mantle interface. The edifices of the two main volcanoes on Hawaii—Mauna Loa and Mauna Kea—stand over 4000 m above sea level and more than 10,000 m above the ocean floor.

The volcanism on Hawaii is dominantly basaltic, and we recognize a series of three typical stages of volcanic activity. Evidence for the earliest stage of volcanism is the buildup of huge volcanoes above sea level, such as Mauna Loa (although work on Loihi indicates that there is an earlier submarine stage characterized by lavas with different chemical compositions). This first stage is called the *shield-building stage*. The fluid nature of the basaltic lava results in the building of enormous shield volcanoes (∞ see Chapter 4) with very gentle slopes (Fig. 1a).

Lavas may be erupted from lava lakes that occupy circular depressions, or calderas, which are distinguished from smaller craters by their large size; alternatively, lavas may be erupted from fracture systems (rift zones) that run down the flanks of the volcano. Most of the material is erupted relatively quietly as lava flows of both *pahoehoe* and *aa* types (Fig. 1b). Pahoehoe lava is distinct in having a smooth, ropy texture on the surface of the flow, whereas aa lava has a rough, jagged surface. The differences in flow types reflect differences in lava viscosity and flow rate on eruption. Occasionally, the top and sides of a lava flow may solidify, leaving the lava to flow through a lava tube (Fig. 1c). The

(a)

FIGURE 1 Volcanic features of Hawaii: (a) Mauna Loa, a shield volcano; a view across the large craters or calderas of the broad flat summit toward Mauna Kea in the distance. The "bumpy" appearance of Mauna Kea is due to numerous small cinder cones. (b) Smooth, ropy *pahoehoe* lava overlying rough, jagged *aa* lava. (c) A lava tube. (d) Diamond Head, Oahu. This is a small pyroclastic volcano related to a cinder cone formed during the posterosional stage of volcanism.

lavas on Hawaii often flow several miles downhill, eventually reaching the ocean. Here, they add material to the shoreline, which constantly increases the overall size of the island.

Mauna Kea is a somewhat older volcano and, accordingly, shows a more advanced stage of volcanism—the *alkalic stage*. The lavas have a different composition from those involved in the shield-building stage. In particular, the lavas contain higher concentrations of alkali elements, such as sodium and potassium. The alkalic composition of the magmas results from lower amounts of partial melting in the mantle, which concentrates incompatible elements such as sodium and potassium into the melt. The alkali stage represents much smaller volumes of lavas, many of which are erupted from small subsidiary volcanoes or cinder cones, rather than from rifts or calderas (Fig. 1a). The lower magma volumes and the lower degrees of partial melting are both consistent with the mantle source underlying the volcano's being somewhat cooler than the source that supplied the voluminous basalts of the shield volcanoes.

The third stage of volcanism, the *posterosional stage*, is present on the island of Oahu, which lies 100 km to the northwest of Hawaii. Volcanism on Oahu was most intense about 5 million years ago when the island must have been much like Hawaii is today. Since then, the island has been eroded considerably—a factor that accounts for its rather irregular coastline. The posterosional stage represents the last hiccups of volcanic activity, following a significant period of time during which rocks of the earlier shield-building and alkalic stages suffered erosion. Many of the igneous rocks from the posterosional stage are pyroclastic. They form small volcanoes with craters such as Diamond Head (Fig. 1d).

Studies of other ocean islands formed through volcanic activity related to hot spots (see Fig 11.14) show similar patterns of volcanic activity. Typically a voluminous shield-building stage, which builds the volcano above sea level, is followed by a later alkali stage.

(b)

(c)

(d)

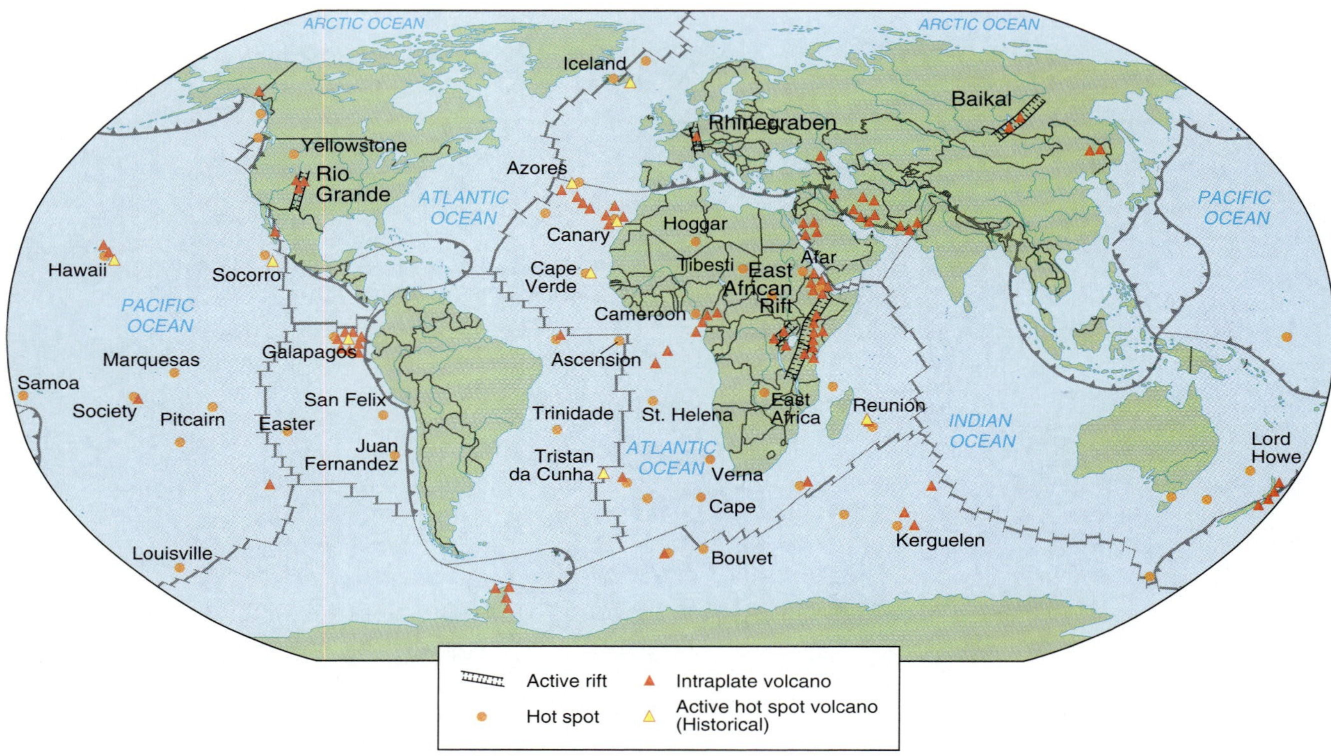

FIGURE 11.14 Global distribution of intraplate volcanism, continental rifts, and inferred hot spots.

this is so, then the locations of mantle plumes and the associated magmatism are independent of plate margins, which are simply features of the lithosphere at the top of the mantle. Most plumes intercept the lithosphere in the interior of a plate, although one may fortuitously reach the top of the lithosphere at the location of a plate boundary. Note that some plumes are close to divergent margins (such as along the Mid-Atlantic Ridge; see Fig. 11.16) because the rifting that led to the formation of the divergent margin was ultimately caused by the plumes. Thus, plumes may influence the locations of divergent plate boundaries, but the converse is not true.

Intraplate magmatism is commonly associated with mantle plumes. Plumes transport hot material from deep within the mantle and are a critical element in mantle convection within Earth.

The origin of plumes is still being explored. Much of the argument is related to questions of whether convection encompasses the whole mantle or is confined to separate upper and lower mantle convection regimes. We know from seismic studies that a discontinuity exists some 660 km within Earth (∞ see Chapter 5). This discontinuity may represent a thermal boundary and may therefore be a source of plumes. Alternatively, plumes might actually originate at the core–mantle interface. This interpretation is consistent with geochemical observations indicating that lavas from intraplate hot-spot volcanoes are quite distinct from lavas derived from the shallow mantle at mid-ocean ridges. The chemical characteristics of magmas generated from plumes have been used to argue that many plumes are actually derived from subducted lithospheric slabs, which may heat up after long storage times in the mantle, perhaps at their final resting place, the core–mantle boundary.

When rising plumes impinge on the lithospheric plates, the resulting hot spots cause them to upwarp. With the recent advances in satellite technology, we can measure the lithospheric upwarp, which is usually too subtle to detect easily on the ground. Plumes that cause large-scale swells of the lithosphere, such as are observed around Hawaii, are interpreted to be derived from the core–mantle boundary. This boundary is argued to be the only place in Earth's interior that has a sufficient temperature and composition contrast across it to sustain a long-lived plume, transporting hot material upward at high rates. These plumes also appear to be fixed in position relative to the lithosphere. This suggests they originate deep in the mantle.

A picture is now emerging of a dynamic convecting mantle in which the upward flow of hot material is dominated by inflow at ridges but with a secondary contribution from deep-seated plumes. The corresponding downflow of cold material is supplied by the subduction of cool, oceanic lithosphere back into the mantle (Fig. 11.16).

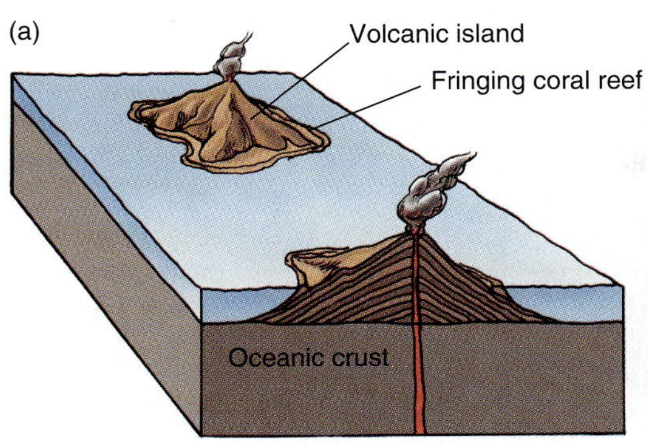

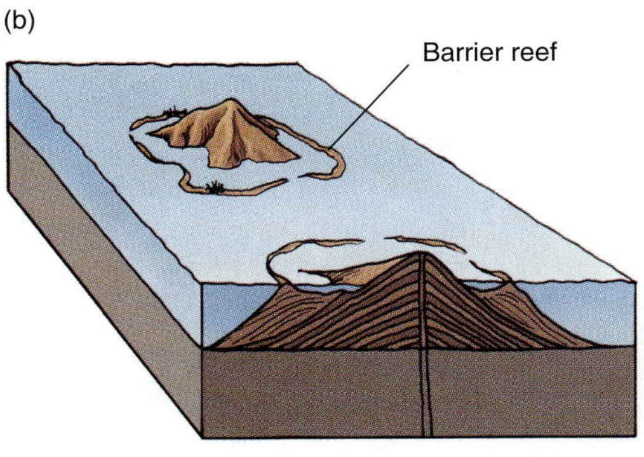

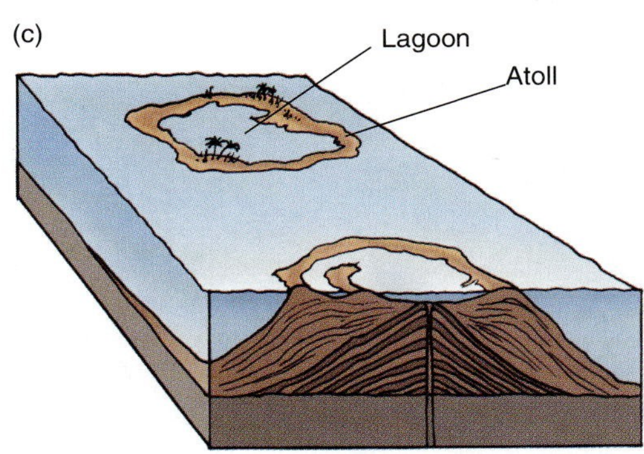

FIGURE 11.15
Formation of (a) a fringing reef (St. Eustatius), (b) a barrier reef (Bora Bora), and (c) an atoll (Palau Islands, Micronesia) in response to subsidence of a volcanic oceanic island. The photographs shown are examples of these types of reefs.

Across the globe, the plumes are relatively fixed in position through time, while the lithospheric plates move over them. Consequently, the location of the hot spot on the lithospheric plate will change as the plate moves, and the plume will leave a hot-spot trail, or a linear string of volcanoes (Fig. 11.17) (∞ see Chapter 6). If sufficient volcanic material is produced, the volcanoes will form islands. Many ridgelike topographic highs on the ocean floor, however, are entirely or mostly submarine and are also interpreted to be hot-spot trails. These features are known as **aseismic ridges**. This is to distinguish them from spreading ridges, which are true plate margins and are associated with earthquakes. They are particularly common in the Atlantic Ocean, where the age of an aseismic ridge apparently decreases toward the active mid-ocean ridge (Fig. 11.18). These relations are easy to understand when we realize that new oceanic lithosphere is actually being formed at the spreading ridge, so the lithosphere is effectively moving away from the ridge and passing over the hot spot, which leaves its trace as an aseismic ridge. The fact that the hot spots are apparently located very close to the spreading ridge itself probably reflects the role that hot spots played in the initial breakup of the continental lithosphere to form the Atlantic Ocean.

Plumes may arise from a range of depths in the mantle. Those originating at the core–mantle boundary are an important mechanism of heat loss from the core.

Figure 11.14 shows that many intraplate volcanic centers are clearly related to rifts in the continental lithosphere. With time, these rifts might develop into plate boundaries if new oceanic lithosphere begins to form along the rift. Then, of course, the volcanism that was part of the original rifting is no longer located within the plate, since it is now at a plate margin. This scenario reemphasizes the way in which plate boundaries and plate-tectonic dynamics are continually changing with time. The original hot spot may persist; in regions like Iceland, we can see the effects of superimposing a hot spot and a divergent plate boundary. Here, the extra heat provided by the hot spot produces a much greater volume of magma than elsewhere along the Mid-Atlantic Ridge. Consequently, the pile of volcanic rocks is much greater than elsewhere along the ridge and has built up above sea level to form the island of Iceland. In contrast to most of the oceanic crust, which averages a thickness of 7 km, the crust beneath Iceland is 24 km thick.

Hot Spots, Continental Flood Basalts, and Rifting

The question of how continents rift apart to form new ocean basins was examined earlier (∞ see Chapter 8). We suggested that rifting may result from a weakening and thinning of the lithosphere by rising plumes. Plumes are generally associated with topographic uplift that, if not actually driving continental breakup, probably plays a significant role in determining where the lithosphere fractures. Where a number of weak spots are in alignment, the continent may be effectively "torn along the dotted line." We argued in Chapter 8 that the combination of fractured crust, topographic drive (ridge push), and slab pull or suction is needed to form a successful rift. The close association between continental rifting and hot spots is underscored by the occasional occurrence of continental flood basalts (∞ see Chapter 6), which are associated in time and space with continental rifting. Rifting above a plume increases the decompression melting effect, enhancing the overall volume of magma produced. Flood-basalt provinces are believed to be the former locations of hot spots generated by large mantle plumes (Fig. 11.19a). Dating of the basalts generally shows that they were erupted during rifting and just before the formation of a new spreading plate boundary (Fig. 11.19b).

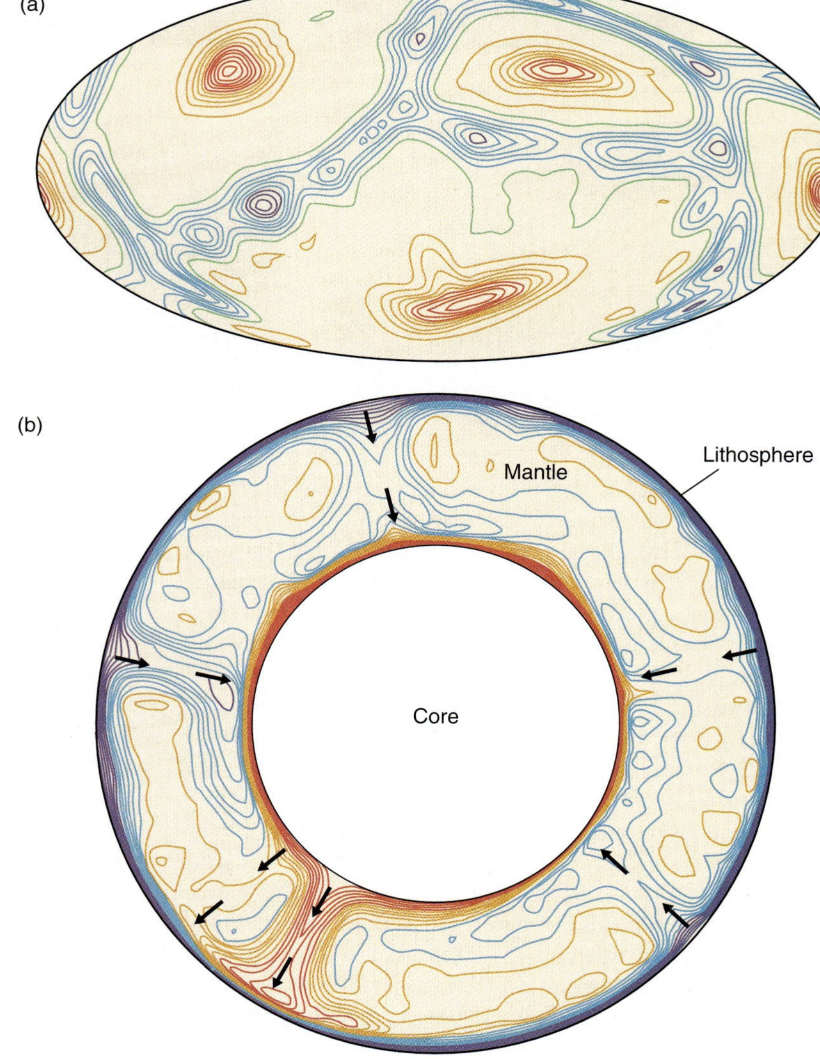

FIGURE 11.16 How mantle convection may work within Earth—a computer model. The mantle is considered to be a viscous fluid, heated from below. (a) A map view of hot (red) and cold (blue) regions that correspond to upwelling and downwelling, respectively. (b) A cross-sectional view. Plumes rise from the core–mantle boundary to form hot spots. The downwelling limbs of convection cells are subducted oceanic lithosphere.

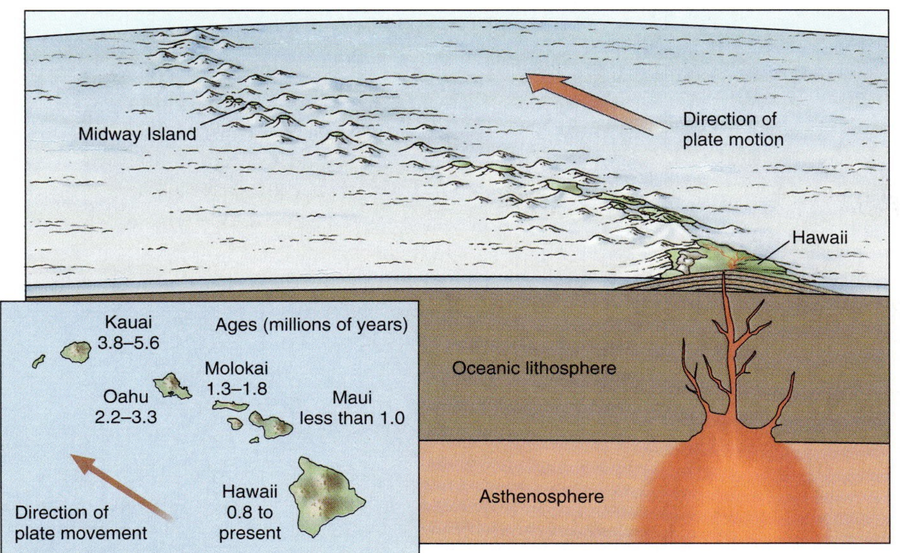

FIGURE 11.17 Formation of a hot spot trail by movement of a lithospheric plate over a fixed, deep-mantle plume. The inset shows a map of the Hawaiian island chain with ages shown in millions of years. Toward the northwest—in the direction of plate motion—the islands become older. We expect the next island to appear to the southeast of the island of Hawaii (see Focus 11.3).

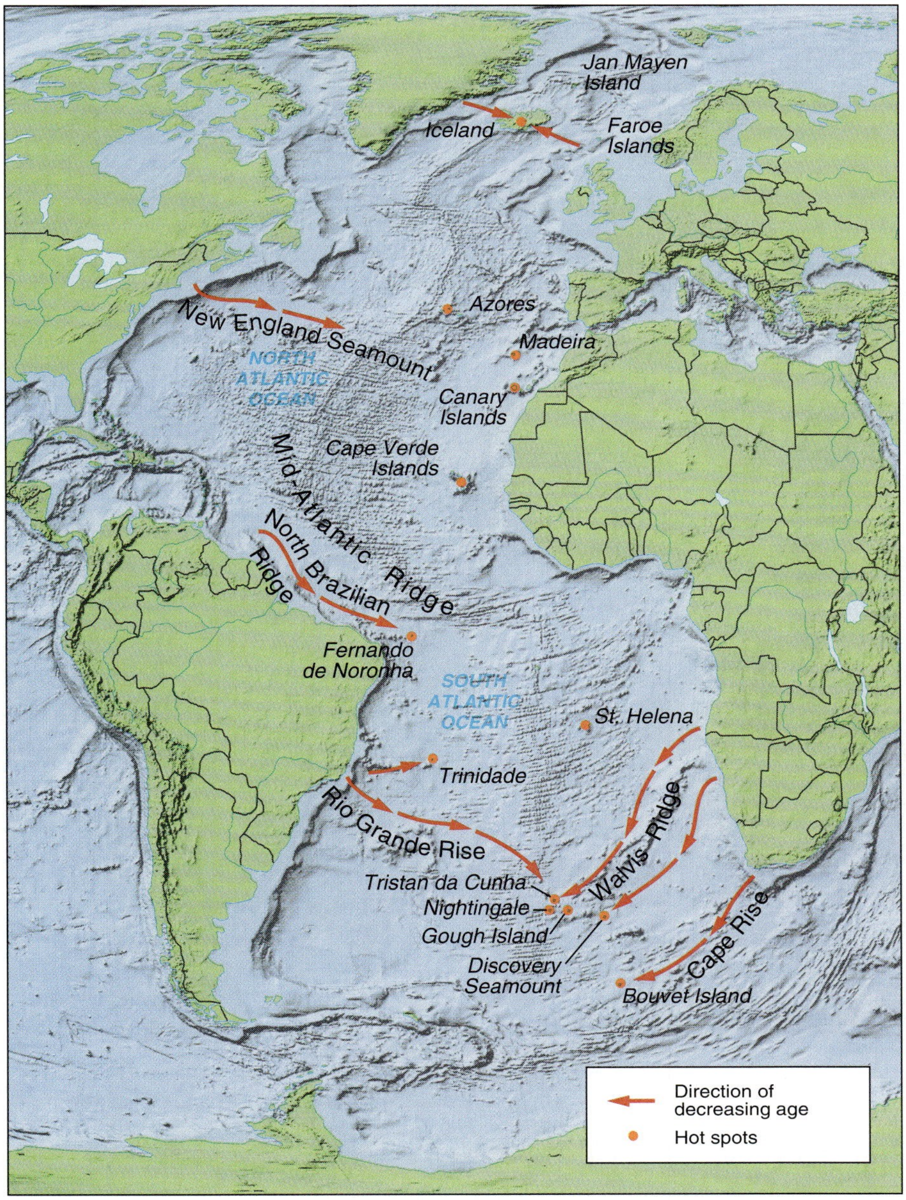

FIGURE 11.18 Aseismic ridges in the Atlantic Ocean. These are interpreted as hot spot trails formed as the Atlantic lithosphere spread away from the Mid-Atlantic ridge. Volcanic islands are also shown. Note that many are at the ends of the aseismic ridges. (Arrows point in the direction of decreasing age of the volcanic material that makes up the ridge.)

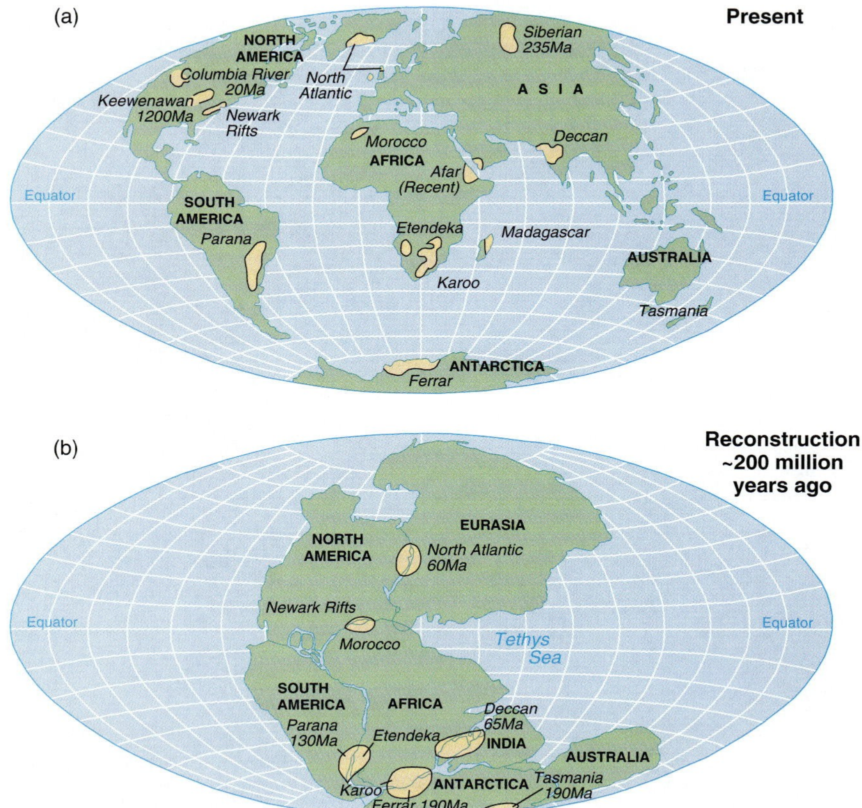

FIGURE 11.19 (a) Map showing the locations of flood-basalt provinces. Ages given are millions of years (Ma). (b) The location of continents is shown before the most recent phase of rifting; the correlation between many of the flood-basalt provinces that are now separated by ocean basins is clear. The Keewenawan and Siberian basalts, in (a), were not generated by the rifting apart of Pangaea (they are older), nor were the Columbia River basalts (they are younger).

FIGURE 11.20 (a) The Deccan flood basalts, forming the Western Ghats, which is an immense escarpment along the northwest coast of India. The thick sequence of basaltic lavas forming the layers here were all erupted 65 million years ago in under 5 million years. (b) Map of India showing distribution of the Deccan flood basalts (yellow).

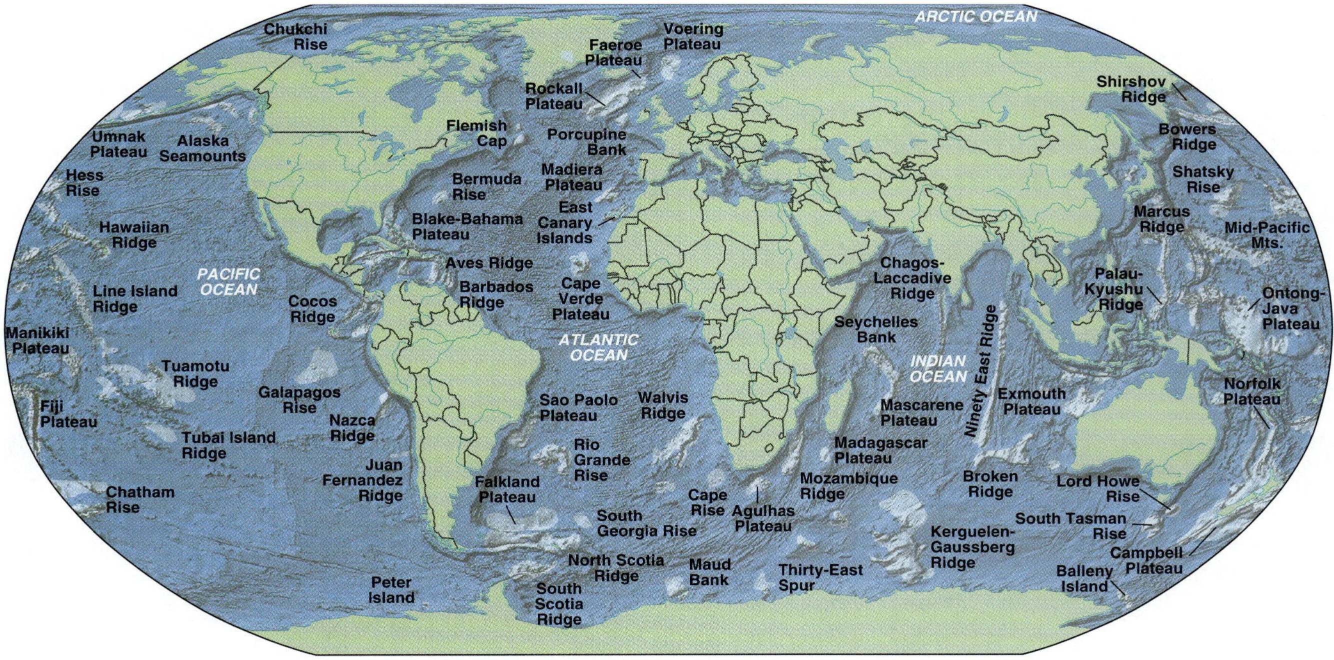

FIGURE 11.21 Distribution of oceanic plateaus and aseismic ridges.

The Deccan flood basalts that form the Western Ghats of India (a huge west-facing escarpment; Fig. 11.20) comprise over 1 million km^3 of basalt that probably erupted 69–64 million years ago within a period of 5 million years. These basalts are related to the separation of Madagascar from India and the eventual formation of the Indian Rise spreading center. This represents a major flux of magma to Earth's surface—far greater than is currently observed even at mid-ocean spreading ridges.

Interestingly, the time at which the Deccan flood basalts erupted (about 67 million years ago) corresponds to the timing of the much-discussed extinction of the dinosaurs. As an alternative to extinction from meteorite impact (∞ see Chapter 2), there has been speculation that the enormous outpourings of basalt and associated gases may have caused a drastic change in Earth's climate, leading to global warming (or cooling), which, in turn, decimated the dinosaurs' food supplies and eventually led to their demise. The most likely culprit to cause a major change in climate would be sulfur dioxide, which is a major constituent of the gases released from basaltic magmas and which can absorb solar radiation if injected into the upper atmosphere (∞ see Chapter 15).

The *cause* of the Deccan flood-basalt volcanism remains elusive though. It may be related to the early stages of a mantle plume, which is currently located at Reunion Island in the Indian Ocean (see Fig. 11.14). It has also been suggested that the Deccan flood basalt formed in response to an enormous meteorite impact that caused widespread melting of the mantle.

Rifting of continents may be triggered by mantle plumes. The initial stages of plume activity might also be accompanied by voluminous basaltic outpourings that form continental flood basalts.

Geologists used the locations of flood-basalt provinces to support the theory of continental drift—the precursor of plate tectonics (Chapter 6). You can clearly see that, when ocean basins are closed up, many of the flood-basalt provinces on opposite margins match up (Fig. 11.19). More recent studies have shown that they can also be matched by ages and by geochemical characteristics. The flood basalts of eastern Greenland match the igneous rocks of the North Atlantic province in northwest Scotland and Ireland, although the latter is more deeply eroded to expose some of the intrusive roots of the volcanoes. In the southern Atlantic, the Parana basalts of eastern Brazil and the Etendeka basalts of Namibia can be closely correlated in age and composition. Many of these flood-basalt provinces are related to the breakup of Pangaea (see Fig. 11.19) (∞ see Chapter 6). However, there are exceptions. The Keewenawan basalts of central North America are not located at a continental margin and were erupted more than 1 billion years ago. They occupy an ancient failed rift, stretching southwest from the Great Lakes, and they represent an attempt to rift apart the North American continent, which never progressed sufficiently to produce a new ocean basin. Although outcrops are found around the shores of the

Great Lakes, the Keewenawan basalts are largely buried beneath younger platform sediments of the American Midwest, and their extent has been determined largely through geophysical methods (see Figure 5.20). Other flood-basalt provinces, such as the Columbia River basalts of the western United States, are not obviously related to rifting at all.

Some submarine oceanic plateaus, such as the Ontong Java Plateau in the southwestern Pacific, are also the result of tremendous basaltic outpourings (Fig. 11.21 on the previous page). These submarine plateaus are very poorly studied and remain an enigma. The rate of basaltic production during the formation of the Ontong Java Plateau may well exceed that of any other volcanic province on Earth. It appears that mantle plumes rising beneath a continent commonly cause the continental lithosphere to rift apart, whereas if the plume rises beneath the stronger oceanic lithosphere, it will not cause rifting, but it will generate oceanic plateaus. As the effect of the plume decreases, it leaves hot spot trails such as the aseismic ridges of the Atlantic Ocean (see Fig. 11.18) or the Hawaiian Islands (see Fig. 11.17) on the oceanic lithosphere.

Unraveling the effects of plumes, both on the continents and oceans, is a new frontier in intraplate geodynamics.

SUMMARY

- Plate interiors are tectonically quiet today; they are not typically associated with active deformation, earthquakes, or volcanoes. What is now a plate interior, however, could have been a plate margin in the past.
- Most of the continental crust comprises older crystalline (igneous and metamorphic) basement covered by a thin veneer of younger sediment.
- Age patterns and analogy with present-day plate-tectonic processes suggest that much of the continental crust has grown by lateral accretion.
- A passive continental margin represents what was originally one side of a continental rift. The crust grades from fully continental on land to fully oceanic—that is, from less dense to more dense material.
- Much sediment accumulates on continental shelves. The extent of the continental shelves (and therefore the amount of sediment that can accumulate) will depend on sea level, which rises and falls through geologic time.
- Although earthquakes are less likely to occur within plates rather than at plate margins, they can be very powerful when they do occur. They are commonly located at ancient plate-tectonic scars or result from intense crustal collision.
- Intraplate magmatism is commonly associated with mantle plumes. Plumes transport hot material from deep within the mantle and are a critical element in mantle convection within Earth.
- Plumes may arise from a range of depths in the mantle. Those originating at the core–mantle boundary are an important mechanism of heat loss from the core.
- Rifting of continents may be triggered by mantle plumes. The initial stages of plume activity might also be accompanied by voluminous basaltic outpourings that form continental flood basalts.

KEY TERMS

peneplain, 275
seamount, 275
craton, 276
shield, 277
platform, 277
lateral accretion, 278
transgression, 284
regression, 284
aseismic ridge, 293

QUESTIONS FOR REVIEW AND FURTHER THOUGHT

1. What processes might cause global sea level to change?
2. What factors may make an earthquake in the eastern United States more damaging than an earthquake at the western margin of the North American Plate?
3. The oldest oceanic crust is about 200 million years old; however, the average age of continental crust is greater than 1 billion years. Why is this?
4. Why is it that passive continental margins are largely underwater (the continental shelves), yet, when they originally formed during continental rifting, they were several hundred meters above sea level?
5. Why do we think that most intraplate magmatism is the result of mantle plumes?
6. How does an active plate margin become part of a plate interior? Can you give an example of a current plate boundary that may soon be incorporated into a plate?
7. What is the relationship among plumes, flood basalts, and passive continental margins?
8. The Roman ruins shown in Figure11.8d have been studied extensively to determine the history of relative uplift and subsidence. The base of the building is thought to have been originally built in the second century B.C. at 6 m above sea level. It had subsided to 6 m *below* sea level by the eleventh century. What was the average yearly subsidence rate over the period between 200 B.C. and A.D. 1100?

12

The Formation of Sediment: Weathering

Satellite view of the Himalayan mountains. Monsoons moving from India (right) provide heavy rains, weathering and eroding the mountains and generating enormous volumes of sediment.

Overview

- Weathering includes all processes that modify and alter the physical and chemical character of rocks exposed at Earth's surface.
- Physical weathering is the mechanical breakdown of rocks by mechanisms such as fracturing, freeze-thaw breakage, or root wedging. Chemical weathering is the reaction among minerals, water, and oxygen.
- With equivalent climatic conditions, the rock type exerts a major control over the type and extent of weathering.
- The time available for weathering to occur is a critical control on the process, which in turn, is controlled by tectonic activity.
- Sedimentary processes separate and isolate individual elements based on the behavior of each in water solution. Soluble elements move to the oceans, and insoluble elements remain on land.
- Climate controls the weathering processes. Arid and arctic climates are dry; hence, chemical weathering is ineffective there, but is very effective in warm and wet tropical climates.

Introduction

There is a famous old instructional film entitled *Why Do We Still Have Mountains?* At first glance, the title seems silly, but it is truly a profound question: Why are there mountains? It is a simple matter to measure the amount of river-borne sediment flowing out of mountain ranges and, from that measurement, it can be shown that the actions of weathering and erosion proceed so rapidly that the mountains should be quickly—in the geologic sense of time—reduced to nearly flat plains. Thus, mountain building must occur continuously or all continents would have been reduced to flat plains long ago.

Erosion (∞ see Chapter 13) comprises the processes that wear down mountains and transport the detritus to a site of deposition. *Weathering* includes all processes that modify and alter the physical and chemical character of rocks exposed at Earth's surface. Two types of weathering are distinguished: physical (mechanical) weathering and chemical weathering. **Physical weathering** may be defined as the breakdown of rocks by mechanical methods, including such processes as repeated freezing and thawing of water in cracks or tree-root wedging. Indeed, mechanical weathering is a process by which a mountain or a massive outcrop can be reduced to small fragments (Fig. 12.1). The mineralogical composition of the smaller rock fragments is much the same as the original rock, and thus it is primarily the fragment sizes that have been reduced. Size reduction definitely plays a major role in weathering. Smaller mineral grains are more chemically reactive because water and oxygen have greater access to grain surfaces where the reactions occur.

Chemical weathering is a reaction between mineral grains and water—typically acidic or alkaline water rather than pure water—in the presence of an oxygen-rich atmosphere. Three principal types of reactions are associated with chemical weathering: oxidation, hydration, and hydrolysis. *Oxidation* is the addition of oxygen atoms to cations in a mineral, or the loss of one or more electrons from an ion or atom. During *hydration*, water molecules are added to a mineral and are incorporated into the crystalline solid. Hydration is not a common process in weathering. In *hydrolysis*, the mineral reacts with water and is generally decomposed or altered by that reaction. Oxidation and hydrolysis are the more common reactions in weathering.

Weathering is defined as: *The response of a mineral assemblage that had been in equilibrium within the lithosphere to conditions at or near Earth's surface. These responses may be physical or chemical, but they occur at low temperatures in the presence of water, air, and organic material. The final result of weathering is a reconstitution into a new mineral assemblage that is in equilibrium with conditions at Earth's surface.*

In this chapter, the term *equilibrium* will be used occasionally. Equilibrium indicates a dynamic or static balance between opposing forces or reactions (∞ see Chapter 3). Chemical equilibrium is a balance between opposing chemical reactions. When water is placed in contact with a slightly soluble material, say calcium carbonate, a small amount of the calcium carbonate will dissolve in the water. When the water has dissolved all the calcium carbonate it can under the given conditions of pressure, temperature, acidity, and so on, it is in chemical equilibrium.

We can identify some general characteristics of chemical reactions in the context of different weathering environments. These are:

1. Chemical reactions tend to proceed more rapidly at higher temperatures.
2. For efficient reaction, the reactants must be brought together quickly and easily, and the products must be removed. In nature, water generally supplies the reactants to the mineral surfaces and flushes away the reaction products.
3. The smaller the reacting grains, the more rapidly the chemical reactions proceed to completion.

All these factors play a role in the chemical weathering process. Local climate controls the average temperature of the reactions and the supply of water for reaction. Grain size of the mineral reactants depends largely on the process of mechanical weathering (disintegration) of rocks, as well as abrasion and breakage during transportation. The length of time available for weathering

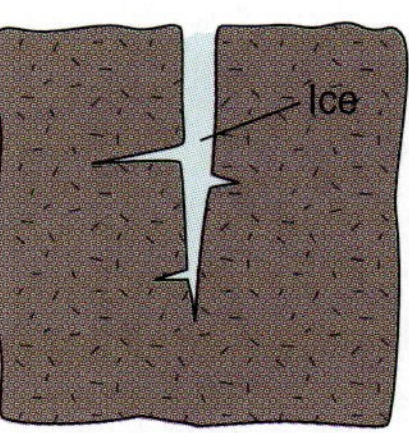
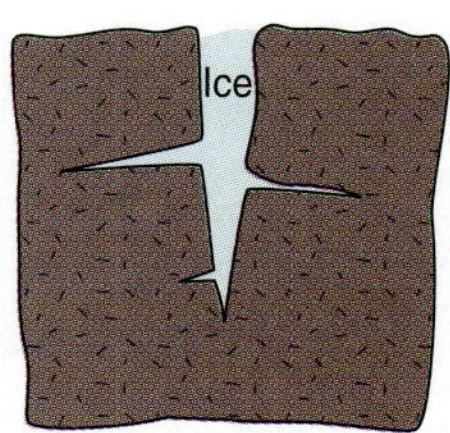

FIGURE 12.1 Progressive joint opening in a block of granite due to frost wedging. An exaggerated view.

reactions depends on the rate of erosion, and therefore the rate of uplift or subsidence. If erosion or deposition occurs rapidly, then weathering reactions will be interrupted because sediments will be buried and removed from the weathering environment; if erosion or deposition occurs slowly, then weathering reactions can proceed for a longer time.

Physical weathering consists largely of the breakage of rocks into smaller fragments. Chemical weathering reactions are controlled by the availability of water, by temperature, and by the time needed for the reactions to proceed.

Physical Weathering: The Disaggregation of Rocks into Sediment

Physical weathering breaks rocks into smaller fragments. *Root wedging* and *frost wedging* are particularly effective mechanisms for breaking rocks apart—especially if cracks or fractures exist in the rock. Plant roots growing in cracks can physically push apart blocks of rock. The same effect is achieved by water that freezes in cracks in the rock (Fig. 12.1). As water freezes to ice, it expands in volume by approximately 9 percent. If water freezes in the small confined cracks of a rock, the force of this expansion can exceed the strength of rocks. Repeated cycles of freezing and thawing cause small cracks in rocks to open progressively wider, allowing water to percolate even deeper into the rock and repeat the process.

The production of progressively smaller rock fragments or even individual mineral grains by mechanical weathering can, by itself, produce sediment. However, its real importance is probably in the increase in surface area available for chemical reactions (chemical weathering).

Effects of Joints and Fractures on Massive Rocks

As we saw earlier (∞ see Chapter 7), most rocks contain a system of joints—fractures or cracks that cross-cut the rock in a regular and repeating fashion. Whether the joints originate through stress release as the rock is brought closer to Earth's surface, or through contraction on cooling from a magma, they are surfaces of weakness that are readily exploited by weathering processes.

Joints commonly form at approximately right angles to one another in three dimensions, and they divide the outcrop roughly into cubes; the size of the cube is determined by the spacing between joints. At the corners of each cube, the rock has the greatest exposure to mechanical and chemical breakdown. This means that the intersections between joint planes (the corners of the cubes in Fig. 12.2a) are weathered more rapidly than are the surfaces (the faces of the cubes), which eventually results in

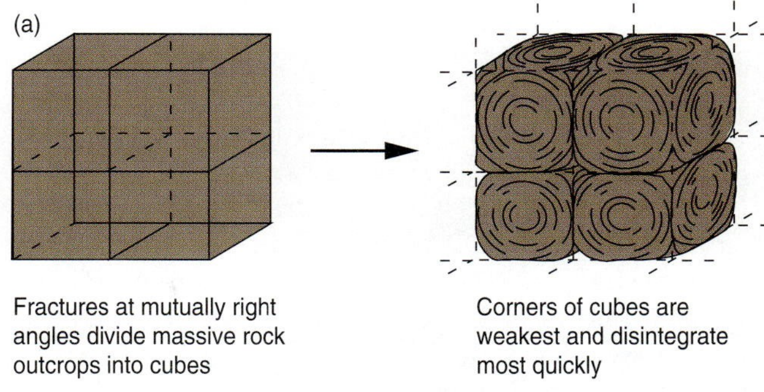

FIGURE 12.2 Spheroidal weathering. (a) The mechanism by which spheroidal weathering forms. It usually breaks down at the corners of mutually orthogonal joints. (b) Spheroidal weathering in both granite (top) and volcanic rocks.

FIGURE 12.3 The breakdown of granite from a massive outcrop to sand grains. (a) Fractured granite on a ledge, broken by frost and ice wedging. (b) These boulders are resting downslope from the outcrop in (a), and ice wedging has begun to break the boulders apart by cracking along grain boundaries. (c) This granite boulder has crumbled to sand.

the formation of crude spheres. This characteristic shape resulting from weathering gives the process its name: *spheroidal weathering* (Fig. 12.2). The spalling off of shells of weathered rock during spheroidal weathering is aided by the outward force exerted by volume increases in the weathered minerals relative to the original minerals. It is clear from Figure 12.2b that jointing helps to focus weathering by providing access to water and chemicals in solution that can react with minerals in the rock. Later in this chapter, we will discuss chemical processes that cause the change from an outcrop of fresh rock to one that is profoundly weathered.

Mechanical Disintegration of Solid Rock into Grains

As is the case with larger joints, most rocks are also fractured on a smaller scale. Minute cracks allow water and plant rootlets to penetrate deeply into the rock. For rocks with relatively coarse grain sizes, such as granite or coarse sandstone, the grain boundaries are preferential sites for water invasion. Typically, a granitic rock is moderately coarse grained: an average grain size 5 to 10 mm. During weathering, small amounts of water penetrate large cracks first; water then flows into increasingly small spaces until it eventually can move along individual crystal boundaries. Frost wedging along these grain boundaries can be a major factor in mechanical disintegration and can cause granitic rocks to disintegrate quickly, grain by grain (Fig. 12.3). For this process to be effective, repeated freeze-thaw cycles are required, and this, in turn, is dependent on climate.

An average granite may have approximately 30 percent quartz, 35 percent potassium feldspar, 20 percent plagioclase feldspar, and 15 percent mafic minerals (biotite or hornblende). After the solid granitic rock has been subjected to repeated cycles of freezing and thawing, it is reduced to a sand that is nearly identical mineralogically to the original granite (Fig. 12.3c). Thus, although the coherence of the rock has been dramatically changed, its composition is largely unaltered. The particle size of the sand produced in this way is controlled by the original grain size of the rock.

Most rocks have pervasive sets of joints or cracks that are pathways for water penetration. Repeated cycles of freezing and thawing cause cracking, breaking, and crumbling of massive outcrops to become sand that has a mineralogy nearly identical to the original rock.

Chemical Weathering: Oxidation, Hydration, and Hydrolysis of Minerals

Over time, rocks that had been in equilibrium at some elevated pressure and temperature within the lithosphere may be brought to Earth's surface. This results in interactions with the atmosphere, with water (the hydrosphere), and with biological activity (the biosphere). By moving rocks from the conditions under which they formed at depth, the overall tendency of weathering processes acting on the rock is to bring the mineral assemblage into equilibrium with this new set of conditions. Rocks formed at very high temperatures or pressures will be more strongly and more quickly affected than will rocks formed at conditions near those at Earth's surface.

Depending on your perspective, weathering can be viewed as either a destructive or a constructive process. In terms of the minerals originally present in the rocks, weathering appears to be destructive. But weathering also represents a *reconstitution* of minerals that are out of equilibrium with their environment into a new assemblage of minerals that *are* in equilibrium with their environment. In this sense, it is a constructive process. For example, a common constituent in many igneous rocks is biotite. Biotite contains iron and magnesium within its crystal structure. Exposed to air and water, the iron atoms within the biotite structure lose an electron—oxidize—from Fe^{2+} to Fe^{3+}, which on combining with oxygen, forms Fe_2O_3, which is stable in air.

On exposure at Earth's surface, a rock that had been in equilibrium with pressure and temperature conditions at depth is usually far from its former equilibrium condition. Weathering is a change from the original mineral assemblage into a new assemblage of minerals that are in equilibrium with conditions at Earth's surface.

How Water Reacts with Silicate Minerals

We are all familiar with iron's rusting when it is left outside; in fact, this is an illustration of weathering—it is an oxidizing reaction and it is a reaction with water. Iron is refined in a high-temperature blast furnace with agents to remove oxygen. Molten iron is produced and cooled—a process roughly analogous to the formation of igneous rocks—and formed into a variety of implements. Pure iron is formed at high temperatures and in the absence of oxygen and water, where it is stable. When a lawn mower, a car, or an old tractor unprotected by paint is left out all winter in the presence of air and water, the iron reacts with the oxygen and water in the atmosphere and forms a new material that is much more stable under these conditions than was the pure iron. We know that new material as iron rust (Fig. 12.4).

FIGURE 12.4 Iron reacts with oxygen in the air to form iron oxide as seen on this rusted iron tractor.

In a simplified form, we can write the oxidizing reaction as:

$$\underset{\text{Iron}}{4Fe} + \underset{\substack{\text{Oxygen}\\\text{(from air}\\\text{and water)}}}{3O_2} \rightarrow \underset{\text{Iron oxide}}{2Fe_2O_3}$$

Igneous minerals display much the same response as iron, except that chemical reactions in silicate minerals are generally more sluggish than reactions involving pure metals.

What happens to the individual minerals in a granitic rock during weathering? Factors controlling the weathering response of minerals include the types of chemical bonds, the mineral structure, and the temperature and pressure at which the mineral originally crystallized (∞ see Chapter 4). In general, minerals that crystallize at higher temperatures are less stable during weathering, and minerals that crystallize at lower temperatures are more stable during weathering.

For instance, quartz is one of the last minerals to crystallize from a granitic magma. Quartz is composed entirely of silica tetrahedra covalently bonded in a strong three-dimensional network. Compounds that are covalently bonded are usually insoluble in water, whereas ionically bonded compounds tend to be water-soluble. Thus, quartz tends to be relatively stable (insoluble) during weathering (Fig. 12.5). Biotite crystallizes at a somewhat higher temperature than does quartz and contains both ionically and covalently bonded components in its structure and so weathers more rapidly.

You may recall (∞ see Chapter 3) that biotite is composed of extensive two-dimensional sheets of covalently bonded silica tetrahedra. The sheets are held together by weak electrostatic (van der Waals) bonds with potassium (K^+) in the spaces between the mineral lattice sheets (Fig. 12.6). Iron (Fe^{2+}) and magnesium (Mg^{2+}) are ionically bonded within the lattice. Ionic bonds are selectively soluble in water; thus, iron and magnesium cations can be removed from the mineral structure during weathering, leaving unfilled spaces inside the crystal lattice. This type of reaction is termed *hydrolysis*. In reality, water in the atmosphere in the form of rain dissolves carbon dioxide to form a weak acid called *carbonic acid* (H_2CO_3). This acid provides hydrogen ions (Focus 12.1) that replace the potassium, iron, and magnesium ions within the crystal lattice. The hydrogen ion, however, is much smaller than the cations it replaces (Fig. 12.6). The result is that the ions in the new mineral structure are not properly stacked together, and the unsupported mineral structure collapses.

It is possible to measure the acidic or basic (alkaline) character of a solution using the pH scale. In fundamental terms, pH is inversely related to the availability of hydro-

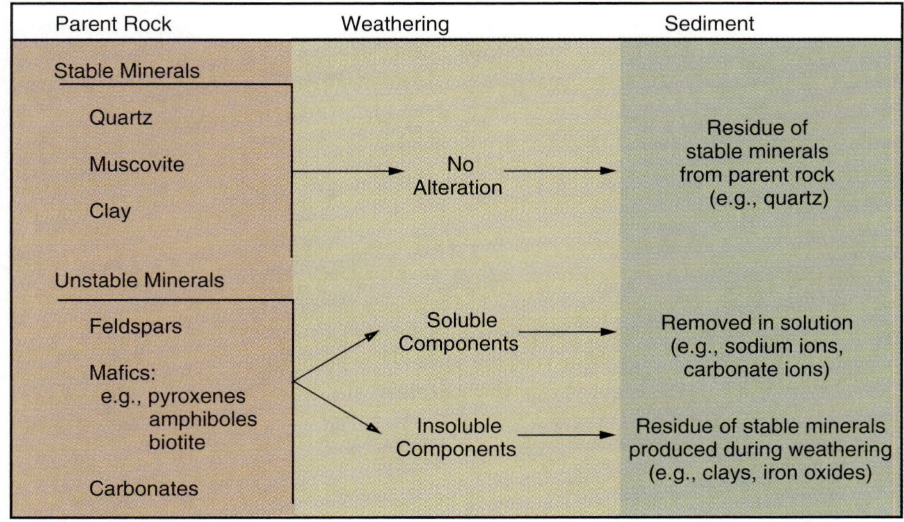

FIGURE 12.5 The alteration products of minerals in a rock that has undergone extensive weathering. Stable primary minerals are unchanged after extensive weathering; unstable minerals undergo major changes in form and mineralogy or may go partially or completely into solution.

gen ions in a solution; hydrogen ions (H^+) are responsible for the acid character of a solution, and hydroxyl ions (OH^-) for the alkaline, or basic, character of a solution. A pH scale, which ranges in value from 0 to 14, is used as a measure of the acidity or alkalinity of a solution. A solution with a pH value of less than 7 is acidic; a value greater than 7 indicates that a solution is basic (alkaline). A solution with a pH value of 7 is defined as being neutral—neither basic nor acidic. Pure distilled water has a pH value of 7.

The strongest acids, such as concentrated hydrochloric acid (HCl), have a pH value close to 0, indicating that there are few HCl molecules in the solution; the acid is almost entirely *dissociated*: It is a solution of discrete H^+ and Cl^- ions. A weak acid will not completely dissociate into its component ions in solution; there will be molecules of the acid in the solution. An example of a weak acid is carbonic acid (Focus 12.1). When carbon dioxide is dissolved in soft drinks, carbonic acid is formed. As a result, some soft drinks have pH values of about 2. Acetic acid (CH_3COOH), the compound responsible for the sour flavor of vinegar, is also a weak acid.

In basic solutions, hydrogen ions cannot exist as free cations but must combine with oxygen to form the reactive hydroxyl anion (OH^-). The strongest bases, such as sodium hydroxide (NaOH), have a pH value of 14. Seawater is slightly alkaline, with a pH value of about 8.

Weathering occurs in areas where a parent rock is being disintegrated by a combination of physical processes, such as freeze-thaw fracturing or root wedging, and chemical attack. This chemical attack is carried out during periodic moistening by rain. Rainwater that is in equilibrium with the carbon dioxide (CO_2) in the atmosphere has an acidic pH value of 5.7. The pH of most rainwater, however, is usually less acidic and has a higher pH, greater than 6.5.

Feldspars are more difficult to weather than are micas, because, like quartz, they have a framework of three-dimensional bonds (∞ see Chapter 3). But cations of potassium (K^+) in the case of potassium feldspar, or sodium (Na^+) and calcium (Ca^{2+}) in the case of plagioclase, reside in the crystal lattices of feldspars. It is an oversimplification to say that potassium, sodium, and calcium are removed from the feldspar structure in the same manner that biotite loses iron and magnesium, but the effect is similar. Again, unfilled spaces in the crystal lattice are left after the sodium, potassium, or calcium is removed. The crystal lattice is no longer stable, and it collapses; after some recrystallization, clay minerals are formed from the residue.

Ionically bonded cations in minerals selectively react with water and are dissolved into solution. Removal of the cations leaves the mineral lattice unsupported, and it collapses.

How Water Reacts with Carbonate Minerals

Carbonate rocks, principally limestone ($CaCO_3$) and dolomite [$CaMg(CO_3)_2$], are very common, and these rocks weather easily when exposed at Earth's surface. Most carbonates are sedimentary in origin, and therefore we might suspect that such rocks would be in equilibrium with conditions at Earth's surface. In fact, as long as they remain in contact with seawater, which is nearly saturated with calcium carbonate (see Focus 12.2), the minerals are in equilibrium. However, if carbonate rocks are exposed to rainwater and biota, they will react. The same processes involved in the weathering of silicate rocks are also involved in the weathering of carbonate rocks. So that chemical reactions may continue, there must be a continuing supply of fresh reactants, and the products of the reactions must be removed. Flushing with abundant new rainwater supplies fresh reactants and removes the dissolution products. Thus, the primary weathering feature of carbonates is dissolution (Focus 12.1).

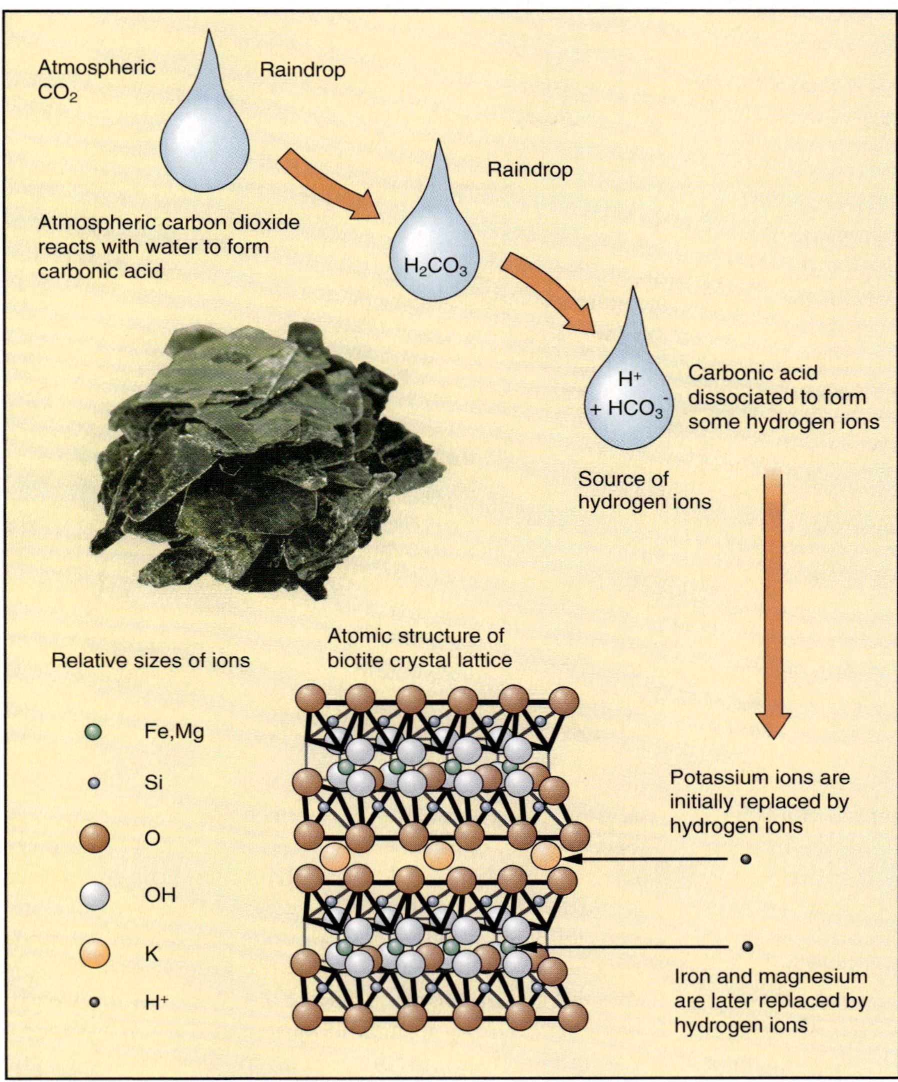

FIGURE 12.6 Atmospheric carbon dioxide dissolves in rainwater to form carbonic acid. This is a weak acid, only partly dissociating to form hydrogen ions (H^+) and bicarbonate ions (HCO_3^-). The hydrogen ions from the acid that are available preferentially react to replace the potassium ions in the biotite lattice; iron and magnesium are more protected within the lattice and therefore are replaced by hydrogen ions more slowly.

Weathering Features in Carbonate Terranes Several distinctive weathering features are formed where carbonate rocks are broadly exposed at the surface. These features are primarily the result of the tendency for carbonate rocks to dissolve rather than disaggregate, or "crumble," on weathering. Exposures of carbonate rocks are usually fractured in much the same way that we described earlier for igneous rocks. These fractures, cracks, and joints in otherwise solid rock allow water to flow through the outcrop, dissolving the limestone. With a continuous supply of water, these small features interconnect so the water can move downward or horizontally, dissolving the carbonates and widening the openings in places. This dissolution results in tunnels, cavities, or caverns in the carbonate rocks.

As the cavities increase in size by dissolution of the carbonate, the cavity becomes unstable, and the roof collapses, further enlarging the caverns upward. Water percolating downward or within the groundwater reservoir dissolves the carbonate rocks above the cavern as it flows through the joints, fractures, and cracks. For this reason, the water that drips into underground limestone caverns is saturated with calcium carbonate. As the water evaporates, the beautiful columns, stalagmites, stalactites, and other formations are deposited from the carbonate carried in water solution (Fig. 12.7).

This process can be understood in terms of the dissolution reaction given in Focus 12.1. When the drop of water is exposed to the air in the cavern, evaporation increases the concentration of calcium and carbonate ions. This drives the reaction to the left, causing precipitation of solid calcium carbonate.

There are many more solution caverns in limestone terranes than just those that are tourist attractions; most are hidden, though. The entrance to most caverns is little more than a slightly widened joint or crack at the surface. However, some caves do have spectacular entrances that open into vertical shafts. Most caves are never discovered or explored; some, however, are found catastrophically.

Focus On 12.1 ACIDS AND MINERALS

When an acidic solution reacts with a mineral, the hydrogen ions exchange for the cations in the mineral. This reaction consumes the hydrogen ions, thus using up the acid, and releases cations from the mineral lattice. An example of this process can be illustrated by the reaction of carbonic acid with calcium carbonate: Carbonic acid formed as carbon dioxide in the atmosphere is dissolved by rainwater, and calcium carbonate is the main mineral in limestone.

H_2CO_3	+	$CaCO_3$	$\rightleftharpoons$	$Ca(OH)_2$	+	$2CO_2$
Carbonic acid	+	Calcium carbonate		Calcium hydroxide	+	Carbon dioxide
(in rainwater)		(in limestone)		(in solution)		(gas)

The exchange of hydrogen ions for cations in any other compound or mineral is broadly similar. Acidic water attacks minerals by exchanging hydrogen ions from the acid for cations in the mineral lattice. These cations are most commonly sodium (Na^+) and potassium (K^+), from silicate minerals, ions that form soluble hydroxides when in water solution (NaOH and KOH). Weak acids are not efficient donors of hydrogen ions. Weak acids, however, are much more common in nature than are strong acids. Naturally formed carbonic acid is the predominant acid involved in the weathering process.

In the mid-continent of North America, the abundance of limestone at or near the surface results in common features called **sinkholes**. A sinkhole is a depression on Earth's surface above a shallow cavern where the cavern roof has collapsed (Fig. 12.8a). Where there is an abundance of sinkholes, a distinctive dimpled landscape called **karst topography** is formed (Fig. 12.8b).

A peculiar type of karst topography, called *tower karst*, is beautifully exposed in parts of China. In this case, much of the carbonate has been dissolved and removed in solution, leaving the remnants that form the towers (Fig. 12.8c).

Weathering Products and Solubility

Weathering occurs most effectively in wet climates. After rainwater has percolated through the rocks, the surfaces of individual mineral grains in a rock are covered with a thin layer of water. This water is slightly acidic owing to the absorption of carbon dioxide from the atmosphere, causing the mineral to react chemically: acidic water adheres to the mineral grain, and the hydrogen ions in the water replace cations in the mineral lattice; in turn, the mineral cations react with the water. Acid is "used up" in this process because the hydrogen ions substitute for cations in the mineral lattice and these cations are released into solution. If the exchangeable cations in the mineral are sodium or potassium, then the pH of the water increases. The pH may reach values as high as 11 because these cations form strong bases.

If the exchange of hydrogen ions for cations in the mineral lattice is to continue, reaction products must be removed and a continuing supply of hydrogen ions must be available. Continued weathering depends on renewal of the acid; that is, additional hydrogen ions are needed to further the weathering process.

Equilibration with the atmosphere is not an adequate source of acid to account for all the observed weathering processes. Biologic processes are also important in weathering. Oxidation of organic matter in the soil by microscopic organisms produces carbon dioxide. Thus, the content of carbon dioxide in the soil may be as much as 50 percent higher than are atmospheric levels. Moreover, feeder roots—the tiny, hairlike rootlets covering the surfaces of larger plant roots—are a potent source of hydrogen ions; pH values as low as 2 have been measured immediately adjacent to the feeder roots of plants. A single tree might have many miles of feeder roots, and thus the chemical influence of plants on mineral weathering, and on soils, can be significant.

In summary, the weathering process of hydrolysis is a replacement (**leaching**) of the cations within a mineral lattice by hydrogen ions. The most soluble ions remain in solution and ultimately are carried by streams and rivers to the oceans (Focus 12.2); less soluble ions may be precipitated along the route, forming new rocks and minerals.

Development of Soils

Different soil types are the result of the specific conditions of weathering and of the rock types that are exposed. Precise definitions of "soil" depend upon the purpose of the definition and who is doing the defining.

(a)

(b)

(c)
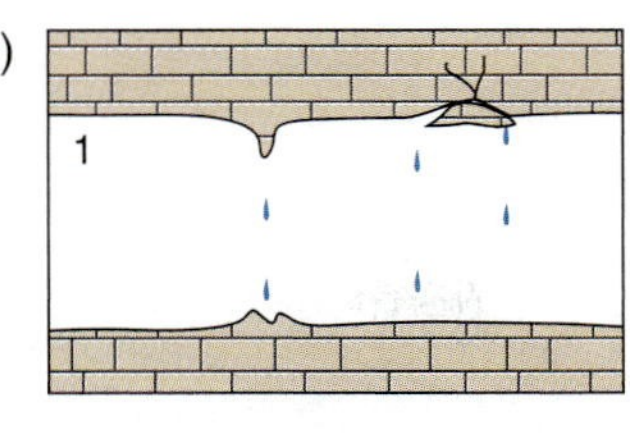

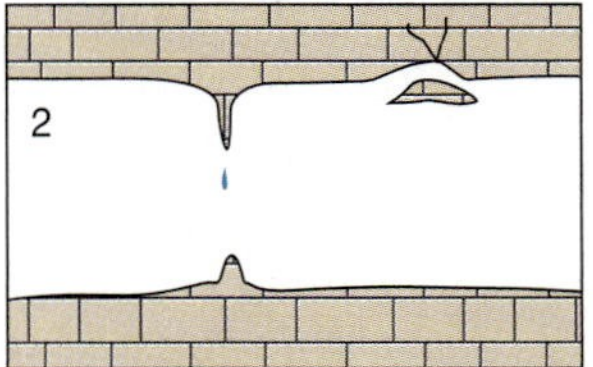

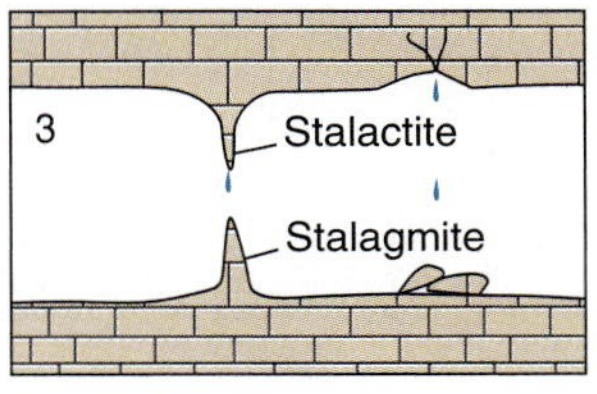

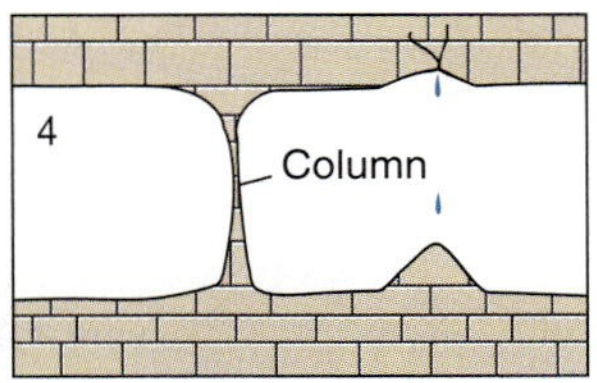

FIGURE 12.7 Photographs of cave features. (a) A solution-widened joint that represents an entrance to a limestone cavern. (b) Stalagmite, stalactites, and columns in a carbonate cavern. (c) A sketch representation of the creation of a column from stalactites and stalagmites. Once the column is complete, water may continue to run down it, depositing the fluted textures on the column shown in part b. Also shown in the illustration is roof collapse, causing new drip points for water entry into the cave.

Soil scientists have detailed classification schemes for soils that take into account mineralogy, cohesiveness, iron and aluminum contents, and other factors. Farmers have a large number of descriptive names for types of soil that they find important to distinguish. We define soils as mixtures of unaltered to partially decomposed organic matter, mineral constituents that have been altered to some degree by exposure to both Earth's surface conditions and water.

This definition of soil excludes loose mineral and rock material such as mechanically disaggregated rock, volcanic ash, sand, glacial moraine, and even lunar "soil," which may rest directly on a solid rock surface. These materials are more properly referred to as *regolith*. Although regoliths are generally sterile, biological processes are very active in soils. Bacterial oxidation of the soil organic matter produces carbon dioxide, which, as we discussed earlier, combines with water to form carbonic acid (H_2CO_3). In soil this additional carbon dioxide enhances further mineral alteration.

A soil contains partially decomposed organic material, weathered rock debris, and water.

Characteristics of Soils Soil formation differs with specific climatic regimes with characteristic vegetation and soil-moisture conditions. Details of soil classification are somewhat complex, and we will not offer a complete treatment here. A well-established descriptive scheme

(a)

(b)

(c)

FIGURE 12.8 Karst topographic features. (a) A catastrophic sinkhole in Florida. This feature appeared suddenly and quickly grew, swallowing cars and buildings. (b) Typical sinkhole topography in a limestone region of Kentucky. There is no integrated drainage system; instead, the landscape is dominated by small depressions. (c) Tower karst from the Kwangsi region, Yangshuo, China.

divides well developed soils into six zones, known as **soil horizons**, which are called **O, A, E, B, C**, and **R**, on the basis of composition, texture, and color (Fig. 12.9).

The uppermost zone, the O horizon, is the region of plant litter and humus accumulation—it was called the O horizon for the organic matter in the zone. Below the O horizon lies the A horizon, a mixture of organic and mineral matter. Beneath the A horizon is the E horizon, which stands for "eluvial" and is characterized by the loss of iron, aluminum, and clay and is usually whitish or gray in color. **Eluviation** refers to the process of washing out fine soil components or what most of us call "dirt." The O, A, and E horizons are subject to the leaching of soluble components by water percolating downward. These dissolved components are carried deeper into the soil, where they may be precipitated in the B soil horizon. Where it exists, the B horizon represents a zone of accumulation for components leached out of the overlying soil horizons. Directly below the B horizon is the C horizon, which typically consists of disaggregated parent material—regolith or bedrock—that is slightly weathered. The R horizon below the C horizon is unaltered regolith or underlying bedrock. The C and R soil horizons are regions where the most active chemical weathering takes place. It is in these zones that this type of weathering will be enhanced by the downward percolation of water, which will be acidic or basic, depending on the extent of the B zone and depth of root penetration.

Residual Soils Some deposits of pure quartz in the soil environment are termed *residual deposits*, meaning that they are the residue left behind after all else either has dissolved out of the parent rock or has abraded away during the process of erosion and transportation. The cation in quartz is silicon, which is entirely covalently bonded to four oxygens to form the silica tetrahedron (∞ see Chapter 3). For this reason, quartz is largely insoluble in water. However, as discussed earlier, the cations can be dissolved from feldspars, resulting in a collapse of the feldspar crystal structure. Removal of cations from minerals causes the iron oxide and clay content to increase with extended weathering. If erosion and transportation have not occurred, extensive weathering of a rock consisting of quartz and feldspar would produce a sediment that is a mixture of clay (the residual aluminum compound from

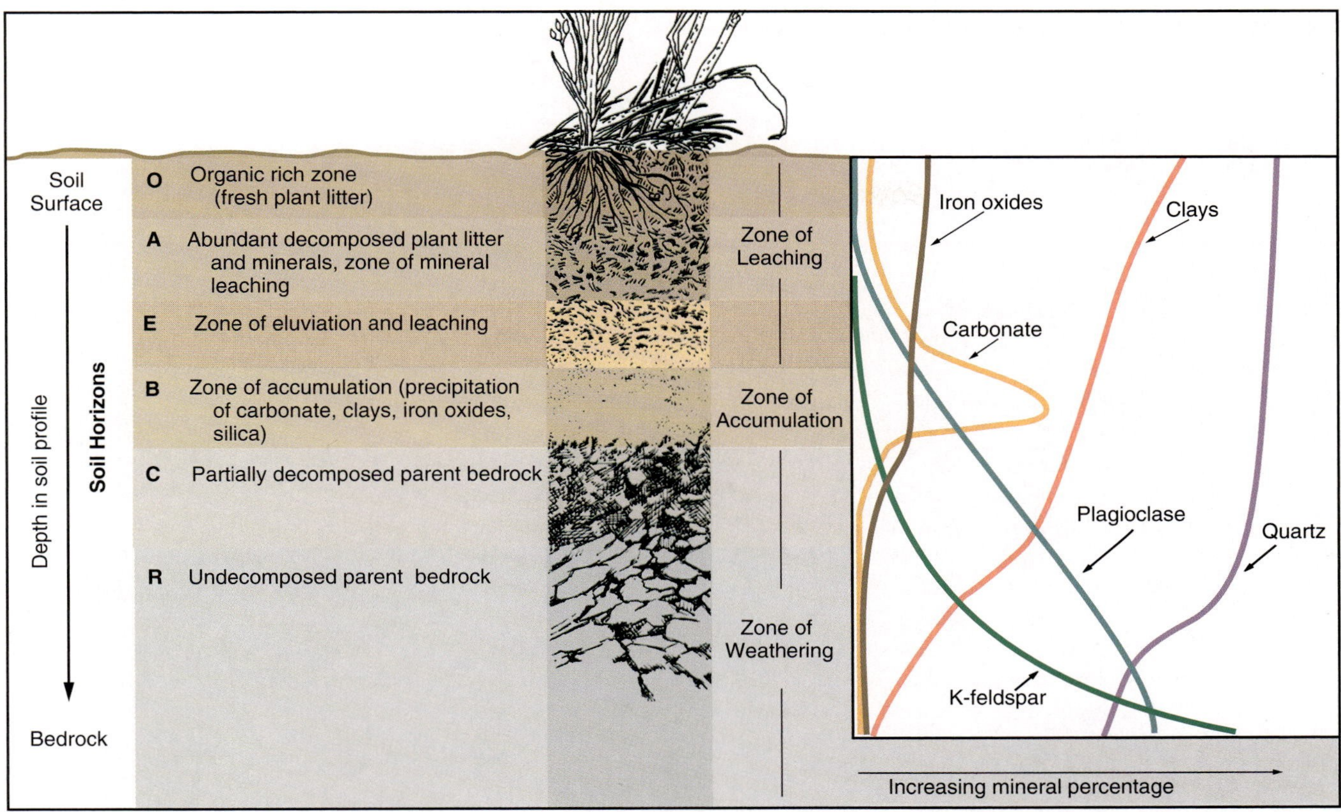

FIGURE 12.9 An idealized illustration of soil zones and characterizing features: The uppermost zone, the O *horizon*, is the region of plant litter. The uppermost parts of the soil are the A *and* E *horizons*. The O, A, and E horizons are subject to leaching. The B *horizon* is often called a zone of accumulation because chemical precipitation is the important mechanism in its formation. The C *horizon* consists of disaggregated parent rock that is usually slightly to strongly weathered; the underlying R *horizon* is unaltered bedrock. The diagram to the right illustrates mineralogical changes that may occur with progressive weathering of a granitic rock outcrop. Compare the alteration shown with the relative stabilities of minerals in weathering illustrated in Figure 12.5.

the breakdown of feldspar) and quartz. Table 12.1 summarizes the effects and the major products of weathering.

Mafic minerals are the least stable, and quartz is among the more stable of all minerals in the weathering environment. The amount of quartz in a soil resting on top of its parent rock appears to decrease with depth toward the unaltered parent rock (Fig. 12.9). This occurs because the percentage of quartz increases with more prolonged weathering. Quartz is not actually accumulating in the soil; rather, its relative increase is due to the

TABLE 12.1 Products of Weathering

MINERAL	COMPOSITION	WEATHERING PRODUCTS
Quartz	SiO_2	SiO_2 (no change)
Feldspar	$KAlSi_3O_8$ $NaAlSiO_3O_8$ $CaAl_2Si_2O_8$	Hydrous aluminum silicates (clay minerals) such as kaolinite $Al_2Si_2O_5\ (OH)_4$ and K, Na^+, Ca^{2+} in solution
Muscovite	$KAl_2(AlSi_3O_{10}\bullet 2H_2O)$	Muscovite (no change)
Ferromagnesian minerals, e.g.,		
Biotite	$K(Mg,Fe)_3(AlSi_3O_{10})(OH)_2$	Hydrous aluminum silicates (clay minerals) plus soluble Mg^{2+}, Ca^{2+}, Fe^{2+}
Olivine	$(Mg,Fe)_2SiO_4$	
Hornblende	$Ca_2Na(Mg,Fe)(Fe,Al,Ti)_3$ . . . . . . $Si_8O_{22}(OH, Cl, F)_2$	
Carbonates	$CaCO_3$, $CaMg(CO_3)_2$	Soluble bicarbonates (+ Ca^{2+} + Mg^{2+} in solution)

Focus On 12.2 WHY IS THE OCEAN SALTY?

The sedimentary fate of elements is fundamentally tied to their behavior in water solution. For example, the element boron virtually always remains in aqueous solution, so the ultimate sink for boron is the oceans, or deposits resulting from the complete evaporation of water. Rock salt, composed entirely of sodium chloride (NaCl), weathers completely by dissolving in water. If the water carrying the weathered products of rock salt joins a river that flows to an ocean, then the sodium and chloride in solution would eventually be carried out to sea.

Other soluble components ultimately reach the oceans as well; thus, the oceans are the final repository of all highly soluble elements. This might imply that the oceans are gradually becoming saltier. In fact, estimating the amounts of salts delivered to the oceans by rivers and measuring the saltiness of the oceans formed the basis of one of the earlier scientific estimates of the age of the Earth. The age estimate of 100 million years was too young by a factor of 45, but this mistake is easily forgiven. At the time the calculation was made, it was difficult to estimate the quantity of material deposited at the bottoms of the oceans, the amount of material taken up by organisms living in the oceans, the amount of nutrients incorporated into the biomass of forests, or the amount of material added to oceans by mechanisms other than stream flow. For example, the chemical interchange between volcanic rocks and ocean water within the hydrothermal circulation cells developed at the mid-ocean spreading ridges (∞ see Chapter 8) must have a major influence on the chemistry of seawater, but just how much is the subject of modern research today.

Evidence for the composition of seawater in the geologic past is based primarily on the sequence of salts deposited by evaporation of the ancient sea. If modern ocean water is evaporated, the sequence of salts that crystallize out are, from first (least soluble) to last (most soluble): calcite ($CaCO_3$), gypsum ($CaSO_4 \bullet nH_2O$), a variety of mixed salts, halite (NaCl), which precipitates after about 90 percent of the water volume has already evaporated, and lastly sylvite (KCl).

To illustrate the relatively high concentration of carbonate in seawater, a simple evaporation experiment can be performed. Place a pan of seawater (or sea salt dissolved in a pan of water) in a low-temperature oven—the best temperature would be about 60°C (about 140°F on your oven control). Carbonate begins to crystallize during evaporation of seawater after the volume of water has been reduced by about 50 percent. Other salts precipitate with further evaporation; for example, sodium chloride (halite, NaCl) precipitates only after approximately 90 percent of the water has evaporated away (Fig. 1).

From old sedimentary deposits, we can deduce the same *sequence* of salt crystallization that can be seen in the evaporation of modern ocean water. This indicates that the *relative* proportion of dissolved ions (Table 12.1) has not changed substantially over time, but it tells us nothing about how much water had to evaporate to produce the crystallized salts—in other words, the *concentration* of salts in the ancient oceans.

Using a bit of indirect reasoning, however, we can deduce that the ionic concentrations in seawater have not changed over the past several hundred million years. We can estimate the salinity of ancient oceans from fossil carbonate shells. The modern oceans are nearly saturated with respect to calcium carbonate, which means that seawater has dissolved as much carbonate as is possible.

loss of other, more unstable minerals from the soil, which causes the appearance of an upward increase in the amount of quartz (Fig. 12.9)

Chemical Separation During Weathering

Sediment Composition A common difficulty in dealing with sedimentary products concerns the great amount of information that must be visualized all at once. Diagrams are useful because they allow us to see the relationships among different components—something often very difficult to do with tables of numbers. A very useful type of diagram is a triangular plot, in which compositional data can be plotted for three variables at once, as illustrated in Figure 12.10. Each apex of the triangle represents 100 percent of one component, with diminishing percentages of each component plotted as lines parallel to the side of the triangle opposite the appropriate apex. Five examples are plotted in Figure 12.10.

For example, point 4 has a composition of 50 percent C, 25 percent A, and 25 percent B. The lines of dimin-

That calcium carbonate is the most nearly saturated of all salts dissolved in ocean water is almost certainly the reason many marine animals and plants use it for their shells and skeletal framework. Modern marine organisms that secrete carbonate shells may show no dissolution features on their shells, even long after the organisms have died. For instance, delicate, needlelike spines on microscopic marine fossils recovered from sediments deposited in shallow marine waters are often perfectly preserved, suggesting that seawater and the carbonate shells are close to chemical equilibrium. We saw earlier (∞ see Chapter 1) that organisms with carbonate shells first appeared approximately 600 million years to 500 million years ago. If the seas of 400 million years ago, for example, were much less saline than are the oceans of today, we might expect to see abundant evidence that fossil carbonate shells were being dissolved by the seawater in which they lived. Although it is rare to find perfectly preserved fossils in rocks that old, perfect fossils have occasionally been found. Thus, we conclude that the ancient oceans must have had approximately the same concentrations of dissolved constituents as do today's oceans. It is also possible that seawater has had a similar composition and concentration of salts for a somewhat longer time than 400 million years. However, in more ancient times (greater than about 2 billion years), the oceans may have differed considerably from those of today. Nearly pure silica deposits interbedded with iron oxides are characteristic of that time period.

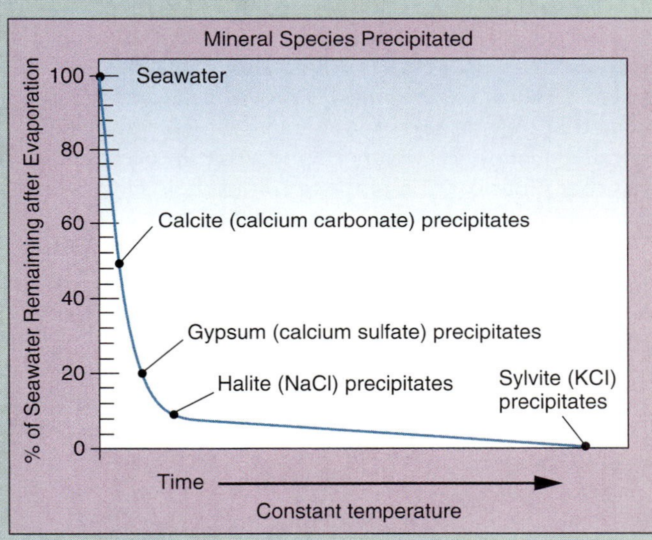

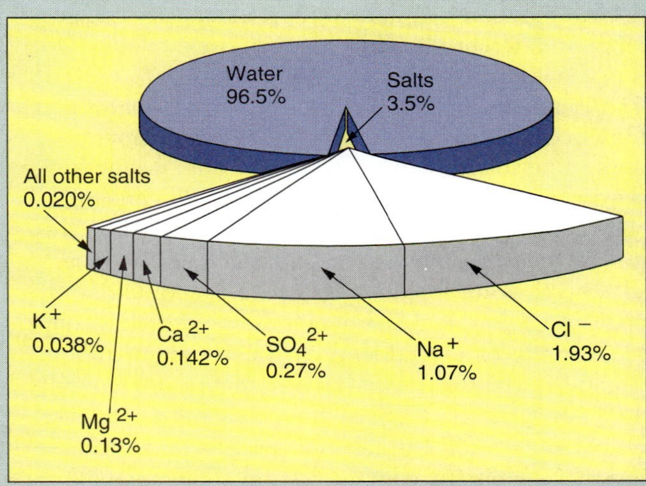

FIGURE 1 Precipitation of the major salts with evaporation of seawater, and the proportions in which dissolved ions (salts) are present in water.

ishing C content are parallel to side AB of the triangle, and there is 0 percent of C along the AB edge of the triangle. Triangular plots can be used to group or categorize sedimentary rocks based on any three components or any three combinations of components. Recall (∞ see Chapter 4) that sedimentary rocks may consist of both mineral grains and substantial amounts of sand-sized rock fragments. The apices of the triangle may represent igneous, metamorphic, or sedimentary rocks, or they may represent characteristic minerals of a specific rock type.

This type of diagram can be used to examine the modification of newly formed sedimentary material from an igneous parent rock (Fig. 12.11). Silica (SiO_2) is plotted at one apex; alumina (Al_2O_3) at a second apex; and calcium oxide (CaO), magnesium oxide (MgO), and iron oxide (FeO) are combined and plotted at the third apex. Gabbro (basalt) and granite (rhyolite), representing the range of common igneous rocks, plot quite close together on the diagram—more or less in the center of the triangle (Gb and Gr in Fig. 12.11). Most other igneous

FIGURE 12.10 A triangular diagram in which three variables are plotted simultaneously. The examples contain the following proportions of components A, B, and C.

Point 1: 0% A and B, 100% C
Point 2: 25% A, 50% B, and 25% C
Point 3: 60% A, 0% B, and 40% C
Point 4: 25% A, 25% B, 50% C
Point 5: 40% A, 50% B, and 10% C

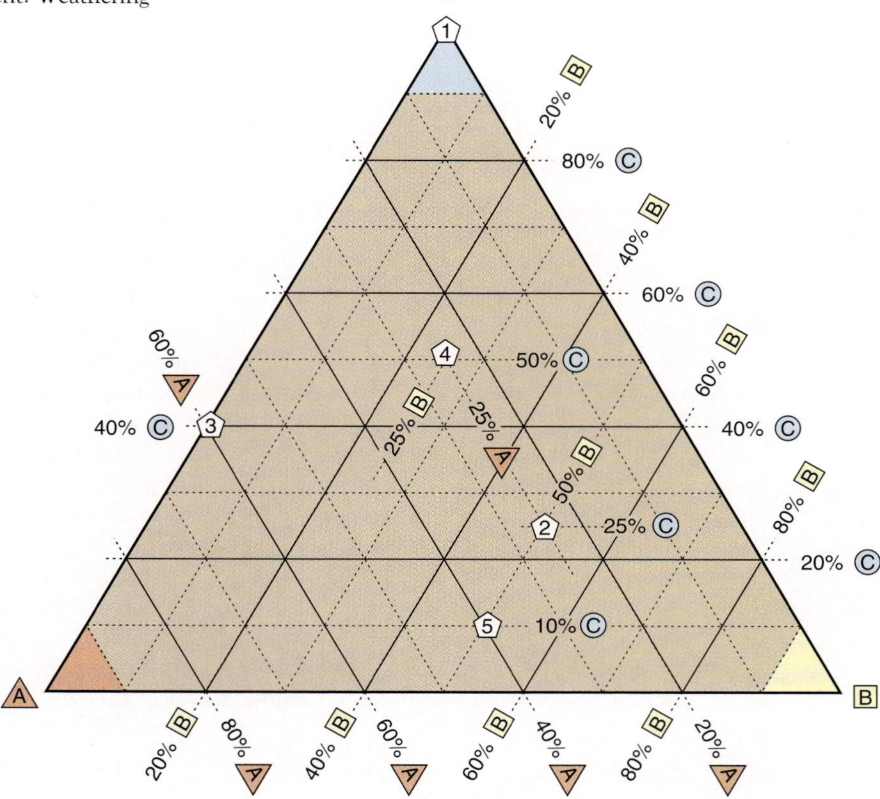

rocks will plot between or near these examples. The percentage of difference in silica between these two lithologies is about 30 percent, from approximately 75 percent in granitic rocks to 45 percent for gabbros. Gabbros and granites contain approximately equivalent amounts of alumina, and the concentrations of calcium, magnesium, and iron combined vary from about 10 to 25 percent.

The compositions of sediments, when plotted on this same diagram, show much greater ranges. A pure quartz sandstone plots at the silica apex because it contains only quartz (SiO_2) (Fig. 12.11). Pure limestone, which is composed entirely of the mineral calcite (calcium carbonate), plots at the CaO apex because there is no alumina (Al_2O_3) or silica (SiO_2) in calcium carbonate. However, the field shown for natural limestone is large because limestones commonly contain detrital components such as sand or clay.

The sedimentary processes of weathering, transportation, and sedimentation are capable of completely separating and isolating elements. Sedimentary differentiation is far more efficient at chemical modification than is magmatic differentiation. The separation of elements, as represented on triangular diagrams, plots as a migration of the original composition toward one of the apices (Fig. 12.12).

In general, the segregation of elements can be observed in the products of weathering: The coarser, clastic fraction (sand) migrates toward the SiO_2 apex; the finer clastic fraction (mud and clay) migrates toward the Al_2O_3 apex; and the composition of the solutions migrates toward the CaO-MgO-FeO apex. The diagram illustrates the concept of sediment maturity, which we introduced in Chapter 4: A sandstone that plots near its parent igneous rock would be called *immature*; a sandstone that plots near the SiO_2 apex of the triangle would be called *mature*. As weathering reactions proceed, products that plot near one of the apices of the compositional triangle are considered to be mature.

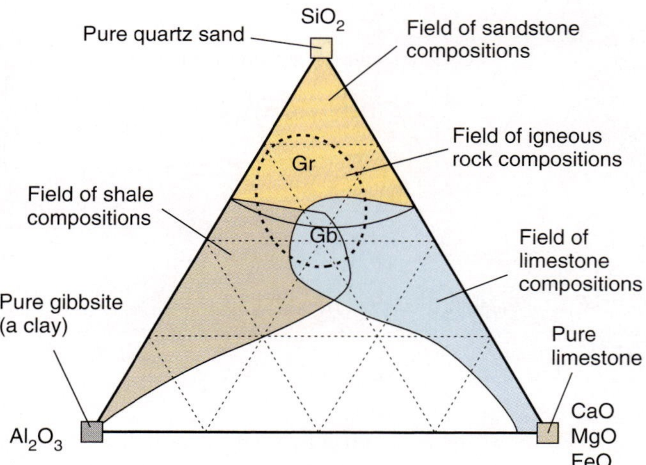

FIGURE 12.11 Common sediments plotted on the SiO_2, Al_2O_3, (CaO, MgO, FeO) triangular diagram. Location of the igneous rocks, gabbro (Gb) and granite (Gr) define the range of most igneous rocks. This plot illustrates that sedimentary processes separate and isolate elements more effectively than do igneous processes, such as fractional crystallization or partial melting that are responsible for the entire field of igneous rock compositions. Note, for example, that pure quartz sandstone plots at the SiO_2 apex, and shale plots toward the Al_2O_3 apex.

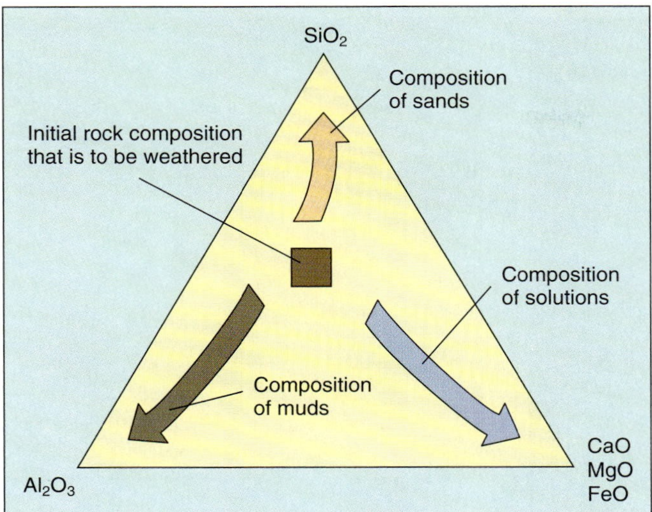

FIGURE 12.12 Sediment maturity triangle. During the processes of weathering, the reaction products migrate away from the composition of the original rock toward the apices of the compositional triangle with increased processing. Sandstones tend to migrate toward the SiO_2 apex. Pore waters and river compositions, generated during weathering, carry the most soluble elements ultimately to the oceans. The insoluble residue accumulates at the Al_2O_3 apex. When the composition of the weathered product reaches one of the apices of the maturity triangle, it is very mature.

Sedimentary processes—weathering, transportation, and sedimentation—can effectively separate and isolate chemical elements.

Sediment Reprocessing The reprocessing of sediments—repetition of the processes of weathering, erosion, transportation, and deposition—will result in a sediment that has been more completely weathered than is possible in a single cycle. Let us consider reprocessing of sediment derived from granite. In rivers, disaggregated mineral grains from the granite are continually deposited, weathered, reeroded, and redeposited as the sediment load moves downstream. The mafic minerals are continually subjected to chemical change and are likely to have been completely altered. The feldspars might be either altered into clay minerals or broken along cleavage planes into very small sizes. Magnetite, a common accessory mineral, alters to rust, but quartz remains unaltered. Granites have grain sizes that are in the general range of medium to coarse sand. The unaltered quartz remains approximately that size, the broken feldspars are finer, and the clay mineral grains are even smaller. After successive episodes of weathering, erosion, transportation, and deposition, the granite has been substantially altered—both chemically and physically.

For chemical weathering to be effective, the chemical components reacting on mineral grain surfaces must be continually replenished. Grain abrasion occurring through grain-to-grain collisions during erosion and transportation is a common surface-renewal mechanism that exposes unweathered cleavage or fracture surfaces. Consequently, reprocessing of sediments may weather a rock thoroughly and segregate its constituent chemical components, as shown in Figure 12.12.

Effect of Lithology on Weathering

We have discussed the weathering of carbonate rocks, which proceeds primarily by dissolving the carbonate in water and transporting the ions away from the site of dissolution. Weathering of silicate minerals depends on replacement of one ion for another, usually the hydrogen ion for a mineral lattice cation. It may seem obvious that in a given climate, weathering reactions and the extent to which weathering occurs will be strongly influenced by the rock type. It is not unusual to have several different rock types (lithologies) in a small area.

Consider a succession of sandstone, shale, and limestone that has been intruded by a granitoid pluton. The pluton itself will weather differently from the contact metamorphic rocks immediately adjacent to it, and both will weather differently from the unmetamorphosed sediments away from the pluton. If the three rock types were placed together, by faulting or folding, their different weathering characteristics would profoundly alter the character of the landscape. This difference in weathering characteristics is called *differential weathering*.

Statues moved from one location to another often provide excellent examples of differential weathering. Figure 12.13 illustrates this point. The tall spire is the Obelisk of Theodosius, made of granite and carved in Egypt approximately 1500 B.C. In A.D. 390 the obelisk was moved to what is now Istanbul, Turkey, and a marble pedestal was carved at that time to hold the obelisk. Istanbul is on the Mediterranean coast, where wet winters are common. There are obvious differences in the weathering of these two rock types after approximately 1600 years in Istanbul. Carvings on the original granite obelisk are still sharply defined and clear, whereas carvings on the marble pedestal are now faded, with the fine detail obliterated because of solution. Thus, although the granite carvings are nearly 1900 years older than the marble carvings, they have suffered far less weathering.

Weathering and Plate-Tectonic Setting

Up to this point, we have considered the influences of climate and temperature and we have only briefly alluded to the role of tectonic activity. A quick review of the earlier chapters in this book will serve as a reminder that major geologic differences exist between continents and ocean basins and between plate interiors and different types of plate margins. It should therefore be no surprise that weathering is also profoundly affected by the plate-tectonic setting.

FIGURE 12.13 The tall spire is the Obelisk of Theodosius carved of granite in 1500 B.C. The obelisk was placed on a newly carved marble pedestal in A.D 390 . Obvious differences are apparent in the weathering of these two rock types after approximately 1600 years in Istanbul, Turkey. Carvings on the granite obelisk are sharply defined and clear (photo on left), whereas the detail on the more recent marble carvings is now faded (photo on right). Although the granite carvings are nearly 1900 years older than the marble carvings, they have suffered far less weathering.

If we compare two areas with similar climates and bedrock geology, a tectonically quiet region will be more extensively weathered than will a tectonically active region. A key factor in weathering is the time available for the processes to occur. The amount of time available is a function of the rate of tectonic uplift. If uplift is rapid, then relatively unweathered detritus is quickly exposed, eroded, and transported away. In contrast, if the rate of uplift is slow relative to the rate of weathering, then the weathered zone—the soil profile—can become very thick indeed. Thus, the depth of weathering is inversely proportional to the rate of erosion, and therefore also to the steepness of local topographic slopes.

By far the most effective agent of erosion is running water. Slowly moving water that is flowing down a gentle slope is less effective than rapidly moving water that cascades down a steep slope. Thus, steep slopes promote rapid erosion, and steep slopes are, in turn, directly related to tectonic uplift.

Depth of weathering is inversely proportional to the rate of erosion. Rapid uplift causes rapid erosion, which does not allow sufficient time for weathering reactions to proceed.

Shield and Platform Areas The interiors of most continents are usually composed of very old continental crust surrounded by progressively younger material toward the continental edges (∞ see Chapter 11). Because the interior is far from plate margins, the fragments of old crust are commonly flat and low-lying, and may be periodically submerged by shallow seas. The absence of tectonic activity has permitted erosive processes to plane off the landscape to a nearly flat or smoothly sloping surface, often with wide and sweeping rivers flowing across the plains. The absence of tectonism in continental interiors enhances the likelihood that sediment will be reprocessed and that weathering will have proceeded as far as the prevailing climate permits.

Passive Continental Margins Along passive margins, sedimentation is generally a more important process than erosion, because such margins are undergoing active subsidence. A mature passive margin is generally characterized by little topographic relief: broad, low-lying coastal plains, abundant lagoons, estuaries, and tidal flats. The shoreline is typically lined by chains of barrier islands, such as are found along the coasts of Texas and North Carolina.

Because passive margins tend to be dominated by sedimentation, weathering of fresh bedrock is far less important in these areas than in other tectonic settings. Weathering of sediments that are being transported or that have already been deposited is an important and ongoing process. This tendency for sediments to be reprocessed produces more mature sediments than those found at active plate margins.

Weathering of sediments is somewhat different from the weathering reactions occurring in fresh bedrock; sediments are already disaggregated and partly weathered. Thus, weathering along passive margins may represent a second or third stage of reaction. By this, we mean that the sedimentary particles may have suffered repeated episodes of weathering, transportation, and deposition since their derivation from a parent rock. The most soluble components may already have been removed from the mineral grains, leaving behind only the most resistant components (see Fig. 12.5). The effectiveness of weathering in these cases reflects the balance between the rate of sedimentation and the climatically influenced weathering reactions. If the sedimentation rate is high, this secondary weathering will not have sufficient time to be effective. If the sedimentation rate is slow and the climate is appropriate for weathering, then all but the most stable minerals will be severely weathered. It is left to the transportation processes to separate the different grain sizes.

Passive margins are dominated by sedimentation. Weathering of sediments that are being transported or that have already been deposited is an important and ongoing process. This reprocessing of sediments produces mature sediments.

Active Continental Margins Active margins contain features that are distinctive to this tectonic setting and that strongly influence weathering. Magmatism at convergent margins adds fresh volcanic or plutonic material to the edge of the continent, and rapid tectonic uplift characterizes active plate margins, whether the margin is convergent or transform. These two features combine to make weathering along active margins quite different from cratonic or passive margin settings. The uplift of active margins creates steep slopes for streams issuing from small drainage basins; these streams actively cut into the freshly exposed rock. Rapid erosion into these steep drainage basins generates an accumulation of largely unweathered sediment downstream. Moreover, because these sediments have suffered less weathering than those along passive margins, they tend to contain abundant sand-sized fragments of the rocks (lithic grains) that are being weathered and eroded. As a result, sediments along active margins are coarser and less mature than are sediments derived from the craton or from passive margins.

Fresh lavas and pyroclastic rocks extruded along active convergent margins, and their intrusive equivalents, contain materials that are not at chemical equilibrium with conditions at Earth's surface. Highly reactive phases, such as volcanic glass, may be present in the volcanic deposits as well as minerals that are readily soluble, such as olivine or pyroxene. As discussed earlier (∞ see Chapter 9), convergent margin volcanoes are characterized by frequent explosive eruptions, which produce pyroclastic deposits of ash and larger fragments, rather than more durable lava flows. This material is rapidly weathered and eroded (Fig. 12.14). Surface water easily percolates through this volcanic debris and, depending on the climate, can quickly reduce it to soil. Water is also vented as steam during eruptions. This magmatic water is

(a)

(b)

FIGURE 12.14 Comparison of weathering on two active volcanoes. (a) An intensely weathered and dissected tropical volcano, Mt. Pelée, Martinique, and (b) an essentially unweathered volcano, located in a more arid climate zone, Licancabur volcano, Chile.

typically highly acidic and promotes alteration of the fresh volcanic rocks almost as soon as they are erupted.

Rapid uplift characteristic of active margins does not permit extensive weathering, so sediments that are deposited along active margins tend to be coarser and less mature than sediments deposited along passive margins.

The Role of Climate in Weathering

As described in Chapter 6, it appears that global climatic belts have been relatively stable with respect to Earth's poles, and that the poles have been fixed with respect to the Sun throughout geologic time. There is no doubt that Earth's climate has fluctuated from temperatures much warmer to much colder than those of today. However, Earth has had stable climatic zones because the climate is directly related to the angle at which the Sun's light strikes our planet. Rainfall maxima at the equator and 60° north and south latitudes, and rainfall minima at 30° north and south latitudes and at the poles. Vegetation and climate vary with latitude, which, in turn, exerts a strong influence on weathering and soil formation (Fig. 12.15).

Arid Environments

Desert Environments The term "arid" indicates the absence of water. Arid regions can occur at almost any temperature, although most people associate arid regions

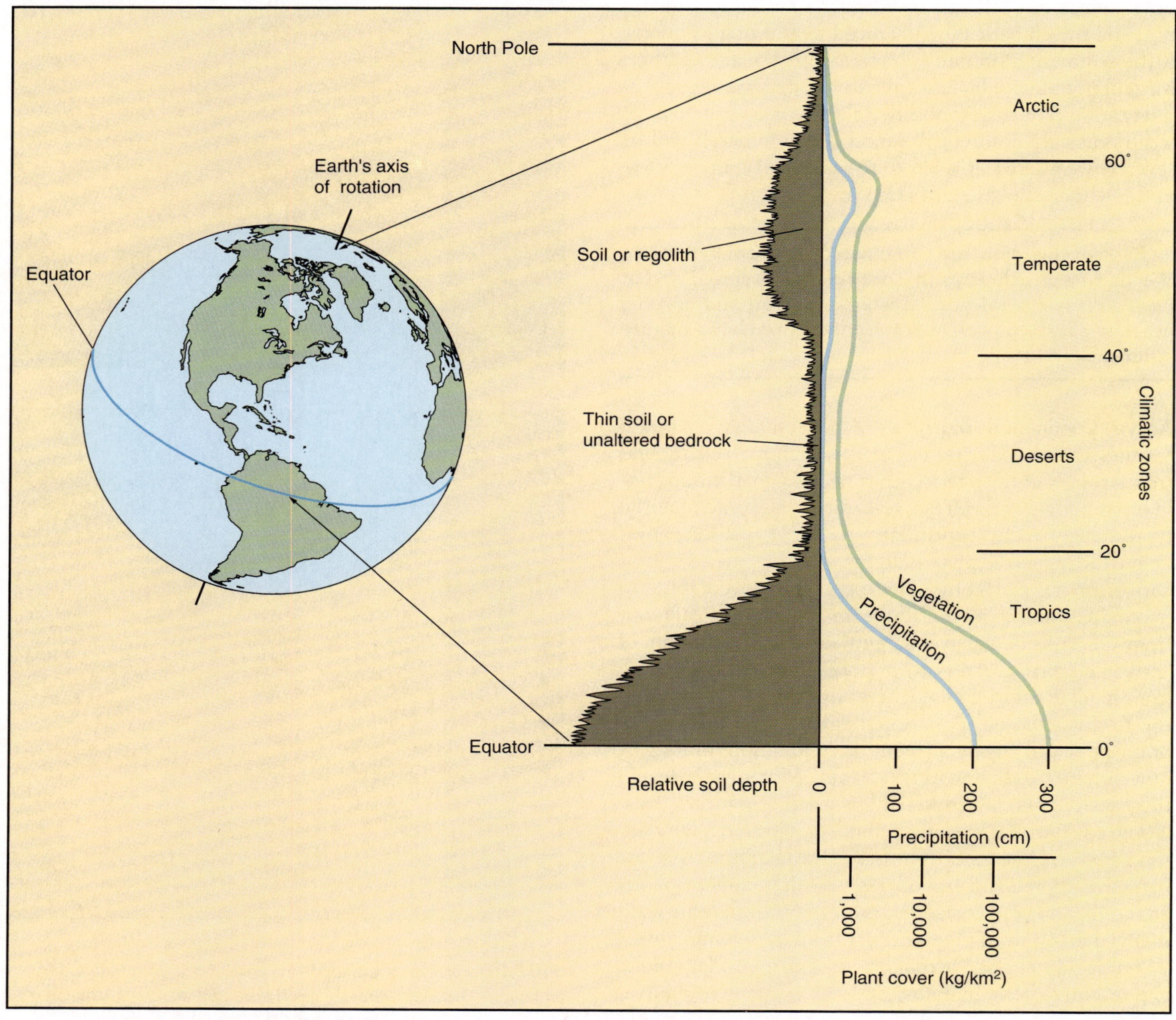

FIGURE 12.15 Relationship between precipitation and vegetation from the North Pole to the equator. Vegetation and climate vary with latitude, which, in turn, exerts a strong influence on weathering and soil formation.

with hot deserts. Weathering in arid climates is impeded by lack of water for the chemical reactions to proceed. Most desert regions on Earth experience at least some rain. Also, during the coolest part of the day, there is likely to be some condensation out of the air onto the rocks. This small amount of water is important for the reactions that do occur in arid environments.

Perhaps the most striking climatic feature of a desert is the persistent wind, which sometimes approaches speeds that can only be described as violent. Sand picked up and carried in the wind severely abrades everything in its path; cars may be stripped of their paint and wooden fence posts abraded or even cut through. This sandblasting also affects exposed rocks, constantly stripping away the outer portions and exposing fresh surfaces to the small amounts of moisture that are intermittently present. Angular faces may be abraded on the surfaces of larger stones so that they look faceted (Fig. 12.16); these are called **ventifacts** (from the Latin *ventus* for "wind" and *facies* for "face").

Quartz grains found in deserts are commonly etched or frosted, and the etching is similar in appearance to decorative etched glass. Close inspection of the grains in a sand dune clearly shows their frosted character. This frosting has been attributed by some geologists to repeated collisions with other grains during wind transport, and there seems to be little doubt that this process contributes to frosting, but chemical weathering effects are also important. It is enlightening to compare sand grains from a beach with grains from desert sand dunes. The grains are immediately distinguishable in your hand; the beach grains are polished and shiny, whereas the dune sand grains are dull.

FIGURE 12.16 Ventifacts—wind-faceted rocks—exhibit desert varnish: a thin, dark brown coating of an iron and manganese clay.

Figure 12.17 shows photographs of quartz grains taken by a scanning electron microscope—one from a beach and the other from a desert sand dune. Quartz grains from the beach are subjected to repeated and violent collisions during surf action, and these grains show distinctive conchoidal fractures resulting from those collisions (Fig. 12.17a). These fractures are small, so that the grains appear highly polished.

Scanning electron microscopic observations of the dull frosted quartz grains from desert sand dunes show some collision fractures, but also the distinctive texture of chemical etching (Fig. 12.17b). Quartz, which is stable in acid solutions, does dissolve a little in alkaline solutions. Observations of frosted quartz sand grains have shown that the frosting is due to slight dissolution (etching) by alkaline (higher pH) solutions that occur on the grain surfaces. In some limestone-depositing regions, persistent winds may blow quartz grains offshore where they

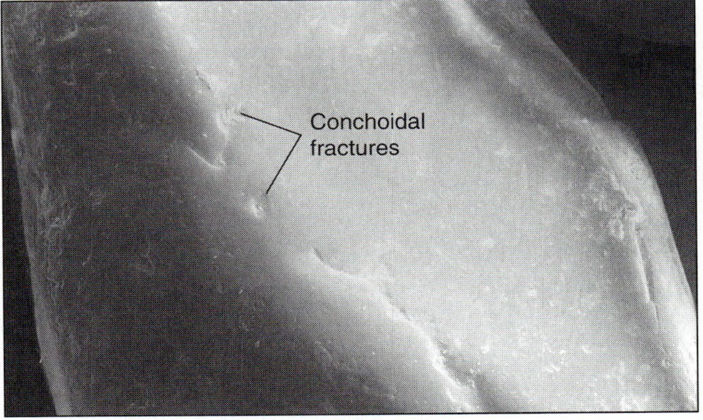

(a)

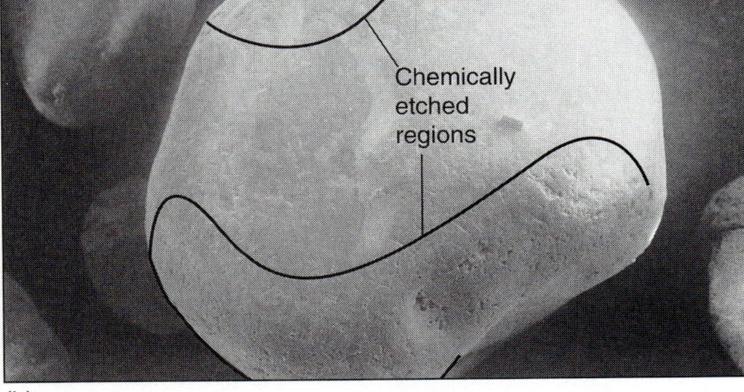

(b)

FIGURE 12.17 Photographs of quartz grains taken with a scanning electron microscope. (a) A quartz grain from a modern beach showing clear conchoidal fractures on its surface that are the result of grain-to-grain collision during the high turbulence of wave action. Adjacent to the fractures are polished surfaces that result from the continual rubbing of the sand grains together during sand movement. (b) A quartz grain from a modern sand dune showing both a conchoidal fracture from a grain-to-grain collision and areas that have been chemically etched.

are incorporated into the carbonate sediment. Grains recovered from limestones are always frosted because the water between carbonate grains in limestones characteristically has a pH value of approximately 8.5. This is sufficiently alkaline for quartz to be slowly dissolved by the pore water in carbonate rocks.

A distinctive dark brown to black coating on rocks is formed in deserts by a somewhat similar mechanism. This coating, called **desert varnish**, is caused by the slow dissolution of cations out of the rocks lying on the ground surface. Most of the cations leached from the rocks are removed, but iron and manganese remain on the rock surface and combine with alumina to form a type of clay mineral. One thing is clear, desert varnish is formed only on the exposed rock surfaces. Pick up a stone in the desert that is heavily coated with desert varnish and you will find that the bottom of the stone is uncoated (Fig. 12.16).

Overall, processes of etching in deserts are not effective mechanisms that can cause wholesale modification of minerals (chemical weathering). The primary modification that occurs in arid climates is the physical breakdown of rocks by mechanical weathering and abrasion. The common desert extremes of cold nights and hot days cause alternating cycles of contraction and expansion, which mechanically break rocks as surely as ice wedging, but much more slowly. The wind carries sand that can abrade rock exposures, but only to a minor extent when compared with the action of water (∞ see Chapter 13).

In desert latitudes on some continents, such as Australia, the central areas are arid; this deters weathering. In continental areas outside of desert latitudes, climates are wetter and exhibit a number of common features. Winter and summer temperatures and rainfall may vary widely in higher latitudes, but show virtually no seasonal differences at low latitudes. Plate interiors are also populated with vegetation types that are controlled by local climate. Each vegetation type in combination with climatic conditions influences the soil pH and weathering in its own peculiar fashion (Fig. 12.18).

Chemical weathering in desert climates proceeds slowly because of the absence of water or wet and dry cycles.

Arctic Environments Although the climate in arctic regions is harsh, air temperatures are so low that the humidity approaches 0 percent, and for that reason, the amount of precipitation is often small. In Yellowknife, located in the Northwest Territories of Canada, the winter temperature can plunge as low as −50°C, but the total snowfall may be less than 1 m. Thus, although the temperature is cold, the precipitation is so low that these regions are truly arid deserts. The air is simply too dry for snow to form in the winter.

During the rest of the year, fluid water may be present, although it may freeze in shadows or at night. Beneath the shallow thaw layer on Earth's surface is the **permafrost** zone, a permanently frozen zone that may extend hundreds of meters in depth. Permafrost prevents water percolation downward into the soil, resulting in a surface zone that, though wet in the summer, contains

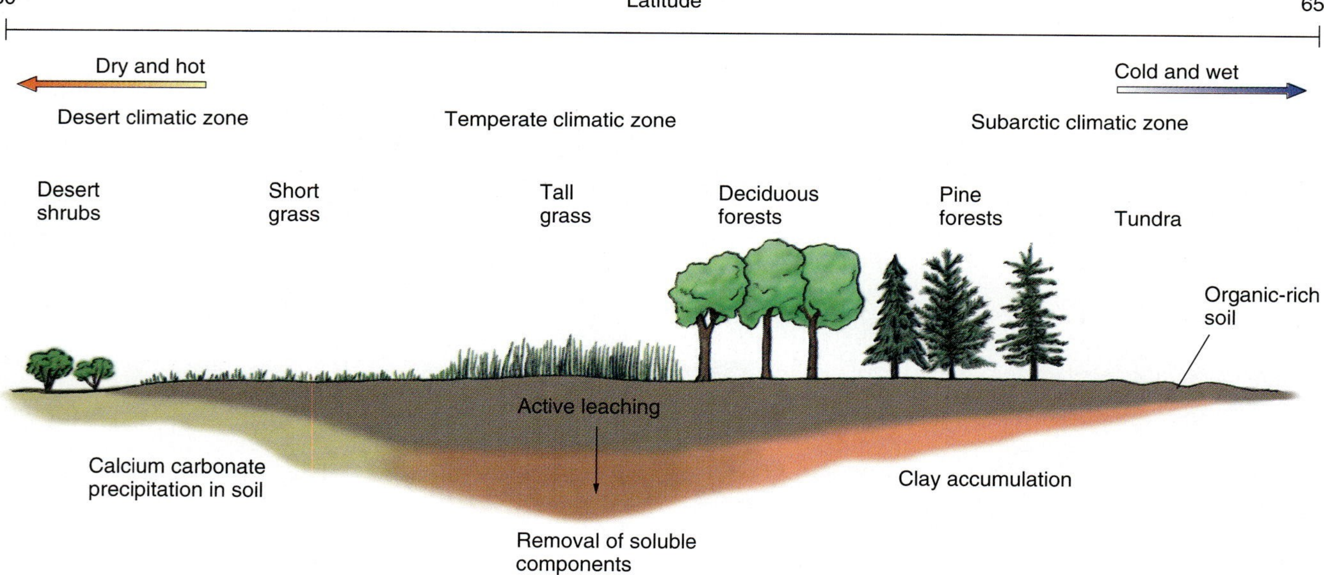

FIGURE 12.18 Relationship among climates, plant growth types, and soil processes. Differences in soil types with latitude correspond to differences in climates. In the cold and wet northern forests, which are primarily pine, the organic-rich soil zone is thin. Abundant rainfall causes some alteration of minerals by the leaching of elements from the upper parts of the soil, but the cold temperatures tend to inhibit this process. In contrast, in the temperate-climatic belt, the organic layer is thick and the alteration of minerals by leaching is effective, creating a thick soil horizon.

stagnant water, which is ineffective at weathering (Fig. 12.19). In this type of environment, chemical breakdown of minerals occurs either very slowly or ineffectively. The major chemical action that does occur on rocks in arctic regions is the result of biological activity.

(a)

(b)

FIGURE 12.19 Two types of tundra. (a) Water-saturated tundra with abundant lakes (Quebec, Canada). The ground is matted with sphagnum moss and is water-saturated everywhere. (b) Arid tundra on Spitsbergen, Norway. The tussocks (approximately 15 cm across and 10 cm high) are caused by frost heave. The ground is not saturated except where springs rise to the surface.

Lichens, an intimate association between algae and fungi, are abundant in this climatic region and are able to etch shallow pits into rocks. But, in terms of rock weathering to produce soils, these lichen-etched pits constitute a very minor effect.

The mechanical breakdown of rocks occurs primarily as a result of the expansion of water during freezing. Small cracks are wedged open by the freeze-thaw process, and during warm periods, the water percolates deeper into the rock outcrop. The next cycle of freezing widens the cracks, ultimately resulting in the formation of a type of soil. This is not a true soil, but rather a regolith—fragmented parent rock with minor mineralogical changes from chemical weathering reactions and little or no organic content.

In subarctic regions, there are vast flat plains covered with as much as several meters of water-saturated moss, called sphagnum moss, with an underlying zone of permafrost. These plains are the **tundra** (Fig. 12.19). The plant matter in the tundra is in a state of preservation rather than a state of decay. It is water-saturated and the water is very acidic—with pH values as low as 2. Under these conditions, cations may be leached from silicates in the rocks and replaced by hydrogen ions. These reactions tend to occur very slowly, partly because of low temperatures, but mainly because the water that saturates the tundra is stagnant. Without water movement allowing continual renewal of the reactants and removal of reaction products, weathering slows and ultimately ceases.

Chemical weathering in arctic climates is ineffective because of the prevailing low temperature and the stagnant waterlogged condition during the wet period.

Temperate Environments

Probably the most complex weathering cycles occur in temperate climatic belts because these regions are the most diverse. Rainfall varies widely, as does temperature, but the year is separated into a well-defined cool, wet period (winter) and a well-defined hot, dry period (summer). Temperate climatic belts boast an abundance and variety of plants and exhibit all other variables that collectively determine the extent and rate of weathering. Soils developed on the mid-continent prairies are different from soils developed in northern pine forests. Part of the reason lies in the different nature and preservation of organic matter in the soils, but such factors as climate, soil moisture, types of vegetation, and parent rock also influence the final product (Fig. 12.18).

Soils of the boreal (northern subpolar) pine forests tend to be thin, in part because the cold winter temperatures retard weathering reactions for a substantial portion of the year. In the mid-latitudes, the climate is warmer and rainfall is more seasonal, with distinct wet and dry periods during the year. As a result, the soil pH tends to be high because there is insufficient water to wash the cations out of the soil (recall that the leached cations form bases). The increased pH of these soils permits the B horizon to form. Weathering of the bedrock proceeds, so there is a clear distinction between unweathered bedrock (the R horizon) and a zone of weathered and disaggregated bedrock (the C horizon).

Organic acids from the decomposing plant litter of the O horizon percolate downward, leaching soluble components. These dissolved components are carried deeper into the soil, where pH conditions, the amount of oxygen and organic matter present, and other parameters differ from the top of the soil. The most common materials that form within the B soil horizon are either carbonate or clay, rarely silica. If carbonate dominates, the zone is called **caliche** or **calcrete** (Fig. 12.20). If clay is dominant, the zone is called **hardpan**; and if silica is the major component precipitated in the B soil horizon, the zone is called a **silcrete**. These terms are normally applied when the soil contains a sufficiently high concentration of the precipitate to be hardened or solidified. Iron is soluble under acidic conditions, but it is insoluble under alkaline conditions and is removed from solution as a precipitate. The iron accumulates with the clay-rich layer (the hardpan), causing it to have a distinctive orange or red color; farmers and gardeners are well acquainted with this layer because it is aptly named: It is hard and forms an impermeable "pan."

Calcretes form only in semi-arid regions that have a strong moisture deficit; that is, where potential evaporation strongly exceeds rainfall. The net evaporation results in precipitation of calcium carbonate from solution. Thus, the occurrence of calcrete in the geologic record is of particular interest because these deposits provide climatic information from sediments that might otherwise be difficult to interpret.

Complex soil profiles are most common in temperate climates. Abundant plant growth and seasonal rain separated by hot, dry periods encourage weathering and leaching of surface soils.

(a)

(b)

FIGURE 12.20 Carbonate soil deposits. (a) Veins of soil carbonate (caliche) in arid region. (b) Soil carbonate in northern Chile forming the white mottling in the deposit in the foreground.

Tropical Environments

Whereas chemical weathering reactions in arctic and arid environments are largely ineffective, these reactions can be carried to near-completion in the tropics. Tropical climates are characterized by warm temperatures all year round and large amounts of rainfall, often causing the rivers to flood surrounding areas. The biological productivity of both plants and animals is high, and there is intensive microbial and fungal reworking of organic matter in soil. Roots and animals turn over and churn the soil, constantly exposing fresh surfaces to biological or chemical attack. In many tropical regions, the rain forest does not cover the entire area; a good example is the Brazilian basin surrounding the Amazon River. Here, vast tracts of land are covered by grasslands that are locally called *savannas*. The forested areas tend to be slightly elevated above the level of these savannas, which preferentially flood during the rainy season.

All the necessary elements for extensive weathering reactions are present: warm temperatures, plenty of moving water to react with the minerals and flush out the reaction products, alternating wet and dry periods, abundant plant growth, and microbial activity. In the extremes of tropical weathering, nearly all reactive cations are removed from the soil minerals, leaving behind a residue of the least reactive material, which is mostly alumina with minor amounts of silica. The silica and alumina recrystallize into a structure similar to the mica structure, with alternating layers of alumina and silica. This is a stable configuration exhibited by the class of minerals called clays. These deposits are termed **bauxite** (∞ see Chapter 16) if the soil is relatively pure aluminum oxide and silica, and **laterite** if the soil contains sufficient oxidized iron (Fe^{3+}) to color it red. The concentrations of alumina in bauxite and iron in laterite are sufficiently high that these residual soils are mined for their metal content (∞ see Chapter 16).

Conditions found in tropical weathering systems are strongly acidic owing to abundant plant roots, abundant microbial activity (producing large amounts of carbon dioxide), and vast quantities of organic acids released during plant decay. In fact, many rivers in the rain forests are so full of dissolved organic acids that the waters are stained dark brown to black; the Rio Negro is so named because the water is black with water-soluble organic components (Fig. 12.21). The river water is darkly colored by soluble organic compounds for precisely the same reason that strong tea or coffee is dark in color. The rainforest soils are thick, but contain virtually no soluble cations or anions that are plant nutrients, for example, magnesium, calcium, potassium, and nitrate. Virtually all the nutrients are tied up in the biomass. Consistent with this observation, rivers that drain such areas have relatively low concentrations of dissolved ionic constituents (iron, calcium, potassium).

FIGURE 12.21 The Rio Negro (with high concentrations of dissolved organic components) seen here as it joins the muddy Amazon, in Brazil.

If deforestation occurs—for instance, by a fire—the rivers immediately begin to carry substantial concentrations of dissolved ionic constituents out of the forested area. The burning removes the nutrients (calcium, potassium, nitrogen, phosphorus, and others) from the biomass and leaves them as a water-soluble residue in the ashes. Runoff from the burned area can result in a permanent loss of these nutrients from the forest system. The nutrients cannot be replenished because the soil has already been completely leached of nearly all soluble cations and anions. However, if the burned area is flat and without much runoff, the nutrients remain in the ground in a water-soluble form that can be taken up by a new generation of plants.

The organic acids tend to make the soil water strongly acidic. We do not expect silica to dissolve in an acid environment, yet tropical soils are depleted in silica. Moreover, under acidic conditions, alumina should be mobile, yet tropical soils are enriched in alumina. Thus, a gap exists in our understanding of how weathering functions in these tropical environments.

Tropical climates are ideal for extensive weathering: warm temperature, abundant water, alternating wet and dry periods, abundant plant growth, and microbial activity.

Living with Geology

Acid Rain

Coal-fired electrical-generation plants produce vast quantities of ash and combustion gases. Ash produced from these plants is easily trapped in bags, but the combustion gases are not so easily trapped. Combustion is defined as oxidation, so complete combustion of carbon produces carbon dioxide:

$$C + O_2 \rightarrow CO_2$$

We have discussed the influence of atmospheric carbon dioxide on weathering processes. Beyond weathering, concern has been expressed by many scientists regarding the increasing concentration of CO_2 in the atmosphere from burning fossil fuels, such as coal or oil, and the role that the resulting increase in CO_2 may play in global warming (∞ see Chapter 15). Although this might be a serious problem in the future, an immediate problem emanates from coal-fired power plants, and that problem is acid rain. How is acid rain associated with coal-fired power plants?

Plant material is metamorphosed to form coal. After plant detritus is buried, it is converted progressively to peat, lignite, bituminous coal, and, finally, to anthracite coal (∞ see Chapter 16). This transformation to progressively higher *grades* of coal is due to increased temperature and pressure with increasingly deep burial. Constituents of the buried plant matter lost during the transformations from one form to the next are water, methane (CH_4), hydrogen sulfide (H_2S), ammonia (NH_3), and carbon dioxide (CO_2). If taken to the extreme, the end product of this alteration would be graphite (pure carbon), but this is quite rare in rocks.

Bituminous coal that has been mined from the eastern Appalachian coal belt contains substantial concentrations of sulfur and nitrogen. When this coal is burned, the sulfur and nitrogen in coal are oxidized to SO_x and NO_x. (The $-O_x$ is used here to indicate that the combustion gases are not pure compounds, but are complex mixtures of sulfur and nitrogen oxides.) When the SO_x and NO_x are released into the atmosphere, they combine with water to form sulfuric acid (H_2SO_4) and nitric acid (HNO_3). Unlike the weak carbonic acid that is naturally present in rainwater, both are strong acids, and both are powerful oxidizing agents. These compounds are the cause of acid rain.

Most coal-fired power plants are in Ohio, Indiana, Pennsylvania, Illinois, Missouri, West Virginia, and Tennessee. Prevailing wind patterns carry the combustion gases from these power plants and industrial facilities toward the northeastern United States and into Canada. Across a broad swath of the northeastern United States and southeasternmost Canada, the pH value of rain derived from this cloud of windblown pollutants might be as low as 4. Most lakes and rivers in this region have too little carbonate in solution to neutralize the acid, and the forest soils are naturally acidic because of the microbial production of CO_2 released during the breakdown of plant matter. Thus, the pH of the air and of many lakes and rivers is decreasing (becoming more acidic) and is having major deleterious effects on life at all levels, including humans.

The action of acid rain or acid fog on the geological and biologic environment is analogous to the natural processes involved in weathering, except carried to an extreme. Prevention of acid rain is not mysterious; it is a matter of economics and logistics. In the United States, legislation is already in place governing coal-fired power plants built after 1978. Plants built before that date are not required to meet these air pollution standards, yet those older plants are responsible for approximately three-quarters of the SO_x emissions and one quarter of the NO_x emissions. The issue is not science; it is the cost of retrofitting the older facilities, which, of course, would be passed on to consumers in the form of higher electric bills.

SUMMARY

- Physical weathering consists largely of the breakage of rocks into smaller fragments. Chemical weathering reactions are controlled by the availability of water, by temperature, and by the time needed for the reactions to proceed.
- Most rocks have pervasive sets of joints or cracks that are pathways for water penetration. Repeated cycles of freezing and thawing cause cracking, breakage, and crumbling of massive outcrops to become sand that has a mineralogy nearly identical to the original rock.
- On exposure at Earth's surface, a rock that had been in equilibrium with pressure and temperature conditions at depth is usually far from its former equilibrium condition. Weathering is a reconstitution of the original mineral assemblage into a new assemblage of minerals that are in equilibrium with conditions at Earth's surface.
- Ionically bonded cations in minerals selectively react with water and are dissolved into solution. Removal of the cations leaves the mineral lattice unsupported, and it collapses.
- A soil contains partially decomposed organic material, weathered rock debris, and water.
- Sedimentary processes—weathering, transportation, and sedimentation—can effectively separate and isolate chemical elements.
- Depth of weathering is inversely proportional to the rate of erosion. Rapid uplift causes rapid erosion, which does not allow sufficient time for weathering reactions to proceed.
- Passive margins are dominated by sedimentation. Weathering of sediments that are being transported or that have already been deposited is an important and ongoing process. This reprocessing of sediments produces mature sediments.
- Rapid uplift characteristic of active margins does not permit extensive weathering, so sediments that are deposited along active margins tend to be coarser and less mature than sediments deposited along passive margins.
- Chemical weathering in desert climates proceeds slowly because of the absence of water or wet and dry cycles.
- Chemical weathering in arctic climates is ineffective because of the prevailing low temperature and the stagnant waterlogged condition during the wet period.
- Complex soil profiles are most common in temperate climates. Abundant plant growth and seasonal rain separated by hot, dry periods encourage weathering and leaching of surface soils.
- Tropical climates are ideal for extensive weathering: warm temperature, abundant water, alternating wet and dry periods, abundant plant growth, and microbial activity.

KEY TERMS

physical weathering, 300
cheamical weathering, 300
sinkhole, 306
karst topography, 306
leaching, 306
soil horizons, 308
eluviation, 308
ventifact, 317
desert varnish, 318
permafrost, 318
tundra, 319
caliche, 320
calcrete, 320
hardpan, 320
silcrete, 320
bauxite, 321
laterite, 321

QUESTIONS FOR REVIEW AND FURTHER THOUGHT

1. What role does water play in the weathering process?
2. The tropics are regions of extreme weathering where most soluble cations have been removed from the soil profile. How can you rationalize the absence of plant and animal nutrients in the soils with the lush growth common in tropical rain forests?
3. Why are weak acids more important in the weathering process than strong acids?
4. What arguments can be advanced to suggest that the oceans have had approximately the same composition for several hundreds of millions of years? What are the assumptions that you make in these arguments?
5. Suggest a process, or series of processes, by which a pure quartz sand might be derived from a granite source area.
6. How does the efficiency of separation of elements in magmatic processes compare with that of sedimentary processes? Explain your answer.
7. How is weathering in arctic and desert environments more or less similar?
8. Why is lunar "soil" actually regolith, rather than a true soil?
9. The most complex soil zonation occurs in temperate climatic zones. Why?
10. What are the three most important controls on weathering?
11. Compare weathering processes on active margins with those on passive margins. Take into account possible differences in the bedrock being weathered, but consider the climates to be similar.

13

Sedimentary Systems: Transportation and Deposition of Sediments

Satellite view of Betsiboka River in Madagascar, colored brown by a sediment, which it carries to the ocean.

Overview

- Clastic sediments are mechanical mixtures of grains brought to the site of deposition from various source areas by a variety of transport mechanisms.
- Sediments are transported in running water and in air, mud, or ice. The characteristics of a deposit are a reflection of the transporting medium.
- The gradient of a stream adjusts to carry the water and sediment load to the stream's base level, maintaining a balance between deposition and erosion.
- Biological activity prevents the preservation of sedimentary features in shallow marine environments.
- Biochemical (mainly carbonate) sediments originate from plants and animals that use carbonate for their houses and skeletons.
- Active continental margins exhibit faulting, volcanism, and plutonism; they are steep and often characterized by small sedimentary basins.
- Passive continental margins are flat and subsiding, with minor tectonic activity, characterized by voluminous sediment accumulation.

Introduction

Sedimentary deposits tell a detailed story if we are able to interpret them. Much of our knowledge of past times and environments comes from ancient sediments that have been preserved. Each process acting on a sedimentary environment tells us whether the climate was warm or cool, whether certain rivers were rushing torrents or slow and muddy. Fossils and fossil fragments give us clues as to the conditions under which the organisms lived. The state of preservation of the fossil fragments provides information regarding abrasion, dissolution, or compaction.

Several factors were introduced that control the nature and distribution of sediments and sedimentary rocks (∞ see Chapter 4). By far the most voluminous sediments in the geologic record are detrital or clastic sediments; biochemical and chemical sediments are less abundant. Detrital sediments are usually mixtures of minerals and rock fragments derived from several sources; this detritus, along with the rocks formed from it, constitutes a mechanical mixture (∞ see Chapter 4). Such sediments can range from deposits composed of a single mineral, such as quartz, to diverse mixtures of different minerals and types of grains.

The mineralogy of igneous and metamorphic rocks is limited in the sense that, for a given composition, only certain minerals can exist together in chemical equilibrium (∞ see Chapter 3). The rules of phase chemistry do not limit the mineral constitution of sediments; any two minerals can be deposited next to each other. The only controls on sediment composition are the physical process of mixing and the original composition of the source rock.

As a general rule, then, clastic sediments are detrital mixtures of just about anything that has been eroded and then deposited. There are some limitations, though. For example, it is unlikely that a chemical sediment, such as halite (NaCl), could be eroded by a stream and the halite grains subsequently transported downstream to be deposited in a lake as a clastic sediment. Halite is highly soluble in water and would immediately dissolve. Other chemical sediments, however, do form clastic deposits. The famous sand dunes forming the white sands of New Mexico are made of gypsum; in northern Norway, there is a sequence of 400–million-year-old sediments, deposited in the intertidal zone, where the clastic grains are anhydrite.

The mineralogic diversity of sediments poses something of a conceptual problem in naming them, classifying them, and discussing them in general. We can simplify the problem with the realization that sediments are almost always composed of just two general types of constituents: (1) grains, such as sand and gravel, that were physically transported to the site of deposition (**detrital** components); and (2) components that originated at the site of deposition (**authigenic** components). Authigenic minerals are chemical precipitates that occur either between the detrital grains as a cement, as individual crystals between sediment grains, or as entire sediment sequences such as evaporites.

Sediments are both mixtures of mineral grains that have been carried to the site of deposition and minerals that have precipitated (crystallized) in the sediment after deposition.

Transportation of Sediments

For this discussion of transportation and sedimentation, we must think about some of the characteristics of flowing fluids in nature—particularly air and water. Some materials move by *laminar flow*, meaning that the flow is even and steady. The fluid moves in parallel streamlines with very little mixing from one parcel of fluid to the next (Fig. 13.1a,b). An example of laminar flow might be the movement of glycerin flowing across a gently sloping table. A streak of color in the glycerin might be stretched, but would not be otherwise deformed. *Turbulent flow* is neither an even nor a steady flow; it is swirling and chaotic. In turbulence, there is mixing within the fluid from one position in the flowing medium to another (Fig. 13.1c). This turbulence in natural flows is crucial for erosion and sediment transport. Even along river courses of seemingly quiet flows, turbulence is well developed. Quiet reaches of the Colorado, Ohio, Delaware, and Mississippi rivers are turbulent, which is why there is so much sediment in their waters. The water of the Colorado River is red because of the suspended sediment in the water.

Turbulence simply indicates that small parcels of fluid differ in flow direction from the general flow. We have all seen leaves swirling in gusts of wind; this phenomenon is common around buildings or behind hills or trees, and is the result of turbulence. In rivers or streams, the water flow is not uniformly moving in a single direction at a single velocity (Fig. 13.1c). If measured carefully and on a sufficiently small scale, it is easy enough to show that portions of the fluid are flowing at rates that may be much faster or much slower than the average rate. Further, they may flow in different directions—upstream, downstream, up, down, or sideways. The average of all velocities and directions defines the general flow.

For a given amount of water in a stream, the stream velocity will be affected by the gradient of the stream bed and by the shape of the stream channel. The rougher the channel, the more friction there will be; velocities of shallow streams flowing through a rocky channel will be less than those of streams in a deep, smooth channel. As well as the small-scale random velocity effects of turbulence, variations exist in velocities within the channel itself. Water flows fastest where it is farther away from the effects of friction or drag on the sides and bottom of the channel.

The amount of water flowing in a stream at a given location is called the stream *discharge*, and is simply the cross-

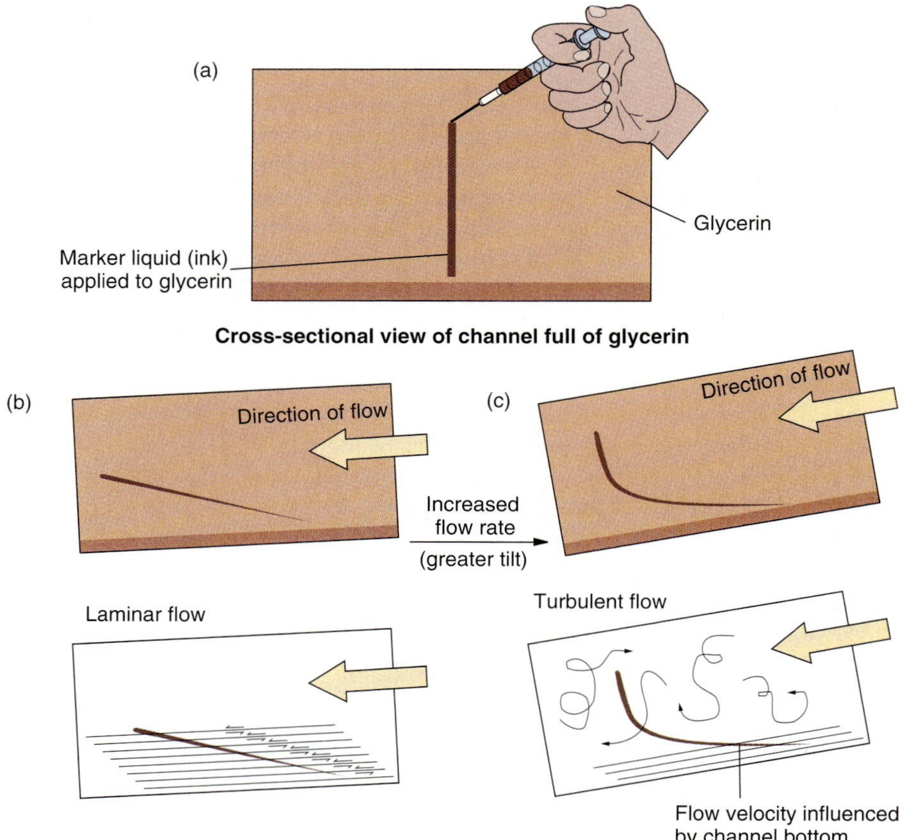

FIGURE 13.1 A comparison between turbulent and laminar flow. (a) In an artificial channel filled with glycerin (a viscous liquid), a liquid (ink) is injected into the column to serve as a marker for flow conditions. In one experiment, the end of the channel—right side—is raised slightly to cause flow. In the other experiment, the right side of the channel is elevated sufficiently to cause rapid flow. (b) In laminar flow, the fluid particles move downstream without mixing. Flow velocity steadily increases upward into the flowing fluid, as shown by the transformation of the vertical line of dye into a straight but inclined line during flow. (c) In turbulent flow, there is irregular motion within the fluid, with mixing and exchange of mass and momentum between different layers of fluid. Flow velocity slightly above the channel bottom is constant owing to the extreme mixing caused by the turbulence.

sectional area of the stream multiplied by the flow velocity at that location. Clearly, discharge has an influence on both stream and flow mechanics and on sediment transport. In a given climate, discharge is directly related to the size of the drainage basin that collects water for the stream.

The two independent parameters in streams and rivers are the discharge and the average sediment size furnished for transport. The other major features of the stream—that is, depth, width, slope, and sediment load—are dependent not only on discharge and sediment grain size but also on each other. For example, if the slope changes, there will be adjustments in stream depth, width, and sediment load transported. Should there be an abrupt narrowing or widening along a stream channel, the flow velocity must be adjusted: If a stream course flows through a narrow gorge composed of very hard rock difficult to erode, the velocity through this narrow region will increase to maintain the discharge delivered upstream of the narrow reach. In the same way, widening of a reach along a stream will cause a decrease in flow velocity. With changing conditions, the stream is constantly modifying its slope, velocity, and cross section to accomplish transport of sediment—with the least expenditure of energy.

Sediment Transport in Air and Water

Sediment transport depends on the turbulence of the transporting medium—either wind or water. Turbulence, in turn, greatly depends on flow velocity: The *greater the flow velocity, the greater the turbulence.* Sediment particles are picked up by the turbulence of wind or water and either carried along with the flow or dropped. Why is a particle carried or dropped? The answer is that transport represents a trade-off between the turbulence of the flow that keeps material in suspension, and the tendency of the material to settle out of the flow. This trade-off is described by the *settling velocity.* The tendency for grains to sink through a fluid depends on the grain size, which is described by the average grain diameter, and the grain density (determined by the type of mineral) (∞ see Chapter 3). For small grains (smaller than about 0.2 mm for settling in water), the **settling velocity** is very sensitive to size; for larger grains, the settling velocity is sensitive to shape. Transport of sediment in air obeys the same principles. However, corresponding particle diameters will be much smaller for similar grains settling in air.

If the settling velocity of a grain is so slow that the small upward "gusts" of turbulence are sufficient to keep the grain suspended in the transporting medium, then the grain is transported by **suspension**. If the settling velocity of the grain is high relative to the turbulence, then the grain may be regularly or only occasionally bumped along; it is being transported by **traction**. In most flows, some grains will have an intermediate settling velocity. These grains may be periodically picked up by a turbulent swirl, carried briefly in a flow, then settle out of the flow. Such grains are **saltating** (Fig. 13.2). The latter two transport

(a)

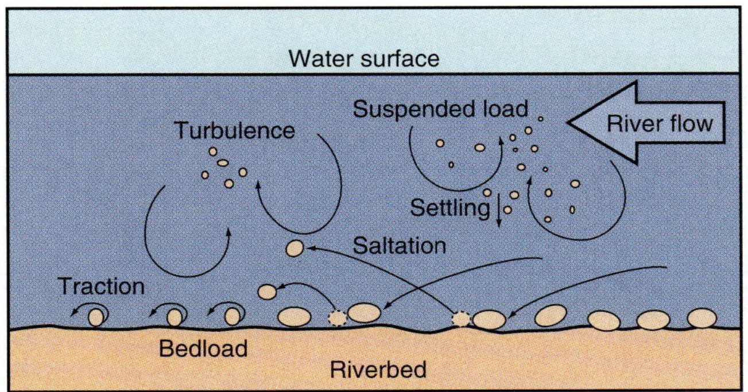

(b)

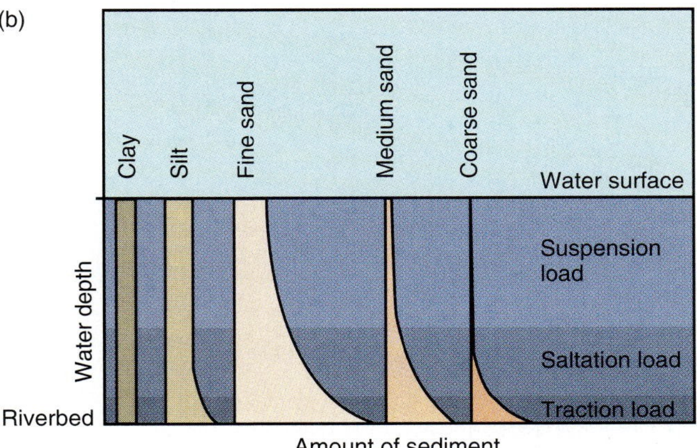

FIGURE 13.2 Movement of sediment in a river. (a) The suspended, saltation, and traction sediment loads, and the fluid forces acting on the sediment grains are shown. (b) Sediment concentration with height above riverbed. The sediments identified represent grain sizes: Clay is defined as grains smaller than 2 μm, silt ranges in size from 2 to 62 μm, fine sand from 62 to 250 μm, (0.25 mm), medium sand from 0.25 to 0.5 mm, coarse sand from 0.5 to 2 mm. Sediment-size ranges normally transported in the different loads are indicated.

mechanisms make up what is called the **bedload**—that is, the sediments transported as part of the channel bed. Finally, material can also be transported as dissolved ions in solution in the river water.

The sediment load carried by a stream is distributed throughout the water column. The fine-grained sediment is distributed about equally throughout the flow, whereas coarser materials are concentrated near the bottom (Fig. 13.2b).

Imagine a stream channel where the sediment on the bottom is a mixture of grains from very small (0.001 mm) to moderately large (10 mm or even 100 mm). If the channel is filled with flowing water, the turbulence picks up the smaller grains into suspension; some of the larger grains move by saltation or traction, and some of the grains are too large to move at all. The smallest grains are carried along suspended in the water; they move at essentially the same rate as the flow of water. These particles remain in suspension until the water slows or stops moving. Grains transported along the bottom of the flow by either saltation or traction move at much lower velocities than that of the water flow. Movement of these grains is influenced by the flow velocity; if the velocity is increased very slightly and larger grains saltate, then more and larger grains move by traction. If the flow is decreased slightly, the reverse effect occurs. The ability of a stream to transport sediment can be described by two terms: **capacity**, which is the maximum amount of sediment that can be transported, and **competence**, which is the maximum size of sediment clast that can be transported. Capacity is related to the total flow of the stream—the volume of water passing a point in a given time, or discharge. Competence depends on the stream's velocity or, more strictly, the square of the velocity. If the velocity doubles, it can carry a clast four times as large. Typical streams carry sediments a fraction of a millimeter in size. Football-sized boulders require rushing torrents.

Sediment grains are transported in suspension, by saltation, and by traction. The settling velocity of a grain and the extent of turbulence determine the transport mechanisms governing the movement of any grain.

Sediment Transport in Mud and Ice

Sediment transport is also influenced by the **viscosity** and density of the transporting fluid. Fluids of low density and viscosity, such as air or water, can transport only a narrow range of small grain sizes. The settling velocity of larger grains is so rapid that the grains quickly fall out of the low-viscosity transporting medium. In high-viscosity fluids, however, such as thick mud or ice, the settling velocity of grains is so slow that even large grains may remain suspended in the medium for long periods of time.

In addition to the viscosity, the density of the transporting medium is equally important. Imagine a very dense fluid, such as mercury. This element is a liquid metal under conditions at Earth's surface. Mercury has a density nearly 14 times that of water; even very dense mineral grains such as magnetite (the density of magnetite = 5.5 g/cm^3) will float on the surface of a dish of mercury. If there were a river of mercury, any rock or mineral encountered by this river would be picked up and carried along, never settling out. Although rivers of mercury do not exist, there are some natural flows that are dense and have very high viscosities.

Thick suspensions of mud can flow down river channels or overland. The density and viscosity vary, depending on how diluted the mud is with water; thick mudflows have densities significantly greater than that of water and a viscosity that may be many hundreds of times greater than the viscosity of water. Thus, concentrated mudflows can transport automobile-sized rocks for long distances. The turbulence within the flow is great, and both the density and the viscosity of the flowing mud are so high that the settling velocity of large boulders in the fluid is very low; they remain in suspension for some time (see Focus 13.1).

Focus On 13.1 CONTROLS ON SETTLING VELOCITIES OF GRAINS

The settling of sediment particles in water is a fundamental process governing the transport of grains in the water column and sorting by grain size or mineralogy. Two equations were derived many years ago to describe the settling of particles in fluids—Stokes's law and Newton's law. These "laws" are old and well established, but large departures from theory are known. Stokes's law states that the settling velocity for small particles varies with the square of the grain size. If the grain size doubles, the settling velocity increases by a factor of 4. This relation predicts that the force of gravity on the grain, minus the buoyancy of the grain in water, equals the viscous resistance of particle movement through the water. Another way to think about this is that the gravitational force acting on the grain must do physical work to separate the fluid so that the grain can move through it. If the grain that is settling is small, flow around it is smooth and laminar.

Newton's law is valid for larger grain sizes; it states that the settling velocity varies with the square root of the particle diameter. If the grain size doubles, the settling velocity increases by 1.4, which is the square root of 2. This relationship indicates that larger particles cause a large drag to be encountered by the particle as it impacts the fluid and forces its way through, creating substantial turbulence behind as it settles through the fluid.

These equations do not accurately predict sediment grain-size distributions and mineral (actually density) segregations that occur naturally. In 1933, W.W. Rubey, a U.S. Geological Survey geologist working on sediment transport in rivers, published a series of experiments accompanied by theoretical studies in which he determined the actual settling characteristics of different minerals in water. Rubey showed that a 0.055-mm grain of galena (density = 7.5 g/cm^3) settles through water at the same velocity as a grain of quartz 0.117 mm in diameter; this result is predicted by Stokes's law. Experiments with smaller grains of these minerals showed that the same ratio in settling velocities is consistently maintained.

Experiments with progressively larger grains of these same minerals, however, exhibited increasingly large departures from the settling velocities predicted by Stokes's law. At even larger grain sizes, the settling velocities conform to those predicted by Newton's law. In the grain-size interval between the sizes where Stokes's law and Newton's law correctly predict the settling velocities, however, the particles obey neither law. A large range of grain sizes did not comply with the predictions.

The broader significance of Rubey's experiments is that Stokes's law and Newton's law do not merge (Fig. 1). The implication is that a complete understanding of the behavior of sediment grains

Glaciers move very slowly, but their movement is a flow similar to that of a slow-moving mudflow. Ice has a very high viscosity and therefore has the ability to carry exceedingly large rocks for long distances. The base of most glaciers, other than those in Antarctica and Greenland, is near the melting point of ice. Therefore, meltwater from the glaciers can flow downward into fractures in the rock outcrops and, upon freezing, break the rock apart. By this action, rocks beneath the glacier may be picked up and carried along by the glacier as it moves downhill. Glaciers moving over bare rock outcrops crush the rocks and grind them into an exceedingly fine powder *(rock flour)*. Subglacial streams carry the rock flour out from under the glaciers and into progressively larger streams. Such rivers are very cloudy and commonly have a peculiar light green color caused by the high concentration of rock flour in the water.

The general motion of glacial ice is that of laminar flow. The absence of turbulence within the glacier prevents the upward movement of debris into the ice. For that reason, both the rocks plucked from the outcrops beneath the ice and the fine rock flour remain at the base of the flowing ice. Thus, glaciers are primarily clear ice with some rocky debris at the base and relatively small amounts of debris on the upper surface.

Sedimentary Structures Formed During Transportation

In a controlled environment, it is possible to perform experiments and monitor the results; this is generally more difficult to do in natural streams. As the flow velocity of water or air increases over a bed of loose sand, the grain motion forms systematic irregularities on the surface

in water has not yet been accomplished. Until then, we must content ourselves with the empirical relations that Rubey derived.

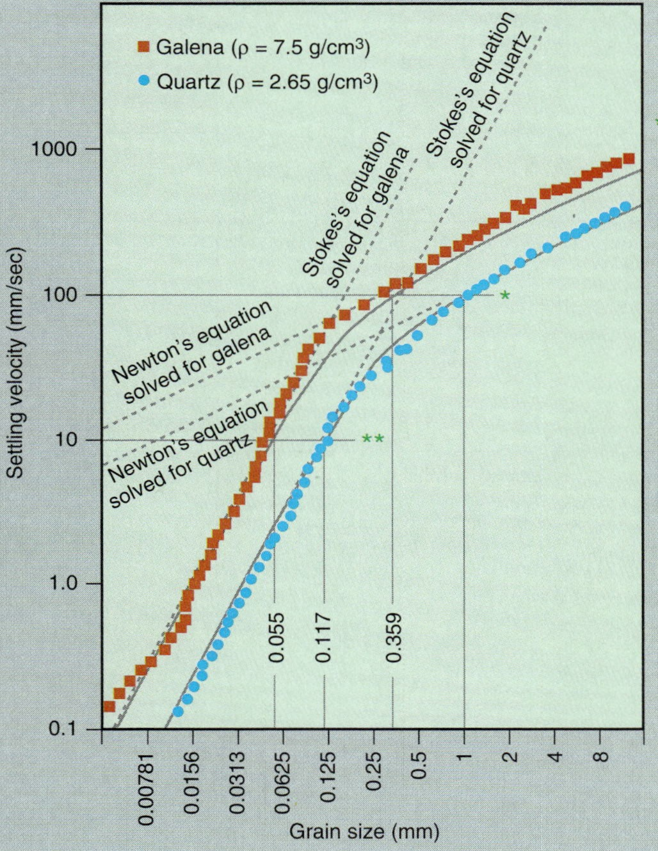

FIGURE 1 Experimentally determined settling velocities of quartz and galena grains in water. The vertical and horizontal lines separate regions of transition between Stokes and Newtonian behavior.

of the sand, called **bedforms**. Bedforms are present where any natural flowing fluid transports sediment across a loose bed. They are seen on sand dunes, in stream beds, on beaches, and at the bottom of the ocean. A flowing stream cannot control the amount of water supplied to it, nor can it control the mineralogy, angularity, or grain size of the sediment delivered to it. However, interaction between the flowing water and the bed of loose sand does influence the flow velocity, the slope, the width, the average depth of a channel, and the amount of sediment transported. Bedforms are elements of roughness on the stream bottom that are generated in response to the flow, and, in turn, serve to control the flow velocity.

Several types of bedforms are commonly observed (Fig. 13.3). Both the shape and the type of bedform are directly related to flow velocity and secondarily related to the grain size of the sediment. Under conditions of relatively slow flow, ripples and dunes form. These features in the sediment are somewhat like asymmetrical waves, and the difference between them is size: If the length of the wave is small (a few centimeters), it is a ripple; if the wavelength is several meters, it is a dune. Sediment movement on both ripples and dunes is characterized by grains rolling up the gentle upstream face and sliding down the steeper downstream face where they are deposited. Deposition on the downstream side of ripples and dunes causes these features to migrate slowly downstream. These bedforms are used to classify flow conditions into the *lower flow regime*.

As flow velocity is increased, dunes begin to diminish in height and ultimately wash out. At this point, the interaction between the flowing fluid and the loose sediment (*fluid shear*) causes particles on the stream bottom to be in motion several grain diameters deep. At still higher flow velocities, undulations again form on the bottom.

FIGURE 13.3 Sediment transport in flowing water. Water flow in a channel shows a variety of bedforms at varying flow velocities. In the lower-flow regime, ripples and dunes are the primary features; standing waves and antidunes are present for upper-flow-regime conditions.

Increasing flow velocity ↓

Flow Regime	Characteristic	Example
Ripple	Water depth is nearly uniform and is unrelated to sediment ripple height. Deposition is on downstream face of ripple.	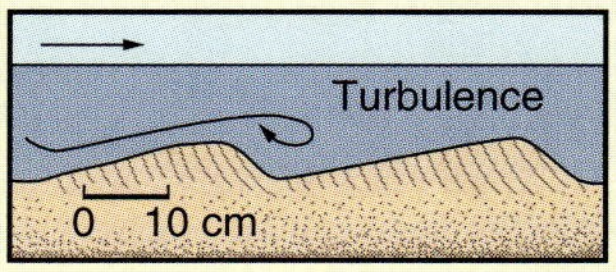 Turbulence; 0 10 cm
Dune	Surface boils; water depth is at its minimum over the dune crest; water depth approximately equal to dune height. Deposition is on downstream face of dune.	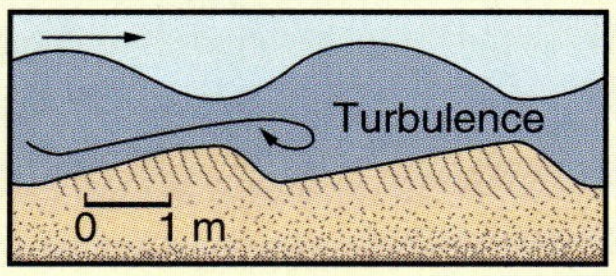 Turbulence; 0 1 m
Plane (flat) bed	Flow is sufficiently rapid that ripples cannot form or are planed off. Sediment deposited on horizontal layers.	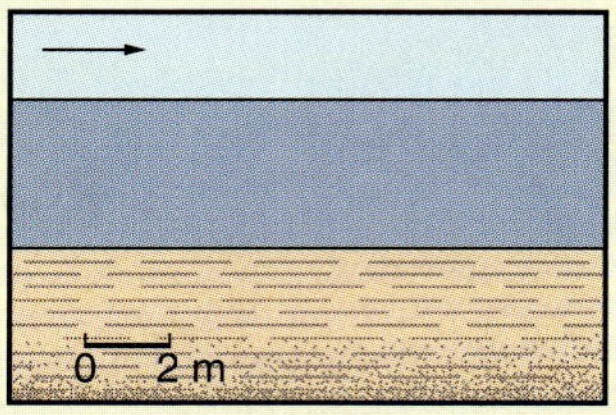 0 2 m
Antidune/ standing wave	Large rhythmic waves on water surface breaking intermittently. Deposition is on upstream face of wave and downstream face is eroded.	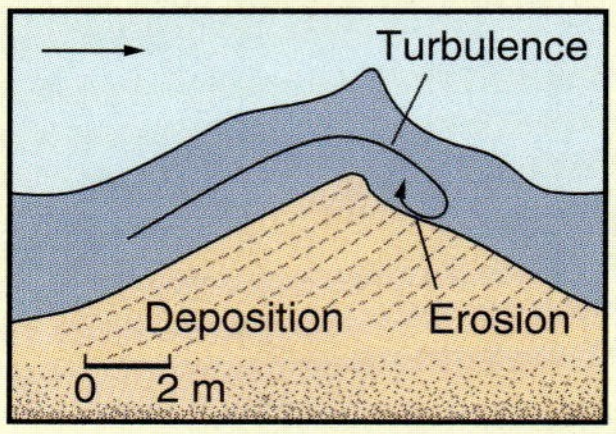 Turbulence; Deposition; Erosion; 0 2 m

These undulations, called *standing waves* and *antidunes*, are nearly sinusoidal in form and may sometimes migrate upstream. Sediment is deposited on the upstream face of the bedform, but is eroded from the downstream face because of the intense flow turbulence located there. These bedforms are used to classify flow conditions into the *upper flow regime*. These forms are hidden underwater in streams and rivers, but are well displayed in experiments (see Fig. 13.3).

Grain Sorting

Let us combine a couple of the concepts that we have discussed so far in this chapter—namely, individual grains of different sizes or densities settle through water at different velocities, and sand grains of different sizes will be picked up or transported differently according to the density, viscosity, and velocity of the transporting fluid. Most areas undergoing erosion—those that supply sediment—provide a broad mixture of grain sizes and minerals. During transport, the grains may be segregated to varying degrees by size, by shape, and by density (mineralogy). If the flow velocity of a stream (or a wind) is held constant, particles larger than a certain size will not be moved, and particles smaller than a certain size will remain in suspension and will not settle out of the flow. If this process is repeated several times, grains of similar size, shape, and mineralogy will become segregated from grains that are different and will be deposited together. If the grains that have accumulated together are similar in size, shape, and mineralogy, the sediment is said to be well sorted. The degree of sorting ranges from *very well sorted*, in which the grains are of the same mineralogy as well as the same size and shape, to *very poorly sorted*, in which there is little segregation by size or mineralogy. Statistical measures are commonly applied instead of these adjectives to describe sorting.

One type of quantitative measure for sorting is illustrated in Figure 13.4. On this diagram, the grain-size distributions of three sediments are plotted. These are determined by plotting the *grain-size distribution*—that is, the proportion (percentage) of each grain size against the grain size. A well-sorted sediment will have a very narrow range in sizes, expressed as a sharp, narrow curve on

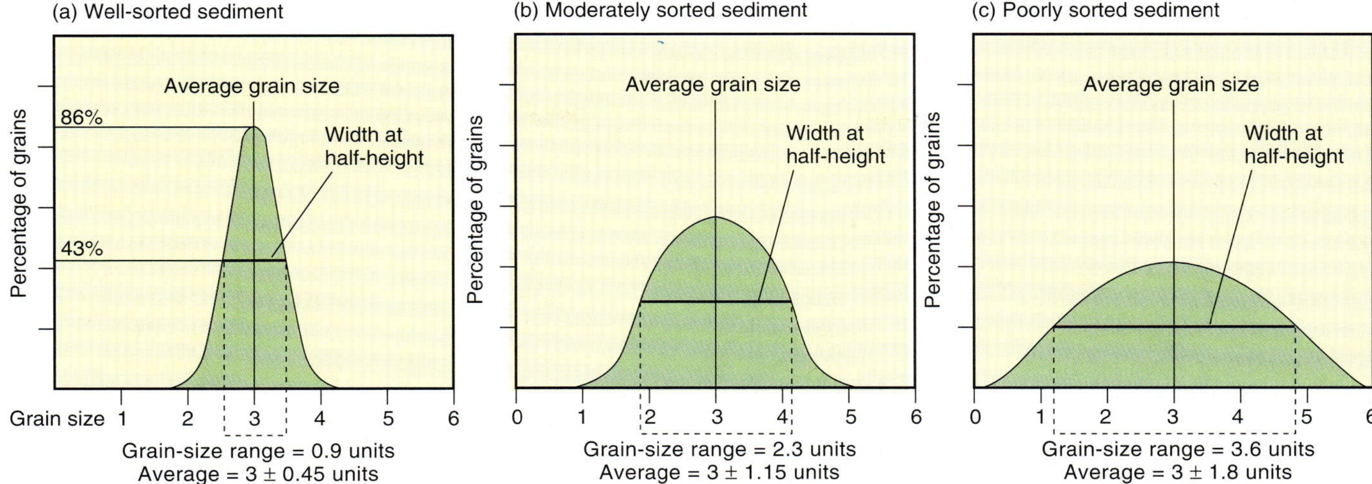

FIGURE 13.4 Sorting in three sediments displayed as "frequency plots," using grain-size distributions of fictitious samples. A frequency plot is a continuous line that plots the sediment grain size against the percentage of that grain size in the total sample. The sediment shown in (a) is very well sorted, in (b) it is moderately sorted, and the sample is poorly sorted in (c). Sorting is shown here as width at half-height, and the grain-size ranges for each level of sorting—well sorted, moderately sorted and poorly sorted sediments—are indicated using arbitrary size units. Note that for this measure of sorting to apply, the curve needs to be roughly symmetrical.

the diagram. The more poorly sorted the sediment is, the wider will be the range in size—expressed as a wider curve on the diagram. Several statistical calculations are used to express sorting quantitatively, but we can make the point easily if we adopt the measurement of *width at half-height*: the width of the curve at exactly half of the maximum percentage level for each sample. The average grain size is at the center of the distribution along the horizontal (x) axis, and sorting is expressed as a departure from the average grain size, which can be expressed as a plus/minus (±) variation in the measurement (Fig. 13.4).

Flow and Sediment Transport in Natural Rivers

Rivers provide us with a natural laboratory in which to examine sediment transport and the effects of other influences. A fundamental property of streams that enables us to evaluate their behavior is the stream profile. The profile of a stream is measured around every curve and bend along the length of a stream course. The stream length is then plotted against the change in elevation along that distance. Let us first look at an ideal stream condition, then at some of the perturbing influences that alter the ideal situation.

Graded Streams Streams are constantly cutting or filling and modifying their slopes, velocities, and cross sections so as to accomplish the imposed work with the least expenditure of energy. The end result of this process is a **graded stream**, and its profile is known as a graded stream profile. This profile represents the gradient of the stream course, which is generated so that the stream is able to transport all of the sediment supplied to it, maintaining a balance between erosion and deposition. Once this condition of equilibrium is established, the overall graded profile does not change through time.

For a stream that is near grade, the profile is a smooth curve; it is steeper near the head of the stream and more shallow near its mouth (Fig. 13.5). The elevation at the mouth of a river is called its **base level**. The worldwide base level is sea level, of course, but there are also local base levels. The local base level for a particular river might be a lake or the level of a larger trunk stream into which it flows. Sea level is known to have fluctuated somewhat over geologic time (∞ see Chapter 11). Also, the elevation of local base levels can be altered suddenly and even catastrophically. We will examine some effects of different perturbations on river processes.

A graded stream is in equilibrium; it has the exact slope necessary to transport the water and sediment supplied to it.

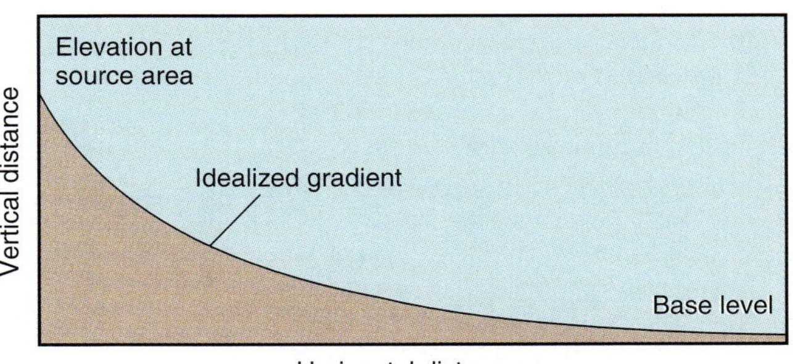

FIGURE 13.5 A graded stream profile. This profile is ideally one in which the stream is able to transport the amount of sediment and water supplied to it, maintaining a balance between erosion and deposition.

Departures from Grade *Uplift:* If uplift has occurred after a stream has reached or is close to reaching its graded profile and if the uplift takes place slowly enough, then stream erosion can keep pace with the uplift. The stream's ability to erode has been renewed because the stream slope has steepened; this is called *rejuvenation*. If uplift is too rapid for erosion to keep pace, then the course of the stream will be diverted around the uplift.

Change in Base Level: In the modern world, the most common way that the base level of a stream is affected is by the construction of a dam. A dam and its associated lake change the base level of the inflowing streams to the level of the lake. The activity of the river is dramatically affected as a result. The stream bed above the lake must be adjusted to a graded profile, which causes deposition upstream (Fig. 13.6a). Below the dam, a new graded profile will also form.

Before construction of the dam, the river had been loaded to capacity with sediment; afterward, the sediment is deposited as a delta where the river enters the lake (Fig. 13.6a). Below the dam, the water flowing out of the lake now lacks a sediment load (Fig. 13.6a). This is a very different situation from the one prior to dam construction and, in fact, is a condition far from equilibrium. The stream below the dam tries to reestablish a graded profile by eroding its channel. After construction of the Hoover Dam, the channel of the Colorado River was altered as far as 550 km downstream from the dam.

Natural causes of changes in base level exist also. The sudden draining of a lake dramatically lowers the local base level of streams feeding into the site of the former lake. Lakes may be formed for a variety of reasons, but they often develop because a stream is suddenly blocked; for example, a lava flow or a landslide may fill the stream valley, thereby damming the river. In the case of a landslide, the lake is likely to be short-lived because the landslide debris may be loose and easily washed away. Breaching of the natural dam is likely to be catastrophic (see Focus 13.2).

During the ice ages (from 2 million years ago up to about 10,000 years ago), sea level was alternately raised and lowered, depending on how much seawater was frozen as glacial ice (∞ see Chapter 11). Rivers that had approximately graded profiles to sea level were forced to adjust their grades because of the fluctuations in sea level. Examples of this influence abound. The Hudson River in New York cut a deep canyon across the continental shelf directly offshore of the mouth of the river during a low stand in sea level. The Mississippi River cut a deep channel during that same time. With the rise in sea level beginning about 10,000 years ago as the glaciers retreated, the river backfilled its channel to accommodate its grade. Today, the bed of the Mississippi River is approximately 20 feet higher than the city of New Orleans but levees protect the city from flooding.

Change in River Length: Clearly, stream length, stream slope (gradient), and sediment load are closely interrelated, as are other channel features such as width and depth. If the length of a river is changed, then all of

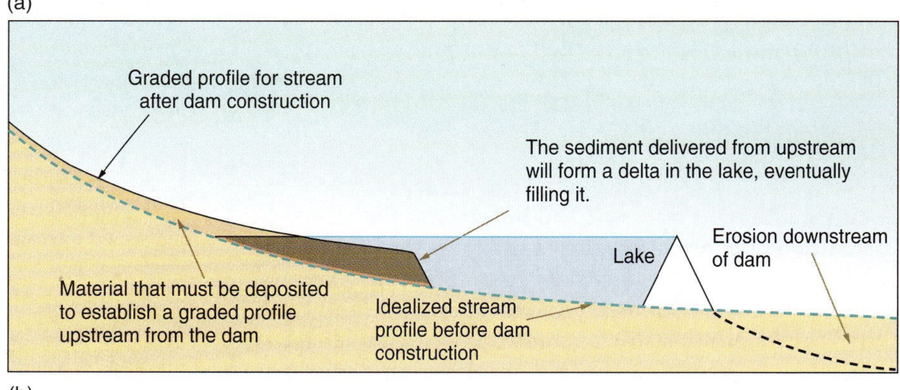

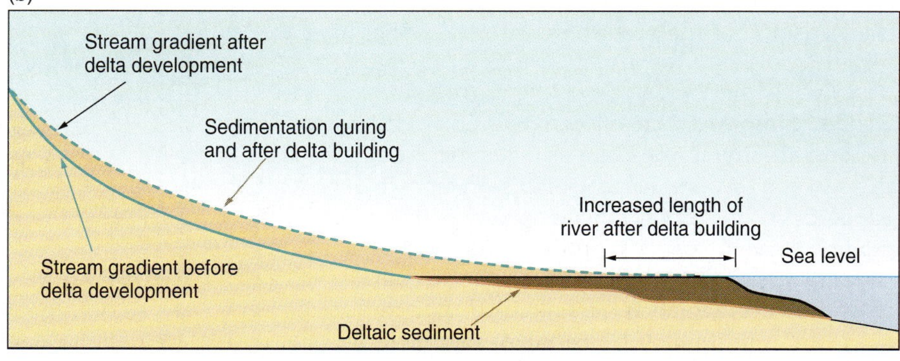

FIGURE 13.6 Stream erosion or sedimentation as a result of natural or human-caused changes on base level length of a river. (a) Diagram showing the influence of stream grade that the construction of a dam and its lake might have on a stream. (1) Material upstream from the lake must be deposited to establish a graded profile. (2) Below the dam, a new graded profile will form. (b) Deposition of a delta lengthens the stream and requires that it backfill a considerable part of its length in an attempt to regenerate a graded profile. The backfill may be small, but in some cases, such as the Mississippi River, the backfill is significant.

Focus On 13.2 THE GREAT PLEISTOCENE FLOODS OF WESTERN NORTH AMERICA

From approximately 30,000 years ago until 12,000 years ago (during a period known as the Pleistocene [see Appendix II]), there was a lake in western Utah that has been given the name of Lake Bonneville. The lake covered approximately 50,000 km^2 during the time that continental glaciers advanced across much of the North American continent. The level of Lake Bonneville rose and fell periodically; it usually peaked when the glaciers melted back (deglaciation), and the meltwaters flowed into a series of connected topographic basins that formed the lake. The spill point for the lake was at its north end, where it flowed into what is now Idaho, through a low pass between higher mountains. The dam that held back the water was a lava flow. As spillage over the top of the lava dam began, the weathered basalt was quickly eroded to form a deep gorge.

Although Lake Bonneville overflowed intermittently, there were two particularly catastrophic floods that flowed overland into Idaho, then along the course of the Snake River to the Columbia River. One of these floods occurred 30,000 years ago and released 600 km^3 of lake water through a narrow gorge at a maximum rate of 0.5 km^3/hr. Floodwaters moved down the Snake River gorge, forming spectacular erosion features and resulting in a deposit of poorly sorted basalt boulders and sand. This deposit of approximately 1 km^3 of flood debris blocked the canyon, briefly impounding the floodwater to a depth of as much as 100 m. Enormously high flow velocities were required to transport boulders—as large as 3 m in diameter—that were found in this gravel deposit along the canyon. A second flood occurred 14,500 years ago, catastrophically dropping the lake level by 100 m. This second flood was at the time of the most recent major deglaciation.

These major catastrophic floods from the overflow of Lake Bonneville generated a peculiar topography known as the *channeled scablands* of southern Idaho and eastern Washington, which cover an area larger than the state of Maryland. Many distinctive topographic features characterize this area: vast interconnecting channels and abandoned canyons; dry waterfalls and water-polished, rock-strewn areas that were undoubtedly huge rapids; gravel deposits that resemble river channel bars except that they are hundreds of times larger. When the area was initially investigated, it was suggested that the features were caused by a huge flood. This proposal met with great skepticism from the established scientific community because geologists of that time did not understand that enormous catastrophic events could play an important role in the shaping of Earth's surface. Instead, the prevailing attitude held that processes operating on a daily basis over vast stretches of time were the most important influences shaping Earth's features. However, unusual events may be preserved. With subsequent work, this proposal has indeed been shown to have been correct, and we now know that unusual events are important in the geologic record.

The climate warmed after the glacial retreat that began 11,800 years ago. Lake Bonneville slowly evaporated to lower water levels and higher salt contents. After a certain level of salt concentration was reached, salt began to precipitate out of the lake water as the water receded, leaving behind the Bonneville Salt Flats. The Great Salt Lake is a small remnant of this once gigantic Lake Bonneville.

these interrelated features must readjust to the new slope. If the river lengthens, the gradient is reduced; if the river is shortened, the gradient becomes steeper.

Let us first consider the effects of an increase in the length of a river. The most common mechanism for lengthening a river is the construction of a delta (Fig. 13.6). As the river lengthens, the gradient of the stream becomes out of equilibrium with its length, so it must backfill its channel to reestablish an approximation of grade. Depending upon the size of the river and the increase in length, this backfilling may be extensive or minor. The delta of the Mississippi River itself has progressed seaward approximately 160 km (Fig. 13.7a), necessitating major backfill along the river channel to reestablish a graded profile.

A great deal of work has been done to map former courses of the Mississippi River and the ancient deltas, including many drill holes for the recovery of sediment cores, and a great many radioactive dating analyses using ^{14}C to date shells and wood in the sediment. As a result of this work, the ancient river courses are well documented, as are the ages that the river occupied each course.

Figure 13.7b is a plot for two ancient courses of the Mississippi, labeled A and B, determined by scientists at Louisiana State University. The ancient Mississippi abandoned river course A because river course B had a steeper gradient, and water tends to run down the steepest hill. The process of one river diverting the flow from another is known as *stream capture*, and is a nearly continuous process where the channels of low-gradient streams such as the Mississippi River are abandoned in favor of channels with steeper slopes.

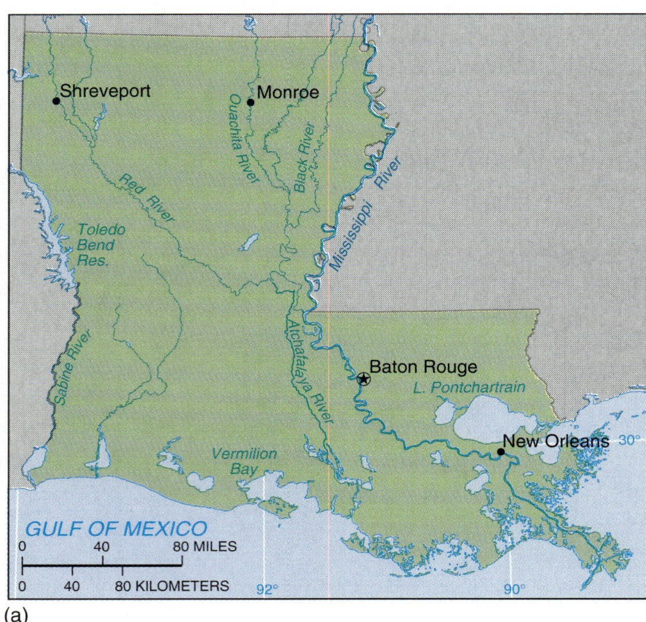

(a)

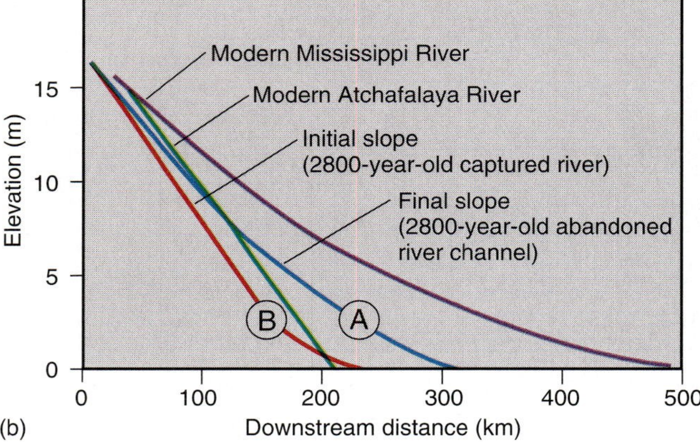

(b)

FIGURE 13.7 Modern and abandoned courses of the Mississippi River and the effects on the development of the Mississippi delta. (a) A map view of the modern river courses. (b) Graded profiles of ancient abandoned courses of the Mississippi. Approximately 2800 years ago, the Mississippi abandoned river course A because river course B had the steeper gradient. Carbon-14 dating of shells and wood has indicated that ancient river course A is older than ancient river course B. Note the modern Atchafalaya River has a stronger gradient advantage over the Mississippi than ancient river B had when it captured river A.

If the modern courses of the Mississippi River and the Atchafalaya River are compared, the Atchafalaya River has a much steeper gradient and therefore is a more desirable course for flow (Fig. 13.7b). Since the mid-1950s, however, the U.S. Army Corps of Engineers has maintained the flow in the main channel of the Mississippi. It is possible, though, that some day there will be a catastrophic jump in which the main flow from the Mississippi River is suddenly taken over by the Atchafalaya River—probably during the flood stage from a major storm.

The Influence of Groundwater on River Flow Groundwater is discussed later, as a resource (∞ see Chapter 16), but there is a fundamental relationship between river flow and groundwater that should be noted here. Depending on the local climate, the upper part of the ground contains only slight amounts of water, but it is saturated at some greater depth. This means that the pore space between grains is completely filled with water. The upper boundary of this zone of saturation is called the *water table*. This water table more or less follows topography—although the undulation of the water table is much more subdued than is the topography. Where the water table meets the land surface, water flows out of the ground—for example, at springs. **Perennial rivers** flow year round, and the water table meets the land surface approximately at the river level. During times of low river flow, water is added to the river from the groundwater reservoir, helping the river to maintain its flow. In perennial streams, then, the sediment load is continually diluted by the contribution of groundwater to the river. Thus, the concentration of sediment in the river water remains constant or decreases downstream. In arid environments, rivers do not flow year round (**ephemeral rivers**) because the groundwater table is far below the stream bottom. As the temporary flow in dry creek beds moves downstream, water is lost from the flow by soaking into the stream bed. Therefore, in contrast to perennial rivers, the sediment concentration continually increases downstream, with the mud becoming denser and more viscous the farther it flows.

Deposition of Sediments

Sediment deposition simply means that the transporting process is no longer capable of carrying its burden. Grains of sediment entrained by high winds are deposited when the wind velocity slackens; in the same manner, deposition occurs in rivers and oceans as the transporting velocity diminishes. In this section, we will discuss depositional environments and the sediments that characterize those environments from the land to the sea; we begin with rivers and lakes and then proceed into the marine realm from the shoreline out into deep water. Figure 13.8 provides a summary of the context of each depositional environment with respect to the others. Some environments may be isolated,

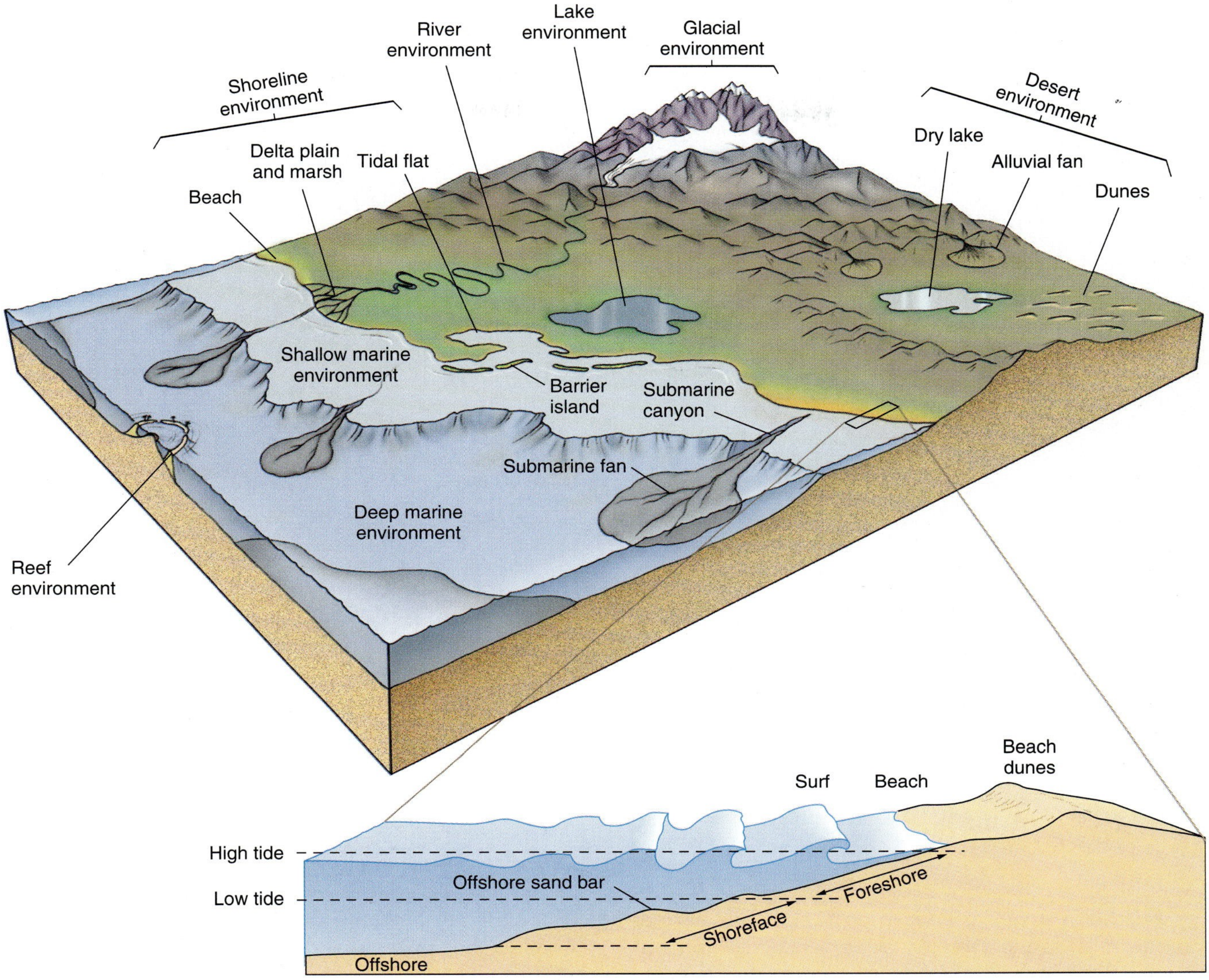

FIGURE 13.8 Different sedimentary environments and the relationships among the environments. It is unlikely that a location could be found that would exhibit all the features shown in such a small area. The box indicates the nearshore marine environment, shown in detail below.

but most show a strong degree of interconnection. The meaning of these connections is that changes in one environment inevitably lead to changes in other environments. Whatever these changes might be, they will be reflected in the sediments that are deposited, or the *depositional record*.

River Environments

Sediments cannot be deposited during the erosive phase of a stream or river. In the depositional phase, the stream is said to be *aggrading*, which means that active sedimentation is occurring on the stream bottom and along the banks. Thus, the highly picturesque mountain streams tumbling over large rocks in narrow canyons with beautiful waterfalls will leave no sedimentary record of their presence. Aggrading streams generally move slowly and flow over flatter landscapes; this is a necessary condition for sediments to be deposited and preserved. River (fluvial) deposits can be generally categorized by the mechanics of deposition—whether by fallout from suspension or by deposition of the bedload.

The mechanics of river sedimentation and the characteristics of the resulting sediment depend on both the type of river and its flow stage. The flow stage of a river indicates its flow conditions—whether it is in a low-flow stage or in flood (high-flow stage). Although there is some variety, *braided* and *meandering* streams are the main types of flow for which sedimentary records exist. A meandering stream follows a single sinuous channel of looplike bends (Fig. 13.9). A braided stream is divided into many strands that constantly shift position (Fig. 13.10).

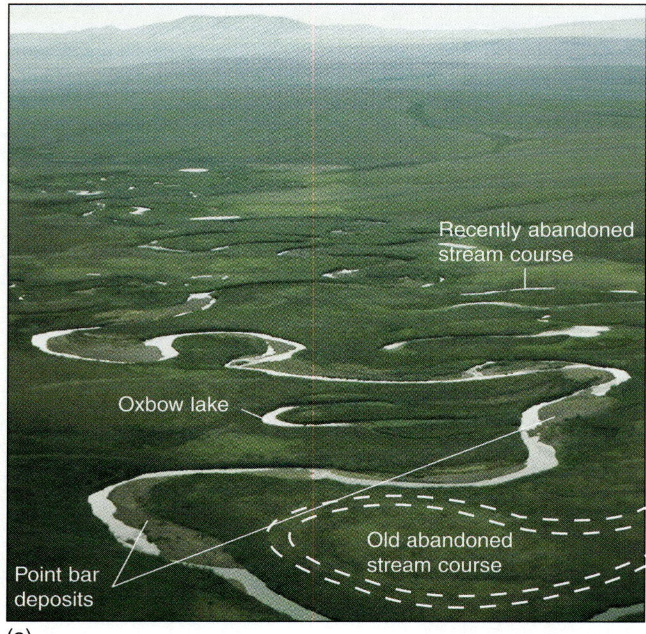

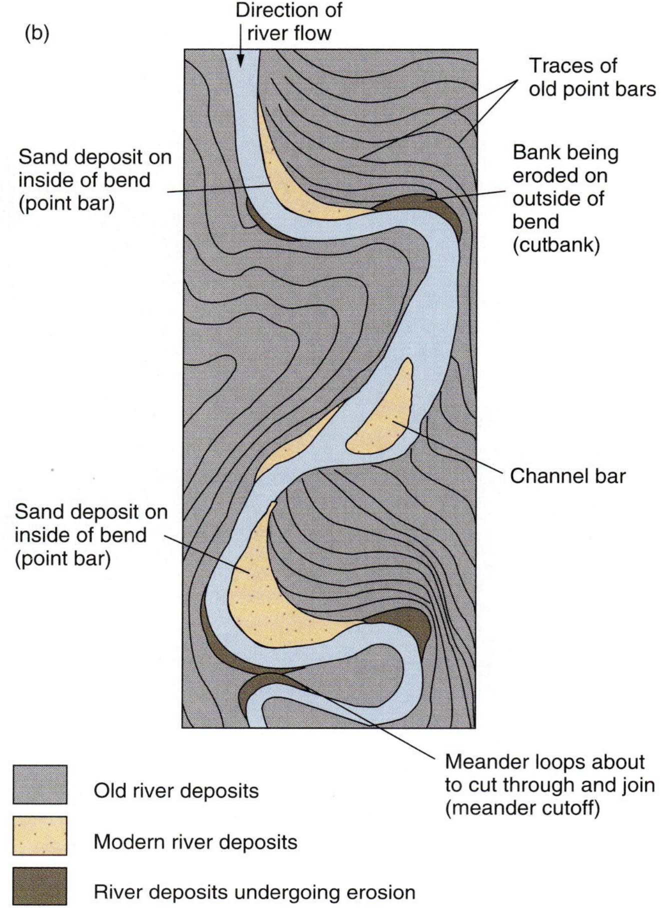

FIGURE 13.9 Features of meandering rivers and their floodplains. (a) The Serpentine River, Alaska, is a strongly meandering river; old, abandoned stream courses with associated point bar and channel bar deposits, and oxbow lakes can be clearly seen. (b) A sketch that illustrates how meanders develop. Sand is deposited on the inside of each meander loop, and is preserved as the meander sweep moves along. Near the bottom of the figure, a meander loop is about to be cut off, to form a new channel and an oxbow lake.

Whether a stream is braided or meandering depends on the relationship between slope (gradient) and discharge (flow rate). Rivers flowing on steeper slopes are more likely to be braided, and greater discharge favors braided streams. Braided streams also tend to form with coarser sediments in the bed. Meandering streams tend to form with finer sediments in the bed, flatter slopes, and lower discharges. The valleys are flat **floodplains** containing a single river channel. Braided streams comb rapidly back and forth across their valley. Often, the entire valley is the stream bed, called a braid plain (Fig. 13.10). Because of rapidly shifting stream locations, the valley floor is kept quite flat and often shows only sparse vegetation or is altogether barren. Mud is carried out of the river system in the suspended load; the main sediment type in braided streams is bedload transported by saltation and traction. To deposit the sands of the bedload, the stream channel must move laterally away, leaving behind a layer of deposited channel sand. The sand deposits left in the geologic record reflect the braided channel system.

Meandering river valleys have several component parts, and the stream processes are more diverse. The topography of a meandering river valley has significant relief; the river is confined to a channel, and the floodplain is shielded from flooding by the slightly elevated border of the channel. The location of the main channel is held in place by mud in the banks and by vegetation along the banks. As water flows around a meander bend, it moves faster on the outside of the meander curve and more slowly on the inside of the bend. The relationship among flow velocity, erosion, and deposition results in erosion on the outside curve of a meander, and deposition on the inside. These deposits are constantly building outward into the river channel, forming deposits called *point bars* (see Fig. 13.9). Erosion on the outside of the meander bend causes the meanders to form tighter and tighter curves until, eventually, the river cuts through the intervening land. By doing so, a short section of the river is abandoned, forming a small curved lake that occupies the former meander bend, called an *oxbow lake* (named after the U-shaped collar of an ox yoke).

Rivers such as the Mississippi do not follow a straight line downslope but instead travel in great meandering loops that slowly move downstream as well as back and forth across the river floodplain. Reading Mark Twain's writings of his time as a riverboat pilot on the Mississippi River gives one a vivid impression of how changeable the

FIGURE 13.10 The Ninzina River (Alaska) is a braided stream. Note the very flat valley bottom, and notice that the flowing water is subdivided into many strands. These channels move from place to place constantly, keeping the floodplain flat and free of vegetation.

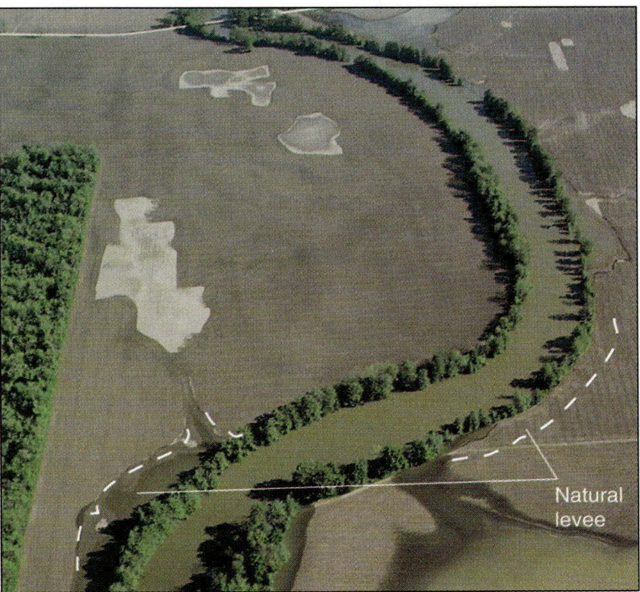

FIGURE 13.11 A natural levee showing its elevation above the normal flow stage river level, and the abundant plant growth along the river. These levees may block access of tributary streams to the main trunk stream, forcing the tributaries to flow parallel to the trunk stream some distance downstream to their confluence.

river course is from day to day, and how the sediment movement had to be watched carefully to prevent a riverboat from running aground.

During flood stage, the flow is strongly turbulent and carries a maximum load of suspended sediment, which is generally coarser than the suspended load during normal flow stages. Grain sizes in suspension range from clay through silt and fine sand (0.002 mm to 0.25 mm), and may continue up to medium sand grains depending upon the extent of turbulence. When the high-flow stage reaches the top of the riverbank, it overflows onto the surrounding floodplain. Once the flood water overflows the riverbank, it is no longer confined to the channel and the water spreads out with a decrease in flow velocity. This decrease in velocity results in a sudden reduction in carrying capacity, so the coarsest sediment being transported is immediately deposited on the riverbank adjacent to the channel. This process, repeated many times over, creates small elevated embankments (natural levees) parallel to the channel (Fig. 13.11). In some cases, the levee blocks tributaries from joining the main river, forcing them to flow parallel to the main stream for some distance. Tributary systems that are redirected in this manner are called Yazoo tributaries. Subsequent to overtopping the levee, the fine-grained suspended load is carried out onto the surrounding flood basin, where flow ceases. Over time, even the finest sediment settles out of the standing water—by fallout from suspension. The floodplain, as the name suggests, contains sediments deposited by repeated floods that leave standing water in the low areas.

A braided stream is divided into many strands that constantly shift position. These streams have a coarser sediment bedload and the valley floor is quite flat. Meandering streams have a finer sediment bedload and a single river channel.

Lake Environments

Lake, or lacustrine, sediments are not widespread but are important beyond their preserved record because of the fossil deposits they often contain. Many of the modern lakes of North America originated as glaciers retreated approximately 8000 to 12,000 years ago. The Great Lakes occupy low areas gouged out by continental glaciers, and the Great Salt Lake is a small remnant after the evaporation of a huge freshwater lake that existed during the last glacial episode (Focus 13.2). At the height of the most recent glacial advance, the western United States was the site of many lakes that existed briefly, then dried up, leaving precipitates. Some of the lakes filled desert basins. It may be surprising to learn that Death Valley was filled to a depth of more than 60 m by a freshwater lake.

Perhaps the most important lacustrine (lake) deposit in North America is the Green River shale. This sediment was deposited in a gigantic inland lake that covered parts of Utah, Colorado, and Wyoming approximately 60 million years ago. The local climate at that time was not very different from today. As rainfall and temperature varied, so did the level of the lake; it ranged from a full, freshwater body of water through phases of partial to com-

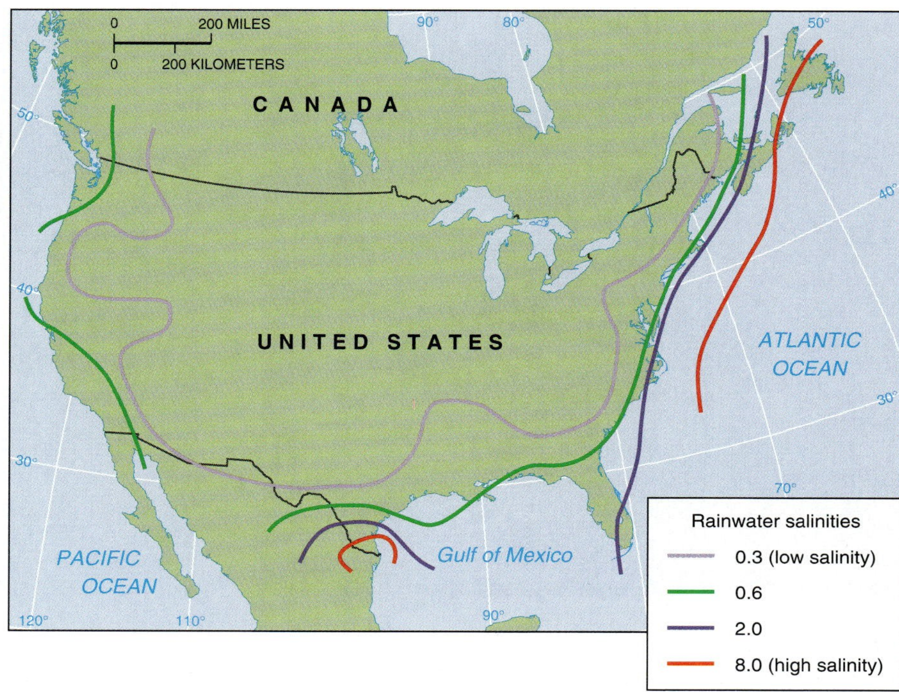

FIGURE 13.12 Salinity of rainwater, using chloride concentrations as a tracer of seawater contamination. Note that the concentration contours (arbitrary units) approximately follow the coastline.

plete desiccation with salt deposition. Salts were deposited not only as separate layers, but salt crystals also grew in the pore spaces within the sediment, displacing and distorting the surrounding sediment as they grew.

Occasionally, the lake contained large quantities of algae and thus supported very high biological productivity. The preserved lacustrine deposits from those times of high productivity are the famous Green River oil shales. There is sufficient organic matter present in these shales that some chunks of it will burn almost like coal.

The "life cycle" of most lakes is brief in the geologic sense of time. Lakes quickly evaporate in arid climates, leaving a deposit of salts. In temperate and tropical climates, lakes become choked with vegetation, which serves to catch and retain sediment. The sediment can be clastic detritus or organic matter. Extensive swamps may border a lake—particularly in areas of low topographic relief. Biological fixation of carbon—the conversion of carbon dioxide (CO_2) to organic matter through the process of photosynthesis—is called *productivity*. In warm, moist climates, productivity is high and these lakes and swamps may be sites of thick deposits of organic matter to be later transformed into petroleum, oil shale, peat, or coal.

Surprisingly, much of the salt from lakes located in arid regions originates from the oceans. Sea salt is present in the atmosphere because at the crest of breaking ocean waves, small droplets of water containing salt are cast into the air. If the droplets are small enough, they remain in the atmosphere. Sodium and chloride and sulfate concentrations have been measured for rainwater in different parts of North America, and these measurements show vanishingly small concentrations in the center of the continent, with concentrations increasing toward the coasts (Fig. 13.12). Contours of salt concentrations in rainwater parallel the coastlines, indicating that sea water is the source of salt in rainwater.

Lakes in temperate climates experience the full range of climatic conditions, from hot summers, when lake temperatures may become quite warm, to freezing winters, when the lake may freeze over. The climatic extremes make temperate lakes the most diverse in their nutrient and biological makeup.

Temperate and tropical lakes go through a definite life cycle, which includes a *primitive* phase, a *mature* stage, and a *waning* stage (Fig. 13.13). The primitive stage of a lake involves the initial filling of the basin with water and its invasion and habitation by organisms. Primitive lakes are usually very clear (Fig. 13.13a) and poor in nutrients. One or more of these stages may be omitted or shortened but all can be observed.

Once settled by organisms, the lake evolves into a mature stage, in which productivity, the fixation of organic carbon and the consumption of organic matter, and the oxidation of organic carbon, ultimately reach a balance. The margin of the lake, both above and below the water level, becomes populated with a variety of plants, depending on local conditions. The plants serve as sediment traps, enhancing sedimentation on the margins and within the lake basin. As sedimentation proceeds, the lake becomes shallower, with plants progressively covering more of the lake area. This results in an increased rate of sedimentation. The waning stage of the lake occurs as it eventually progresses to a shallow pond surrounded by

(a)

(b)

FIGURE 13.13 Two phases in the lacustrine (lake) cycle. Primitive, mature, and waning lacustrine stages occur in many different settings. (a) Lake Louise, British Columbia, is a primitive lake. It is characterized by few nutrients in the water, with little algal or bacterial growth, and hence has very clear water. (b) The waning (or mature) stage of a lake in New York State. It is nearly filled with sediment and organic detritus, as suggested by the plants growing out of the water in the middle of the lake.

marsh (Fig. 13.13b). Ultimately, the lake vanishes and is replaced by a meadow.

Sediments deposited in lakes often show a distinctive suite of sedimentary structures—for example, symmetrical ripple marks that suggest oscillating wave activity. If the lake is deep, the only sedimentary structure may be thin parallel laminations that indicate deposition in still water and an absence of burrowing organisms. Particulate organic matter content indicates the level of biological productivity—both within and around the lake. A particularly distinctive feature of lacustrine sediments, called *varves*, occurs when there are strong seasonal differences in biological productivity and sedimentation in the lake. Varved sediments are paired beds: one light colored and the other dark (Fig. 13.14). These are especially pronounced if the lake is covered by ice for part of the year. These two parts of the bed represent an annual cycle of higher and lower organic productivity (dark and light layers, respectively), or times of lesser and greater sediment influx (dark and light layers, respectively). Varved sediments have been used for dating in much the same way as tree rings have been used to date annual growth variations in the lifetime of a tree.

FIGURE 13.14 Thinly bedded sediments deposited in a glacial lake in Massachusetts. These varved clays comprise dark layers (little sediment influx during winter) and light layers (greater sediment influx during summer when more melt water is produced).

Desert Environments

Sedimentation in a desert setting often occurs as a result of a storm accompanied by major flooding. Desert rainstorms are sudden and violent. These storms can dump huge amounts of water that may exceed the rate at which water can soak into the dry regolith, resulting in little penetration of water. Thus, much of the rainfall runs off directly into channels, creating flash floods that are highly erosive and cut deep, steep-sided channels (Fig. 13.15).

(a)

(b)

FIGURE 13.15 A flash flood and its effects in a desert environment. (a) The Santa Cruz River, near Tucson, Arizona, in flood. The water is sediment-laden, highly turbulent, and rapidly moving. (b) After the flood, the waning water reworked the sediment into a braided network of small channels. Note that the small building on the left side of photo (a) is missing in photo (b).

FIGURE 13.16 An alluvial fan in Death Valley, California; the fan shape of this particular feature is obvious. Alluvial fans are formed partly by mudflows, which are later reworked by small streams. Small stream courses are clearly evident on the surface of this fan.

Because the slope of a desert channel may be very steep, the water flowing in such a channel will be turbulent and moving with great speed; it will therefore be highly erosive. As a result, these flows are laden with sediment, forming a thick and muddy suspension. In deserts the water table is usually far below the level of the channel. As floodwaters move down a channel, water is continually lost from the flood by soaking into the walls and bed of the channel. As the water is lost, the muddy suspension gets thicker, denser, and more viscous. Its erosive power therefore increases and, although its velocity may decrease, the capacity for carrying rocks and boulders increases because both the viscosity and the density of the mudflow increase.

In desert mountains, channels may spill out onto flat plains, leaving large fanlike deposits of sediment at the mouths of canyons (Fig. 13.16). These **alluvial fans** build outward by the addition of material. As a stream flows out of a narrow mountainous channel onto a desert plain, the flow velocity suddenly decreases, with a consequent loss of carrying capacity. This loss of carrying capacity causes sediment deposition. As one portion of the alluvial fan builds upward and outward, the stream gradient across the fan at that location is reduced, so the stream shifts its direction and other parts of the fan begin to receive sediment. Thus, the fans are kept more or less symmetrical as they develop (Fig. 13.16).

Sediments of alluvial fans are mixtures of massive mudflow deposits, rockfall, and landslide deposits. Transport distances are usually short. The uppermost parts of some deposits may be partly reworked and redeposited by small streams that flow over the fan surface immediately after storms. Thus, alluvial fan deposits are poorly sorted and display mixtures of sediments ranging in size from mud to house-sized boulders. Bedding in these deposits is usually poorly developed.

Perhaps someone who has never visited a desert might imagine it to be endless stretches of sand dunes. In fact, although sand dunes are spectacular, they usually comprise only a small portion of most deserts. In mountainous regions of deserts, fields of sand dunes occur primarily in the lee of mountains or in topographic bowls where winds are

(a)

(b)

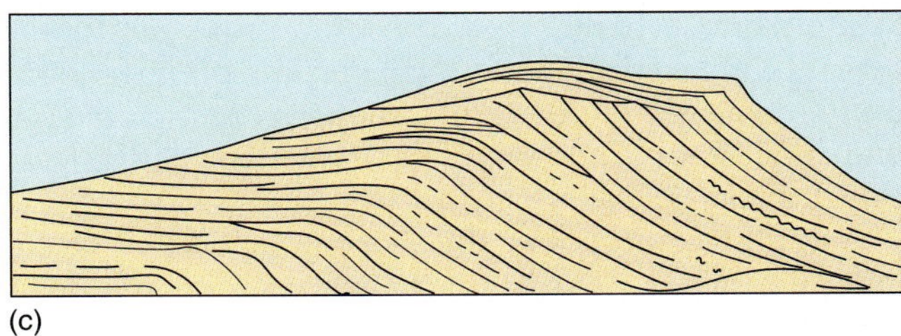

(c)

FIGURE 13.17 Sand dune features. (a) A view across the crest of a large dune. Sand is transported by traction up the shallow slope to the left and the sand slides down the slip face. The dune causes a disruption in the wind around it, so wind directions are variable as shown by ripples in the sand. (b) A view of the slip face on the front of the dune shown in (a). Note that the primary sand transport is by mass movement down the slip face; also note that the wind blows nearly perpendicular to that direction at this location. (c) Internal cross section through dune, showing cross-bedding.

forced to circle and therefore deposit the blowing sand as the wind velocity decreases (Fig. 13.17). Most of the sand that forms the dunes is winnowed material from the barren surface of the desert. In the great deserts of the world, such as the Sahara in northern Africa, the Kalahari in southern Africa, or the Gobi in China, the area covered by sand dunes occupies only a few percent of the total desert area. Much of the material eroded from the desert surface is transported away. The finest dust can be transported enormous distances by winds and may be deposited in continental regions as *loess* or carried out into the oceans where it settles to the sea floor as red or brown "clays."

Alluvial fan deposits form where streams flow out of narrow mountainous channels onto a desert plain, resulting in a sudden decrease in the flow. The remaining areas of desert regions are mostly rocky, with small areas of sand dunes.

Shorelines

The overall transport of sediments is from continents to oceans. On the continents, sediment is transported by water, wind, or ice. In the oceans, water is the transporting medium. Sediment is eroded and deposited by wave action and by oceanic currents. The interface between continents and oceans—the shoreline—therefore represents an important contrast in sedimentary processes.

Deltas The velocity of a river abruptly decreases upon meeting the ocean, so the sediment load is immediately deposited in the relatively quiet oceanic environment. Deltas are usually triangular, low-lying areas that may extend oceanward as well as some distance inland. The delta plain is the geographic delta, with low-lying marshes, bays, and *distributaries* (channels that branch off of the main river in a downstream direction and serve to distribute the water across the delta); the *delta platform* is a prism of sediment offshore from the geographic delta that is largely underwater.

Deltas migrate seaward because of the sediment supplied to them by the rivers (Fig. 13.18). Sediments deposited far offshore in the marine environment are thinly bedded clays and silts that become increasingly thinner with distance from shore. Sediments deposited on the immediate front of the delta are not deposited horizontally but on a slight slope. Then, as the delta is overridden by the river and the delta plain, there is an upper sequence of sediments deposited in those environments. Reduced to its simplest components, a delta con-

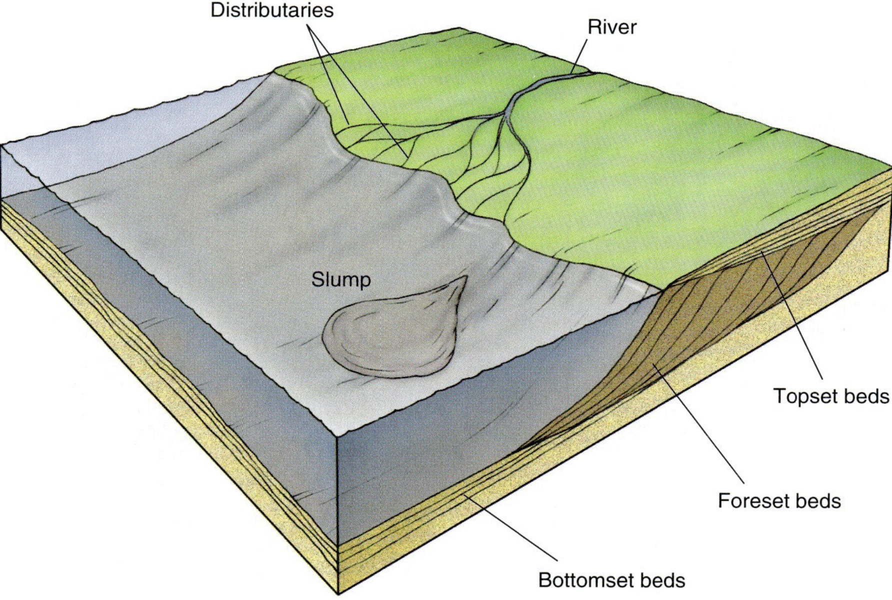

FIGURE 13.18 Structure of a delta, illustrating topset beds (1), foreset beds (2), and bottomset beds (3) are evident.

sists of three general types of beds: a lower horizontal sequence of beds called the *bottomset beds*, an upper horizontal sequence of beds called the *topset beds*, and a set of inclined beds deposited on the front slope of the delta called the *foreset beds* (Fig. 13.18).

Several factors control the shape of deltas and determine the nature of sedimentary processes on the delta platform. Marine currents, waves, and tides serve both to disperse the sediments brought to the mouth of the river and transport the sediments to other locations—either along the shoreline or out into deeper water. The balance between wave activity and the amount of water and sediment flowing out of the river, and the sediment grain size, are secondary controlling mechanisms on the delta form.

There are two extremes in delta morphology: a wave-dominated delta (Fig. 13.19a) and a river-dominated delta (Fig. 13.19b). In the former case, the coastline is accessible to strong wave activity and the shape of the

(a)

(b)

FIGURE 13.19 (a) A false-color satellite image of a wave dominated delta on the east coast of Mozambique. Beach ridges were constructed by waves and suggest strong marine redistribution of deltaic sediment by wave and marine currents. (b) A false-color satellite image of modern Mississippi delta, a delta dominated by river processes, with little wave or current erosion on the seaward edge of the delta.

delta is the result of wave erosion off the front of the delta. Sediments delivered to the coast will be redistributed by wave action and transported offshore by shallow marine currents. If the wave activity is strong, but wave and river processes are in balance, then the delta tends to be arcuate in shape. If wave activity is weak and the river is large, river processes may dominate. River-dominated deltas extend well beyond the local shoreline, forming irregular shapes that may resemble a bird's foot (Fig. 13.19b). Such deltas are not very common; a more common type is the wave-dominated delta, such as that shown in Figure 13.19a.

The flow stage of a river controls the sedimentary processes that operate on deltas. If a river is in low-stage flow, all of the water is guided out to sea by the active distributaries; sedimentation occurs in these channels and immediately offshore from them. The delta plain and other intervening areas of the delta receive no sediment. When a river floods, however, the delta plain is flooded and the areas between distributaries receive sediment.

Deltas form when the river velocity decreases upon meeting the ocean and the sediment load is immediately deposited. Deltas are shaped by the interaction among the river and marine currents, waves, and tides that disperse the sediments.

FIGURE 13.20 Emergence runoff features on tidal flats. An emergent tidal flat showing ripples formed from tidal currents. As the ripples emerge above water level, they form channels that guide late-stage emergence runoff. (Austfjorden, Spitsbergen, Norway)

Tidal Flats Low-relief coastlines are often very muddy, with wide areas of mud exposed during low tide and covered by a thin veneer of water during high tide. Because these regions are subjected to the constant ebb and flood (lowering and rising) of the tides, the tides themselves continually deposit sediment and then rework the sediment that has already been deposited. Tidal currents almost always flow in and out approximately perpendicular to the coastline. Sedimentary features created as a result of tidal action will be oriented either perpendicular to the shoreline or, in the case of tidal current ripples, parallel to the shoreline. Even more important to the sedimentary processes is the fact that the tidal flats are flooded and then emerge from that flood twice a day. This periodic flooding and emergence has a profound influence on sedimentary and biological processes of the flats.

The zones created by differences in sediment transport mechanisms on tidal flats are shown most graphically by variations in sediment grain sizes. Sand dominates the farthest offshore portion of the tidal flat because grain movement is by bedload transport in the tidal current. The sediment that is closest to shore consists of mud because the fine sediments can fall out from suspension in the still water at slack tide. Thus, differences in the sediment transport mechanism from one part of the tidal flat to another create a zonation ranging from coarse-grained sediment in the lowest part of the tidal flat to the finest sediments inshore.

During emergence from the flooded stage, sediments on the tidal flat are modified by water flow and by biological activity, such as burrowing by organisms. During emergence runoff, ripples may be washed out or rounded. Emergence from the flooded state creates unique sedimentation processes because the sedimentary structures themselves guide the water flow during different stages of emergence. As the ebb flow proceeds, larger sedimentary features, such as sandbars, emerge first and the topography of that tidal flat then begins to exert a control on the direction of water flow. The more the tide recedes, the more important this effect becomes until small ripples direct the last stages of water runoff (Fig. 13.20).

Tidal flats range from coarse-grained sediment in the lowest part of the tidal flat, where sands are transported as bedload by strong currents, to fine muds inshore, where sediments fall out of suspension as the tidal current stops.

In arid coastal areas of low clastic input, tidal flats and coastal lagoons form a zone where evaporites are deposited. Algal and bacterial mats are also found in these areas. Tidal flats dominated by evaporites are called *sabkhas*—a phonetic transliteration of the Arabic word for salt flat. The surface of these areas is constantly winnowed and eroded by wind. The land surface elevation is controlled by the level of the groundwater table. Periodic storms bring seawater onto the flat, and subsequent intense evaporation concentrates this water to salinities approximately 10 times that of normal seawater. Precipitation of gypsum ($CaSO_4 \bullet 2H_2O$) and aragonite ($CaCO_3$) occurs in the spaces between carbonate grains as this saline water evaporates. In some areas, so much material is precipitated that the originally deposited grains are forced apart by the chemical sediments. Anhydrite ($CaSO_4$) may form by diagenesis as a secondary replacement of the original gypsum. This is a dehydration reaction, removing water from the crystal structure of gypsum. Diagenesis also leads to the formation of dolomite, $CaMg(CO_3)_2$, by replacement of some calcium by magnesium in aragonite.

Shallow Marine Environments

Beaches Although beaches are really features of the shoreline, it is more appropriate to discuss them within the shallow marine environment. Beaches are the uppermost part of a continuous succession of sedimentary deposits that range from the beach into deeper water where offshore sediments accumulate (Fig. 13.10). Beaches occupy the same tidal position as do tidal flats. So, why is it that muddy tidal flats occur in some areas and sandy beaches in others?

The most distinctive characteristic of a beach is the surf, caused by wave action. Waves are caused by wind blowing over the oceans. The energy of the blowing wind is transferred to undulations, which move over the water's surface. To understand why a wave breaks, we must consider its motion. Any particle in a body of water will follow a circular motion as a wave passes through the water (Fig. 13.21). If you watch a floating object on the waves away from the shore, you will see it bob up and down and move to and fro as the waves pass beneath it. The circular path has a characteristic depth, which is about half a wavelength and called the *wave base*, beneath which the wave has no effect on the water. If you swim in the ocean and dive beneath the waves, you feel a strong surge during the passage of a wave overhead; this surge is the wave motion under water. When the wave approaches the shore, the water depth decreases and the bottom of the ocean becomes the wave base. At this point, friction at the bottom of the wave slows it down, causing the wave to become "top-heavy" as the back catches up with the front, and topple over, breaking on the shore (Fig. 13.21).

Within the intertidal zone (the *foreshore*, Fig. 13.8), sediment is moved with every wave that breaks on the beach. Reach down and grab a handful of beach sediment and examine it. Several distinctive characteristics are easily observed with the unaided eye: Most beach sediments are quite well sorted by size, there is virtually no mud, the grains tend to be highly polished, and they are well rounded. These features are easy to explain using the dynamics of the beach system.

The crashing surf churns the sand several centimeters deep. This movement winnows all mud and other fine particles out of the sediment and carries this fine debris away in suspension. Fine-grained mud cannot remain where there is vigorous surf. With each breaking wave, the sediment grains rub vigorously against one another. In this way, each grain polishes every other grain it contacts. This action also rubs the sharp edges and corners off each grain. Thus, every grain tends to be rounded. The slope of the beach surface is, in fact, a bedding plane, so beach deposits show nearly planar beds, dipping shallowly toward the sea.

Sediment grains in the beach and on foreshore environments travel great distances up and down the beach, but they also are subjected to transport along the coast. As each wave of the surf crashes into the sediment, grains are moved inland during *upwash* and then seaward during *backwash*. The grains, however, will not follow their own tracks precisely. If waves come in at a slight angle to the coast then the sediment grains on the beach will be washed along the path of the incoming wave. But the force of gravity causes the water and the sediment grains to move directly down the beach slope back toward the sea in a direction perpendicular to the shore. Thus, there

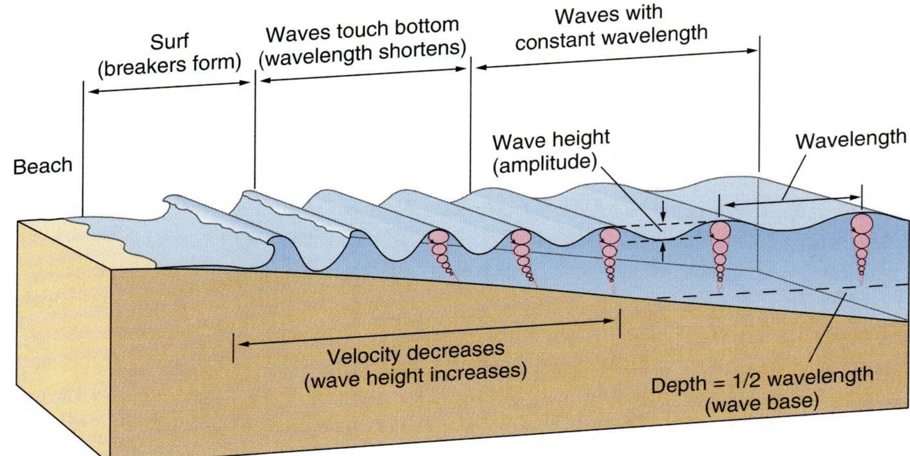

FIGURE 13.21 Ocean waves. The circular oscillatory motion decreases in amplitude with depth. Near the shore, bottom drag causes waves to steepen and topple over to form breakers in the surf zone.

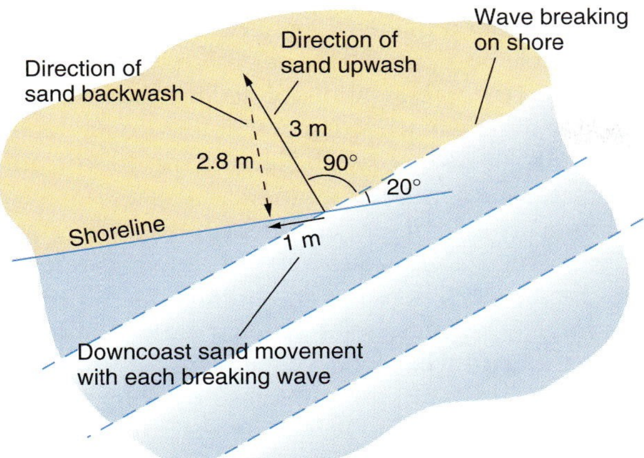

FIGURE 13.22 Mechanism of longshore sediment movement. Waves impinge on a beach at an angle. Sand movement follows the direction of the wave in the upwash, whereas the sand movement during backwash is directly downslope—different from the upwash direction. In this way, sand is moved parallel to the coast; the direction of sand migration is determined by the orientation of the waves striking the coast.

is an angle between the direction of upwash and that of backwash (Fig. 13.22). Similar effects occur in the underwater currents in the surf zone. Waves obliquely incident on a beach generate a current along the surf zone that transports sediments disturbed when the waves break. Both these effects transport sediments along the coast.

In the example shown in Figure 13.22, for every meter that the grain is moved downcoast, it may actually travel up the beach 3 m with the incoming wave and slightly less than 3 m back down the beach with the backwash. Thus, every grain moves a total of nearly 6 m for each meter it travels parallel to the shoreline. This clearly influences the amount of abrasion that each grain receives and accounts for the highly polished nature of beach sand.

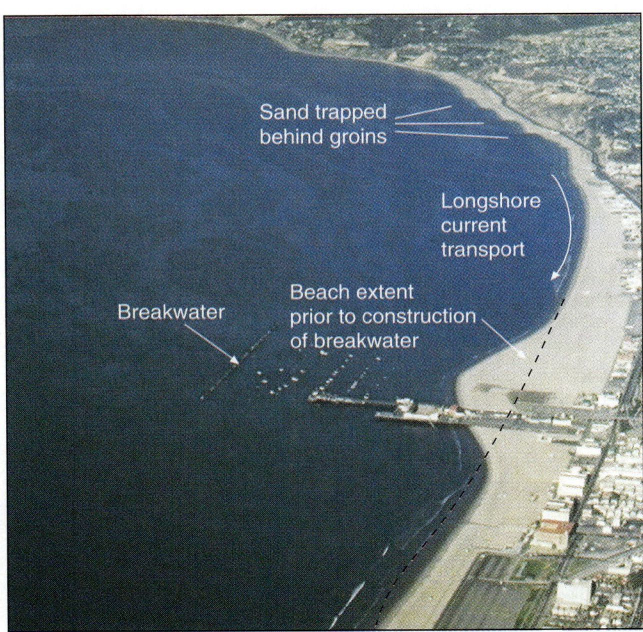

FIGURE 13.23 Longshore drift and the influence of structures on sand migration. The Santa Monica pier and breakwater as they appeared about 1948. A part of the beach is protected from incoming waves, and for that reason, longshore sand drift is stopped. Waves are deflected around the breakwater, causing beach erosion downcoast from the breakwater.

The movement of coastal sediment parallel to the shoreline is called *longshore drift*, and it represents an important mechanism by which sediment is moved from one location to another along the beach. This movement, however, only occurs near the surf zone, and is not effective elsewhere. Along a given coastline, the prevalent winds are usually consistent in direction, and this determines a dominant direction of longshore drift.

Downcoast transport of sediment is easily identified. A variety of structures are built along coasts, including harbors, breakwaters, jetties, and groins (small walls or breakwaters that are built out from shore). These structures will influence "down current" sediment movement because they each interfere with the waves cresting and breaking on the beach. A longshore current turns into the relatively still water created when a jetty is built to create a protected anchorage. As the current moves into the still water of the harbor, it slows; having lost its sediment-transporting ability, it deposits sand in the entrance to the harbor. A breakwater prevents surf from reaching the shore, effectively stopping longshore drift (Fig. 13.23). Down current from the breakwater, breakers are deflected into the still water (refracted in the same manner as seismic waves (∞ see Chapter 5) and erode the beach. A groin standing perpendicular to the coast can function as a sediment dam, stopping longshore sediment transport.

Beaches represent the junction between land processes and marine processes. Wave action causes winnowing of fine particles, polishing and rounding the sediment grains on the beach. Longshore sediment transport can result if waves approach the coast at an angle.

Subtidal Sediments Sediments below the tidal range, but shallower than the reach of the deepest storm wave, are considered to be within the *shoreface* environment (Fig.13.8). This zone is the richest biological habitat along clastic coastlines because of the ample light and abundant organic detritus in the water. Sediments deposited in this environment are generally extensively mixed by animals moving over and through the sediment, a process referred to as *bioturbation* (Fig. 13.24). Within the shoreface zone, the surge from surface wave motion touches the bottom, which may cause the current action to move large volumes of sediment. After deposition, individual layers of sand and mud are mixed by organisms; for this reason, shoreface sediments are often poorly sorted. It is not uncommon to find thin layers of shell debris in shoreface sediments; these almost certainly represent storm debris.

In several shallow marine environments, we must consider the effects of marine organisms that either make their homes in or on the sea floor or eat the sediment and digest available nutrients. These animals have a profound influence on the preservation of sedimentary features such as bedding. Sediments that are moved around rapidly by currents or have no available nutrients for the animals, such as sand bars, are not satisfactory places for most marine animals to build a permanent home. Therefore, these sediments will show prominent bedding features. If oxygen, food supply, housing, and salinity are all within the appropriate tolerances for resident marine animals, then the sediments will become populated. Under normal marine conditions, delicate laminations are obliterated by animal activity. Mixing of sediment by bioturbation is often so thorough that all other sedimentary structures are destroyed, and the sediment is left with a peculiar, mottled texture.

FIGURE 13.24 Sedimentary structures formed by the activity of animals that live in sediment. Burrows in marine muds of a deltaic sequence that was overridden by delta distributary channel sandstone. (Etchegoin Formation, Coalinga Anticline, California) The sands filled the burrows of living clams, killing them; most of these burrows have fossils in them. The depth of burrowing is approximately half a meter; the hammer handle is 50 cm long.

Well-bedded shallow marine sediments usually indicate that the sediments were lifeless. Preservation of delicate primary bedding features indicates that conditions important for animal life were not present in a given location. Reasons for this might include: *bottom stagnation*, resulting in depletion of oxygen or an accumulation of hydrogen sulfide—either of which is poisonous to life; conditions that are too dry or too saline for abundant life; *rapid deposition*, in which burrowers cannot keep up with the rate of sediment accumulation; or *continual shifting* or *coarse sediment*, either of which inhibit burrowing animals.

Sediments considered part of the *offshore* environment are those that are deposited in quiet water below the wave base. Except for the quiet rain of sediment falling out of suspension from the water column, the only way sediments can reach the offshore zone is by massive transport during large storms or by catastrophic events similar to underwater landslides, which we will discuss later in this chapter. Far from shore, or in very deep water, the rate of sediment accumulation is very slow, ranging from a few millimeters to a few centimeters per thousand years, depending on the distance from shore.

Carbonate-Depositing Environments Carbonate-depositing environments are among the most scenic areas in the oceanic world. Flourishing life comprises a wide variety and diversity of species, including fish, coral reefs, and abundant sea grasses. The water is amazingly clear because these areas are devoid of silicate muds. Marine organisms make carbonate shells by taking advantage of the fact that, above depths of approximately 5500 m, the water is nearly saturated with calcium carbonate. The solubility of calcium carbonate (calcite) increases in ocean water with increasing pressure. Knowing that the oceans are saturated or supersaturated with respect to carbonate caused sedimentologists to believe that most carbonate rocks were direct chemical precipitates from seawater. It was not until the mid- to late 1950s that sedimentary structures such as cross beds (∞ see Chapter 4) and channel features in carbonate rocks were accepted as evidence of the transported nature of most carbonate sedimentary rocks. Thus, most carbonate rocks (limestones) are clastic rocks in the sense of being transported grains, influenced and winnowed by currents.

A fundamental difference exists between sediments made of silicate minerals and those made of carbonate, even though they may be deposited in similar marine basins under similar conditions of current action. Silicate minerals are almost always derived from outside the basin and are transported to the site of deposition. Carbonate grains almost always originate within the depositional basin. The sources of these grains are the organisms that have carbonate shells and structures. Generally, these organisms live in shallow water because many require sunlight for photosynthesis. We use the term biochemical sediments (∞ see Chapter 4) to indicate that the carbonate detritus has an intimate connection with biological processes. The most abundant growth of carbonate-secreting organisms is within a few degrees of latitude of the equator, partly because the high sun angle is required for photosynthesis and partly because the warm water is suitable for biological precipitation of carbonate. In such shallow water, as the animals and plants die, skeletal carbonate is broken into smaller grains by wave action and then transported by marine currents to the site of deposition.

Carbonate sediment originates from shells and carbonate-secreting plants. Carbonates are clastic sediments in the sense that they have been eroded, transported, abraded, and sorted.

Limestone Grains The grains that make up carbonate sediments, and ultimately rocks, vary in size from entire shells to sand-sized shell fragments and to finer lime mud. Carbonate or lime mud is defined as mud by grain size, but the mineral composition is actually pure calcium carbonate (either calcite or its polymorph aragonite). In the larger size range, along with shell and coral fragments, regular, ovoid grains of carbonate occur, known as *ooids*. Ooids form by inorganic accretion of concentric layers of carbonate around an initial nucleus—commonly a detrital sand grain or a tiny shell fragment. They typically form in high-energy environments where the grains are rolled around by currents, in contrast to lime mud, which forms in quieter environments such as lagoons or deeper water. If ooids eventually are the dominant component in a rock, it is referred to as an *oolitic limestone*.

Many carbonate-secreting organisms live in shallow water because they are filter feeders; they guide water through a feeding apparatus that acts like a sieve to filter out their food. For filter feeders, the violent currents of the surf environment are perfect; they obtain fresh food with every wave. The disadvantage of this environment, is that they must build strong fortresses of carbonate such as coral reefs for protection from the wave action.

Coral growth is typically confined to warm, clear shallow tropical water such as now exists in the Bahamas and off the northeast coast of Australia. The environmental sensitivity of corals makes fossil corals particularly useful in the interpretation of ancient environments and climates (∞ see Chapter 6, Figure 11.15).

The reef environment is characterized by an abundant and diverse fauna, including those organisms that secrete massive carbonate skeletons to form the structural frame of the reef. Animals that build the corals of the carbonate reef mass make up only a small percentage of the total animal and plant population of the reef tract. Within the crevices of the reef are quiet microenvironments inhabited by a great many other species. Strong current activity and winnowing action, particularly during storms, may break off bits of coral and tear up carbonate-secreting algae and other organisms. The crashing surf quickly grinds these fragments into sand-sized grains, transporting away any finer material that may be produced. Sediments deposited within this environment contain a great many fossil fragments.

There is an ecological zonation within the reef–lagoon complex that is controlled by both the depth of light penetration and the wave energy. Both the light and the wave energy influence the nutrient supply and the degree of destruction of the reef mass. This zone thus represents not only an area in which a large and diverse biota thrives, yielding diverse sedimentary fabrics, but it also represents the "carbonate factory" supplying carbonate detritus to the sedimentary regimes of the associated offshore and onshore regions.

Deep Marine Environments

Three dominant types of sediments exist in the mid-oceanic deep sea: red or brown clay, siliceous ooze, and carbonate ooze. Red and brown clays occur on ocean bottoms near the 30° latitude zones; that is, within the same latitude range as the world's deserts. These sediments are thought to be material that has been blown out to sea by strong winds from deserts such as the Sahara. But why are the clays colored red or reddish brown? Organic matter contained between and around sediment grains causes sediments to develop a dark gray color. If no organic matter is present, the sediments are light colored. During weathering, most mineral grains develop a fine surface coating of oxidized iron (possibly $Fe(OH)_3$). Because these grain coatings are oxidized, the sediment develops the color of iron rust, which is reddish brown.

Despite the name, the tiny particles that make up deep-sea clays are not actually clay minerals (∞ see Chapter 3), but rather fine clay-sized grains of other minerals. The settling velocity of clay-sized mineral grains in seawater is sufficiently slow that even slight currents can keep it suspended. Thus, marine currents are able to move the suspended clays far from their site of origin. Sediments deposited in the deep sea mix with shells of microscopic marine organisms that live in the uppermost layers of the ocean (plankton), and upon dying, sink to the bottom. As with clay, this organic matter settles very slowly. But these dead algal cells represent food for many small animals called zooplankton, so nearly every particle of organic debris that settles through the oceanic water column has passed through the digestive processes of many organisms, including bacteria. Nearly all organic components are consumed and never reach the sea bottom. Thus, the sediments on the deep sea floor are red or brown in part because they contain no organic matter, and in part because they consist of oxidized mineral fragments that originated in the desert regions of the world.

Near the equator, sediments on the ocean bottom are primarily carbonate and result from the accumulation of single-celled organisms (foraminifera) that secrete carbonate shells. Deposition of masses of these tiny shells, called *tests*, creates a carbonate ooze (fine mud). Foraminifera grow abundantly in equatorial waters, so the bottom sediments are composed almost entirely of the tests of these organisms. Chalk is one example of a sediment deposited in the deep sea, but entirely derived from plankton. The chalk sediment at the famous White Cliffs of Dover in England is composed almost entirely of microfossils called *coccoliths*, the internal carbonate remains of single-celled algae. To give you an idea of the size of an individual coccolith, the periods on this page are about 0.2 mm in diameter; about 2000 coccoliths would fit nicely onto each period. Yet the accumulation of these tiny microfossils is responsible for many thick chalk deposits around the world.

Siliceous oozes are deposited in narrow bands immediately north and south of the equator and are primarily composed of the shells of two types of organisms: diatoms and radiolarians. These organisms use silica to build their tests, so they require a plentiful source of silica. Although silica is nearly insoluble in water, a small amount is in solution. The form of silica that diatoms and radiolarians secrete is not quartz, but a quasi-crystalline substance containing water in its structure. On burial, the silica from both types of organisms is unstable; it dehydrates and eventually converts to a type of finely crystalline quartz called *chert*. If conditions are not appropriate for the conversion to chert, the silica remains unaltered.

Carbonate oozes occur in equatorial sediments; siliceous oozes are deposited in narrow bands north and south of the equator. Both types result from the accumulation of the shells of microscopic organisms. Fine-grained clays derived from the continents are red or brown after deposition owing to oxidation and to the lack of organic matter.

Sedimentation on Continental Shelves and Transport to the Deep Ocean Continents are topographically elevated relative to ocean basins and therefore represent sources of sediment. The ultimate repository of that sediment must be the ocean basins because they are the lowest regions on Earth's crust. Sediment accumulated in shallow water on the continental shelves can be transported by **turbidity currents** down the continental slopes into the deep water of the continental rise and the abyssal plain. Turbidity currents are landslides of suspended sediment, typically with a bulbous head and a long tail. The head is caused by flow resistance by the substratum and the surrounding water. Sediment is supplied from the body of the flow into the head, providing the energy to overcome this resistance.

Coastlines with narrow continental shelves usually occur along convergent and transform plate margins and are incised by many submarine canyons (Fig. 13.25). Typically, the heads of these canyons are located not far from the mouths of rivers.

Slopes of the submarine canyons are so steep that divers have observed sand flowing down the canyons. This is a grain-by-grain movement similar to the downslope movement of sand grains on the fronts of sand dunes. This grain flow is erosive—the sand flowing down the canyons is the canyon-cutting mechanism. But why are these canyons associated with rivers?

In locations where a river course has suddenly shifted to a new outlet into the sea, it has been observed that the sediment is deposited in the form of an extended lobe directed more or less in a line offshore. With a narrow

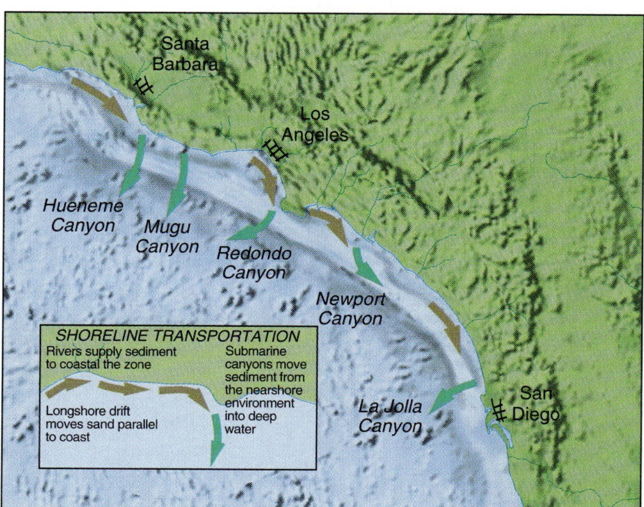

FIGURE 13.25 Distribution of submarine canyons along the southern California coast. Each submarine canyon is supplied by sand from a specific region of the coastline, usually fed by rivers. The inset map shows such an area. The canyons, in turn, remove the sediment from the nearshore system and transport it into the deep sea.

continental shelf, this lobe of sediment reaches the shelf edge and begins to flow over the edge and down the continental slope. Once the sand begins to flow downslope, the process of canyon cutting has started.

Headward erosion allows the submarine canyons to migrate steadily landward until the canyon heads are sometimes only a few hundred meters offshore.

Sediment migrates along the coast by the process of longshore drift. Because the heads of the submarine canyons reach into shallow water and even approach the surf zone in some cases, sediment that is transported by longshore currents falls into and is trapped in the submarine canyons. This process drains the sediment supply out of the coastal migration system and into the submarine canyons (Fig. 13.26). As more sediment accumulates, the sediment pile enlarges and ultimately fails, creating a landslide down the canyon. As the debris moves down the steep canyon, it erodes the sides and bottom of the canyon and ultimately becomes a mudflow. As with mudflows on land, these submarine mudflows (turbidity currents) have a higher density and viscosity than the surrounding seawater, so they are able to transport large amounts of sediment. Depending on the amount of mud incorporated into the current, the transported sediment can be very coarse indeed, including pebbles or cobbles. The downslope flow generates momentum, and as long as the increased density is maintained, turbidity currents are able to flow down the continental slope and across a flat abyssal plain for many tens to hundreds of kilometers until the momentum of the flow is exhausted.

How fast do turbidity currents travel? This question was answered following an earthquake in 1929, which triggered a turbidity current off the eastern coast of Newfoundland, causing breaks in transatlantic communication cables on both the continental slope and rise off the Grand Banks of Newfoundland. The sequence of cable breaks extended from cables located high on the continental slope to those farther down. The times at which communications were interrupted indicated the times at which the sequence of different cables broke, which led to the first direct measurement of a turbidity current's speed at 45 to 60 km per hour.

Turbidity current deposition depicts a catastrophic event in that each bed represents a single depositional episode. A turbidity current can deposit a lobe of sediment, called **turbidite** that is more than a meter thick, but the length of time required for this deposition might be only a few minutes. The base of a turbidite shows evidence for scouring and erosion. Individual deposits of each turbidite are characterized by graded bedding. This is a direct reflection of the mechanism of sedimentation from the turbidity current. As the current loses velocity, the sediment can no longer be transported and it settles out. The settling velocity of heavy grains, which tend to be concentrated in the front of the flow, is greater than that of the fine material. Coarser grains therefore settle out before finer particles.

The overall characteristics of the sediments deposited by turbidity currents vary greatly with respect to position relative to the submarine canyon. Immediately adjacent to the outlet of the submarine canyon, a large fan—very much like an alluvial fan—is deposited. Like alluvial fans, the upper reaches of this fan contain coarser sediment and the lower regions contain finer sediment. Flows sweeping out of the submarine canyon erode deep channels in the upper part of the fan, but these channels disappear farther downslope. In the channelized upper portion of the fan, turbidity currents may spill over the

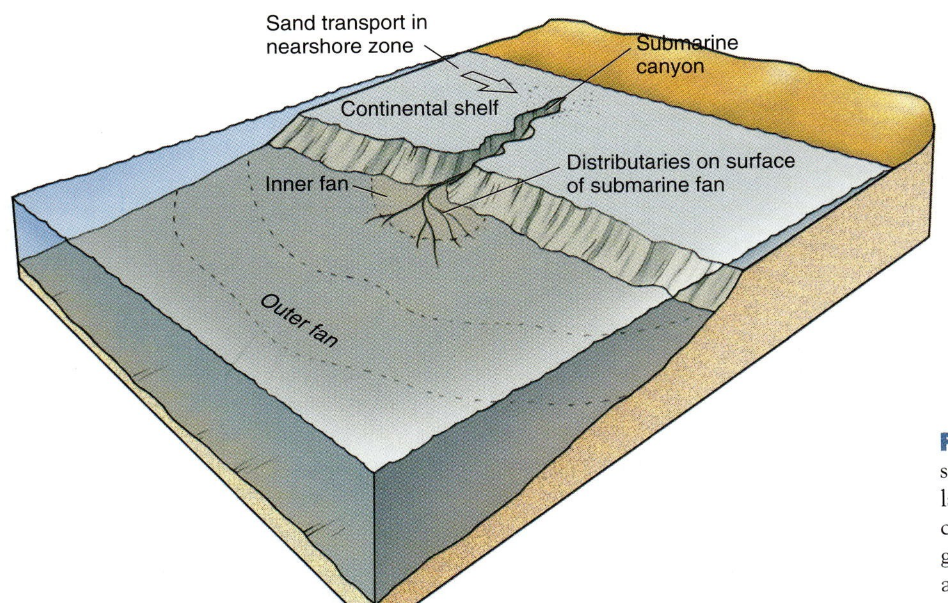

FIGURE 13.26 Transport of sediment to a submarine canyon causes landslides in the form of turbidity currents to flow down the canyon, generating channels (distributaries) and submarine fans. Finer sediments are found on the outer fan.

channel banks, depositing fine sand, silt, and clay between channels. Although this activity occurs in the deep marine environment, in some respects it is similar to overbank flooding of a river.

The heads of submarine canyons are close to shore, so that sediment carried by longshore currents may be trapped within them. After sufficient sediment has accumulated in the canyon, a landslide in the canyon might trigger a turbidity current that can carry sediment out of the shallow marine zone into the deep sea, forming a submarine fan.

Tectonic Control on Sedimentary Processes and Depositional Environments

Convergent Margins Sedimentary processes reflect their tectonic setting. Erosional areas along active continental margins are steep, and sediments are deposited into deep, steep-sided basins. Such basins are usually formed by faulting so that subsidence can be very rapid; the basins are commonly small and irregularly shaped. The processes that fill the basins with sediment are primarily the result of turbidity current action.

Along convergent continental margins, volcanic and plutonic activity is characteristic and uplift is rapid. The steep hillsides created by uplift are rapidly eroded, with little time available for weathering; thus, sediments deposited in adjacent areas are filled with fresh minerals and rock fragments derived from newly exposed volcanic and plutonic terranes. Greywackes are typical of the immature sedimentary rocks of convergent plate margins (∞ see Chapter 9). These sediments are deposited in fore-arc basins on the continental shelf offshore from the magmatic arc.

Divergent Margins In its "youthful" stage during continental rifting, a divergent margin is the setting for active extensional tectonics, including normal faults, dipping toward the newly forming ocean basin, and faulted basins, which may drain inward or may open to the ocean and be periodically inundated. The characteristic sedimentary deposits along rifted margins initially show coarse terrestrial sediments deposited by landslides, rockfalls, and steep streams (alluvial fan deposits). As subsidence continues, the alluvial deposits are followed by lake or shallow marine sediments. The marine sediments deposited in the isolated rift basins may typically contain thick salt deposits depending on the climate. The high heat flow may lead to volcanism, which, in turn, provides sediment with volcanic detritus.

As a rift develops into maturity, the plate boundary is located at an oceanic spreading ridge, and the continental edge is left as a passive margin. Passive margins tend to be characterized by low topographic relief, generally slow subsidence, and, except in the earlier stages of development, an absence of volcanic activity. Sediments deposited on the continental shelves of passive margins are compositionally and texturally mature. Instead of mainly being transported immediately into deep water by turbidity currents, as at active margins, passive margin sediments move slowly across the continental shelf in response to marine currents, and they accumulate along the continental slope and continental rise.

Transform Margins A number of distinguishing features characterize transform margins, the most prominent of which is the absence of igneous activity. However, these margins are very active seismically, with shallow earthquake hypocenters. Motion on the faults is more or less pure strike-slip, with the movement distributed across a number of parallel fault lines. The faulting has a strong influence on stream drainage patterns, the lateral movement commonly causes streams to be progressively offset, and the source areas for sediments dislocated some distance from the depositional basins. Sedimentary basins are typically small and usually bounded on both sides by faults. Because of the fault boundaries, the basins are generally fragmented by fault movement. If the sedimentary sequences survive at all, they will survive as wedges of sedimentary rock caught between faults.

Fault movement causes rapid local uplift and subsidence and creates abrupt discontinuities among stream sizes, drainage patterns, or topography on the two sides of the fault trace. Streams on one side of a fault may not have created a drainage basin and a sediment distribution system. Rocks of different hardnesses might be juxtaposed on opposite sides of a fault, yielding different river-channel characteristics, plant communities, or soil and slope features. Thus, sediments derived from one side of a fault might be simply dumped locally in small basins on the opposite side of the fault.

Active-margin sediments form thick deposits of immature sediments with a high concentration of igneous components. Passive-margin sediments tend to be compositionally and texturally mature. Transform margin sediments are immature and diverse in composition.

Living with Geology

Damming a Wild River

Any discussion of the construction of a dam across a "wild" river usually pits conservationists against power companies and farmers. One side stresses the need to conserve for future generations; the other stresses the needs of the modern population and economy.

Prior to the construction of a dam on the Trinity River near Mt. Shasta in northern California, the river was a typical youthful mountain stream; its bed was rocky and strewn with boulders, and only minor brush grew along the river edge. It was a clear, cold stream with a large resident population of trout and annual spawning runs of steelhead and salmon. The stream bed was kept free of sediment and brush by vigorous floods derived from the spring thaw in the high mountains nearby. This mountainous region was a wilderness, with little industry except logging. Downstream, however, was a growing population and economy. Energy needs of the area grew, and the spring floods became more than just a nuisance—they were a threat to safety. A dam was built to control the floods, conserve the water for later use in irrigation, and as a hydroelectric facility.

After construction of the dam, the spring floods were contained by the new Trinity Lake. Control of the river meant that regular water delivery could be provided, and irrigation for farms and ranches downstream was a boon to agriculture. Power generated from the hydroelectric plant is completely nonpolluting and provides a substantial proportion of the energy needs of the area. But there are no spring floods to clean the river of sediment, weeds, and brush. Instead of a clear, fast-moving, cold stream, the Trinity River has become sediment-laden, slower moving, and warmer. It is less hospitable to the prized game fish that formerly populated the river. There are no longer any salmon or steelhead runs up the Trinity River.

Trinity Lake, formed behind the dam, is a popular recreational area that attracts many people for boating, water skiing, and fishing. Agriculture and the cities and towns downstream benefit from the reliable supply of water and power. The lake has an excellent resident population of trout, and because of the good fishing, the surrounding forest supports a growing population of osprey—a hawk that many conservationists considered on the brink of extinction.

SUMMARY

- Sediments are both mixtures of mineral grains that have been carried to the site of deposition and minerals that have precipitated (crystallized) in the sediment after deposition.
- Sediment grains are transported in suspension, by saltation, and by traction. The settling velocity of a grain and the extent of turbulence determine the transport mechanisms governing the movement of any grain.
- A graded stream is in equilibrium; it has the exact slope necessary to transport the water and sediment supplied to it.
- A braided stream is divided into many strands that constantly shift position. These streams have a coarser sediment bedload, and the valley floor is quite flat. Meandering streams have a finer sediment bedload and a single river channel.
- Alluvial fan deposits form where streams flow out of narrow mountainous channels onto a desert plain, resulting in a sudden decrease in the flow. The remaining areas of desert are mostly rocky, with small areas of sand dunes.
- Deltas form when the river velocity decreases upon meeting the ocean and the sediment load is immediately deposited. Deltas are shaped by the interaction among the river and marine currents, waves, and tides that disperse the sediments.
- Tidal flats range from coarse-grained sediment in the lowest part of the tidal flat, where sands are transported as bedload by strong currents, to fine muds inshore, where sediments fall out of suspension as the tidal current stops.
- Beaches represent the junction between land processes and marine processes. Wave action causes winnowing of fine particles, polishing and rounding the sediment grains on the beach. Longshore sediment transport can result if waves approach the coast at an angle.
- Carbonate sediment originates from shells and carbonate-secreting plants. Thus, carbonates are clastic sediments in the sense that they have been eroded, transported, abraded, and sorted.
- Carbonate oozes occur in equatorial sediments; siliceous oozes are deposited in narrow bands north and south of the equator. Both types result from the accumulation of the shells of microscopic organisms. Fine-grained clays derived from the continents are red or brown after deposition owing to oxidation and to the lack of organic matter.
- The heads of submarine canyons are close to shore, so that sediment carried by longshore currents may be trapped within them. After sufficient sediment has accumulated in the canyon, a landslide in the canyon may trigger a turbidity current that can carry sediment out of the shallow marine zone into the deep sea, forming a submarine fan.
- Active margin sediments form thick deposits of immature sediments with a high concentration of igneous components. Passive-margin sediments tend to be compositionally and texturally mature. Transform margin sediments are immature and diverse in composition.

KEY TERMS

detrital, 325
authigenic, 325
settling velocity, 326
suspension, 326
traction, 326
saltation, 326
bedload, 327
capacity, 327
competence, 327
viscosity, 327
bedform, 329
graded stream, 331
base level, 331
perennial river, 334
ephemeral river, 334
alluvial fan, 340
turbidity currents, 348
turbidite, 349

QUESTIONS FOR REVIEW AND FURTHER THOUGHT

1. What is the difference between detrital minerals and authigenic minerals? Can you think of at least one criterion that might be used to distinguish these two types of minerals?
2. Why is turbulence in flowing fluid important to sediment transport?
3. Bedforms are irregularities in loose sediment under a flowing fluid (either wind or water). Describe the differences between bedforms that are characteristic of slowly flowing fluids and those that are characteristic of fluids flowing at higher velocities.
4. The headwaters of rivers are steeper than lower reaches of the same river. Why is this always the case?
5. What might happen in a river that flows through an area that has been recently burned or logged? Over what time span would you anticipate this effect to last?
6. It is not uncommon in deserts to find large boulders much farther away from the front of a mountain range than you might expect them to roll due to gravity alone. Can you come up with a hypothesis to explain this observation? How would you test your hypothesis?
7. Some marine animals move across soft sediment or use it as housing or as a place to search for food. What effects might the biota have on sediments? Describe the clues you might look for to explain these effects.
8. Most carbonate sediments are clastic in the strict sense of the term. Explain this assertion.
9. Explain the mechanism of turbidity-current sedimentation. Why can turbidity currents transport relatively coarse sediment?

14 Geomorphology: The Study of Landforms

Arid climate and flat-lying sandstone layers produce the characteristic canyon and butte scenery of the southwestern United States, such as this view of Canyonlands, Utah.

Overview

- Landforms vary in character from region to region in response to changes in climate and geology.
- Natural agents responsible for shaping Earth's surface are rain, wind, rivers, glaciers, ocean waves, and gravity.
- Surface material may move downslope under the influence of gravity.
- Weathering, erosion, and deposition of surface materials are generally controlled by the dynamic actions of water in its different forms (rain, rivers, ice, oceans), as described by the hydrologic cycle.
- Landscapes constantly change (evolve) over time.
- Because plate tectonics affects topography and geology, it also influences geomorphology.

Introduction

We are all familiar with the magnificent array of scenery on Earth's surface. It varies from the ice-covered wastes of the polar regions to dry deserts to steamy jungles. *Geomorphology* is the study of the varied landforms on the surface of Earth. Even within the United States, striking contrasts exist between the spectacular arid and rocky scenery of the Southwest, the glaciated alpine mountains of the West, and the wooded rolling hills of the Northeast. What is responsible for these varied landscapes?

Three primary influences affect the type of scenery that may be produced at a given point on Earth's surface. The first is the local geology—the types of rocks and structures that make up the land. The second is the local climate. You know from Chapters 6 and 12 that climate varies with latitude; it is warm and wet at the equator and cold and dry at the poles. Climate also varies as a result of atmospheric circulation. The arid deserts of the world, for instance, are located in two subtropical belts approximately 30° north and south of the equator, where air originally heated at the equator descends and forms a hot, dry, high-pressure region.

Time is the third major influence on landforms. We have learned that Earth's continents are moved around on tectonic plates, slowly changing position on the surface of the globe. Accordingly, the local climate will change with time, and so will the landscapes that are formed. As a result, continents now located in tropical regions may show evidence of past glaciations, whereas deposits believed to have formed in deserts may now be exposed in the arctic (∞ see Chapter 6).

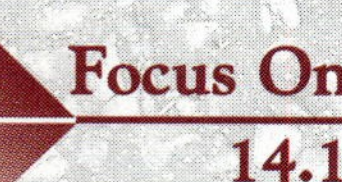

Focus On 14.1 Fractals and Scale Invariance in Geology

When a geologist takes a photograph of a rock outcrop, a familiar object such as a hammer or a pen is included in the photo to identify the scale of the photo. Folds, for example, occur over a vast range of sizes and, without such a scale, folds in a hand sample could be misidentified as huge folds associated with mountain building. Topography, which is largely formed by weathering and erosion, looks similar on the scale of the surface of a grain or on the scale of the surface of the planet. This behavior of patterns that exhibit the same general appearance irrespective of the scale is called *scale invariance*. It arises because the laws of chemistry and physics have no intrinsic scale. Consider Newton's third law: force = mass × acceleration. There is nothing in this law about size; it applies equally well to the atom as it does to bodies the size of stars. Thus, when a force associated with physical weathering breaks rock of uniform composition, fracturing occurs on all scales. The weathering products range in size from dirt to huge boulders, leaving behind a similar size distribution of erosion features such as ridges, channels, gullies, and valleys. Such scale-invariant patterns are called *fractals*, and the ratio of the number of elements at one scale divided by the number at another scale is called the *fractal dimension*. The fractal dimension is a measure of the preponderance of small-scale phenomena over large-scale phenomena.

Numerous geologic phenomena are distributed as fractals. Some important ones include rock fragments, faults, folds, earthquakes, volcanic eruptions, topography, and river networks. A good example is illustrated in Figure 10.8, which shows a model of a sheared block of clay, compared with a large strike-slip fault in Iran. The distribution of fractures is remarkably similar despite the fact that the scale differs by a factor of nearly a million! There are, however, a greater number of events or features at the small scale than at the large scale. If rock fragments from a weathered outcrop are passed through a series of sieves with progressively smaller holes and the number of fragments passing through each sieve is counted, one obtains many more sand grains than stones, more stones than cobbles, more cobbles than boulders, and so on. Similarly, there are more small earthquakes than large ones, more small volcanic eruptions than big ones, more gullies than valleys, more small streams than rivers.

To calculate the fractal dimension of, for example, sieved rock fragments, the number (N) with dimension greater than r is found to satisfy the relation

$$N = C / r^D$$

where D is the fractal dimension and C is a constant that depends on the total number counted.

Landforms are controlled by the local geology and climate, and by time.

In this chapter, landscapes are divided into a number of distinct geomorphologic environments, although we must recognize that such a subdivision is rather arbitrary. There are rarely abrupt changes in scenery as you drive across a continent; rather, the changes are more gradual in nature. These geomorphologic environments are discussed in terms of the two processes that shape landforms: *erosion* and *deposition*. For a discussion of the fundamental principles of weathering, erosion, and deposition in sedimentary systems, refer back to Chapters 6, 12 and 13.

It is important to point out that erosion, rather than deposition, will always be the dominant process on balance, because we are discussing regions that are on dry land. Remember that the forces of weathering and erosion (rain, wind, rivers, glaciers, and so on) are constantly working to wear down Earth's surface on all scales (see Focus 14.1) and to transport the eroded sediment ultimately into the ocean basins. Erosion is therefore the primary factor that sculpts the land surface—for instance, cutting cliffs along coastlines and steep canyons in deserts. Nevertheless, deposition does occur locally. Sand dunes in deserts, moraines in glacial environments, and floodplains in river valleys, provide a secondary modification of the landscape. These deposits, although not very common in the overall geologic record, are important because they are the evidence of past climates and environments on different continents.

Before examining different geomorphologic environ-

Generally, the logarithm of N is plotted against the logarithm of r and the slope of these data gives D. For the case of sieved rock fragments, studies show that $D = 2.6$. In this case, the equation tells us that the ratio of 1-cm fragments to 2-cm fragments is

$$\frac{N(1\text{ cm})}{N(2\text{ cm})} = \frac{C/1^{2.6}}{C/2^{2.6}}$$

$$2^{2.6} / 1^{2.6} = 6$$

Similar calculations can be made for undulations in topography, sizes of folds, rivers, numbers of eruptions, or numbers of earthquakes. The Gutenberg-Richter relation (Figure 7.15), which is derived from a plot of the number of earthquakes above a given magnitude versus magnitude, is a fractal distribution. The slope of the plot is the fractal dimension (D), which has a value of about 1 for earthquakes.

The fractal behavior breaks down if either material properties or the physical law governing the process changes. For example, the size of plate boundaries would limit the size of the largest fault or shear zone and we would not expect scale invariance for larger structures. At the other end of the size scale, below the size of atoms, fractures have no meaning and different physical laws apply. Also, some geologic processes act to destroy fractality. Processes such as sedimentary sorting may have an effect similar to sieving, with the result that in a given area, one size of sediment grain may dominate—an effect similar to obtaining just one sieved fraction of sediment. For example, in a stream bed, sand or gravel particles might have similar sizes depending on the flow properties of the stream. Smaller fragments have been carried away, whereas larger fragments have been left behind at the source of erosion (∞ see Chapter 13). Taken together, all the erosion products in the sedimentary system are fractal, but samples from the stream bed alone would appear to be nonfractal.

The study of fractals is a new and a rapidly developing field. The fractal dimension has proved very successful in providing a quantitative measure of such physical features that hitherto had appeared random. Modern research is involved in determining what basic physical and chemical interactions give rise to the fractal dimension for the different phenomena, and what limits the scale length range over which fractality is exhibited.

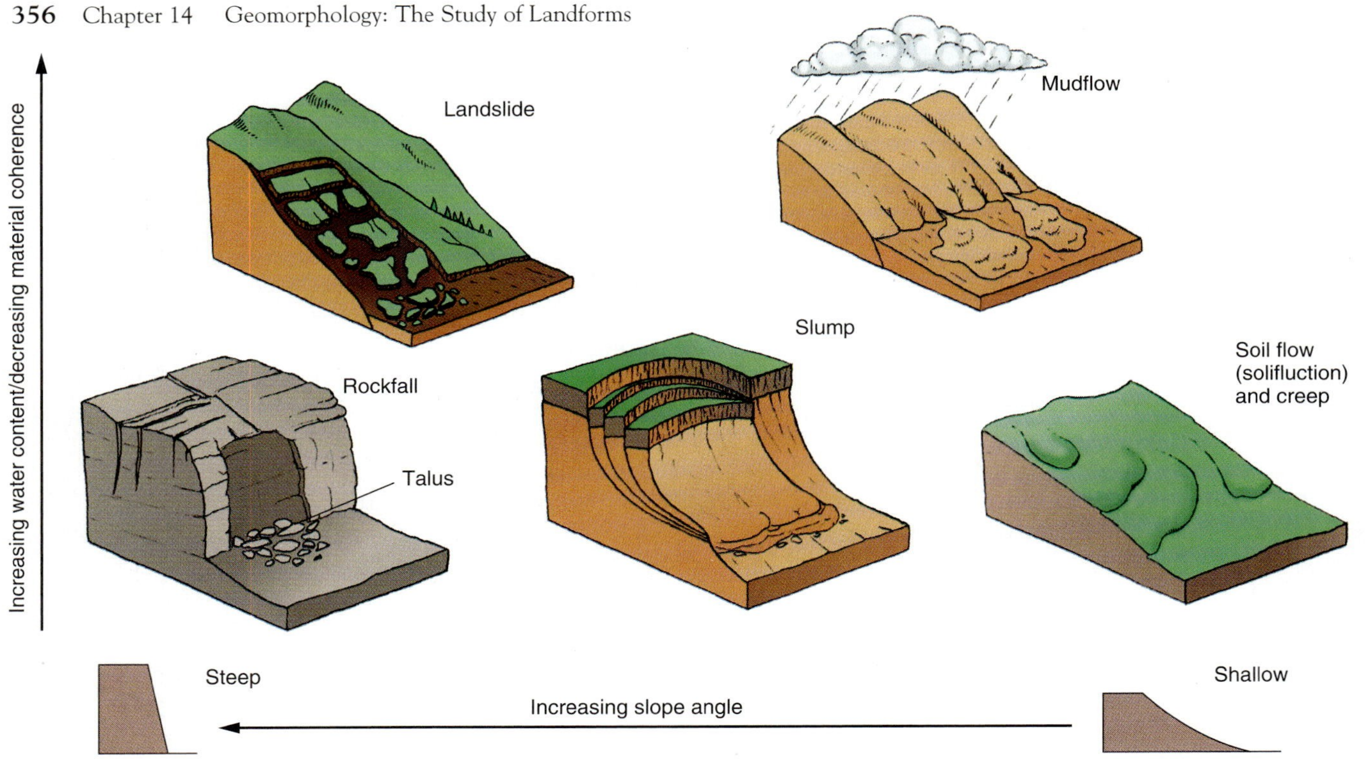

FIGURE 14.1 Different types of mass wasting. The most important factors influencing mass wasting are the steepness of the slope, the coherence of the material on the slope, and the amount of water involved. The most likely form of mass movement on a steep slope is a fall; creep dominates on a shallow slope. The style of mass wasting on intermediate slopes depends on the water content of the soil.

ments, we must explore the roles of gravity (mass wasting) and running water (the hydrologic cycle) in shaping the planet's surface.

Mass Wasting

We appreciate that water and ice will move downhill and transport sediment. Solid material may also move downslope by itself, in response to gravity. **Mass wasting,** the term applied to this downslope movement of material on Earth's surface, can take many different forms, depending on the nature of the movement. In simple terms, we can divide gravity-driven processes into falls, slides or slumps, flows, and creep (Fig. 14.1). The mechanism of downslope movement depends on the slope of the hillside, the coherence of the material, and the water content. Focus 14.2 looks briefly at the forces acting on a hillslope to explain quantitatively how mass wasting occurs.

The Importance of Water in Mass Wasting

Water is an important agent in the process of mass wasting. Water not only acts to weaken materials on a slope but it also adds weight and lubricates existing planes of weakness. Under normal conditions, a hillslope may be stable, meaning that the friction holding the soil on the hillside is greater than the forces tending to pull it downslope. If we look at the same hillside after a protracted rainy period, we will find that the balance of forces has changed. The friction is less because the water "lubricates" the sliding surface (Fig. 14.2).

Small amounts of water can strengthen soils. Because of surface tension, the water gives the soil a great deal of cohesion. If the soil is too dry, it is easily eroded. Too much water forces the soil grains apart and reduces cohesion to near zero. You can demonstrate these effects with a simple sandbox experiment. Make a castle of dry sand. It will not stand very high. If you dampen the sand, you can form a steep castle because of the increased cohesion. Form a depression in the top and add water until the sand becomes *waterlogged*, and it will eventually collapse. If you try to rebuild the same hill using the waterlogged sand, you will find that the wet sand is unstable on any slopes except a very slight one. As a consequence of the potential weakness of waterlogged hillsides, the most effective method to prevent or control landsliding is to drain the water out by drilling horizontal boreholes into the slope.

Falls

Rockfalls occur in regions of very steep slopes, where it is possible for rocks to fall free, such as off cliff faces. If material is loosened by rain, plant-root wedging, frost wedging, or by earthquakes, then it may simply fall from

THE PHYSICS OF MASS WASTING

Consider an object lying on a sloping surface. The main forces acting on it are gravity, which pulls the object directly downward toward the center of Earth and friction, which holds the object in place. The gravitational force is the weight of the object, which is dependent on its mass:

$$W = mg$$

where W = weight, m = mass of the object, and g = acceleration due to gravity. Let the surface make an angle θ with the horizontal (Fig. 1).

We can resolve the gravitational force W into two vectors at 90° to each other (see Focus 6.4): one acting perpendicular to the slope—$W \cos\theta$—and one parallel or tangential to it—$W \sin\theta$. You see that the tangential vector pulls the object downhill. But there is a component of force preventing such movement; this is the friction f acting at the surface of contact between the object and the slope. Friction is proportional to the perpendicular force on the slope, with the constant of proportionality μ called the coefficient of friction, that is:

$$f = \mu W \cos\theta \qquad (1)$$

For the object to be stationary, the frictional force up the slope must equal or exceed, the tangential force down the slope, or Force down the hill ≤ Friction:

$$W \sin\theta \le f \qquad (2)$$

Using Equation 1 for f, we find:

$$W \sin\theta \le \mu W \cos\theta$$

Dividing both sides by $W \cos\theta$, and noting that $\sin\theta/\cos\theta = \tan\theta$:

$$\tan\theta \le \mu$$

The object will slip when the tangential force is greater than the friction, or when $\tan\theta > \mu$.

From this equation, you can see that you could get the object to move by either increasing the angle of slope (which will also make the tangential component of the resolved vector greater) or by decreasing the frictional force (μ); for instance, by lubricating the surface of the slope.

You can easily verify this yourself using a coffee cup and a table! Put a coffee cup (or any other suitable object) on a table, then tilt the table by raising one side slowly. At a particular critical angle, the cup will begin to slide. At greater angles of slope, sliding will occur; at lesser angles, the cup is stable. If you lubricated the surface—for instance, by adding a film of soapy water—you would see that the angle at which the cup begins to slide is decreased. The same principles apply to soil and rocks on a hillside that may move due to mass wasting.

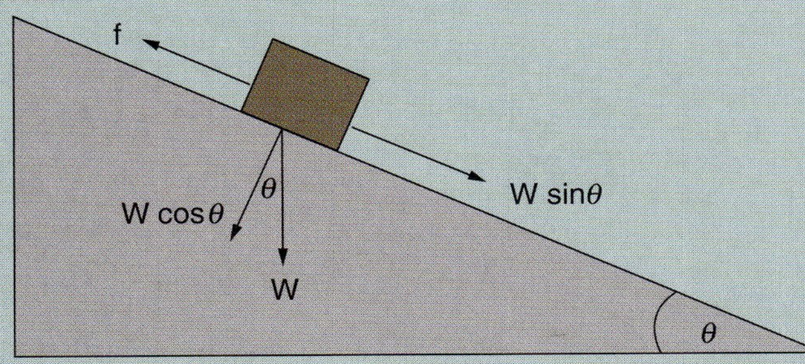

FIGURE 1 A block resting on a plane inclined at angle θ to the horizontal has a weight W. The block is pulled down the hill by a force $W\sin\theta$, which must be resisted by an equal frictional force of $\mu W\cos\theta$ *up* the plane.

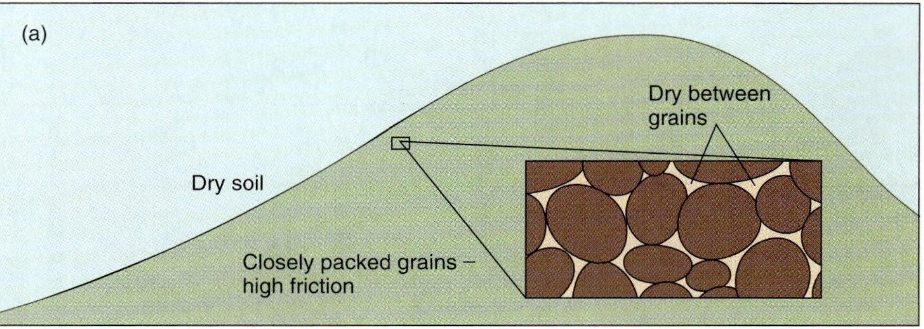

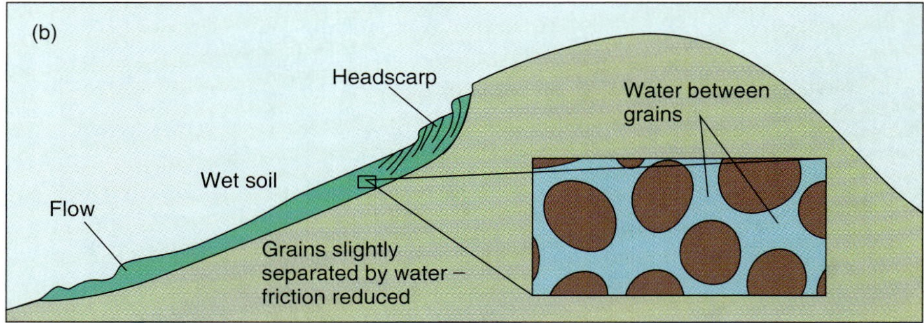

FIGURE 14.2 The effect of water in mass wasting. (a) The hillslope is dry. (b) The same hillslope is wet (after rain) and the spaces between soil particles are filled with water. Individual particles on the hillslope will slide more easily relative to each other as the water supports their weight better by lubricating the loose soil. Friction is lessened and the result is a landslide or slump.

the cliff face. The accumulation of broken material at the base of steep slopes or cliffs is known as **talus**. As the rocks fall and roll downslope, the larger rocks tend to roll farthest. For that reason, talus slopes are crudely sorted in that the largest rocks tend to be the greatest distance from the cliff face simply because they have greater momentum, and because of their size, they are less likely to become lodged among smaller ones, hence tend to override them. The smaller particles accumulate near the base of the cliff (Fig. 14.3). The maximum stable inclination that a loose pile of material can sustain is called the *angle of repose*. This angle varies with the moisture content, the size and sorting of the particles, and their angularity, but is typically up to about 30°.

Slides and Slumps

In contrast to a fall, where material plummets freely through the air, the material involved in a **slide** maintains contact with the slope down which it moves. As with rockfalls, landslides can move very rapidly and can have disastrous consequences (Fig. 14.4). However,

(a)

(b)

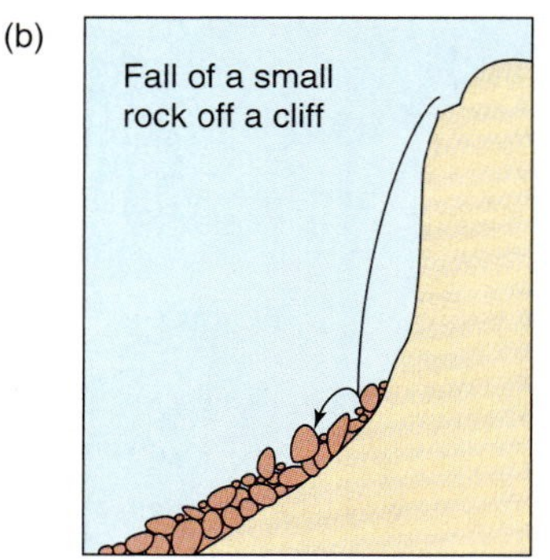

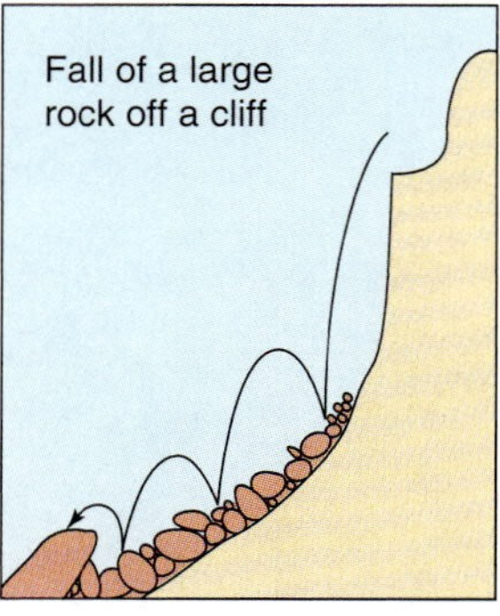

FIGURE 14.3 (a) Rockfall-generated talus slopes in Glendalough, Ireland. Note the very large boulders at the base of the talus slope. (b) Illustrations of the contrast between small fragments falling on a steep talus slope and large fragments falling on an identical slope. The small fragments stop at the top of the talus slope or quickly become lodged in the interstices between larger blocks; larger fragments will roll down the slope.

FIGURE 14.4 A typical landslide in Lost Trail Creek.

(a)

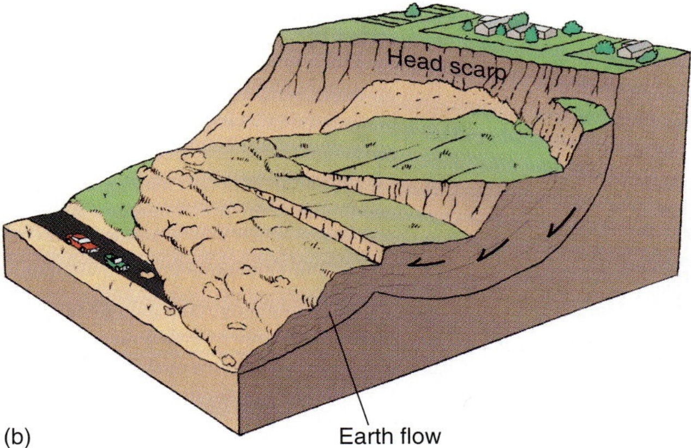

(b)

FIGURE 14.5 (a) Slump in roadcut clearly showing the head scarp, the side slip surfaces, and the toe of the slump. The cause was a combination of an oversteepened slope, too much moisture, and cohesionless soil. (b) Illustration of the structure of a slump.

slides also may move with imperceptible slowness, with movement occurring episodically.

One common type of slide is called a **slump**. These are interrupted landslides, which means that they do not slide far. Slumps have a characteristic "scoop" shape, with the top of the slump marked by a head scarp where the slump has detached itself from the upslope material (Fig. 14.5). The toe is at the bottom of the slump, comprising deformed material that has been pushed downhill in bulldozer like fashion by the slump, overriding the hillslope below. The toe of the landslide may be so unconsolidated that it flows. The motion of a slump is a rotational one rather than a simple slippage parallel to the slope surface; this rotation causes a subsurface scoop or spoonlike shape. It is not necessary for slumps to move far to cause serious damage.

Flows

During a **flow**, the material in motion—rock, mud, or soil—is mixed with enough water so that the mixture behaves more like a viscous liquid than a solid; it is fluidized. These flows occur in much the same way as landslides, but the soil–water mixture flows as a mudflow and commonly deposits large volumes of mud downhill or downstream. Usually such flows occur suddenly, and, depending on the amount of material in the flow, the results can be catastrophic (Fig. 14.6a). A small area on a hillside may become so saturated with water that it loses all cohesion and spontaneously flows downslope.

Mudflows associated with volcanoes are known as **lahars**. Volcanoes characteristically have steep sides and are often associated with large accumulations of loose pyroclastic material. Furthermore, many volcanoes loom high above the surrounding land, so that their summits are affected by voluminous amounts of rain or snowfall (∞ see Chapter 15).

In arctic regions, a somewhat similar process occurs, where soil flows downhill but very slowly. These flows form distinct lobes with a shape rather like that of a mudflow lobe (Fig. 14.6b), and they occur during the short arctic summer. The mechanism is straightforward. The uppermost layer of soil thaws and becomes water-saturated. Immediately below the water-saturated soil zone—generally within one meter—is a layer of permanently frozen soil, called permafrost. The water-saturated soil has no strength, and this mud flows downslope. These flows are sometimes called *mud glaciers*, but the name given to the process is **solifluction** (from the Latin *solum*, meaning "soil" and *fluere*, "to flow"; hence, "soil flow").

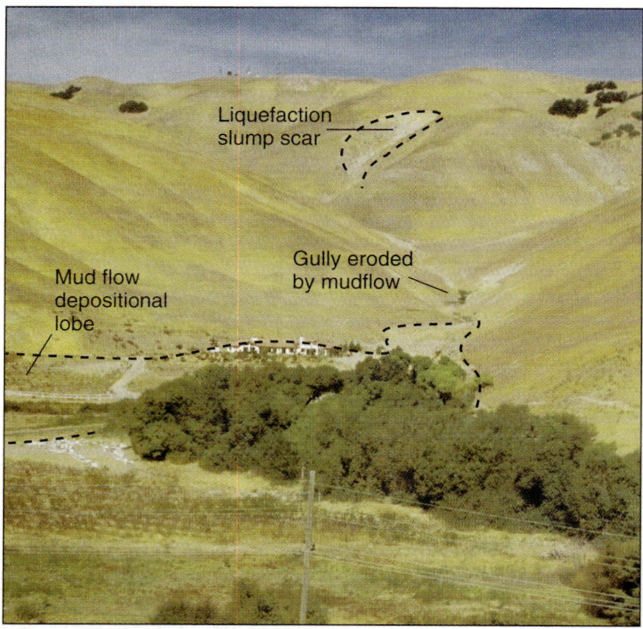

(a)

(b)

FIGURE 14.6 Examples of mass flows (a) A water-saturated mudflow triggered by a liquefaction landslide deposited the flat area on which the house is built; southern Tehachapi Mountain, California. (b) Solifluction lobes in Rocky Mountain National Park, Colorado. Three solifluction lobes are outlined, but there are many in the photograph.

Soil Creep

Loose soil on hillsides has the tendency to move downslope, either as a mass or as individual particles. Even when there is strong and cohesive bedrock at the surface, the entire soil profile that extends from a few centimeters to more than a meter in depth may be slowly moving downslope under the force of gravity, a process called **soil creep** (Fig. 14.7).

Why does the soil move downhill? Most soils expand as moisture is absorbed. Each soil particle moves outward perpendicular to the slope as the water content of the soil increases. As the moisture content decreases, the soil contracts. But during shrinkage, the path of each particle is pulled vertically downward by gravity, so each soil grain moves slightly downhill from its original position (Fig. 14.7a).

As the soil moves downslope, fence posts may be pushed over until they stand out from the hillside nearly horizontally. In the same way, trees are also pushed over. However, trees do not continue to lean downslope because they have an inherent tendency to grow vertically upward. Thus, although a tree is continually pushed over by soil creep, the top of the tree continually corrects this by pointing its growing tip upward. The result is that the trunk becomes bent, and as this process continues over time, the bend in the tree trunk becomes more extreme (Fig. 14.7b).

Mass wasting is the downhill movement of material under the influence of gravity. Material can fall, slide, or flow, and rates can be catastrophically fast (rockfall, landslide) or extremely slow (creep).

Shaping the Surface: The Hydrologic Cycle

Although mass wasting can significantly modify slopes and transport material over short distances, there is a limit to its influence on the landforms that we see on Earth's continents. Mass wasting also affects the surfaces of other planets in our Solar System, for all have a significant gravitational force acting on any surface slope. The principal agents shaping Earth's surface are actually wind and water—particularly water, in the form of rain, rivers, glaciers, and waves. The role of water in determining the physical characteristics of the planet's surface cannot be overstressed. Earth is unique in the Solar System in this respect. The hydrosphere—liquid water on the surface and water vapor in the atmosphere—not only influences climate and geomorphology but also supports organic life, which itself may affect landforms.

The water of the hydrosphere (∞ see Chapter 1) is found in a number of different reservoirs. By far the most important are the oceans, which account for more than 97 percent of Earth's water. The freshwater reservoirs can be divided into land ice (ice caps and glaciers), groundwater (subterranean water found in the pore spaces of rocks and soils), rivers, and lakes. The importance of water as a resource will be discussed in Chapter 16. You may recall (∞ see Chapter 1) that the constant transfer between the various reservoirs is referred to as the *hydrologic cycle*. In the hydrologic cycle, a major process is evaporation—mainly of ocean water—to form clouds. In turn, this evaporation causes precipitation of either rain or snow, depending on latitude and elevation. Rainfall may either seep into the soil (*infiltrate*) to join the groundwater reservoir, which flows underground and can

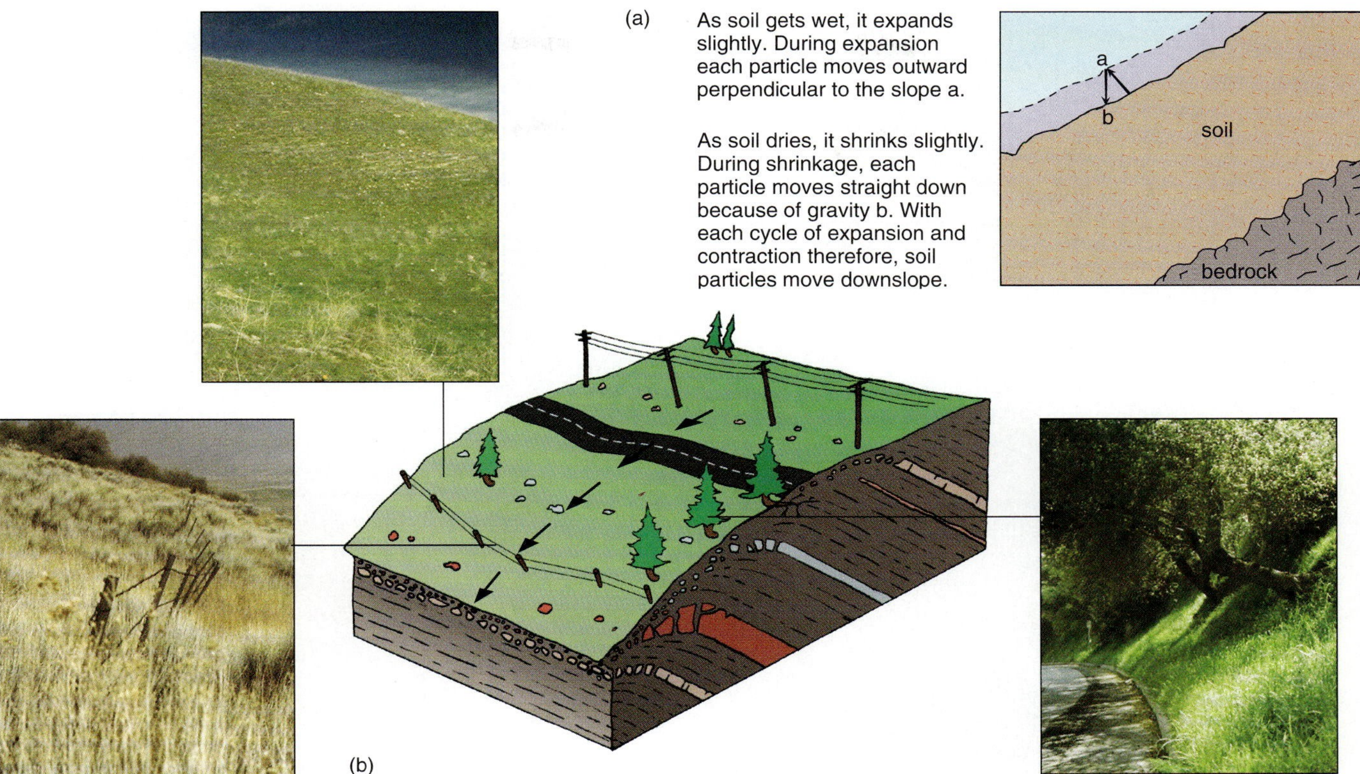

FIGURE 14.7 Soil creep: The slow downhill movement of soil and loose material. (a) The mechanism of soil creep. (b) Possible effects of soil creep. (Bottom left) Originally upright fence posts leaning outward from hillside as a result of upper surface of soil moving downslope. (Top left) Soil terraces that result from downhill soil movement. (Bottom right) Bent trees that have resulted from soil creep leaning the tree outward, as with the fence posts, followed by the tree growing vertically upward.

exit to the surface at springs, or run off and collect in rivers and lakes. Snow may accumulate to form glaciers, which help to shape cold highland regions as these glaciers move slowly downhill. Ultimately, all of the water on or just below the land surface will return to the oceans or will evaporate back into the atmosphere.

We have introduced the characteristics of sediment transport and deposition in moving fluids, such as water (∞ see Chapter 13). For our purposes in this chapter, it is important to appreciate that moving water—either as rivers, glaciers (rivers of frozen water), or waves—represents an enormous source of energy and is the principal means of erosion, sediment transport, and deposition shaping the planet's surface.

You may recognize a similarity between the hydrologic cycle, and the rock cycle (∞ see Chapter 4), which illustrates the ways in which rock types may be transformed with time. Both cycles consist of components that are linked by energy-driven processes (Fig. 14.8). The energy driving the rock cycle is mainly the internal heat of the planet—the same energy that provides the forces responsible for plate tectonics. In contrast, the energy that drives the hydrologic cycle is external to the planet; that is, solar energy interacting with the hydrosphere. The hydrologic and rock cycles are closely linked in that the interaction between rocks and the hydrologic cycle leads to weathering, erosion, and sedimentation, producing the varied landforms on Earth's surface.

The hydrologic cycle works on a shorter time scale than does the rock cycle, and therefore its effects are much easier for us to appreciate given our relatively short lifetimes. It is possible for the hydrologic cycle to turn over in a few days or years. For example, water evaporated from the oceans may form clouds that are blown overland. The clouds release their water as rain. After the rain, streams collect the water, which may run immediately back into the ocean, thereby completing the cycle in just a few days.

The rock cycle generally operates over much longer time scales. Consider a sediment that has been deposited in the marine environment along a convergent margin. It will be covered by succeeding layers of sediment so that it is buried ever more deeply. Diagenesis will lithify the sediment to form a sedimentary rock (∞ see Chapter 4). Later, the rock body that was our original parcel of sediment may be uplifted, weathered, and eroded, forming detrital grains that are then deposited as another sediment. This process, which you will recognize as part of

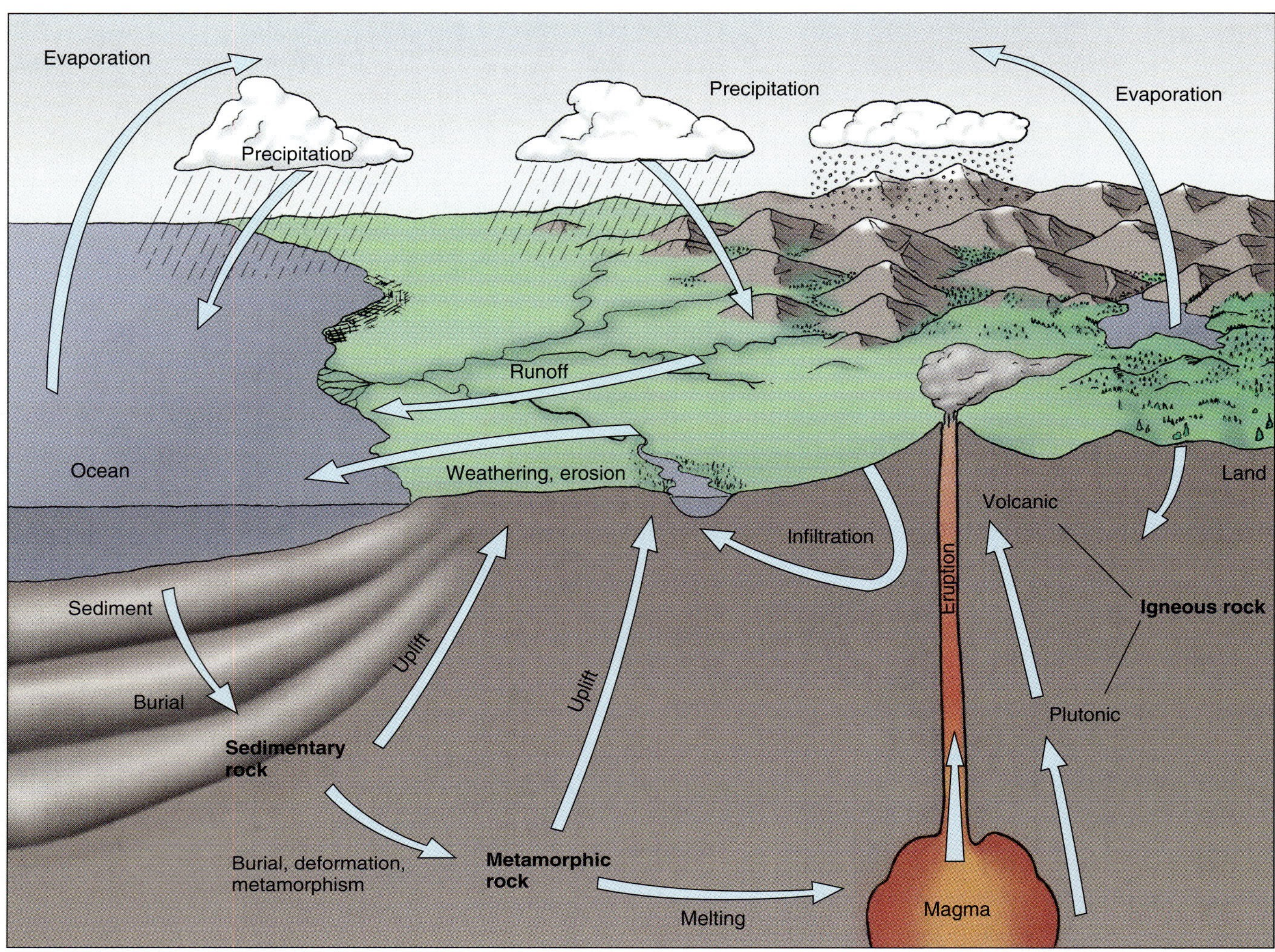

FIGURE 14.8 The rock cycle and the hydrologic cycle are closely linked. Note that they both share surface processes—the weathering and erosion that are consequences of the hydrologic cycle make sediment from solid rock. Notice also that water is actually cycled through the rock cycle. Water dissolved in the mantle is outgassed at volcanic eruptions, and water contained in minerals of the ocean crust is returned to the mantle through subduction.

the rock cycle (see Fig. 14.8), might take a few million years or more to complete.

The following sections will examine landforms that characterize the surface of the continents; we will see how they are related to climate (the hydrologic cycle) and geology (the rock cycle).

Fluvial Geomorphology

One of the most common landforms is the valley. This feature is carved by small streams and rivers, and takes a wide variety of forms—from wide broad plains, such as the lower Mississippi Valley, to steep-sided canyons, such as the Grand Canyon of the Colorado River (Fig. 14.9). The shape of a river valley depends on many factors such as the climate, the slope of the land surface, and the nature of the soil or rock over which the river flows. The valley will change both in location (depending on whether it is near the source or the mouth of the river) and with time.

We have discussed the downstream slope of rivers and streams and developed the concept of a steady state gradient called *grade* (∞ see Chapter 13). The slope of a valley along the length of a stream is steepest near its source and most gentle near the mouth. At the same time, the shape of the valley (the cross-sectional profile) also changes downstream. It is steep and V-shaped near the source; farther downstream, it is much more open, with a wide, flat bottom. The shaping of stream valleys is due to a combination of erosion by flowing water and mass wasting. By itself, the stream could only erode a nearly vertical notch. Mass wasting—slumps, landslides, falls, and soil creep—widens the upper portion of the valley to form the characteristic V shape of stream canyons (Fig.

14.10). Near the source, the stream is a bubbling torrent. The steep slope means that it will erode its bed and carry material away from the uplands. The valley floor is uneven, rocky, and often characterized by waterfalls. Toward the mouth, where the gradient is much more gentle, the stream begins to deposit much of the sediment that it has transported from upland regions. The river lays down a carpet of sediment at the bottom of its valley. Because the valley is wide and flat, the river will tend to flood its own valley frequently and will thus build up its own banks or levees (∞ see Chapter 13).

Geomorphic Age

The geomorphology of a region in a continental interior can be described as youthful, mature, or old age, based on the proportion of upland surface, valley slope, and valley bottom. A *youthful landscape* is characterized by having large tracts of relatively flat, uneroded upland; a small proportion of the area consists of valley slopes, and the bottoms of the few valleys are filled with rivers (Fig. 14.11a, page 366).

Mature landscapes have undergone much more stream erosion; there are few upland surfaces remaining, and most of the area consists of valley slopes (Fig. 14.11b). The stream bottoms are widened by lateral movement of and erosion by the rivers, with the development of river floodplains. Eventually the area is reduced to a flat plain, over which rivers slowly meander, depositing extensive veneers of sediment upon the floodplain as the channels shift laterally.

An *old-age landscape* possesses no upland area and consists mainly of valley bottoms with broad river floodplains; very little of the area consists of valley slope (Fig. 14.11c). The rivers flood often; oxbow lakes and swamps are common. These stages in the development of a given landscape through time often occur simultaneously at different sites downstream along a river valley. The headwaters of a river system may be youthful, whereas the mouth of the river may be at the mature or old-age geomorphic stage. Thus, geomorphic age is local, reflecting erosional maturity and not simply the age of the river.

What happens when this landscape evolution is interrupted by a change in the base level of the river systems due, for instance, to tectonic uplift or a global fall in sea level? The increase in potential energy triggered by the relative decrease in the river's base level causes rejuvenation (∞ see Chapter 13).

If the river channel is well established, then it will be cut deeply (incised) into the landscape. In the case of a river that has reached a late meandering stage, incised meanders will form. Recall that meanders develop because of lateral shifting of a stream's course over that flat floodplain. If the gradient is increased, the tendency of the stream will be to cut downward rather than sideways, but the meandering course that has already been established will be maintained. Many of the rivers of the Colorado Plateau region are clearly incised. The Colorado Plateau was originally close to sea level, so rivers meandered over wide, flat floodplains. Uplift began about 5 million years ago and the river channels were incised. The ultimate expression of such rejuvenation is the Grand Canyon, which has been cut down through nearly 1850 m (about 6000 feet) of rock strata in the last 5 million years as the plateau has risen (see Fig. 14.9). A stream system that preserves characteristics developed on a preexisting land surface is called an *antecedent stream*.

Landscapes evolve through time as erosion reduces relief. At any time, tectonic forces may interrupt the process, causing uplift and rejuvenation of the erosive agents.

Effects of Climate

Distinct differences exist in the features, or forms, of valleys in different climatic settings. In arid climates, mass wasting of the valley sides is limited and steep-walled canyons result. The lack of water, and the consequent absence of vegetation, prevents the formation of soil on the valley walls. Occasionally, the valleys will widen by rockfalls. The stream bed is usually dry for most of the year in arid desert regions, with erosion restricted to periods of flash flooding. In more temperate climates, the valley sides will degrade to more modest slopes, supporting soil and vegetation.

The year-round (perennial) streams that characterize temperate climates have valleys controlled to a greater extent by mass wasting (∞ see Chapter 13). The zone of groundwater saturation more or less parallels the topography in areas of significant rainfall, sloping downhill to intersect the land surface at the level of the river. Thus, rain infiltration increases the soil moisture content to saturation far more rapidly than happens in arid climates, where groundwater is deep underground. Erosion is a direct function of the relation between infiltration and runoff; the greater the runoff, the greater the erosion. However, greater rainfall also causes a thicker vegetative cover. Vegetation inhibits erosion by holding the soil in place and increasing the amount of infiltration relative to runoff. Although the hillsides in areas of higher rainfall are more heavily vegetated than those in deserts, landsliding may occur when the stream banks are steepened by undercutting (see Fig. 14.10).

Drainage Patterns

Runoff water from rainfall collects in tiny streams (rivulets), which, in turn, collect into larger and larger streams, or tributaries; these eventually flow into the main (trunk) streams. All of the streams that join together are known collectively as a drainage system and they

FIGURE 14.9 Some examples of typical valley features in North America. (a) Yellowstone River, Yellowstone National Park, Wyoming. The typical cross section of a youthful stream valley such as this is V-shaped. (b) Meandering stream southeast of Bismarck, North Dakota. Meander loops, natural levees, and oxbow lakes can all be seen as the river channel migrates over a wide, flat flood plain. (c) Delaware River, New Jersey. The open, wide, flat-bottomed valley is typical of the lower reaches of a river. (d) The Grand Canyon, Arizona, an example of rejuvenation in which tectonic uplift caused the meandering Colorado River to cut a canyon nearly 2000 m deep—the same process that resulted in the incised meanders shown in Figure 14.9e. (e) San Juan River, Colorado Plateau, Utah. Incised meanders are formed when a meandering river is uplifted to increase its potential downcutting power. Downcutting takes over as a dominant form of erosion (rather than lateral shifting of the channel) but the original meandering channel shape is maintained.

occupy a **drainage basin**. The term *basin* is appropriate since all of the rivers run downhill from a topographic high, or *drainage divide*, which separates one drainage basin from another. Drainage basins become larger in size through time as the individual stream erodes back into upland regions. This process is known as *headward erosion*. Figure 14.12 shows the drainage basins of some of the world's major rivers. Notice that a relationship exists between the size of the drainage basins and the plate-tectonic environment. We will discuss this relationship further at the end of the chapter.

The streams in a drainage system define a drainage pattern—an arrangement of stream courses. In homogeneous terranes that lack oriented rock structures, stream drainage patterns tend to be *dendritic*, or treelike. In other words, a map of the stream courses that make up a river system resembles a branching tree (Fig. 14.13a). The streams flow from the smallest outer branches into the larger main trunk.

In some regions, though, the underlying bedrock structure has a strong influence on the drainage pattern. A tilted sequence of rocks with alternating soft and resistant layers creates alternating ridges and valleys. The streams will be confined largely to the soft rock, only occasionally cutting through the ridges. This will lead to a *trellis* drainage pattern (Fig. 14.13b). The trellis pattern is formed from parallel streams running along the valleys and feeder streams flowing at near-right angles to the main valley streams running down or cutting through the valley walls. Similar effects can be observed if the bedrock is strongly jointed (∞ see Chapter 7). The joints, which typically form a systematic geometric pattern, serve as lines of weakness in the rock and are preferentially eroded, so that the stream channels describe a *rectangular* pattern, directly reflecting the joint pattern.

A third type of drainage pattern, a *radial* pattern, is found around large isolated hills such as volcanoes (Fig. 14.13c). In this pattern streams flow directly downslope away from the high point on the hill. In contrast to the dendritic drainage pattern in which streams converge because they are flowing into a basin, when streams flow down an isolated hill in a radial drainage pattern, they diverge. Note, however, that radial drainage patterns tend to be local and on a smaller scale (the scale of individual mountains) than are typical dendritic patterns, which may occupy a drainage basin approaching the size of a continent (see Fig. 14.12). Thus, while a series of hills might be locally drained by radially arranged streams, these may ultimately all flow into a drainage basin that has dendritic characteristics.

GLACIERS

The landscape that results from shaping by glaciers is distinctive and in many cases spectacular (Fig. 14.14, page 368). We can see glaciers at work in high mountain ranges such as the Alps, the Rockies, and the Himalayas. Much of the rugged topography, the bare rock, and the steep cliffs in these places has been shaped primarily by ice. However, because of erosion, there is little likelihood that evidence for mountain glaciers will be preserved long in the geologic record.

Less obvious are the effects of glaciations in lowland areas, such as the Great Lakes region of the United States. Nevertheless, at various times throughout geologic history, ice sheets have covered huge landmasses, eroded vast areas, and, at the same time, left extensive deposits of sediment. Effects of these large-scale glacia-

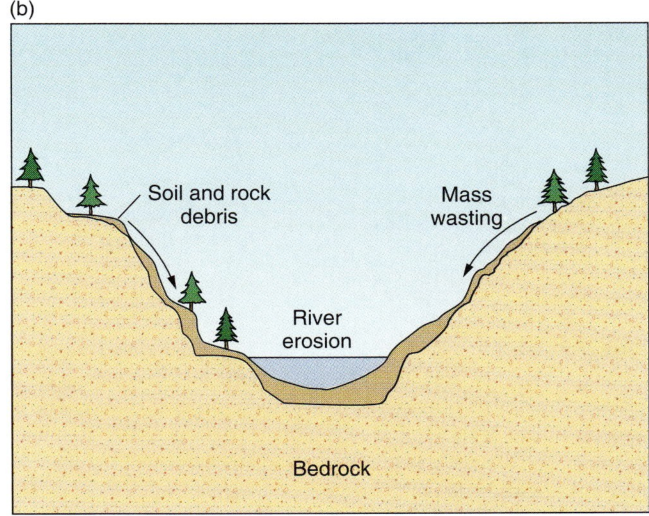

FIGURE 14.10 The role of mass wasting in the formation of valleys. If the valley is cut by stream erosional processes only, the valley will be a narrow slot that is essentially the width of the stream. Most material, however, is not strong enough to support vertical cliffs that would be cut by stream erosion, and mass wasting from the valley sides widens the valley to a familiar V-shape. The resultant valley is therefore shaped by both stream erosion and mass wasting.

(a) Youthful landscape

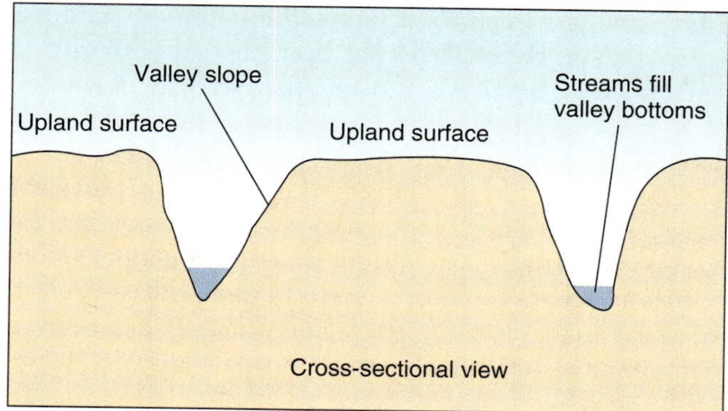

(b) Mature landscape

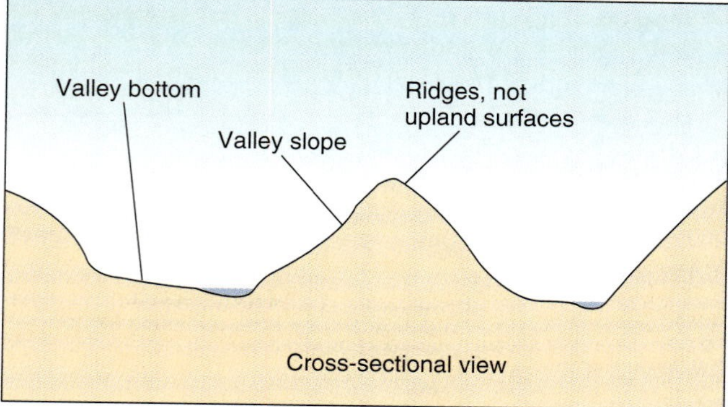

(c) Old-age landscape

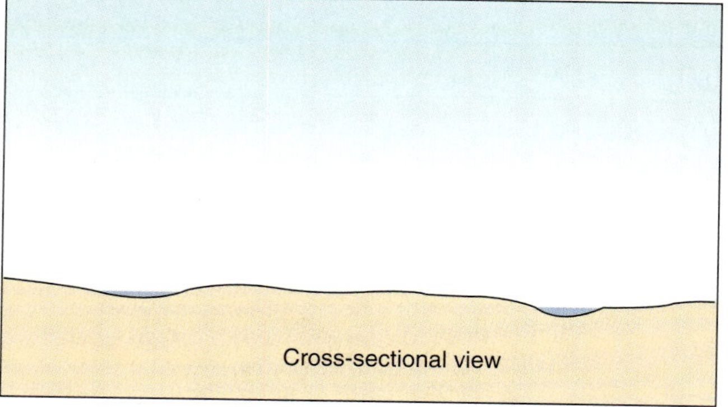

FIGURE 14.11 The evolution of a landscape due to river erosional processes. (a) Youthful topography is characterized by broad upland plains incised by a few steep-walled canyons. The streams flowing in these canyons often completely fill the valley bottoms. (b) Mature topography consists mainly of valley sides and walls, with little undissected upland remaining. The stream may not necessarily occupy the entire valley bottom. (c) Old-age topography shows no upland and only minor remnants of valley slopes. Both lateral stream erosion and mass wasting have reduced the landscape to a nearly flat plain.

tions in the past are therefore more important than those of the current glacial action in mountainous areas.

The periods during which polar ice extended to much lower latitudes than normal are known as *ice ages* (see Focus 14.3). As we saw earlier (∞ see Chapter 6), there have been several periods of extensive glaciation—the effects of which are preserved in the geologic record. The "ice age" you have probably heard about was a period of extensive glaciation that began about 2 million years ago (∞ see Chapter 15). The great expansion of ice did not occur in one episode, but rather as several ice advances with intervening interglacial periods when a warmer climate prevailed. Our present warm climate began about 10,000 years ago; it is not clear whether this climate moderation represents an end to the ice ages or is simply another interglacial period. Analyses of dust and pollen recovered from recent ice-core drilling from the large tract of ice called the Greenland ice sheet provide evidence that the onset of glacial climate conditions was very rapid—taking between 10 and 100 years. The actual formation of ice sheets and the melting back of the ice sheets (deglaciation) required longer periods of time—as much as a few thousand years.

Geologists and climatologists work together to understand why ice ages occur (see Focus 14.3). Efforts to model climatic variations have been increased recently with our realization that the planet's climate is a very complex and fragile system. Short-term climatic changes, such as the frequently mentioned "global warming," might well be due to human influences. Understanding these can help us understand the larger-scale variations such as those that occurred during ice ages. Records of past climates that we have obtained so far indicate that our present climate is quite moderate in comparison to the extremes of the past 2 million years. There are strong indications that the climate of Earth has had much colder average temperatures than those of today.

What Is a Glacier?

Ice currently accumulates in cold regions of Earth near the poles and in the highest mountains. Large tracts of ice—ice sheets—cover high-latitude continental regions such as Greenland and Antarctica. Glaciers are moving ice. They might be mountain glaciers or they might be the more extensive continental glaciers, formed as ice sheets move slowly down toward the coast. Currently, with the exception of polar regions, glaciation is largely restricted to the high mountains. During the last ice age, as recently as 18,000 years ago, continental glaciers advanced over much of North America and Europe (Fig. 14.15).

In a mountain valley, a glacier can be likened to a river of ice. Like a river of water, it starts high in the mountains where snow collects and is changed into denser ice. Snowflakes originate as delicate, feathery crystals of ice. With time, if the snow does not melt but is buried beneath additional snowfall, the original crystals turn into a granular material called *firn*. With further compaction, the firn turns into solid ice. Although ice is a solid, under pressure it can deform and flow plastically.

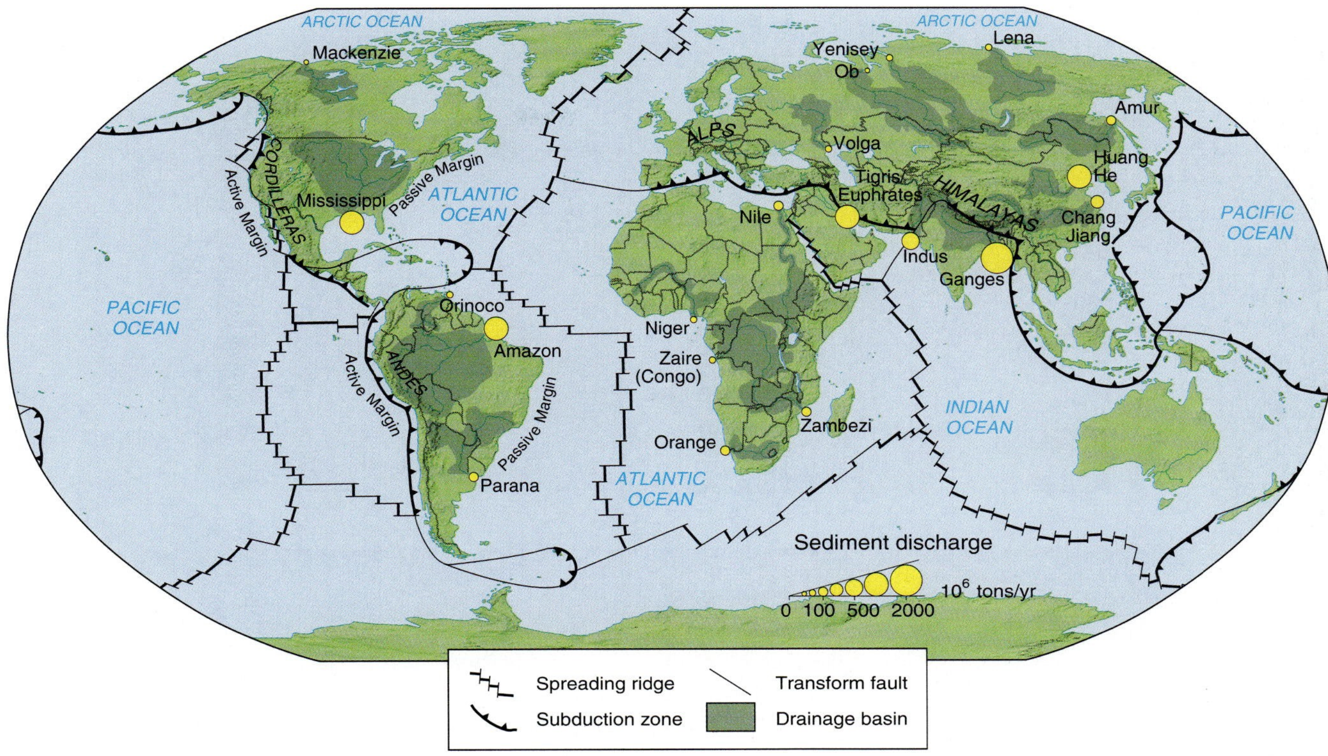

FIGURE 14.12 The world's greatest drainage basins relative to plate-tectonic setting. Sizes of circles indicate annual amount of sediment transported to the ocean by each system.

Unlike wind and water, which generally exhibit turbulent flow (∞ see Chapter 13), glacial flow is laminar. The flow rate of a glacier is very slow. If you were to stand and watch a glacier, it would appear to be motionless. If, however, you were to place a straight row of stakes across the glacier and return to observe the markers a year or two later, you would see that the stakes were no longer in a straight line, but bowed downhill (Fig. 14.16a, page 374). The stakes trace the slow plastic flow of the glacier. It flows more rapidly at its interior than at its base or sides, which is why the stakes are moved farther in the center, and define a curve.

In addition to internal flow, mountain glaciers in temperate climate zones may also *slide* downslope on the contact between the ice and the underlying surface (Fig. 14.16b). The sliding is facilitated by the high pressure existing at the base of the glacier owing to the overlying weight of ice. Why does this cause the glacier to slide downhill? Unlike most other solids, the density of ice is less than that of liquid water. This means that the denser form of H_2O, liquid water, is stable at higher pressure (see Focus 3.4); thus, ice will melt under pressure. Pressures at the base of a glacier are usually high enough that the ice melts and lubricates the glacier; it moves downhill over the ground on a thin layer of water. These same principles enable a skater to glide over the surface of an ice rink. The weight of an ice skater is borne on a very narrow blade, so that the entire weight of the per-

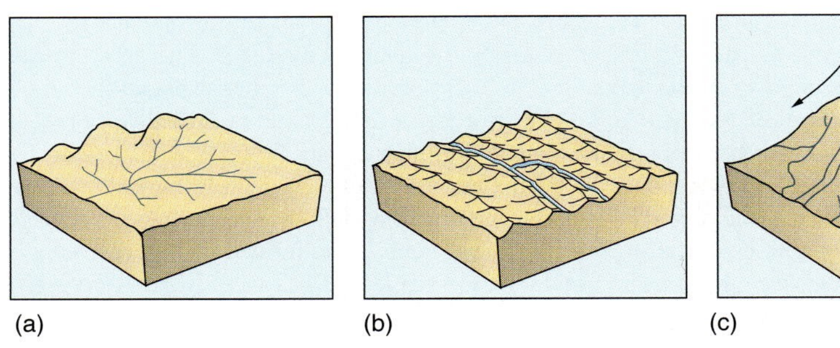

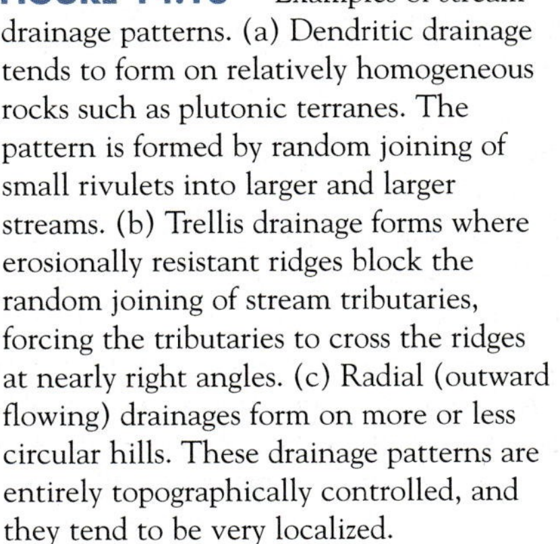

FIGURE 14.13 Examples of stream drainage patterns. (a) Dendritic drainage tends to form on relatively homogeneous rocks such as plutonic terranes. The pattern is formed by random joining of small rivulets into larger and larger streams. (b) Trellis drainage forms where erosionally resistant ridges block the random joining of stream tributaries, forcing the tributaries to cross the ridges at nearly right angles. (c) Radial (outward flowing) drainages form on more or less circular hills. These drainage patterns are entirely topographically controlled, and they tend to be very localized.

FIGURE 14.14 Examples of typical glacial features in North America. (a) Glacier with outwash plain (Glacier Bay, Alaska). Note the accumulations of debris along the sides of the glacier, which are lateral moraines. The debris ridges along the middle of the glacier are medial moraines, formed as two glaciers converge from tributary valleys, merging one of their respective lateral moraines. The debris at the terminus of the glacier forms a terminal moraine. Glacial streams rapidly rework sediment from the moraines and from subglacial streams in outwash deposits in the foreground. (b) Glacial striations in Central Park, New York. The scratches formed parallel to the direction of glacier flow as rocks in the base of the glacier were dragged across exposed rock surfaces. (c) Icefall from a cirque, Angel Glacier, British Columbia. The cirque is a bowl-shaped depression caused as accumulating ice moves away from the source of a glacier. (d) Typical features of glacial erosion, Wind River Range, Wyoming. The U-shaped valley was carved by a glacier moving downhill away from the viewer. It has melted and left a wide flat floor now occupied by lakes. The pointed peaks (horns) and ridges (arretes) were formed by the intersection of steep backwalls of cirques (see Fig. 14.14d), which were progressively eroded by accumulating ice. (e) Drumlins east of Rochester, New York. These ridges form parallel to the flow direction as glaciers move across previously deposited till.

son is in contact with the ice in a few square centimeters. Therefore, the pressure at the point where the skate meets the ice is enormous. For instance, a 55-kg (120-lb) ice skater with skate blades a few mm wide exerts a pressure of over 1000 kg per square meter on the surface of the ice. The skater slides on a thin layer of water that melts when the pressure is applied, and refreezes as the pressure of the ice skate is removed.

You may have noticed from Figure 14.16a that, even though the glacier is flowing downhill, as is clear from the displaced stakes, the end (terminus) of the glacier does not move down an equivalent distance—in fact, it retreats as it melts back. What processes are affecting the terminus of a glacier? As glaciers move down valleys, they merge with glaciers from adjoining valleys, just like rivers joining together, to form increasingly larger glaciers. Unlike rivers, however, there is an upper limit to the size a glacier can actually reach. Ice has a peculiar property that is important in glacial processes—the solid ice can actually vaporize rather than melt. Ice cubes left open in a freezer gradually shrink in size. They are not melting, but vaporizing—the proper term applied to the vaporizing of a solid is **sublimation**. Because glaciers move slowly, they either sublimate or melt as they progress downslope, depending on the climate. In areas where glaciers are barely able to sustain themselves, such as in the Alps of southern Europe, there is substantial melting at the terminus. Meltwater from the terminus forms rivers and lakes downstream from these glaciers. However, in colder regions farther north, such as Norway or Alaska, there is more sublimation at the ends of glaciers than true melting. This is clearly shown by the distinct lack of rivers flowing off the ends of glaciers. If a glacier does reach the sea, or a lake, mass may also be lost at the terminus as blocks "calve" off the front of the glacier to form icebergs.

The upper portion of a glacier is the *zone of accumulation*, where the glacier is growing, and a lower region, known as the *zone of ablation*, is where the glacier is decreasing in size as a result of melting, sublimation, and calving of icebergs. Thus, even though the body of a glacier is continually moving downhill, the terminus may actually be retreating due to excess ablation, and receding back up the valley (Fig. 14.16a). The division between the growth and shrinkage of a glacier occurs at the lower limit of accumulation of fresh snow, which is called the *firn line*. In a colder climate, the firn line is at a lower elevation. At the onset of an ice age, global cooling causes a change in the balance between accumulation and ablation, and the increased accumulation leads to an expansion of glacial ice to both lower elevations and lower latitudes.

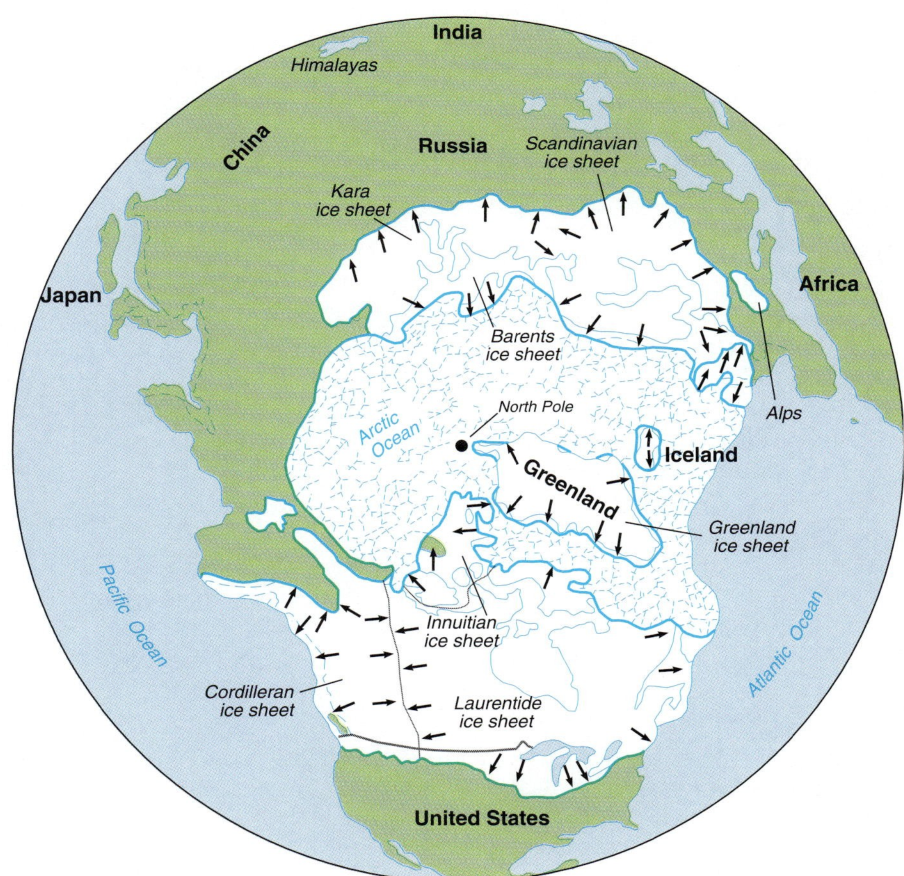

FIGURE 14.15 Map showing the extent of glaciation during the most recent ice age.

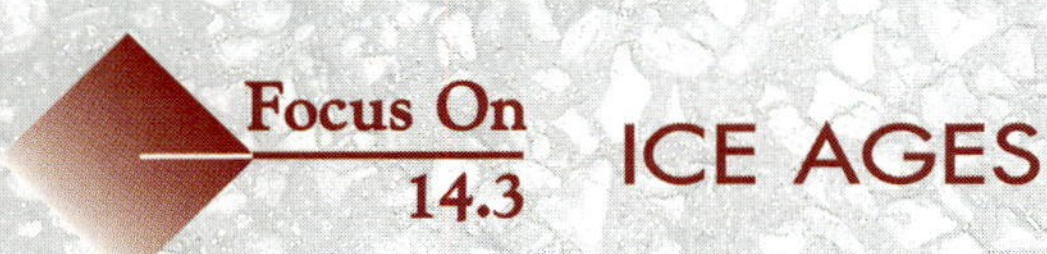

ICE AGES

CLIMATE CHANGE AND THE TIMING OF GLACIATIONS

As discussed in Chapter 12, climate is influenced by latitude and topography. To this list, we can add ocean currents and ocean-surface temperatures. In fact, the oceans exert a strong influence on climate; the latitude of London, England, is approximately 52°N; Edmonton, Alberta, Canada, is 53°N; and Moscow, Russia is about 56°N. Edmonton and Moscow both suffer harsh winters, yet the climate of London is mild because of the moderating influence of the ocean. A warm oceanic current flows from the Gulf of Mexico northward past Britain. As a result, the winter pack ice rarely reaches farther south than about 81°N, yet on the other side of the Atlantic, between Ellesmere Island and Greenland, there is no warm current and it is not uncommon to have winter pack ice extend as far south as 72°N.

We have discussed glaciations in the context of continental positions, with glaciations occurring at polar latitudes (∞ see Chapter 6). We showed evidence recorded in the rocks that the major glaciation of 300 million years ago (Fig. 1a) was a polar glaciation, and that a continent (Gondwanaland) was positioned over the South Pole. We cannot say whether there was a major ice cap at the North Pole of the time; there was no landmass there to record it. However, with modern chemical techniques, we can analyze the shells of oceanic plankton that secrete carbonate shells and interpret the data in terms of global temperature. How can we do this?

As discussed earlier (∞ see Chapter 2), elements may have more than one naturally occurring isotope. Carbon, for example, has two stable (nonradioactive) isotopes, ^{12}C and ^{13}C, and oxygen has three stable isotopes, ^{16}O, ^{17}O, and ^{18}O. However, ^{17}O is usually not measured because its low abundance makes it difficult to analyze accurately, but ^{16}O and ^{18}O can be routinely measured. Carbon and oxygen are particularly useful because these elements are incorporated in organic tissue and shells (carbonate), and oxygen is bonded with hydrogen to form water. We will confine our discussion here to water and oxygen isotopes.

Consider two molecules of water (H_2O), each with a different oxygen isotopic composition: $H_2{}^{16}O$ and $H_2{}^{18}O$. At a given temperature, the molecules with light oxygen ($H_2{}^{16}O$) vibrate faster and have a higher vapor pressure than $H_2{}^{18}O$ molecules. When water evaporates, the water vapor is enriched in ^{16}O because molecules of $H_2{}^{16}O$ preferentially evaporate, and the water left behind is enriched in ^{18}O. How is this useful in a determination of the timing of ice ages?

For rain or snow to fall, water must evaporate from the oceans. That water vapor is enriched in ^{16}O, leaving the ocean water slightly enriched in ^{18}O. During the most recent ice ages, a substantial volume of the oceans evaporated and ultimately precipitated onto the land surface to form glacial ice, which was enriched in ^{16}O relative to the ocean water. Because so much of the light isotope is trapped in glacial ice during an ice age, the ocean water left behind is significantly enriched in ^{18}O.

Marine organisms make their shells from calcium carbonate. The oxygen in the carbonate has an isotopic composition that reflects the oxygen isotope composition of the ocean water from which it is precipitated and the temperature at which it formed (the composition is not exactly the same, but for a given shell formed at a given temperature the composition is directly proportional to that of the ocean). During glaciations, the shells of animals living in the ocean are enriched in ^{18}O, reflecting both the higher ^{18}O content of the oceans and the lower temperatures of the ocean water. Thus, measurement of the oxygen isotopic composition of shells grown at different times permits us to determine when the oceans were affected by the buildup of large volumes of glacial ice (Fig. 1b). Calibration with modern lime-secreting planktonic organisms enables us to estimate the ocean water temperatures during the past.

In the past few years, the U.S. National Science Foundation, in cooperation with the Danish government, has conducted a drilling project in central Greenland, the goal of which is to retrieve a complete core through the ice sheet. Simultaneously scientists from the University of Bergen have retrieved an ocean bottom core from a region off the southwest coast of Norway. Together, these two sets of cores provided us with some major surprises. Work to date on these samples has shown that the climate in the northern regions shifted from "mild" to "glacial" much more rapidly than we had imagined. Some of these climatic shifts occurred in under 10 years, and most in far less than 100 years (Fig. 1c).

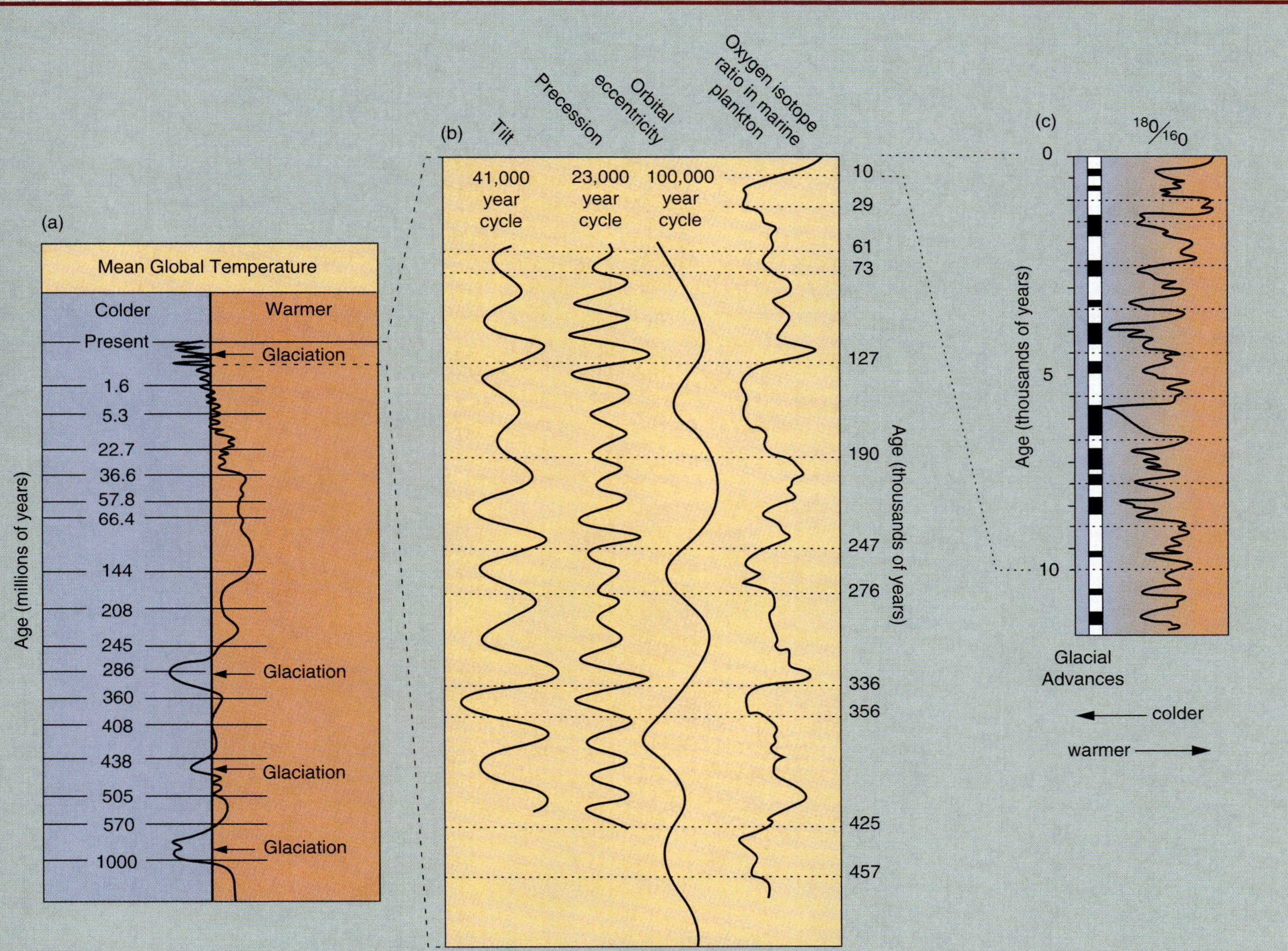

FIGURE 1 (a) Variations in mean global temperature with time, for the last billion years. Note that the time scale is nonlinear, and that the apparent increased "spikiness" of the curve toward the present day is merely a reflection of better resolution in the more recent geologic record. (b) Oxygen isotope ratios recorded in marine plankton over the last 500,000 years. Scale is arbitrary, but lower values (to the left) reflect warmer ocean temperatures and smaller ice mass, and vice versa. The calculated variation in tilt, precession, and eccentricity are plotted immediately to the left of the oxygen isotope curve. When the three curves on the left are arithmetically superimposed, the resulting complex variation closely follows the oxygen isotope curve, which in turn is believed to reflect variations in mean global temperature. (c) Oxygen isotope ratios over the last 10,000 years in an ice core from Greenland. In this case the lower $^{18}O/^{16}O$ ratios (peaks to the left) reflect colder temperatures and correspond closely to the ages of known glacial advances, shown as shaded bars.

CAUSES OF GLACIATIONS

Although we do not know exactly what causes glaciations, many theories have been advanced.

Orbital Variations (Milankovitch Cycles)

During the last million years, oxygen isotope ratios in deep-sea cores show cyclic variations through time, which are interpreted to reflect glacial episodes about once every 100,000 years. Other smaller, periodic fluctuations appear in the climatic record with periods of 26,000 and 41,000 years.

These cycles in climate correspond with changes in the geometry of Earth's orbit around the Sun, which, in turn, affect the amount of sunlight incident on Earth. The Yugoslavian (Serbian) astronomer Milutin Milankovitch first recognized the correlation between orbital parameters and climate, now referred to as *Milankovitch cycles*. The shortest period (26,000 years) is related to the precession of Earth's axis (∞ see Chapter 2). The 41,000-year period is related to the variation in tilt with respect to the orbital plane. The longest variation (100,000 years) is thought to be due to changes in the shape of Earth's orbit (Fig. 2). Variations in energy from the Sun caused by changes in orbital geometry are small—about 10 percent at the shorter periods and much smaller than this at the 100,000-year period. Understanding their effects on global temperatures requires a study of many different yet interrelated factors that are in sensitive balance.

Large-scale fluctuations in the oxygen isotope ratios of marine plankton through time appear to vary systematically with the calculated Earth's orbital eccentricity (Fig. 1b). To examine the oxygen isotopic temperature curve in more detail, we can use a mathematical technique called Fourier analysis, discovered by the French mathematician J.B.J. Fourier. His theorem states that any curve, no matter how complex, can be reproduced by combining a series of regular simple harmonic curves.

Using this technique on the oxygen curve, two other harmonic cycles with periods of 41,000 and 23,000 are identified. Variations in Earth's tilt and Earth's precession are calculated by astronomers (Fig. 1b) to have predominantly these frequencies. Superimposing the effects of these three astronomical variations gives a very strong correlation with the oxygen isotope ratio.

The congruity between global temperatures and the astronomical variations identified by Milankovitch does not prove a cause-and-effect relationship with glaciation, but it certainly is a striking coincidence.

Although the Milankovitch parameters show a good correlation with the long-term variation in global temperatures, the recent realization that the onset of a glaciation may occur over 10 to 100 years suggests that major climatic shifts occur far more quickly than 26,000 years or 40,000 years. The recent ice age has affected Earth in only the last 2 million years. Although ice ages have affected the planet at other times in the geologic past (for instance, 300 million years and 600 million years ago), these events are not regularly spaced in time. It would appear that the orbital effects may moderate or influence the rhythms of glacial advances but cannot be the ultimate cause of glaciations. The climate variations caused by Milankovitch cycles must be superimposed on broader controls such as the configuration of the continents and the effects of atmospheric circulation.

Volcanic Eruptions

Another theory suggests volcanic dust or gases, such as sulfur dioxide in the atmosphere, might account for rapid cooling of the climate. A large increase in volcanic activity would reduce the intensity of sunlight falling on Earth and might cool the climate sufficiently to trigger an ice age. However, two conditions must be met for volcanic activity to influence worldwide climate. The eruption must be sufficiently energetic that material is ejected into the stratosphere, and the eruption must occur in the tropics in order for the particulate material to be efficiently distributed around the globe. The eruptions of Krakatoa (1883) in Indonesia and Mt. Pinatubo (1991) in the Philippines both meet these criteria. They produced minor global cooling for two or three years, but they did not trigger glacial advances. Given that even larger eruptions may affect climate more significantly, we would expect to find huge volumes of volcanic material associated with geologic evidence for glaciations in the geologic record if the two were linked. Although there is a hint of a slight increase in volcanic activity toward the beginning of the recent ice age no compelling link has yet emerged.

Solar Activity

Solar activity, particularly sunspot activity, varies in 11-year and 22-year cycles, and these are weakly correlated to global climate on Earth. Annual records of sunspot numbers extend back into the late seventeenth century, and the most striking feature of this historical record is the Maunder Minimum. This is a period lasting from about 1645 to 1715, when virtually no sunspots were observed. This has no particular geologic significance by itself except for the fact that between the years 1250 to 1740, the climate of Europe was unusually cold—so cold that is has been termed the Little Ice Age.

Sunspots are cooler regions on the surface of the Sun, and an absence of sunspots should correlate to a slight increase in solar luminosity. Such an increase might enhance water evaporation, thereby increasing cloud cover. Sunlight reflects off clouds much more effectively than off ground or water, and the additional cloud cover might have a global cooling effect. The paucity of observational data on sunspot numbers prior to 1645 is unfortunate, but is it only coincidence that a time of unusually low sunspot activity correlates with at least part of the Little Ice Age?

Tectonic Activity

Aside from the movement of plates into polar regions, tectonic activity may have other effects on climate. The positions of plates strongly control oceanic currents, and these in turn have profound climatic influences. However, tectonic motions are imperceptibly slow compared to climatic shifts documented in the Greenland ice cores and the Norwegian sediment cores.

One interesting idea discussed earlier (∞ see Chapter 6), is that the recent ice age was a consequence of the collision between India and Asia, which formed the Himalayas. In this case, the plate configuration does not significantly influence oceanic circulation; rather, the Himalayan range affects atmospheric circulation patterns, which in turn affect climate.

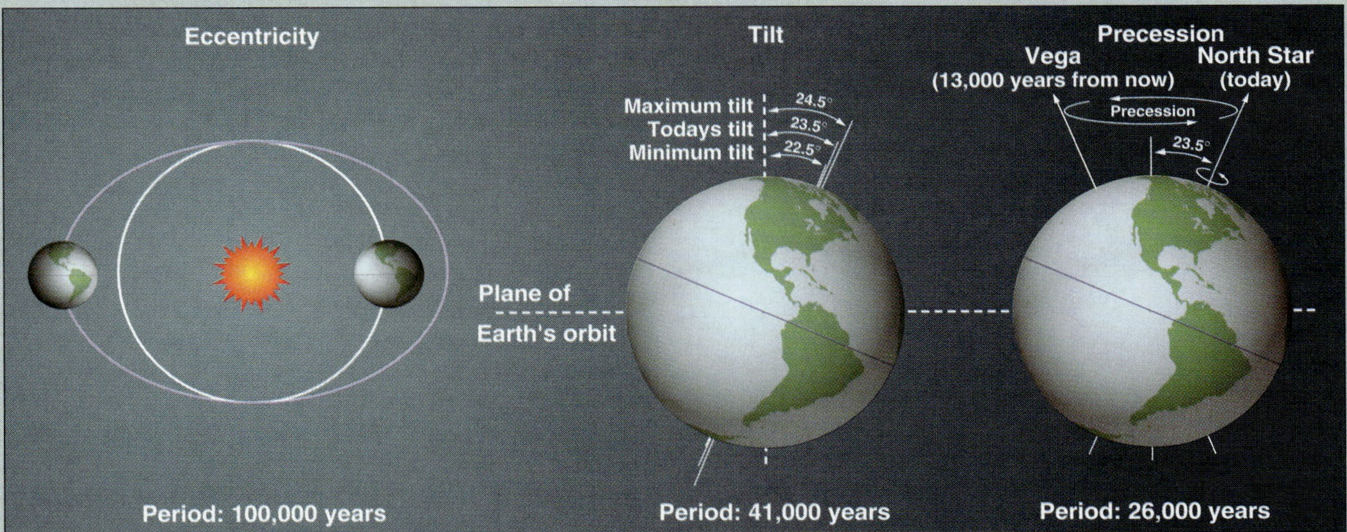

FIGURE 2 Schematic illustration of the causes of Milankovitch cycles, through variations in orbital eccentricity (greatly exaggerated), tilt, and precession of Earth's spin axis.

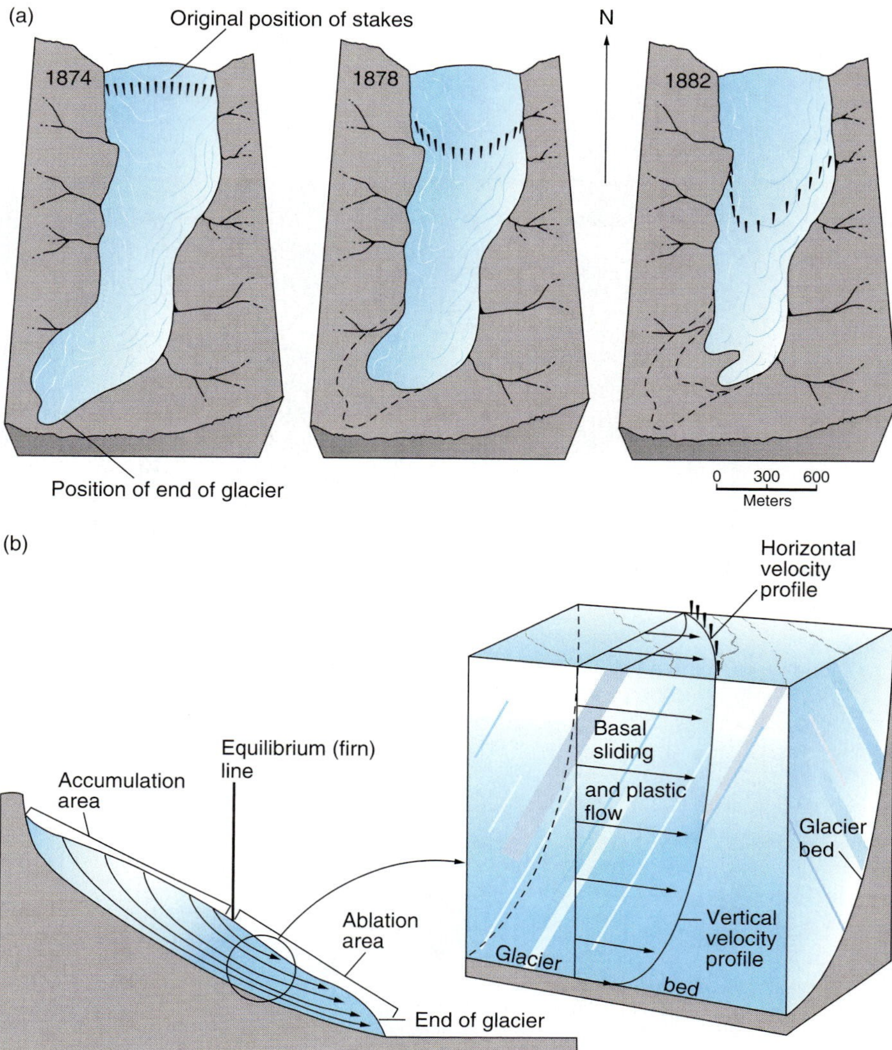

FIGURE 14.16 Mechanism of flow in glaciers. (a) Demonstration of slow flow by observing the distortion of a straight line of stakes driven into the ice of the Rhone glacier. After eight years, the row of stakes has moved approximately halfway down the length of the glacier. Note that the terminus of the glacier has retreated, even though the ice is moving down the glacier. (b) Cross section of a glacier showing flow lines and the effect of basal sliding. Note that the flow lines are parallel, indicating laminar flow. The length of the arrows indicates the total amount of flow throughout the body of the glacier over a given time period.

Glacial Erosion and Deposition

Glaciers can be very powerful agents of erosion, depending on the climate in which they are found. As the ice moves over the land surface, it picks up loose fragments of bedrock and incorporates them into the glacier—a process known as *plucking*. Consequently, the glacier is not simply a "river of ice"; it can also contain rocks and dirt. These rock fragments act as an abrasive, scraping the land surface as the glacier moves over it. Thus, the action of a glacier might be compared with that of a giant file scraping against the rocks as it flows downslope.

Glacial erosion begins at the source of a glacier (see Fig. 14.14c). Large volumes of ice accumulate in depressions on the sides of mountains. The ice plucks away at the headwalls of the depression to form a bowl-shaped feature, or **cirque**. As with rivers, the glacier will erode headward, cutting back the walls of the cirque. Cirques may cut back into each other, forming a sharp knife-edge-like wall between two cirques called an **arête**. If more than two cirques intersect, a pointed peak, or *horn*, is formed (see Fig. 14.14d). The Matterhorn of the Swiss Alps is probably the most famous example of this feature, but glacial horns are plentiful in every glaciated mountain range.

Most glaciers will follow preexisting river valleys; a pattern of glacial erosion is often superimposed on a preexisting valley system. During glaciation, a glacier occupies the entire valley and supports the walls. Valleys that have been eroded by glaciers have characteristic steep sides and wide, flat bottoms, and are called *U-shaped valleys*, which aptly describes their shape in cross section (see Fig. 14.14d). In contrast, rivers occupy only the very bottom of their valleys, and, as we have seen, the valley walls are typically modified by mass wasting. The shape of a valley that has been eroded by a river is more like the letter V. As a result of glaciation, the steep sides of a U-shaped valley may truncate the valleys of smaller tributaries. Following glaciation, streams in the tributary valleys are stranded high on the sides of the valley, forming *hanging valleys*. In the most spectacular instances, the stream pours over a vertical cliff into the valley as a waterfall. The deep, wide valley floor is commonly also the site of many lakes after the glaciers have retreated (see Fig. 14.14d). Deep, U-shaped glacial valleys that reach the coast may become inundated by the sea, forming spectacular inlets known as *fjords*.

If the glacier moves over hard rock surfaces, it may leave scratches or *striations*. The striations that we commonly see on rock surfaces in high latitudes are evidence that glaciers once existed there (see Fig. 14.14b). Furthermore, the striations are parallel to the direction that the glacier traveled. Thus, by studying the orientations of glacial striations over large regions, we can determine the directions of ice flow during the ice age.

In mountainous regions, glacial erosion predominates and glacial deposits are scarce and transient. As with rivers, however, extensive deposition occurs in the lowland reaches of the system. These deposits are typically reworked rapidly by rivers, wind, and mass wasting, and are therefore poorly represented in the geologic record. At the current time, glacial deposits are unusually well represented in lowland regions of the Northern Hemisphere simply because the last ice age was so recent.

Glacial deposits, whether left directly by glaciers or subsequently reworked, are referred to as *drift*. This curious term was adopted by early European geologists who recognized that much of the loose sediment material covering portions of the continent had been transported considerable distances from their source. They incorrectly surmised that it had "drifted there, perhaps on ice rafts in ancient floods."

Drift material deposited directly from glaciers is termed *till*: a very poorly sorted accumulation of detritus ranging in size from rock flour to boulders. This debris has been transported both within the body and on the surface of the glacier, where talus has fallen onto the glacier by mass wasting off the valley walls. Till deposits commonly form ridges or mounds known as **moraines**. In mountain glaciers, the highly angular talus debris is transported on the glacier surface as elongate moraine ridges along the sides of the glacier. Moraines protect portions of the glacier from the atmosphere, and therefore from sublimation. As a result they stand as high ramparts above the surface of the glacier (see Fig. 14.14a). If two glaciers converge, these *lateral* moraines will coalesce to form a *medial* moraine—a debris ridge that runs down the middle of the glacier surface. If the glacier retreats, the moraine deposits may be left behind, both at the terminus of the glacier (terminal moraine) and along its sides.

Beneath and adjacent to high latitude arctic and temperate glaciers are large rivers, usually moving very swiftly and laden with sediment. Subglacial streams may be braided, moving back and forth under the glacier. Glaciers in temperate climates are near the melting temperature of ice and, for that reason, water flows in tunnels throughout the glacier. Rivers crisscross the tops of these glaciers, and lakes may accumulate between the terminus of retreating glaciers and the terminal moraine. Glaciers in colder climates such as the arctic, are formed from ice that is well below its melting point. Even so, there is water flowing in through the sides of the glaciers as well as streams flowing from beneath them. These processes erode and redeposit the morainal material left by a retreating glacier. By this mechanism of reworking, large angular rocks that originally fell on the glacier as talus may become rounded. This accounts for the puzzling observation that moraines may contain rounded boulders, despite being very poorly sorted.

Both mountain and continental glaciers probably never make a single advance; instead, they advance, retreat, readvance, continually pulsing forward or backward. In so doing, they push up mounds of previously deposited moraine, then retreat back from the pile, only to perhaps readvance entirely over the mounds earlier created. In lowland regions, extensive deposits of moraine may exist that were laid down during the ice ages. In places, glaciers flowing over the moraine reshaped it into groups or fields of elongate mounds called **drumlins** (see Fig. 14.14c). These features are rather unspectacular from ground level but, from the air, they are seen to be teardrop-shaped and have their steepest slope on the upstream side (relative to the direction of ice flow). Drumlin fields, like glacial striations, have been used to determine the flow directions of continental glaciers.

Glaciers move through a combination of internal plastic flow and sliding at their base. They scrape and pluck the rock surfaces over which they move and transport unsorted debris both internally and on their surfaces.

In addition to the deposits left directly by the glaciers, there is a good deal of sediment that results from reworking of glacial till by streams and wind. These are referred to as *outwash deposits* (deposited by streams, Fig 14.14a) and *loess* (deposited by wind).

As a glacier melts, it releases the sediment that it contains and produces meltwater. Not surprisingly, much of the sediment is transported by meltwater streams both issuing from the end of a glacier and, in some cases, flowing beneath the glacier. The reworking by water results in drift deposits that, in contrast with till, are crudely stratified and a little better sorted. In the case of widespread continental ice sheets, meltwater streams produce broad outwash plains. In the case of subglacial streams, the sediments are deposited in the stream channel. After the glacier retreats, these deposits are left, forming a sinuous ridge, which connects to the fan-shaped deposit of a meltwater delta, formed as sediment-laden streams flow into meltwater lakes.

The huge continental ice sheets that existed during the ice ages undoubtedly modified the weather in a major way. Cold air flowing away from the glaciers may have generated strong winds. The fluvial deposits left in front of the glaciers after a major glacial retreat are vulnerable to wind winnowing and erosion. The sediments are loose and abundant, and the climate immediately following the ice age was cold and arid. Silt was winnowed from these floodplains and blown long distances from the glaciers to accumulate in thick deposits across Asia, northern Europe, and the northwestern and midwestern United States. Like the fine, windblown deposits originating from the world's deserts today, these deposits are called loess, and are particularly important because they produce fertile agricultural soils.

Deserts

As with the glacial environment, the desert environment is defined by climate. The scarcity of water that serves to define desert regions gives rise to unique landforms (Fig. 14.17), although the basic principles of erosion, sediment transport, and deposition are the same in deserts as anywhere else on the continents.

Distribution and Origin of Deserts

A desert is a part of the world that is literally deserted because of extremes in the environment. Although we may think of deserts as being hot and sunny, desert regions are principally characterized by low rainfall. There are other arid regions on Earth—not all of which are hot. For example, the dry, ice-free valleys of Antarctica are true deserts. In fact, these Antarctic valleys contain lakes of very high salinity; their high salinity is due to the fact that evaporation exceeds the flow of meltwater into the lakes. The existence of strong evaporation in such a cold place may seem paradoxical, but the humidity of continental polar regions is very low; thus, evaporation rates are high.

Most of the deserts of the world are located in two belts adjacent to the tropics at 30°N and 30°S of the equator. These are regions that receive little rainfall owing to their location on the downflowing limits of atmosphere circulation cells (∞ see Chapter 6). Desert conditions can also be caused by rainshadow effects.

Erosion and Deposition in Desert: Effects of Water and Wind

The general lack of water in deserts limits its role as an agent in shaping the land surface. Nevertheless, water has an important role in the erosion and deposition of sediments (∞ see Chapter 13). A sudden downpour commonly results in flash flooding, where ephemeral streams scour deep channels in the desert surface. Because the erosive energy generated by the flash flooding is great and because the slopes of desert mountains are generally steep, erosion cuts down quickly to bedrock. This often happens without leaving the slightest veneer of sediment. As erosion proceeds, these violent flood channels erode down to grade, in the same fashion as do perennial streams. The difference is that the grade is established on bedrock; this eroded bedrock surface is called a *pediment surface*. When the floodwater in a canyon leaves the mountains, it quickly loses power and will drop its sediment load at the mouth of the canyon, depositing alluvial fans. The water that flows beyond the alluvial fans may collect in small, short-lived (ephemeral) lakes. These *playa lakes* quickly evaporate, leaving behind encrustations of salts or evaporites (see Fig. 14.17a).

A more constant, if less effective, agent in deserts is the wind. Sand grains are carried by the wind in a manner similar to that of rivers (∞ see Chapter 13). More commonly, however, grains move by bouncing off the desert floor or by rolling along it. Wind that is carrying sand acts as an abrasive on the surfaces over which it blows. Rocks on the desert floor can become polished by this process. Even strong winds are incapable of moving rocks that are 2 to 5 cm in size, but sand grains of 0.5 mm or so are easily picked up and carried away. As smaller particles are removed by wind in a process known as *deflation*, the remaining rocks settle slightly and nestle more closely together. The surface that remains is covered by a type of armored coating consisting of closely spaced polished and faceted rocks often referred to as *desert pavement* (Fig. 14.18). *Blowouts* form when the surface of the desert is disturbed and becomes more susceptible to erosion by prevailing winds. They take the form of dimple-shaped depressions ranging from meters to kilometers (Fig. 14.17e). Wind eddies are thought to remove fine particles in a process similar to deflation.

Sand deposits are relatively uncommon features even in the driest deserts, covering only about 30 percent of the total desert area (∞ see Chapter 13). Barren rocks and desert pavement are by far the most common desert landscape. However, depending on sediment supply, surface topography, and wind strength and direction, sand may be deposited to create dunes. The shapes of sand dunes vary with wind direction and the abundance of sand. If the sand is abundant and the wind blows consistently from a single direction, transverse dunes are formed that have the appearance of large ripples in the desert surface, aligned perpendicular to the wind direction. If the sand is in short supply, Barchan dunes form (Fig. 14.17c)—crescent-shaped dunes with tips pointing downwind. If, in contrast, the wind blows from various directions with various velocities, dune shapes vary from starlike configurations to irregular shapes.

Vegetation can have profound influence on desert morphology, in some cases (Fig 14.17b) preserving sand from erosion.

The action of desert streams is important but episodic. Wind is a more constant climatic factor, but in either case deposition is the exception rather than the rule in deserts.

FIGURE 14.17 Examples of typical desert features in North America. (a) A saltpan in Death Valley, California. This is an example of a playa lake, where evaporite deposits are left when water evaporates from the basin floor. (b) Indian corn field, Death Valley, California. Here, deflation has occurred—the loose sand has been blown away to expose the root clumps of these plants where the sand is protected from removal. (c) Sand dunes near Yuma, Arizona. If the sand is abundant and the wind blows from a single direction, transverse dunes are formed that have the appearance of large ripples. If the sand is in short supply, barchan dunes, such as the ones shown here, form. They are crescent-shaped dunes with tips pointing downwind. (d) Wash in dry gully, Death Valley, California. These features form by periodic flash flooding when torrents and mudflows wash sediment from the mountains during storms. (e) A small blowout formed when the surface of the desert is disturbed, in this case by the presence of a plant and becomes more susceptible to wind erosion.

(a)

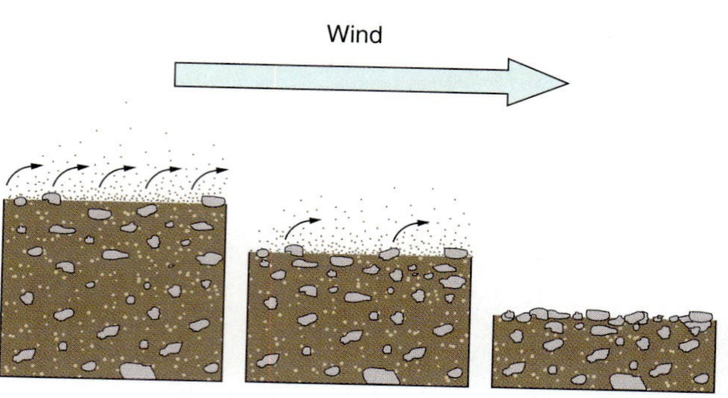

(b)

FIGURE 14.18 Formation of desert pavement. (a) Desert pavement from the southwestern United States, with scattered, polished rocks lying on the floor. (b) Drawing depicts how such pavements form by progressive removal of loose sand and dust (deflation).

COASTS

Coastlines are a dynamic interface between land and sea. They change position and form through time (Fig. 14.19). On the time scale of a single year, the coast at one location can vary from being barren of sediment to forming a sandy beach (Fig. 14.20). Over thousands of years, shorelines have moved inland or regressed from land by several kilometers.

Waves and Tides

The two main agents responsible for erosion and deposition along coasts are waves and tides. Tides are caused by the gravitational pull of the Sun and especially the Moon (Fig. 14.21). The surface of the oceans will be raised by gravitational attraction—the highest tides (spring tides) are caused when the Sun and Moon are aligned and their gravitational attractions act in concert. Occasionally, tides may themselves cause erosion through what is known as a tidal bore, which occurs where there is a channel such as a river estuary into which the rising tide is forced, causing a wavelike surge that can erode the channel into which it flows. The main effect of tides, however, is to determine the range in height over which waves will act.

At the shore, waves expend their energy by breaking (∞ see Chapter 13), and the resulting surf is an important agent of coastal erosion and transport. Wave size is related to the strength of the wind blowing, and to the distance over the water that the wind blows. These two factors are interrelated. A strong wind blowing a short distance over water might cause large waves, or a moderate wind blowing over a longer distance might cause waves of equal or greater size. As a general rule, waves generated by winter storms have a longer wavelength and a higher amplitude than do waves generated in the summer. If there are no barriers, waves generated by storms can travel thousands of miles across the ocean with essentially no decrease in energy. For this reason, in the summer, some south-facing coastlines in the Northern Hemisphere may be strongly affected by winter storms occurring in the Southern Hemisphere.

Coastal Erosion and Deposition

At any given time, the geomorphology of a coastline is determined by a complex interplay between the geology of the coastline (for example, whether the outcropping rocks are soft or hard) and the processes of wave erosion and deposition. Waves acting on a coastline will tend to straighten out the coast. This could be regarded as a sort of equilibrium condition similar to that of rivers, glaciers, and wind trying to erode continents down to a flat surface. Where promontories exist, they are attacked by the waves. Wave refraction due to differences in depth close to shore will focus the wave energy onto protruding headlands increasing erosion there. At the same time, material is deposited in the adjacent bays to form beaches. Waves generally approach the coastline obliquely; consequently, sediment is transported along the coast by longshore drift (∞ see Chapter 13).

The force of breaking waves can be immense, particularly during storms. Air and water are compressed into cracks in the rock, forcing them apart and eventually breaking the rock. Where the rock is resistant, vertical cliffs will be cut. As the cliffs are worn back, a flat surface of rock called a wave-cut platform is left at the base of the zone of erosion (Fig. 14.22). Note that there is no talus at the base of the cliff shown in Figure 14.22. This means that the cliff is being actively eroded by the surf, and the sediment produced by this erosion is transferred offshore by the surf, and downcoast by longshore currents. Where the resistance of the rock varies, some parts

FIGURE 14.19 Some examples of typical coastal features in North America. Note in particular the contrast between the emergent Pacific coast and the submergent Atlantic coast. (a) Outer Banks, Duck, North Carolina. An example of a coastal barrier island, which separates the Atlantic Ocean (left) from a quiet water lagoon. (b) Natural Bridges State Park, California. Erosion by wave action created this sea arch, which, upon collapsing, would form sea stacks such as the one on the right. (c) Bolivia Point, California. An example of a wave-cut platform at the base of a cliff. The platform forms in response to the concentration of erosion power at the level of the surf zone, that gradually erodes the cliff. The flat surface at the top of the low cliffs marks a former beach level that has been uplifted. (d) The California coast at Big Sur. The steep shoreline is typical of emergent coastlines.

(a)

(b)

FIGURE 14.20 Seasonal movement of sand on a beach. The two photos were taken from approximately the same location in summer (a) and winter (b), and they show a dramatic change in the amount of sand present. (La Jolla, California)

of the coast will be eroded back faster, leaving small embayments separated by headlands or promontories. The force of the waves will then be concentrated on the promontories, and may actually cut through the headlands to create sea arches (see Fig. 14.19b). With further erosion, the arches will collapse to leave isolated columns of rock called *sea stacks*.

The main deposit formed at the shore is the beach. Material deposited on beaches usually consists of cobbles and sand, which come from two main sources. One is the erosion of the coastline by waves, and the second, volumetrically more important, is the material transported to the shore by rivers. Whether sand accumulates on a beach or is eroded off the beach is related to the wavelength and amplitude of the local waves. Larger waves with longer wavelengths, such as occur during winter storms, erode the beach and transfer the sediment offshore beyond the surf zone. The sediment is deposited there in ridges, parallel to the shoreline, and are called *offshore bars*. In contrast, short-wavelength, low-amplitude waves move the sediment that had accumulated as offshore bars shoreward again and back onto the beaches. Thus, in the absence of a strong longshore current, sediment moves from the beach to the bar and back again (see Fig. 14.20).

Where strong longshore currents do exist, sediment is transported along the coast. Occasionally, where the beach turns sharply landward from the direction of longshore drift, this drift will build a ridge of sand out to sea. This forms a special type of beach called a *spit*.

Waves expend energy by breaking on the shore. The resulting surf erodes the coastline. The eroded material, together with material supplied from inland, is deposited mainly on beaches.

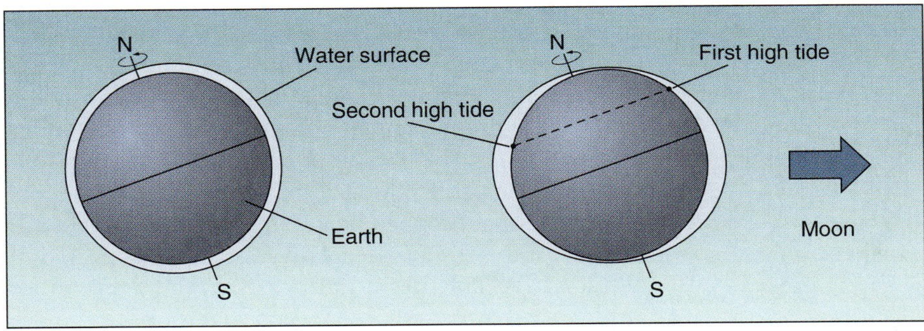

FIGURE 14.21 Tides are formed by the gravitational pull of the Moon. Depending on the position of the Moon's orbit relative to Earth, the tidal bulge resulting from the Moon's pull on the oceans may be inclined to the equator. The gravitational pull of the Sun has a similar but less marked effect.

Emergent and Submergent Coasts

A glance at a map of any coastline will quickly convince you that few have attained the straight equilibrium condition described earlier. Most coastlines are highly convoluted, with numerous promontories, bays, and inlets. One reason for this is that the interplay between the varied geology of the coastline and the coastal processes of erosion and deposition seldom reaches equilibrium. Irregular coastlines can also be formed as they gradually rise or sink relative to sea level. As discussed in earlier (∞ see Chapter 11), sea level may change globally in response to widespread glaciations or to alterations in ocean-ridge spreading rates. On a regional or local scale, sea level may appear to change as a result of tectonic activity or isostatic adjustments.

Rising coastlines (falling relative sea level) are called emergent coasts. Most of the western (Pacific) coast of the United States is emergent because of active tectonics. Such coasts tend to be dominated by erosion rather than deposition and are often characterized by cliffs (see Fig. 14.19d). A record of uplift may be preserved in *raised terraces*—topographic platforms marking successively older beach levels upward from the current sea level (see Fig. 14.19c). In contrast, most of the eastern (Atlantic) coast of the United States, which is a passive continental margin, is sinking; it is a submergent coastline. Such coastlines are characterized by flat coastal regions with drowned river estuaries (see Fig. 14.19a).

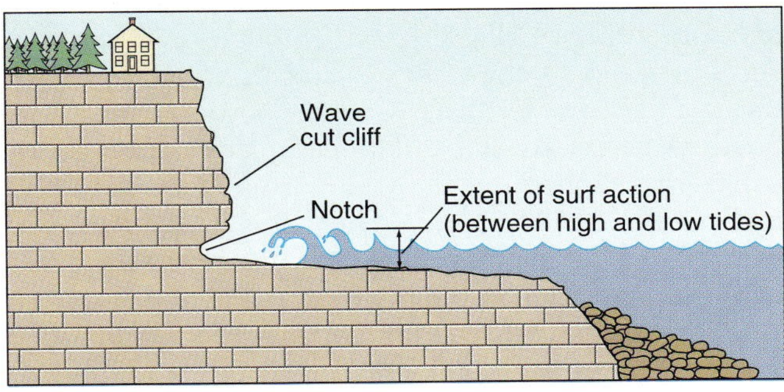

FIGURE 14.22 Formation of a wave-cut platform. The surf erodes a notch in the base of the cliffs. Eventually the cliff above the notch collapses, and the talus may be removed so that the process gradually causes the cliff line to retreat. The photograph shows a wave-cut platform at Robin Hood's Bay, England.

Many geomorphologic features along coastlines reflect emergence or submergence of the shore.

Controls on Geomorphology at Passive and Active Continental Margins

We have discussed passive and active continental margins from a plate-tectonic standpoint. However, it is important to note that the plate-tectonic environment will also affect the geomorphology of landscapes, such as coastlines and river drainage basins. This can be readily appreciated by comparing the eastern (passive) and western (active) continental margins of North and South America (Figs. 14.19 and 14.23).

Active continental margins are generally characterized by deformation and uplift, forming tectonically emergent coasts. Cliffs and raised beaches are common features of such coastlines. Plate convergence also leads to the formation of a volcanic arc near the edge of the continent. This mountain chain will form a topographic high fairly close to the shore, and rivers arriving at the coastline will have flowed down steep gradients for short

FIGURE 14.23 Plate-tectonic control on drainage-basin characteristics. The South American continent is bounded to the west by an active (convergent) margin. River systems are constrained to be relatively short and steep by the subduction-related Andean mountain range. On the eastern side of the Andes, much larger, lower-gradient river systems drain the passive continental margin. The Maipo and Amazon river drainages are emphasized to make the comparison.

distances. Streams that flow out of tectonically active areas have small but very steep drainage basins. The distinctive characteristic for rivers that flow into the Pacific Ocean from Alaska to Chile is their young geomorphic age—the steep incline and limited extent of their drainage basins. The Maipo River draining the western Andes in southern Chile serves as a good illustration (Fig. 14.23). In these rapidly uplifting regions, streams cannot erode deeply enough or rapidly enough to establish a well-linked network of rivers. Continued uplift in these tectonically active regions thwarts stream integration, keeping the rivers small and steep and often confined to deep canyons. The steep terrain results in the rapid downslope movement of rocks and gravel (mass wasting) into the rivers. This material is quickly removed owing to the erosive power of the steep rivers.

The sediment supplied to the shore by these rivers will be rapidly transferred to deeper waters as the continental shelf at active margins is narrow (< 50 km) and drops sharply into the topographic trench that forms at the site of subduction. If sediment is supplied rapidly, it will accumulate in the trench and be actively deformed as a result of subduction to form an accretionary wedge (∞ see Chapter 9).

Because passive margins are not active plate margins, there tend to be no young mountain ranges associated with them, so they are characterized by subdued topography. In terms of geomorphic age, these are typically mature or old landscapes. Large stream drainage networks develop in such tectonically inactive areas. The current drainage network of the Amazon River, for instance, covers roughly a quarter of the entire land area of South America (Fig. 14.23). Through time, the predecessor rivers of the Amazon and its tributaries have traveled back and forth across a large portion of northern South America and have flattened many of the continent's topographic irregularities. With no tectonic activity to renew topographic relief, this large area was reduced to a flat plain by the erosion caused by the back-and-forth migration of the river. After a period of time, a kind of topographic equilibrium has been established where the slope is low and erosion is more or less in balance with deposition. In such areas, erosion occurs slowly with long-term sediment storage along the banks and islands of the river. The results are extensive weathering and thick soils along and between the rivers.

Most of the sediment from these drainage systems accumulates, at least temporarily, on the continental shelf. The shelf at a passive margin may be quite extensive, depending on the relative position of sea level—the extent of the shelf determining how much and how quickly sediment is recycled into the deeper ocean. During times of low sea level, such as during an ice age, the shelf is exposed and sediment is eroded and transferred closer to the edge of the continental slope, allowing for more rapid transfer to the abyssal plain.

The relationship between drainage systems and tectonic environment can be appreciated by examining the locations of the world's greatest rivers (see Fig. 14.12). The large drainage basins of the Amazon, Mississippi, and Zaire rivers all coincide with passive continental margins. In many places, passive margins are crossed by failed rifts that formed at the time of continental rifting (∞ see Chapter 8) and are now the sites of major rivers. The lower reaches of the Amazon and Mississippi occupy failed rift segments as does the Niger on the other side of the Atlantic. Notice that, despite their smaller size, the Tigris/Euphrates and Ganges rivers deliver enormous masses of sediments to the oceans. These rivers drain the tectonically active Alpine-Himalayan zone of continental collision, where rapid uplift and consequent extremely high erosion rates provide plentiful sediment.

Many geomorphologic features are clearly not preserved for long. But those that are enable us to establish a clear link between geomorphology and plate tectonics. We see evidence in the rock record of extensive ancient rivers that drained now worn-down mountain ranges, evidence for past glaciations and for ancient deserts, all of which helps us to piece together the plate-tectonic puzzle back through geologic time.

The sizes of drainage basins, the steepness of slopes, and the types of rocks vary according to the plate-tectonic environment, and will therefore influence geomorphology.

SUMMARY

- Landforms are controlled by the local geology, climate, and time.
- Mass wasting is the downhill movement of material under the influence of gravity. Material can fall, slide, or flow, and rates can be catastrophically fast (rockfall, landslide) or extremely slow (creep).
- Landscapes evolve through time as erosion reduces relief. At any time, tectonic forces can interrupt the process, causing uplift and rejuvenation of the erosive agents.
- Glaciers move through a combination of internal plastic flow and sliding at their base. They scrape and pluck the rock surfaces over which they move and transport unsorted debris both internally and on their surfaces.
- The action of desert streams is important but episodic. Wind is a more constant climatic factor, but in either case deposition is the exception rather than the rule in deserts.
- Waves expend energy by breaking on the shore. The resulting surf erodes the coastline. The eroded material, together with material supplied from inland, is deposited mainly on beaches.
- Many geomorphologic features along coastlines reflect emergence or submergence of the shore.
- The sizes of drainage basins, steepness of slopes, and types of rocks vary according to plate-tectonic environment, and will therefore influence geomorphology

KEY TERMS

mass wasting, 356
talus, 358
slide, 358
slump, 359
flow, 359
lahars, 359
solifluction, 359
soil creep, 360
drainage basin, 365
sublimation, 369
cirque, 374
arete, 374
moraine, 375
drumlin, 375

QUESTIONS FOR REVIEW AND FURTHER THOUGHT

1. Suppose an elderly relative asked you to accompany her to her childhood home in the hills. Upon arrival, you find the house in good condition, but the fence posts are leaning away from the hillside. Your relative attributes this to cattle rubbing against the fences. What alternative explanation can you offer?
2. Consider a mature river system that suddenly experiences tectonic uplift. What geomorphologic features will replace the broad flat region as a result?
3. Ephemeral streams in arid climates erode steep-walled, narrow canyons; perennial streams in rainy climates erode V-shaped canyons; and glaciers erode U-shaped valleys. Can you think of an explanation for these differences?
4. This chapter suggests that the effects of mountain glaciers are probably unimportant in the geologic record. Defend or contradict this statement.
5. Will mass wasting be a more effective process in a mountainous tropical region or in lowland desert? Why?
6. From Figure 14.16, calculate the maximum yearly rate at which the Rhone glacier advanced between 1874 and 1882. What is this rate in meters per *day*? Do you think you would be able to see the glacier moving?

15 Geologic Hazards and the Environment

A collapsed house in Pacific Palisades, California. The steep slope in the foreground gave way due to shaking in the 1994 Northridge earthquake.

Overview

- During earthquakes, the potential for damage is controlled by earthquake size, building design, and the local geology.
- Pyroclastic flows, mudflows, and ashfalls pose the greatest danger to property and people in regions of active volcanism.
- Urbanization affects stream flow because the ground cover causes increased runoff, which may cause flooding.
- Landslides are common in tectonically active regions where rapid uplift of weak or layered rocks gives rise to unstable slopes.
- Oil spills, despite accounting for only a small fraction of oil added to the oceans, pose an environmental hazard until diluted.
- Fossil fuel burning increases carbon dioxide in the atmosphere, which may increase the degree of solar warming. Chlorofluorocarbons in the atmosphere destroy ozone, which absorbs damaging ultraviolet solar radiation.

INTRODUCTION

Geologic processes affecting human safety and habitations are diverse and are dependent on local conditions. Many geologic hazards, such as earthquakes and volcanoes, are the direct result of plate tectonics; others, which can be indirectly related to plate tectonics, are the result of either steep topography, high amounts of rainfall, or even arid environments. In addition to these natural hazards, humans play an increasingly important role in controlling the environment in which we live. Environmental issues such as global warming, pollution, and the ozone hole will be matters of concern well into the twenty-first century.

EARTHQUAKES

Building Failure in Earthquakes

Few natural events strike as much fear into people as do earthquakes. Even our expressions for Earth's solidity indicate that we expect it to be solid and unshakable. Some examples include *terra firma, having one's feet on the ground*, and *being firmly grounded in a subject*. When an earthquake strikes, there is no refuge. What had previously been solid now ruptures and shakes with such violence that buildings collapse.

Structures straddling a fault that ruptures Earth's surface may be literally torn in two during an earthquake. Most damage, however, occurs near the fault because of the radiated seismic waves. The fault undergoes a jerking motion as the rupture travels along it, radiating a complicated pattern of waves that reverberate as they bounce off various geologic interfaces in the crust, such as the bottom of a sedimentary basin or the base of the crust, and then are reflected back again from Earth's surface. The combination of jerking action and reverberation accounts for the oscillatory motion of the ground seen in the record of a seismogram. The primary factors controlling ground motion at a particular location during an earthquake are the seismic energy released by the earthquake, the distance of the site from the earthquake epicenter, and the response of the surficial geology to bedrock motion beneath the building location.

Three characteristics determine the potential damage to buildings as a result of ground shaking: amplitude, frequency, and duration.

The *earthquake-wave amplitude* (∞ see Chapter 5) is half of the difference in height between the crest of a wave and the preceding or following trough. Water waves provide a good analogy: Surfers hope for high-amplitude waves for the best ride. During earthquakes, building damage increases with increasing wave amplitude.

Frequency content of the earthquake is the second characteristic determining potential building damage. Earthquakes differ not only in magnitude but also in the types of waves generated and the energy carried by each wave type. This is a key characteristic because buildings and surficial geologic deposits respond to shaking in a resonant manner, amplifying the motion at particular frequencies.

Buildings have a natural frequency of vibration, rather like a plucked guitar string. If pushed, they will sway, and a tall building will take a longer time per swing than will a shorter building of similar construction. When the seismic waves have the same frequency as the building, each successive oscillation causes it to sway with an increasingly larger amplitude until the building fails. Such behavior, known as **resonance**, is the principle used when pushing someone on a swing. A series of well-timed, small pushes causes a person sitting on the swing to rise higher and higher; if you were to close your eyes and push without synchronizing with the swing, the amplitude would not build up and the person would not get much of a ride. The frequency of your pushes must match the natural frequency of the swing. As a rule of thumb, engineers estimate that the resonant period of a building is 0.1 sec times the number of stories in the building. A 100-story building will be affected by seismic waves having a period of 10 secs (0.1 Hz), whereas a one-story building will be affected by 0.1-sec waves (10 Hz).

In 1985, a large earthquake occurred off the coast of Mexico. The seismic waves traveled several hundred miles to Mexico City, which sits on an old lake bed. Because the lake bed is composed of much more compliant sediments than the surrounding rock, the waves were amplified and the bed vibrated like a bowl of jelly. Those buildings with resonant frequencies similar to the frequencies of the incoming waves were destroyed. Each oscillation of the building was pushed synchronously by the seismic waves. Nearby buildings—either taller or shorter but having different resonances—survived (Fig. 15.1).

The *duration of shaking* is the third important characteristic influencing earthquake damage to buildings. Building failure depends on the cumulative number of cycles that flex the building. The more times that a building or structure flexes, the more likely it is to fail. The rupture of a fault travels at speeds of approximately 3 km/sec. Thus, the duration of a magnitude 6 earthquake with a fault length of 18 km is about 6 sec. In contrast, a magnitude 8 earthquake with a fault length of 400 km would have a duration of 133 sec. The actual duration of shaking is longer than the duration of faulting because the former depends on seismic-wave propagation effects such as reflections from geologic structures. Such effects have smaller amplitudes but can increase the duration by more than a factor of 10.

Building failure is linked to the earthquake-wave amplitude, the frequency content of the waves, and the duration of shaking.

FIGURE 15.1 Damage to high-rise buildings during the 1985 Mexico City earthquake. Buildings with a natural resonance equal to the frequency of shaking fared much worse than did neighboring buildings with different frequencies of vibration.

Note that in many earthquakes, the greatest damage is caused by fire rather than by the shaking itself. Fires are caused by ruptured gas mains or electrical shorts. After the great 1906 San Francisco earthquake, fires raged out of control for three days. When the fires initially broke out, it was discovered that the water mains had burst, making it difficult to extinguish the flames. Similar problems were encountered in controlling fires in the Marina District in San Francisco 83 years later during the 1989 Loma Prieta earthquake. Auxiliary pumps were used to pump water from San Francisco Bay.

Effects of Geology on Building Survival

The physical properties of the ground material upon which a building stands may modify seismic waves traveling through it, possibly increasing or decreasing the amplitude of specific seismic waves. Thus, the building site exerts an important influence on the response of a structure to earthquakes. Tests on different soil and mud types have been conducted in areas where housing tracts have been constructed on soils of questionable strength. Such tests have shown that soft mud, for example, cannot transmit intense, high-frequency motion but can amplify low-frequency motion.

When landfill is placed on mud, such as the San Francisco Bay mud, the landfill is subjected to seismic shaking as modified by the mud underlying the fill. Building damage in San Francisco during the 1989 Loma Prieta earthquake was primarily in areas where landfill had been placed over bay mud. Ironically, the landfill consisted of rubble from buildings destroyed in the great 1906 earthquake. Earthquakes can also readily trigger catastrophic slope failures such as rock slides, landslides, and debris flows, which will have disastrous consequences for any structures built in the vicinity.

Liquefaction

In some instances, earthquake damage is more extreme in areas that are some distance from the epicenter than it is at the epicenter. Several factors may be responsible for this, including loose-fill building sites, which lead to a phenomenon called **liquefaction**. Liquefaction occurs when the ground becomes fluid from the shaking of an earthquake. It occurs in soils where the water table is near the surface. Over time, the soil grains settle so that they are in solid-solid contact with each other. These frictional contacts give the foundation its rigidity. When subjected to severe shaking by earthquake waves, however, the grains move apart slightly and groundwater flows in across the contacts, thereby lubricating them. The foundation acts like a liquid, and structures rooted in it experience severe damage (Fig. 15.2).

In the disastrous Kobe earthquake in Japan that occurred on January 17, 1995, liquefaction of reclaimed land in Osaka Bay caused as much as 3 m subsidence. Gasoline storage tanks may pop out of the ground as a

FIGURE 15.2 Liquefaction of the foundation soils of a building in the Niigata earthquake, Japan, June 16, 1964. Although the structures were undamaged, they were unusable.

(a)

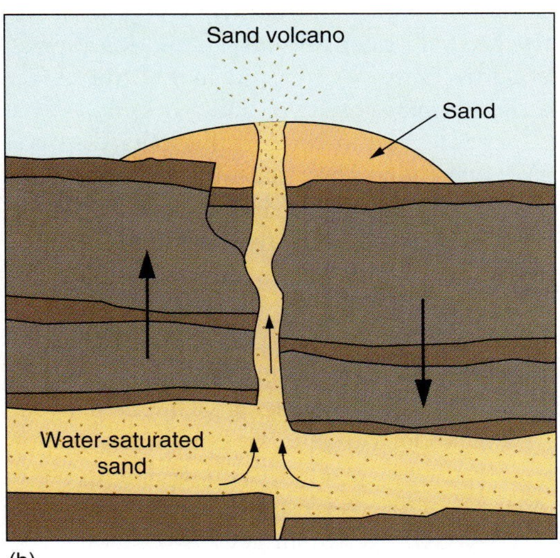

(b)

FIGURE 15.3 (a) Photograph of mud volcanoes. (b) Principle of a mud volcano or sand volcano. When mud or sand becomes liquefied in an earthquake, pressures build up, forcing the liquefied material through a fissure to form a miniature volcano.

consequence of liquefaction, creating very dangerous conditions. Earthquake-hazard maps take into account the depth of both the sediments and the water table in assessing potential risk. In general, safer locations are found on hills where the water table is low and the hard rock is not easily shaken. An extreme case of liquefaction can generate sand or mud volcanoes. Liquefied soil under an impermeable surface cap of hardened, dry soil builds up sufficient pressure to break through to the surface. The soil then erupts like a miniature volcano (Fig. 15.3).

Effects of Construction Techniques and Materials on Building Survival

In earthquake-prone areas, buildings must be designed to withstand substantial lateral forces imposed by earthquake shaking. Buildings are conventionally designed to withstand the considerable stresses caused by the weight of the structure—very few buildings collapse spontaneously under their own weight. During an earthquake, however, the building is shaken from side to side. In general, though, buildings are not designed to withstand a force of similar magnitude to that of gravity that is applied from the side, and the sideward force from earthquakes can reach values as large as or greater than that from gravity.

Why are some buildings not earthquake resistant? Building codes specify design requirements that are able to withstand the lateral forces created by earthquakes. The intent is to design structures that will not collapse during a major earthquake. The seismic provisions of building codes are not intended to prevent all damage to buildings, but should at least allow residents of the structures to live through the earthquake. In San Francisco, there are approximately 160,000 buildings in the city; more than half were built before earthquake standards were written into building codes in 1933. Furthermore, building-design codes lag behind our knowledge of the effects of earthquakes on buildings. It is the nature of such legislation to be a compromise between what is known about building safety and the economics of construction. For example, it may be possible to design and build a completely earthquake-resistant structure, but the cost might be so prohibitively high that it is impractical to build it.

Many structures have survived major earthquakes in cities such as San Francisco and remain in use after refurbishment. These structures may not be safe in the next earthquake. By evaluating construction materials and techniques, engineers can make a fairly accurate prediction of the extent of damage that building will incur during an earthquake. An unreinforced brick building is likely to collapse during a moderate earthquake. A reinforced building that is strongly attached to its foundation with roof strongly affixed to the walls is likely to survive with cosmetic damage at most. Brick or stone facings on walls, unless strongly attached to the building, are likely to fall off during an earthquake. Decorative parapets over building entries can be deadly. These commonly fall during earthquakes, reaching the ground at about the same time occupants of the building reach the entry in their rush to exit the structure.

In some developing countries, the typical mode of construction is adobe, which is a combination of sun-dried mud and straw. Many of the earthquake deaths caused by such a structure are due to lateral shearing of the walls, which causes the structure to collapse. A magnitude 7 earthquake, which might claim tens of thousands of lives in one country, may claim less than a hundred in a nation utilizing modern engineering practices.

The 1988 magnitude 6.9 Armenian earthquake claimed 25,000 lives, due largely to inadequate building construction (apartments built from concrete slabs), while the magnitude 7.1 Loma Prieta earthquake the following year in California (Focus 7.2) claimed only 62 lives even though the population densities around the hypocenters at Loma Prieta and Armenia were similar.

Earthquake damage depends on the earthquake source as well as on the foundation material and building resonances relative to the seismic-wave oscillations. Zones with a shallow water table in loose fill are prone to liquefaction, whereas buildings with solid rock foundations are safest.

Tsunamis

Tsunamis (after the Japanese word for "great wave") are ocean waves caused by sudden movements of the sea floor such as earthquakes, volcanic eruptions, or sediment slumps. In contrast to wind-driven waves that have a wavelength of dozens of meters and oscillate at a period of about 10 sec, tsunamis have wavelengths of hundreds of kilometers and oscillate over several minutes. Even though they have comparable amplitudes to wind-driven waves of several meters, they are not observed out at sea because they are so spread out and take so long to oscillate. On reaching shallow coastal water, the speed of the wave decreases. The part of the wave in the deep water catches up with that at the front, causing the wave to pile up, reaching heights of 20 m or more.

Flooding occurs as the water moves rapidly inland. Their description as "tidal waves" is a misnomer, however, because tsunamis have nothing to do with gravitationally driven tides. Tsunamis travel undiminished across the world oceans and therefore can present a hazard from an earth movement at a distance of thousands of kilometers. The tsunami from the great 1960 Chilean earthquake caused damage in Hawaii and Japan.

Tsunamis generally originate at convergent margins marked by submarine trenches, such as the Aleutian trench, the Japanese trench, and the Chile-Peru trench. The subduction that occurs at these trenches is not smooth but is marked by a series of jolts in the form of earthquakes. Large earthquakes can trigger sudden upward or downward movement of the sea floor over many hundreds of kilometers. The movement acts like a plunger, generating waves in the ocean. The great Alaskan earthquake of 1964 is a good example. At the time of the earthquake, a region of the sea floor approximately 1000 km long along the trench and 250 km wide buckled with an amplitude of about 10 m. This enormous push on the ocean resulted in a series of great waves traveling outward from the faulted margin.

A seismic sea-wave warning system in Honolulu monitors Pacific earthquakes, and water-level gauge readings on islands in the Pacific transmit data directly to the warning center. Once a major earthquake has been registered, the scientists issue warnings for times that the tsunami will hit Pacific rim coasts. Tsunamis travel at speeds of about 800 km per hour in the deep oceans. Thus, a warning can be issued many hours in advance.

Earthquake Strength

Mercalli Intensity We saw in Chapter 7 that the most common way to express the strength of an earthquake is to use the Richter magnitude scale that was designed for the earthquakes measured in California. The concept was soon extended to worldwide earthquakes using either body waves or surface waves (∞ see Chapter 7). In either case, we refer to magnitudes determined from seismic wave amplitudes as Richter magnitudes. However, the magnitude scale does not convey information about damage. For example, a magnitude 7 earthquake at a depth of 600 km causes negligible damage. An earthquake of this size at a shallow depth in a populated area can be catastrophic. As mentioned in Chapter 7, the Mercalli intensity is a method for expressing the effective strength of an earthquake at the surface and is based on the damage observed and other effects such as landslides. It was devised in 1902 by the Italian seismologist Giuseppe Mercalli. The Mercalli intensity scale ranges from I to XII; note that Roman numerals are used to distinguish the Mercalli scale from the magnitude scale (∞ see Chapter 7). Table 15.1 presents a general summary that includes the names applied by early observers.

Richter magnitude is an instrumental measurement obtained using a seismometer, whereas Mercalli intensity is based on observation of the area affected by an earthquake. After about 60 years of comparing magnitude and Mercalli intensity, scientists have been able to correlate these two methods to some extent. Using this correlation and eyewitness accounts of historical earthquakes, we can determine an approximate magnitude for earthquakes of long ago. Thus, when you read of a magnitude applied to an earthquake that occurred before the invention of the seismograph (late in the nineteenth century) the magnitude is derived from this correlation.

For a large earthquake, the intensity of damage is generally maximum in an area near the epicenter and decreases at greater distances. Contours can be drawn that separate regions experiencing the same intensity of shaking. These lines are called *isoseismals*. The Mercalli intensity for the event is the largest intensity on the isoseismal map (Figure 1 of Focus 11.2 shows isoseismal lines for the New Madrid, Missouri, earthquake of 1811).

Richter Magnitude and Moment-Magnitude Scales The Richter magnitude of an earthquake depends on the maximum displacement as recorded on a standard seismometer at a given distance (∞ see Chapter 7).

TABLE 15.1 Mercalli Intensity Scale

MERCALLI INTENSITY	CHARACTERISTIC EFFECTS	APPROXIMATE RICHTER MAGNITUDE
I—Instrumental	Not felt	1
II—Just perceptible	Felt by only a few persons, especially on upper floors of tall buildings	1.5
III—Slight	Felt by people lying down, seated on a hard surface, or in the upper stories of tall buildings	2
IV—Perceptible	Felt indoors by many, few outside	3
V—Rather strong	Generally felt by everyone; sleeping people may be awakened	4
VI—Strong	Trees sway, chandeliers swing, bells ring, some damage from falling objects	5
VII—Very strong	General alarm; walls and plaster crack	5.5
VIII—Destructive	Felt in moving vehicles; chimneys collapse; poorly constructed buildings seriously damaged	6
IX—Ruinous	Some houses collapse; pipes break	6.5
X—Disastrous	Obvious ground cracks; railroad tracks bent; some landslides on steep hillsides	7
XI—Very disastrous	Few buildings survive; bridges damaged or destroyed; all services interrupted (electrical, water, sewage, railroad); severe landslides	7.5
XII—Catastrophic	Total destruction; objects thrown into the air; river courses and topography altered	8

Richter magnitude was designed for the moderate earthquakes that are common in California; however, a problem arose. For the very largest earthquakes (greater than Richter magnitude 8), the maximum amplitude of the displacement does not increase systematically with the size of the event. The largest earthquakes generated readings as if they were composed of a series of smaller earthquakes sequentially strung end to end across the countryside. Thus, the amplitudes of the waves were closer to that of smaller earthquakes strung end to end. The seismograms have longer durations, but the amplitudes do not increase for larger events.

The solution to this problem was to use the *moment*, which can be estimated from a seismogram by summing the amplitudes. Recall that moment is the product of slip times area times elastic modulus (∞ see Chapter 7). Clearly, this quantity increases for increasingly larger earthquakes. A simple formula (Focus 15.1) is used to convert moment to *moment magnitude*, denoted as M_w. The Richter magnitude scale and the moment magnitude scale give approximately the same values for earthquakes up to about magnitude 7. Currently, newspaper reports of earthquakes provide moment magnitude rather than Richter magnitude. The largest Richter magnitudes are in the high 8s. In contrast, the largest moment magnitudes are in the 9s; the great Chile earthquake of 1960 had a moment magnitude of 9.7 but a Richter magnitude of only 8.5 (Focus 15.1). No earthquake having a moment magnitude greater than 10 has occurred since scientists started taking such measurements.

From a hazard point of view, the fact that the amplitude does not proportionately increase with increasing earthquake size is an important finding for engineers who design earthquake-resistant buildings. It means that the shaking forces from larger earthquakes do not increase without limit. The force exerted on a building is given by its mass times the acceleration caused by the ground motion. Maximum accelerations generally are never much greater than 1 *g* for earthquakes in the range from magnitude 6 to 9. However, durations increase for larger magnitudes, which has to be taken into account in the building design because extended shaking can cause material failure such as metal fatigue.

The Mercalli intensity scale for earthquakes is defined from observing the damage and related effects generated by the shaking. The Richter magnitude, or *strength*, of an earthquake is instrumentally measured using a seismometer. *Moment magnitude* is used for large earthquakes.

Volcanic Hazards

Volcanic eruptions are among the most spectacular natural phenomena, but they also possess enormous capacity for destruction. Volcanic hazards may be immediate (the direct effects of the eruption), short term (such as flooding and *lahars* resulting from the accumulation of pyroclastic material), or long term (such as changes in climate that may result from the injection of ash and gases from volcanic eruptions into the atmosphere). We have learned in earlier chapters that different types of volcanoes and therefore eruptions are associated with different plate-tectonic

Focus On 15.1

CALCULATION OF MOMENT MAGNITUDE

We defined *moment* (∞ see Chapter 7) as the product of slip, area, and elastic modulus:

$$M_o = \text{slip} \times \text{area} \times \text{modulus} \qquad (1)$$

Seismograms for distant seismometers are used to obtain a value for M_o by summing up amplitudes. Then, to calculate the moment magnitude, M_w, from the moment, we use

$$M_w = 0.69 \log_{10} M_o - 6.4 \qquad (2)$$

We can use the Chilean earthquake of May 22, 1960, as an example. The Richter magnitude was found to be M = 8.5 (Fig. 1). However, the moment (M_o) determined from the seismograms was found to be 2×10^{23} Newton meters. If we substitute this into Equation 2, we find a moment magnitude of 9.7.

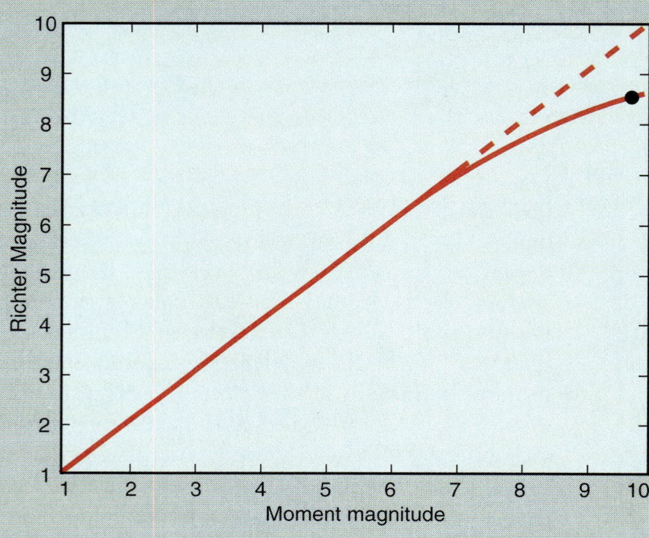

FIGURE 1 Comparison of Richter magnitude and moment magnitude, showing that Richter magnitude gives lower estimates for larger events. The dot shows the Richter (M = 8.5) and moment (M_w = 9.7) magnitudes for the great Chilean earthquake of 1960. Moment magnitude is regarded as the better estimate of the size of a large earthquake.

DIRECT CALCULATION OF MOMENT

Instead of using seismograms to calculate moment, suppose that we could make field observations of the slip and fault length and could estimate the depth of the seismic event from the aftershock distribution. We could then calculate the moment directly. We can again use the Chilean earthquake as an example.

Given that the modulus (∞ see Chapter 5) is 10^{11} Pa, length = 900 km = 900×10^3 m, width = 100 km = 100×10^3 m, and slip = 22 m, we can find the moment by substituting into Equation 1:

$$\text{Moment} = 10^{11} \times 900 \times 10^3 \times 100 \times 10^3 \times 22 = 2 \times 10^{23} \text{ Newton meters.}$$

You can see that the dimensions of a magnitude 10 earthquake would probably require a fault length greater than 1000 km—a fearsome prospect.

settings. Intraplate (hot spots or plumes) and divergent margin magmatism can produce voluminous outpourings of lava, but the eruptions tend to be relatively gentle. In contrast, volcanoes along convergent plate margins tend to erupt explosively, and sometimes without warning.

Types of Hazards

As with earthquakes, volcanic hazards can generally be divided into those directly associated with the eruption itself (lava flows, volcanic blasts, pyroclastic flows, and

FIGURE 15.4 A basalt lava flow from Kilauea, Hawaii, surrounded and destroyed the National Park Service Visitors' Center at Wahalua in 1989.

ashfall), and secondary effects triggered by the eruption (such as lahars, landslides, and gas emissions).

Primary Effects Lava tends to flow quite slowly and, as a result, people can get out of the way. The same is not true for property. So, while lava flows pose only a minor threat to human life, they may cause considerable damage to buildings in their path, which are incinerated (Fig. 15.4). The paths followed by lava flows can often be predicted based on the topography of the volcano and the direction taken by previous flows. Populations in the path of a flow can be evacuated. Occasionally, attempts are made to divert the flow of lava. Earthen barriers constructed to protect houses from lava at Mt. Etna in Sicily have been moderately successful. Lava from the 1973 Heimaey eruption on Iceland was prevented from blocking the entrance to the local harbor by spraying the lava continuously with water to solidify it. This action saved the major livelihood of the local population—fishing. Though the eruption destroyed much of the town, it actually improved the harbor and the town is now heated by geothermal energy from the hot lava flow.

Pyroclastic flows present a far more serious hazard. They comprise mixtures of hot gas, ash, and larger volcanic fragments (∞ see Chapter 9), and can move at speeds greater than 150 km per hour (100 mph). During the 1980 eruption of Mount St. Helens, a motorist attempted to escape a pyroclastic flow by driving at speeds exceeding 90 mph and, even so, was overtaken by the flow. Fortunately, the driver escaped because he was near the edge of the flow, but his testimony is startling. The speed, heat, and sheer force of such flows enable them to destroy life and property in a matter of seconds. Some 26,000 residents of the town of St. Pierre, Martinique, were wiped out by pyroclastic flows when Mt. Pelée erupted in 1902 (see Focus 9.2). The casts of people and animals of Pompeii that were preserved in pyroclastic deposits from the eruption in A.D. 79 of Mt. Vesuvius in Italy indicate that they were killed very suddenly. Many appear to have been knocked down by the flow in the street. Others died covering their faces from the searing hot, suffocating ash and gases (Fig. 15.5).

Explosions accompanying pyroclastic eruptions can also be rapid and deadly. The lateral blast at Mount St. Helens (see Focus 9.3) generated sufficient force to mow down a forest for a distance of more than 12 km from the volcano, overturning machinery weighing 10 tons or more. In 1993, a small summit explosion at Galeras volcano, Colombia, killed a group of volcanologists participating in an international workshop studying volcanic hazard prediction. The suddenness of the tragedy underscores the fact that some volcanic eruptions continue to elude prediction.

One of the most spectacular manifestations of explosive volcanic eruptions is the enormous column of hot ash and gas rising from the volcano to heights commonly exceeding 10,000 m. Large eruption columns in remote areas may pose a serious threat to air traffic (∞ see Chapter 9). Ashfall also poses a serious hazard. Ash accumulating on the roofs of buildings can cause them to collapse. In the eruption of Rabaul volcano in New Guinea in the fall of 1994, the population was successfully evacuated from the town, but many of the buildings were subsequently destroyed by collapse under the weight of accumulating ash. Ash accumulates as a blanket across all of the land downwind from a volcanic eruption. The long-term effects of this may also be hazardous—killing crops, obstructing communications routes such as roads and railways, and clogging streams, to cause flooding.

Secondary Effects Among secondary volcanic hazards, lahars pose the greatest threat to life and property. Lahars

FIGURE 15.5 Casts of victims of the A.D. 79 eruption of Mt. Vesuvius. Pyroclastic flows overwhelmed victims at Pompeii.

are volcanic mudflows, common on many volcanoes because of the combination of loose, unconsolidated (pyroclastic) material, the availability of water (as melted snow, streams, or high rainfall), and the presence of steep slopes.

The hazards from lahars are particularly acute during an eruption, when pyroclastic material is rapidly erupted and channeled into river valleys where it mixes with water. These volcanic mudflows can transport large boulders and other debris because mud has a greater carrying capacity than does water (∞ see Chapter 13). The chief hazard to human habitations and populations from these mudflows arises if the flows reach river courses that lead into towns located near the volcanoes. At Mount St. Helens, some of the greatest damage in the 1980 eruption occurred as a result of mudflows (see Focus 9.3).

In 1985, a lahar on the volcano Nevado Del Ruiz in Colombia caused 22,000 fatalities. Heat from the volcano's renewed activity melted ice at the summit of the volcano. Hot water mixed with mud and raced down a canyon to engulf the town of Armero. Volcanologists had observed the warning signs of earthquake activity and minor eruptive activity. In addition, a lahar had also killed about a thousand people in this area 145 years earlier. A warning was issued that this sequence might be repeated. The town of Armero, the largest that was destroyed, was built on an ancient mudflow from the volcano. However, poor coordination between officials and residents, as well as the inevitable uncertainty in the prediction, led to a tragedy that might have been averted by evacuation of threatened towns and villages.

More recently, the eruption of Mt. Pinatubo in the Philippines (1991) was accompanied by devastating effects from lahars. Powerful though the eruption was, most of the population had been evacuated from the area immediately affected. The combination of accumulating pyroclastic material with the arrival of a typhoon bringing heavy rainfall, however, triggered widespread lahars that traveled tens of kilometers downslope from the volcano, washing away buildings and bridges and causing widespread flooding (Fig. 15.6a). Even today, several years after the eruption, the drainage system around the volcano is still choked with unconsolidated ash, which continues to be washed down in lahars.

Volcanic landslides (termed *sector collapse*) have now been recognized to be a common feature of many volcanoes at convergent margins (∞ see Chapter 9). The sector collapse at Mount St. Helens in 1980 produced an enormous debris avalanche. In this case, the slide was triggered by the eruption itself; in other cases, it seems that the slide is simply triggered by the gravitational instability of the steep volcano flanks (∞ see Chapter 14). Although the actual area affected by such a debris avalanche may be limited, its effect on local drainage can be profound.

Sector collapses have the potential of causing sudden and devastating destruction of populations located close to convergent margin volcanoes (Fig. 15.6b). Ocean island volcanoes, such as Hawaii (∞ see Chapter 11), may also collapse despite the rather gentle slopes that characterize them. In this case, it is the constant injection of magma into the volcanoes that acts to wedge large sections of the island away. Giant landslides have occurred in the past around the Hawaiian Islands, leaving landslide deposits on the surrounding sea floor. In such cases, it is perhaps the tsunamis that would be generated by such an enormous mass of land sliding into the sea that would cause a greater threat to human life than the landslide itself.

Finally, emission of volcanic gases at fumaroles or during eruptions may also be hazardous. Plants may die as a result and cause a break in the food chain. In today's world of rapid communication and transportation, this might not ultimately pose a serious threat of starvation to humans, but food shortages as a consequence of volcanic eruptions in the past have been serious. Most notably, the eruption of Laki in Iceland caused widespread famine and death.

Volcanic Hazards in the United States

For most of the highly populated regions of the United States, volcanoes do not pose a direct hazard. The Hawaiian volcanoes, over the Hawaiian plume (∞ see Chapter 11)are an ever-present danger, but their eruptive style is rather quiet. Fissure and crater eruptions cause basaltic flows to cascade down the slopes of the mountains. These are "well-behaved" volcanoes, and their activity follows a fairly predictable pattern (∞ see Chapter 7). Houses and entire towns may be overwhelmed by the flows, but generally with minimal loss of life. In contrast to the quiet character of the Hawaiian

(a)

(b)

FIGURE 15.6 (a) Destruction caused by lahars from Mt. Pinatubo in the Philippines. (b) Avachinsky and Koryaksky volcanoes in Kamchatka, Russia. The city of Petropavlosk in the foreground has more than 100,000 inhabitants and is built on debris avalanche deposits from the volcanoes. Debris avalanches that formed as a result of sector collapse pose a major threat to cities in the proximity of steep-sided volcanoes such as these.

volcanoes, subduction-related volcanoes of North America's western coast are explosive and dangerous. The major volcanic hazards of North America are associated with the convergent western margin of the continent in northern California, and in Oregon, Washington, and Alaska. Several of these volcanoes are near sizable population centers: Mount Shasta is near Redding, California; Mount Hood is near Portland, Oregon; and Mount Rainier is near Seattle, Washington.

When volcanoes of the Cascades arc erupt, the hazard is very real, as Mount St. Helens indicated in 1980. Even this eruption was extremely small in size compared with some of the enormous caldera-forming eruptions known to have occurred in the geologic past. Fortunately, a large caldera-forming eruption such as occurred in Long Valley, California, 750,000 years ago has not happened in recorded history.

In the interior of the western United States, relatively recent volcanism is associated with either plumes or rifting. The Rio Grande Rift, an incipient continental rift, may ultimately open to create an ocean basin millions of years in the future. The volcanism associated with the Valles caldera on the edge of the rift presents a long-term caldera threat—although the last major eruption occurred over a million years ago. Huge caldera-forming eruptions have also occurred within the past 2 million years at Yellowstone National Park, possibly related to a mantle plume rising beneath the continent. Although caldera-forming eruptions are catastrophic, they occur rarely (see Fig. 9.9).

Volcanoes at subduction zones pose the greatest common volcanic threat to humans. Explosive activity produces pyroclastic flows, ashfall, and lahars.

Flooding

Flooding is a normal part of the development of streams and stream valleys and poses a major hazard to populations living on low ground near waterways. Major factors controlling the frequency of stream flooding are the amounts of rainfall or snow pack. Areas hit frequently by hurricanes suffer disastrous floods as a result of the torrential rains accompanying the storms. Tsunamis trig-

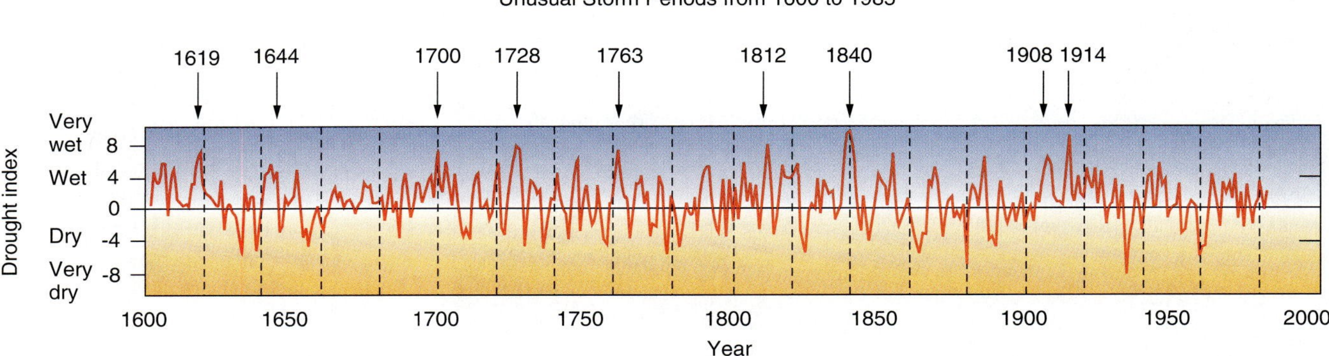

FIGURE 15.7 Diagram of storm frequency in northeastern Nevada. Unusually wet periods are flagged by arrows and are defined as consisting of abnormally high rainfall at least three winters in succession. Nine unusually wet winters over 380 years yield an average interval between wet years of approximately 40 years, although the actual interval between wet years varies from 6 to 68.

ger flooding from the ocean. Floods caused by the collapse of a dam are quite rare but, nevertheless, can be devastating.

You may have heard such terms as a "50-year flood" or a "100-year flood" in the media. About every 50 or 100 years or so, an exceptionally large rainfall inundates an area. This does not mean that these exceptional events occur at regular intervals; they are more random than that. In any given region, the intensity of such storms tends to remain about the same when averaged over a period of several years. Thus, whereas there is some variation on a year-to-year basis, most storms fall within an average range of wind speed and rainfall intensity. Figure 15.7 shows the rainfall data from northeastern Nevada over the past 385 years. This figure clearly shows that, although the unusual storms do not occur with exact regularity, major storms in this area do have an average cycle of about 40 years.

Like earthquakes and volcanic eruptions, storms have many more small events than large ones. Large events are determined by the scale of the system. For earthquakes, it is the size of plate margins. For storms, it is the size of the hydrosphere–atmosphere system. Why a given system exhibits a large event rather than many small events is the subject of much present research (∞ see Focus 14.1).

We have discussed the rainwater runoff sequence that starts with the infiltration of dry soil (∞ see Chapter 14). Initially, most of the rainwater soaks into the dry soil with minor runoff, but the soil eventually becomes saturated and total runoff begins. The more torrential the rainfall, the more rapidly this sequence of events occurs. Runoff eventually reaches rivers; this causes the rivers to run at high flow during times of heavy rainfall. During low rainfall periods, the rivers flow at their lowest stage. Perennial streams depend on contributions from the groundwater to maintain their flow during times of low flow (Fig. 15.8). Thus, the amount of soil infiltration and, in turn, the extent of groundwater discharge are both directly related to river flow during low-flow stages.

Land use, particularly urbanization, strongly affects peak-flow characteristics, total runoff, and the water quality of a stream. A principal factor governing the year-round nature of stream flow is the percentage of drainage area that is impervious to water infiltration and percolation. In urban or suburban areas, impervious surfaces (asphalt, concrete, roofing, and so on) are areas of rapid runoff with no water infiltration. With no infiltration, the rainfall flows directly to the rivers, resulting in higher high-stage flows (Fig. 15.8b). Lack of water percolation into the ground lowers the groundwater level; thus, between rainy seasons, there will be little or no groundwater contribution to the rivers. This results in a low river flow that is lower than it was prior to urbanization. In extreme cases, urbanization causes streams to dry up.

Urbanization near rivers or streams causes higher flood stages and lower low-flow stage than before settlement. Immediate runoff during storms without water percolation into the ground causes higher floods. Lack of percolation decreases groundwater contribution to river flow between rainy seasons.

Incidental Pollution and Flood Runoff

Water quality is also affected by urbanization. Storm wash from streets carries waste oil products, partially burned soot from cars and trucks, and rubber scrapings into the storm drains and, ultimately, to rivers and lakes. In many cities, the major sewage-treatment plants run at design capacity for handling and treating municipal wastewater. Sewer accesses exist in the streets and gutters; these invariably leak water into the sewers during heavy rains. Usually there is no way to separate the leakage from the sewage, so the wastewater treatment plants may be overloaded during storms. Therefore, during nearly every heavy rainstorm, these treatment plants can be overloaded with amounts of water that are two or three times their treatment capacity. This additional water bypasses

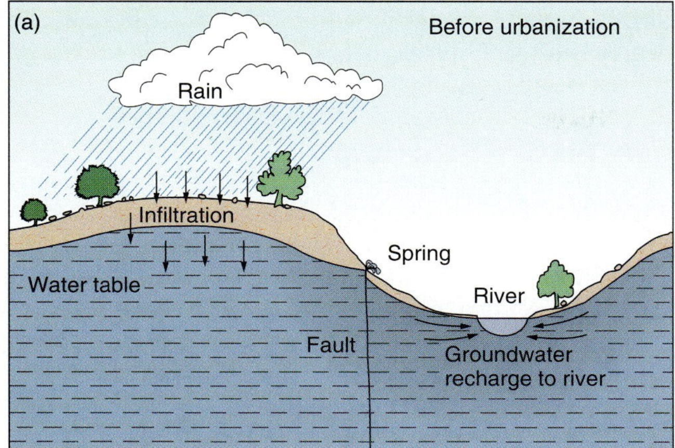

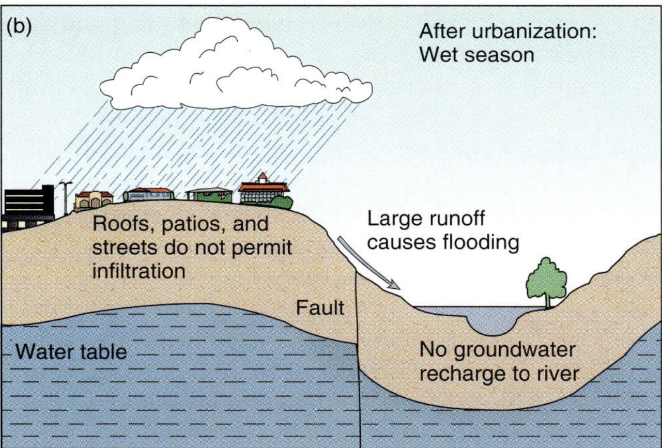

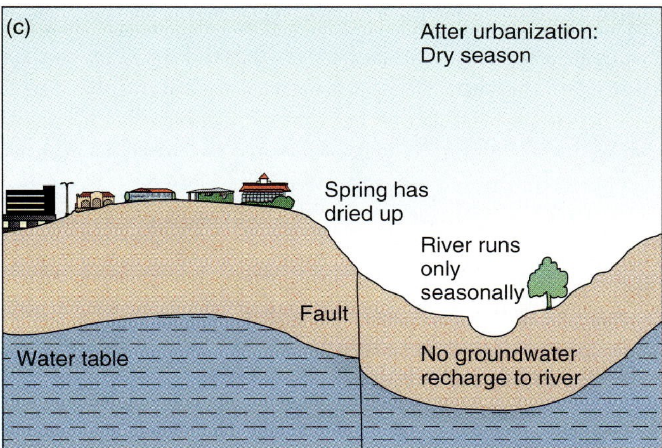

FIGURE 15.8 Effects of urbanization on infiltration and runoff of rainfall. (a) Before urbanization, the water table is high; in this case, dammed by a fault that causes a spring to flow. Groundwater recharges the river. (b) After urbanization, in the wet season, rapid runoff causes flooding and does not replenish the water table. (c) In the dry season after urbanization, the spring and river are dry.

treatment systems, causing a massive release of water that is untreated and polluted. In coastal cities, the excess water flows out into the ocean, creating health hazards and resulting in closures of beaches and fisheries.

Inland cities and towns utilize waterways as both a supply for drinking water and a sewage disposal system. Even treated sewage contains dissolved constituents that can act as nutrients for algal growth and may alter the balance of organisms in the stream. Runoff from lawns and gardens carries fertilizers and various pesticides. As a result, it is not uncommon for the water quality of these waterways to be seriously degraded after urbanization or suburbanization.

Stream Channelization

From an engineering perspective, some streams may be inconveniently located, and efforts to relocate such streams to another channel are sometimes attempted. If a meandering stream is straightened, the slope of the new channel will be much steeper than that of the original, natural channel. As the stream strives to achieve an equilibrium profile, severe erosion will occur in the upper reaches of the new channel, whereas substantial flooding and sediment deposition will take place in the lower reaches.

Building Design and Floodplains

Municipalities vary greatly in their susceptibility to flooding, and even areas that are subject to frequent flooding might not take rudimentary precautions to prevent property damage or loss of life. Streams and rivers usually have well-defined floodplains, which are relatively flat areas inundated with water when stream flow overtops the stream banks. Unfortunately, river floodplains are often wonderful housing sites: The ground is relatively flat, requires little grading, and the soil is rich in humus and nutrients. Old and often beautiful trees cover these areas, slightly elevated above the river level. With advancing urbanization and its attendant impervious cover creating sudden runoff during rainstorms, these low-lying areas stand in danger of severe flooding when unusually heavy rains do occur. Regardless of this potential danger, houses, businesses, and roadways are often built on floodplain surfaces.

Homes and businesses constructed on such sites are subject to periodic flooding based on the frequency of heavy storms in the particular area. A few municipalities have passed ordinances to minimize death and property loss from flooding. Areas that receive floodwaters every few years—a 10-year flooding cycle, for example—may be zoned for parks and open space. Any structure in these areas must be built so that damage by floodwaters is minimized. This type of construction requires that the buildings be erected on pillars, with walkways, landscaping, and parking areas in lower areas. Unfortunately, too few cities or counties have such ordinances. If an area does not experience frequent major flooding, then it seems to be more likely that the danger and damage resulting from such flooding will be overlooked.

LANDSLIDES

Of all geologic hazards affecting humans, landslides and slope stability problems over the long term fall second only to earthquakes in financial loss. The reason for this is quite simple: If given a choice, people prefer to build a house on a hill with a good view. As a general rule, anything built on a slope tends to move downslope. Architects, engineers, and builders must evaluate the source of potential slope instability and take appropriate measures to prevent sliding. In this section, we use landslide in its broadest sense to describe all types of mass wasting.

Geology of Landslide Damage

As we have learned (∞ see Chapter 14), slope stability is strongly influenced by the nature of the underlying bedrock. Granites and well-cemented limestones, for instance, are stronger than clays, mudstones, and salt. Even a structure constructed on solid bedrock, however, may show a tendency to move downhill. One major control on site stability is the orientation of rock beds relative to the hillside. On a flat surface, mudstones may provide a perfectly adequate foundation for a house, but such substrates are more likely to fail on a hillside.

If sedimentary layering or fractures in massive rock are approximately parallel to the hillslope, the site is weak. The bedding or fracture surfaces are weak planes that may slip, particularly if the weight of a house or swimming pool is placed on top of them. Even deep and strong foundations are not very helpful under these conditions because the bedding surfaces or fractures may lie deep within the hillside below the base of a foundation. In wet weather, groundwater lubricates fractures or bedding planes, which reduces frictional resistance and can cause slippage.

FIGURE 15.9 Slump in Daly City, California.

Of the many types of downslope mass movement, only a few such as rockfalls are likely to pose risks to human life. Slumps and earth flows (∞ see Chapter 14) represent the primary dangers to human-made structures (Fig. 15.9). These two processes grade into each other. A slump largely maintains its cohesiveness, sliding downhill more or less as a unit for some distance, whereas an earth flow behaves more like a liquid.

Causes of Landslides

The steepness of the slope is one of the most significant factors contributing to landslides. This is obviously modified by the type of rock on which the structure is built, the thickness and type of soil, and the climate. Given that steeper slopes are more likely to fail by landsliding, it should be clear that any modification of the land that results in steeper slopes increases the danger of landsliding. The most common modification of this type is undercutting the base of a slope, which decreases the stability of the hillside.

Undercutting is practiced for several reasons, including construction of access roads to homes in a new subdivision or to increase the number of available lots in a hillside housing development. Each housing site is flattened by cutting into the uphill portion of a lot, and the fill dirt is then dumped on the downhill side. Building codes require that houses be constructed on "cut" rather than "fill" because fill is known to be less stable. Thus, the problem of stability becomes compounded for each lot situated farther downhill as there is an artificially steepened slope of unstable fill dirt above each house (Fig. 15.10). Although it is true that an entire subdivision is not likely to fail, one or more of the homeowners in a hillside subdivision might arrive home some rainy afternoon to discover the kitchen or bedroom filled with their uphill neighbor's yard.

Increased water content increases the weight of wet soil on a steep hillside and tends to decrease the friction holding the soil in place (∞ see Chapter 14). This effect is heightened by the construction of undrained, impervious walls, which prevent the hillside from draining naturally and reducing its slope. If walls are desirable, it is essential that the slope be adequately drained so that excess water will not accumulate. Buildup of a large amount of water behind a high wall may cause it to collapse catastrophically.

Prevention of landslides in housing developments involves a number of procedures: drainage of slopes and walls; reducing the slope, if possible; planting the slope with deep-rooted trees or shrubs to consolidate the soil if

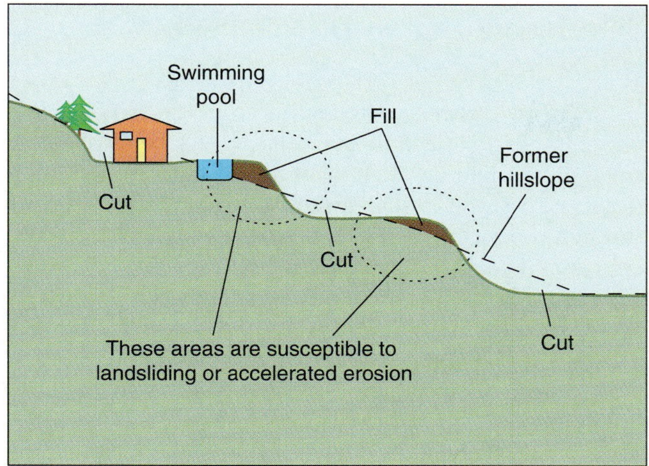

FIGURE 15.10 Homesites on hillside communities are cut into the slope on the uphill side and filled on the downhill side of the lot. Access roads are also cut into the sides of the hill. In each case, the slope is steepened, and the likelihood of landsliding or soil erosion is increased.

slope reduction is not possible. If walls are needed to keep your neighbor's yard where it belongs, pilings and buttresses, in addition to drainage, are worth the investment.

Buildings on steep hillsides are inherently unstable. Landscape modifications, such as excavating the downslope side of the hill, heavy watering of the slope, and high retaining walls without proper drainage may make a structure even less stable.

Coastal Erosion Problems

A beach may vary as much as 3 to 5 m in height because the shoreline undergoes seasonal variations depending on the height and wavelength of the incoming surf (see Fig. 14.20). Winter storms in the open ocean tend to generate large waves with long wavelengths, and summer winds generate smaller waves with shorter wavelengths. The winter storm waves erode the beach, depositing the sand on offshore bars (see Fig. 14.20). Generally, this seasonal sand migration does little damage. However, towns, cities, and vacation homes adjacent to the coastline occasionally suffer the damage from major storms. We have discussed the dynamics of sand transport along beaches (∞ see Chapter 13).

Sediment emptied into the sea by rivers and streams, and then migration along the coast under the influence of longshore drift, is the primary replenishing agent for beaches. Interference with this process, such as the construction of breakwaters and groins that inhibit longshore drift, will result in diminishment or complete disappearance of beaches.

Very few rivers in the United States are without dams, weirs, or other impediments to sediment transport. Most sediment is trapped in lakes formed along the rivers behind dams, and it no longer reaches the coast; the net effect is sediment starvation of the beaches. As transportation has increased in efficiency in the past few decades, construction of homes and recreational housing adjacent to particularly scenic coastal areas has increased. "The closer to the water the better" has been the developer's guiding principle. As a result, small changes in the sand budget for local areas along a coast may have a profound effect on the near-water buildings. Lessening of sediment delivery from a single river may cause the shoreline to move landward, with the result that the near-water buildings no longer have the safety buffer of a zone of sand between violent storm waves and the buildings.

Surface Subsidence

Subsidence Due to Fluid Withdrawal

An important geologic hazard in flat areas is subsidence. In areas underlain by thick sedimentary sequences, part of the load of the sediment pile is carried by contact forces across grain-grain boundaries; the remaining part is carried by fluids within the pore spaces among the grains. This latter portion may be borne two ways: by buoyancy or by the pressure developed in a confined fluid. With the former mechanism, each sediment grain displaces a volume of water (or another fluid in the sediment, such as oil) equal to the volume of the particle itself. The weight of water displaced represents the buoyant force exerted on each grain. This buoyant force is less than the total weight of the grain. For example, because quartz has a density of 2.65 gcm^{-3} and water has a density of 1.0 gcm^{-3}, slightly more than one-third of the weight of each quartz grain will be supported by its buoyant force.

The other fluid mechanism that supports the weight of the overlying sedimentary sequence is the pressure developed in a confined fluid. If the water or oil is confined, the fluid is nearly incompressible, so that it can support the entire weight of overburden in much the same way that hydraulic systems in machinery can support massive weights (for example, a car jack).

This system is more or less in equilibrium. If the fluid is withdrawn from the sediment, however, then the equilibrium is destroyed. For instance, major subsidence has been recorded in parts of California where large quantities of groundwater have been removed for irrigation by pumping from deep wells. Near the California town of Mendota, a U.S. Geological Survey benchmark dropped 7.6 m (25 ft) during approximately 30 years of intensive farming and irrigation.

FIGURE 15.11 Near-surface subsidence in the San Joaquin Valley of California, caused by the rapid withdrawal of irrigation water. This type of subsidence cannot be restored because the removal of water allows a closer packing of the sediment grains.

There are several areas in California's San Joaquin Valley (Fig. 15.11) where this type of subsidence has occurred. Given the location and the slow rate of subsidence, no major damage was incurred, so this type of subsidence is more a nuisance than a hazard. However, the city of Long Beach, California, faced a serious situation from withdrawal of petroleum from the Wilmington oil field. In the period between 1928 and 1968, petroleum extracted from an area near the Long Beach docks caused up to 8.5 m (28 ft) of subsidence (Fig. 15.12). Several of the docks were awash at high tide and parts of that port city were in danger of being flooded by the sea. The operating company managing the Wilmington Oil Field began to inject water at a slightly greater rate than the rate that petroleum was being produced; this stopped the subsidence and caused a small amount of rebound. Although petroleum is still produced at Wilmington, no further subsidence has occurred since this water injection program was initiated.

Subsidence is caused by withdrawal of underground fluids. Production of large volumes of irrigation water and petroleum has caused major subsidence in some areas.

Subsidence Due to Dissolution of Underlying Rocks

You may have either visited a limestone cavern or seen a photograph of one and marveled at its beauty. These caves are formed by the flow of freshwater through limestone. Limestone is slightly soluble. As the water flows through the rock, channels are slowly widened until caverns are created (∞ see Chapter 12).

It should be apparent that a large underground cavity is not a stable situation for the ground above. Collapse of the roof is the rule rather than the exception. In fact, most caves collect large heaps of rubble on their floors, testifying to the periodic collapse of the ceilings. Occasionally, a type of archway forms, making the ceiling at least temporarily stable. It is not unusual for collapsing caves to progressively work their way up to Earth's surface. When this happens, the landscape develops a peculiar pattern of small (dozens of meters) depressions called sinkholes. Generally, sinkholes pose no significant danger; they are usually only a few meters deep and rarely open downward into major caverns. Occasionally, however, collapse of these underground caves is so rapid that a wide cavern opens to Earth's surface. In Winter Park, Florida, a sinkhole began to develop in 1981 that was large enough to swallow several houses. Eventually, of course, the surface hole stops widening as the collapse of the underlying cavern becomes complete (Fig. 15.13).

POLLUTION

Our surface environment can be affected by artificial introduction of material and chemicals. Increasingly, geologists are called upon to evaluate such environmental effects to establish criteria for the control and limitation of pollution to safeguard water supplies, wildlife, and even human life. Dealing with pollution involves the logistics

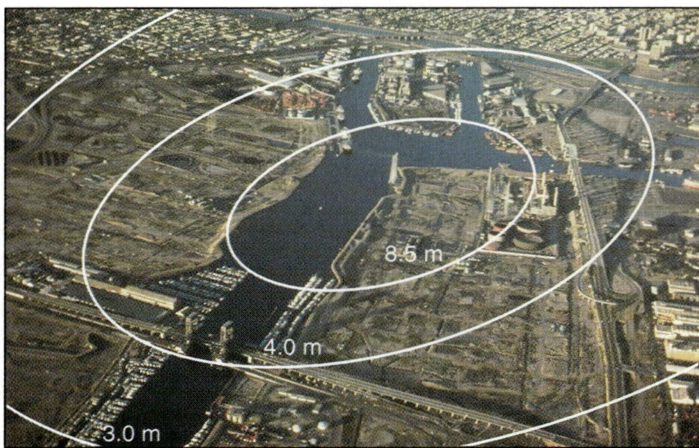

FIGURE 15.12 An oblique view of the city of Long Beach, California, with contours of subsidence from 1928 to 1968 (in feet) superimposed. Subsidence was caused by withdrawal of petroleum from the Wilmington oil field.

FIGURE 15.13 A surface sinkhole caused by the collapse of an underground limestone cavern in Winter Park, Florida, 1981. These sinkholes may expand to damage or destroy property, but most occur away from settled areas. Although the roofs of all limestone caverns are subject to collapse, not all cause sinkholes that reach Earth's surface.

of cleanup, the organisms affected, the environments that are polluted, and the degradation time of the pollutants. We will deal here with only a few types of pollution, and some of the remediation techniques employed.

Legislation exists to regulate the dumping of chemical and biological materials; the strictest controls exist for the most toxic materials, including radioactive waste, pesticides, and metals, such as mercury. Long-lived radioactive wastes or chemical and biological warfare agents are certainly dangerous by almost any standard and are carefully controlled. No biological agents or processes can consume or alter these materials into harmless byproducts. The result is that they accumulate in the environment.

Some "trace metals" that have not been prohibited can be incorporated into living organisms and accumulate there; mercury, for example is a dangerous nerve toxin. Examples of biologically active materials that may be dumped are arsenic, nickel, and zinc. In addition, strong evidence indicates that lead, chromium, and cadmium accumulate in living organisms, including humans, and cause poisoning. Near sewage outfalls, concentrations of these metals may be high, and there is some evidence that mucus-coated fish living in these areas develop skin diseases and chancres. However, the serious question of whether petroleum should be included in the same category as radioactive wastes or biological warfare agents is appropriate to ask.

TABLE 15.2 Petroleum Inputs into Oceans

SOURCE	INPUT (MILLION METRIC TONS/YR)	
Transportation losses[a]	2.13	
Loading and unloading		1.83
accidents (oil spills)		0.30
Offshore production losses[a]	0.08	
Coastal city contributions[b]	0.8	
Refineries		0.2
Other (wastewater, etc.)		0.6
Atmosphere[b]	0.6	
Offshore oil seeps[c]	0.6	
Urban runoff (storm drains)[c]	0.3	
River runoff	1.6	
TOTAL	6.11	

[a] The accuracy of the estimate is high.
[b] The accuracy of the estimate is adequate.
[c] The accuracy of the estimate is poor.

Petroleum in the Marine Environment

The amount of petroleum that enters the ocean each year is estimated to be about 6 million metric tons (Table 15.2). Oil spills account for only 5 percent of this total. Oil is lost in transportation when the ballast in an empty tanker is off-loaded. Empty oil tankers are not very seaworthy. The ship must carry ballast on its return trip; the ballast most commonly used is seawater. Near the reloading port, the seawater is off-loaded; that is, it is pumped overboard. Some of the oil is pumped into the sea with the ballast. We have relatively few ways to determine the influence and effects of oil spills on any specific environment. We can conduct follow-up studies after an oil spill. In addition, we can perform experiments in which a small amount of oil is purposely spilled in a given environment, or we can study the organisms near active natural oil seeps. There are severe constraints on the first two methods, but studying known seep areas has been particularly useful.

Natural Petroleum Degradation

Several variables profoundly influence the effects of an oil spill. The type of product spilled, the air and water temperature, exposure to wave energy and therefore oxygen, and the location of spillage—whether near or far offshore—are the main variables.

Petroleum is a natural material and therefore natural mechanisms serve to dilute or to decompose it. The most toxic components of petroleum are the *volatile* components—compounds that evaporate at low temperatures—and the components that are water-soluble.

In contrast to crude petroleum, refined products or artificially synthesized compounds are alien to the natural world; for this reason, few natural mechanisms, or biological pathways, decompose them. The chief mechanisms that serve to render crude petroleum *relatively* harmless in the marine environment are bacterial degradation and evaporation. Some bacterial strains degrade oil, meaning they utilize petroleum for food. Because of the environmental hazard from oil spills, scientists have developed engineered bacteria to carry out this task for use in regions where the bacteria do not occur naturally.

Spills in the Open Ocean

When crude petroleum is suddenly introduced into the open seas, oil-degrading bacteria begin to reproduce, utilizing the petroleum as a nutrient. If the water and air are cold, the growth rate of the bacteria is slow, and evaporation of the lower molecular-weight portions of the oil is impeded. If the ocean and air temperatures are warm, both processes proceed more rapidly.

Oil floats on the water surface, so animals and plants at the sea–air interface are most severely affected by spills. Seabirds are the most seriously affected; the coating of oil on their feathers destroys their capability to maintain warmth by fluffing their feathers. In the early days of oil-spill cleanup (late 1960s), most birds that were cleaned by kindhearted people died anyway. Washing the birds with solvents that removed the oil also removed the water-repellency of their feathers; thus, the affected birds died of hypothermia. It became clear that the birds must be cared for until their natural waterproofing secretions recoat their feathers. Survival rates for treated birds are now moderately good.

Animals and plants that live in or below the water column, rather than on the surface of the water, are largely unaffected by oil spills. Fish and lobster catches off the east and west coasts of North America have been normal following major oil spills and the biological productivity not affected.

Eventually, an oil slick is thoroughly degraded into a hard tarlike residue, largely or completely harmless to seagoing organisms of all kinds; in fact, some floating pieces of tar have been found with barnacles attached to them. Wave action eventually breaks these pieces of tar into progressively smaller pieces, until the tar is invisible to the unaided eye. The tar remains as an inert floating substance for a considerable time, but there is no evidence that it is harmful to any marine plant or animal.

Much of the petroleum in the ocean evaporates, and much of it is rapidly degraded by bacteria. Eventually, the oil becomes a hard, inert block of tar.

Oil Washed Onshore

A different scenario occurs when spilled oil washes ashore along wave-washed coastlines. Agitation with seawater causes petroleum to form a foamy emulsion called "chocolate mousse" by workers in the field because the appearance of the emulsion is exactly like that of the confection. Incoming waves and tidal action carry the light, foamy emulsion to shore, where it is stranded in the intertidal zone (Fig. 15.14). The emulsion covers everything: rocks, plants, animals, and sand. If the slope is sufficiently steep, the light emulsion drains down the beach as the tide goes out, but it is washed back up with the next tide.

The high dissolved oxygen content in the water due to the wave or current activity permits rapid oxidation and bacteriological degradation of the spilled oil. Response time of these bacteria is directly related to the frequency that the area is exposed to petroleum. If petroleum is a relatively common part of the environment, as it is near active oil seeps, then the response time of oil-degrading bacteria is rapid; if an area has never been exposed to petroleum, the bacterial response will be slower.

The situation is very different in environments protected from wave action, although bacterial activity may be high. Water exchange in marshes, estuaries, and some bays is slow, and the oxygen concentration in the sediment is very low. Plants and invertebrates in bays and marshes are quickly killed by suffocation or toxic reactions. Thus, the effects of the petroleum might be lasting. A large salt marsh in West Falmouth, Massachusetts, was subjected to a small spill of heating oil in 1969. Shellfish remained quarantined for over five years, and clear chemical traces of the petroleum contamination exist to this day.

Larger animals use bays and estuaries as mating, nesting, or nursery areas. In Baja, California, Scammons Lagoon is renowned as a nursery for the Pacific gray whale. In the Alaskan arctic, near Prudhoe Bay, two small bays serve as the nesting sites for nearly the entire global population of northern phalaropes, a small aquatic bird. In both areas, there have been near misses: A large diesel fuel spill (1.9 million gallons) occurred in 1957, during the season that the whales were rearing their calves. Luckily, the tanker went aground a few inlets distant from Scammons Lagoon. Similarly, a small spill of crude oil occurred in Alaska near the bird site. Had these spills happened in the vulnerable bays, the global populations of these two animals would have been severely damaged, and the animal populations might have been driven to extinction.

The Greenhouse Effect and Global Warming

A greenhouse is a glass-enclosed structure used to keep plants warm in cool climates by trapping the Sun's radiation (Fig. 15.15). On Earth, the surrounding atmosphere

FIGURE 15.14 Stranded oil emulsion (mousse) draining off a beach in Britanny, France, at low tide. With the next high tide, the oil will again be carried back up the beach.

plays the role of the greenhouse glass. In fact, this "greenhouse effect" is a most important one; without it, Earth would be too cold to sustain life. Gases that increase the greenhouse effect are called "greenhouse gases." Carbon dioxide is the most important of these. Burning of fossil fuels is thought to have increased the amount of carbon dioxide in the atmosphere, possibly causing global warming. Geologic processes are sources and sinks of greenhouse gases. This section examines the physical and geologic processes involved in the greenhouse effect.

How the Greenhouse Effect Works

The Sun transmits its heat to Earth via electromagnetic radiation. To understand the physical processes involved with the greenhouse effect, we will first discuss electromagnetic radiation (see Focus 15.2), which includes radio and TV waves, light, and ultraviolet (UV) and infrared radiation, among others. How is electromagnetic radiation generated?

All moving charges generate electromagnetic radiation in the form of waves. When electrons move to different orbits within atoms, the movement generates visible light waves with wavelengths in the range from 0.4 to 0.7 micrometers (Fig. 15.16). The visible portion of electromagnetic radiation is perceived either as white light when it is the superposition of radiation carrying all the colors, or as a spectrum (∞ see Chapter 2) of colors when white light is passed through a prism. The spectrum of colors appears as red, orange, yellow, green, blue, indigo, and violet (easily remembered as ROYGBIV); the longest wavelengths are red and the shortest are violet.

Wavelengths beyond the violet end of the spectrum and thus beyond the range of human vision are called *ultraviolet* radiation. At the other extreme of the spectrum are wavelengths called *infrared* radiation. Most heat is transmitted by infrared waves generated by the oscillatory motions of molecules. The temperature of a body is proportional to the average motion of its atoms and molecules. When you warm yourself at a fire, the oscillating molecules of the flames generate infrared waves that travel to your skin. There, the oscillating electric and magnetic fields cause the molecules in the outer layers of your body to oscillate more rapidly. They, in turn, cause molecules farther within your body to increase their oscillations; the net effect is an increase in warmth.

FIGURE 15.15 The principle of the greenhouse effect. Light energy from the Sun consists of short wavelengths in the visible range of the electromagnetic spectrum that pass through the glass of a greenhouse. Reflected energy from plants and the interior of the greenhouse has longer wavelengths (in the infrared range) that do not pass through the glass but reflect back, causing heating.

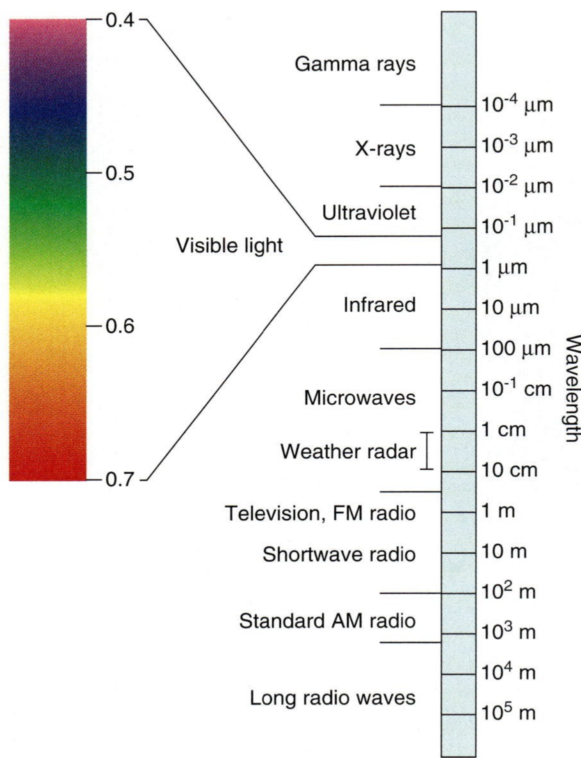

FIGURE 15.16 The electromagnetic spectrum. Energy is carried through space as electromagnetic waves. Some familiar forms of electromagnetic waves include visible light, X-rays, and radio waves. Ultraviolet waves cause sunburn. Infrared waves carry heat. Gamma rays are generated by radioactivity.

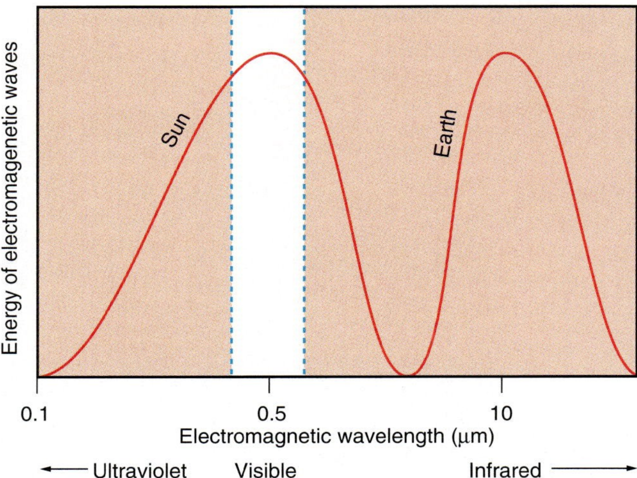

FIGURE 15.17 Energy emitted from the Sun and Earth as a function of wavelength. Most of the electromagnetic radiation emitted by the Sun is in the visible wavelength range of the electromagnetic spectrum, consisting of short wavelengths. In contrast, the radiation emitted from Earth is in the infrared range, at longer wavelengths.

Radiation is made up of electromagnetic waves originating from oscillating electric charges. Light is visible electromagnetic radiation. Ultraviolet and infrared radiation are not part of the visible spectrum and cause sunburn and heating, respectively.

The electromagnetic radiation transmitted to Earth by the Sun keeps our planet warm. The strength of the solar radiation varies as a function of the light's wavelength. The peak amplitude occurs in the visible range of wavelengths (Fig. 15.17) at about 0.5 µm. This radiation is absorbed by Earth and retransmitted. The amount retransmitted equals the amount absorbed; otherwise Earth would heat up indefinitely. The peak amplitude in the transmitted radiation is in the infrared range of wavelengths at about 10 µm (Fig. 15.17). Both absorption and retransmission occur at the atomic or molecular level. For example, an atom on Earth may absorb radiation by exciting an electron, causing it to jump to a higher energy shell. As this electron returns to its lower shell, it releases some of the absorbed radiation in the form of light. This action confers a characteristic color upon the material containing this atom. Air molecules preferentially absorb the blue portion of the white light of the Sun and reradiate it in all directions, causing the sky to appear blue. The reradiated blue light comes to our eyes from all directions in the atmosphere.

The atoms or molecules on Earth absorb the visible radiation from the Sun and become more agitated, or warmer. They then transmit infrared radiation. Most of the Sun's radiation is converted in this manner; to an extraterrestrial observer, Earth would appear to emit more infrared than visible radiation. In a greenhouse, the glass allows the Sun's visible radiation to enter. This radiation is absorbed by the plants and retransmitted as infrared radiation. The glass does not transmit this infrared radiation but reflects it back into the greenhouse, causing the greenhouse to heat up. For visible radiation, the glass acts like a window; for infrared radiation, it acts like a mirror.

On Earth, visible radiation from the Sun passes through the atmosphere to reach Earth, but the atmosphere is much more effective at absorbing infrared wavelengths. A large fraction of the Sun's incoming radiation on reaching the surface of our planet is converted to infrared radiation, which is then absorbed by the greenhouse gases (mainly water and carbon dioxide) in the atmosphere. The atmosphere is not heated directly by the Sun; rather, it is heated by Earth, which explains why the atmosphere becomes colder at higher altitudes. Even though mountaintops have less atmosphere above them absorbing the Sun's rays, they are much colder than are valleys.

The Sun radiates visible light, which is converted to infrared radiation at Earth's surface. The infrared radiation leaving Earth is absorbed and reflected back by Earth's atmosphere—the same effect as occurs with the glass in a greenhouse.

Focus On 15.2 ELECTROMAGNETIC RADIATION

Radiation is transmitted in the form of electromagnetic waves. Some electromagnetic waves are harmful to humans, but most are not. A radio or television station generates radio waves, which carry programs to radio and TV receivers. Electromagnetic waves are generated by a transmitter connected electrically to an antenna, and that causes electric current to oscillate from one end of the antenna to the other. Electromagnetic waves radiate out from the antenna in a manner analogous to the way in which waves of water radiate out when a pebble is thrown into a pond (∞ see Chapter 5).

Electric currents consist of electrons in motion. A stationary electron has an electric field; a moving electron generates a magnetic field. The oscillating electrons in an antenna generate oscillating electric and magnetic fields, or electromagnetic fields, which radiate away, carrying information about their oscillation rate and amplitude.

Electromagnetic fields can occur at many frequencies and wavelengths, depending on the source of excitation. In a vacuum, the waves travel at the speed of light: 3×10^8 m/sec. Their velocity is given by the product of the frequency (number of cycles per second, Hz) and wavelength (distance between peaks, m):

$$\text{velocity} = \text{frequency} \times \text{wavelength}.$$

You have probably tuned your radio receiver to a given frequency; for example AM1010 on your radio corresponds to a frequency of 1010 kHz. We can use the equation just introduced to determine the wavelength:

$$3 \times 10^8 = 1{,}010{,}000 \times \text{wavelength}$$

which gives a wavelength of 297 m.

Infrared radiation has a frequency of about 10^{13} Hz corresponding to a wavelength of 3×10^{-5} m. Visible light has wavelengths of 4×10^{-7} to 7.5×10^{-7} m, whereas ultraviolet light has even smaller wavelengths of 10^{-8} m. Both X-rays and gamma rays are even smaller. Small wavelengths can be absorbed by living tissue and harm the tissue. Long wavelengths, such as radio waves, pass right through such tissue and have no effect. Radios work inside buildings because the waves pass right through the walls. In contrast, smaller wavelengths of light do not enter. Absorption of infrared waves causes warming. Ultraviolet waves cause sunburn—and skin cancer. Large doses of X-rays and gamma rays can also cause cancers.

The atmosphere is divided into the **troposphere** and the **stratosphere**. The troposphere is the lowest zone of the atmosphere, ranging from 0 km to 12 km above Earth. The troposphere contains the clouds, and is sometimes called the *weather-sphere*. The stratosphere lies above the troposphere in the height range 12 km to 50 km above Earth. It is stratified, in contrast to the well-mixed gases of the underlying troposphere. (The composition of the atmosphere is shown in Table 15.3.) Water vapor ranges from 0 to 4 percent of the atmosphere by volume. This gas absorbs about five times more radiation than all the other gases combined; it accounts for the warmer temperatures present in the lower troposphere, where it is most highly concentrated.

TABLE 15.3 Composition of the Atmosphere

Nitrogen	N_2	78.08	
Oxygen	O_2	20.95	
Argon	Ar	0.93	
Carbon dioxide*	CO_2	0.035	(1992)
		0.07?	(2092)

* Carbon dioxide contents for 1992 (measured) and 2092 (estimated) are given.

Without the greenhouse effect, Earth would possess an average temperature of −100°C, rather than its more temperate average of 15°C. The atmosphere of Venus—a hundred times more dense than Earth's atmosphere—is composed almost entirely of carbon dioxide. The greenhouse effect on Venus gives rise to temperatures of 400°C at the surface; this is too hot for liquid water and hot enough to melt lead. In the absence of the greenhouse effect, the temperature on Venus would be well below freezing (−50°C).

Evidence for Greenhouse Effects

The effect of the human-induced increase of atmospheric carbon dioxide on our climate has been the focus of many research projects. Burning of fossil fuels such as coal and petroleum (which are mainly carbon) produces carbon dioxide (CO_2) gas. Evidence shows that the injection of industrial carbon dioxide into the atmosphere has changed its average composition and may even have had an effect on temperature. Although there is general agreement on the former effect, many different views abound on the latter suggestion. Since 1860, the amount of carbon dioxide in the atmosphere has increased from about 290 parts per million (ppm) to 350 ppm—a 20 percent increase! During the same period, temperatures on the continents in the Northern Hemisphere, where the measurements were made, are estimated to have increased by 0.6°C (Fig. 15.18a). Are these two changes related?

In an attempt to answer this question, carbon dioxide levels have been estimated for the last 160,000 years from an ice core from Antarctica (Fig. 15.18b). Each new layer of ice laid down contains samples of the atmosphere at that time in pores. These preserved samples of ancient atmospheric content have provided the longest record of variations in the atmospheric composition available.

The correlation observed in Figure 15.18 suggests that temperature is related to carbon dioxide levels in the atmosphere. More recently, measurements at Mauna Loa Observatory in Hawaii have recorded about a 20 percent increase in carbon dioxide levels between 1957 and the present. The rate at which we are polluting the atmosphere is increasing. In fact, experts anticipate that global concentrations of carbon dioxide will double by the end of the twenty-first century. Climate models that take the greenhouse effect into account predict that the global temperature will increase by several degrees.

In addition to greenhouse gases, the atmosphere contains microscopic particles of organic matter, spores, pollen, and seeds that have the opposite effect. Dust is injected into the upper atmosphere from volcanic eruptions, from upwelling wind currents, and even from meteors burning up as they encounter the braking effects of the atmosphere. Water droplets in clouds condense on atmospheric dust. These particles reflect the Sun's radiation, causing cooling.

The small percentage of carbon dioxide (350 ppm) in the atmosphere, even today, is a far cry from the virtually complete carbon dioxide atmosphere of Venus. In fact, Earth's oceans contain 50 times as much carbon dioxide as does its atmosphere. Some scientists postulate that, as the level of carbon dioxide in the atmosphere increases, it will be buffered by a greater absorption by the oceans,

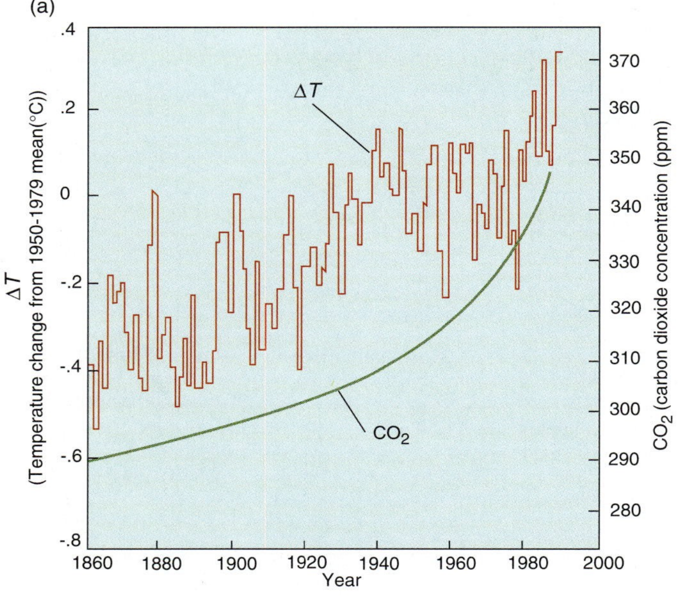

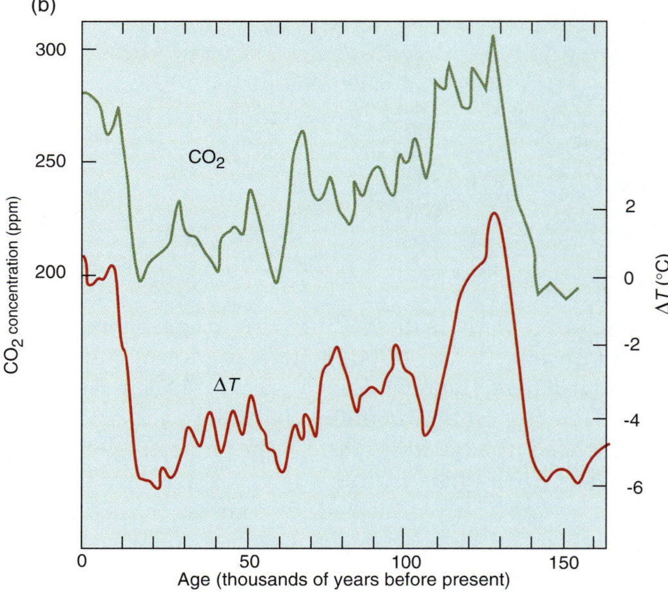

FIGURE 15.18 (a) Variations in temperature and the amount of carbon dioxide in the atmosphere since 1860. The temperature has increased by about 0.6°C while the carbon dioxide content of the atmosphere has increased by 20 percent. (b) Variations of carbon dioxide and temperature as a function of time for the last 160,000 years as determined from the Vostok ice core in Antarctica. The correlation suggests these two factors are related, but their relationship is still unresolved.

both by dissolution and an increase of biotic activity. However, others argue that if the temperature increases before this happens, the oceans may instead release carbon dioxide, which is less soluble at higher temperatures, and would thereby accelerate the greenhouse effect. Further study of this delicate balance is required to reach a definitive conclusion.

In addition to carbon dioxide, industry is releasing many other gases and particles into the atmosphere. At this stage, models of Earth's climate, called Global Circulation Models (GCMs), predict a significant warming of our planet if carbon dioxide emissions are not controlled. It is also understood, however, that such models must take into account the effects of other emissions, some of which (sulfate particles) promote climatic cooling rather than warming. Some scientists argue that carbon dioxide in the atmosphere is already absorbing about as much radiation as it can. Any more carbon dioxide will not affect the total absorption. Others argue that as carbon dioxide increases, a small increase in cloudiness will occur that will not be noticeable; however, such an increase in cloud cover may cause a cooling trend that would completely cancel the warming effect.

The 0.6°C warming trend that has occurred since 1860 (Fig. 15.18a) may simply be due to an increase in solar output and may have little to do with any coincidental increases in the level of carbon dioxide. It was noted that a striking correlation existed between the undulations in the temperature curve with time and the frequency of sunspots (Fig. 15.19). This new result suggests that climate may also be controlled by small variations in solar activity.

Much of this discussion reveals that correlations between two variables, such as carbon dioxide and temperature, do not necessarily mean that they are causally related. It is a necessary condition but not a sufficient one to prove such a relationship. Details of the processes operating must be understood to make a convincing argument. However, we must be cautious, for an increase in the global temperature of just a few degrees could cause melting of glacial ice, which, in turn, would trigger a rise in sea level and flooding of coastal regions. Even though the actual mechanism that has caused the 0.6°C temperature rise since 1860 is debatable, the 20 percent increase in the carbon dioxide content to reach its highest level in the last 160,000 years is uncontested. Humans have made their mark on the atmosphere; if temperatures start rising as a result, it will take a long time to reverse the present situation.

Geologic Feedback

Various geologic processes are sources and sinks of atmospheric carbon dioxide. During the ancient history of Earth, a balance has existed between the volcanic degassing of carbon dioxide and the absorption due to weathering. Indeed, it is thought that the early atmosphere of Earth was primarily carbon dioxide and that the transition to an oxygen-rich atmosphere happened about 2 billion years ago—at a time when there was a significant increase in life-forms as determined from the fossil record. The interplay of geologic conditions causing this transition is not clear.

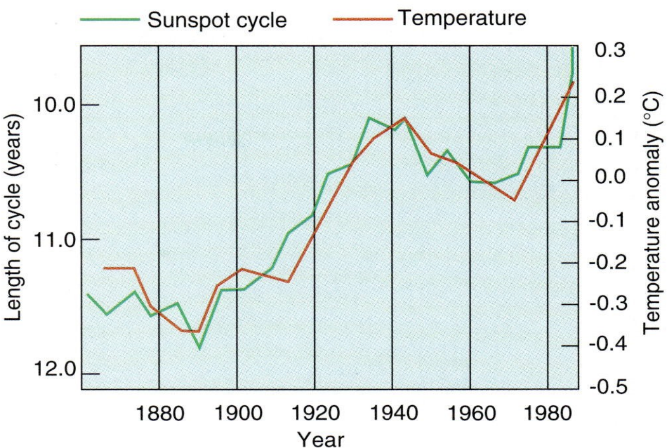

FIGURE 15.19 Variations in sunspots and temperature as a function of time. This graph suggests that the recent warming trend may be related to small fluctuations in solar activity rather than to the increase in greenhouse gases.

We saw that weathering absorbs carbon dioxide from the atmosphere, forming carbonate rocks (∞ see Chapter 6). The uplift of the Himalayas caused changes in the Asian climate with an associated increase in weathering. It has been speculated that absorption of atmospheric carbon dioxide may have given rise to global cooling and glaciation, perhaps followed by a reduction in weathering and reversal of the trend.

If the carbon dioxide effect were to cause global warming, melting of the polar caps would cause flooding of the continents. However, the resulting shallow seas would be ideal for the production of limestone, which would extract carbon dioxide from the atmosphere. The interplay between life and the carbon dioxide content of the atmosphere involves many feedback mechanisms that can be elucidated with systematic analysis of the geologic record.

The Ozone Hole

We know that too much exposure to the Sun's ultraviolet (UV) radiation causes sunburn. In extreme doses, it causes skin cancer and cataracts. Ultraviolet radiation is also harmful to small plant organisms called *phytoplankton*, which are important to the oceanic food chain. About 50 percent of the incident UV radiation from the Sun is absorbed by the atmosphere; 30 percent is absorbed in the upper atmosphere (the stratosphere) by ozone. Ozone is therefore a natural sunscreen, operating in the atmosphere at heights between 10 and 50 km, with maximum activity at 30 km above Earth.

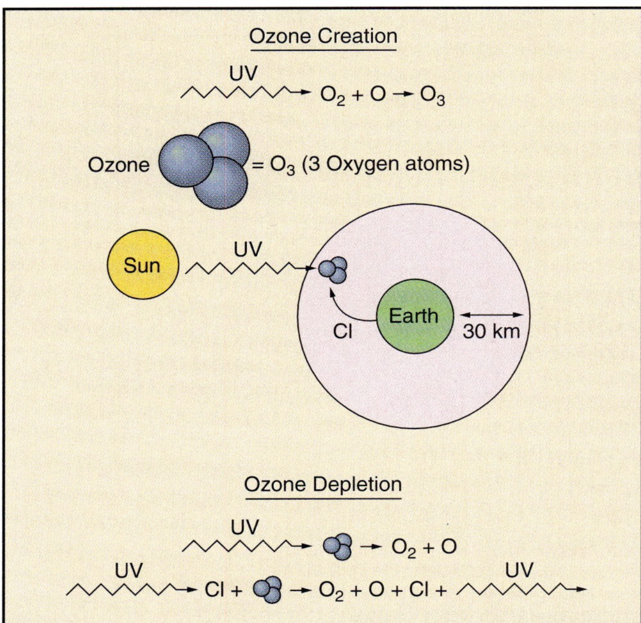

FIGURE 15.20 Ozone creation and depletion. Ozone—a combination of atomic and molecular oxygen—is created from the Sun's ultraviolet radiation (UV). Once formed, ozone is highly effective in absorbing ultraviolet radiation. However, in the presence of chlorine from chlorofluorocarbons, ozone reverts back to atomic and molecular oxygen and the ultraviolet radiation is no longer absorbed, as shown in the lower set of reactions.

Ozone has the chemical formula O_3; it is triatomic oxygen (a molecule composed of three oxygen atoms bonded together). It is a light blue gas of oxygen with a sweetish smell that you may detect in the vicinity of an electric discharge—for example, near an electric generator or a motor. In nature, ozone is formed by the interaction of UV rays from the Sun with atmospheric oxygen molecules in the stratosphere (Fig. 15.20). Ozone is extremely effective at absorbing UV radiation. However, if chlorine gas is present (for example, due to the release of chlorofluorocarbons into the air), the reverse reaction occurs; that is, the UV radiation converts the ozone back to oxygen molecules and chlorine. Once the layer of ozone is gone, the UV radiation it absorbs can pass directly through the atmosphere to reach Earth's surface. A 1984 National Academy of Sciences report states that a 1 percent decrease in ozone will cause a 2 percent increase in ultraviolet radiation reaching Earth's surface.

Scientists estimate that a 1 percent decrease in the protecting layer of ozone surrounding Earth can cause a 6 percent increase in skin cancers. Recently, depletions of the ozone layer have been measured over Antarctica (Fig. 15.21) and scientists have sought the reason for such a depletion (see Focus 15.3). A leading candidate is the injection into the atmosphere of chlorofluorocarbons—gases used as refrigerants or as propellants in aerosol packs.

As with carbon dioxide, measurements of the atmospheric content indicate that the percentage of chlorine has steadily increased since the Industrial Revolution. However, establishing a causal relationship between the antarctic ozone holes and the release of chlorofluorocarbons is still the subject of ongoing research. Chlorofluorocarbons are extremely unreactive in the lower atmosphere, which accounts for their widespread use in industrial processes. However, when these gases rise into the upper atmosphere, strong UV radiation in sunlight separates them into their constituent atoms, the most important in this context being chlorine.

Figure 15.20 shows that UV radiation interacting with ozone creates diatomic and monatomic oxygen. Without chlorine, the production and depletion rates of ozone are about equal. However, in the presence of chlorine, less radiation is needed for the depletion reaction. This shifts the reaction balance and more ozone is destroyed than created, while the UV radiation passes through the atmosphere to Earth. The chlorine acts as a *catalyst* for the reaction—it increases the rate of a chemical process—and remains free after the reaction to promote further reactions. It has been estimated that one chlorine atom ultimately eliminates many thousands of ozone molecules. As a result, ozone is rapidly converted to oxygen, leaving that part of the atmosphere unprotected.

Ozone in the upper atmosphere absorbs the Sun's ultraviolet radiation, thereby protecting Earth from harmful levels of this radiation. Chlorine atoms from chlorofluorocarbons accelerate the breakdown of ozone to atomic and molecular oxygen, creating an ozone hole through which the Sun's ultraviolet radiation passes.

Volcanoes and Atmospheric Pollution

During eruptions, volcanoes inject solid particles and gases into the atmosphere (Fig. 15.22). Particles may remain in the stratosphere for months to years, much longer than the days or weeks that such particles remain in the troposphere, where they rain back onto Earth. Volcanoes also release chlorine and carbon dioxide, which are important to the ozone and greenhouse issues that we have discussed. The main products injected into the atmosphere from volcanic eruptions, however, are volcanic ash particles and small drops of sulfuric acid in the form of a fine spray known as an *aerosol*.

Most chlorine released from volcanoes is in the form of hydrochloric acid, which is washed out in the troposphere. This is in contrast to the fate of chlorine, tied up in chlorofluorocarbons, which pass through the troposphere to the stratosphere where ultraviolet radiation breaks them down and releases chlorine.

Volcanoes also emit carbon dioxide, but at 0.1 percent of the current rate of release by humans. During times of giant volcanic eruptions in the past, the amount of carbon dioxide released may have been enough to affect the

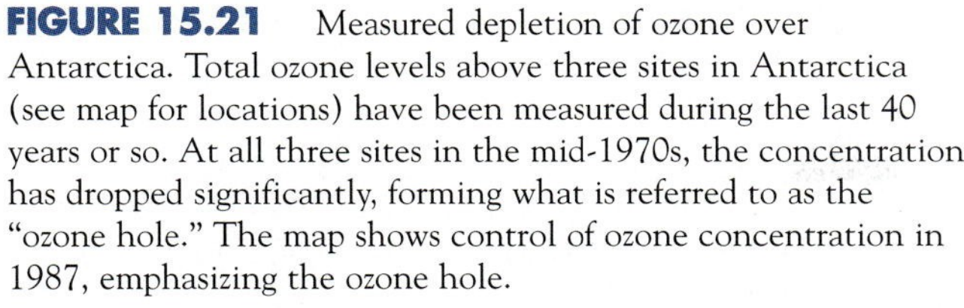

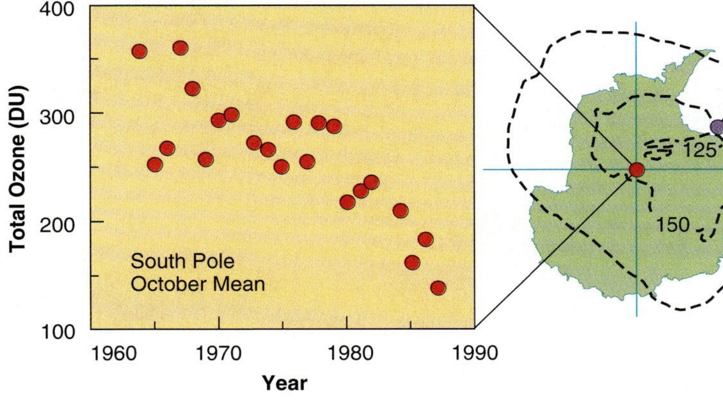

FIGURE 15.21 Measured depletion of ozone over Antarctica. Total ozone levels above three sites in Antarctica (see map for locations) have been measured during the last 40 years or so. At all three sites in the mid-1970s, the concentration has dropped significantly, forming what is referred to as the "ozone hole." The map shows control of ozone concentration in 1987, emphasizing the ozone hole.

climate. At such times, however, the reflection of sunlight back into space by volcanic products in the atmosphere, such as fine ash and aerosols, appears to have a more important role in the short term (years) than the greenhouse effect. In general, global temperatures are cooler for a year or two after a major eruption.

Volcanoes and Climate

One of the interesting indirect influences of plate tectonics on the biosphere and atmosphere is the effect of volcanic eruptions on climate. The magnitude of such possible effects was emphasized following the relatively small 1982 eruption of El Chichón in Mexico. This eruption is thought to have affected the atmosphere—from unusually colorful sunsets to cooler than normal temperatures several years after the eruption.

A large-magnitude pyroclastic eruption such as a caldera-forming event can be expected to eject huge volumes of fine ash high into the stratosphere, where it may remain for several years, carried around the globe by strong air currents in the upper atmosphere. The presence of this ash will increase the opacity of the atmosphere; that is, it will reduce the amount of sunlight reaching Earth's surface. Accordingly, Earth's surface and climate will become cooler. Various other atmospheric effects may be observed. Particularly noticeable is an increase in the intensity of sunsets. The beautiful watercolor renditions of sunsets in London, England, in November 1883 (see Fig. 15.22) illustrate the effects of upper-atmosphere ash in the months following the cataclysmic August 1883 eruption of Krakatoa, some 11,000 km away in Indonesia.

It is more difficult to appreciate the significant effects that large basaltic eruptions may have on climate. We have learned that it is silicic eruptions that tend to be explosive and produce pyroclastic deposits. In contrast, basaltic eruptions are more gentle, giving rise to fluid lavas. However, basaltic eruptions are commonly accompanied by the emission of many gases—in particular, sulfur dioxide (SO_2). Although invisible, these gaseous aerosols may act just like particulate ash in the stratosphere, reducing the amount of sunlight passing through the atmosphere. It was SO_2 that principally affected the climate following the 1982 El Chichón eruption. More recently, the 1990 eruption of Mt. Pinatubo injected a large SO_2 plume into the atmosphere, which had a small but significant effect on the global climate during the following years.

Enormous basaltic eruptions are known to have occurred in the geologic past and may have pumped huge volumes of SO_2 and other gases into the air. The seventeenth-century eruption at Laki, a basaltic volcano in Iceland, produced gas emissions that triggered cooling and poor crop production for the next year. It caused the deaths of 25 percent of the Icelandic population through starvation. It has even been suggested that vast basaltic eruptions in western India 65 million years ago, which produced the Deccan plateau (∞ see Chapter 11), may have reduced the sunlight available for the photosynthesis of plants and may have cooled the climate sufficiently to contribute to the extinction of the dinosaurs.

Aerosols from volcanoes also appear to have enhanced ozone depletion. Understanding the effects of human-

Focus On 15.3 THE ANTARCTIC OZONE HOLE

The main difference between the arctic and antarctic stratospheres is that temperatures in winter are lower in the antarctic (–90°C) than they are in the arctic (–60°C). The arctic is buffered by the temperature in the ocean beneath the ice sheet. A polar vortex (a circulating wind) forms, which circulates Antarctica. High in this vortex, the extremely low temperatures cause clouds to form, which are thought to be important for the chemical reactions that break up chlorofluorocarbons and deplete ozone. In the arctic, higher temperatures cause the vortex to be much weaker and shorter-lived, and high-level clouds do not form. As a result, there is less degradation of the chlorofluorocarbons. This difference is supported by the measurement of free chlorine in the arctic, which is 1/20 of the chlorine concentration in the antarctic.

At a 1987 conference held in Montreal, Canada, by the United Nations Environment Program (UNEP), 24 nations signed a protocol to limit the use of industrial chlorofluorocarbons. They agreed to freeze the consumption of chlorofluorocarbons at 1986 levels on July 1, 1989; to reduce the consumption of chlorofluorocarbons by 20 percent by mid-1993; and to reduce consumption by an additional 30 percent by mid-1998. However, the ozone hole in Antarctica occurred with about 3 parts per billion (ppb) of chlorine in the atmosphere, and the accord permits this concentration to increase to 6 or 7 ppb. It has been suggested that it will require an almost total ban on chlorofluorocarbon release to return the atmosphere to its preindustrial purity.

You may have noticed products marked "ozone friendly," "no chlorofluorocarbons used," or "nitrous oxide used as a propellant." Such changes are the result of international concern over the fate of the ozone layer, as expressed in the Montreal protocol.

FIGURE 15.22 Watercolor renditions by William Ascroft of the spectacular sunsets November 1883 observed in London, England, caused by ash in the upper atmosphere from the eruption of Krakatoa in Indonesia.

made pollution and natural pollution must take into account both the interdependence of the many different chemical reactions involved and the dynamics of atmospheric mixing. It is not a simple problem; many feedback mechanisms are involved, such as increased cloud formation, both in the stratosphere and troposphere, as well as unknown reactions involving catalysts and solar radiation. Global warming and ozone depletion from human activities will continue to be prime subjects of research as industrialization continues to increase across the globe.

Volcanic activity injects carbon dioxide, sulfur dioxide, and chlorine into the atmosphere. For large eruptions, greenhouse effects are less important than the cooling effect from aerosols reflecting sunlight.

SUMMARY

- Building failure is linked to the earthquake-wave amplitude, the frequency content of the waves, and the duration of shaking.
- Earthquake damage depends on both the earthquake source and on the foundation material and building resonances relative to the seismic-wave oscillations. Zones with a shallow water table in loose fill are prone to liquefaction, whereas buildings with solid rock foundations are safest.
- The Mercalli intensity scale for earthquakes is defined from observing the damage and related effects generated by the shaking. The Richter magnitude, or *strength,* of an earthquake is instrumentally measured using a seismometer. *Moment* magnitude is used for large earthquakes.
- Volcanoes at subduction zones pose the greatest common volcanic threat to humans. Explosive activity produces pyroclastic flows, ashfall, and secondary lahars.
- Urbanization near rivers or streams causes higher flood stages and low-flow stage lower than before settlement. Immediate runoff during storms without water percolation into the ground causes higher floods. Lack of percolation decreases groundwater contribution to river flow between rainy seasons.
- Buildings on steep hillsides are inherently unstable. Landscape modifications, such as excavating the downslope side of the hill, heavy watering of the slope, and high retaining walls without proper drainage, make a structure even less stable.
- Subsidence is caused by withdrawal of underground fluids. Large volumes of irrigation water and petroleum have caused major subsidence in some areas.
- Much of the petroleum in the ocean evaporates, and much of it is rapidly degraded by bacteria. Eventually the oil becomes a hard, inert block of tar.
- Radiation is made up of electromagnetic waves originating from oscillating electric charges. Light is visible electromagnetic radiation. Ultraviolet and infrared radiation are not part of the visible spectrum; they cause sunburn and heating, respectively.
- The Sun radiates visible light, which is converted to infrared radiation at Earth's surface. The infrared radiation leaving Earth is absorbed and reflected back by Earth's atmosphere—the same effect as occurs with the glass in a greenhouse.
- Ozone in the upper atmosphere absorbs the Sun's UV radiation, thereby protecting Earth from harmful levels of this radiation. The chlorine atoms from chlorofluorocarbons accelerate the breakdown of ozone to atomic and molecular oxygen, creating an ozone hole through which the Sun's UV radiation passes.
- Volcanic activity injects carbon dioxide, sulfur dioxide, and chlorine into the atmosphere. For large eruptions, greenhouse effects are less important than is the cooling effect from reflecting sunlight from aerosols.

KEY TERMS

resonance, 385 **liquefaction**, 386 **troposphere**, 403 **stratosphere**, 403

QUESTIONS FOR REVIEW AND FURTHER THOUGHT

1. What types of buildings are most likely to suffer serious damage in an earthquake? What types of buildings are least likely to suffer serious earthquake damage? Explain your reasoning.
2. Unconsolidated sediment is a poor foundation material for a building. Why is that true?
3. Explain why pyroclastic volcanic eruptions are so dangerous. Are pyroclastic flows the only major danger from volcanic eruptions? Can you name others?
4. If given the choice, would you rather farm on the erosional terrace of a river or would you think this might be a good place to build a home? Explain your reasoning.
5. Near-surface subsidence takes several forms. Can this process represent a hazard to a town? Explain.
6. How much oil goes into the world's oceans each year and what percentage of this comes from oil spills?
7. What environments are the most sensitive to oil contamination? What environments are not very sensitive to oil contamination? Explain the differences.
8. Why does the sky appear blue?
9. What is the difference between the radiation absorbed by Earth from the Sun and that emitted from Earth?
10. What is the greenhouse effect, and what problems would we face if it did not operate on Earth?
11. How does chlorine affect the balance of creation and depletion of ozone in the atmosphere?
12. In 50,000 years, do you think Earth will be hotter or colder than it is now, based on historical evidence? Give reasons and estimate roughly what the temperature change might be.

16 EARTH'S RESOURCES

Beyond metals and fossil fuels, water and heat are perhaps Earth's most important natural resources. Here, they combine to form geysers at Rotorua, New Zealand.

OVERVIEW

- Economically important elements and compounds do not occur in sufficient concentration in most rocks to be extracted for profit.
- The forces of mantle convection and plate tectonics bring magma to Earth's surface. The tectonic forces also bury organic matter to form fossil fuels and exhume buried reservoirs, rendering them accessible to mining.
- Water heated by magma bodies dissolves minerals, which precipitate out as valuable ores upon cooling. The water in contact with magma becomes superheated and can be used to drive steam turbines to generate electrical energy.
- Coal and peat are formed by the burial of many types of vegetation, including mosses, ferns, algae, and trees.
- Petroleum results mainly from burial and alteration of marine microorganisms.
- Fresh water is arguably Earth's most important natural resource. Most potable water is contained as groundwater in fractures and pores in the crust.

Introduction

Plate tectonics provides a framework for exploration geologists to relate mineral or hydrocarbon deposits to the tectonic events and structures associated with their formation. Many mineral deposits can be directly related to magmatism at plate margins, but others are associated with magmatism in plate interiors. Hydrocarbon deposits are found in sedimentary basins formed by plate motions. The formation of deposits that depend on biological processes—coal and hydrocarbons—is also dependent on climate; climate, in turn, depends on plate morphology and paleo-latitude. Water is not only a valuable resource but is also important to deposition of most minerals, migration of hydrocarbons, and the production of geothermal energy. Relationships between plate tectonics and resources are still being developed and require an understanding not only of current plate tectonics and associated mineral-forming processes, but also of past tectonic processes and plate positions over the millions of years during which the resources formed.

Mineral Resources

Rocks that come directly from the mantle, such as basalts, may contain economically important elements such as lead, copper, zinc, and iron; however, concentrations of these metals are so small that it would require too much energy to extract them for the process to be profitable. There are, though, natural processes taking place in the crust that concentrate important elements into ores. An **ore** is a mineral or an aggregate of minerals from which a valuable constituent—particularly a metal—can be mined or extracted profitably. In this chapter, we will examine ore-forming processes—many of which are related to plate tectonics.

Minerals and Society

The first metals used by humans were probably gold and copper. These metals are found today in the nearly pure metallic, or native, state. But most metallic ores occur as compounds; that is, they are found in chemical combination with other elements (∞ see Chapter 3). Important examples include various metals bonded to the element sulfur, to produce metal sulfides: iron (pyrite), lead (galena), zinc (sphalerite), mercury (cinnabar), and copper (covellite, chalcocite, and chalcopyrite). Compounds in which oxygen or hydroxide is bonded to a metal—oxides or hydroxides of metals—are also important: as ores of iron (hematite and magnetite) or aluminum (bauxite).

During the Stone Age, the time between about 1 million and 4000 years B.C., early humans used stones such as flints and obsidian to make tools (Fig. 16.1). These rocks break with a conchoidal fracture (∞ see Chapter 3) and can therefore be used to make cutting edges such as arrowheads and scrapers. Production of metals, however, required a higher level of development. To extract metals, ores must be processed. This process is called *smelting*, which involves melting the ores mixed with certain materials that separate the metal from the compound. Before 4000 B.C., early humans learned to extract copper from copper sulfide by melting the copper sulfide with charcoal. By 3000 B.C., silver, tin, lead, zinc, and other metals were also being extracted and combined to form alloys such as bronze (copper and tin), brass (copper and zinc), and pewter (tin and other metals such as lead, copper, or antimony). Thus, the Bronze Age began, a time when bronze replaced flint and other stones as the chief material for weapons and tools (Fig. 16.1).

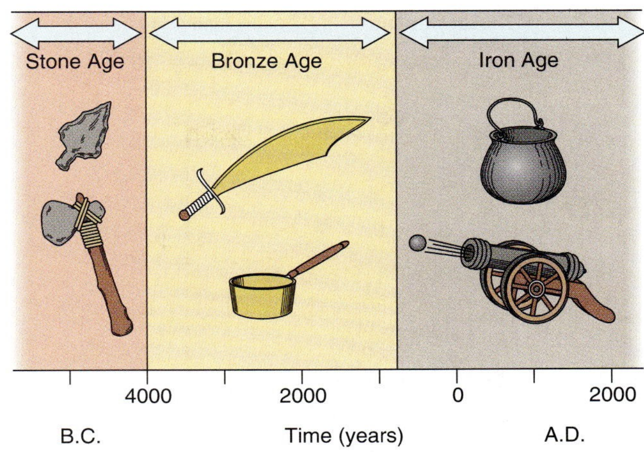

FIGURE 16.1 Progressive utilization of raw materials by early humans.

It was not until about 800 B.C. that the Bronze Age was followed by the Iron Age, a time characterized by the introduction of iron metallurgy in Europe. The first iron used is believed to have come from meteorites, but these are rare. Iron occurs in the crust in silicate minerals (∞ see Chapter 3) and as iron oxide and is more abundant in Earth's crust than many other metals, but it is much more difficult to extract from ore than is copper. Smelting iron involved combining iron ore with charcoal in a furnace operating at very high temperatures to form metallic iron and releasing carbon dioxide.

Ores are minerals from which valuable constituent compounds can be profitably extracted. Smelting is the process by which metals are extracted from ores.

Mineral Deposits

We found that (∞ see Chapters 1–3), owing to differentiation, the composition of the crust is markedly different from Earth as a whole (see Fig. 3.18). Oxygen, silicon, aluminum, iron, calcium, potassium, sodium, and magnesium compose 99 percent of Earth's crust; the remaining 1 percent comprises all the remaining elements. Consequently, many elements that we rely on for indus-

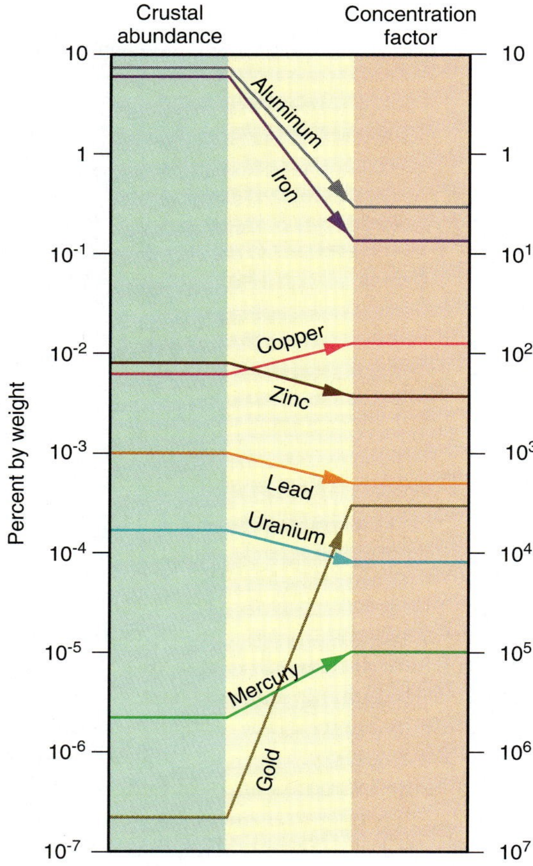

FIGURE 16.2 Factors by which natural processes must concentrate metals to form ores. The concentration factor represents how much more abundant a metal must be in an ore deposit compared with its typical crustal concentration or abundance in the crust. For instance, typical crustal concentrations of copper are 0.0058 percent. To be an economically viable ore, concentrations must be a factor of 80 to 100 times greater (about 0.6 percent).

trial purposes are extremely scarce. Gold, for instance, exists in minute quantities measured in parts per billion of the continental crust, and silver is measured in parts per 10 million.

A process of concentration is required for economically important elements or compounds to form ores. For common metals such as aluminum, the concentration need only be about four times the typical abundances in the crust. Rarer metals such as uranium and gold may have ores that are thousands of times more concentrated than the typical abundances in Earth's crust (Fig. 16.2). For a given metal occurring in an ore deposit, we can talk about the grade of the deposit. High-grade ores have high concentrations; low-grade ores have low concentrations (although concentrations that are still economical to mine). If an ore is very low grade, such as the ores of copper, which may contain less than 1 percent of the metal, huge volumes of rock must be mined to obtain a small amount of metal. Some of the most important ores and ore minerals and their economic uses are listed in Table 16.1.

A number of different ore-forming processes have been recognized, the most important of which are: magmatic processes; hydrothermal processes; sedimentary processes; weathering processes.

We will discuss these processes briefly and emphasize the main sources of our most important ores. If we understand the processes that generate and then concentrate minerals, we can increase our chances of finding new deposits by knowing where to look.

Magmatic Processes

We learned earlier (∞ see Chapter 4) how elements can be separated by magmatic differentiation. We also discovered that the most common minerals found in igneous rocks are silicates. Although silicates do contain appreciable concentrations of many important metals, such as iron and aluminum, it is very difficult and not economically worthwhile to extract these metals from the silicates. However, some economically important minerals do occur naturally in igneous rocks and may be concentrated by processes such as crystallization.

Chromium is a metal concentrated in this way. The metal is used in manufacturing steel, and in chrome plating to protect metal surfaces. It is extracted from chromite, an oxide of chromium, which crystallizes from certain mafic to ultramafic magmas. Much of the world's chromite comes from a single enormous igneous intrusion, called the *Bushveld Complex*, in South Africa (Fig. 16.3). Smaller deposits are also commonly associated with mid-ocean

FIGURE 16.3 Dark chromite layers in the Bushveld complex, South Africa, formed by crystal settling from a chrome-rich magma.

TABLE 16.1 Origin and Use of Some Important Ores

NAME	FORMULA	OCCURRENCE	USE
Azurite	$Cu_3(CO_3)_2(OH)_3$	Upper (oxidized) zones of copper veins	Copper used in electronics
Barite	$BaSO_4$	Found in many ores, near hot springs, in veins, or as masses in limestones, also stratiform deposits	Used (powdered) in drilling muds, paints, and as an indicator for X-rays of the digestive tract
Bauxite	Variably hydrated Al-oxides	Common residual of clay deposits in tropical and subtropical regions; formed by weathering	Commercial source of aluminum
Cinnabar	HgS	Hydrothermal deposits	Principal ore of mercury, used in thermometers
Chalcocite	Cu_2S	Generally massive hydrothermal zones; porphyry copper deposits	Ore of copper
Chalcopyrite	$CuFeS_2$	Generally massive hydrothermal zones; porphyry copper deposits	Principal ore of copper
Chromite	$(FeMg)(CrAl)_2O_4$	Accessory mineral in basic and ultrabasic igneous rocks, precipitated in magmas	Principal ore of chromium, used in stainless steel and chrome plating
Covellite	CuS	Hydrothermal deposits	Ore of copper
Diamond	C	Kimberlite pipes and placer deposits	Jewelry, cutting, drilling tools
Fluorite	CaF_2	Accessory mineral deposited in lakes and hydrothermal veins	Principal ore of fluorine, used in toothpaste and water supply treatment
Galena	PbS	Common in veins and massive ores; associated with sphalerite, in stratiform deposits	Principal ore of lead, used in batteries
Gold	Au	Hydrothermal vents, placer deposits, igneous veins	Jewelry, electronic circuits, aerospace alloys, dentistry
Hematite	Fe_2O_3	Found in igneous, sedimentary, and metamorphic rocks	Principal ore of iron, used in most construction
Laterite	Variably hydrated Fe-oxides	Residual soil formed from tropical weathering	Ore of iron
Malachite	$Cu_2CO_3(OH)_2$	Upper (oxidized) zone of sulfide deposit, associated with azurite	Ore of copper, used in wire, ornamental objects, roofing
Marcasite	FeS_2	Common as nodules and concretions in sedimentary rocks and in hydrothermal zones	Ore of sulfur, used in making sulfur dioxide and sulfuric acid
Pyrite	FeS_2	Most widespread and abundant of the sulfide minerals, occurring in all kinds of rocks	Ore of sulfur, used in sulfur dioxide and sulfuric acid
Quicksilver	Hg	The name for mercury as a native mineral; hydrothermal deposit	Thermometers
Sphalerite	ZnS	Commonly associated with galena in hydrothermal veins and stratiform deposits	Ore of zinc, used for galvanizing

ridge magmatism. Chromite is very dense, so it sinks rapidly in the magma to accumulate as layers—much as sedimentary grains settle out of a fluid (∞ see Chapter 13).

Another mineral that owes its origin to igneous processes is diamond, one of the world's most valuable natural materials. It is the hardest mineral known and therefore has important industrial applications as an abrasive and as a cutting material. More important economically, diamond is a rare and valuable gemstone. The primary occurrence of diamonds is in kimberlite, an igneous rock derived from deep within the mantle. Kimberlite rocks are rare, occurring as small pipelike bodies (Fig. 16.4a) in intraplate regions, exclusively on continents and generally far away from other types of tectonic-related igneous activity such as arcs, rifts, and hot spots. This rock type is really a mixture, or hybrid, of a small-degree melt from the mantle mixed with fragmented mantle material picked up on the way to the surface. The disaggregated material is comprised of both xenoliths (lumps of ripped up mantle wall rock) and

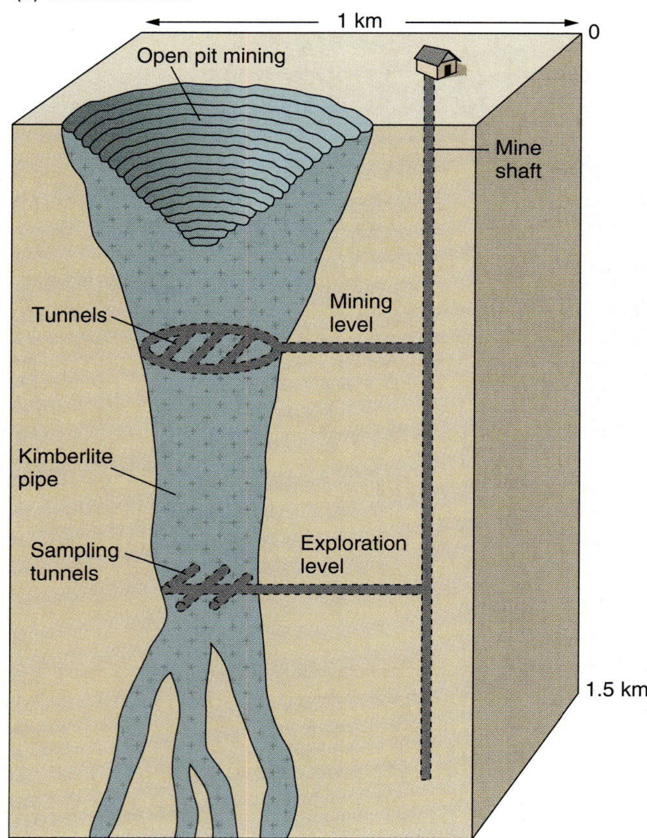

FIGURE 16.4 (a) Kimberlite pipes bring diamonds to the surface in rocks called kimberlites. They are thought to originate at depths of greater than 150 km because diamonds only form at pressures associated with these depths. (See Figure 2.18.) (b) Kimberlites consist of scattered and very rare diamond crystals among many different types of minerals picked up as the magma travels through the kimberlite pipe to the surface. (c) Worldwide distribution of kimberlite provinces. The size of the circle represents the relative sizes of the deposits. Greatest deposits occur in central and southern Africa and Russia. Note the association between diamond occurrences and ancient cratons.

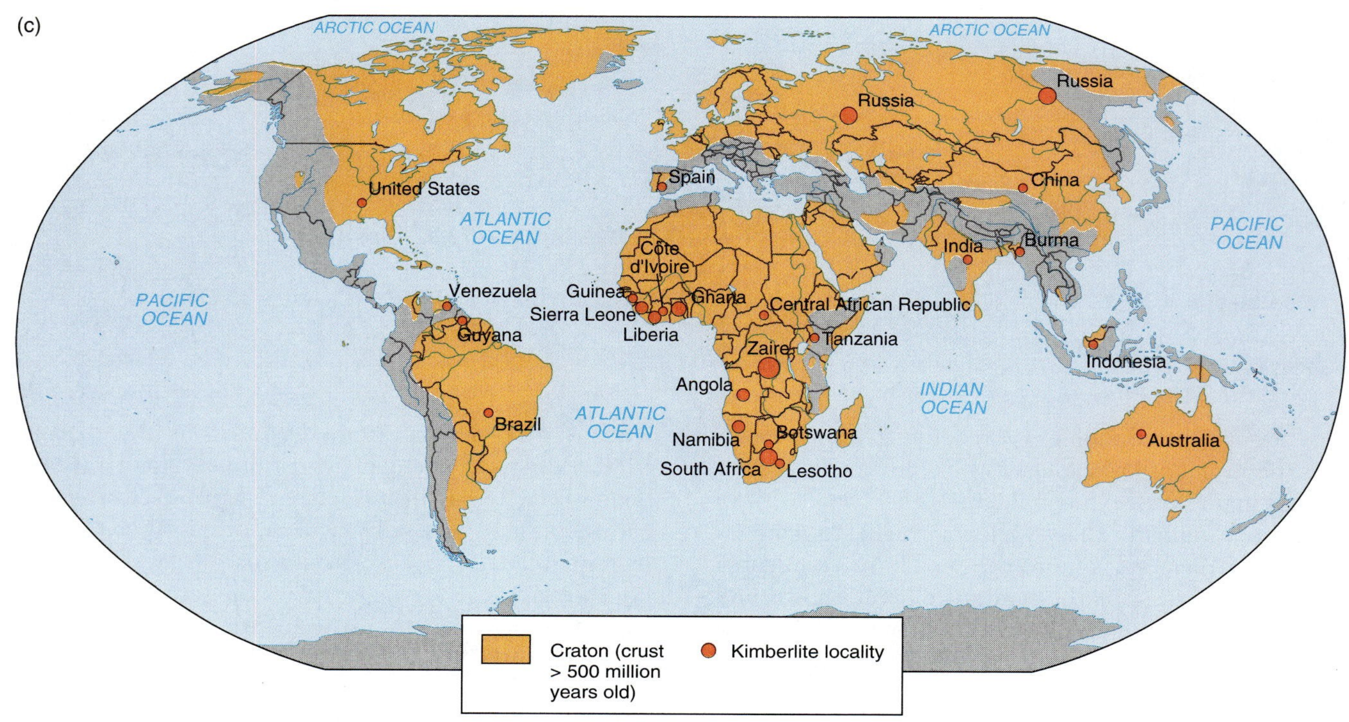

xenocrysts (individual crystals like olivine, pyroxene, and garnet). Diamonds are generally found as isolated xenocrysts (Fig. 16.4b); more rarely, they are found included in xenoliths.

The principal kimberlite diamond fields are shown in Figure 16.4c. Note that most kimberlites occur in Archean cratons—areas of very old, stable lithosphere (∞ see Chapter 11). Examples include the South African, Siberian, and Australian cratons. This relationship with cratons has an important significance. Recall (∞ see Chapter 4) that diamonds are high-pressure polymorphs of carbon formed at pressures corresponding to depths greater than about 150 km within Earth (Fig. 16.4a). Further, the lithosphere is only this thick (150 to 250 km) in cratonic areas, so these are the only places where diamond is stable in the lithosphere. Radioisotope dating of minerals included in diamonds recovered from kimberlite pipes in South Africa yields ages of about 3 billion years, which indicates that the diamonds and the lithosphere in which they reside are very ancient indeed.

Kimberlite is mined by open-pit operations near the surface and by underground shafts. Only about 5 g of diamond are found in every 100 tons of kimberlite ore. Extraction of diamonds from the ore makes use of the unique mineral properties of diamond. Heavy minerals (including diamond) are first separated from the crushed rock and then are passed along on a greased belt washed with water jets. Diamond, which does not wet, sticks to the belt while the other minerals are flushed away. Alternatively, the heavy minerals can be passed through an X-ray beam, which makes the diamonds fluoresce, so they can be easily picked out.

Hydrothermal Processes

Many metal ores are formed in **hydrothermal zones**—regions in the upper crust through which hot water has circulated. In ocean hydrothermal zones, the water may be seawater that penetrates into deep cracks. On land, the water is provided by rainfall and is referred to as *meteoric water*. The water is commonly heated by magmatic intrusions, which causes it to expand and hence rise, to be replaced by cooler water. The net result is that a convective circulation is set up for the water contained in the pores and cracks in the rock (Fig. 16.5). The hot water dissolves minerals in both the magma and the surrounding rocks. The resulting mineral solution is called a **brine**. The solubility of most minerals is high at high temperatures and decreases as the temperature drops. As the solution rises in fractures and cools, the dissolved solids precipitate out of solution and are deposited on the walls of the fractures as veins. At times, the deposition completely clogs up the fractures and new fractures form elsewhere. Many metals, such as copper, gold, and silver, are deposited in hydrothermal systems associated with volcanoes and igneous intrusions.

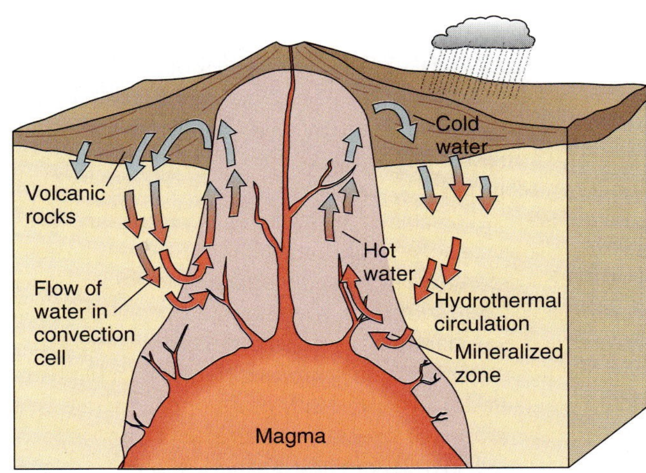

FIGURE 16.5 Convection of water in a hydrothermal zone within a volcano. Groundwater is heated by magma, causing the water to rise in a convection cell. This water carries dissolved minerals from the magma and the surrounding rock. When the water in the convection cell cools, it precipitates out minerals.

The world's most important copper ores are porphyry coppers—low-grade copper deposits in which the copper minerals occur in disseminated grains and veinlets, rather than in discrete veins. A porphyry-type igneous rock is one that contains large crystals embedded in a fine-grained groundmass (the texture is porphyritic; hence, they are called *porphyry copper*; ∞ see Chapter 4). A shallow intrusion acts as a heat source to drive hydrothermal circulation through the cool, solid outer part of the intrusion, and through the surrounding wall rocks. Copper deposited from the hydrothermal fluids fills what were tiny pore spaces among the crystals of the porphyry. At times, mercury deposits of cinnabar or of quicksilver are associated with porphyry coppers. Smaller deposits of gold, silver, tin, lead, and molybdenum are also found with porphyry coppers. Many porphyry copper deposits are concentrated around the rim of the Pacific Ocean, having been formed by magmatism associated with subduction.

Silver is an element similar to copper in many ways. Silver-bearing minerals are found in hydrothermal veins principally in andesitic and rhyolitic rocks. Most of the world's silver is found in the great mountain chain that runs down the western edge of the Americas, where subduction has generated andesitic and rhyolitic volcanoes.

We saw earlier (∞ see Chapter 8) that hydrothermal circulation is also an important process along oceanic spreading centers. Pools of hot brines have been discovered on the sea floor along mid-ocean ridges, especially where transform faults intersect the central ridge. Such pools have also been found in the Red Sea; they contain brines of zinc, copper, and lead—all of which have potential commercial value. The metals are thought to have formed by a thermally induced circulation of seawater through volcanic rocks.

Hydrothermal processes have been observed directly from submersibles on a number of mid-ocean ridges as black smokers and white smokers. Black smokers (Chapter 8) are towering mineral deposits on the ocean floor that discharge mineral-loaded water. The discharge from black smokers precipitates iron- or zinc-rich minerals. White smokers are similar but have lower temperatures and precipitate white particles of barium sulfate. Eventually these deposits become covered by sediments as the ocean basin continues to form. However, they may reappear in ophiolite complexes (Chapter 8) as a result of tectonic collisions.

Oceanic subduction zones are associated with deposits of massive sulfides of zinc, lead, and copper, which are known as Kuroko-type ores, after their site of discovery in Japan. Although details of their formation are not clear, these ores are thought to have been deposited in a shallow marine environment when rifting split the axis of the island arc to form a back-arc basin (Chapter 9). The rift filled with ocean water, which reacted with the magma of the arc and generated the ores—in a process presumed to be similar to that occurring along other spreading ridges. Ancient deposits of Kuroko type may have been tectonically incorporated into the continents following island arc continental collisions. Examples include the Rio Tinto deposit in Spain, the Umm Samiuki deposit in Egypt, and the Buchan mine in New Brunswick, Canada.

Some minerals become concentrated by crystallization from magmas. Diamonds are brought to the surface in kimberlite pipes. Hydrothermal systems contain mineral-laden water heated by intrusions of magma. Minerals precipitate as the water cools.

Sedimentary Processes

Sedimentary processes include direct precipitation from seawater or concentration by grain sorting during transport and deposition.

Stratiform deposits are found in sedimentary rocks, usually shales, in which the mineralization occurs in (stratalike) layers. The ores are commonly sulfide deposits, but also include native copper deposits. A famous example, the Kupferschiefer, or copper shale, averages only 25 cm in thickness, but extends throughout much of northern Europe. The copper it contains has been mined continuously since the fourteenth century.

A common sulfate that is precipitated is barium sulfate, or barite, which is used in drilling for oil and gas. Because of its high density, finely ground barite is added to clay and water to form a drilling mud, which is pumped down the drill stem. Drilling mud has the dual purpose of cooling the drill and acting as a seal if high pressures are encountered that otherwise would generate a gusher.

A large percentage of the world's stratiform lead-zinc deposits occurs in limestones and dolomites that have undergone little tectonic disturbance since their deposition. Because these deposits are common in the Mississippi Valley, they are referred to as "Mississippi Valley-type ore deposits." It is thought that brines rising from deep, compacting sedimentary basins are channeled through the porous carbonate rocks where they react with a carbon-rich residue such as petroleum. The carbon-rich residue is required to chemically reduce the metals that are in aqueous solution in an oxidized condition to a reduced condition, which is insoluble and so they precipitate. The deposits form by diagenetic growth of lead or zinc sulfides in limestone. The most common sulfide minerals that are precipitated include galena, the chief source of lead; sphalerite, the main source of zinc; tetrahedrite, a silver ore; and, at times, pyrite used to extract iron and manufacture sulfuric acid. One of the principal uses of lead is in the manufacture of car batteries. The principal use of zinc is in galvanizing, a process in which zinc is used to coat a metal that otherwise would corrode.

One interesting and potentially valuable form of sedimentary mineralization in the deep ocean is the precipitation of manganese-rich sediments and manganese nodules. Some parts of the ocean floor are covered by **manganese nodules**, which are complex mixtures of iron and manganese oxides and hydroxides along with minor amounts of other metals, such as copper, cobalt, and nickel (Fig. 16.6). The nodules are roughly spherical, ranging in size from several to more than 10 cm in diameter. The manganese and other metals that are in the nodules are thought to be mainly derived from the weathering of

FIGURE 16.6 Manganese nodules are irregular potato-shaped lumps of manganese-rich minerals that carpet some parts of the deep ocean floor.

rocks and sediments on land. The metals are transported in rivers and streams to the oceans. Another source of the metals in manganese nodules is the hydrothermal and volcanic vents along the mid-ocean ridges. The metals precipitate onto a central nucleus of rock or shell material in onion-like layers to form the nodules. Where they are concentrated, they have value as sources of manganese and other metals used in the production of iron alloys.

Banded iron formations, one of the world's most important iron-ore deposits, are enigmatic ancient sedimentary rocks consisting of layers of iron oxide separated by layers of silica in the form of chert (Fig. 16.7). Banded iron formations are found on all the continents, and all were deposited between about 2 billion and 3 billion years ago. Over the same time interval, large amounts of gold and uranium were also deposited. What conditions were responsible for this period of peak deposition and why are they not present today? It has been estimated that when banded iron formations accumulated, the atmosphere contained less oxygen and more carbon dioxide. Such a difference in the atmosphere could have increased the acidity of rainwater as well as seas, rivers, and lakes when compared with present values. An increase in acidity would have provided excellent conditions for dissolving iron and transporting it to lakes and shallow seas where it subsequently precipitated. There is much debate on the origin of the banding; cyclical periods of evaporation or climate changes are two possibilities.

Sedimentary processes may also concentrate minerals by density sorting in placer deposits. **Placers** are accumulations of economically important minerals along stream beds and offshore sandbars. As a consequence of gravitational settling, the denser minerals preferentially settle out into these deposits, whereas other lighter minerals are washed away by waves or rivers (∞ see Chapter 13).

FIGURE 16.7 Banded iron formation (layers of iron oxide separated by chert); an important ore of iron.

Native gold, varying from small flakes to nuggets, is found in placer deposits. Gold is a soft, malleable metal that is corrosion resistant. In addition to its ornamental use, half of the world's annual production of gold is used in modern industry for electronic components (it is an excellent electrical conductor, even better than copper), aerospace alloys, and in dentistry. Gold is primarily deposited in veins from hydrothermal systems, but when these veins are exposed at the surface, they are subjected to weathering and erosion like any other rock. Unlike most metals, gold does not react to form new compounds and it remains in the native state through weathering and erosion. Gold is concentrated in streams where flowing water washes away less dense sand grains and leaves the gold behind, concentrated behind barriers. These alluvial placers may contain gold in grains that are 1 mm to 2 mm in diameter. Prospectors panning for gold attempt to discover the source of the placer deposits—the main vein, or mother lode—by panning progressively upstream. The process of panning itself relies on the same principles as placer concentration: Lighter minerals are carried off in the water swirled in the pan, which leaves the denser minerals, such as any gold, behind in the sediment.

The greatest known gold deposits in the world are the Witswatersrand conglomerate deposits of South Africa. These sediment beds are thought to have been deposited in a shallow marine basin. Ancient rivers ran into the marine basin and formed large deltas. The gold is found in conglomerates in the deltas as clastic fragments, presumed to have been carried by the streams that deposited the conglomerate pebbles.

Placers containing diamonds can be found in regions where diamond-bearing kimberlite pipes have been eroded and the resulting sediment transported away. The high density of diamond accounts for its occurrence in these deposits. The beach sands of Namibia contain diamonds derived from erosion of kimberlites in nearby South Africa, and the sands are mined as diamond placer deposits.

Sedimentary processes concentrate minerals by precipitation or by grain sorting during transport and erosion.

Weathering Processes

The process of weathering can also concentrate certain elements sufficiently to form ores. In equatorial regions, weathering is extremely effective because both temperature and humidity are high (∞ see Chapter 12). As a consequence, very thick soils are developed. As part of the soil-forming process, minerals react with water, and many of the constituent elements are removed in solution. Some minerals are, however, relatively insoluble and resist being removed by dissolution. They are effectively left behind, and in this way concentrated to form what is known as a residual ore deposit.

Bauxite and laterite are tropical soils formed by this process, and they are mined as ores for aluminum and iron respectively.

Prospecting for Mineral Deposits

Most of the sedimentary rocks or soil that cover Earth's land have no value as ores. The rich mineral deposits that crop out at the planet's surface have largely been discovered. To find mineral deposits covered by sedimentary layers requires indirect methods. For example, a geologic interpretation of the tectonic environment in which the underlying rocks formed can improve the chances of detecting underground ore bodies. Geochemical analysis of groundwater in contact with an underground ore body can sometimes detect increased concentrations of dissolved trace elements associated with an ore body. In some cases, ore bodies are magnetic, and their presence can be detected by magnetic surveying; in other cases, an ore body conducts electricity well and its presence can be discerned by electrical measurements made at Earth's surface. Gravity measurements can also detect the presence of ores—particularly if they are more or less dense than the surrounding rocks.

Uranium ores are unusual because they are radioactive. The classic picture of the prospector looking for uranium depicts someone carrying a Geiger counter to measure radioactivity. Such counters detect radioactive emissions electrically and a readout gives the number of emissions per second passing through the sensor. If the radioactivity is buried too deep, the emissions are absorbed by the overlying strata. Thus, Geiger counters are only sensitive to the radioactivity contained in the upper 20 or 30 cm of crust. Deeper deposits may pass undetected, and other exploration methods must be used to discover them.

Another way to detect uranium is to search for radon gas, which is emitted during the decay of uranium (^{235}U, ^{238}U), and thorium (^{232}Th). Radon is chemically unreactive and thus escapes rising upwards from uranium deposits that lie at greater depths than those at which a Geiger counter can detect radioactivity. The gas is radioactive and can be detected by a Geiger counter. Detection of radon in homes is important to recognize because radon is known to cause cancer.

Nonmetallic Economic Deposits

Various deposits that we rarely consider as valuable are highly sought after by industrialized society. Among these, limestone is used for building stone and cement; sand and gravel are utilized in road materials and concrete aggregates; and the clay *kaolinite* is used in paint and as a coating to make magazine covers glossy. It is used as filler in modern athletic shoes, in chocolate to make it more heat resistant, and is fired into ceramic to make the finest china dishes. The clay *montmorillonite* is used in paint and for drilling mud. Other deposits can be equally valuable, of course, but this is a reasonable sampling of this type of resource.

A wide variety of stones can be used for interior decoration but, for exterior facing, the rock must be strong and reasonably weather resistant. Limestone and granites that are used for building facings need to be massive, unbedded, and essentially without joints. One of the best-known locations to quarry limestone building stone is in south central Indiana. Nearly all the buildings on the campus at Indiana University are made of the local building stone, and one of the local quarries supplied the facing stone for the Empire State Building. The limestone used in making cement must have a particular composition that includes a small amount of silica. Processing plants are commonly constructed at sites where limestone containing the proper composition is found in sufficient quantities to quarry. During processing, the limestone is finely crushed, fired to a temperature in excess of 900°C, and then reground to a fine powder. The firing in a large kiln converts the limestone ($CaCO_3$) to calcium oxide. When water is added to a mixture of calcium oxide and silica, these chemicals combine to form a calcsilicate. Although concrete appears to harden in a matter of hours, the reaction between these two chemicals occurs in a solid, and therefore is very slow; it actually takes years for concrete to reach its maximum strength.

Aggregate for concrete has special requirements. Angular aggregate makes stronger concrete than do rounded pebbles, but crushing and sieving the material costs money, which raises the cost to the consumer. Hard, dense rocks, such as basalt or a massive, unfoliated metamorphic rock, are stronger than soft rocks for the aggregate; moreover, the aggregate must not react with water or with the curing cement. Gravel for road material is less restrictive. Again, angular gravel is preferred over rounded gravel for roads because tires cling better to it in wet weather. The relative hardness and composition is really unimportant for road material. Both gravels in concrete and road aggregate are usually obtained from stream beds, where sieves are erected to segregate the material by size. Cobbles and boulders may be crushed, and this crushed aggregate is mixed with the smaller grains of the aggregate and adds strength to the aggregate if it is to be used for concrete.

Clays are usually mined from lake deposits or terranes that have undergone intense chemical weathering. Kaolinite forms during the weathering of feldspars in subtropical regions. For the uses that we have listed, very pure kaolinite is required, and such deposits are rare. Montmorillonite has a very peculiar property when first dried, then dampened. The mineral lattice absorbs water between layers of the lattice, causing the lattice to swell by a factor of approximately three. In this condition, ionic charges exist on specific parts of the mineral lattice. These charges weakly bond with water and with other nearby

Focus On 16.1 URANIUM ORE AND NUCLEAR ENERGY

Uranium deposits are found in such a wide variety of geologic settings that it is difficult to generalize about their origins. The average concentration of uranium in Earth's crust is about 2 parts per million (ppm). Rich uranium ores have concentrations of 20,000 ppm. The ores must be enriched to give uranium with a concentration sufficient for use in nuclear reactors. Uranium is a valuable radioactive metal used as a nuclear fuel and is concentrated naturally in rocks in several different ways. The large size of uranium atoms prevents them from being easily incorporated into early crystallizing phases in a cooling magma. The uranium becomes concentrated in the final melts and fluids, and in the rocks derived from them; that is, in silica-rich rocks like granites.

Uranium may occur concentrated in veins as an important ore mineral called *pitchblende,* which is a form of uranium oxide, or uraninite (UO_2). Metamorphic uranium deposits form at the interface between molten igneous rocks and the rocks they intrude, or they form from remelting of rocks deep in the crust. Detrital uranium minerals are thought to collect in stream channels and to be deposited in much the same way as gold is deposited. The high densities of both gold and uranium minerals cause both metals to collect in placer deposits. One of the most famous of these uranium placer deposits occurs in the Witswatersrand of South Africa, which is not only the world's foremost gold producer but also accounts for 10 percent of the world's uranium production. Finally, uranium may be leached by groundwater and precipitated in sedimentary deposits, or it can be taken up by plants and animals, concentrated and preserved in wood or bone.

The radioactive decay of uranium releases energy used in nuclear reactors to produce electricity or in nuclear weapons. Let us examine the nuclear processes involved. Uranium may decay by nuclear fission, a rare form of radioactive decay. When fission occurs, a nucleus breaks into two large and fairly equal fragments, releasing neutrons (∞ see Chapter 2). Some nuclei undergo fission spontaneously. For example, ^{238}U (an isotope of uranium with mass number 238) undergoes spontaneous fission, but such fissions are very rare and most uranium decays by emitting alpha particles. Some other nuclei, notably ^{235}U (and ^{239}Pu), undergo fission if they are bombarded with neutrons. Because the fission process itself produces neutrons, if the fissionable material is sufficiently concentrated (about 3 percent) to ensure that many of the emitted neutrons hit other active nuclei, then a chain reaction occurs, releasing vast amounts of energy. An uncontrolled chain reaction leads to a nuclear explosion.

Controlled fission is achieved in nuclear reactors by adjusting the concentration of neighboring atoms by inserting moderators, which are materials that absorb neutrons and thus control the chain reaction. If this is not managed carefully, reactor meltdown can occur, which is what happened in the Chernobyl nuclear reactor in Ukraine in 1985.

Uranium fuel contains several percent of ^{235}U. The uranium in uranium ores is composed mainly of the ^{238}U isotope; the fissionable ^{235}U isotope makes up only 0.7 percent of the uranium in the ore. Thus, even the richest ores—at 20,000 parts of uranium per million—are stable and are not about to undergo an atomic explosion in nature. However, one exception has been reported. A natural fission event appears to have occurred at Oklo in the Gabon Republic of West Africa 2 billion years ago. It is now a uranium mine. In scattered pockets within the deposit, the ^{235}U reached sufficient relative concentrations to become critical (about 3 percent of the total uranium). The isotope of ^{235}U decays about six times faster than does the ^{238}U isotope. Two billion years ago, ^{235}U represented a greater proportion of uranium ores. Natural concentration of the ores at that time was more likely to produce a fissionable material than today—when the concentration is only 0.7 percent.

clay grains, causing the clay–water mixture to solidify. When shaken, however, the bonds are disrupted and the mixture becomes liquid again. This curious property of liquefying when shaken and solidifying when left alone is called *thixotropy*. In drilling water or oil wells, pieces of rock are broken off the bottom of the hole by the drill bit; circulating mud picks up these drill chips and carries them up and out of the hole. The use of montmorillonite in the mud causes it to stop circulating away from the drill bit and to stiffen on its way to the surface. The thixotropic fluid (drilling mud) solidifies and prevents the drill chips from sinking back to the bottom of the hole.

The same property that causes thixotropy allows another use for montmorillonite. When the mineral is heated to approximately 1000°C, it partially fuses and suddenly expands like popcorn. The result is a lightweight, but strong, aggregate. It is so light that it floats on water, but when mixed with cement of the right composition, it makes a very strong concrete. The resulting concrete weighs less than one-third that of normal concrete, but it is nearly as strong. When used in buildings, much less steel is needed to support the weight of the structure, resulting in large savings in construction costs.

Minerals and Plate Tectonics

The previous discussion of ore-forming processes illustrates how plate tectonics is an important factor in determining the distribution of economically important mineral deposits. Wherever magmatic processes occur, minerals are generated—either by direct concentration of crystal phases or by the secondary role of magmas in providing heat for hydrothermal circulation. Most magmatic processes occur at plate margins. Thus, plate tectonics provides a reference frame for understanding mineralization. Applying the plate-tectonic scheme to the search for mineralized zones not only requires an understanding of present tectonic margins but also a piecing together of ancient margins, many of which are now contained within the continents (∞ see Chapter 11). Some mineral resources exhibit an affinity for certain plate margins because of the conditions found there, such as temperature, pressure, geologic structure, and fluid flow (Fig. 16.8). This is not to say, however, that these ores are exclusively found at such margins, and ores of some elements (such as uranium; see Focus 16.1) may form in many different ways.

Mineral ores are formed in the oceanic lithosphere as it moves away from the mid-ocean ridge, where hydrothermal processes operate. Massive sulfide deposits, such as marcasite, a polymorph of pyrite; chalcopyrite, the most important ore of copper; and sphalerite, an ore of zinc, are found in the upper crust on top of or within the pillow lavas of crustal layer 2 (∞ see Chapter 8). Other large, sulfide-rich deposits have been found in ancient fragments of oceanic crust (ophiolites) that have been obducted onto the continents. The ophiolite deposits on the island of Cyprus have been a rich source of copper sulfide, which has been mined since ancient times. In fact, the word copper comes from the Greek *cyprus*.

As we saw earlier (∞ see Chapter 9), subduction is commonly associated with magmatism—the formation of vol-

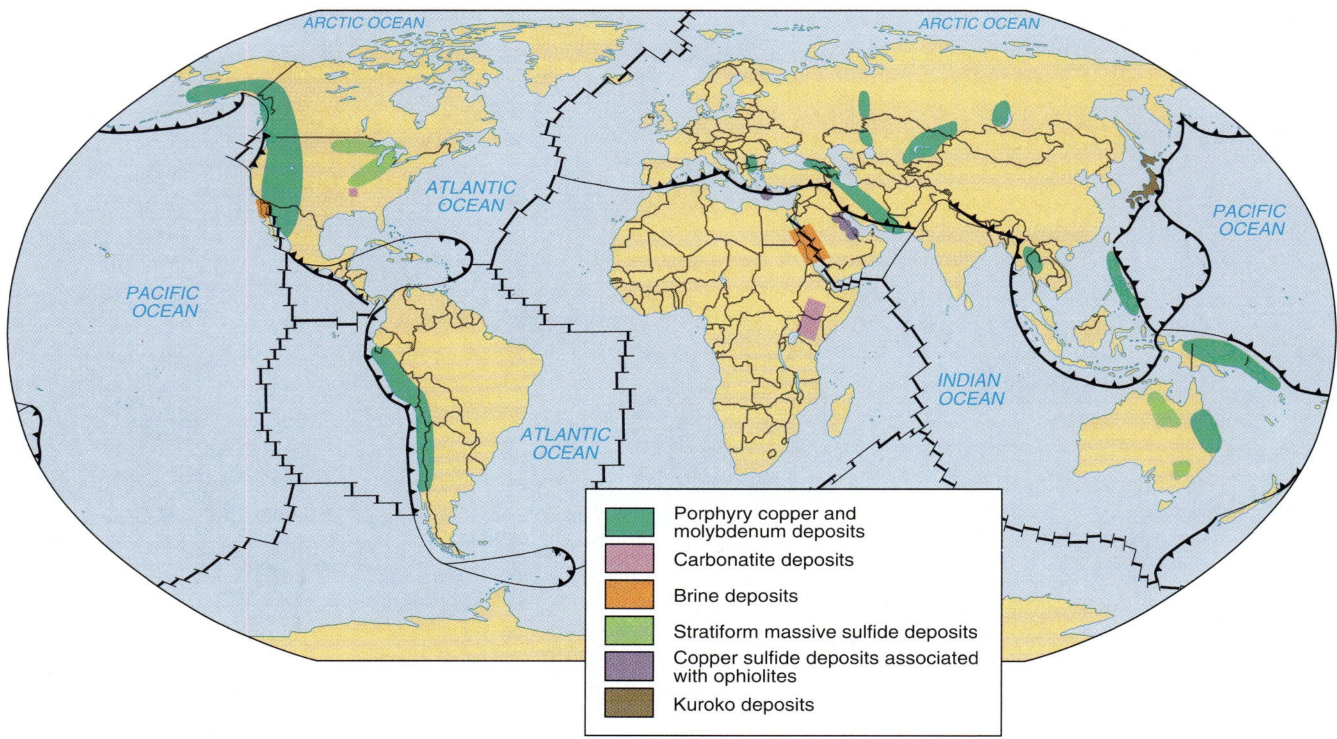

FIGURE 16.8 Some of the important mineral deposits of the world and their relation to plate tectonics. The main role of plate tectonics is the control of magmatism, which may be a primary source of ore minerals or may simply provide the heat necessary for ore-depositing hydrothermal systems. Most of the porphyry copper belts are associated with present or ancient subduction activity. Major belts are found along the west coasts of the Americas. Recent carbonatite deposits (East Africa) and brine deposits (Red Sea) are associated with rifting. Sulfide deposits associated with subduction of sea floor are found in Cyprus and in Oman. Deposits related to back-arc basin formation called Kuroko-type ores are found in northern Japan. In contrast, stratiform sulfide deposits such as the Mississippi Valley-type ore deposits and deposits in Australia are formed in interior basins away from plate margins.

canoes and huge belts of intrusive plutons. Granite belts occur in arc regions of Andean-type (continental) subduction zones such as in Southeast Asia and along the western margins of the American continent. Porphyry copper deposits are perhaps the most important ores found here (Fig. 16.8). For example, along the western edge of South America, the long and narrow nation of Chile provides a significant proportion of all the world's copper, boasts the world's largest copper mine at Chuquicamata, and has an economy that is strongly dependent on the price of copper.

Where transform faults intersect mid-ocean ridges, the crust is highly fractured and permeable. Fracturing allows the downward propagation of seawater, which is heated and returns mineral-laden water to the surface. In effect, transform faults serve to facilitate fluid flow, although the hydrothermal circulation is primarily due to the heat associated with a divergent spreading center. The brine pools of the Red Sea, discussed earlier, fall into this category, as do brines found in the Salton Sea region of California. The Romanche fracture zone of the equatorial Atlantic has been found to contain iron sulfide concretions that may have originated by this mechanism.

The initial rifting of continents, as is currently occurring in East Africa, may be associated with the formation of some magmatic deposits. An unusual magma type called *carbonatite* may be erupted or intruded at volcanoes associated with continental rifting. Dikes of carbonatite are found east of Hot Springs, Arkansas. Carbonatites are igneous rocks consisting largely of calcium carbonate or dolomite. They frequently crystallize minerals with high concentrations of rare metals such as phosphorus, niobium, tantalum, uranium, thorium, zirconium, and rare-earth elements, which can be used as ores. Erosion and dissolution of the carbonatites result in deposits of sodium bicarbonate (baking soda), sodium chloride, and sodium fluoride as found in the lakes of the East African Rift system.

Certain types of mineral deposits are associated with particular plate-tectonic environments, where magmatism, fluid flow, or sedimentary processes are suitable for ore concentration.

Fossil Fuels

It has been said that organic matter equivalent in quantity to Earth's weight has been created by living creatures since life originated on this planet. Much of this matter decays rapidly when the organism dies, and is recycled into the atmosphere as carbon dioxide. In certain cases, however, the remains of an organism are preserved from immediate decomposition after death, allowing for the formation of fossil fuels.

When wood is used as a fuel, carbon and hydrogen burn; that is, they combine with oxygen and liberate heat. Fossil fuels are carbohydrates that contain carbon and hydrogen in higher concentrations than are present in wood, making them much more efficient fuels. Fossil fuels occur in three familiar forms: coal, oil, and natural gas. Less familiar fossil fuels include forms of shale oil, tar sands, and peat.

Although the familiar forms of fossil fuels are quite diverse, ranging from an opaque solid to a yellowish liquid to a colorless gas, they share a similar origin as organic matter trapped in sedimentary rock. The form of fossil fuels depends on the type of organic matter originally deposited; coal comes from tree trunks, leaves, stems, and other plants deposited in freshwater swamps; oil and gas are mainly generated from phytoplanktonic organic matter deposited in marine basins. Formation of economically important fossil fuels from organic matter depends on subsequent alteration by bacterial decay and the action of temperature and pressure resulting from burial.

At very shallow depths, biochemical changes occur in the deposited material, including bacterial decay and the growth of fungi and other microorganisms. A primary product of this activity is the release of methane gas (CH_4). Coal damp, a foul and poisonous gas that pollutes the air in mines, consists mainly of methane (with traces of hydrogen sulfide), as does swamp gas. At greater depths, microbial activity diminishes. Increased temperatures and pressures drive off volatiles and water and cause cracking or degradation of complex hydrocarbon molecules into smaller ones.

Peat

All natural fuels, including wood, peat, coal, oil, gas, and shale oil, come from living matter. Green plants and many bacteria grow by photosynthesis, a process in which carbon dioxide, water, and light energy from the Sun interact to produce carbon compounds. These carbon compounds include carbohydrates such as cellulose, starch, and sugar. During this process, oxygen gas is released into the air (Fig. 16.9a); this is the reason for the presence of oxygen in the atmosphere.

When organisms die, the organic matter decays, and the process reverses. Mircroorganisms use the oxygen from the atmosphere to decompose the carbon-containing compounds in these organisms. This process releases carbon dioxide back into the atmosphere (Fig. 16.9a). If all organic matter were completely decomposed in this fashion, little free oxygen would exist. Burial processes, however, protect a proportion of dead organisms from decay because the microorganisms causing decomposition no longer have access to oxygen. In particular, waterlogged vegetation, such as is found in bogs and marshes, does not undergo complete oxidation. The incompletely decomposed products form layers of peat in peat bogs.

Peat forms from a variety of vegetation ranging from algae and mosses to trees. The climate for peat formation must be humid and the conditions such that the production rate is greater than losses due to oxidation. Preservation requires rapid burial. Today, peat-forming areas occur in both temperate and cold regions, as well as

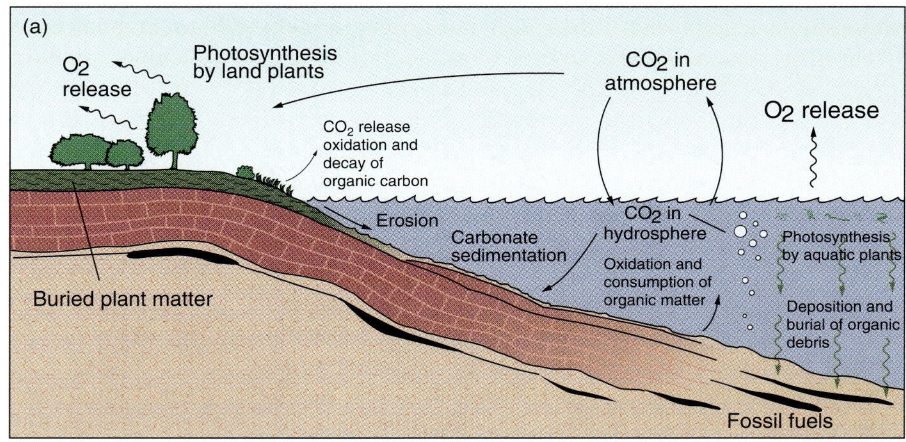

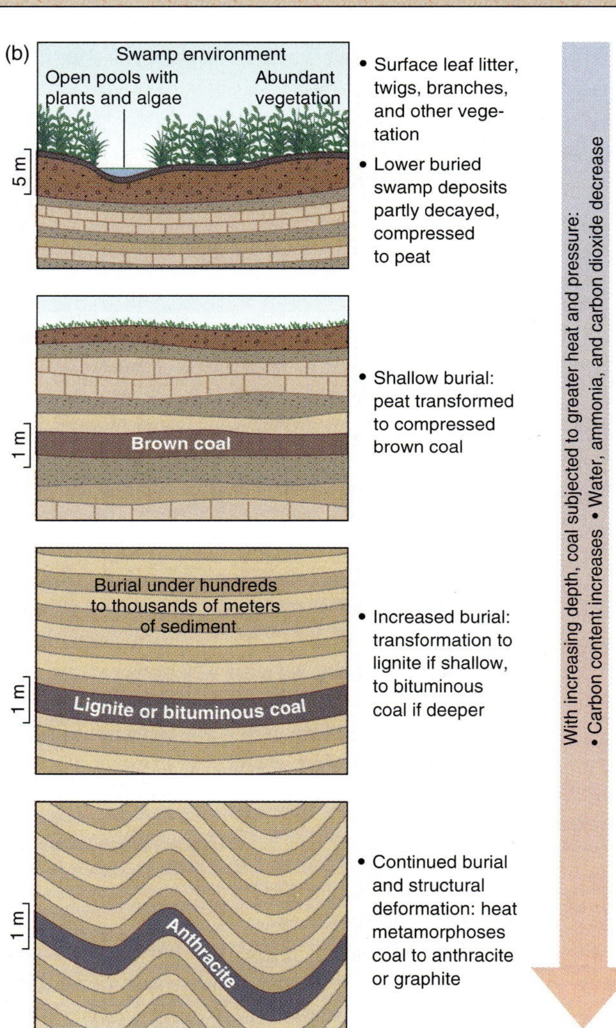

FIGURE 16.9 (a) The carbon cycle: Photosynthesis converts carbon dioxide in the atmosphere to plant matter, releasing oxygen into the atmosphere in the process. Plant matter subsequently decays, returning carbon dioxide to the atmosphere. Some plant matter is buried by sediments to be preserved as fossil fuels. (b) Formation of coal. Decaying vegetable matter in a swamp environment is buried by sediments and is thus protected from further decay. Deeper burial causes heating and increased pressure, which drive off volatiles. If sufficient temperatures and pressures are reached, a high-grade coal such as anthracite is formed.

the tropics. Cold regions include the arctic, Ireland, Scandinavia, Alaska, and Canada, where abundant rainfall promotes rapid plant growth but cooler temperatures retard bacterial decay.

The many thousands of shallow lakes formed in glaciated regions have become peat bogs as these lakes gradually evolved to form marshes (∞ see Chapter 13). Peat also forms in shallow lagoons and lakes of low-lying coastal plains, floodplains, and deltas. Tropical regions that provide ideal conditions for peat formation include the densely forested swamps of the Ganges River and other such deltas. Along the northeast coast of Sumatra, peat has formed in swamps to a depth of 9 m. In North Carolina and Virginia, peat is found to a depth of 2 m over an area of 2400 km^2 on the coastal plain, including an area called the Dismal Swamp. Peat is commonly dug up from the ground, dried, and used as a fuel for heating homes in regions such as Scotland and Ireland.

Burial of organic material protects it from decay by oxidation. Peat forms from the burial of a wide range of vegetation, from mosses to trees.

Coal

When peat is compacted by burial beneath layers of sediments, pressures and heat at depth drive out the water and gases in a manner that progressively enriches the peat with carbon until it finally forms coal (Fig. 16.9b). The *rank* (quality) of coal is a measure of carbon content. Rank depends on the amount of volatiles (mainly CO_2 and NH_3) and moisture that have been removed from the coal; the rank varies from *lignite*, which contains only about 30 percent carbon; *bituminous coal*, which has about 87 percent carbon; to *anthracite*, which contains over 94 percent carbon (Fig. 16.10). Rank increases with time and depth of burial. It has been estimated that it takes about 5 m thickness of peat to generate 1 m of coal.

The peak period of coal formation was approximately 300 million to 285 million years ago (Fig. 16.11). Suitable land plants for peat formation were not in existence until about 350 million years ago. By 300 million years ago, a rich flora had developed. The chief coal-making vegetation was tree ferns, and they were the size

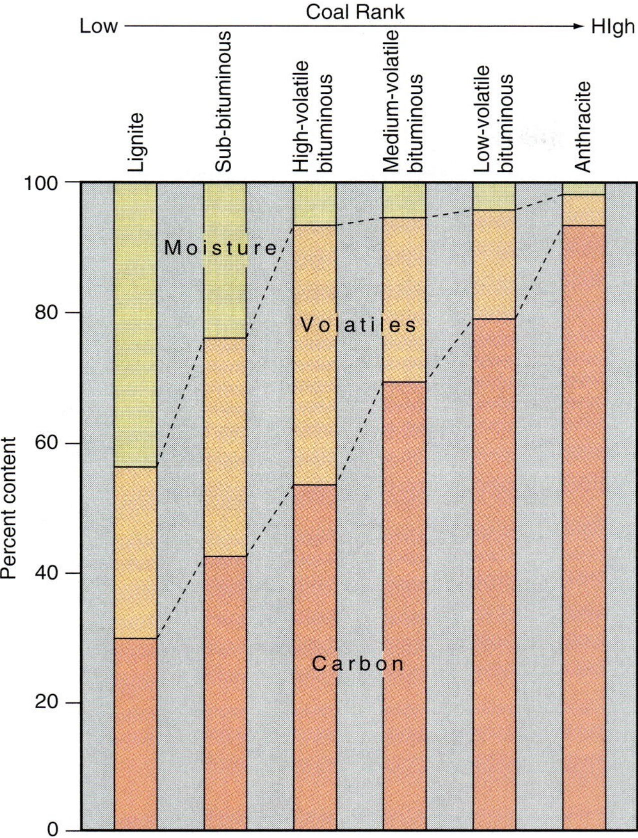

FIGURE 16.10 Rank variation of coals showing the increase in quality, or rank, from lignite with 30 percent carbon to anthracite, which contains over 94 percent carbon. The more carbon, the better the coal is as a fuel.

of tall forest trees. Many plant remains such as leaves, stems, and tree trunks are visible to the naked eye in coal. However, most plant remains decay almost immediately in the first stages of burial, leaving no trace of their original forms.

Coalfields Extensive coalfields were deposited approximately 300 million years ago along a belt running through North America (bordering the Appalachians), the British Isles, Europe, and parts of Asia (Fig. 16.12 and Focus 16.2). At those times, plate motion had brought this region close to the equator. The climate then was humid and supported abundant plant growth. Earth movements associated with collision tectonics formed extensive subsiding basins in which vegetation was buried and converted to coal seams. At later times, vertical motions caused upfolded portions of the crust to expose coal seams at the surface. Underground continuations of the exposed coal-fields are currently being mined beneath the strata.

According to estimates made by the U.S. Geological Survey, about 7.64 trillion metric tons of coal remain unexploited in the United States. To date, we have used only 1.7 percent of available reserves. Trillions of tons of

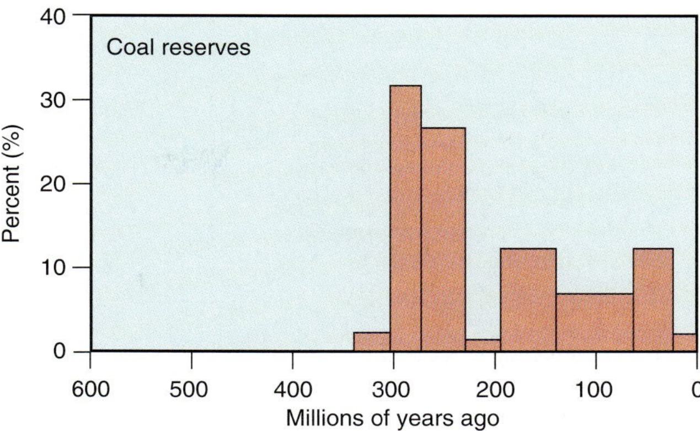

FIGURE 16.11 Worldwide distribution of coal reserves as a function of geologic time of coal formation.

additional coal exist, but lie either so deep or in such thin beds that mining it is not commercially viable. The northern Plains States of the United States have huge reserves of low-grade coal (lignite). The upper midwest and Appalachians have large reserves of higher rank coal (bituminous and anthracite), but most of this coal is high in sulfur content, which causes environmental problems when used (∞ see Chapter 12), and to convert the coal to energy (electricity) requires more water than is available in that area. The use of coal as a fuel has tapered off during the last century because it is more expensive than oil and gas, and because of the environmental problems associated with it (∞ see Chapter 12). Smoke from coal combustion, including noxious sulfurous gases, causes pollution of the atmosphere and acid rain. Earlier in this century, coal smoke in London, England, was responsible for "pea soup" smog—a dirty fog that reduced visibility to less than a meter. This fog was also quite toxic. The sulfur and nitrogen contained in bituminous coal formed sulfur dioxide and nitric oxide, resulting in acid fog. When inhaled, the sulfuric and nitric acids in the fog corroded mucous membranes and destroyed lung tissue. Many deaths directly resulted from these dense fogs.

Additional environmental issues concern the serious disposal problems associated with the coal ash that remains after burning coal; this residue amounts to several percent of the original weight. Strip mining can disfigure the countryside unless sufficient funds are allocated to reclamation, and underground mining is dangerous, claiming many lives every year. For these reasons, despite coal's abundance, petroleum has become the fossil fuel of choice. As petroleum reserves become depleted, however, we must anticipate developing environmentally acceptable methods of mining and burning coal; it is one of the largest energy resources we have.

Formation of basins leads to burial of vegetation, eventually turning it into coal. Subsequent tectonic motions may expose the buried strata, enabling the coal to be mined.

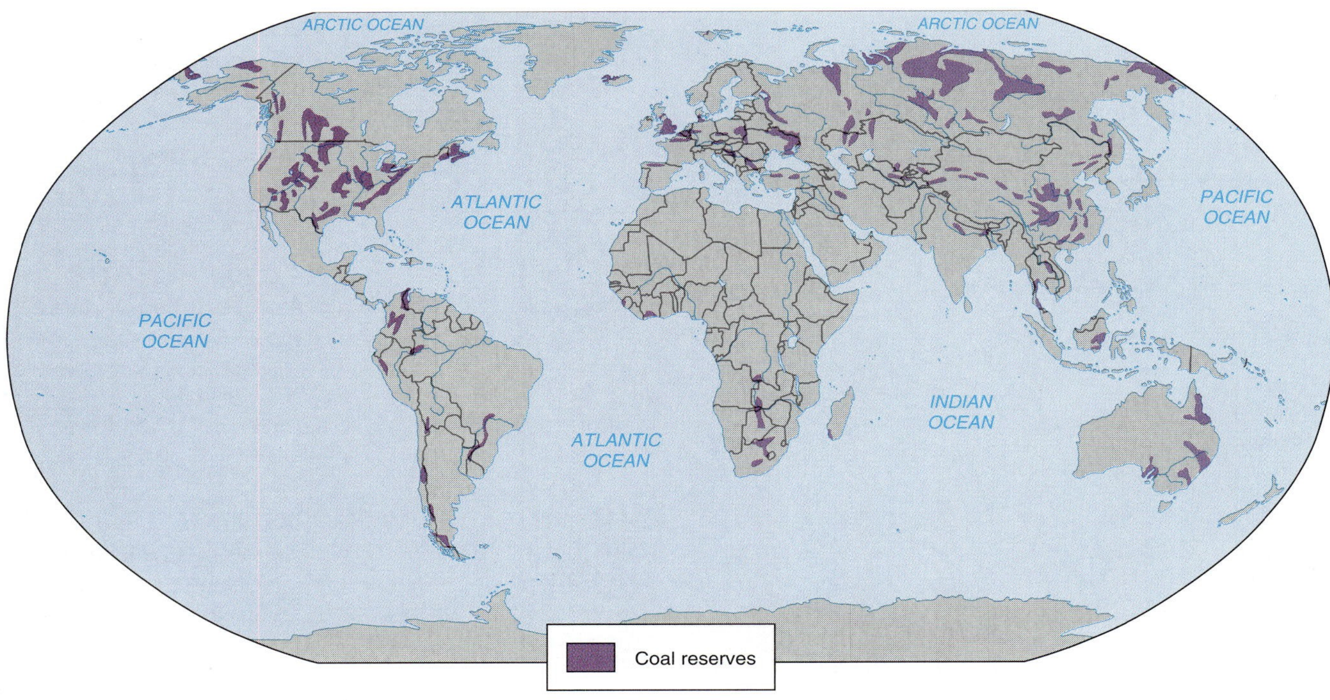

FIGURE 16.12 Geographical distribution of the most important coal occurrences on Earth.

Petroleum

The term *petroleum* describes all hydrocarbons (gaseous, liquid, or solid) found in rocks. The word is derived from the Greek *petra*, meaning rock, and the Latin, *oleum*, meaning oil. In common usage, petroleum refers to all liquid hydrocarbons such as crude oil.

Natural gas is composed mostly of methane (99 percent) with small amounts of ethane, propane, and butane, as well as other gases. The mixture heptane, octane, and nonane (C_7H_{16} to C_9H_{20}), along with many other similar hydrocarbons, is what we know as gasoline. Gasoline is not a natural product; its composition has been modified from the natural petroleum components in a refinery by a variety of techniques including the breaking of molecular chains or rings, re-forming them into another geometry, and adding or subtracting hydrogen to particular molecular types.

Oil Formation Oil forms almost exclusively from organic matter trapped in marine sediments. Some oil forms in lake sediments. Natural gas forms both in marine and nonmarine rocks. Oil is not found associated with peat or coal, illustrating the difference between the two sources of organic matter—marine and nonmarine.

The main contributors to oil are minute marine organisms called *plankton*, which thrive in coastal waters. For this organic material to be preserved as it accumulates on the sea floor, oxygen supply in the sediment layers must be insufficient to oxidize it all. In addition, anaerobic bacteria (bacteria that do not require oxygen to live) may release methane in these locations, and this action can contribute to the formation of petroleum. Hydrocarbons in significant concentrations are not found in newly deposited muds; they are found at depths greater than 1 m beneath the ocean floor in materials older than 3000 to 9000 years.

With continued sedimentation, organic matter is buried. At depths of up to several kilometers, compaction of the sediments expels water. At a temperature of 150°C and pressures found at several kilometers depth, the organic material is altered to *kerogen* and liquid petroleum. Kerogens are large complex molecules formed during the same process that creates petroleum, but kerogen is generally believed to represent a stable end product of organic alteration.

Petroleum forms from organic matter trapped in marine sediments. Burial of organic matter to depths of several kilometers expels water; heat at depth transforms organic molecules into petroleum.

Distribution of Oil Reserves Figure 16.13 shows the distribution of the world's oil and gas as a function of geologic time. Most of these resources formed in the last 200 million years—well after the breakup of Pangaea. Why is this so?

The amount of marine planktonic organisms in the world has generally increased over time; known reserves reflect this increase. Sedimentary rocks less than 200 million years old contain 83 percent of the world's oil, whereas only about 20 percent is contributed by older rocks. The lack of ancient oil may be due to causes other

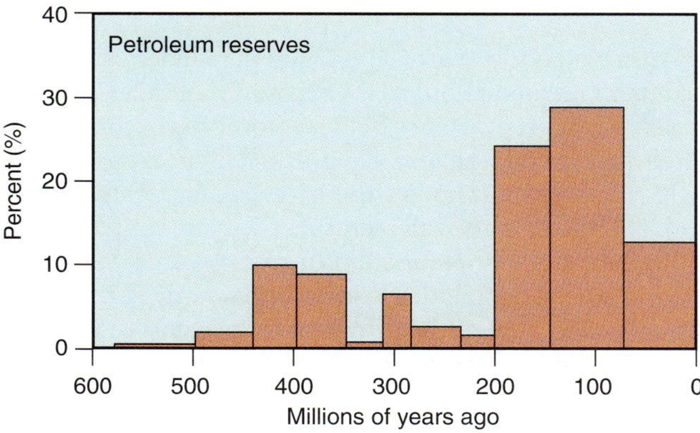

FIGURE 16.13 Distribution of known petroleum (oil and gas) generation through time, showing that most of the petroleum was generated between 65 million and 200 million years ago.

than scarcity of life-forms. The older a rock, the more likely it is that the oil will have been converted to other forms due to metamorphism. Furthermore, with time, tectonic processes and erosion are more likely to have disrupted reservoirs and dispersed trapped oil.

Tectonic setting is one of the major influences on the distribution and hydrocarbon potential of thick sedimentary sequences. Geologic processes acting at divergent, transform, and convergent margins determine basin morphology, rock types, heat flow, and rates of sedimentation, subsidence, and uplift, which, in turn, control generation and pooling of hydrocarbons.

Formation of Oil Reserves We now examine tectonic conditions favorable for the generation and accumulation of petroleum. Four steps comprise the process: source, maturation, expulsion-migration, and entrapment.

Source The initial step in the development of favorable source rocks for petroleum is the formation of a sedimentary basin in a marine environment with conditions favorable for high organic productivity. This may occur in an extensional tectonic setting or in basins associated with convergent margins. Sedimentary rocks containing marine organisms are referred to as *thermally immature source rocks*, which are predominantly shallow marine shales and limestones, but also include nonmarine and deltaic shales.

Maturation When a sedimentary basin forms, organic matter is oil-like but is not petroleum, nor is it sufficiently concentrated in source rocks to be commercially useful. As basins deepen, pressures and temperatures rise, which changes organic matter by a process called *maturation*. During maturation, the large organic molecules are broken into smaller components that comprise petroleum and that flow more readily in the rock strata.

Expulsion-migration Increasing pressures due to burial and tectonic stresses expel the matured oil, which migrates along pathways in more porous rocks such as sandstones. This process is rather like squeezing water out of a sponge.

Entrapment An **oil trap** is a geologic structure consisting of reservoir rock sufficiently porous and permeable to contain an economically significant amount of oil. The oil is prevented from escaping by an overlying impermeable cap rock. Traps can form in a variety of ways (Fig. 16.14), including differences in rock type and a range of geological structures such as folds and faults.

Traps might be the result of changes in depositional environment. When sea level rises, organic deposition occurs in shales (source rocks). Then, when sea level falls, these shales are overlain by sandstones (reservoir rocks) (∞ see Chapter 11). Unlike the shales, the sandstones are porous and can store the oil in the pore spaces. When sea-

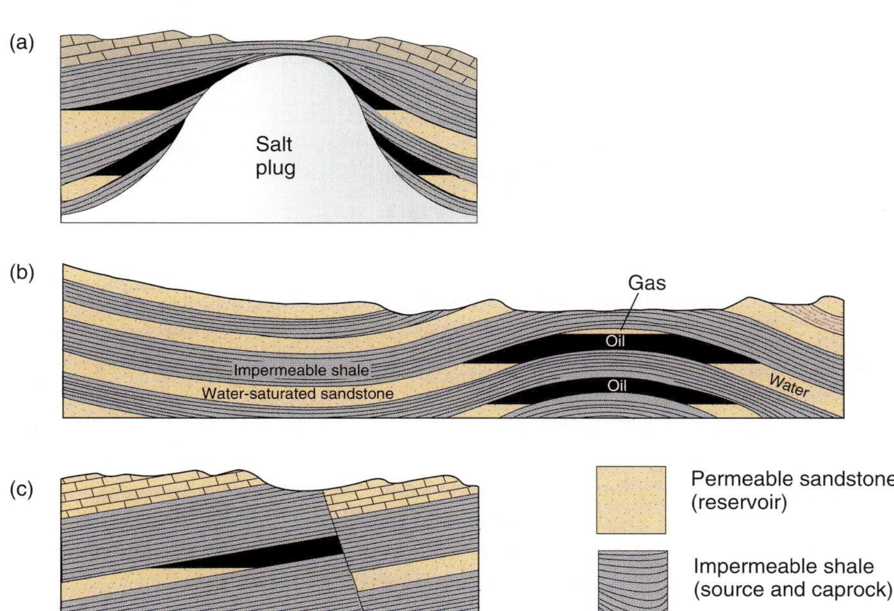

FIGURE 16.14 Different types of oil traps. In each case, low-density maturated oil produced from an organic-rich source rock (shale) moves upward into a permeable reservoir rock (sandstone), and is trapped by the geometry of the reservoir relative to impermeable rocks such as shale and evaporite. (a) Salt dome. The structure is similar to that in part (b), but folding is due to the rise of a salt dome. Oil collects against the dome, which is formed from impermeable salt.
(b) Anticline trap. Oil collects in the upper part of the fold hinge.
(c) Faulted trap. Oil collects in the upper part of a tilted sandstone layer, and is trapped by impermeable shale on the other side of the fault.

water evaporates, evaporites remain (∞ see Chapter 4). Additional rises and falls may cover this evaporite layer with shales and sandstones, which have higher densities. The light evaporites tend to float up through the denser sedimentary overburden. Salt *diapirs* rise in localized zones that push up to form salt domes. The up-arched strata of a salt dome make excellent traps for oil (Fig. 16.14a). Evaporites and shales can also act as seals above sandstones. Tectonic stresses can also generate traps. Folding generates anticlines that may trap oil in a reservoir (Fig. 16.14b) in the core of a fold, beneath an impermeable folded layer. Tilting and faulting can seal off permeable layers at their upper end, producing reservoirs (Fig. 16.14c).

Oil (like water) flows through porous sedimentary rock. If the anticline is capped by an impermeable rock (a rock through which water or oil cannot flow), over millions of years, as oil and water flow through, the oil will accumulate to form an inverted "pool" (Fig. 16.14). Such pools may have a sufficient concentration of oil to be worth drilling. The space in the rock that contains the oil is typically only a few percent of the total rock volume because the oil is contained in the pores. Thus, the pool is not a liquid pool but more like liquid saturating a sponge. An example of an oil reservoir in Texas is shown in Figure 16.15.

Formation of a petroleum reservoir requires the following: source rocks containing marine organisms, burial to achieve maturation, expulsion and migration of maturated oil, and structures suitable for entrapment.

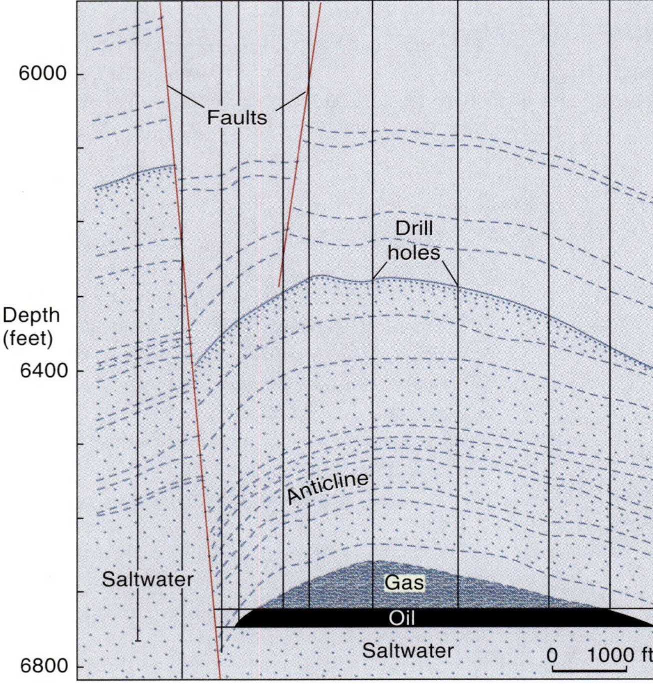

FIGURE 16.15 A cross section through the Amelia oil field of Texas, showing an anticline oil trap.

Oil and Plate Tectonics Oil is found in sedimentary rocks. Sediment accumulation occurs in basins that differ in character between various plate-tectonic environments. Production of oil from organic material contained in these sediments will depend on the pressure-temperature-time path followed by the sediments as they are buried in the basin, which is also a function of plate-tectonic environment. At continental rifts, heating occurs early in the basin formation because of the proximity of magmas to the surface. Oil maturation may therefore occur within a few tens of millions of years. In contrast, basins formed at convergent margins are generally displaced from the magmatism and so are cooler and take longer to heat up. Maturation can take much longer, more than 100 million years.

Figure 16.16 shows the major areas of sedimentary rocks on the continents and on continental shelves, and highlights those areas where large oil and gas fields have been found. Only about 10 percent of the major sedimentary provinces contain the vast majority of the world's known oil. The largest reserves are found in the Middle East, which contain nearly half of the total known amount of oil.

Continental margins begin as continental rifts. Young continental rifts that are landlocked have not been associated with large accumulations of hydrocarbons because of the nonmarine environment of the sedimentary units. As rifting continues from continental to oceanic, shallow seas develop, in which source rocks form. Continental shelves and rises then develop as the shallow seas turn into ocean basins.

Weathering of continental rocks generates sediments transported by rivers to the oceans. The sediments may accumulate on continental shelves that are subsiding because of cooling of the lithosphere as it moves away from the mid-ocean ridge. High heat flow from the cooling lithosphere promotes maturation of organic sediments. Sediments are often underlain by salt deposited by evaporation of seas during initial rifting stages. This salt is important in the later formation of certain types of oil traps, as it may rise up through the oil-bearing sediments to form salt domes. As a result of these effects, sedimentary successions that are many kilometers in thickness are common at passive continental margins, such as the gulf coast of North America. These settings are good places to search for oil (Focus 16.2, page 432).

Oil Exploration

In oil companies, research groups model the geologic history of a basin of interest. They ascertain the composition of the rocks, when they were formed, and their history. For example, they try to determine the relationship between sandstone reservoirs and shale source rocks and whether there has been tectonic movement that would produce traps overlain by rocks that would act as seals. Of particular interest is the sea level and thermal history of the basin and the history of sediment burial, which determines the pressure history. Mechanical models are con-

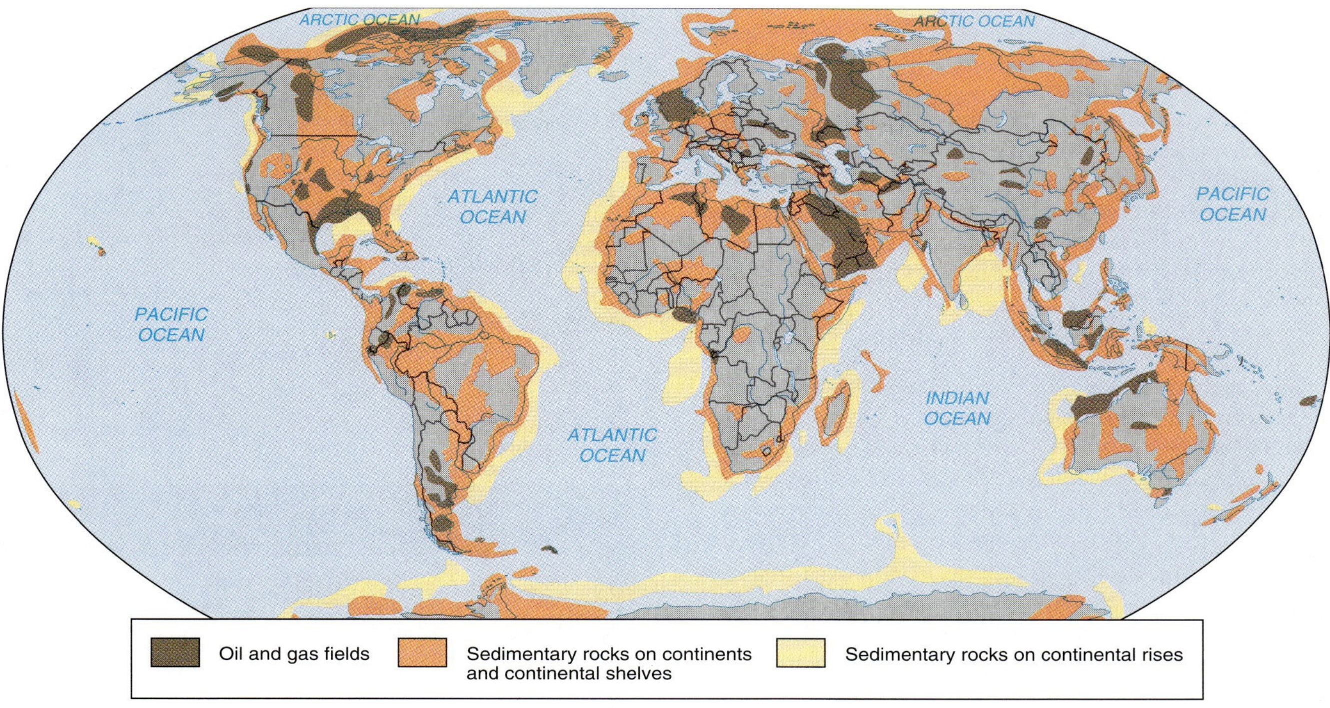

FIGURE 16.16 Distribution of the world's sedimentary rocks; large oil and gas fields are indicated.

structed of the basin tectonics; for example, in an extensional setting, models of continental extension are made that result in graben formation and transition to a passive margin; in a convergent setting, models are made of the formation of a sedimentary wedge and folding and thrusting. Thermal models describe heat supplied by hot rocks from the asthenosphere, and the transition to sea-floor spreading, and the heat flow associated with burial.

Such models give a pressure–temperature history of the sediments. Once this analysis has been completed, estimates can be made as to the depth range at which the rocks were maturated and whether it is economical to drill to these depths. If maturated rocks are too deep, the recovery process could be too expensive. If sufficient time has not occurred for maturation, oil may not have formed. If the rocks have been overmaturated, the oil may have been converted to methane. The depth range in a particular basin where economical recovery of oil is expected is referred to as the "oil window." Predicting oil maturation is an example of a practical application of the theory of plate tectonics, which provides a framework for testing models of hydrocarbon generation.

Sedimentary rocks suitable for generation of oil are deposited in basins from most tectonic settings. Heat flow from magmatism accelerates maturation.

Seeps An oil seep is a location where oil has migrated from a reservoir to leak out at Earth's surface. Over 50 mapped oil-seep areas are off the southern California coast alone. We can find oil seeps along beachfronts, and we can even detect methane bubbling to the surface in coastal ponds. These are classic indications that fractures are providing passageways to the surface for oil and gas to escape from their reservoirs. In the Santa Barbara channel on the California coast, a line of offshore oil rigs marks the fold axis of an oil-bearing anticline beneath the ocean floor, the source of many of the seeps.

The first oil wells were drilled in Pennsylvania, in regions exhibiting this type of surface evidence for underground reservoirs. In the early twentieth century, the importance of oil for vehicles led to systematic geologic exploration for oil. Initially, structures that were interpreted to be potential oil traps from the surface geology were drilled. However, as these sources were exhausted, hidden underground structures such as anticlines or salt domes that had been covered by sediments, forests, or oceans were eventually explored using geophysical methods.

Seismic Prospecting Because oil occupies the pores of a rock, which are just a few percent of the rock's volume, geophysical signals generated by the oil itself are generally too small to be observed. Instead, geophysical methods are designed to search for buried geologic structures that experience has shown act as traps for oil. By far the most common method used to explore such hidden structures has been seismic prospecting. Seismic waves generated by an explosion (or shot) at Earth's surface reflect back to the surface from boundaries between different rock layers within Earth, like echoes reflecting from a cliff (∞ see Chapter 5). The timing of the reflection, or echo (Fig. 16.17), depends on the depth of the layer from which the wave reflects. This reflection time is measured on a seismogram (Fig. 16.17b). If the seismometer and shot are placed side-by-side on the surface, the velocity of seismic waves in the rock is used to determine the depth of the layer from the formula

$$\text{depth} = \text{time}/2 \times \text{velocity}.$$

The travel time of a reflection is divided by two because the measured travel time comprises a two-way travel time down to the layer and back to the surface. If enough reflections are measured at different points across the countryside, the variation in depths of the reflecting rock layers can be mapped. This method has been used to reveal the locations of salt domes, anticlines, and faulted traps. In some cases, the oil reservoir itself has been detected.

The amount of reflected energy depends on the contrast in the *impedance* of the rock either side of a reflecting boundary. Impedance is the product of density and velocity of the rock. If the rock pores contain gas and oil, the overall density of the rock might be quite low. The most extreme contrast will be detected when comparing the properties of the gas-filled region with the properties of surrounding regions where the pore space contains cement and other fluids. In rare cases, this contrast results in enhanced reflections in a seismic record over the reservoir called "bright spots" because of their high amplitudes (Fig. 16.17c).

Modern methods of seismic exploration can reveal the three-dimensional structure below Earth's surface. Seismic surveys are carried out both on land, sea and ice (Fig. 16.18a). On land, thousands of seismometers are laid out on the surface in a grid (Fig. 16.18b) and explosions are detonated throughout this array. As an alternative to explosions, some surveys apply vibrations to Earth's surface using **vibroseis** trucks. These trucks lower a weight of several tons onto the ground. The rear and front wheels are raised. Then, using a series of hydraulic pumps, vibratory pressure is exerted upon the weight, which passes as waves into the ground. These vibrations serve as a source of seismic waves to image underground structures. Reflections from the volume of rock below the array are recorded and processed to give a focused image of the structure.

In a three-dimensional seismic survey, a given seismic station receives reflected signals that arrive from all angles (Fig. 16.18b). Signals are converted into reflections from layers directly beneath the station; each blip marks a boundary beneath it. All common boundaries are correlated and their depths can be contoured, just like a topographic map. Modern imaging techniques represent the various rock strata in different colors. The surveyed volume is represented as a cube, three sides of which are

(a)

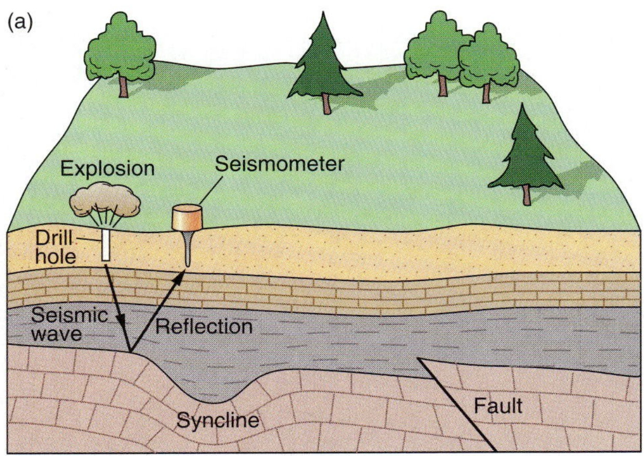

(b)

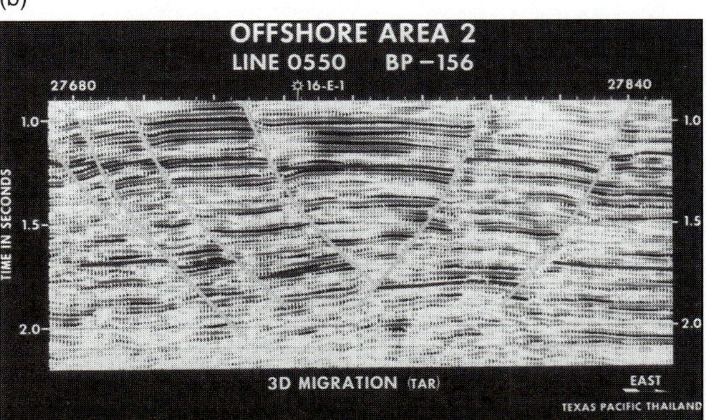

(c)

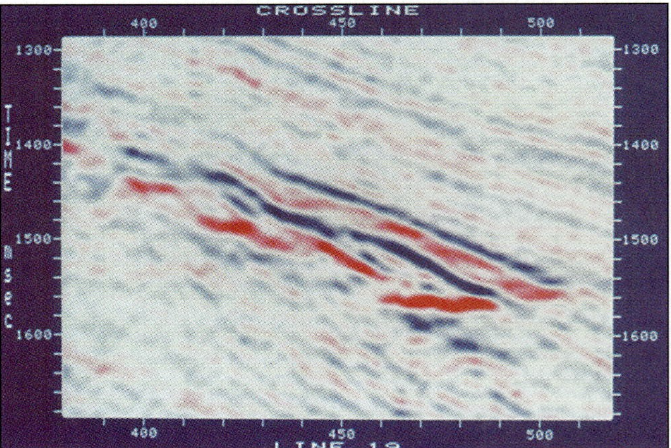

FIGURE 16.17 (a) Reflected energy from a series of sedimentary layers, a syncline, and a fault will be detected by a seismometer. (b) A seismic section. Each vertical line represents a seismogram where the shot and seismometer are placed side-by-side on the surface. The energy reflected from each layer generates a blip on the seismogram. The dark lines running across the page are made up of seismic reflections, or blips, from buried layers that return to seismometers on the surface at about the same time. They can be used to trace reflections across the countryside and to follow curves such as synclines and anticlines or offsets due to faulting. (c) Bright spots in a reflection record may be indicators of a major change in the seismic properties of an underground layer and may indicate the presence of a layer filled with oil or gas.

(a)

(b) 3–D Seismic Survey
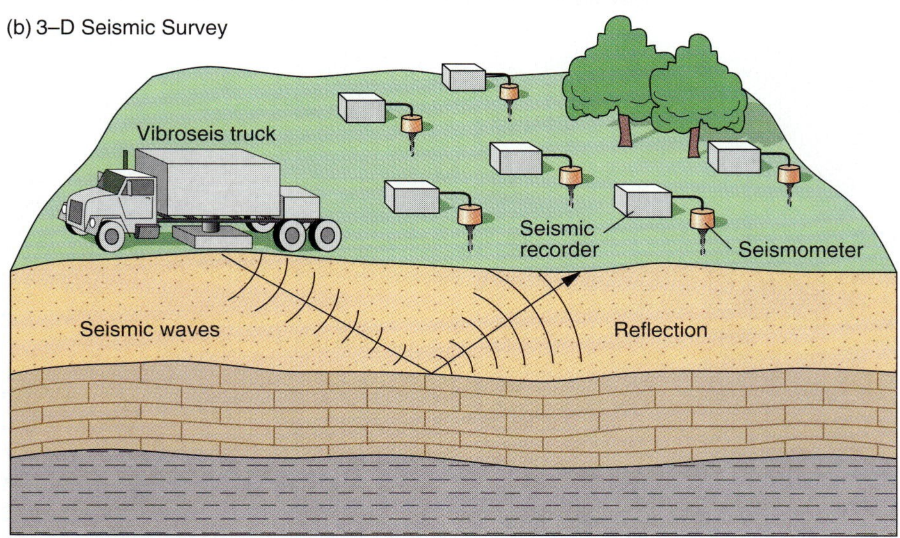

FIGURE 16.18 (a) Installation of seismic stations. (b) The field layout for imaging underground structures using distributed seismometers and a vibroseis truck as a source of seismic waves.

visible in a perspective view on a color computer monitor. Using the computer, this cubic representation of the rock strata can then be examined in detail; hypothetical slices through the volume are generated by removing layers from one of the surfaces and noting how the structure changes. For example, Figure 16.19 illustrates the three-dimensional data obtained from a volume where the development of a salt dome and the positions of a dipping fault and a syncline are both revealed.

Gravity and Magnetic Prospecting Geophysicists can also discover underground structures that might be oil or gas traps by measuring the small changes in gravity and magnetic fields that these structures cause. Because of the low density of the salt, a salt dome will produce a change in the Earth's gravitational field that can reveal the dome's presence (Fig. 16.20). An igneous intrusion will also cause local anomalies in Earth's geomagnetic and gravitational fields because the intrusion is generally more magnetic and denser than its surroundings. Before conducting expensive seismic surveying, oil-company geophysicists perform less expensive gravity and magnetic surveying as a means of reconnaissance. Although such methods have been found to be essential in the search for reserves, no method has proved to be foolproof for recognizing underground oil other than drilling, the most expensive method of all.

Oil seeps provide evidence of oil at depth within Earth. Seismic prospecting—in which waves are reflected from underground geologic structures—can reveal potential oil traps. Gravity and magnetic surveying are used as reconnaissance methods to characterize oil-bearing basins.

(a)

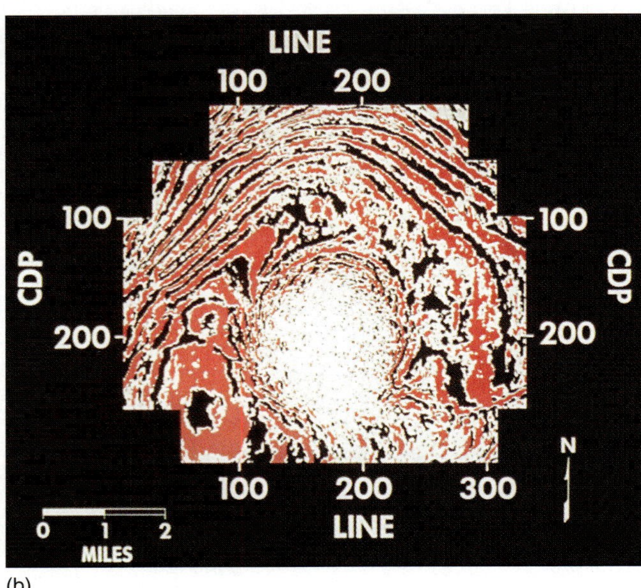

(b)

FIGURE 16.19 (a) and (b) A three-dimensional seismic survey records all reflections beneath a surveyed area. The volume of data can be used to view the three-dimensional variation of the geologic structure by using a computer to portray different slices in depth or laterally across the image.

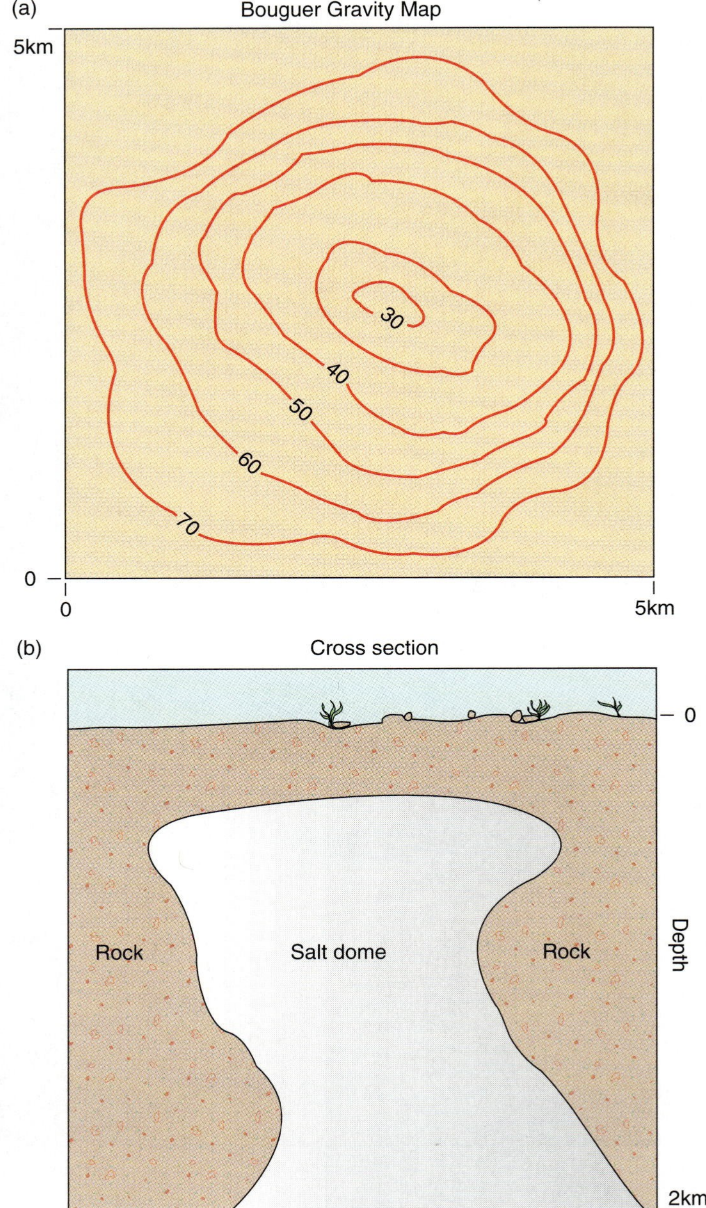

FIGURE 16.20 (a) Gravity anomaly over a salt dome shown in cross section. Because salt domes are of lower density than the surrounding rock, the gravity is less above them. Gravity contours are in milligals. (b) The shape of the salt dome has been determined from drilling, confirming the interpretation of the gravity in (a).

Oil Recovery

Because of the huge pressures that have squeezed the oil out of shale into sandstone, the oil reservoir itself can also be under great pressure. When it is drilled, it can form a gusher—an oil fountain that spews oil hundreds of meters into the air. To prevent such gushers, dense barite muds are pumped into the well at the time of drilling. The oil well must be capped and pressure valves used to regulate the flow. In some cases, the oil pressure is not high and pumps must be used to bring oil to the surface from the reservoir. Even with pumping, however, about 70 percent of the oil is not recovered from a typical oil field. Secondary recovery methods are used to tap this remaining oil, such as flooding the reservoir with water or steam to flush it out. Another method involves pressurizing oil wells until the reservoir rocks fracture. Then sand is pumped in to hold the cracks open, allowing the trapped oil to flow more readily.

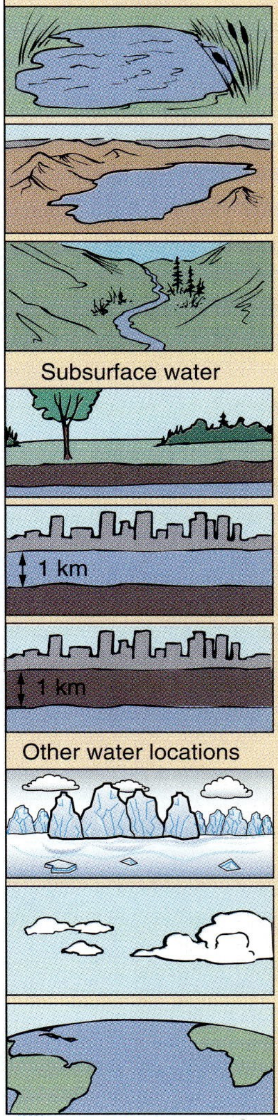

Location	Water volume (liters)	Water volume (percent)
Surface water		
Fresh-water lakes	125×10^{15}	0.009
Saline lakes and inland seas	104×10^{15}	0.008
Average in rivers and stream channels	1×10^{15}	0.0001
Subsurface water		
Vadose water (includes soil moisture)	67×10^{15}	0.005
Groundwater within depth of 1km	4170×10^{15}	0.3
Groundwater (deep lying)	4170×10^{15}	0.3
Other water locations		
Ice caps and glaciers	$29{,}000 \times 10^{15}$	2.1
Atmosphere	13×10^{15}	0.001
World oceans	$1{,}320{,}000 \times 10^{15}$	97.2

FIGURE 16.21 Distribution of water on Earth. (Compare with Figure 1.11.)

WATER

Perhaps the most essential natural resource is fresh water since it is needed to sustain life. Most of Earth's surface water (97 percent) resides in the oceans; 2.1 percent is held in the polar ice caps and glaciers (Chapter 1, see Fig. 1.11). Potable water—water suitable for drinking—makes up only 0.65 percent of the total. The distribution of water in various forms and locations on Earth is shown in Figure 16.21. Most of the potable water is accounted for by groundwater, which is stored in the pores and fractures in rocks. At greater depths within Earth, the pressure of the overlying rock causes pores and cracks to close, and almost complete closure occurs at a depth of about 10 km. The greatest water storage, therefore, lies near the surface. Figure 16.21 shows that nearly 50 percent of all groundwater is stored in the upper 1000 m of Earth. This is about 4000 times greater than the amount in all rivers and 20 times greater than the amount in all rivers and lakes.

Groundwater is stored in a variety of rock types. A groundwater reservoir from which water can be extracted is called an **aquifer**. We can effectively think of an aquifer as a "deposit" of water (See Focus 16.3). Extraction of water from an aquifer depends on two properties of the aquifer: **porosity** and **permeability** (Fig. 16.22a, page 434). Between sediment grains are spaces that can be filled with water. This pore space is known as porosity, and is expressed as a percentage of the total rock volume. Porosity is important for water-storage capacity, but for water to flow through rocks, the pore spaces must be connected. The ability of water, or other fluids, to flow through the interconnected pore spaces in rocks is termed permeability. Fractures and joints have very high permeability. In the intergranular spaces of rocks, however, fluid must flow around and between grains in a tor-

Focus On 16.2 HOW PLATE TECTONICS REDISTRIBUTES MINERAL AND ORGANIC RESOURCES

Various hypotheses have been proposed for the relationship between plate tectonics and resources. One is that during a continental collision, fluid-filled sediments (mostly saline brines) rich in minerals and organic substances are squeezed out of the collision zone by the forces of colliding plates. The mineral-laden brines may travel hundreds of kilometers inland into foreland basins (Fig. 1). Fluids rising from such depths would carry heat. They leach minerals from the rocks as they pass and, as they cool, precipitate mineral ore bodies. The heat maturates organic substances into hydrocarbons or increases the rank of coal deposits. A huge volume of sediments is buried in the collision zone to depths beyond practical drilling. Thus, as the continents collide, a large amount of brines would be expelled.

One example of the tectonic control exerted upon resources is found in the Appalachian region of eastern North America. Tectonic reconstructions of the plate motions (see Fig. 9.19) indicate that 500 million years ago, a great ocean similar to the Atlantic existed that was called Iapetus. It separated two continents that were similar to modern North America (plus Greenland) and present-day Africa (plus

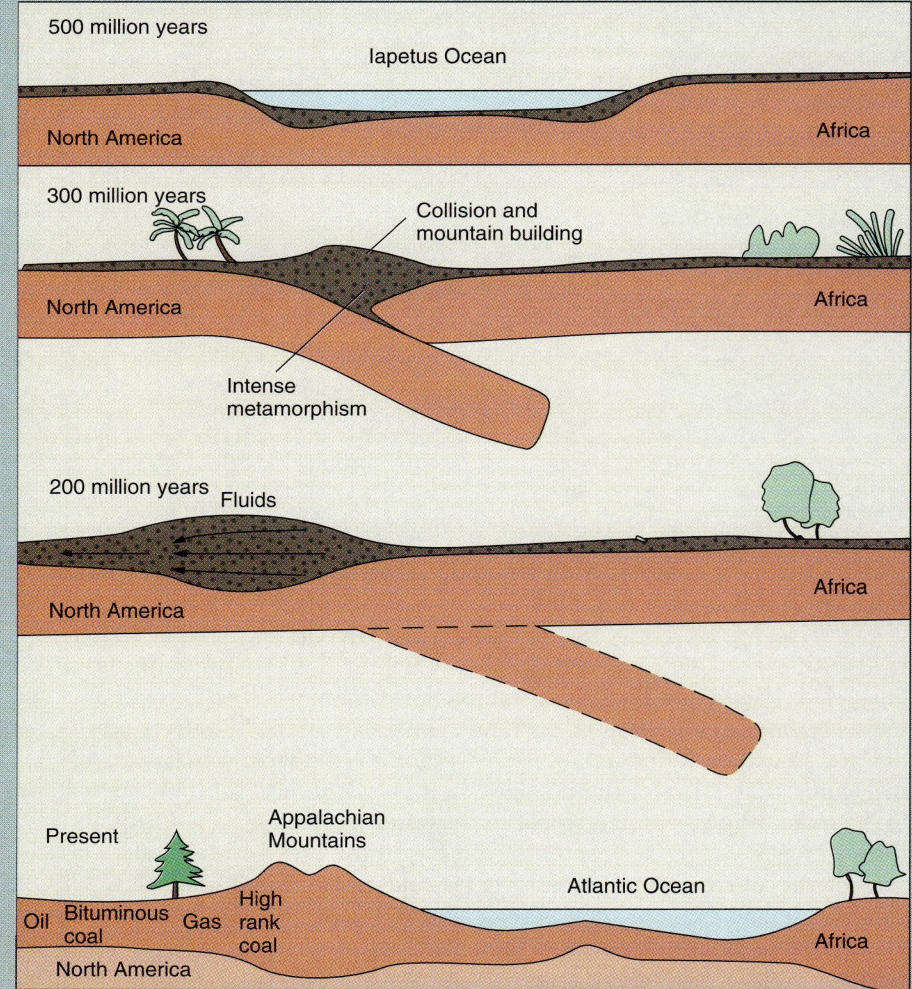

FIGURE 1 With the collision of Africa and North America about 300 million years ago, sediments were piled up on the future North American continent. The forces of the collision squeezed brines westward, which gave rise to coal of higher rank near the collision zone.

Europe). Initially, this sea was spreading and sediments were deposited at the continental margins in basins or deltas as rivers drained the continents. The continental sediments included organic matter and nutrients for marine organisms, which also contributed to the sedimentary deposits.

Then Iapetus started to close. The sea floor began to subduct under the African continent. Africa, acting like a giant bulldozer, scraped sediments and seamounts from the subducting oceanic plate. When North America collided with Africa, its leading edge is thought to have descended beneath Africa. The huge wedge of sediments was pushed over the basins and deltas of the North American continental margin, to form the Appalachian Mountains. The sediments were porous, containing mineral-rich fluids in the pores. Under the great pressures associated with burial of the overriding wedge, the fluid would have been expelled. Africa subsequently split away from North America, to form the modern Atlantic Ocean basin. However, telltale signs of the collision are thought to be seen in the distribution of resources around the Appalachians.

Many oil wells are found near the Appalachians in Pennsylvania and Ohio, as well as in nearby states. Although the distribution of oil reservoirs appears to be complex, perhaps even random, there is one trend that may be explained by past plate motions. Gas is more prominent near the Appalachian tectonic zone in the east and oil is more abundant in the west. Higher maturation (formation of gas) in the eastern regions is consistent with the migration of hot fluids from the collision zone, which cool as they migrate west.

Further support for this hypothesis is found in the coal deposits in eastern Pennsylvania, which are classified as high-rank clean anthracite, whereas coal in western Pennsylvania is bituminous. In other words, the eastern coal has been subjected to greater metamorphism by heat. Although these differences can be explained by deeper burial of the eastern deposits and subsequent exhumation, they can also be explained by hot brines from the collision zone metamorphically increasing the grade of the deposits.

In other regions affected by continental collisions, hydrocarbon migration effects are seen. In Wyoming, Oklahoma, and Alaska's North Slope, hydrocarbons can be shown to have consistently migrated away from the collision zone. This is in contrast to areas unaffected by nearby collisions, such as the gulf coast region, California, Michigan, the Williston basin of Canada, Montana, and North Dakota, where there is little migration of hydrocarbons and this migration is not in one particular direction.

The Middle East oil fields contain the largest amount of known oil in the world. They occur in one of the largest sedimentary basins on the planet. About 15 million years ago the Tethys ocean floor between Arabia and Asia was consumed and the two continents collided. The collision formed the Zagros mountains and a huge basin of deep sediments called the *Mesopotamian basin* lying on the southwest side of the Zagros. The oil is found in this basin in a linear region extending from southern Turkey to Oman.

Why are the Middle East oil fields so large? This extraordinary concentration of petroleum is thought to be due to the coexistence of all the conditions necessary for oil accumulation. Abundant source rock covered by shallow seas has been deposited on the Arabian Plate for over 100 million years, in conditions ideal for formation of marine life. Thick sequences of reservoir rock were also deposited over evaporites, which have formed salt dome traps in areas such as the Persian Gulf. The continental collision exerted a compressive stress, causing maturated oil at depth to migrate into the reservoir rock. The stresses also faulted and folded the sediments, to form structures such as anticlines that have acted as large-scale structural traps.

These studies are in a preliminary phase. They rely on unraveling the motions of the plates and piecing together the history of paleo-sedimentation and fluid flow. Vast sedimentary sections lie buried beyond the reach of the drill stem. Expelled brines, however, have left their mark in the mineral and hydrocarbon deposits squeezed into shallower strata. A more complete understanding of these phenomena will help reveal valuable deposits of natural resources now hidden from view.

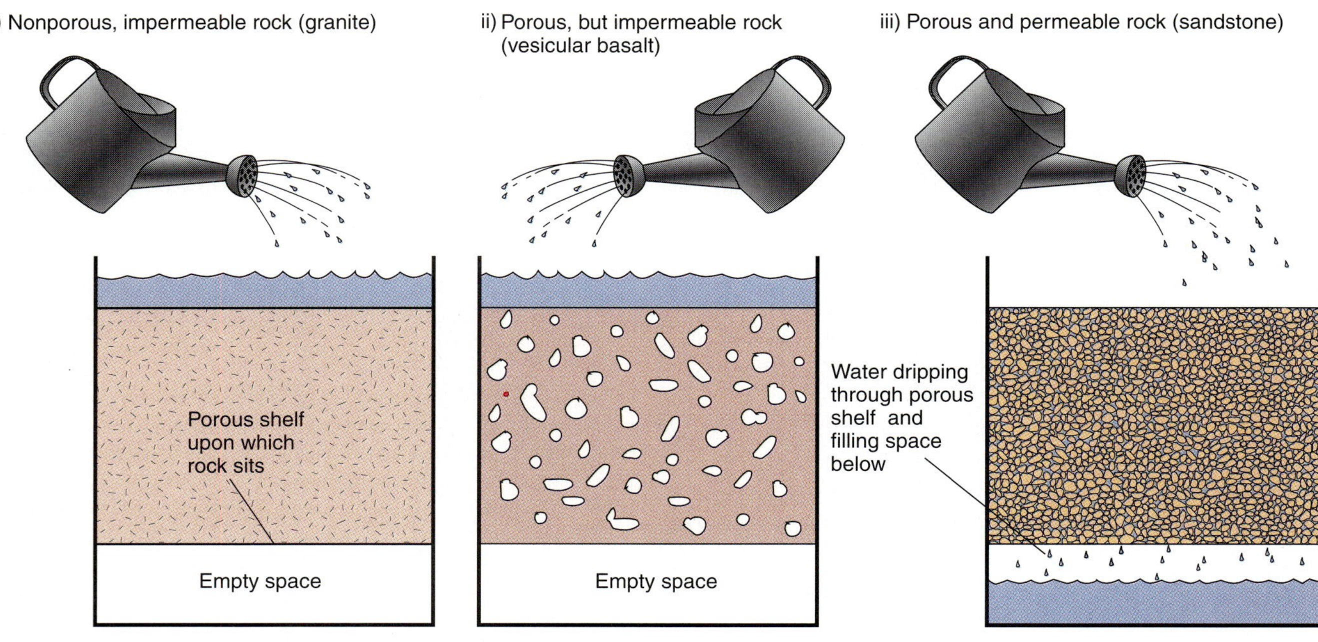

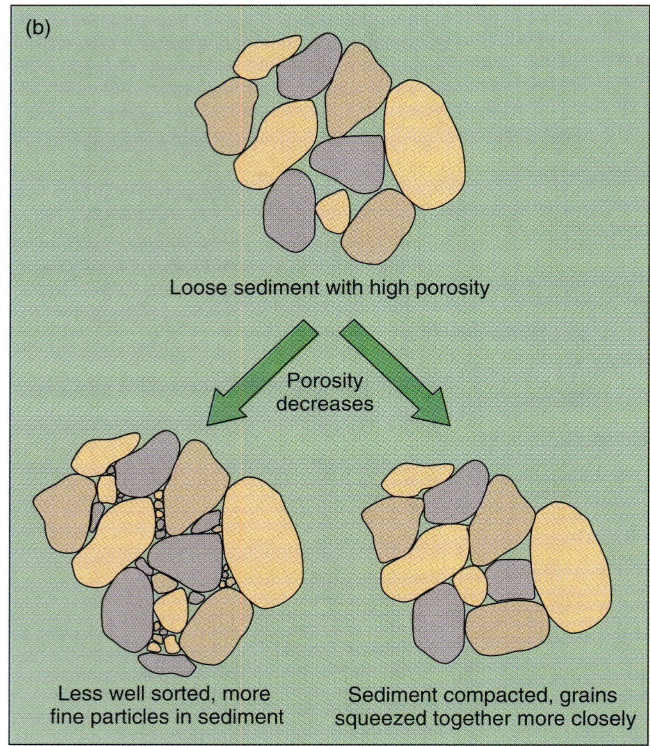

FIGURE 16.22 (a) The concepts of porosity and permeability. (b) Dependence of porosity on sediment sorting and compaction.

tuous path; this winding path causes a resistance to flow. The rate at which the flowing water overcomes this resistance is related to the permeability of the rock. Both porosity and permeability are influenced by sediment sorting and compaction. The more poorly sorted or the more tightly compacted a sediment, the lower its porosity and permeability (Fig. 16.22b). Sedimentary rocks—the most common rock type near the surface (∞ see Chapter 4)—are also the most common reservoirs for water because they contain the greatest amount of empty space that can be filled with water. Sandstones generally make good aquifers while finer-grained mudstones are typically impermeable. Impermeable rocks are referred to as **aquicludes**. Igneous and metamorphic rocks are more compact, commonly crystalline, and rarely contain spaces between grains. However, extensive fracturing may occur in such rocks, and if the fracture system is interconnected, even igneous and metamorphic rocks may act as groundwater reservoirs.

The **water table** is the underground boundary below which all the cracks and pores are filled with water. In some cases, the water table reaches Earth's surface where it is expressed as rivers, lakes, and marshes. Typically, though, the water table may be tens or hundreds of meters below the surface. The water table is not flat but usually follows the contours of the topography (Fig. 16.23a). This is a consequence of the slow rate of movement of the groundwater, which prevents the water table from attaining the equilibrium geometry of a horizontal plane. Above the water table is the **vadose zone**, through which rain water percolates. Water in the vadose zone drains down to the water table, leaving a thin coating of water on mineral grains behind. The vadose zone also supplies the roots of plants with water.

Most of Earth's potable water is contained in groundwater. Water-filled fractures and pores are located in a region below the water table that extends to about 1000 m in depth.

Focus On 16.3 EARTH'S WATER: WHERE DID IT COME FROM?

Earth is the only terrestrial planet with an abundance of accumulated water on its surface. Only Earth satisfies all the conditions to form and maintain a wet planet:

1. Earth is sufficiently large that degassing of water from its interior occurs more rapidly than subsequent loss to space.
2. The planet is at the right distance from the Sun, so that average temperatures are not above the boiling point of water or below its freezing point.
3. Earth's temperature is sufficiently moderate to allow the evolution of organisms that are able to remove carbon dioxide from the atmosphere, thereby preventing greenhouse heating such as that occurring on Venus.

Every year, about 15 km^3 of new rock is erupted onto Earth's surface. This rock is derived from mantle rock by melting, and it contains about 0.5 percent water (stony meteorites contain about the same amount). If this amount of water is liberated from the magma before solidification, about 0.075 km^3 of new or "juvenile" water is released to the atmosphere and oceans annually. At this rate, the total amount of water released into our oceans and atmosphere during Earth's lifetime (4.5×10^9 years) is 0.34×10^9 km^3; that is, 25 percent of the current volume of the oceans (1.4×10^9 km^3). This rate of water release is an overestimate because some of this water is retained in rocks that are subducted back into Earth. The release rate was faster in the past, when Earth's radioactive heat content was higher and mantle convection was more vigorous. Most of the water was released by early degassing of the planet. Formation of the oceans has been an ongoing process ever since Earth formed.

Given that if Earth's mantle when it formed had, on average, the same concentration of water as do stony meteorites (0.5 percent), then Earth has only released 1/16 of its water to form the oceans since its formation. The remainder is still locked up in the mantle.

Because the surface of the groundwater table is not flat, but rises and falls in elevation more or less in concert with the topography, groundwater is affected by gravity in the same fashion as surface water. Groundwater flows downhill to topographic lows. If the water table intersects the land surface, groundwater will flow out onto the surface at springs, either to be collected there or to subsequently flow farther along a drainage (Fig. 16.23a). Groundwater commonly collects in stream drainages, but may remain entirely beneath the surface of dry stream beds in arid regions. In particularly wet years, short stretches of an otherwise dry stream bed may have flowing water because the water table rises sufficiently to intersect the land surface (Fig. 16.23b).

If aquifers are exposed at higher elevations than the surrounding terrain, water may enter the groundwater system there (recharge the aquifer) and fill the zone nearly to the elevation of the recharge area. If the aquifer is surrounded by impermeable rock, the groundwater is at a pressure related to the level of the water table at the recharge area (Fig. 16.23c). When a hose is nearly filled with water and the ends of the hose are elevated, the water level at each end will always be at precisely the same elevation. If one end is lower, water will run out of the hose. This is the principle of a siphon, used, for instance, to extract gasoline from the gas tank of an automobile, a process that requires the gas to flow uphill out of the tank and then down to a container placed below the level of gas in the tank. In the geologic example, if the water table in the recharge area is higher than the level of a well drilled into the aquifer, water will spontaneously flow from the well. This is called an **artesian** well because water flows out of the well bore without pumping.

Should groundwater meet a subsurface obstacle that blocks its downhill flow, such as a fault or a geologic contact with an impermeable rock (an aquiclude), it will be dammed in much the same way as a surface stream is blocked by a dam. Faults are commonly marked by lines of springs where groundwater is pooled behind the impermeable fault gouge; the water level rises behind the "dam" until it reaches the land surface, where it emerges as a spring. Should groundwater encounter a highly permeable zone, such as a highly fractured volume of rock, a sudden drop in the level of the water table will occur—much like a waterfall.

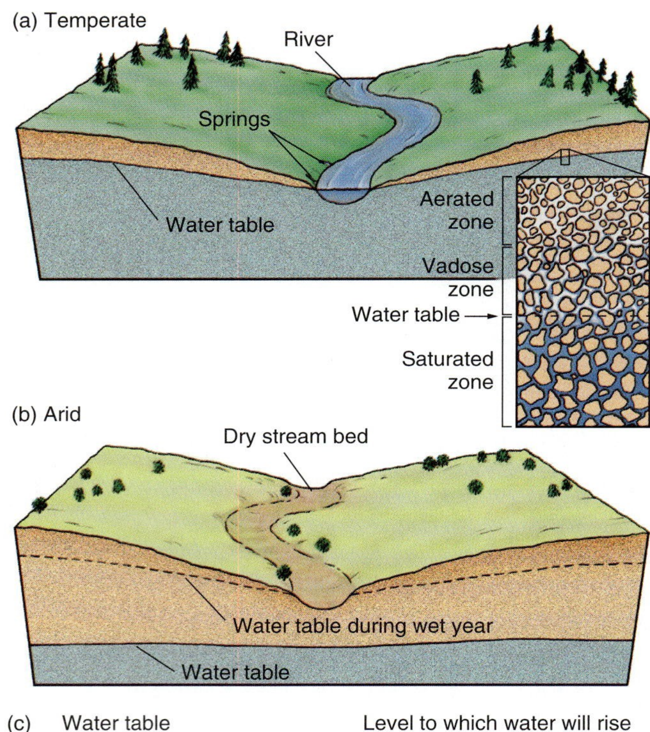

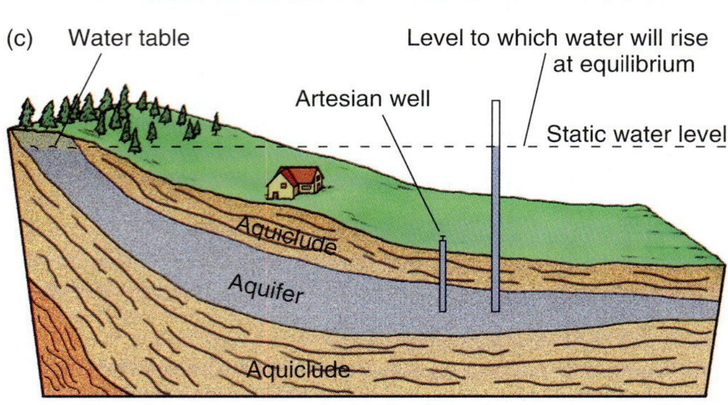

FIGURE 16.23 (a) Water table in a temperate environment. (b) Water table in an arid environment. (c) Aquifer in an artesian basin, and the principle of an artesian well.

Rates of groundwater movement in the subsurface vary widely. In large fracture systems, such as those that feed the subsurface rivers flowing in some limestone caverns, the water may flow at rates of a few meters per second. In contrast, water in intergranular spaces of poorly sorted or strongly compacted sediment may flow only a few millimeters or a few centimeters per year.

Mining Water The actual water absorbed by agricultural crops is a small fraction of that used to irrigate these crops. Most of the water is absorbed by the ground or is returned to the atmosphere by evaporation or transpiration—the evaporative loss of water by plants through leaves and stem. About 81 percent of the fresh water used in the United States is used for irrigation of crops. If a region is irrigated by pumping water out of the ground and this water is subsequently removed in the form of evapotranspiration or uptake by crops, the water table may eventually become depressed faster than it is replenished by rainfall. When the water table becomes too depressed, wells run dry. This is an especially serious problem in the cotton and grain fields of the high plains of western Texas. In the arid southwest, alfalfa is grown as a cash crop and to enrich the soil. The roots of this plant extend downward 2 to 3 m, and it requires a great deal of water to thrive. Excessive pumping of wells for irrigation water may cause the groundwater table to decrease, making it expensive to drill and produce the water.

Why is the aquifer not replenished as rapidly as it is pumped from the ground? The answer lies in the slow rate at which groundwater typically moves, which may be far less than the rate at which it is removed by pumping. Rapid pumping from wells sunk below the water table will deplete the aquifer in the vicinity of the well, and form a *cone of depression* around the well. The local level of the water table may drop to the point where the well runs dry (Fig. 16.24a).

Rapid pumping may also lead to contamination problems by disturbing the geometry of the local water table. One small coastal town in northern California pumps its water from wells. The ocean is nearby, and because seawater is denser than fresh water, the potable water in the aquifer is underlain by saltwater from the ocean. This town is supplied by 5 water wells, each producing approximately 3000 gallons of water per minute, amounting to roughly 4 million gallons per day. That amount of pumping affects the groundwater table. If the wells are pumped too rapidly, the underlying saltwater may be sucked upward into the well, permanently contaminating it (Fig. 16.24b). If this should occur, the well must be abandoned and a new one drilled.

Replenishing water In cases where the removal of groundwater is greater than replenishment, water can be replaced by flooding a region with water from elsewhere. Experiments in the San Jacinto basin of southern California have used water from the California aqueduct to replenish groundwater depleted during a recent drought (1986–1992). The California aqueduct is the canal that carries water from melting snow in the Sierra Nevada range approximately 500 km to the city of Los Angeles. Persistent pumping in the San Jacinto basin had lowered the level of the water table by 30 m, causing wells and boreholes to run dry. When water was plentiful during the winter runoff, the California aqueduct was diverted into the basin and the water table depth was restored in a limited area, with plans to replenish the supply over a number of years. The alternative—to build large reservoirs—is impractical in this populated area. This experimental program involves geologists who identify rock types and perform mapping, geophysicists

(a) i) Pumping begins

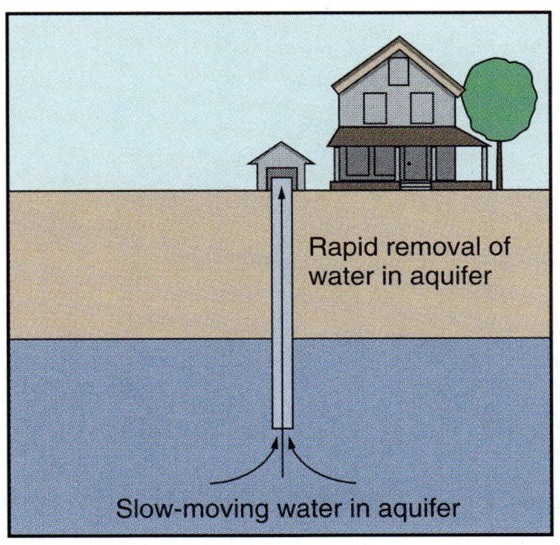

ii) Pumping causes draw-down

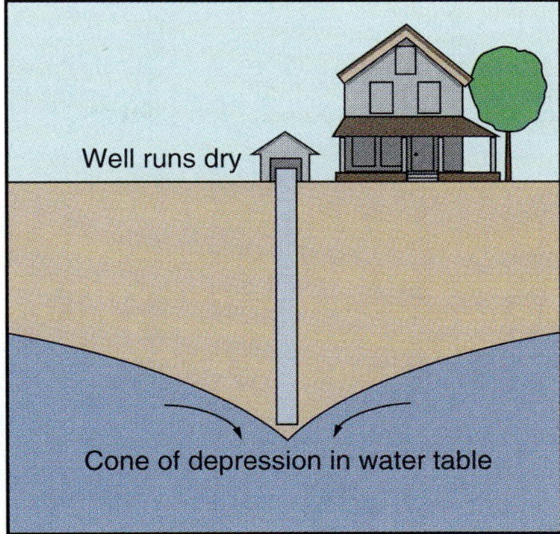

(b) i) Water well at rest

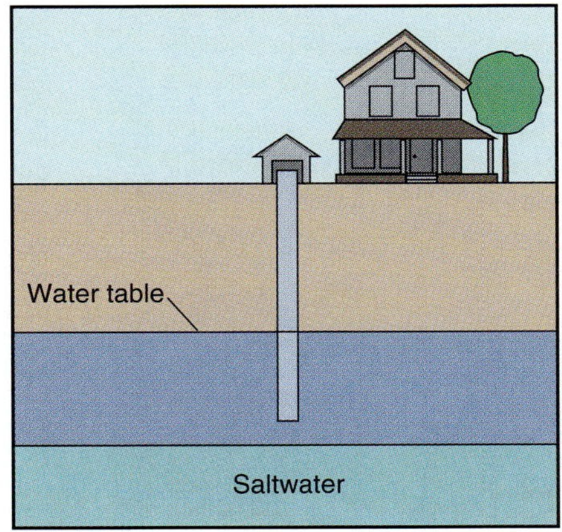

ii) Water well pumping rapidly

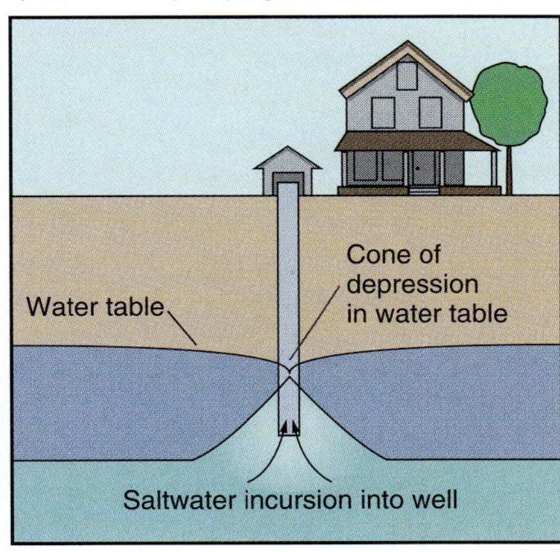

FIGURE 16.24 (a) Rapid removal of water from an aquifer causes a cone of depression in the water table. (b) Saltwater incursion into a well in which rapid pumping results in a cone of depression that intersects upwarp of salt water lying beneath fresh groundwater.

who probe and map the deep structure of the basins, geohydrologists who model the water flow during the replenishment, and geochemists who test for water quality. Excessive removal of water from an aquifer may also have severe consequences for the local land surface. The water in the aquifer actually supports some of the weight of the overlying material. Pumping this water out can cause subsidence of the surface. Collapse of the reservoir also prevents the effective replenishment of the reservoir. Subsidence in the San Joaquin Valley, California; and Houston, Texas, has been attributed to groundwater withdrawal. As key regions become more populated, such replenishment of resources may become critical.

Exploring for Water

Even in deserts, groundwater can be found by deep drilling. But drilling to such depths can be impractical. Remote sensing of the water table depth can improve the chances of discovering a shallow aquifer. The exploration for underground water reservoirs takes advantage of the fact that water is a better electrical conductor than is rock.

As a result, water-saturated rock situated below the water table conducts electricity much better than the dry rock above, allowing detection of the water by electrical surveys. Electric-resistivity sounding is a common method used to determine the depth of the water table. *Resistivity* is a property that describes the resistance of a material to electrical current flow. Two electrical probes connected to a voltage source are placed in the ground, and the amount of current that flows for a given voltage provides a measure of the resistance. The current flows to a depth approximately equal to the distance between the probes (Fig. 16.25). By increasing the distance between the probes, deeper regions can be sounded. Once the separation between probes is greater than the depth of the water table, the electrical resistance drops and the current increases. In this way, both the depth to the water table and the amount of water present can be estimated. Other electrical methods based on this principle are also utilized.

In cases where the removal of groundwater is greater than replenishment, water can be replaced by flooding a region with water from elsewhere. The exploration for underground water reservoirs takes advantage of the fact that water is a better electrical conductor than is rock.

Focus On 16.4 RESOURCES FROM SPACE

What is the likelihood that we will be able to mine resources from space? At present, the expense of collecting and returning resources to Earth makes space mining a doubtful proposition. But if colonization of the Moon or of other planets is contemplated for the future, it is useful to evaluate the materials that could be mined on site. We saw in Chapter 2 that the building blocks of the universe are essentially the same: the elements of the periodic table generated by the big bang and supernova explosions. When the Solar System condensed out of the primeval nebula, the first step in the concentration of elements into compounds (that are useful to humans) took place. Iron cores separated from rocky mantles. This was then followed by asteroidal collisions that sent iron meteorites crashing to Earth to be used by our Stone Age ancestors.

We will probably never mine the gaseous planets. However, it is conceivable that ore bodies have formed on the other terrestrial planets, the Moon, and asteroids. Over time, the planets have convected and degassed as they cooled. Hydrothermal circulations may have occurred as the planets released water. Planetary mineralization offers new challenges in space science. On a hot planet like Venus, where temperatures at the surface are above 400°C, convection of water in a hydrothermal zone is unlikely to occur at present. However, conditions may have been different in the past on Venus and on other planets. Mars is thought to have had running water (Fig. 1) and may even have groundwater circulation today, which could possibly produce ores to be mined for human habitation.

It is unlikely that the terrestrial planets, other than Earth, possess accumulations of hydrocarbons, because the evidence so far available suggests that life has not formed on them. Some asteroids may be sources of iron and nickel, and perhaps even other minerals. Planetary mineralization may have occurred in ways not even suspected on Earth, given the different evolutionary histories of the planets. In the same way that early miners discovered placer deposits of gold nuggets in the Yukon or California, or the diamonds in the placers or kimberlite pipes of Africa, exploration of the planets may reveal near-surface concentrations of precious minerals, the mechanism of their formation not well understood even after their discovery and exploitation.

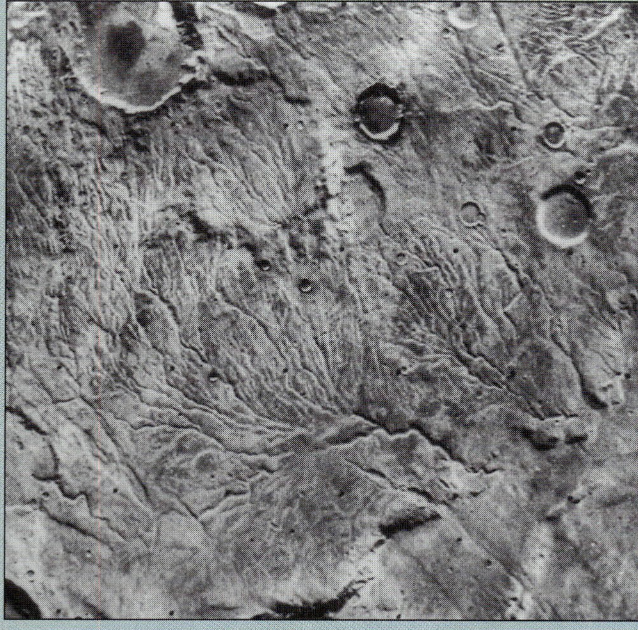

FIGURE 1 The surface of Mars showing channels that may once have carried water.

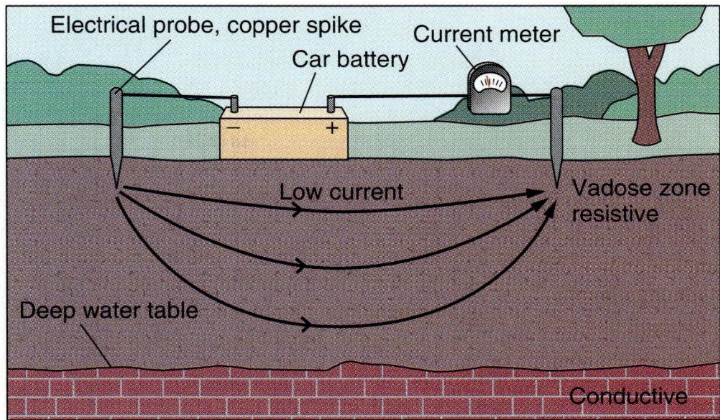

FIGURE 16.25 Water saturated rock conducts electrical current better than does dry rock. Resistivity sounding is a tool to search for water. A voltage is connected to two electrodes implanted in the ground and the resistance is calculated by measuring current (which is inversely proportional to resistance). By increasing the separation between the electrodes, deeper regions are sampled by the current. A series of such measurements can be used to determine resistance as a function of depth. A sudden decrease with depth indicates the presence of the water table.

Superheated Water and Geothermal Energy

On average, the temperature of the upper crust increases with increasing depth at a rate of approximately 20°C per kilometer. This is called the *geothermal gradient* (∞ see Chapter 4). In the lower crust and mantle, the geothermal gradient decreases. At a depth of about 100 km, the temperature is about 1000°C and the temperature slowly continues to rise with depth in the mantle until it reaches about 6000°C at Earth's inner core. Thus, the mantle is extremely hot. If we could tap this **geothermal energy**, it would provide a clean source of energy for much of Earth's population. The geothermal heat energy per unit area flowing out of Earth is 1/3000 of the energy from the Sun that reaches Earth. The total amount of geothermal energy is 44×10^{12} watts. As a comparison, the total energy used by humans is about 1×10^{12} watts.

Geothermal energy is too widely distributed over the planet to be a useful energy source in most parts of the world. At divergent and convergent plate margins and hot spots, however, hot rocks from magmatic intrusion and eruptions are found close to the surface. Geothermal gradients in these locations can be many times the worldwide average. Water seeping down into Earth in these regions becomes superheated (above the boiling point), forming geysers, or hot springs. This heat can be tapped by geothermal wells drilled into a heated aquifer (Fig. 16.26). Superheated water and steam tapped by drilling can drive turbines, which, in turn, generate electricity. In dry underground areas, it is not so easy to extract heat energy. The Los Alamos National Laboratory in New Mexico has conducted experiments in which water is injected into boreholes in hot, dry rocks in volcanic zones, for the purpose of extracting the heat energy in the form of steam. After the steam has been used to drive a turbine, which, in turn, drives a generator, it cools and condenses to water, to be reinjected once more.

Locations of geothermal power plants are shown in Figure 16.27. These plants are found chiefly along the plate boundaries bordering the Pacific, in Iceland, and in the central Mediterranean. The total global output is about 4733 megawatts or about 0.4 percent of the world's needs.

Superheated water and steam from aquifers can be used to drive turbines that generate electricity. This geothermal energy is most important at plate margins where magmatism provides increased heat flow.

FIGURE 16.26 A geothermal power station in Kenya.

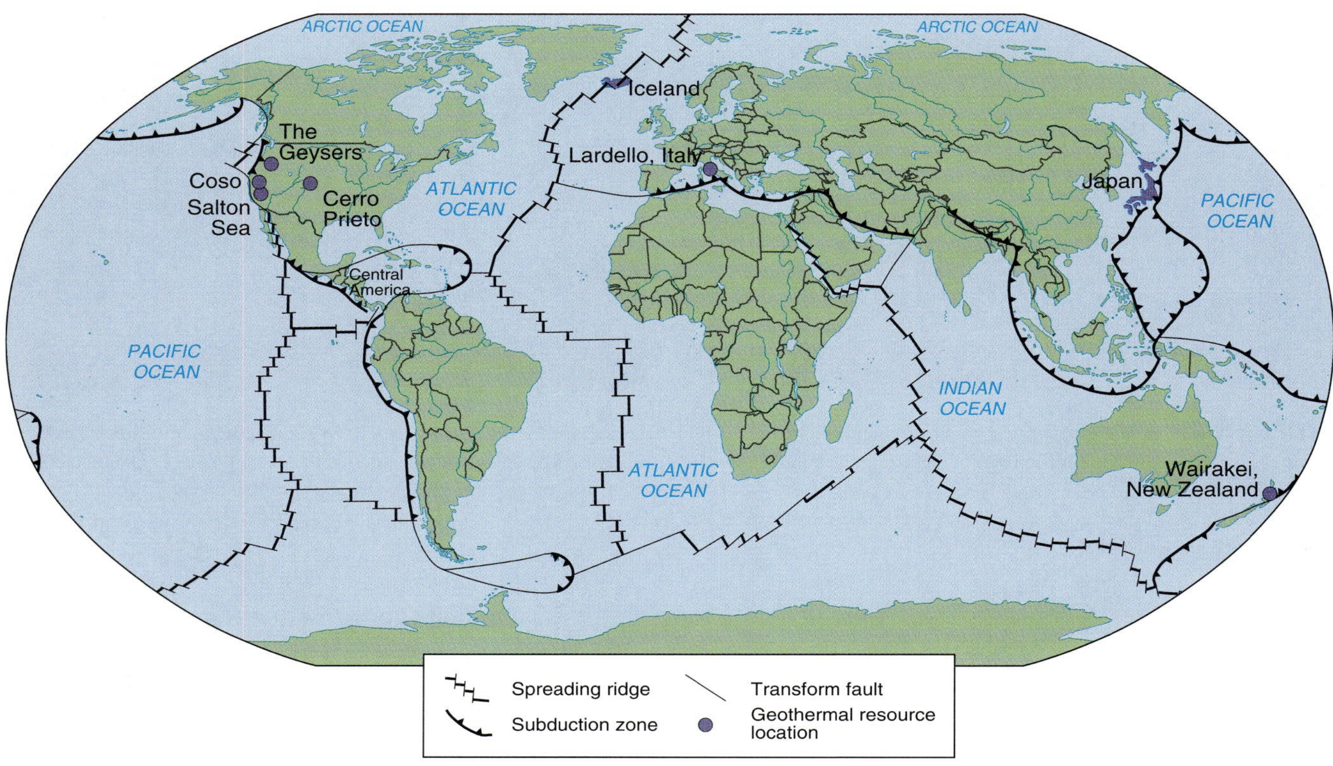

FIGURE 16.27 Geothermal resources of the world showing their proximity to plate margins where magmatism provides increased heat flow.

SUMMARY

- Ores are minerals from which valuable constituents can be profitably extracted. Smelting is the process by which metals are extracted from ores.
- Some minerals become concentrated by crystallization from magmas. Diamonds are brought to the surface in kimberlite pipes.
- Hydrothermal systems contain mineral-laden water heated by intrusions of magma. Minerals precipitate as the water cools.
- Sedimentary processes concentrate minerals by precipitation or by grain sorting during transport and deposition.
- Certain types of mineral deposits are associated with particular plate-tectonic environments, where magmatism, fluid flow, or sedimentary processes are suitable for ore concentration.
- Burial of organic material *protects* it from decay by oxidation. Peat forms from the burial of a wide range of vegetation, from mosses to trees.
- Formation of basins leads to burial of the vegetation, eventually turning it into coal. Subsequent tectonic motions may expose the buried strata, enabling the coal to be mined.
- Petroleum forms from organic matter trapped in marine sediments. Burial of organic matter to depths of several kilometers expels water; heat at depth transforms organic molecules into petroleum.
- Formation of a petroleum reservoir requires the following: source rocks containing marine organisms; burial to achieve maturation; expulsion and migration of maturated oil; structures suitable for entrapment. Sedimentary rocks suitable for generation of oil are deposited in basins from most tectonic settings. Heat flow from magmatism accelerates maturation.
- Oil seeps provide evidence of oil at depth within Earth. Seismic prospecting—in which waves are reflected from underground geologic structures—can reveal potential oil traps. Gravity and magnetic surveying are used as reconnaissance methods to characterize oil-bearing basins.
- Most of Earth's potable water is contained in groundwater—water-filled fractures and pores located in a region below the water table that extends to about 1000 m in depth.
- In cases where the removal of groundwater is greater than replenishment, water can be replaced by flooding a region with water from elsewhere. The exploration for underground water reservoirs takes advantage of the fact that water is a better electrical conductor than is rock.
- Superheated water and steam from aquifers can be used to drive turbines that generate electricity. This geothermal energy is most important at plate margins where magmatism provides increased heat flow.

KEY TERMS

ore, 411
hydrothermal zone, 415
brine, 415
stratiform deposit, 416
manganese nodule, 416
banded iron formation, 417
placer, 417
oil trap, 425
vibroseis, 428
aquifer, 431
porosity, 431
permeability, 431
aquicludes, 434
water table, 434
vadose zone, 434
artesian, 435
geothermal energy, 439

QUESTIONS FOR REVIEW AND FURTHER THOUGHT

1. Why do we not mine basalts for ore?
2. Describe hydrothermal circulation and how it gives rise to ore bodies.
3. What process forms manganese nodules and where are they found?
4. Where would you search for gold and why?
5. How are diamonds formed?
6. What is the difference between coal and peat?
7. Describe the origin of petroleum and compare it with the origin of coal.
8. Describe reservoir rocks, source rocks, and structural traps.
9. Why is there very little oil from ages greater than 200 million years ago?
10. What methods are used to search for oil?
11. What is secondary recovery of oil?
12. How is groundwater used to generate electrical energy?
13. How long would it take a seismic wave to reflect from an oil reservoir at a depth of 2300 m if the velocity of seismic waves in the region is 3 km/sec?

Appendix
Conversion Tables

TEMPERATURE CONVERSION

Temperature is measured in degrees Celsius, a scale in which the interval between the freezing and boiling points of water is divided into 100 degrees, with 0°C/32°F representing the freezing point and 100°C/212°F the boiling point. So:

$$\frac{5}{9}°\text{C (Celsius)} = 1°\text{F (Farenheit) or } °\text{C} = \frac{(°\text{F}-32°)}{1.8} \text{ or } °\text{F} = (°\text{C} \times 1.8) + 32°$$

CONVERSION OF MASS

	Grams (g)	Kilograms (kg)	Pounds (lb)	Ounces (oz)
g	1	1,000	453.6	28.35
kg	0.001	1	0.4536	2.835×10^{-2}
lb	2.205×10^{-3}	2.205	1	6.25×10^{-2}
oz	3.527×10^{-2}	35.27	16	1

Multiply units in the column heads by the figures in the table to convert to units at left. (For example, to convert pounds to kilograms, multiply the number of pounds by 0.4536.)

CONVERSION OF LENGTH

	Meters (m)	Yards (yd)	Centimeters (cm)	Inches (in)	Feet (ft)	Kilometers (km)	Miles (mi)
m	1	0.9144	1×10^{-2}	2.54×10^{-2}	0.3048	1×10^{3}	1,609
yd	1.094	1	1.094×10^{-2}	2.778×10^{-2}	0.3333	1,094	1,760
cm	100	91.44	1	2.54	30.48	1×10^{5}	1.609×10^{5}
in	39.37	36	0.3937	1	12	3.939×10^{4}	6.336×10^{4}
ft	3.281	3	3.281×10^{-2}	8.333×10^{-2}	1	3,281	5,280
km	1×10^{-3}	0.144×10^{-4}	1×10^{-5}	2.54×10^{-5}	3.048×10^{-4}	1	1.609
mi	6.214×10^{-4}	5.682×10^{-4}	6.214×10^{-6}	1.578×10^{-5}	1.894×10^{-4}	0.6214	1

Multiply units in the column heads by the figures in the table to convert to units at left. (For example, to convert kilometers to feet, multiply the number of kilometers by 3,281.)

CONVERSION OF AREA

	Acres	Square Inches in^2	Square Centimeters cm^2	Square Feet ft^2	Square Meters m^2	Square Miles mi^2	Square Kilometers km^2
acres	1	6.27×10^{-6}	2.471×10^{-8}	3.296×10^{-5}	2.471×10^{-4}	640	247.1
in^2	6.272640×10^{6}	1	0.1550	144	1,550	4.015×10^{9}	1.550×10^{9}
cm^2	4.047×10^{7}	6.452	1	929	1×10^{4}	2.59×10^{10}	1×10^{10}
ft^2	4.356×10^{4}	6.944×10^{-3}	1.076×10^{-3}	1	10.76	2.788×10^{7}	1.076×10^{7}
m^2	4.047×10^{3}	6.452×10^{-4}	1×10^{-4}	9.290×10^{-2}	1	2.590×10^{6}	1×10^{6}
mi^2	1.562×10^{-3}	4.25×10^{-9}	3.861×10^{-11}	3.587×10^{-8}	3.861×10^{-7}	1	0.3861
km^2	4.047×10^{-3}	6.452×10^{-10}	1×10^{-10}	9.290×10^{-8}	1×10^{-6}	2.590	1

Multiply units in the column heads by the figures in the table to convert to units at left. (For example, to convert square meters to square feet, multiply the number of square meters by 10.76.)

CONVERSION OF VOLUME

	Cubic Meters (m^3)	Cubic Yards (yd^3)	Cubic Centimeters (cm^3)	Cubic Inches (in^3)	Cubic Feet (ft^3)
m^3	1	0.7646	1×10^{-6}	1.639×10^{-5}	2.832×10^{-2}
yd^3	1.308	1	1.308×10^{-6}	2.143×10^{-5}	3.704×10^{-2}
cm^3	1×10^{6}	7.646×10^{5}	1	16.39	2.832×10^{4}
in^3	6.102×10^{4}	46,656	6.102×10^{-2}	1	1,728
ft^3	35.31	27	3.531×10^{-5}	5.787×10^{-4}	1

Multiply units in the column heads by the figures in the table to convert to units at left. (For example, to convert cubic inches to cubic centimeters, multiply the number of cubic inches by 16.39)

II

Appendix

Geologic Time Scale

Eon	Era	Period	Epoch	Million Years*	Life Events	Earth Events
Phanerozoic	Cenozoic	Quaternary	Holocene	0.01	Humans	Rifting in East Africa, Baikal, Rhine Graben, Rio Grande; Ice ages
			Pleistocene	1.64		
		Tertiary	Pliocene	5.2		
			Miocene	23.3		Himalayas form
			Oligocene	35.4	Mammals	
			Eocene	56.5		
			Paleocene	65	Extinction of dinosaurs	North Atlantic rifting starts; Rocky Mountains form
	Mesozoic	Cretaceous		145.6		South Atlantic rifting starts
		Jurassic		208		Central Atlantic rifting starts
		Triassic		245		Final assembly of Pangaea
	Paleozoic	Permian		290	Reptiles	Appalachian Mountains form
		Carboniferous	Pennsylvanian / Mississippian	362.5		
		Devonian		417		
		Silurian		443	Land plants, Fishes	
		Ordovician		495	Shelled organisms	
		Cambrian		545		
Precambrian	Proterozoic			2500	Early multicelled organisms	
	Archean			4000	First primitive life	Oldest rocks
				4550	Age of Earth	Heavy meteorite bombardment

*Note scale is strongly non-linear.

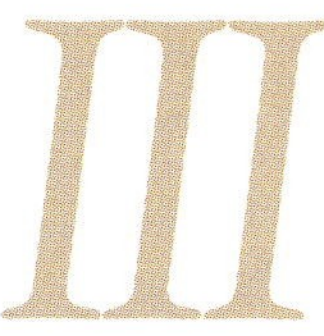

Appendix

The Periodic Table of The Elements

Periodic Table of the Elements

Elements in Earth's crust and their abundances

Element	Na	K	Ca	Mg	Fe	Al	Si	O
%	2.1%	2.3%	2.4%	4%	6%	8%	28%	46%

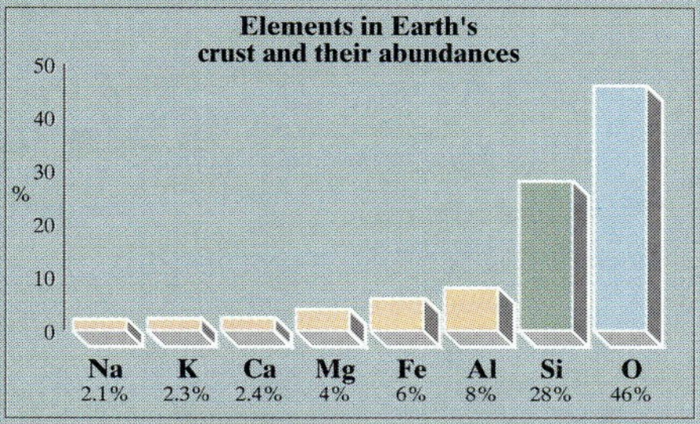

Key: 24 — Atomic number; **Cr** — Element symbol; 51.9961 — Atomic mass; Chromium — Element name

PERIOD	GROUP 1 1A	2 2A	3 3B	4 4B	5 5B	6 6B	7 7B	8 8B	9 8B	10 8B	11 1B	12 2 B	13 3A	14 4A	15 5A	16 6A	17 7A	18 8A
1	1 **H** 1.00794 Hydrogen																	2 **He** 4.00260 Helium
2	3 **Li** 6.941 Lithium	4 **Be** 9.01218 Beryllium											5 **B** 10.811 Boron	6 **C** 12.011 Carbon	7 **N** 14.0067 Nitrogen	8 **O** 15.9994 Oxygen	9 **F** 18.998403 Fluorine	10 **Ne** 20.1797 Neon
3	11 **Na** 22.9898 Sodium	12 **Mg** 24.3050 Magnesium											13 **Al** 26.9815 Aluminum	14 **Si** 28.0855 Silicon	15 **P** 30.9738 Phosphorus	16 **S** 32.066 Sulfur	17 **Cl** 35.4527 Chlorine	18 **Ar** 39.948 Argon
4	19 **K** 39.0983 Potassium	20 **Ca** 40.078 Calcium	21 **Sc** 44.9559 Scandium	22 **Ti** 47.88 Titanium	23 **V** 50.9415 Vanadium	24 **Cr** 51.9961 Chromium	25 **Mn** 54.9381 Manganese	26 **Fe** 54.9381 Iron	27 **Co** 58.9332 Cobalt	28 **Ni** 58.69 Nickel	29 **Cu** 63.546 Copper	30 **Zn** 65.39 Zinc	31 **Ga** 69.723 Gallium	32 **Ge** 72.61 Germanium	33 **As** 74.9216 Arsenic	34 **Se** 78.96 Selenium	35 **Br** 79.904 Bromine	36 **Kr** 83.80 Krypton
5	37 **Rb** 85.4678 Rubidium	38 **Sr** 87.62 Strontium	39 **Y** 88.9059 Yttrium	40 **Zr** 91.224 Zirconium	41 **Nb** 92.9064 Niobium	42 **Mo** 95.94 Molybdenu	43 **Tc** (98) Technetium	44 **Ru** (98) Ruthenium	45 **Rh** 102.906 Rhodium	46 **Pd** 106.42 Palladium	47 **Ag** 107.868 Silver	48 **Cd** 112.411 Cadmium	49 **In** 114.818 Indium	50 **Sn** 118.710 Tin	51 **Sb** 121.75 Antimony	52 **Te** 127.60 Tellurium	53 **I** 126.904 Iodine	54 **Xe** 131.29 Xenon
6	55 **Cs** 132.905 Cesium	56 **Ba** 137.327 Barium	57 ***La** 138.906 Lanthanum	72 **Hf** 178.49 Hafnium	73 **Ta** 180.948 Tantalum	74 **W** 183.85 Tungsten	75 **Re** 186.207 Rhenium	76 **Os** 186.207 Osmium	77 **Ir** 192.22 Iridium	78 **Pt** 195.08 Platinum	79 **Au** 196.967 Gold	80 **Hg** 200.59 Mercury	81 **Tl** 204.383 Thallium	82 **Pb** 207.2 Lead	83 **Bi** 208.980 Bismuth	84 **Po** (209) Polonium	85 **At** (210) Astatine	85 **Rn** (222) Radon
7	87 **Fr** (223) Francium	88 **Ra** 226.025 Radium	89 **†Ac** 227.028 Actinium	104 **Db** (261) Dubnium	105 **Jl** (262) Jollothium	106 **Rf** (263) Rutherfordium	107 **Bh** (262) Bohrium	108 **Hn** 186.207 Hahnium	109 **Mt** (268) Meitnerium	110 (269)	111 (272)	112 (277)						

Series														
* Lanthanide series:	58 **Ce** 140.115 Cerium	59 **Pr** 140.9077 Praseodymium	60 **Nd** 144.24 Neodymium	61 **Pm** (145) Promethium	62 **Sm** 150.36 Samarium	63 **Eu** 151.965 Europium	64 **Gd** 157.25 Gadollnium	65 **Tb** 158.925 Terbium	66 **Dy** 162.50 Dyprosium	67 **Ho** 144.24 Holmium	68 **Er** 144.24 Erbium	69 **Tm** 144.24 Thulium	70 **Yb** 144.24 Ytterbium	71 **Lu** 144.24 Lutetium
† Actinide series:	90 **Th** 232.0381 Thorium	91 **Pa** 231.0359 Protactinium	92 **U** 238.0289 Uranium	93 **Np** 237.048 Neptunium	94 **Pu** (244) Plutonium	95 **Am** (243) Americium	96 **Cm** (247) Curium	97 **Bk** (247) Berkelium	98 **Cf** (251) Californium	92 **Es** (252) Einsteinium	92 **Fm** (257) Fermium	92 **Md** (258) Mendelevium	92 **No** (259) Nobellium	92 **Lr** (260) Lawrencium

Atomic masses are relative to carbon-12. For certain radioactive elements, the numbers listed (in parentheses) are the mass numbers of the most stable isotopes. The metals are ▪ and the nonmetals are ▪. Metalloids are indicated by ▪. The noble gases are ▪.
Names for elements 104 to 109 are still being debated and elements 110, 111, and 112 have not yet been named.

IV

APPENDIX
COMMON MINERALS AND THEIR PROPERTIES

Mineral	Formula	Color	Hardness	Distinguishing Characteristics
Silicates				
Quartz	SiO_2	Commonly colorless, impure varieties are white (milky quartz), pink (rose-), gray (smoky-), purple (amethyst)	7	Conchoidal fracture, good crystal shapes, hexagonal cross-section
Olivine	$(Mg,Fe)_2SiO_4$	Green-yellow	6.5–7	Glassy luster, no cleavage
Plagoiclase felspar	$(Na,Ca)(Al,Si)_2Si_2O_8$	White to gray (some varieties show bluish iridescence)	6	Two cleavages at 90°; may have fine striations on one cleavage surface
Alkali feldspar	$(Na,K)AlSi_3O_8$	White, sometimes stained pink	6	Two cleavages at 90°
Pyroxene	$(Ca,Mg,Fe)_2Si_2O_6$	Green to brown or black	5–7	Two good cleavages at 90°
Amphibole	$(Na,Ca)_2(Mg,Al,Fe)_5Si_8O_{22}(OH)_2$	Brown to green to black	5–6	Two cleavages at 120°
Biotite (mica)	$K(Mg,Fe)_3AlSi_3O_{10}(OH)_2$	Brown to black	5.5	Excellent cleavage in one direction, forming parallel sheets
Muscovite (mica)	$KAl_3AlSi_3O_{10}(OH)_2$	Colorless to pale brown	3–2.5	Excellent cleavage in one direction, forming parallel sheets
Epidote	Hydrous Ca, Al, Fe silicate	Yellow to green	6–7	Vitreous luster may resemble olivine. One good cleavage
Chlorite	Hydrous Fe, Mg, Al silicate	Green	2–2.5	Usually occurs as aggregates of small grains with poor crystal shape
Talc	$Mg_3Si_4O_{10}(OH)_2$	White	1–1.5	Greasy feel (commonly referred to as soapstone)
Clay	Hydrous Al silicate with Ca, Na, Ca, Fe, Mg	White (can be colored by impurities)	1–2	Forms soft, compact earthy masses of microscopic grains, earthy odor when moist
Garnet	$(Ca,Mg,Fe)_3(Al,Fe)_2Si_3O_{12}$	Variable; commonly red to purple brown	7	Well-developed symmetrical crystal, no cleavage. Dense
Sulfides				
Pyrite	FeS_2	Brassy yellow	6–6.5	Metallic luster, cubic or octahedral crystals, greenish black streak
Chalcopyrite	$CuFeS_2$	Brassy yellow (lighter than pyrite)	3.5–4.5	Metallic luster, softer than pyrite
Sphalerite	ZnS	Reddish brown	3.5	Resinous to vitreous luster (unlike metallic luster of most metal sulfides). Six perfect cleavages
Galena	PbS	Silvery gray	2.5	Metallic luster, cubic crystals, very dense. Three perfect cleavages at 90°

Mineral	Formula	Color	Hardness	Distinguishing Characteristics
Oxides				
Hematite	Fe_2O_3	Red to dark gray	5.5-6.5	Dark red streak
Magnetite	Fe_3O_4	Black	6	Strongly magnetic
Ruby / Sapphire	Al_2O_3	Red (ruby), Blue (sapphire)	9	Both gemstones are varieties of aluminum oxide (corundum), distinguished and used for its hardness
Sulfates				
Gypsum	$CaSO_4 \cdot 2H_2O$	Colorless	1.5-2.5	Forms glassy, clear crystals or tubular or fibrous crystals. One perfect cleavage
Barite	$BaSO_4$	White	2.5-3.5	Pearly luster, distinctive high specific gravity. Two cleavages at 78°
Anhydrite	$CaSO_4$	Colorless to white	3-3.5	Cleavage in two directions
Carbonates				
Calcite	$CaCO_3$	Colorless, impure varieties colored	3	Effervesces in weak acid, cleaves into rhombohedra. Two cleavages at 78°
Dolomite	$(Ca,Mg)CO_3$	White to pink	3.5-4	Effervesces in weak acid if powdered (less readily than calcite). Two cleavages at 78°
Azurite	$Cu(CO_3)_2(OH)_2$	Azure blue	4	Effervesces in HCl
Malachite	$CuCO_3 \cdot Cu(OH)_2$	Bright to dark green	3.5-4	Effervesces in HCl
Halides				
Fluorite	CaF_2	White, yellow, green, or purple	4	Vitreous luster, cubic crystals may be fluorescent in ultraviolet light. Four perfect octahedral cleavages
Halite	NaCl	Colorless to white or pale gray	2.5-3	Perfect cubic cleavage, salty taste, soluble in water
Sylvite	KCl	Colorless to white	2	Perfect cubic cleavage, salty taste (more bitter than halite), soluble in water
Native elements				
Diamond	C	Colorless	10	Brilliant (adamantine) luster
Graphite	C	Dark gray	1-2	Greasy to the touch, marks paper
Sulfur	S	Yellow	1.5-2.5	sulfurous smell

V

APPENDIX
EARTH STATISTICS

The Earth

- Equatorial radius: 6378 km
- Polar radius: 6357 km
- Surface area: 5.1006×10^8 km^2
- Volume: 1.083×10^{12} km^3
- Mass: 5.974×10^{24} kg
- Average density: 5142 kgm^3
- Mean distance to Sun: 1.496×10^8 km
- Mean distance to Moon: 3.844×10^5 km

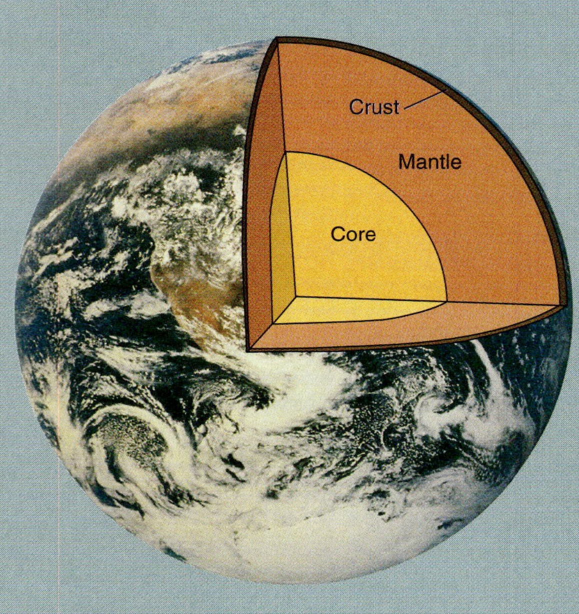

Surface

- 71% ocean and 29% land
- Average land elevation: 875 m
- Average ocean depth: 3800 m

The Oceans

Surface areas
- Indian Ocean: 7.343×10^7 km^2
- Pacific Ocean: 16.624×10^7 km^2
- Atlantic Ocean: 8.252×10^7 km^2

Crust

Continental Crust
- Area: 1.479×10^8 km^2
- Average thickness: 35 km
- Average density: 2700 kg/m^3
- Volume: 4.437×10^9 km^3
- Mass: 1.198×10^{22} kg

Oceanic Crust
- Area: 3.621×10^8 km^2
- Average thickness: 7 km
- Average density: 2850 kg/m^3
- Volume: 2.535×10^9 km^3
- Mass: 7.225×10^{21} kg

Mantle

- Average thickness: 2883 km
- Average density: 4405 kg/m^3
- Volume: 9.08×10^{11} km^3
- Mass: 4.00×10^{24} kg

Core

- Average thickness: 3471 km
- Average density: 11,132 kg/m^3
- Volume: 1.75×10^{11} km^3
- Mass: 1.95×10^{24} kg

Glossary

aa Blocky and fragmented basalt lava flow.

accretionary wedge An accumulation of sediment at the trench of a convergent margin, supplied from both the volcanic arc and the subducting plate.

acicular habit A characteristic needle-like crystal shape.

acidic solution An aqueous solution containing an excess of hydrogen ions.

active continental margin A continental margin that is also a plate margin, either represented by a subduction zone or a transform fault.

active rift A rift that results from active movement of material in the mantle; for example, upwellings of hot material in mantle plumes.

adamantine luster Brilliant, reflective "sparkling" appearance to the mineral, such as diamond.

alluvial fan Large fanlike deposit of sediment deposited at the mouth of a canyon, typically in desert environments.

alpha (α) decay Spontaneous radioactive decay involving ejection of an alpha particle (a helium nucleus comprising two neutrons and two protons) from the nucleus of a radioactive isotope.

alumina A term for the oxide Al_2O_3.

amorphous Without form; in mineralogy, a lack of the orderly, repetitive arrangement of atoms that characterizes crystalline structure.

amplitude The height of a wave crest above a mean or average value, or the depth of the adjacent trough below the average value.

angle of repose The maximum angle that loose, cohesionless material on a slope can remain stationary.

anhydrous Without water; in geology, this term usually refers to a mineral or to magma that contains essentially no water.

anion A negatively charged ion.

antecedent stream A stream course established prior to an uplift. During uplift, the water course erodes downward as the mountains rise.

anthracite coal The highest grade of coal, containing more than 90 percent carbon.

anticline A fold in layered rocks in which the opposing sides of the structure dip away from the fold axis, giving it a convex upward form.

aphanitic An igneous rock texture in which the crystals are too small to see with the unaided eye.

aquiclude An impermeable sedimentary bed that acts as a barrier to the flow of groundwater.

aquifer A horizon of rock that can transport and store water.

arc-trench gap The linear distance on Earth's surface from the subduction zone (commonly the oceanic trench) to the volcanic arc.

arrete A sharp rocky wall analogous to a knife-edge, which has resulted from the backward erosion of two glaciers.

aseismic creep Motion along a fault that does not cause earthquakes.

aseismic ridge As lithospheric plates moves over a hot-spot, a linear string of volcanoes is produced. If sufficient volcanic material has been erupted, the volcanoes will form ridge-like topographic highs on the ocean floor.

assimilation Incorporation into magma of wall rock or another magma.

asthenosphere The portion of Earth beneath the lithosphere that yields plastically to stress. Its upper boundary ranges from approximately 5 km at spreading ridges to more than 100 km under continents.

atmosphere The gaseous envelope that surrounds Earth.

atom The smallest unit of an element that retains the properties of that element. Atoms combine together to form molecules. An atom consists of a central nucleus containing a given number of positively charged protons and uncharged neutrons, with negatively charged electrons orbiting the nucleus.

atomic mass The mass of an element measured in atomic mass units.

atomic number The number of protons in an atom. The atoms of each element are defined by a unique atomic number. The simplest atom is the hydrogen atom, which has a single proton and a single electron.

aulacogen Narrow sediment-filled and fault-bounded basin, formed as the failed arm of a rift system.

authigenic Describing minerals or cement formed in sediments after deposition.

auxiliary plane A plane perpendicular to the true fault plane.

axial plane An imaginary surface that bisects a fold, that is, it cuts through the fold axis for each bed.

axis The locus of points of maximum curvature on a folded surface, or, in mineralogy, an imaginary line that passes through the center of a crystal to which the different faces and angles of the crystal may be referred.

back-arc basin A basin that is similar to a mini-ocean basin, where sea-floor spreading takes place behind a magmatic arc at a convergent margin.

back arc The side of a volcanic arc away from the trench and subduction zone.

banded iron deposit A deposit of sedimentary rocks in which chert and iron oxide alternate.

basalt A dark-colored rock with a mineral composition of calcic plagioclase, pyroxene, perhaps olivine, and iron ore minerals. The term is usually reserved for extrusive rocks; it is the fine-grained equivalent of gabbro.

basement The bulk of the continental crust, composed of crystalline igneous or metamorphic rocks, generally overlain by younger, stratified sedimentary rocks.

batholith A large body of igneous rock that has intruded into crustal rocks.

bathymetry The topography of the ocean floor.

bauxite A deeply weathered soil, usually found in tropical climates, in which silica, and most soluble ions have been flushed out by abundant water to leave an ore of aluminium.

bed A unit of sediments in a stratified sequence bounded top and bottom by a surface known as a bedding plane.

bedload Material that is transported by flowing water or wind that remains in contact with the bed.

bedform A sedimentary structure formed by flowing water or wind over a bed of loose sediment.

beta decay Spontaneous radioactive decay involving ejection of an electron (a beta particle) from a neutron in the nucleus of a radioactive isotope.

big bang A theory explaining the origin of the universe whereby all matter was involved in a gigantic explosion about 15 billion years ago.

bioclastic A term that generally describes carbonate sediments composed of shells or other carbonate fragments that once were parts of plants or animals.

biosphere A general term that includes all living organisms on Earth.

bituminous coal A type of coal containing at least 15 percent volatile components.

black smoker A submarine hot spring near spreading ridges, which emits sulfide-laden water.

bladed habit A characteristic flat and wide, but thin and elongate crystal shape, that develops when conditions are appropriate for large crystals, or aggregates of crystals, to grow.

blocking temperature The temperature (about 100°C below the Curie point) below which a mineral's permanent magnetism becomes locked in the direction of the ambient field.

blueschist A metamorphic rock, invariably associated with subduction complexes, that is formed under high pressures but relatively low temperatures. It contains sodium-rich amphiboles that are distinctively blue.

body wave Seismic waves that travel through the body of Earth; P and S waves are body waves.

botryoidal habit A characteristic clumping of approximately spherical forms of a mineral, giving the general appearance of a bunch of grapes.

Bouguer correction A calculated change in gravity measurements that corrects for topography and density variation at Earth's surface; it shows gravity measurements at a uniform elevation as if all mountains had been scraped off and all valleys were filled with material of the same density as the surroundings.

Bowen's Reaction Series A hypothetical sequence of mineral crystallization during solidification of a molten silicate melt; named after the experimental petrologist N. L. Bowen who succeeded in simulating natural crystallization processes experimentally.

breccia A sedimentary rock formed from angular rock fragments

brine Water that has been enriched in salts; mainly chlorides.

brittle deformation Deformation of geologic media characterized by brittle behavior, such as fracturing or faulting.

brittle-ductile transition A depth at which the behavior of geologic media changes from brittle breakage to plastic flow.

bulk modulus A measure of the amount a given material will decrease in size but increase in density when placed under increasing force per unit area.

calcrete (or caliche) A soil zone formed within the "B" soil horizon characterized by precipitation of carbonate which may either be between soil grains or replacing soil mineral grains.

caldera A broad circular volcanic depression formed when an eruption removes large volumes of magma from a magma chamber, causing its roof to collapse.

capacity The maximum sediment load that can be transported by a stream with a given water flow and slope.

carbonate A variety of minerals in which the anion is CO_3^{-2}. Examples are calcite, aragonite, and dolomite.

cataclastic metamorphism Rock deformation characterized by crushing or granulation so that the brittle minerals have been broken, flattened, or strung out.

cation A positively charged ion.

cementation Precipitation of a mineral between loose grains of sediment, causing the sediment to become compact and hard.

centrifugal force A force directed outward from the axis of rotation in a rotating system.

chemical bond A force connecting atoms together into a molecule.

chondrite A meteorite that contains chondrules.

chondrule Small, rounded droplets that are thought to have condensed from the original solar nebula from which the Solar System derives.

cinder cone A conical volcanic structure composed of pyroclastic fragments (vesicular cinders or scoria) that accumulate near the vent. They usually form as a result of fountaining activity in basaltic volcanoes.

cirque An amphitheater-like area, surrounded by high, steep-walled cliffs, that is the major area feeding ice to a glacier.

clast A general term for a sedimentary fragment without regard to its composition, roundness, or size.

clastic sediment A general term for a sediment whose constituents have been transported either by wind or water.

clay A specific mineral class that has the general composition of hydrous aluminum silicates; also a general term for grains that are less than 0.004 mm in diameter.

cleavage In mineralogy, some crystals have preferential alignments along which they tend to break. These planes are always parallel to a possible crystal face; also, a metamorphic fabric in which platy minerals are aligned more or less parallel, causing the rock to preferentially split along planar surfaces that are formed by that mineral alignment.

coal Accumulated vegetative matter that has been transformed by pressure and chemical reactions to a relatively high carbon content, and variable concentrations of volatiles (methane, carbon dioxide, water).

collage tectonics The assembly of many small crustal fragments added to a continent at various times in the past. Each terrane must have a different history and origin from its adjacent blocks.

columnar jointing Contraction fractures usually in lava or pyroclastic flows that are due to cooling and perpendicular to the cooling surfaces

comet A small body made up of mainly ice and dust, that follows an elliptical orbit around the Sun.

compatible element Elements that have ionic radii and charge that permits them to substitute into the common rock forming minerals without significant distortion of the crystal structure.

competence The largest sediment clast that can be transported by a stream.

composite volcano See stratovolcano.

compound A substance composed of at least two different types of atoms that is held together by chemical bonds.

compression A stress that forces the molecules of a material closer together.

concentric folding Layered geologic media that has been deformed into gentle wave-like shapes called folds. In a concentric fold, the bent layers all have a common center of curvature.

conchoidal fracture A series of smoothly curving fractures (appearing shell-shaped) in glasses and minerals without cleavage; for example, quartz.

conduction A process of heat transport from hotter to cooler regions by contact so that the heat is carried internally from atom to atom.

confining bend A bend or curve in a fault trace such that movement along the fault tends to pinch the two sides of the fault together.

conglomerate A poorly sorted sedimentary rock that contains rounded or partially rounded clasts greater than about 4 mm set in a finer matrix.

continent A major landmass; continents stand above ocean basins because the crust is much thicker, averaging approximately 30 km in contrast to oceanic crust, which is approximately 7 km thick.

continental breakup The large-scale disintegration of continents by plate tectonic forces.

continental drift The motion of continents across ocean basins. (Note: Continental drift is not synonymous with plate tectonics.)

continental lithosphere Lithosphere that has a component of continental crust in it. Such lithosphere has a slightly lower density than oceanic lithosphere, but the crust is much thicker. For this reason, continents rise above sea level.

continental margin The edge of a continent, usually defined geographically as the 200 m bathymetric contour. Only at active margins does the continental margin and the plate margin coincide.

convection Movement of a material as a result of density variations caused by heating.

convergent plate margin A plate margin in which two plates are converging, or moving toward one another, with one plate subducting beneath the other.

core The inner part of Earth, having a radius of approximately 3400 km. The core is divided into a solid portion, which extends out to a radius of about 1200 km, and a liquid portion that is approximately 2270 km in thickness. The core is thought to be composed mainly of iron.

covalent bond A type of chemical bond in which two atoms are linked together by sharing an electron pair.

cover the younger, more stratified part of the crust that overlays the igneous or metamorphic basement rocks. The cover is typically sedimentary or metamorphosed sedimentary rocks.

craton One of two major parts of continental interiors. The craton is the older stable core, to which younger mountain belts are accreted.

cross-bedding Repetitive, inclined laminations confined to one or more beds in sediments or sedimentary rocks.

crust Earth's outer rocky layer, comprising approximately 0.7 percent of the mass of Earth, and extending to a depth of about 30 km beneath the continents and to an average depth of 7 km beneath the ocean floor. The crust is largely made up of minerals containing the elements calcium, aluminum, magnesium, iron, silicon, sodium, potassium, and oxygen. (Note: Crust is not synonymous with lithosphere.)

crustal root Crustal rocks that extend to unusual depths.

crystal form All the faces of a crystal that have a similar position relative to the elements of symmetry; loosely used to designate the general shape of a crystal.

crystal habit The characteristic shape (form) of a mineral crystal.

crystal An orderly repetitive arrangement of atoms, or groups of atoms, that may have an outward expression as crystal faces.

Curie point The temperature above which a magnetic material loses its magnetism. This temperature is well below the material's melting point.

décollement A detachment of sedimentary strata, usually almost horizontal, that separates an overlying deformed zone from an underlying zone.

deformation A general term that describes the mechanical bending or breaking of material.

degree of symmetry The relative symmetry of one mineral crystal compared to another mineral.

delta Usually triangular, low-lying areas of sediment deposition that extend oceanward from a river mouth.

dendritic habit The tendency for a crystal or group of crystals to grow in a branching pattern.

density Mass per unit volume.

depth of compensation The depth at which columns of rock (or rock plus ocean) standing on equal areas will have the same weight, regardless of surface topography.

desert varnish A thin dark film formed of iron oxide and manganese oxide on the surface of rocks exposed in the desert for long periods of time.

detrital Describing sediments or sedimentary grains that are transported into the site of deposition.

devitrification The process by which glass, which is unstable, may crystallize.

diagenesis The chemical and physical changes that occur in a deposited sediment in response to increasing pressures and temperatures due to burial as well as the effects of fluids.

diapir An intrusive body that has risen through overlying rocks because it is less dense than the overlying column of rocks.

differential weathering Different rock types weather at different rates and by different reactions. Differential weathering is a major influence in the generation of topography.

differentiation On a planetary scale, the separation of a planet into a core, mantle, and crust. In the transformation of igneous melts from an initial composition to another composition, differentiation refers to partial melting of an original rock, or to the separation of crystals that have formed early from the remainder of the melt.

dike A sub-vertical planar igneous rock body that cuts across the rock into which it is intruded. Compare with sill.

dip The angle measured from the horizontal that defines the inclination of a discrete layer in a rock body.

direct ray The ray drawn for a wave that travels by the most direct path from a seismic source to a seismometer.

dispersion The seismogram from an earthquake changes shape with time (distance) traveled because seismic waves of different frequencies travel at different speeds through Earth.

divergent margin A divergent plate margin forms where a plate splits and the two pieces move away from each other, leaving a rift where new oceanic crust is formed.

Doppler effect An apparent shift in the frequency of a wave due to the relative motion of the source and the receiver.

drainage basin An area of converging streams that eventually leads to the same trunk stream.

drumlin A teardrop-shaped hill caused by glaciers flowing over previously deposited moraine, reshaping it into groups or fields of elongate mounds.

ductile A material property indicating that irreversible deformation is due to plastic flow rather than to elastic or brittle deformation.

duplex structures A braided pattern of several fault traces at the point where a strike-slip fault bends or curves.

dynamo A device that generates electrical current.

earthquake cycle A cycle in which plate motion causes increasing elastic strain to accumulate across a locked fault, an earthquake releases that elastic strain, and the rocks on each side of the fault then adhere, re-establishing the friction to start another cycle.

earthquake Violent and rapid shaking due to fault slip in Earth. Seismic energy is released suddenly; this energy was stored as elastic strain immediately prior to the earthquake.

elastic limit The limit of stress that a material can tolerate, above which permanent deformation occurs.

elastic modulus The ratio of stress to strain in a material. For example, a spring is stretched by applying a force. The ratio of the force to the stretch is a measure of the elastic modulus of the spring.

elastic rebound The sudden release of elastic strain across a locked fault due to an earthquake.

elasticity The extent to which a material can return to its original shape after being subjected to a stress.

electron A negatively charged particle, approximately 1/1900 the mass of a neutron, that orbits an atomic nucleus.

element A substance composed of just one type of atom. The atoms of an element all have the same number of protons and electrons, and thus, the same atomic number.

***en echelon* structure** A series of approximately parallel linear features (faults, fold hinges etc.) that are slightly offset from one another.

ephemeral river A stream that flows only in direct response to rainfall in the immediate vicinity.

epicenter The position on earth's surface immediately above the hypocenter of an earthquake.

epicentral angle The angular distance formed at Earth's center between the source of an earthquake and a seismic station.

epirogenic uplift An intraplate uplift that is commonly very broad in regional coverage, and may produce uplifts of as much as hundreds of meters.

equilibrium A state of balance.

erosion Movement of rock and mineral particles by the action of flowing fluids, including wind and water, moving mud or ice, and downslope transport due to gravity.

escape velocity The minimum velocity needed to escape the gravitational pull of a planet; the actual velocity varies according to the size and density of each planet.

evaporite A mineral deposit formed by the evaporation of a body of water.

exfoliation Shedding of thin sheets or plates of rock, usually curved, from a solid rock core.

exotic terranes The individual blocks that are accreted in collage tectonics. The term exotic terrane is used to emphasize that the origins of these small continental fragments are unknown.

extrusive Magma that has erupted at the surface from a volcano or fissure. Compare with intrusive.

fabric A term that describes how mineral grains are arranged relative to one another, including grain shape, grain size, and the distribution of grain sizes in rocks. Compare with texture.

failed rift See aulocagen.

fault A brittle fracture across which relative movement of rocks has occurred.

fault creep Slow, nearly continuous movement across a fault that does not permit the buildup of elastic strain necessary for an earthquake.

fault gouge Finely ground rock caused by the crushing of two sides of a fault.

fault plane The surface along which offset occurs. Usually thought of as a flat surface, but fault planes may be curved or irregular.

fault plane solution The orthogonal planes formed by plotting the first movement measured by seismometers on an imaginary sphere surrounding the fault, that separates positive and negative motion. One is the true fault plane and the other the auxiliary plane.

fault scarp A small ledge or cliff generated as a result of fault movement.

fault surface see fault plane

fault trace A line that defines the intersection of the fault with the land surface.

felsic A term used to describe a light colored rock that is rich in feldspar.

fibrous habit A flexible, thin, elongate fiber-like shape that characterizes some crystals.

first motion The very first movement of the ground in an earthquake, up or down, compression or extension.

flood basalts Huge outpourings of basalt that are believed to be a result of large-scale melting of the upper mantle due to rising mantle plumes immediately prior to continental breakup.

floodplain The portion of a river system adjacent to the channel that is covered by water when the river overflows its banks.

flow The movement of fluid from one place to another; or, irreversible deformation of solid media, cauiing it to behave more like a liquid.

flower structure A region along a confining bend of a large fault where multiple faults are formed, which are forced upward and outward, bending into a flower petal-like geometry.

focus See hypocenter.

fold A deformational structure in rock that is due to flexure.

foliation A fabric, generally in metamorphic rocks, in which platy minerals are aligned more or less parallel to each other and perpendicular to the direction of maximum compression.

footwall The block that lies under the fault.

fore-arc basin The basin formed between an arc and an accretionary wedge, in which sediments commonly accumulate.

fore arc The side of a volcanic arc that faces toward the trench and subduction zone.

fossil Preserved evidence of a pre-existing organism.

fractional crystallization Differentiation of magma based on the separation of liquid and crystals because minerals crystallize at different times during crystallization. Compare with Bowen's Reaction Series.

fracture A material property (the way in which a mineral or rock breaks other than along a cleavage plane) or a joint or a crack in a rock.

free-air gravity Gravity measurements that are corrected for topography to show values at a constant elevation.

frequency The number of wave crests, or peaks, in a wave passing by a given point per unit time.

gabbro An igneous rock—the intrusive equivalent of basalt, comprising typically olivine, pyroxene and plagioclase in a phaneritic texture.

geode A cavity or hole in a rock that is partially or completely filled with minerals.

geodimeters Instruments that measure horizontal distances by timing laser pulses between the instrument and a reflector.

geodynamo The source of Earth's magnetic field. It is thought to be the result of convection in Earth's liquid outer core.

geologic time scale A relative time scale based on fossil content. Radiometric dating has permitted fine resolution of the Periods and Ages on this scale, but fossils remain the primary defining tool.

geophysics The study of Earth using the fundamental principles and methods of physics.

geotherm A curve representing the variation of temperature with depth.

geothermal energy Energy tapped from hot rocks beneath Earth's surface.

geothermal gradient The rate of increase in temperature with depth in Earth. Near Earth's surface, the average geothermal gradient is about 30° C per kilometer.

glacier A large mass of perennial ice, recrystallized from snow, that has moved downslope due to gravity or radially outward due to its own weight.

glass An amorphous, noncrystalline material with no order to the arrangement of silica tetrahedra, anions, and cations. Natural glasses are mostly volcanic in origin, caused by sufficiently rapid cooling that prohibited crystallization. Compare with crystal.

Global Positioning System A navigation and location system developed by the military. A receiver on the ground receives signals from several of the 24 satellites in Earth orbit for accurate location in three dimensions.

gneiss A banded or lineated metamorphic rock typically consisting of alternating concentrations of platy or linear minerals (for example, biotite or amphibole) with quartz and feldspar.

graben A down-dropped block between two normal faults of opposing dips, caused by extension, and commonly expressed as a valley at the surface.

graded bedding A continuous variation in grain size from the base of a sedimentary bed to the top. Normal grading consists of coarser grains at the base and finer grains at the top. Reverse grading can occur also, with the coarser grains at the top of the bed.

graded stream A stream close to a steady state condition where the stream transport capability is closely matched to the material supplied for transportation.

grain size There is a formal classification for sediment grain sizes. The classification in commonest usage is called the Wentworth Size Scale: clasts <0.004 mm are classified as clay sized grains, >0.004 mm to 0.063 mm are silt grains, >0.063 mm to 2 mm are sand grains, >2 mm to 4 mm are granules, >4 mm to 64 mm are pebbles, >64 mm to 260 mm are cobbles, and > 260 mm are boulders.

gravimeter A sensitive instrument used to measure the force of Earth's gravitational field.

gravity The force of attraction of two masses to each other.

gravity anomaly A departure from the gravity field expected if the Earth were homogeneous. Caused by variations in the density of subsurface rocks.

greywacke A poorly sorted sandstone containing angular fragments of quartz, feldspar, clay minerals and a variety of rock fragments. Common at convergent plate margins.

groundwater Water in the uppermost parts of the crust, where it is available for use by society.

Gutenberg-Richter relation A graph of the logarithm of the number of earthquakes above a given magnitude against the magnitude of each earthquake which shows, for example, that there are 10 times more earthquakes of magnitude 5 than there are of magnitude 6. This relationship holds both globally and regionally.

habit The characteristic form that minerals exhibit when conditions are appropriate for large crystals, or aggregates of crystals, to grow.

hackly fracture A fracture surface that is rough and uneven, characterized by many small sharp corners and points.

half-graben An asymmetrical down-dropped block bounded on one side by a normal fault and a tilted ramp on the opposite side.

half-life The time required for half of the atoms of a given radioactive isotope to decay to a daughter isotope.

hanging wall The block that lies above a fault.

hardness The resistance of a mineral to scratching. See Mohs Hardness Scale.

hardpan A hard impervious soil layer cemented by clays. Hardpan accumulates within the "B" soil horizon.

harmonic tremor A distinctive seismic signal that accompanies underground magma movement; a sustained oscillatory seismic motion usually at a dominant, fixed frequency.

heat of radioactivity The energy released during radioactive decay, which heats up the surrounding atoms.

hinge The region of curvature in a fold.

horn An irregular three- or four-sided pyramidal peak formed at the intersection of three or more cirques.

hot spot trail A trail of extinct volcanoes that marks the movement of a lithospheric plate over a stationary hot spot.

hot spots Regions of unusually high heat flow due to upwelling within the mantle.

hydration The chemical combination with water. (Note: This is not equivalent to a reaction with water.)

hydrologic cycle The continual transfer, or cycle, of water among the oceans, clouds, rain, groundwater, lakes, rivers, and glaciers.

hydrolysis A reaction with water, in geology usually associated with weathering reactions.

hydrosphere A general term encompassing all water above and immediately beneath Earth's surface, including oceans, groundwater, rivers, lakes, glaciers and water vapor.

hydrothermal circulation Circulation of water within the crust, driven by a heat source, such as hot magma.

hydrothermal zone A region in which hot rocks cause the circulation of hot water away from the site and cold water into the site.

hypocenter The actual point of initial breakage on a fault.

igneous rock A rock that originated as a melt.

ignimbrite A rock formed by the deposition from a pyroclastic flow. The rock consists largely of ash and pumice, with a few crystals.

IGRF International Geomagnetic Reference Field consists of a worldwide compilation of magnetic field measurements.

immature sediment Textural immaturity is a lack of sorting, so that a sediment may have clay-sized grains together with much coarser material. Mineralogic immaturity is the occurrence in a sediment of highly reactive minerals in a weathering environment together with very stable minerals. May also indicate abundant rock fragments in a sediment.

impermeable A body of rock through which a gas or fluid cannot pass.

incompatible element Elements that have ionic radii and charge that inhibit their substitution into the common rock-forming minerals without significant distortion of the crystal structure.

inelasticity Nonelastic behavior under stress; normally implies plastic behavior.

inert Unreactive.

inner core A solid spherical volume at the center of the Earth, approximately 1200 km in radius, composed mainly of iron.

intraplate earthquake A large earthquake that occurs within a plate. The New Madrid earthquakes of 1811 and 1812 are examples of such earthquakes.

intraplate volcanism Volcanic activity located away from plate margins, commonly associated with plumes and hot spots (example Hawaii).

intrusive Magma that has risen into the lithosphere and solidified there below the surface to form an intrusive body such as a pluton, dike or sill. Compare with extrusive.

ion An electrically charged atom that has gained or lost one or more electrons.

ionic bond A type of chemical bond in which the electrical force of attraction between positively and negatively charged ions is the primary bonding energy.

island arc A magmatic arc on oceanic crust.

isostasy The concept whereby Earth's lithosphere is "floating" in gravitational balance upon the asthenosphere.

isostatic rebound The rise, or rebound, of the continental crust to conditions of isostatic equilibrium after a weight has been removed from it (such as melting of ice caps or erosion).

isotopes Atoms of the same element that contain different numbers of neutrons and therefore have different atomic masses.

isotope dating (radiometric dating) The use of naturally occurring radioactive elements and their decay products to determine the absolute age of the rocks containing those elements.

joint A rock fracture across which there has been no differential offset.

karst topography A type of topography found in limestone or dolomite, which is characterized by caves, caverns, and sinkholes.

kimberlite A variety of brecciated periodite that has been brought to the surface from great depth (>150 km).

lahar A volcanic mudflow.

lateral accretion In tectonics, the addition of crustal material to the margins of continents. In sedimentology, depositional processes in streams that cause the sedimentary deposit to grow laterally.

laterite A deeply weathered soil, usually found in tropical climates, in which silica, and most soluble ions have been flushed out to leave iron oxides and alumina rich clays, forming an ore of iron.

lattice In crystallography, the geometric repetitive pattern of atoms or groups of atoms in a mineral.

lava A flow of magma that has reached Earth's surface.

leaching The removal of elements from solid material by dissolution.

left lateral fault Motion on the far side of a strike-slip fault that is directed to the observer's left.

lignite A brownish coal that is the first stage of coalification after peat. Lignites have a high content of volatiles, and low heating capability compared to bituminous coal.

limb The part of a fold that is not curved, or the region on a fold where the curvature is reversing from one direction to the opposite.

limestone A sedimentary rock formed principally of carbonate minerals, most commonly derived from marine plant and animal shells.

liquidus The temperature above which a material is completely liquid. Compare with solidus.

liquefaction The transformation of a water-saturated solid into a liquid, usually as a result of earthquake shaking.

listric fault A curving normal fault in which the curvature is concave upward, commonly becoming near horizontal at depth.

lithification The process of converting loose sediment into rock through compaction or cementation.

lithosphere The outer, cooler, rigid portion of Earth. It ranges in thickness from approximately 5 km at spreading ridges to 100 km under continents. (Note: This term is not synonymous with crust.)

longshore drift The movement of sediment within and near the surf zone that is the result of waves approaching the shore at an angle.

low-velocity zone (LVZ) A thin region in the mantle beneath the lithosphere that is defined by a significant decrease in seismic wave velocity, corresponding to the asthenosphere.

luster A term that expresses how a mineral reflects light, and is related to the smoothness of a crystal surface.

mafic A term used to describe a dark-colored rock with abundant pyroxene, amphiboles or biotite.

magma chamber A region in the crust or upper mantle where magma accumulates.

magma Underground molten rock.

magmatic arc An arc on which a line of volcanoes form; found on an overriding plate that is adjacent to a subducting plate.

magmatic differentiation The evolution of a magma from one composition to another by mechanisms that include fractional crystallization, assimilation, partial melting, and magma mixing.

magnetic declination The difference between the direction that the north arrow of a magnetic compass points and true north.

magnetic inclination The angle of dip at which Earth's magnetic lines of force intersect the surface.

magnetic north The magnetic pole on Earth.

magnetic stripes Areas of different magnetic field strength which appear as "stripes" extending either side and parallel to a spreading ridge, formed by reversals of the geomagnetic field during seafloor spreading.

magnitude In seismology, a scale used to relate the size of one earthquake to another.

manganese nodule A semi-spherical accumulation of primarily manganese dioxide (MnO_2) that forms on the seafloor.

mantle The thickest compositionally-defined layer within the Earth, extending from the base of the crust to the outer surface of the core.

mass The amount of matter present in an object.

mass wasting The downslope movement of material by gravity.

mature sediment Textural maturity is indicated by a high degree of size sorting, so that the sediment has a very narrow size range of grains. Mineralogical maturity is the accumulation as a sediment of highly stable minerals in a weathering environment. This usually means a pure quartz sandstone.

mélange A mixture; used to describe pervasively sheared and folded sedimentary rocks within a portion of the accretionary prism.

metallic bond Sharing of electrons between all the atoms in the solid. The shared electrons can flow between the atoms, which makes metals good electrical conductors.

metallic luster A term that describes the silvery reflective surface of fresh metals.

metamorphic grade The intensity, or degree, of metamorphism to which a rock has been subjected.

metamorphic rock A rock that has been transformed by heat, pressure, and reaction with fluids.

metamorphism The alteration of a rock from one form to another in response to heat, pressure or reaction with fluids.

meteorite Asteroid fragment that reaches Earth's surface.

microplate A relatively small lithospheric plate.

Mid-Atlantic Ridge A topographic ridge that roughly follows the centerline of the Atlantic Ocean from north of Norway to just north of Antarctica.

migmatite A very high grade metamorphic rock, that has begun to melt.

mineral A naturally occurring inorganic solid with a definite (fixed) composition, an orderly arrangement of atoms (a crystal structure), and a characteristic set of physical properties.

mineral assemblage A specific group of minerals that are characteristic of a given rock composition, conditions of formation, or process of formation.

Modified Mercalli Intensity Scale A scale used for estimating the energy released in an earthquake, which is based on observed damage and the size of the affected region.

modulus A quantity that indicates the response of a material to external stress.

Mohorovičić discontinuity (Moho) The boundary between the crust and the mantle.

Mohs hardness scale A standard set of ten minerals used to determine comparative mineral hardness.

molecule A combination of atoms—of either the same type or different types—bonded together.

moment magnitude In seismology, a relatively new measure of earthquake size. Moment is the product of amount of offset multiplied by the area of the fault that moved, multiplied by the elastic modulus.

moraine Poorly sorted debris that has been transported to its site of deposition by the direct action of a glacier.

mother lode A miner's term for the source of a gold placer deposit.

nappe A tight overturned fold that has been sheared off along its lower limb.

native element Any non-gaseous element that is found naturally in nature. Some examples are carbon, sulfur, copper, gold, and silver.

neutron An electrically neutral particle found in the nucleus of an atom.

newton A unit of force (force = mass × acceleration) equivalent to 1 kg m/sec/sec.

normal fault A fault in which the hanging wall moves downward relative to the footwall, in response to extensional stress.

normal mode The natural resonance, or vibrating pattern of Earth.

normal polarity The present magnetic field orientation of Earth.

nuclear fission The process in which a nucleus breaks apart into two or more fragments, releasing vast amounts of energy.

nuclear fusion The process in which two nuclei combine to form a single, larger nucleus, and release a tremendous amount of energy.

oblique slip Movement on a fault that is neither purely up or down dip nor strike slip, but at some angle to the strike and dip of the fault.

obsidian Naturally occurring volcanic glass, most commonly of rhyolitic or dacitic composition.

oceanic lithosphere Lithosphere generated at spreading ridges.

oil trap Geologic structure that prevents oil seeping to the surface due to it's natural buoyancy.

open fold A fold in which the angle between the limbs is very wide.

ophiolite Slivers of ocean floor caught up and deformed between colliding plates. In ancient mountain belts, outcrops of ophiolite indicate the suture zone along which two plates were joined.

ore A concentration of an element or compound in rock, which can be mined for profit.

orogen An elongated region in Earth's crust that has undergone folding; formed at convergent margins, and generally in continent-continent collisions.

orogenesis The process by which mountains are formed.

outcrop An exposure of rock that is not covered by (for instance) sand, soil, or vegetation.

outer core The liquid outer region of Earth's center that is approximately 2270 km thick and is thought to be composed mainly of iron.

overturned fold A fold in which one of the limbs is top-side down, or overturned.

oxidation The loss of one or more electrons from an ion or an atom; the combination of an element and oxygen.

oxide A mineral generally composed of a metal combined with oxygen.

P wave The primary wave of an earthquake; it is the first seismic wave recorded.

pahoehoe Smooth- or ropey- surface textured basalt flow.

paleomagnetism The study of the permanent magnetism of rock.

Pangaea An ancient supercontinent formed from the assembly of most of Earth's continental landmasses some 200–300 million years ago.

partial melting The melting of some minerals in a mixture before others. Mixtures of silicate minerals do not have a single melting point, but a range of temperatures over which melting occurs.

passive continental margin The edge of a continent, corresponding to the junction between continental and oceanic crust where it occurs in the interior of a plate rather than at a plate boundary.

passive rift A rift caused by stresses in the lithosphere generated from remote plate forces that cause it to become extended; the associated faulting results in rift formation.

peat The earliest stage of coal formation from plant debris. The carbon content is low at approximately 60 percent.

pebble Rounded sedimentary clasts that range from 4 centimeters to 6.4 centimeters.

peneplain An extensive, flat erosional landscape.

perennial river A river that flows year round.

peridotite An ultramafic rock, common in the mantle that contains olivine (typically with pyroxenes).

period The time required to complete an oscillation of a single wave; or, a group of elements arranged in vertical columns in the Periodic Table; or, a geologic measure of time.

periodic table of the elements A table of the chemical elements arranged in columns (groups) and rows (periods) according to atomic number, which relates to the physicochemical properties of each element.

permafrost Permanently frozen soil in the arctic and antarctic regions.

pH scale A measure of the acidicity or basicity of a solution, strictly defined as the negative $\log_{10}$ of the hydrogen ion activity.

phaneritic A textural term, applied to igneous rocks, indicating that the crystals in the rock are sufficiently large to be seen with the unaided eye.

phase A substance that can be defined by its composition and physical state.

phase boundary A line on a phase diagram that defines the stability field—the limiting pressure, temperature, or composition—for a mineral or compound.

phase diagram A means of expressing the equilibrium phases that exist relative to pressure, temperature or composition.

phase transition The transformation from one phase to another in response to changing temperature, pressure, or composition, as occurs at a phase boundary.

Phenocrysts Well formed larger crystals in an extrusive igneous rock that give rise to a porphyritic texture.

photon A unit of electromagnetic energy that appears as a burst of light when an electron jumps from an outer shell to an inner shell of an atom.

physical weathering The breaking or granulation of rocks exposed at the surface by such processes as wedging due to ice and seasonal freeze and thaw.

pillow lava Bulbous pocket-like texture, resembling pillows, that may form when lava erupts underwater.

placer A stream deposit in which heavy minerals are concentrated by constant winnowing.

plastic deformation Irreversible rock deformation in which the style of deformation is more ductile or fluid than brittle or elastic.

plate A portion of lithosphere that is distinguished by boundaries and has a consistent motion.

plate motion The movement of a piece of lithosphere.

plate tectonics A theory that interrelates the internal and external processes of Earth, which involves the interaction of lithospheric plates.

platform A region of continental crust where ancient basement is covered by flat-lying sedimentary layers.

plume A column of hot, rising solid mantle material.

pluton A structural term indicating a body of igneous rock, intrusive in character, but restricted in size.

polarity In geology, the term polarity may refer to the magnetic field, normal polarity defined at the present orientation and reversed polarity defined as the opposite of the present orientation.

polymorph A mineral of a given composition that can exhibit more than one crystalline structure.

porosity The proportion of the total volume of a rock that is void space or fluid-filled.

porphyritic A textural term, applied to igneous rocks, indicating a population of crystals substantially larger than the remainder.

porphyroblast A textural term, applied to metamorphic rocks, indicating a population of crystals, substantially larger than the remainder, that have grown as a result of metamorphism.

precession The angular movement of the spin axis of an object around an axis fixed in space.

preservation potential A general term indicating the likelihood that a sequence of rocks or sediments will be preserved in the geologic record.

protolith A precursor to a metamorphic rock.

proton A positively charged particle found in the nucleus of an atom.

pumice Frothy glass that results from expulsion of water and other volatiles as bubbles from a magma as it moves upward.

pyroclastic Fragmented magmatic material that is ejected explosively from a volcano.

pyroclastic fall Pyroclastic material that falls back to the ground following an eruption. Near the vent it can accumulate a considerable thickness (such as at a cinder cone), farther away it may mantle the topography like snow.

pyroclastic flow Pyroclastic material which moves downhill as a hot gas-charged flow, formed commonly when a large eruption column collapses.

radiating habit A form of crystal growth characterized by a center from which thin, elongate crystals appear to emanate.

radioactivity The tendency for unstable nuclei of certain atoms to emit radiation.

ramp The inclined section of a thrust fault where it cuts across bedding.

recumbent fold A fold in which the axial plane is nearly horizontal; that is, one limb of a fold is completely overturned and the other may be approaching horizontal.

red shift The shift in the frequency of light observed coming from a distant star. The shift is to a lower frequency, red, because the star is moving away from Earth at high speed. See doppler effect.

reflection The return, or bouncing back, of waves from a surface.

refraction The bending of light or sound waves passing obliquely from one medium or layer into another due to variations in the velocity of wave travel in these media.

regression A seaward migration of the shoreline.

rejuvenation Relative lowering of the base level of a river, causing downward erosion of the river bed in an attempt to re-establish equilibrium (grade).

resinous luster A reflection from the surface of a mineral that makes it appear slightly translucent and yellowish brown. Named from pine resin, because the appearance is somewhat similar.

resonance The fundamental frequency that an object naturally vibrates if excited.

reverse fault A fault, often steeply inclined, in which the hanging wall moves upward relative to the footwall. Compare with thrust fault.

reverse polarity The reverse of Earth's present magnetic field orientation.

Richter magnitude scale A scale of the magnitude of an earthquake measured on a standard seismometer.

ridge push The outward-directed force on the lithosphere, exerted by uplifted asthenosphere at a divergent plate margin.

rift A long, narrow valley bounded by normal faults on each side.

right lateral fault Motion on the far side of a strike-slip fault that is directed to the observer's right.

rock cycle The geologic cycle that emphasizes the interconnections between processes that produce different rock types. For example, sedimentary rocks may be metamorphosed to metamorphic rocks, or melted to create igneous rocks, and these rocks may be weathered to form sediments and complete the cycle.

rock A hardened aggregate of one or more minerals bound tightly together, either by interlocking grains or by cement.

sag pond A small lake or pond that fills a surface depression, or sag, along a large fault.

saltation Particle movement in which the grains are transported by periodically bouncing along the bed of the flow.

S wave A shear wave occurring after an earthquake, it is known as the secondary wave because it is usually the second seismic wave recorded.

schist A metamorphic fabric that is defined by a preferred alignment of platy minerals such as mica; the alignment may be planar or wavy.

schistosity The plane defined by mineral orientation in a schist fabric.

sea-floor spreading The spreading of the sea floor outward from long submarine mountain ranges, the spreading ridges.

seamount Large submarine mountains, isolated or in linear chains on the sea floor. Many, but not all, seamounts are of volcanic origin.

sector collapse A large gravitational collapse landslide of the flanks of a volcano in reposnse to slope oversteepening. May be associated with magmatic activity, whereby collapse is accompanied by eruption due to pressure release (e.g. Mount St Helens).

secular variation The variation with time of the horizontal component (declination) and vertical component (inclination) of Earth's magnetic field.

sediment A wide variety of materials deposited at Earth's surface by physical processes (such as wind, water, and ice), chemical processes (precipitation from oceans, lakes, and rivers), or biological processes (organisms).

sedimentary rock A rock that consists of hardened sediment.

sedimentary structure A structure within a sedimentary bed formed by depositional processes; includes bedding, cross-bedding, and so on.

seismic station A place where seismometers are located.

seismic wave Any of a variety of waves that result from the sudden release of elastic strain in an earthquake.

seismogram A recording of seismic waves from an earthquake.

seismometer An instrument that detects vibrations in the ground.

settling velocity The terminal velocity that a particle reaches when falling through a fluid.

shale Mud that has been lithified, and that shows preferential breakage parallel to bedding planes (cleavage).

shear A stress that tends to cause adjacent parts of a solid to slide past one another.

sheeted dike complex Magma injected into cracks above a magma chamber, particularly at spreading ridges, where it cools and solidifies to form a nearly continuous tract of sub-parallel dike structures between the chamber and Earth's surface.

sheeting Planar, closely spaced joints.

shield An extensive tract of low-lying, ancient continental crust exposed at the surface; used interchangeably with the term craton.

shield volcano Volcano with shallow slopes, resembling the shape of an upturned shield. Formed from the accumulation of a large volume of very fluid basalt flows.

shoulder uplift The higher elevation of the edges, or margins of continental rift grabens compared with the surrounding plateau.

shutter ridge A hill that is moved by faulting so that it offsets and shuts off the stream flow out of a canyon.

silcrete A soil zone formed within the "B" soil horizon characterized by precipitation of silica (quartz), which may be between soil grains or more commonly replacing soil mineral grains.

silica tetrahedron The fundamental building block of silicate minerals—a silicon atom surrounded by four oxygens in the shape of a tetrahedron. Tetrahedra can be linked together in chains, sheets or networks and bonded to other ions.

sill A thin planar igneous rock body formed by the injection of magma typically between beds of rock in a sub-horizontal orientation. Compare with dike.

silt Sedimentary particles that range in size from 0.004 millimeters to 0.063 millimeters.

siltstone Lithified silt.

sinkhole A depression, several meters to hundreds of meters in diameter, formed by the collapse of an underground cavern.

slab Term applied to subducted lithosphere.

slab pull The force exerted on a plate as it is pulled into a subduction zone by a dense, sinking slab.

slab suction The suction force exerted on an overriding plate at a subduction zone when the subducted slab sinks vertically downward and backward, acting like a paddle sweeping back through the mantle.

slide A portion of a hillside that moves downhill due to gravity, and is generally disrupted to the point that the lowermost portions may form a disaggregated flow.

slump A portion of a hillside that moves as a unit downhill due to gravity.

soil creep A downhill movement of soil generally as individual particles.

soil horizon A vertical layering, or variation of materials that compose soil developed from weathering bedrock. The zones are termed O, A, B, C, and R.

solar wind A constant stream of high speed particles, mostly hydrogen ions, emitted from the Sun.

solid solution A mineral, or a group of minerals, in which the chemical composition varies uniformly so that every proportion between two end member compositions is represented. An example is olivine (end members $FeSiO_4$ and $MgSiO_4$).

solidus The temperature below which a material is completely solid. Compare with liquidus.

solifluction A slowly moving mudflow in polar regions.

solubility The capability of one material to be dissolved in another. In geology we speak, for example, of the solubility of water in magma, or of one mineral in another (solid solution).

sorting A measure of the uniformity of particle sizes in a sediment or sedimentary rock.

specific gravity The ratio of the weight of a given volume of a material to that same volume of water at 4° C.

spectrum The separation of electromagnetic radiation (including light) into its component frequencies.

spreading ridge Topographic feature formed on the seafloor marking the location of a divergent plate margin.

Standard International (SI) system of units A standard system of weights and measures in which the fundamental units are kilograms, meters, and seconds.

strain release Deformation associated with the release of stored elastic strain. An earthquake is a strain release mechanism, as is folding.

strain Any change in shape or size (deformation) of a body when stress is applied to it.

strata Layers in a sequence of sedimentary rock.

stratiform deposit A type of ore deposit that is confined to undeformed sedimentary rock layers.

stratosphere The upper part of the atmsophere, lying above the troposphere.

stratovolcano The classic volcano form with steep, slightly concave-up flanks sloping upward the summit, formed from layers of pyroclastic deposits and occasional lava flows.

strength A measure of the resistance of material to elastic or inelastic deformation.

stress The force applied to an object per unit area.

strike-slip fault A fault along which the dominant motion is parallel to the fault trace (strike).

strike A line defined by the intersection of a horizontal plane and an inclined surface, or the direction of a horizontal line on the surface of an inclined plane.

subduction The descent of one lithospheric plate beneath another.

subduction zone The structure formed at a convergent margin where one lithospheric plate sinks beneath another.

subduction zone polarity The orientation of the subducting plate.

sublimation The process of changing from the solid phase to gas without going through the liquid phase.

supernova The explosion of a massive star, causing collapse of the star and the release of a tremendous amount of energy.

surface wave A seismic wave that is confined to Earth's surface and shallow layers; it does not penetrate into Earth's interior.

suspension Particles dispersed through a fluid but not dissolved in it; or, grains remaining in a flowing medium without settling out.

suture The interface along which two items are joined together—can be mineral grains in a metamorphic rock or even continents following collision.

syncline A fold in deformed layered rocks in which the opposing sides of the structure dip toward the fold axis, giving it a concave upward form.

talus The accumulation of individual broken fragments at the base of a steep slope or cliff.

texture The mineral shape, size, and distribution of grain sizes in a rock. Compare with fabric.

throw The amount of displacement along a fault.

thrust fault A gently inclined fault in which the hanging wall moves upward and over the footwall.

traction Transportation of the coarser fraction of grains that roll and bump along the bottom of the flow. Compare with saltation.

transform fault A fault that ends abruptly at a plate margin; its activity is confined to the interval between the axes of spreading ridges, subduction zones, or a spreading ridge and a subduction zone.

transform margin A plate boundary at which plates slide past each other along a transform fault.

transgression A landward migration of the shoreline.

transpression The alignment of stresses that cause compression on opposite sides of a strike-slip fault during movement, which results in uplift and outward thrust faulting.

transtension The alignment of stresses that cause tension on opposite sides of a strike-slip fault during movement, which results in normal faulting and subsidence.

travel time In seismology, the length of time that it takes a given wave to travel from its source to the recording seismic station.

travel-time curve A graph constructed by successively timing seismic pulses over greater and greater epicentral distances from a given source. The shape of a travel-time curve can be used to determine wave velocity variation as a function of depth.

trench The elongate bathymetric depression that marks the line on the seafloor at which one plate subducts beneath another at a convergent plate margin.

triple junction The junction between three plates.

troposphere The lowest part of the atmosphere, where all weather occurs.

tuff A rock that primarily consists of consolidated ash.

tundra A relatively flat, treeless plain in alpine, arctic, and antarctic regions.

turbidite A sedimentary deposit resulting from a turbidity current.

turbidity current A current in which a denser fluid flows downward and under another, less dense, fluid.

ultramafic A term used to describe igneous rocks comprising only dark minerals (such as pyroxenes and olivine) with no felsic minerals. Most mantle rocks are ultramafic.

unconformity A significant interruption (hundreds to millions of years) in the sedimentary sequence comprising Earth's history.

upright fold A fold in which the axial plane is near vertical.

vadose zone The partially wet, but not saturated, zone immediately above the water table.

van der Waals bonds A very weak electrostatic attractive force between atoms caused by induced polarity.

ventifact A desert rock with one or more faceted surfaces caused by abrasion from blowing sand.

vibroseis A truck which is vibrated hydraulically to provide a source of seismic energy for seismic prospecting.

viscosity A material's resistance to flow.

volatile A substance, usually of low molecular or atomic weight, that is likely to preferentially evaporate or move into the gaseous phase.

volcanic arc An arc of volcanoes on the overriding plate of a subduction zone.

volcanic bomb Large pyroclastic fragment (>10 cm) ejected in the molten state from a volcano.

volcano The morphological surface consequence of magma reaching the surface (see cinder cone, shield volcano, stratovolcano).

volcano deformation cycle A repeated cycle of inflation with magma, eruption, and deflation as the magma is expelled.

Wadati-Benioff zone An inclined seismic zone that roughly corresponds to the location of the subducted plate at a convergent margin.

water table The upper boundary of the zone of saturation in the ground.

wave A repetitive disturbance that travels over a surface or through a medium.

wave base The depth to which surface waves can disturb the subsurface water as a wave passes, approximately one-half the wave length.

weathering The breakdown and decomposition of rocks ar or near Earth's surface.

welded tuff A deposit formed as a result of a very large pyroclastic flow, which has fused together, forming a hard rock.

xenolith A foreign rock fragment included in a magma.

Photo Credits

Chapter 1

CO1 Tom Van Sant/Geosphere Project, Santa Monica/Science Photo Library/Photo Researchers, Inc. 1.1 Worldsat International/Science Photo Library/Photo Researchers, Inc. 1.2 Robin T. Holcomb 1.3 NASA Headquarters 1.4 NASA Headquarters 1.5 NASA Headquarters 1.8 NASA Headquarters 1.9 NASA/Johnson Space Center 1.10 United States Department of Agriculture Natural Resources Conservation Science 1.13 NASA Headquarters 1.14 NASA Headquarters 1.15a Norman Banks/U.S. Department of the Interior, U.S. Geological Survey, David A. Johnston Cascades Volcano Observatory, Vancouver, Washington 1.15b Lowell, Georgia/Photo Researchers, Inc. 1.15c P.W. Wigand/U.S. Geological Survey 1.15d Jon Davidson 1.18 Tom Till Photography 1.20 Edward A. Hay 1.21 NASA Headquarters Focus 1.1 Upper Left NASA/Jet Propulsion Laboratory Focus 1.1 Middle Left NASA Headquarters Focus 1.1 Upper Right Jon Davidson Focus 1.1 Lower Left Breck P. Kent Focus 1.1Lower Right Jon Davidson

Chapter 2

CO2 NASA Headquarters 2.6 Palomar Observatory/California Institute of Technology 2.9 NASA Headquarters 2.15 Kitt Peak National Observatory/Aura, Inc. 2.16 Giraudon/Art Resource. Bayeux, Mus'e de l'Ev'ch'. With special authorization of the City of Bayeux 2.17 Mr. & Mrs. James M. Baker of Lillian, Alabama/James M. Baker 2.19 Peter L. Kresan Photography 2.20 Chip Clark 2.22 NASA Headquarters 2.23 Gregory G. Dimijian/Photo Researchers, Inc. 2.24b 1987 Sky Publishing Corp. All rights reserved. Reproduced with permission. 2.28a NASA Headquarters 2.28b NASA/Johnson Space Center 2.28c NASA Headquarters Focus 2.2.1 Michael Collier Focus 2.2.2 R.W. Girdler, P.T. Taylor, and J.J. Frawley/Tectonophysics

Chapter 3

CO3 Jeffrey A. Scovil 3.1a Roberto do Gugliemo/Science Photo Library/Photo Researchers, Inc. 3.1b Photo Researchers, Inc. 3.2 John Simpson 3.3a–c Jon Davidson 3.4b Breck P. Kent 3.7a The Natural History Museum, London 3.7b Jeffrey A. Scovil 3.12 Geoscience Features Picture Library 3.13 Geoscience/PH 3.14 Breck P. Kent 3.15 Geoscience Features Picture Library 3.16 (inset) Richard Mantonya 3.17 Jon Davidson 3.18 NASA Headquarters 3.19 Jeffrey A. Scovil Table 3.1: 1 Geoscience Features Picture Library 2a–b Ted Reed 2c Martin G. Miller 3a–f Donna Tucker 5 Natural History Museum 6 David Bayless/RIDA Photo Library 7a Chip Clark 7b Animals/Earth Scenes 8a Donna Tucker 8b Jon Davidson 9a The Natural History Museum, London 9b Ted Reed 9c Gem Media/Gem Media Gemological Institute of America 10a–c Jeffrey A. Scovil

Chapter 4

CO4 Donald C. Johnson/The Stock Market 4.1 Breck P. Kent 4.2 Upper Row: Donna Tucker / Breck P. Kent / Ted Reed / Geoscience Features Picture Library 4.2 Bottom Row: Donna Tucker / Donna Tucker / Jeffrey A. Scovil / Donna Tucker / Geoscience/PH 4.3 Jon Davidson (all Photos) 4.4a Dr. John Wolff, University of Texas 4.5 Jon Davidson 4.6b Jon Davidson 4.7a Breck P. Kent 4.7b Photo Researchers, Inc. 4.7c Geoscience Features Picture Library 4.7d John S. Shelton 4.7e Dr. Marge Wilson, University of Leeds 4.9a Bob I. Tilling/U.S. Geological Survey/VGP/Robert Tilling 4.9b Peter J. Mouginis-Mark 4.9c Alexei Ozerov, Institute of Volcanology, Petropavlovsk, Kamchatsky, Russia 4.9d Tom Till Photography 4.10a–b Peter Kresan/Peter L. Kresan Photography 4.11a Peter Arnold, Inc. 4.11b Woods Hole Oceanographic Institution 4.11c R.S. Fiske, Smithsonian Institution 4.11d Tom Till Photography 4.11e Keith G. Cox 4.14 Earth Satellite Corp./Photo Researchers, Inc. 4.15a Jon Davidson 4.15b Martin Bond/Science Photo Library/Photo Researchers, Inc. 4.17a Photo Researchers, Inc. 4.17b Photo Researchers, Inc. 4.17c Martin G. Muller 4.18 Tom Till Photography 4.19 Geoscience Features Picture Library 4.20a Peter Arnold, Inc. 4.20b Geoff Dore/Tony Stone Images 4.21a Peter L. Kresan Photography 4.21b Geoscience Features Picture Library 4.21c Jon Davidson 4.24a Geoscience Features Picture Library 4.24b–c Geoscience/PH 4.24d Donna Tucker 4.24e Jon Davidson 4.25a Breck P. Kent 4.25b Ted Reed 4.26a–b Geoscience/PH 4.26c–d Jeffrey A. Scovil Focus 4.1a–b Jon Davidson Focus 4.2.1–4 Tom Wright/U.S. Geological Survey Focus 4.3.1 Jon Davidson Focus 4.5.1a Jon Davidson Focus 4.5.1b Donna Tucker/UCLA Focus 4.5.1c Jon Davidson Focus 4.5.2a Martin G. Miller Focus 4.5.2b Geoscience Features Picture Library Focus 4.5.2c American Association of Petroleum Geologists Focus 4.5.3a Tom Bean/DRK Photo Focus 4.5.3b Donna Tucker/UCLA Focus 4.5.3c Jon Davidson

Chapter 5

CO5 © P. J. Tackley, 1993. 5.16 After Steeples and Igor 5.17a Woodhouse, Oxford 5.17b Su and Dziewonski, Harvard 5.19 J.R. Heirtzler, NASA/GSFC Laboratory for Terrestrial Physics and J. Frawley, Herring Bay Geophysics 5.20a R.W. Simpson, T.G. Hildenbrand, R.H. Godson, and M.F. Kane/US Geological Survey/Geophysics Unit 5.24b John Sohlden/Visuals Unlimited

Chapter 6

CO6 Chris Johns/National Geographic Society 6.11 JPL/NASA Headquarters 6.12 Tom Van Sant/Geosphere Project, Santa Monica/Science Photo Library/Photo Researchers, Inc. 6.15c NASA/Johnson Space Center 6.19a NASA Headquarters 6.19b Charles Preitner/Visuals Unlimited

Chapter 7

CO7 James Balog/Tony Stone Images 7.3a Peter L. Kresan Photography 7.3b Tom Till Photography 7.3c Tom Bean/The Stock Market 7.4b Martin G. Miller 7.4c Peter L. Kresan Photography 7.4d Peter Mozley 7.4e Breck P. Kent 7.8f1 Breck P. Kent 7.8f2 Martin G. Miller 7.8f3 John S. Shelton 7.9a G.R. Roberts/Photo Researchers, Inc. 7.9b Martin G. Miller/Visuals Unlimited 7.9c Ted Reed 7.13a Tom McHugh/California Academy of Sciences/Photo Researchers, Inc. 7.13c Michael Holford Photographs 7.19 Kerry Sieh/California Institute of Technology 7.22b Brian Lanker

Chapter 8

CO8 Bruce C. Heezen and Marie Tharp. Copyright 1980 by Marie Tharp. Reproduced by permission of Marie Tharp, 1 Washington Ave., South Nyack, NY 10960 8.9 Dudley Foster/Woods Hole Oceanographic Institution 8.10a Rod Catanach/Woods Hole Oceanographic Institution 8.10c Al Giddings/Images Unlimited 8.13 Ken C. MacDonald 8.18 From Wintershell Focus 8.2.3 GSF Picture Library Focus 8.2.5 Chris Johns/National Geographic Society

Chapter 9

CO9 Colin Prior/Tony Stone Worldwide 9.2 & inset NASA/Johnson Space Center 9.4b A. Hasegawa et al./Nature 9.7a Tass/Sovfoto/Eastfoto 9.7b Herb Spannagl/Tongariro Natural History Society 9.8a Jon Davidson 9.8b John D. Cunningham/Visuals Unlimited 9.8c Jon Davidson 9.8d R.I. Tilling 9.8e Mike Dungan 9.9b Meteor Crater Enterprises, Inc. 9.10a Jon Davidson 9.10b Shan de Silva 9.12 Ray Ingersoll, UCLA 9.13 NOAA/NGDC 9.15 Paolo Koch/Photo Researchers, Inc. 9.16b Mike Andrews/Animals/Earth Sciences 9.17 LANDSAT image courtesy of Phillips Petroleum Company, Exploration Projects Section 9.21 Tom Casedavall/U.S. Geological Survey, Denver Focus 9.1.2 Michael Moore/US Geological Survey Focus 9.2.2a Jon Davidson Focus 9.2.2c Jon Davidson Focus 9.3.2a–d Gary Rosenquist/Earth Images

Chapter 10

CO10 Manaaki Whenua Landcare Research 10.6 Ted Reed 10.12b Ted Reed 10.13a Ted Reed 10.13b Ted Reed 10.13c Ted Reed 10.15b Hall, J.K., 1994. Relief Map of Israel and Adjacent Areas (Digital Terrane Model). Geological Survey of Israel 10.17a Ted Reed Focus 10.1.2 Jon Davidson Focus 10.2.2 Ted Reed

Chapter 11

CO11 NASA Headquarters 11.1 LANDSAT image courtesy of Phillips Petroleum Company, Exploration Projects Section 11.8d John S. Shelton 11.15a Jim Wark/Peter Arnold, Inc. 11.15b Jean-Marc Truchet/Tony Stone Images 11.15c Jeffrey L. Rotman 11.20a Keith G. Cox, Department of Earth Sciences, University of Oxford, England Focus 11.3.1a D.W. Peterson/U.S. Geological Survey Focus 11.3.1b Jon Davidson Focus 11.3.1c J. Judd/U.S. Geological Survey Focus 11.3.1d P.W. Francis

Chapter 12

CO12 NASA Headquarters 12.2b (top and bottom) Ted Reed 12.3a–c Ted Reed 12.4 Ted Reed 12.7a Kenneth Murray/Photo Researchers, Inc. 12.7b Tom Bean/DRK Photo 12.8a Aerial Cartographics of America 12.8b John S. Shelton 12.8c Kjell B. Sandved/Visuals Unlimited 12.13a–c Jon Davidson 12.14a–b Jon Davidson 12.16 Mickey Gibson/Animals/Earth Scenes 12.17 Ted Reed 12.19a John Eastcott/YVA/Momatiuk/Animals/Earth Scenes 12.19b Ted Reed 12.19a–b Ted Reed 12.20a Ted Reed 12.20b Jon Davidson 12.21 Will and Deni McIntyre/Photo Researchers, Inc.

Chapter 13

CO13 NASA Headquarters 13.8a William E. Ferguson 13.10 Tom Bean 13.11 Comstock 13.13a Ted Reed 13.13b John Lemker/Animals/Earth Scenes 13.14 Joseph H. Hartshorn 13.15a–b Peter L. Kresan Photography 13.16 Mike Andrews/Animals/Earth Scenes 13.17a–b Michael Collier 13.19a NASA Headquarters 13.19b NASA/Johnson Space Center 13.20 Ted Reed 13.22 John S. Shelton 13.23 John S. Shelton 13.24 Ted Reed

Chapter 14

CO14 John M. Roberts/The Stock Market 14.3a Jon Davidson 14.4 Ernest Wilkinson/Animals/Earth Scenes 14.5a William E. Ferguson 14.6a Ted Reed 14.6b Ted Reed 14.7 (all photos) Ted Reed 14.9 (center) Worldsat International Science Photo Library/Photo Researchers, Inc. 14.9a Wendy Neefus/Animals/Earth Scenes 14.9b John Lemker/Animals/Earth Scenes 14.9c John S. Shelton 14.9d Tom Bean 14.9e Stephen Trimble 14.14 (center) Worldsat International Science Photo Library/Photo Researchers, Inc. 14.14a Breck P. Kent 14.14b Breck P. Kent 14.14c John S. Shelton 14.14d Martin G. Miller 14.14e Martin G. Miller © Martin Miller 14.17 (center) Worldsat International Science Photo Library/Photo Researchers, Inc. 14.17a–b Jon Davidson 14.17c Peter L. Kresan Photography 14.17d Jon Davidson 14.17e Martin G. Miller 14.18a Martin G. Miller 14.19a Peter L. Kresan Photography 14.19b William E. Ferguson 14.19c John S. Shelton 14.19d Russ Kinne/Comstock 14.20a John S. Shelton 14.20b John S. Shelton 14.22 Ted Reed

Chapter 15

CO15 Chromo Sohm/The Stock Market 15.1 François Gohier/Photo Researchers, Inc. 15.2 Photo courtesy of Caltech 15.3 AP Photo/Douglas C. Pizac 15.4 J.D. Griggs/U.S. Geological Survey 15.5 Peter L. Kresan Photography 15.6a Richard Vogel/The Gamma Liaison Network 15.6b Ben Castellana 15.9 Burton A. Amundson/JLM Visuals 15.11 U.S. Geological Survey 15.12 John S. Shelton 15.13 Aerial Cartographics of America 15.14 NOAA HAZMAT 15.22a–c The Natural History Museum, London

Chapter 16

CO16 John Lamb/Tony Stone Images 16.3 Peter L. Kresan Photography 16.4b The Natural History Museum, London 16.6 Institute of Oceanographic Studies/NERC/Science Photo Library/Photo Researchers, Inc. 16.7 Peter L. Kresan Photography 16.17b–c ©1991 American Association of Petroleum Geologists 16.18a Steve Kaufman/DRK Photo 16.19a ©1991 American Association of Petroleum Geologists Focus 16.4.1 NASA Viking Orbiter image #063A09 courtesy Cornell University's Spacecraft Planetary Imaging Facility

Index